PROGRESS IN BRAIN RESEARCH

VOLUME 173

GLAUCOMA: AN OPEN WINDOW TO NEURODEGENERATION
AND NEUROPROTECTION

Other volumes in PROGRESS IN BRAIN RESEARCH

Volume 136: Changing Views of Cajal's Neuron, by E.C. Azmitia, J. DeFelipe, E.G. Jones, P. Rakic and C.E. Ribak (Eds.) – 2002, ISBN 0-444-50815-5.
Volume 137: Spinal Cord Trauma: Regeneration, Neural Repair and Functional Recovery, by L. McKerracher, G. Doucet and S. Rossignol (Eds.) – 2002, ISBN 0-444-50817-1.
Volume 138: Plasticity in the Adult Brain: From Genes to Neurotherapy, by M.A. Hofman, G.J. Boer, A.J.G.D. Holtmaat, E.J.W. Van Someren, J. Verhaagen and D.F. Swaab (Eds.) – 2002, ISBN 0-444-50981-X.
Volume 139: Vasopressin and Oxytocin: From Genes to Clinical Applications, by D. Poulain, S. Oliet and D. Theodosis (Eds.) – 2002, ISBN 0-444-50982-8.
Volume 140: The Brain's Eye, by J. Hyönä, D.P. Munoz, W. Heide and R. Radach (Eds.) – 2002, ISBN 0-444-51097-4.
Volume 141: Gonadotropin-Releasing Hormone: Molecules and Receptors, by I.S. Parhar (Ed.) – 2002, ISBN 0-444-50979-8.
Volume 142: Neural Control of Space Coding, and Action Production, by C. Prablanc, D. Pélisson and Y. Rossetti (Eds.) – 2003, ISBN 0-444-509771.
Volume 143: Brain Mechanisms for the Integration of Posture and Movement, by S. Mori, D.G. Stuart and M. Wiesendanger (Eds.) – 2004, ISBN 0-444-513892.
Volume 144: The Roots of Visual Awareness, by C.A. Heywood, A.D. Milner and C. Blakemore (Eds.) – 2004, ISBN 0-444-50978-X.
Volume 145: Acetylcholine in the Cerebral Cortex, by L. Descarries, K. Krnjević and M. Steriade (Eds.) – 2004, ISBN 0-444-51125-3.
Volume 146: NGF and Related Molecules in Health and Disease, by L. Aloe and L. Calzà (Eds.) – 2004, ISBN 0-444-51472-4.
Volume 147: Development, Dynamics and Pathology of Neuronal Networks: From Molecules to Functional Circuits, by J. Van Pelt, M. Kamermans, C.N. Levelt, A. Van Ooyen, G.J.A. Ramakers and P.R. Roelfsema (Eds.) – 2005, ISBN 0-444-51663-8.
Volume 148: Creating Coordination in the Cerebellum, by C.I. De Zeeuw and F. Cicirata (Eds.) – 2005, ISBN 0-444-51754-5.
Volume 149: Cortical Function: A View from the Thalamus, by V.A. Casagrande, R.W. Guillery and S.M. Sherman (Eds.) – 2005, ISBN 0-444-51679-4.
Volume 150: The Boundaries of Consciousness: Neurobiology and Neuropathology, by Steven Laureys (Ed.) – 2005, ISBN 0-444-51851-7.
Volume 151: Neuroanatomy of the Oculomotor System, by J.A. Büttner-Ennever (Ed.) – 2006, ISBN 0-444-51696-4.
Volume 152: Autonomic Dysfunction after Spinal Cord Injury, by L.C. Weaver and C. Polosa (Eds.) – 2006, ISBN 0-444-51925-4.
Volume 153: Hypothalamic Integration of Energy Metabolism, by A. Kalsbeek, E. Fliers, M.A. Hofman, D.F. Swaab, E.J.W. Van Someren and R.M. Buijs (Eds.) – 2006, ISBN 978-0-444-52261-0.
Volume 154: Visual Perception, Part 1, Fundamentals of Vision: Low and Mid-Level Processes in Perception, by S. Martinez-Conde, S.L. Macknik, L.M. Martinez, J.M. Alonso and P.U. Tse (Eds.) – 2006, ISBN 978-0-444-52966-4.
Volume 155: Visual Perception, Part 2, Fundamentals of Awareness, Multi-Sensory Integration and High-Order Perception, by S. Martinez-Conde, S.L. Macknik, L.M. Martinez, J.M. Alonso and P.U. Tse (Eds.) – 2006, ISBN 978-0-444-51927-6.
Volume 156: Understanding Emotions, by S. Anders, G. Ende, M. Junghofer, J. Kissler and D. Wildgruber (Eds.) – 2006, ISBN 978-0-444-52182-8.
Volume 157: Reprogramming of the Brain, by A.R. Møller (Ed.) – 2006, ISBN 978-0-444-51602-2.
Volume 158: Functional Genomics and Proteomics in the Clinical Neurosciences, by S.E. Hemby and S. Bahn (Eds.) – 2006, ISBN 978-0-444-51853-8.
Volume 159: Event-Related Dynamics of Brain Oscillations, by C. Neuper and W. Klimesch (Eds.) – 2006, ISBN 978-0-444-52183-5.
Volume 160: GABA and the Basal Ganglia: From Molecules to Systems, by J.M. Tepper, E.D. Abercrombie and J.P. Bolam (Eds.) – 2007, ISBN 978-0-444-52184-2.
Volume 161: Neurotrauma: New Insights into Pathology and Treatment, by J.T. Weber and A.I.R. Maas (Eds.) – 2007, ISBN 978-0-444-53017-2.
Volume 162: Neurobiology of Hyperthermia, by H.S. Sharma (Ed.) – 2007, ISBN 978-0-444-51926-9.
Volume 163: The Dentate Gyrus: A Comprehensive Guide to Structure, Function, and Clinical Implications, by H.E. Scharfman (Ed.) – 2007, ISBN 978-0-444-53015-8.
Volume 164: From Action to Cognition, by C. von Hofsten and K. Rosander (Eds.) – 2007, ISBN 978-0-444-53016-5.
Volume 165: Computational Neuroscience: Theoretical Insights into Brain Function, by P. Cisek, T. Drew and J.F. Kalaska (Eds.) – 2007, ISBN 978-0-444-52823-0.
Volume 166: Tinnitus: Pathophysiology and Treatment, by B. Langguth, G. Hajak, T. Kleinjung, A. Cacace and A.R. Møller (Eds.) – 2007, ISBN 978-0-444-53167-4.
Volume 167: Stress Hormones and Post Traumatic Stress Disorder: Basic Studies and Clinical Perspectives, by E.R. de Kloet, M.S. Oitzl and E. Vermetten (Eds.) – 2007, ISBN 978-0-444-53140-7.
Volume 168: Models of Brain and Mind: Physical, Computational and Psychological Approaches, by R. Banerjee and B.K. Chakrabarti (Eds.) – 2008, ISBN 978-0-444-53050-9.
Volume 169: Essence of Memory, by W.S. Sossin, J.-C. Lacaille, V.F. Castellucci and S. Belleville (Eds.) – 2008, ISBN 978-0-444-53164-3.
Volume 170: Advances in Vasopressin and Oxytocin – From Genes to Behaviour to Disease, by I.D. Neumann and R. Landgraf (Eds.) – 2008, ISBN 978-0-444-53201-5.
Volume 171: Using Eye Movements as an Experimental Probe af Brain Function — A Symposium in Honor of Jean Büttner-Ennever, by Christopher Kennard and R. John Leigh (Eds.) – 2008, ISBN 978-0-444-53163-6.
Volume 172: Serotonin–Dopamine Interaction: Experimental Evidence and Therapeutic Relevance, by G. Di Giovanni, V. Di Matteo and E. Esposito (Eds.) – 2008, ISBN 978-0-444-53235-0.

PROGRESS IN BRAIN RESEARCH

VOLUME 173

GLAUCOMA: AN OPEN WINDOW TO NEURODEGENERATION AND NEUROPROTECTION

EDITED BY

CARLO NUCCI
*Ophthalmological Unit, Department of Biopathology and Diagnostic Imaging,
University of Rome "Tor Vergata", Rome, Italy*

NEVILLE N. OSBORNE
Nuffield Laboratory of Ophthalmology, University of Oxford, Oxford, UK

GIACINTO BAGETTA
Department of Pharmacobiology, University of Calabria, Cosenza, Italy

LUCIANO CERULLI
*Ophthalmological Unit, Department of Biopathology and Diagnostic Imaging,
University of Rome "Tor Vergata", Rome, Italy*

ELSEVIER

AMSTERDAM – BOSTON – HEIDELBERG – LONDON – NEW YORK – OXFORD
PARIS – SAN DIEGO – SAN FRANCISCO – SINGAPORE – SYDNEY – TOKYO

Elsevier
360 Park Avenue South, New York, NY 10010-1710
Linacre House, Jordan Hill, Oxford OX2 8DP, UK
Radarweg 29, PO Box 211, 1000 AE Amsterdam, The Netherlands

First edition 2008

Copyright © 2008 Elsevier B.V. All rights reserved

No part of this publication may be reproduced, stored in a retrieval system
or transmitted in any form or by any means electronic, mechanical, photocopying,
recording or otherwise without the prior written permission of the publisher

Permissions may be sought directly from Elsevier's Science & Technology Rights
Department in Oxford, UK: phone (+44) (0) 1865 843830; fax (+44) (0) 1865 853333;
email: permissions@elsevier.com. Alternatively you can submit your request online by
visiting the Elsevier web site at http://www.elsevier.com/locate/permissions, and selecting
Obtaining permission to use Elsevier material

Notice
No responsibility is assumed by the publisher for any injury and/or damage to persons
or property as a matter of products liability, negligence or otherwise, or from any use
or operation of any methods, products, instructions or ideas contained in the material
herein. Because of rapid advances in the medical sciences, in particular, independent
verification of diagnoses and drug dosages should be made

British Library Cataloguing in Publication Data
A catalogue record for this book is available from the British Library

Library of Congress Cataloging-in-Publication Data
A catalog record for this book is available from the Library of Congress

ISBN: 978-0-444-53256-5 (this volume)
ISSN: 0079-6123 (Series)

For information on all Elsevier publications
visit our website at books.elsevier.com

Printed and bound in Hungary

08 09 10 11 12   10 9 8 7 6 5 4 3 2 1

Working together to grow
libraries in developing countries

www.elsevier.com | www.bookaid.org | www.sabre.org

ELSEVIER   BOOK AID International   Sabre Foundation

List of Contributors

N. Amerasinghe, Singapore National Eye Centre, Singapore, Singapore

A.M. Angrisani, Ophthalmology Unit, Association Columbus Clinic, Catholic University of Rome, Rome, Italy

P. Aragona, Section of Ophthalmology, Department of Surgical Specialities, Azienda Ospedaliera Universitaria Policlinico G. Martino, Messina, Italy

T. Aung, Singapore National Eye Centre, Singapore, Singapore

G. Bagetta, Department of Pharmacobiology; Section of Neuropharmacology of Normal and Pathological Neuronal Plasticity, University Center for Adaptive Disorders and Headache (UCHAD), University of Calabria, Arcavacata di Rende (Cosenza), Italy

E. Balestrazzi, Department of Ophthalmology, A. Gemelli University Hospital, Catholic University of the Sacred Heart, Rome, Italy

M. Bari, Department of Experimental Medicine and Biochemical Sciences, University of Rome "Tor Vergata", Rome, Italy

K. Barton, Moorfields Eye Hospital; Department of Epidemiology, Institute of Ophthalmology, University College London, London, UK

S. Brocchini, ORB (Ocular Repair and Regeneration Biology Research Unit), National Institute for Health Research Biomedical Research Centre, Moorfields Eye Hospital NHS Foundation Trust and UCL Institute of Ophthalmology, London, UK; The School of Pharmacy, University of London, London, WC1N 1AX, UK

P. Brusini, Department of Ophthalmology, Santa Maria della Misericordia Hospital, Udine, Italy

N.D. Bull, Cambridge Centre for Brain Repair, University of Cambridge, Cambridge, UK

F. Cantatore, Glaucoma Center, Department of Ophthalmology, University of Bari, Bari, Italy

J. Caprioli, Jules Stein Eye Institute, UCLA, Los Angeles, CA, USA

R.G. Carassa, Department of Ophthalmology, University Hospital S. Raffaele, Milano, Italy

F. Cavaliere, Department of Pharmacobiology, University of Calabria, Arcavacata di Rende, Italy

C. Cedrone, Physiopathological Optics, Department of Biopathology and Diagnostic Imaging, University of Rome "Tor Vergata", Rome, Italy

M. Centofanti, Biopathology Department; Department of Ophthalmology, University of Rome "Tor Vergata"; G.B. Bietti Eye Foundation, IRCCS, Rome, Italy

W.O. Cepurna, Casey Eye Institute, Oregon Health & Science University, Portland, OR, USA

A. Cerulli, Physiopathological Optics, Department of Biopathology and Diagnostic Imaging, University of Rome "Tor Vergata", Rome; Department of Pharmacobiology, University of Calabria, Arcavacata di Rende, Italy

M. Cesareo, Physiopathological Optics, Department of Biopathology and Diagnostic Imaging, University of Rome "Tor Vergata", Rome, Italy

N.G.F. Cooper, Department of Anatomical Sciences and Neurobiology, University of Louisville School of Medicine, Louisville, KY, USA

G. Coppola, Ophthalmology Unit, Association Columbus Clinic, Catholic University of Rome, Rome, Italy

M.T. Corasaniti, Department of Pharmacobiological Sciences, University "Magna Graecia" of Catanzaro, Catanzaro, Italy

M.F. Cordeiro, Glaucoma & Retinal Degeneration Research Group, UCL Institute of Ophthalmology; The Glaucoma Research Group, Western Eye Hospital, London, UK

V. Cozzolino, Department of Diagnostic Imaging and Interventional Radiology, University of Rome "Tor Vergata", Rome, Italy

A. Dahlmann-Noor, ORB (Ocular Repair and Regeneration Biology Research Unit), National Institute for Health Research Biomedical Research Centre, Moorfields Eye Hospital NHS Foundation Trust and UCL Institute of Ophthalmology, London, UK

P.V. Desai, Jules Stein Eye Institute, UCLA, Los Angeles, CA, USA

B. Falsini, Istituto di Oftalmologia, Universita' Cattolica del S. Cuore, Roma, Italy

W. Fan, Department of Anatomical Sciences and Neurobiology, University of Louisville School of Medicine, Louisville, KY, USA

E. Fazzi, IRCCS Fondazione Istituto Neurologico "C. Mondino", Pavia, Italy

A. Ferreras, Department of Ophthalmology, Miguel Servet University, Zaragoza, Spain

F. Ferreri, Section of Ophthalmology, Department of Surgical Specialities, Azienda Ospedaliera Universitaria Policlinico G. Martino, Messina, Italy

G. Ferreri, Section of Ophthalmology, Department of Surgical Specialities, Azienda Ospedaliera Universitaria Policlinico G. Martino, Messina, Italy

P. Ferreri, Glaucoma Center, Department of Ophthalmology, University of Bari, Bari, Italy

R. Floris, Department of Diagnostic Imaging and Interventional Radiology, University of Roma "Tor Vergata", Rome, Italy

P. Fogagnolo, G.B. Bietti Foundation – IRCCS (Istituto di Ricovero e Cura a Carattere Scientifico), Rome, Italy

F.G. Garaci, Department of Diagnostic Imaging and Interventional Radiology, University of Rome "Tor Vergata"; IRCCS San Raffaele Pisana, Rome, Italy

D.F. Garway-Heath, NIHR Biomedical Research Centre for Ophthalmology, Moorfields Eye Hospital NHS Foundation Trust and UCL Institute of Ophthalmology, London, UK; G.B. Bietti Foundation – IRCCS (Istituto di Ricovero e Cura a Carattere Scientifico), Rome, Italy

F. Gaudiello, Department of Diagnostic Imaging and Interventional Radiology, University of Rome "Tor Vergata", Rome, Italy

S. Georgoulas, The School of Pharmacy, University of London, London, WC1N 1AX, UK; ORB (Ocular Repair and Regeneration Biology Research Unit), National Institute for Health Research Biomedical Research Centre, Moorfields Eye Hospital NHS Foundation Trust and UCL Institute of Ophthalmology, London, UK

M. Gliozzi, Department of Pharmacobiological Sciences, University "Magna Graecia"of Catanzaro, Catanzaro, Italy

E. Goodyear, Department of Ophthalmology, University of Montreal, Montreal, Canada

L. Guo, Glaucoma & Retinal Degeneration Research Group, UCL Institute of Ophthalmology, London, UK

N. Gupta, Department of Ophthalmology and Vision Sciences; Department of Laboratory Medicine and Pathobiology, Keenan Research Centre at the Li Ka Shing, Knowledge Institute of St. Michael's Hospital, University of Toronto, Toronto, Ontario, Canada

M.R. Hernandez, Department of Ophthalmology, Feinberg School of Medicine, Northwestern University, Chicago, IL, USA

D.K. Heuer, Department of Ophthalmology, Medical College of Wisconsin Milwaukee; Froedtert & Medical College of Wisconsin, Eye Institute, Milwaukee, WI 53226, USA

G.R. Howell, The Jackson Laboratory, Bar Harbor, ME, USA

A. Izzotti, Department of Health Sciences, Faculty of Medicine, University of Genoa, Genoa, Italy

S.W.M. John, The Jackson Laboratory; The Howard Hughes Medical Institute, Bar Harbor, ME; Department of Ophthalmology, Tufts University of Medicine, Boston, MA, USA

E. Johnson, Casey Eye Institute, Oregon Health & Science University, Portland, OR, USA

T.V. Johnson, Cambridge Centre for Brain Repair, University of Cambridge, Cambridge, UK

P.T. Khaw, The School of Pharmacy, University of London, London, WC1N 1AX, UK

A. Laabich, Acucela Inc., Bothell, WA, USA

D. Lepore, Department of Ophthalmology, A. Gemelli University Hospital, Catholic University of the Sacred Heart, Rome, Italy

L.A. Levin, Department of Ophthalmology, University of Montreal, Montreal, Canada; Department of Opthamology and Visual Sciences, University of Wisconsion, Madison, USA

R.T. Libby, University of Rochester Eye Institute, University of Rochester Medical Center, Rochester, NY, USA

S.A. Lipton, Center for Neuroscience, Aging and Stem Cell Research, Burnham Institute for Medical Research, La Jolla, CA, USA

A. London, Department of Neurobiology, The Weizmann Institute of Science, Rehovot, Israel

A. Ludovici, Department of Diagnostic Imaging and Interventional Radiology, University of Rome "Tor Vergata", Rome, Italy

T. Lukas, Department of Molecular Pharmacology and Biological Chemistry, Feinberg School of Medicine, Northwestern University, Chicago, IL, USA

T. Lupattelli, Interventional Radiology Department, IRCCS Multimedica, Sesto San Giovanni, Milan, Italy

M. Maccarrone, Department of Biomedical Sciences, University of Teramo, Teramo, Italy

R. Mancino, Physiopathological Optics, Department of Biopathology and Diagnostic Imaging, University of Rome "Tor Vergata", Rome, Italy

G. Manni, Department of Ophthalmology; Biopathology Department, University of Rome "Tor Vergata"; G.B. Bietti Eye Foundation, IRCCS, Rome, Italy

K.R. Martin, Cambridge Centre for Brain Repair, University of Cambridge, Cambridge, UK

P. Mazzarelli, Department of Biopathology, Institute of Anatomic Pathology, University of Rome "Tor Vergata", Rome, Italy

F.A. Medeiros, Hamilton Glaucoma Center, University of California-San Diego, La Jolla, CA, USA

H. Miao, Department of Ophthalmology, Feinberg School of Medicine, Northwestern University, Chicago, IL, USA

F. Missiroli, Section of Ophthalmology, Department of Biopathology, University of Rome "Tor Vergata", Rome, Italy

A. Montepara, Glaucoma Center, Department of Ophthalmology, University of Bari, Bari, Italy

J.C. Morrison, Casey Eye Institute, Oregon Health & Science University, Portland, OR, USA

L.A. Morrone, Department of Pharmacobiology; Section of Neuropharmacology of Normal and Pathological Neuronal Plasticity, University Center for Adaptive Disorders and Headache (UCHAD), University of Calabria, Arcavacata di Rende, Italy

R.W. Nickells, Department of Ophthalmology and Visual Sciences, University of Wisconsin, Madison, WI, USA

C. Nucci, Ophthalmological Unit, Department of Biopathology and Diagnostic Imaging, University of Rome "Tor Vergata"; Experimental Neuropharmacology Center, "Mondino-Tor Vergata", Fondazione C. Mondino-IRCCS, Rome, Italy

F. Oddone, G.B. Bietti Eye Foundation, IRCCS, Rome, Italy
N. Orzalesi, Eye Clinic, Department of Medicine, Surgery and Odontoiatry, San Paolo Hospital, University of Milan, Milan, Italy
N.N. Osborne, Nuffield Laboratory of Ophthalmology, University of Oxford, Oxford, UK
M.M. Pagliara, Department of Ophthalmology, A. Gemelli University Hospital, Catholic University of the Sacred Heart, Rome, Italy
V. Parisi, G.B. Bietti Eye Foundation, IRCCS, Rome, Italy
N. Pasquale, University of Rome "La Sapienza", 2nd Faculty of Medicine, Rome, Italy
A. Perdicchi, S. Andrea Hospital, Rome, Italy
S. Pucci, Department of Biopathology, Institute of Anatomic Pathology, University of Rome "Tor Vergata", Rome, Italy
L. Quaranta, Department of Ophthalmology, University of Brescia, Brescia, Italy
A. Quinto, Glaucoma Center, Department of Ophthalmology, University of Bari, Bari, Italy
S. Ranno, Eye Clinic, Department of Medicine, Surgery and Odontoiatry, San Paolo Hospital, University of Milan, Milan, Italy
S.M. Recupero, S. Andrea Hospital, 2nd Faculty of Medicine, University of Rome "La Sapienza", Rome, Italy
F. Regine, Section of Ophthalmology, Department of Biopathology, University of Rome "Tor Vergata", Rome, Italy
B. Ricci, Ophthalmology Unit, Association Columbus Clinic, Catholic University of Rome, Rome, Italy
F. Ricci, Section of Ophthalmology, Department of Biopathology, University of Rome "Tor Vergata", Rome, Italy
R. Ritch, New York Eye and Ear Infirmary, New York, NY, USA
C.E. Riva, Dipartimento di Discipline Chirurgiche, Rianomatorie e dei Trapianti "Antonio Valsalva", Universitá di Bologna, Bologna, Italy
L. Rombolà, Department of Pharmacobiology, University of Calabria, Arcavacata di Rende, Italy
L. Rossetti, Eye Clinic, Department of Medicine, Surgery and Odontoiatry, San Paolo Hospital, University of Milan, Milan, Italy
G. Ruggeri, Glaucoma Center, Department of Ophthalmology, University of Bari, Bari, Italy
R. Russo, Department of Pharmacobiology, University of Calabria, Arcavacata di Rende, Italy
S.C. Saccà, Division of Ophthalmology, St. Martino Hospital, Genoa, Italy
S. Sakurada, Department of Physiology and Anatomy, Tohoku Pharmaceutical University, Sendai, Japan
C. Sborgia, Glaucoma Center, Department of Ophthalmology, University of Bari, Bari, Italy
C.L. Schlamp, Department of Ophthalmology and Visual Sciences, University of Wisconsin, Madison, WI, USA
M. Schwartz, Department of Neurobiology, The Weizmann Institute of Science, Rehovot, Israel
G.L. Scuderi, S. Andrea Hospital, 2nd Faculty of Medicine, University of Rome "La Sapienza", Rome, Italy
M. Seki, Center for Neuroscience, Aging and Stem Cell Research, Burnham Institute for Medical Research, La Jolla, CA, USA
S.J. Semaan, Department of Ophthalmology and Visual Sciences, University of Wisconsin, Madison, WI, USA
S.C. Sharma, Departments of Ophthalmology, Cell Biology and Anatomy, New York Medical College, Valhalla, NY, USA
G. Simonetti, Department of Diagnostic Imaging and Interventional Radiology, University of Rome "Tor Vergata", Rome, Italy
A. Spanò, Ophthalmological Unit, Department of Biopathology and Diagnostic Imaging, University of Rome "Tor Vergata", Rome, Italy

G. Tezel, Department of Ophthalmology and Visual Sciences; Department of Anatomical Sciences and Neurobiology, University of Louisville School of Medicine, Louisville, KY, USA

A. Tosto, Biopathology Department, University of Tor Vergata, Rome, Italy

M. Vetrugno, Glaucoma Center, Department of Ophthalmology, University of Bari, Bari, Italy

X. Wang, Department of Anatomical Sciences and Neurobiology, University of Louisville School of Medicine, Louisville, KY, USA

C. Watanabe, Department of Physiology and Anatomy, Tohoku Pharmaceutical University, Sendai, Japan

R.N. Weinreb, Hamilton Glaucoma Center, University of California, San Diego, La Jolla, CA, USA

S.M. Whitcup, Research and Development, Allergan, Inc., Irvine; Department of Ophthalmology, Jules Stein Eye Institute, UCLA School of Medicine, Los Angeles, CA, USA

Y. Yücel, Department of Ophthalmology and Vision Sciences; Department of Laboratory Medicine and Pathobiology, Keenan Research Centre at the Li Ka Shing, Knowledge Institute of St. Michael's Hospital, University of Toronto, Toronto, Ontario, Canada

L. Ziccardi, G.B. Bietti Eye Foundation, IRCCS, Rome, Italy

Preface

Glaucoma is a family of diseases that includes primary open angle glaucoma (POAG), normal tension glaucoma (NTG), angle-closure glaucoma, secondary glaucoma and glaucoma with onset in infancy. The effect of glaucoma is the death of retinal ganglion cells (RGCs) which causes ultimate loss of vision. The nature of glaucomatous visual field changes and morphologic abnormalities suggests that the pathophysiology reflects injury which occurs at the level of the optic nerve. It remains a matter of debate as to whether this is caused by ischemia at the optic nerve head, blockade of ganglion cell axonal transport, peripapillary atrophy, or changes in the characteristics of the lamina cribosa. Loss of RGCs in glaucoma, especially in POAG and NTG, is gradual with some cells dying ten to twenty years later than others. The challenge for the future is therefore to devise ways of preventing functional RGCs dying once glaucoma is diagnosed.

Glaucoma has traditionally been diagnosed by either elevated intraocular pressure (IOP), optic nerve head atrophy of a defined clinical characteristic, or "cupping" and loss of visual field. The current treatment strategy involves medical, surgical, or parasurgical interventions, all aimed at reducing the IOP. This is now widely acknowledged to be insufficient, as many glaucoma patients progress despite treatment and many patients do not have elevated IOP in the first place.

Over the past three decades much research has been devoted to discovering novel therapeutic strategies to improve the clinical outcome of the vast majority of patients with glaucoma. In particular, numerous studies have aimed at identifying pharmacological agents that can directly attenuate RGC death as occurs in glaucoma. Much has been learned from experimental studies on animal models of glaucoma and from information related to the treatment of other CNS degenerative diseases.

Unfortunately, a large clinical trial stimulated by impressive experimental studies has proved unsuccessful, raising many questions related to neuroprotection strategies for the treatment of glaucoma. It is clear that further progress is necessary before it is possible to translate basic science into clinically effective technologies. Moreover, this goal can only be reached through the development of suitable procedures to monitor disease onset and progression.

In this book, leaders in the field of glaucoma contribute articles which shed light on current ideas about the pathophysiology of the disease where putative risk factors require serious consideration. Suitable sensitive procedures for diagnosis and clinical monitoring of the disease need particular thought in order to demonstrate whether a neuroprotection strategy is possible. Topics include experimental models of glaucoma, clinical protocols for the study of neuroprotective therapies, mechanisms underlying glaucomatous damage to neurons of the retina and other "relay stations" in the visual pathway in the brain, and, last but not least, future insights into ways of preventing neuronal injury in glaucoma patients.

Carlo Nucci, M.D., Ph.D.
Associate Professor of Ophthalmology
Department of Biopathology and Diagnostic Imaging
University of Rome Tor Vergata,
Rome, Italy

Neville N. Osborne B.Sc., Ph.D., M.A., D.Sc.
Professor of Ocular Neurobiology
Nuffield Laboratory of Ophthalmology,
University of Oxford, Oxford U.K.

Giacinto Bagetta, M.D.
Professor of Pharmacology
Department of Pharmacobiology
University of Calabria
Arcavacata di Rende (CS), Italy

Luciano Cerulli, M.D.
Professor of Ophthalmology
Department of Biopathology and Diagnostic Imaging
University of Rome Tor Vergata, Rome, Italy

Contents

List of contributors v

Preface xi

I. Epidemiology and Clinical Assessment of the Disease

1. Epidemiology of primary glaucoma: prevalence, incidence, and blinding effects
   C. Cedrone, R. Mancino, A. Cerulli, M. Cesareo and C. Nucci (Rome, Italy) 3

2. Predictive models to estimate the risk of glaucoma development and progression
   F.A. Medeiros and R.N. Weinreb (La Jolla, CA, USA) 15

3. Intraocular pressure and central corneal thickness
   G. Manni, F. Oddone, V. Parisi, A. Tosto and M. Centofanti (Rome, Italy) 25

4. Angle-closure: risk factors, diagnosis and treatment
   N. Amerasinghe and T. Aung (Singapore) 31

5. Early diagnosis in glaucoma
   D.F. Garway-Heath (London, UK and Rome, Italy) 47

6. Monitoring glaucoma progression
   P. Brusini (Udine, Italy) 59

II. Anatomical and Functional Monitoring

7. Standard automated perimetry and algorithms for monitoring glaucoma progression
   G.L. Scuderi, M. Cesareo, A. Perdicchi and S.M. Recupero (Rome, Italy) 77

8. Short-wavelength automated perimetry and frequency-doubling technology
   perimetry in glaucoma
   P. Fogagnolo, L. Rossetti, S. Ranno, A. Ferreras and N. Orzalesi
   (Rome and Milan, Italy and Zaragoza, Spain) 101

9. Scanning laser polarimetry and confocal scanning laser ophthalmoscopy: technical notes
   on their use in glaucoma
   F. Ferreri, P. Aragona and G. Ferreri (Messina, Italy) 125

10. The role of OCT in glaucoma management
    M.M. Pagliara, D. Lepore and E. Balestrazzi (Rome, Italy) 139

11. Functional laser Doppler flowmetry of the optic nerve: physiological aspects and
    clinical applications
    C.E. Riva and B. Falsini (Bologna and Rome, Italy) 149

12. Advances in neuroimaging of the visual pathways and their use in glaucoma
    F.G. Garaci, V. Cozzolino, C. Nucci, F. Gaudiello, A. Ludovici,
    T. Lupattelli, R. Floris and G. Simonetti (Rome and Milan, Italy) 165

III. Current Therapy

13. Primary open angle glaucoma: an overview on medical therapy
    M. Vetrugno, F. Cantatore, G. Ruggeri, P. Ferreri, A. Montepara,
    A. Quinto and C. Sborgia (Bari, Italy) 181

14. The treatment of normal-tension glaucoma
    P.V. Desai and J. Caprioli (Los Angeles, CA, USA) 195

15. The management of exfoliative glaucoma
    R. Ritch (New York, NY, USA) 211

16. Laser therapies for glaucoma: new frontiers
    G.L. Scuderi and N. Pasquale (Rome, Italy) 225

17. Modulation of wound healing during and after glaucoma surgery
    S. Georgoulas, A. Dahlmann-Noor, S. Brocchini and P.T. Khaw (London, UK) 237

18. Surgical alternative to trabeculectomy
    R.G. Carassa (Milan, Italy) 255

19. Modern aqueous shunt implantation: future challenges
    K. Barton and D.K. Heuer (London, UK and Milwaukee, WI, USA) 263

IV. Experimental Approaches to Model Disease

20. Model systems for experimental studies: retinal ganglion cells in culture
    E. Goodyear and L.A. Levin (Montreal, QC, Canada and Madison, USA) 279

21. Rat models for glaucoma research
    J.C. Morrison, E. Johnson and W.O. Cepurna (Portland, OR, USA) 285

22. Mouse genetic models: an ideal system for understanding
    glaucomatous neurodegeneration and neuroprotection
    G.R. Howell, R.T. Libby and S.W.M. John
    (Bar Harbor, ME, Rochester, NY and Boston, MA, USA) 303

23. Clinical trials in neuroprotection
    S.M. Whitcup (Irvine and Los Angeles, CA, USA) 323

V. Neuroprotection: New Vistas in Pathophysiology

24. Pathogenesis of ganglion "cell death" in glaucoma and neuroprotection:
    focus on ganglion cell axonal mitochondria
    N.N. Osborne (Oxford, UK) 339

25. Astrocytes in glaucomatous optic neuropathy
    M.R. Hernandez, H. Miao and T. Lukas (Chicago, IL, USA) 353

26. Glaucoma as a neuropathy amenable to neuroprotection and immune manipulation
    M. Schwartz and A. London (Rehovot, Israel) 375

27. Oxidative stress and glaucoma: injury in the anterior segment of the eye
    S.C. Saccà and A. Izzotti (Genoa, Italy) 385

28. TNF-α signaling in glaucomatous neurodegeneration
    G. Tezel (Louisville, KY, USA) 409

29. Involvement of the *Bcl2* gene family in the signaling and control of retinal
    ganglion cell death
    R.W. Nickells, S.J. Semaan and C.L. Schlamp (Madison, WI, USA) 423

30. Assessment of neuroprotection in the retina with DARC
    L. Guo and M.F. Cordeiro (London, UK) 437

31. Potential roles of (endo)cannabinoids in the treatment of glaucoma: from intraocular
    pressure control to neuroprotection
    C. Nucci, M. Bari, A. Spanò, M.T. Corasaniti, G. Bagetta, M. Maccarrone and
    L.A. Morrone (Rome, Catanzaro, Arcavacata di Rende and Teramo, Italy 451

32. Glaucoma of the brain: a disease model for the study of transsynaptic
    neural degeneration
    Y. Yücel and N. Gupta (Toronto, ON, Canada) 465

33. Changes of central visual receptive fields in experimental glaucoma
    S.C. Sharma (Valhalla, NY, USA) 479

VI. Neuroprotection: Evidence for Future Strategies

34. Targeting excitotoxic/free radical signaling pathways for therapeutic intervention in glaucoma
    M. Seki and S.A. Lipton (La Jolla, CA, USA) 495

35. Stem cells for neuroprotection in glaucoma
    N.D. Bull, T.V. Johnson and K.R. Martin (Cambridge, UK) 511

36. The relationship between neurotrophic factors and CaMKII in the
    death and survival of retinal ganglion cells
    N.G.F. Cooper, A. Laabich, W. Fan and X. Wang (Louisville, KY and
    Bothell, WA, USA) 521

37. Evidence of the neuroprotective role of citicoline in glaucoma patients
    V. Parisi, G. Coppola, M. Centofanti, F. Oddone, A.M. Angrisani, L. Ziccardi,
    B. Ricci, L. Quaranta and G. Manni (Rome and Brescia, Italy) 541

38. Neuroprotection: VEGF, IL-6, and clusterin: the dark side of the moon
    S. Pucci, P. Mazzarelli, F. Missiroli, F. Regine and F. Ricci (Rome, Italy) 555

39. Rational basis for the development of coenzyme Q10 as a neurotherapeutic
    agent for retinal protection
    R. Russo, F. Cavaliere, L. Rombolá, M. Gliozzi, A. Cerulli, C. Nucci,
    E. Fazzi, G. Bagetta, M.T. Corasaniti and L.A. Morrone
    (Arcavacata di Rende, Catanzaro, Rome and Pavia, Italy). 575

40. 17β-Estradiol prevents retinal ganglion cell loss induced by acute rise of
    intraocular pressure in rat
    R. Russo, F. Cavaliere, C. Watanabe, C. Nucci, G. Bagetta,
    M.T. Corasaniti, S. Sakurada and L.A. Morrone
    (Arcavacata di Rende, Rome, Catanzaro, Italy and Sendai, Japan). 583

Subject Index 591

See Color Plate Section at the end of this book

SECTION I

Epidemiology and Clinical Assessment of the Disease

C. Nucci et al. (Eds.)
Progress in Brain Research, Vol. 173
ISSN 0079-6123
Copyright © 2008 Elsevier B.V. All rights reserved

CHAPTER 1

Epidemiology of primary glaucoma: prevalence, incidence, and blinding effects

Claudio Cedrone*, Raffaele Mancino, Angelica Cerulli, Massimo Cesareo and Carlo Nucci

Physiopathological Optics, Department of Biopathology and Diagnostic Imaging, University of Rome "Tor Vergata", Rome, Italy

Abstract: Certain general conclusions can be drawn from a series of 56 studies on glaucoma prevalence. Even in the most recently published studies the rate of undiagnosed glaucoma is particularly high. Another fairly constant finding is the discrepancy between the clinical and epidemiologic diagnoses of glaucoma. The prevalence of primary open-angle glaucoma (POAG) has been increasing, and this trend is undoubtedly due at least in part to advances in diagnostic technology. The decreasing prevalence of primary angle-closure glaucoma (PACG) is due to the adoption of more stringent criteria for the diagnosis of this form of glaucoma. Prevalence increases proportionately with age for each racial group. African or African origin populations had the highest POAG prevalence at all ages but the increase in prevalence of POAG is steeper for white populations. PACG is commonest in Asian ethnic groups, with the exception of the Japanese. Low-tension glaucoma (LTG) is quite common in the Japanese population. Over 80% of those with PACG live in Asia, while POAG disproportionately affects those of African derivation. Women are more affected by glaucoma. Very few incidence studies have been completed, because the cost of examining large samples is high. There are only two recent studies conducted on persons of African descent in Barbados (West Indies) and on white inhabitants of Rotterdam (Netherlands). Risk of incident glaucoma was highest among persons classified as having suspect POAG at baseline, followed by those with ocular hypertension. No difference in incidence of POAG between men and women was found. The more recent studies which included routine visual-field testing reveal rates of blinding glaucoma <10% in many countries, including those that are developing.

Keywords: glaucoma; prevalence; incidence; blindness

Introduction

The main objectives of an epidemiological study are to describe the frequency of a disease in a given population and to identify risk factors that may be associated with the disease. As with many other diseases, the biggest methodological problem encountered in studies that aim to investigate the prevalence and incidence of glaucoma is the definition of the disease.

Prevalence of glaucoma

Since the 1920s, numerous studies have been conducted to determine the prevalence of glaucoma. In

*Corresponding author. Tel.: +39 0672596144; Fax: +39 062026232; E-mail: cedrone@uniroma2.it

DOI: 10.1016/S0079-6123(08)01101-1

many cases, however, the participants were simply persons with high intraocular pressure (IOP) because in those days *glaucoma* was regarded as synonymous with *ocular hypertension* (HIOP). After the introduction of gonioscopy (Barkan, 1938), two types of glaucoma were distinguished: primary open-angle glaucoma (POAG) and primary angle-closure glaucoma (PACG). Most of the prevalence studies cited above focused on POAG, which — according to the initial findings — affected 2% of individuals over 40 years of age. This figure reflected the number of subjects with HIOP (i.e., IOP > 20 mmHg, a figure that represents the statistical mean of values observed in numerous population studies increased by a sum equal to twice the standard deviation of the mean), which is not always associated with optic neuropathy, a hallmark of glaucoma. It failed to include persons who did have glaucomatous changes in the optic disc and in the visual fields despite the fact that their IOP was statistically normal.

Since 1972, most of the figures cited above have been considered inaccurate representations of the prevalence of POAG because the studies that generated them failed to respect certain methodological requisites (Khan, 1972): (1) the study population must be well defined, and no subgroup should be systematically excluded from the investigation; (2) the criteria used to define persons affected by the disease must be clearly described; (3) the number of subjects in the population who are eligible for enrollment must be specified; and (4) the participation rate, i.e., the percentage of subjects actually examined must be reported (if possible, for each sex and age group).

The first study that was unanimously acknowledged to be well designed and appropriately conducted was the 1963 Ferndale Glaucoma Survey (Hollows and Graham, 1966), which examined 92% of the eligible population of a town in South Wales. POAG was diagnosed in subjects who presented an IOP of > 20 mmHg, cupping of the optic disc, and visual-field defects. This study introduced the concept of routine visual-field testing with the Friedman analyzer (even though this testing was done on only one out of three patients). This approach led to the identification of certain cases of low-tension glaucoma (LTG), half of which were subsequently reclassified as POAG based on IOPs observed at later visits. In addition, the study revealed that the majority of cases of glaucoma are primary and of the open-angle type.

So far, 55 other population-based surveys have been conducted in the populations of an entire town, village, region, or nation or in those selected with a clearly defined random or clustered sampling procedure. The racial or ethnic groups examined in these studies can be broadly grouped into the following categories: Asian (Alsbirk, 1973; Arkell et al., 1987; Hu, 1989; Shiose et al., 1991; Rauf et al., 1994; Foster et al., 1996, 2000; Jacob et al., 1998; Dandona et al., 2000a, b; Metheetrairut et al., 2002; Bourne et al., 2003; Ramakrishnan et al., 2003; Iwase et al., 2004; Rahman et al., 2004; Yamamoto et al., 2005; Raychaudhuri et al., 2005; Vijaya et al., 2005, 2006, 2008a, b; He et al., 2006; Casson et al., 2007); Black (Wallace and Lovell, 1969; Mason et al., 1989; Tielsch et al., 1991; Wormald et al., 1994; Leske et al., 1994; Buhrmann et al., 2000; Rotchford and Johnson, 2002; Ekwerekwu and Umeh, 2002; Rotchford et al., 2003; Ntim-Amponsah et al., 2004; Friedman et al., 2006); Hispanic (Quigley et al., 2001; Anton et al., 2004; Varma et al., 2004); Mixed (Salmon et al., 1993; Sakata et al., 2007); White (Hollows and Graham, 1966; Bankes et al., 1968; Leibowitz et al., 1980; Bengtsson, 1981; Martinez et al., 1982; Gibson et al., 1985; Tielsch et al., 1991; Ringvold et al., 1991; Klein et al., 1992; Coffey et al., 1993; Dielemans et al., 1994; Leske et al., 1994; Hirvela et al., 1994; Giuffré et al., 1995; Ekstrom, 1996; Mitchell et al., 1996; Cedrone et al., 1997; Bonomi et al., 1998; Reidy et al., 1998; Wensor et al., 1998; Kozobolis et al., 2000; Friedman et al., 2006; Sakata et al., 2007; Topouzis et al., 2007). Prevalence rates of POAG and PACG are reported in Tables 1–3.

These studies varied widely in terms of the eye examination methods and case definitions used, in particular the criteria adopted for defining glaucomatous nerve damage. With the advent of automated perimetry, the definitions of glaucoma used in epidemiologic studies improved. With manual kinetic field testing, the outcome of the examination and the reliability of findings were

Table 1. Glaucoma prevalence studies in Asian racial group

Author	Location	Country	Age	% POAG	% PACG
Alsbirk (1973)	Umanaq	Greenland	40+	0.3	4.8
Arkell et al. (1987)	Alaska	USA	40+	0.2	2.7
Hu (1989)	Shunyi, Beijing	China	40+	0.03	1.4
Shiose et al. (1991)	Whole country	Japan	40+	2.6	0.3
Rauf et al. (1994)	Southall, London	UK	30+	2.7	0.0
Foster et al. (1996)	Hovsgol	Mongolia	40+	0.5	1.5
Jacob et al. (1998)	Vellore,	India	30–60	0.4	4.3
Foster et al. (2000)	Tanjong Pagar District	Singapore	40+	1.8	1.1
Dandona et al. (2000a, b)	Andhra Pradesh	India	40+	2.6	1.1
Metheetrairut et al. (2002)	Bangkok	Thailand	60+	3.5	2.5
Bourne et al. (2003)	Rom Klao, Bangkok	Thailand	50–70	2.3	0.9
Ramakrishnan et al. (2003)	Aravind,	India	40+	1.2	0.5
Iwase et al. (2004)	Tajimi	Japan	40+	3.9	ns
Rahman et al. (2004)	Dhaka	Bangladesh	35+	1.9	0.3
Yamamoto et al. (2005)	Tajimi	Japan	40+	ns	0.6
Raychaudhuri et al. (2005)	West Bengala	India	50+	3.0	0.2
Vijaya et al. (2005, 2006)	Tamil Nadu region	India	40+	1.6	0.9
He et al. (2006)	Guangzhou city	China	50+	2.1	1.5
Casson et al. (2007)	Meiktila district	Myanmar	40+	2.0	2.5
Vijaya et al. (2008a, b)	Chennai city	India	40+	3.5	0.9

Note: POAG, primary open angle glaucoma; PACG, primary angle closure glaucoma; ns, not stated.

Table 2. Glaucoma prevalence studies in Black, Hispanic, and Mixed racial group

Author	Location	Country	Age	% POAG	% PACG
Black:					
Wallace and Lovell (1969)	Jamaica	West Indies	35+	1.4	0.35
Mason et al. (1989)	St. Lucia	West Indies	30+	8.8	0.0
Tielsch et al. (1991)	Baltimore	USA	40+	4.7	ns
Wormald et al. (1994)	London	UK	35+	3.7	ns
Leske et al. (1994)	Barbados	West Indies	40+	7.0	ns
Buhrmann et al. (2000)	Kongwa	Tanzania	40+	3.1	0.6
Rotchford and Johnson (2002)	Kwazulu-Natal	South Africa	40+	2.7	0.1
Ekwerekwu and Umeh (2002)	Alum-Inyi	Nigeria	30+	2.1	ns
Rotchford et al. (2003)	Temba	South Africa	40+	2.9	0.5
Ntim-Amponsah et al. (2004)	Akwapim-South district	Ghana	30+	7.7	ns
Friedman et al. (2006)	Salisbury	USA	73+	20.0	ns
Hispanic:					
Quigley et al. (2001)	Nogales and Tucson	USA	40+	2.0	0.1
Anton et al. (2004)	Segovia	Spain	40–79	2.1	ns
Varma et al. (2004)	Los Angeles	USA	40+	4.7	ns
Mixed:					
Salmon et al. (1993)	Mamre	South Africa	40+	1.5	2.3
Sakata et al. (2007)	Piraquara City	Brazil	40+	3.8	0.8

Note: POAG, primary open angle glaucoma; PACG, primary angle closure glaucoma; ns, not stated.

Table 3. Glaucoma prevalence studies in White racial group

Author	Location	Country	Age	% POAG	% PACG
Hollows and Graham (1966)	Ferndale	Wales	40+	0.4	0.1
Bankes et al. (1968)	Bedford	UK	40+	0.8	0.2
Leibowitz et al. (1980)	Framingham	USA	52+	1.6	ns
Bengtsson (1981)	Dalby	Sweden	55–69	0.9	0.0
Martinez et al. (1982)	Gisborne	New Zealand	65–75	3.6	ns
Gibson et al. (1985)	Melton Mowbray	UK	76+	6.6	ns
Tielsch et al. (1991)	Baltimore	USA	40+	1.3	ns
Ringvold et al. (1991)	middle-Norway	Norway	65+	4.0	ns
Klein et al. (1992)	Beaver Dam	USA	43+	2.1	0.1
Coffey et al. (1993)	Roscommon	Ireland	50+	1.9	0.1
Dielemans et al. (1994)	Rotterdam	The Netherlands	55+	1.1	0.0
Leske et al. (1994)	Barbados	West Indies	40+	0.8	ns
Giuffré et al. (1995)	Casteldaccia	Italy	40+	1.2	ns
Hirvela et al. (1994)	Oulu	Finland	70+	3.0	ns
Ekstrom (1996)	Tierp	Sweden	65–74	3.8	ns
Mitchell et al. (1996)	Blue Mountains	Australia	50+	3.0	0.3
Cedrone et al. (1997)	Ponza	Italy	40+	2.5	1.0
Bonomi et al. (1998)	Egna-Neumarkt	Italy	40+	2.1	0.6
Reidy et al. (1998)	North London	UK	65+	3.0	ns
Wensor et al. (1998)	Melbourne	Australia	40+	1.7	0.1
Kozobolis et al. (2000)	Crete	Greece	40+	2.8	ns
Friedman et al. (2006)	Salisbury	USA	73+	8.5	ns
Sakata et al. (2007)	Piraquara City	Brazil	40+	2.1	0.8
Topouzis et al. (2007)	Thessaloniki city	Greece	60+	3.8	ns

Note: POAG, primary open angle glaucoma; PACG, primary angle closure glaucoma; ns, not stated.

dependent on the psychological/physical conditions of the patient and of the examiner. The automated examination allowed objective evaluation of the reliability of the findings, which are generally rated as good when false positive or false negative rates are ≤33% and loss of fixation rates are <20%. Furthermore, it was more sensitive and specific than manual perimetry and thus capable of detecting early visual-field changes that were missed with older methods.

The Dalby Study (1977–1978) deserves mention because it was the first study in which all participants underwent visual-field testing with the automated perimeter developed and validated by Heiji and Krakau, which is the prototype of the Humphrey perimeter (Bengtsson, 1981). As a result, all cases of LTG in the population examined in this study should have been diagnosed. Despite its improved reliability, visual-field testing with an automated perimeter had its drawbacks. For one thing, it prolonged the eye examination by approximately 30 min. Therefore, in many of the less recent studies, visual fields were not tested in all participants because it represented a true "bottle-neck" in the examination scheme. With the recently introduced program based on the Swedish Interactive Threshold Algorithms (SITA), examination times are considerably shorter than those associated with the program based on the Full Threshold Algorithm. In the assessment of the central 30° of the visual field, the SITA Standard and SITA Fast programs reduce test times by 50% and 66%, respectively. More recently, the introduction of perimetry based on the frequency doubling technology (FDT) has allowed increasingly rapid detection of even earlier changes in the visual field (less than 2 min per eye).

These new methods for visual-field testing have drastically reduced the duration of eye examinations in prevalence studies. This factor is particularly important in research settings because the reliability of prevalence figures is inversely proportional to the time required for examination of the total sample. The prevalence of a disease is calculated by dividing the number of cases ascertained at a given time by the number of subjects in the population at that time. If case ascertainment is significantly prolonged — e.g., 2 years — the onset of some of the cases identified near the end of the study may have occurred long after the beginning of the study. In theory, with respect to the initiation of the study, these would represent *incident* rather than *prevalent* cases. Furthermore, by the end of the 2-year ascertainment period, glaucoma might have developed in some subjects who were examined in the early phases of the study and found to be healthy.

Visual-field testing of only some of the participants in a study can obviously lead to underestimations of the prevalence of glaucoma (Mason et al., 1989). However, in a substantial number of subjects who do not have glaucoma, the results of field tests meet the criteria for abnormality. Therefore, use of these results without an examination of the optic disc can lead to overestimation of glaucoma prevalence. A thorough examination of the optic disc is a fundamental part of the three-tiered system of evidence (Foster et al., 2002), which was recently used to categorize glaucoma in almost all recent population-based prevalence surveys (Bourne et al., 2003; Rotchford et al., 2003; Iwase et al., 2004; Rahman et al., 2004; Raychaudhuri et al., 2005; Yamamoto et al., 2005; Vijaya et al., 2005, 2006, 2008a, b; He et al., 2006; Casson et al., 2007; Topouzis et al., 2007; Sakata et al., 2007). The prototype system was discussed by the ISGEO Glaucoma classification working group at the congress of the International Society for Geographical and Epidemiological Ophthalmology held in Leeuwenhorst, the Netherlands, in June 1998.

The rationale underlying this new classification system is that, although the level of IOP is one of the most consistent risk factors for the presence of glaucoma, the concept that statistically raised IOP is a defining characteristic for glaucoma has been almost universally discarded. This is based on several population-based studies that document the typical disc and field damage of glaucoma in people with a statistically normal IOP and, conversely, people with statistically elevated IOP and no evidence of optic neuropathy (Foster et al., 2002). The authors proposed to follow this current convention except for category 3 diagnosis, as detailed below. The diagnosis of glaucoma is made according to the following three categories:

Category 1 (structural and functional evidence): Eyes with a cup-to-disc ratio (CDR) or CDR asymmetry >97.5th percentile for the normal population, or a neuroretinal rim width reduced to <0.1 CDR (between 11 and 1 o'clock or 5 and 7 o'clock) that also showed a definite visual-field defect consistent with glaucoma.

Category 2 (advanced structural damage with unproved field loss): If the subject could not satisfactorily complete visual-field testing but had a CDR or CDR asymmetry >99.5th percentile for the normal population, glaucoma was diagnosed solely on the structural evidence. In diagnosing category 1 or 2 glaucoma, there should be no alternative explanation for CDR findings (dysplastic disc or marked anisometropia) or the visual-field defect (retinal vascular disease, macular degeneration, or cerebrovascular disease).

Category 3 (optic disc not seen. Field test impossible): If it is not possible to examine the optic disc, glaucoma is diagnosed if: (A) The visual acuity is <3/60 and the IOP >99.5th percentile, or (B) The visual acuity is <3/60 and the eye shows evidence of glaucoma filtering surgery, or medical records were available confirming glaucomatous visual morbidity.

POAG is therefore optic nerve damage meeting any of the three categories of evidence below, in an eye which does not have evidence of angle closure on gonioscopy, and where there is no identifiable secondary cause. PACG is optic nerve damage meeting any of the three categories of evidence above, in an eye which has evidence of angle closure on gonioscopy, according to the following classification of primary angle closure (PAC) discussed below.

PAC suspect

An eye in which appositional contact between the peripheral iris and posterior trabecular meshwork is considered possible.

PAC

An eye with an occludable drainage angle and features indicating that trabecular obstruction by the peripheral iris has occurred, such as peripheral anterior synechiae, elevated IOP, iris whorling (distortion of the radially orientated iris fibers), "glaucomfleken" lens opacities, or excessive pigment deposition on the trabecular surface. The optic disc does not have glaucomatous damage.

PACG

PAC together with evidence of glaucoma, as defined above.

Certain general conclusions can be drawn from the studies on glaucoma prevalence shown in Tables 1–3.

A constant finding in these prevalence studies is that most cases of glaucoma had not been previously diagnosed. This was first noted in the Ferndale study, where two thirds of the POAGs and three quarters of the PACGs were undiagnosed (Hollows and Graham, 1966). It is interesting to note that even in the most recently published studies the rate of undiagnosed glaucoma is particularly high: 33% in the United States (Friedman et al., 2006), 57% in Greece (Topouzis et al., 2008), 70% in Spain (Varma et al., 2004), 74% in China (He et al., 2006), 90% in Brazil and in South Africa (Sakata et al., 2007; Rotchford and Johnson, 2002), and from 91% to 98.5% in India (Ramakrishnan et al., 2003; Vijaya et al., 2005).

Another fairly constant finding is the discrepancy between the clinical and epidemiologic diagnoses of glaucoma, which emerged for the first time in the Framingham Eye Study (Leibowitz et al., 1980). In this study, visual-field defects were actually found in less than one fourth of the patients who reported that they had been diagnosed with or were receiving pharmacological treatment for glaucoma. Later, in the Dalby Study, not even one of the 16 patients diagnosed with glaucoma before enrollment had visual-field defects: all had HIOP that was being treated "to prevent or anticipate glaucoma" and all of the cases of glaucoma were discovered only after enrollment (Bengtsson, 1981). This was an interesting public health issue that has unfortunately not been investigated in more recent studies.

In studies in which routine visual-field testing was not performed and IOP was used to define glaucoma, the prevalence of the disease was lower than that reported when routine field testing was used and POAG was not defined by IOP alone.

Methods of examination have changed markedly since the 1960s up to now. Variations in eye examination methods, case definitions, and the prevalence of primary glaucoma complicate the comparison of individual studies, but the most recent ones (see above) are easier to compare because they are abased on the three-tiered system of evidence (Foster et al., 2002). The prevalence of POAG has been increasing, and this trend is undoubtedly due at least in part to advances in diagnostic technology, which have led to earlier and more reliable identification of cases. It is also important to recall that many of the cases assigned to categories 2 and 3 of the classification developed by Foster et al. (2002) were previously described as "probable," "suspected," or "doubtful" before this classification system was introduced. However, it is impossible to exclude the possibility that these increases are due in part to increased life expectancies.

The decreasing prevalence of PACG is due to the adoption of more stringent criteria for the diagnosis of this form of glaucoma. In the past, the presence of a narrow drainage angle with an elevated IOP or peripheral anterior synechiae was sufficient for a diagnosis of PACG and the presence of glaucomatous field defects or optic disc changes were not considered mandatory for the diagnosis of angle-closure glaucoma (Jacob et al., 1998).

Prevalence increases proportionately with age for each racial group (i.e., prevalence increases exponentially with age); the average estimated prevalence in those older than 70 years of age was 6% in white populations, 16% in black populations, and 3% in Asian populations. African or African origin populations had the highest POAG prevalence at all ages but the increase in prevalence of POAG is steeper for white populations; increases with age in black and Asian populations are similar (Rudnicka et al., 2006).

PACG is commonest in Asian ethnic groups, with the exception of the Japanese. Only 8% of all cases of glaucoma occur in this population (Yamamoto et al., 2005), whereas LTG is quite common in the Japanese population and accounts for 92% of all cases of POAG (Iwase et al., 2004).

Over 80% of those with PACG live in Asia, while POAG disproportionately affects those of African derivation. Women are more affected by glaucoma because of their greater prevalence of PACG, as well as their relatively greater longevity (Quigley and Broman, 2006).

In 2010, there will be 60.5 million people with glaucoma, and this rate is expected to increase to 79.6 million in 2020. Seventy-four percent of these affected individuals will have POAG. Fifty-five percent of POAG cases, 70% of PACG cases, and 59% of all glaucoma cases in 2010 will occur in women. Asians will represent 47% of those with glaucoma and 87% of those with PACG (Quigley and Broman, 2006).

The costs of treating glaucoma are considerable: in 1988 the United States healthcare system spends an estimated $2.5 billion a year for this disease ($1.9 billion in direct costs and $0.6 billion in indirect costs) (Glick et al., 1994). A retrospective analysis conducted in 1988 (Kobelt-Nguyen et al., 1998) estimated that the annual cost of treating newly diagnosed open-angle glaucoma was $1055. In Europe, the direct cost of treatment increased by an estimated 86 euros for each incremental step ranging from 455 euro per person year for stage 0 (HIOP only) to 969 euro per person year for stage 4 (severe glaucoma) disease; medication costs ranged from 42% to 56% of total direct cost for all stages of disease (Traverso et al., 2005). If we multiply the number of patients with glaucoma in Europe estimated by Quigley and Broman (2006) — 12 million — by a mean annual cost of 750 euros (for moderate stage 2 glaucoma) we find that the total cost for 2010 is substantial: 9 billion euros.

Incidence of glaucoma

Incidence rates reflect the rate at which a disease develops in a population within a defined time period. Incidence rates for glaucoma reveal more about the natural history and current causes of this disease, and this information is essential if effective corrective measures are to be adopted. Very few incidence studies have been completed, because the cost of examining large samples is high.

The first study on the incidence of glaucoma was conducted in 1966 on inhabitants of Skowde, Sweden, who were 55–64 years old at baseline and had an IOP < 21 mmHg. Over the next 5 years not one new case was observed (Linnir and Stromberg, 1966). However, this finding is not very reliable due to the low rate of participation in the follow-up study and the diagnostic criteria used. The first reliable data are those collected in Dalby, Sweden, by Bengtsson (1989). This study included an initial examination and two follow-up visits over the next 10 years, and the estimated incidence rate was 0.25% in the population aged 55–69 years. This was the first longitudinal study conducted in an acceptable manner with visual-field testing at baseline and at follow-up.

In 1983, given the unreliability of the available data on glaucoma incidence, Leske et al. (1981) and Podgor et al. (1983) decided to estimate the incidence of this disease based on the prevalence data collected in the Ferndale (Leske et al., 1981) and Framingham (Podgor et al., 1983) studies and on survival tables. They calculated 5-year POAG incidence rates of 0.20% at age 55 and 1.1% at age 75. These studies were later repeated in the United States (Quigley and Vitale, 1997) and in the United Kingdom (Minassian et al., 2000).

The medical charts of patients with glaucoma diagnosed between 1980 and 1982 and confirmed by Goldmann perimetry were analyzed to determine the incidence of glaucoma in the Swedish town of Halsingland (Lindblom and Thorburn, 1984). However, the data collected in this study merely reflect the frequency with which glaucoma was diagnosed in the population by the healthcare system. The mean annual frequency of POAG diagnosis reported by these authors ranged from 0.02% in 51–55-year-olds to 0.44% in those aged 91–95 years. The medical histories of the residents of Olmsted County, Minnesota were also used to calculate the incidence of POAG in the United States (Schoff et al., 2001).

However, more reliable data were collected in two recent studies conducted on persons of African descent in Barbados (West Indies) and on white inhabitants of Rotterdam (Netherlands) (Leske et al., 2001; de Voogd et al., 2005). The 4-year incidence of POAG in Barbados was 2.2%. Incidence rates increased from 1.2% at ages 40–49 years to 4.2% at ages of 70 years or more, tending to be higher in men than women (2.7% vs. 1.9%). Of the 67 incident cases of POAG, 52% had higher pressures at baseline. Risk of incident glaucoma was highest among persons classified as having suspect POAG at baseline (26.1%), followed by those with ocular hypertension (4.9%) and lowest in the remaining population (0.8%) (Leske et al., 2001). The 5-year risk of definite POAG was 0.6% among the Rotterdam study survivors aged 55 years and over at base line; the rate rose significantly from 1% at age 60 years to approximately 3% at age 80 years. No difference in incidence of POAG between men and women was found. HIOP at baseline or use of IOP-lowering treatment, gave a three times higher risk for incident POAG. Normal fellow eyes of POAG eyes had a fivefold higher risk for developing POAG compared with fellow eyes of normal eyes (de Voogd et al., 2005). In both studies, most of the incident cases (half of those observed in Barbados, two thirds of those in Rotterdam) had not been previously diagnosed and were not receiving treatment.

Blinding effects of glaucoma

The percentage of cases associated with binocular blindness has been reported only in the more recent studies on the prevalence of glaucoma, and in many of these studies the criteria used to define blindness were not specified; in others the definition of "legal blindness" was used (VA < 20/200), which is less restrictive than that recommended by the WHO (VA < 20/500).

According to a WHO report, in which correction factors were used to determine the prevalence of best-corrected VA in studies using non-WHO definitions, the estimated number of visually impaired people in 2002 was 161 million, and 37 million of these were blind (Resnikoff et al., 2004). With regard to correction factors to determine prevalence according to ICD-10 from different definitions of visual impairment, there were a sufficient number of studies reporting data with both definitions to enable a table of conversion to be calculated. Glaucoma accounts for 12.3% of the cases of best-corrected blindness in the world (4.5 million in the year 2002).

In contrast, in 2000, the number of people with primary glaucoma in the world is estimated to be nearly 66.8 million, with 6.7 million suffering from bilateral blindness (Quigley, 1996). A later study (Quigley and Broman, 2006) estimated that there will be 60.5 million people with primary glaucoma in 2010 and 79.6 million by 2020. Bilateral blindness will be present in 8.4 million people in 2010 and 11.2 million people in 2020. These authors concluded that "the two estimates differ because of methodological issues; blindness prevalence surveys often assign the most treatable disease as the primary cause of blindness. It is often assumed that cataract is more treatable than glaucoma. This leads to underestimation of glaucoma blindness."

It is possible that rates of blinding glaucoma are underestimated because this disease is not a major cause of central visual loss, which is the sole criterion used to define blindness in most of the studies. However, the more recent studies which included routine visual-field testing reveal rates of blinding glaucoma < 10% in many countries, including those that are developing (Table 4). No cases of bilateral blindness caused by glaucoma were found in India, China, and Brazil (Raychaudhuri et al., 2005; He et al., 2006; Sakata et al., 2007). The Los Angeles Latino Eye Study found a rate of 1% of legal blindness due to glaucoma (Varma et al., 2004). Other prevalence studies in India revealed rates ranging from 1.6% to

Table 4. Blinding rate of glaucoma found in the most recent studies

Author	Racial group	Country	Blinding rate (%)
Dandona et al. (2000a, b)	Asian	India	10.2[a,b]
Foster et al. (2000)	Asian	Singapore	7.0
Buhrmann et al. (2000)	Black	Tanzania	6.0
Rotchford and Johnson (2002)	Black	South Africa	11.8
Bourne et al. (2003)	Asian	Thailand	3.7
Ramakrishnan et al. (2003)	Asian	India	1.6
Varma et al. (2004)	Hispanic	U.S.A.	1.0[a]
Rahman et al. (2004)	Asian	Bangladesh	30.0[a]
Raychaudhuri et al. (2005)	Asian	India	0.0
Vijaya et al. (2005, 2006)	Asian	India	3.1
He et al. (2006)	Asian	China	0.0
Friedman et al. (2006)	Black	U.S.A.	5.3[a]
Friedman et al. (2006)	White	U.S.A.	1.3[a]
Sakata et al. (2007)	White & no white	Brazil	0.0
Casson et al. (2007)	Asian	Myanmar	10.9
Vijaya et al. (2008a, b)	Asian	India	2.4

[a]Visual Acuity < 20/200.
[b]Visual Field < 20° around central fixation.

3.1% (Ramakrishnan et al., 2003; Vijaya et al., 2005, 2006; Vijaya et al., 2008a, b). A blinding rate of 3.7% was found in Thailand (Bourne et al., 2003). In the Salisbury Eye Evaluation Glaucoma Study (Friedman et al., 2006), the rate of blindness in black patients with glaucoma was 5.3% (vs. 1.3% in whites), but this difference was not statistically significant. Blinding rate of 6–7% were found in Tanzania and Singapore (Buhrmann et al., 2000; Foster et al., 2000). The highest blinding rate were found only in India, South Africa, Bangladesh, and Myanmar but visual impairment cases (VA < 20/200, and/or VF < 20° around central fixation) were included in Bangladesh, and Myanmar studies (Dandona et al., 2000a, b; Rotchford and Johnson, 2002; Rahman et al., 2004; Casson et al., 2007).

Finally, changes between 1988 and 2000 in the prevalence and main causes of blindness and low vision in Ponza, Italy was investigated (Cedrone et al., 2007). The prevalence of all types of glaucoma among subjects aged 40 years and over in Ponza was quite high (1988: 3.6%; 2000: 4.8%), and this may be a reflection of inbreeding in this island community, but the rate of bilaterally blinding glaucoma (visual field < 10° around central fixation) dropped from 5.4% to 2.5%. It is interesting to note that the mean age of patients with glaucomatous impairment of at least in one eye also increased considerably (1988: 68.2 ± 10.7; 2000: 75.5 ± 6.3, $p = 0.030$), whereas the mean age of the total glaucoma subgroup remained essentially stable (1988: 66.8, 2000: 68.2, $p > 0.05$).

Abbreviations

CDR	cup-to-disc ratio
FDT	frequency doubling technology
HIOP	high intraocular pressure
IOP	intraocular pressure
LTG	low-tension glaucoma
PAC	primary angle closure
PACG	primary angle-closure glaucoma
POAG	primary open-angle glaucoma
SITA	Swedish Interactive Threshold Algorithms
VA	visual acuity
VF	visual field

Acknowledgment

The authors thank Marian Everett Kent, who received payment for her assistance in writing and editing the manuscript.

References

Alsbirk, P.H. (1973) Angle-closure glaucoma surveys in Greenland Eskimos: a preliminary report. Can. J. Ophthalmol., 8: 260–264.

Anton, A., Andrada, M.T., Mujica, V., Calle, M.A., Portela, J. and Mayo, A. (2004) Prevalence of primary open-angle glaucoma in a Spanish population: the Segovia Study. J. Glaucoma, 13: 371–376.

Arkell, S.M., Lightman, D.A., Sommer, A., Taylor, H.R., Korshin, O.M. and Tielsch, J.M. (1987) The prevalence of glaucoma among Eskimos of northwest Alaska. Arch. Ophthalmol., 105: 482–485.

Bankes, J.L., Perkins, E.S., Tsolakis, S. and Wright, J.E. (1968) Bedford glaucoma survey. Br. Med. J., 1: 791–796.

Barkan, O. (1938) Glaucoma: classification, causes and surgical control. Results of microgonioscopic research. Am. J. Ophthalmol., 21: 1099–1117.

Bengtsson, B. (1981) The prevalence of glaucoma. Br. J. Ophthalmol., 65: 46–49.

Bengtsson, B. (1989) Incidence of manifest glaucoma. Br. J. Ophthalmol., 73: 483–487.

Bonomi, L., Marchini, G., Marraffa, M., Bernardi, P., De Franco, I., Perfetti, S., Varotto, A. and Tenna, V. (1998) Prevalence of glaucoma and intraocular pressure distribution in a defined population: the Egna-Neumarkt Study. Ophthalmology, 105: 209–215.

Bourne, R.R., Sukudom, P., Foster, P.J., Tantisevi, V., Jitapunkul, S., Lee, P.S., Johnson, G.J. and Rojanapongpun, P. (2003) Prevalence of glaucoma in Thailand: a population based survey in Rom Klao District, Bangkok. Br. J. Ophthalmol., 87: 1069–1074.

Buhrmann, R.R., Quigley, H.A., Barron, Y., West, S.K., Oliva, M.S. and Mmbaga, B.B. (2000) Prevalence of glaucoma in a rural East African population. Invest. Ophthalmol. Vis. Sci., 41: 40–48.

Casson, R.J., Newland, H.S., Muecke, J., McGovern, S., Abraham, L., Shein, W.K., Selva, D. and Aung, T. (2007) Prevalence of glaucoma in rural Myanmar: the Meiktila Eye Study. Br. J. Ophthalmol., 91: 710–714.

Cedrone, C., Culasso, F., Cesareo, M., Zapelloni, A., Cedrone, P. and Cerulli, L. (1997) Prevalence of glaucoma in Ponza, Italy: a comparison with other studies. Ophthalmic Epidemiol., 4: 59–72.

Cedrone, C., Ricci, F., Nucci, C., Cesareo, M., Macri, G. and Culasso, F. (2007) Age-specific changes in the prevalence of best-corrected visual impairment in an Italian population. Ophthalmic Epidemiol., 14: 320–326.

Coffey, M., Reidy, A., Wormald, R., Xian, W.X., Wright, L. and Courtney, P. (1993) Prevalence of glaucoma in the west of Ireland. Br. J. Ophthalmol., 77: 17–21.

Dandona, L., Dandona, R., Mandal, P., Srinivas, M., John, R.K., McCarty, C.A. and Rao, G.N. (2000a) Angle-closure glaucoma in an urban population in southern India: the Andhra Pradesh eye disease study. Ophthalmology, 107: 1710–1716.

Dandona, L., Dandona, R., Srinivas, M., Mandal, P., John, R.K., McCarty, C.A. and Rao, G.N. (2000b) Open-angle glaucoma in an urban population in southern India: the Andhra Pradesh Eye Disease Study. Ophthalmology, 107: 1702–1709.

Dielemans, I., Vingerling, J.R., Wolfs, R.C., Hofman, A., Grobbee, D.E. and de Jong, P.T. (1994) The prevalence of primary open-angle glaucoma in a population-based study in The Netherlands: the Rotterdam Study. Ophthalmology, 101: 1851–1855.

Ekstrom, C. (1996) Prevalence of open-angle glaucoma in central Sweden: the Tierp Glaucoma Survey. Acta Ophthalmol. Scand., 74: 107–112.

Ekwerekwu, C.M. and Umeh, R.E. (2002) The prevalence of glaucoma in an onchoendemic community in South-Eastern Nigeria. West Afr. J. Med., 21: 200–203.

Foster, P.J., Baasanhu, J., Alsbirk, P.H., Munkhbayar, D., Uranchimeg, D. and Johnson, G.J. (1996) Glaucoma in Mongolia: a population-based survey in Hovsgol province, northern Mongolia. Arch. Ophthalmol., 114: 1235–1241.

Foster, P.J., Buhrmann, R., Quigley, H.A. and Johnson, G.J. (2002) The definition and classification of glaucoma in prevalence surveys. Br. J. Ophthalmol., 86: 238–242.

Foster, P.J., Oen, F.T.S., Machin, D., Ng, T.P., Devereux, J.G., Johnson, G.J., Khaw, P.T. and Seah, S.K. (2000) The prevalence of glaucoma in Chinese residents of Singapore — a cross-sectional population survey of the Tanjong Pagar District. Arch. Ophthalmol., 118: 1105–1111.

Friedman, D.S., Jampel, H.D., Muñoz, B. and West, S.K. (2006) The prevalence of open-angle glaucoma among blacks and whites 73 years and older: the Salisbury eye evaluation glaucoma study. Arch. Ophthalmol., 124: 1625–1630.

Gibson, J.M., Rosenthal, A.R. and Lavery, J. (1985) A study of the prevalence of eye disease in the elderly in an English community. Trans. Ophthalmol. Soc. UK, 104(pt 2): 196–203.

Giuffré, G., Giammanco, R., Dardanoni, G. and Ponte, F. (1995) Prevalence of glaucoma and distribution of intraocular pressure in a population: the Casteldaccia eye study. Acta Ophthalmol. Scand., 73: 222–225.

Glick, H., Brainsky, A. and McDonald, R.C. (1994) The cost of glaucoma in the United States in 1988. Chibret Int. J. Ophthalmol., 10: 6–12.

He, M., Foster, P.J., Ge, J., Huang, W., Zheng, Y., Friedman, D.S., Lee, P.S. and Khaw, P.T. (2006) Prevalence and clinical characteristics of glaucoma in adult Chinese: a population-based study in Liwan District, Guangzhou. Invest. Ophthalmol. Vis. Sci., 47: 2782–2788.

Hirvela, H., Tuulonen, A. and Laatikainen, L. (1994) Intraocular pressure and prevalence of glaucoma in elderly people in Finland: a population based study. Int. Ophthalmol., 18: 299–307.

Hollows, F.C. and Graham, P.A. (1966) Intra-ocular pressure, glaucoma, and glaucoma suspects in a defined population. Br. J. Ophthalmol., 50: 570–586.

Hu, C.N. (1989) An epidemiologic study of glaucoma in Shunyi County, Beijing. Zhonghua Yan Ke Za Zhi, 25: 115–119.

Iwase, A., Suzuki, Y., Araie, M., Yamamoto, T., Abe, H., Shirato, S., Kuwayama, Y., Mishima, H.K., Shimizu, H., Tomita, G., Inoue, Y. and Kitazawa, Y.The Tajimi Study Group, Japan Glaucoma Society. (2004) The prevalence of primary open-angle glaucoma in Japanese: the Tajimi study. Ophthalmology, 111: 1641–1648.

Jacob, A., Thomas, R., Koshi, S.P., Braganza, A. and Muliyil, J. (1998) Prevalence of primary glaucoma in an urban south Indian population. Indian J. Ophthalmol., 46: 81–86.

Khan, H.A. (1972) The prevalence of chronic simple glaucoma. Am. J. Ophthalmol., 74: 355–359.

Klein, B.E., Klein, R., Sponsel, W.E., Franke, T., Cantor, L.B., Martone, J. and Menage, M.J. (1992) Prevalence of glaucoma: the Beaver Dam eye study. Ophthalmology, 99: 1499–1504.

Kobelt-Nguyen, G., Gerdtham, U.G. and Alm, A. (1998) Costs of treating primary open angle glaucoma and ocular hypertension: a retrospective, observational two year chart review of newly diagnosed patients in Sweden and the United States. J. Glaucoma, 7: 95–104.

Kozobolis, V.P., Detorakis, E.T., Tsilimbaris, M., Siganos, D.S., Vlachonikolis, I.G. and Pallikaris, I.G. (2000) Crete, Greece glaucoma study. J. Glaucoma, 9: 143–149.

Leibowitz, H.M., Krueger, D.E., Maunder, L.R., Milton, R.C., Kini, M.M., Kahn, H.A., Nickerson, R.J., Pool, J., Colton, T.L., Ganley, J.P., Loewenstein, J.L. and Dawber, T.R. (1980) The Framingham eye study monograph: an ophthalmological and epidemiological study of cataract, glaucoma, diabetic retinopathy, macular degeneration, and visual acuity in a general population of 2631 adults, 1973–1975. Surv. Ophthalmol., 24(Suppl): 335–610.

Leske, M.C., Connell, A.M., Schachat, A.P. and Hyman, L. (1994) The Barbados eye study: prevalence of open angle glaucoma. Arch. Ophthalmol., 112: 821–829.

Leske, M.C., Connell, A.M.S., Wu, S.Y., Nemesure, B., Li, X., Schachat, A. and Hennis, A.for the Barbados Eye Studies Group. (2001) Incidence of open-angle glaucoma: the Barbados eye studies. Arch. Ophthalmol., 119: 89–95.

Leske, M.C., Ederer, F. and Podgor, M. (1981) Estimating incidence from age-specific prevalence in glaucoma. Am. J. Epidemiol., 113: 606–613.

Lindblom, B. and Thorburn, W. (1984) Observed incidence of glaucoma in Halsingland, Sweden. Acta Ophthalmol., 62: 217–222.

Linnir, E. and Stromberg, U. (1966) Ocular hypertension. A five years study of the total population in a Swedish town, Skovde. Glaucoma Symposium, Tuzing Castle, 187–214.

Martinez, G.S., Campbell, A.J., Reinken, J. and Allan, B.C. (1982) Prevalence of ocular disease in a population study of subjects 65 years old and older. Am. J. Ophthalmol., 94: 181–189.

Mason, R.P., Kosoko, O., Wilson, M.R., Martone, J.M., Cowan, C.L., Gear, C. and Ross-Degnan D. (1989). National survey of the prevalence and risk factors of glaucoma in St. Lucia, West Indies, I: prevalence findings. Ophthalmology, 96, 1363–1368.

Metheetrairut, A., Singalavanija, A., Ruangvaravate, N. and Tuchinda, R. (2002) Evaluation of screening tests and prevalence of glaucoma: integrated health research program for the Thai elderly. J. Med. Assoc. Thai., 85: 147–153.

Minassian, D.C., Reidy, A., Coffey, M. and Minassian, A. (2000) Utility of predictive equations for estimating the prevalence and incidence of primary open angle glaucoma in the UK. Br. J. Ophthalmol., 84: 1159–1161.

Mitchell, P., Smith, W., Attebo, K. and Healey, P.R. (1996) Prevalence of open angle glaucoma in Australia: the Blue Mountains Eye Study. Ophthalmology, 103: 1661–1669.

Ntim-Amponsah, C.T., Amoaku, W.M., Ofosu-Amaah, S., Ewusi, R.K., Idirisuriya-Khair, R., Nyatepe-Coo, E. and Adu-Darko, M. (2004) Prevalence of glaucoma in an African population. Eye, 18: 491–497.

Podgor, M., Leske, M.C. and Ederer, F. (1983) Incidence estimate for lens changes, macular changes, open-angle glaucoma and diabetic retinopathy. Am. J. Epidemiol., 118: 206–212.

Quigley, H.A. (1996) The number of persons with glaucoma worldwide. Br. J. Ophthalmol., 80: 389–393.

Quigley, H.A. and Broman, A.T. (2006) The number of people with glaucoma worldwide in 2010 and 2020. Br. J. Ophthalmol., 90: 262–267.

Quigley, H.A., West, S.K., Rodriguez, J., Munoz, B., Klein, R. and Snyder, R. (2001) The prevalence of glaucoma in a population-based study of Hispanic subjects: Proyecto VER. Arch. Ophthalmol., 119: 1819–1826.

Quigley, H.A. and Vitale, S. (1997) Models of open-angle glaucoma prevalence and incidence in the United States. Invest. Ophthalmol. Vis. Sci., 38: 83–91.

Rahman, M.M., Rahman, N., Foster, P.J., Haque, Z., Zaman, A.U., Dineen, B. and Johnson, G.J. (2004) The prevalence of glaucoma in Bangladesh: a population based survey in Dhaka division. Br. J. Ophthalmol., 88: 1493–1497.

Ramakrishnan, R., Nirmalan, P.K., Krishnadas, R., Thulasiraj, R.D., Tielsch, J.M., Katz, J., Friedman, D.S. and Robin, A.L. (2003) Glaucoma in a rural population of southern India: the Aravind Comprehensive eye survey. Ophthalmology, 110: 1484–1490.

Rauf, A., Ong, P.S., Pearson, R.V. and Wormald, R.P. (1994) A pilot study into the prevalence of ophthalmic disease in the Indian population of Southall. J. R. Soc. Med., 87: 78–79.

Raychaudhuri, A., Lahiri, S.K., Bandyopadhyay, M., Foster, P.J., Reeves, B.C. and Johnson, G.J. (2005) A population based survey of the prevalence and types of glaucoma in rural West Bengal: the West Bengal Glaucoma Study. Br. J. Ophthalmol., 89: 1559–1564.

Reidy, A., Minassian, D.C., Vafidis, G., Joseph, J., Farrow, S., Wu, J., Desai, P. and Connolly, A. (1998) Prevalence of serious eye disease and visual impairment in a north London population: population based, cross sectional study. Br. Med. J., 316: 1643–1646.

Resnikoff, S., Pascolini, D., Etya'ale, D., Kocur, I., Pararajasegaram, R., Pokharel, G.P. and Mariotti, S. (2004) 9 Global data on visual impairment in the year 2002. Bull. World Health Organ., 82: 844–851.

Ringvold, A., Blika, S., Elsas, T., Guldahl, J., Brevik, T., Hesstvedt, P., Hoff, K., Hoisen, H., Kjorsvik, S. and Rossvold, I. (1991) The middle-Norway eye-screening study, II: prevalence of simple and capsular glaucoma. Acta Ophthalmol. (Copenh.), 69: 273–280.

Rotchford, A.P. and Johnson, G.J. (2002) Glaucoma in Zulus: a population-based cross-sectional survey in a rural district in South Africa. Arch. Ophthalmol., 120: 471–478.

Rotchford, A.P., Kirwan, J.F., Muller, M.A., Johnson, G.J. and Roux, P. (2003) Temba glaucoma study: a population-based cross-sectional survey in urban South Africa. Ophthalmology, 110: 376–382.

Rudnicka, A.R., Mt-Isa, S., Owen, C.G., Cook, D.G. and Ashby, D. (2006) Variations in primary open-angle glaucoma prevalence by age, gender, and race: a Bayesian meta-analysis. Invest. Ophthalmol. Vis. Sci., 47: 4254–4261.

Sakata, K., Sakata, L.M., Sakata, V.M., Santini, C., Hopker, L.M., Bernardes, R., Yabumoto, C. and Moreira, A.T.R. (2007) Prevalence of glaucoma in a South Brazilian population: projeto glaucoma. Invest. Ophthalmol. Vis. Sci., 48: 4974–4979.

Salmon, J.F., Mermoud, A., Ivey, A., Swanevelder, S.A. and Hoffman, M. (1993) The prevalence of primary angle closure glaucoma and open angle glaucoma in Mamre, Western Cape, South Africa. Arch. Ophthalmol., 111: 1263–1269.

Schoff, E.O., Hattenhauer, M.G., Ing, H.H., Hodge, D.O., Kennedy, R.H., Herman, D.C. and Johnson, D.H. (2001) Estimated incidence of open-angle glaucoma in Olmsted county, Minnesota. Ophthalmology, 108: 882–886.

Shiose, Y., Kitazawa, Y., Tsukahara, S., Akamatsu, T., Mizokami, K., Futa, R., Katsushima, H. and Kosaki, H. (1991) Epidemiology of glaucoma in Japan — a nationwide glaucoma survey. Jpn. J. Ophthalmol., 35: 133–155.

Tielsch, J.M., Sommer, A., Katz, J., Royall, R.M., Quigley, H.A. and Javitt, J. (1991) Racial variations in the prevalence of primary open-angle glaucoma: the Baltimore Eye Survey. JAMA, 266: 369–374.

Topouzis, F., Coleman, A.L., Harris, A., Koskosas, A., Founti, P., Gong, G., Yu, F., Anastasopoulos, E., Pappas, T. and Wilson, M.R. (2008) Factors associated with undiagnosed open-angle glaucoma: the thessaloniki eye study. Am. J. Ophthalmol., 145: 327–335.

Topouzis, F., Wilson, M.R., Harris, A., Anastasopoulos, E., Yu, F., Mavroudis, L., Pappas, T., Koskosas, A. and Coleman, A.L. (2007) Prevalence of open-angle glaucoma in Greece: the Thessaloniki eye study. Am. J. Ophthalmol., 144: 511–519.

Traverso, C.E., Walt, J.G., Kelly, S.P., Hommer, A.H., Bron, A.M., Denis, P., Nordmann, J.-P., Renard, J.-P., Bayer, A., Grehn, F., Pfeiffer, N., Cedrone, C., Gandolfi, S., Orzalesi, N., Nucci, C., Rossetti, L., Azuara-Blanco, A., Hitchings, R., Salmon, L.R., Buchholz, P.M., Kotak, S.V., Katz, L.M., Siegartel, L.R. and Doyle, J.J. (2005) Direct costs of glaucoma and severity of the disease: a multinational long term study of resource utilisation in Europe. Br. J. Ophthalmol., 89: 1245–1249.

Varma, R., Ying-Lai, M., Francis, B.A., Nguyen, B.B., Deneen, J., Wilson, M.R. and Azen, S.P.Los Angeles Latino Eye Study Group. (2004) Prevalence of open-angle glaucoma and ocular hypertension in Latinos: the Los Angeles Latino eye study. Ophthalmology, 111: 1439–1448.

Vijaya, L., George, R., Arvind, H., Baskaran, M., Paul, P.G., Ramesh, S.V., Raju, P., Kumaramanickavel, G. and McCarty, C. (2006) Prevalence of angle-closure disease in a rural southern Indian population. Arch. Ophthalmol., 124: 403–409.

Vijaya, L., George, R., Arvind, H., Baskaran, M., Ramesh, V., Raju, P., Kumaramanickavel, G. and McCarty, C. (2008b) Prevalence of primary angle-closure disease in an Urban South Indian population and comparison with a rural population. The Chennai glaucoma study. Ophthalmology, 115: 655–660.

Vijaya, L., George, R., Baskaran, M., Arvind, H., Raju, P., Ramesh, V., Kumaramanickavel, G. and McCarty, C. (2008a) Prevalence of primary open-angle glaucoma in an Urban South Indian population and comparison with a rural population. The Chennai glaucoma study. Ophthalmology, 115: 648–654.

Vijaya, L., George, R., Paul, P.G., Baskaran, M., Arvind, H., Raju, P., Ramesh, S.V., Kumaramanickavel, G. and McCarty, C. (2005) Prevalence of open-angle glaucoma in a rural south Indian population. Invest. Ophthalmol. Vis. Sci., 46: 4461–4467.

de Voogd, S., Ikram, M.K., Wolfs, R.C., Jansonius, N.M., Hofman, A. and de Jong, P.T. (2005) Incidence of open-angle glaucoma in a general elderly population: the Rotterdam study. Ophthalmology, 112: 1487–1493.

Wallace, J. and Lovell, H.G. (1969) Glaucoma and intraocular pressure in Jamaica. Am. J. Ophthalmol., 67: 93–100.

Wensor, M.D., McCarty, C.A., Stanislavsky, Y.L., Livingston, P.M. and Taylor, H.R. (1998) The prevalence of glaucoma in the Melbourne visual impairment project. Ophthalmology, 105: 733–739.

Wormald, R.P., Basauri, E., Wright, L.A. and Evans, J.R. (1994) The African Caribbean eye survey: risk factors for glaucoma in a sample of African Caribbean people living in London. Eye, 8: 315–320.

Yamamoto, T., Iwase, A., Araie, M., Suzuki, Y., Abe, H., Shirato, S., Kuwayama, Y., Mishima, H.K., Shimizu, H., Tomita, G., Inoue, Y. and Kitazawa, Y.Tajimi Study Group, Japan Glaucoma Society. (2005) The Tajimi Study report 2: prevalence of primary angle closure and secondary glaucoma in a Japanese population. Ophthalmology, 112: 1661–1669.

CHAPTER 2

Predictive models to estimate the risk of glaucoma development and progression

Felipe A. Medeiros* and Robert N. Weinreb

Hamilton Glaucoma Center, University of California, San Diego, La Jolla, CA 92093-0946, USA

Abstract: In this chapter, we review the motivations and steps involved in the construction and validation of predictive models for the development and progression of glaucoma. We start with a critical review of the literature on the risk factors that have been identified as associated with development and progression of the disease. Subsequently, we review the steps necessary to build and validate a predictive model containing risk factors. We analyze the current models that have been proposed to assess risk of glaucoma development in patients with ocular hypertension and we discuss the potential for creating models to assess risk of progression in patients already diagnosed with the disease. Finally, we discuss some of the limitations of currently available models and the perspectives on this area.

Keywords: glaucoma; ocular hypertension; risk; predictive model

Making predictions is an essential part of health care. Prediction models have been continuously developed in several areas of medicine and their use has significantly contributed to the management of many disorders. The identification of risk factors for development or progression of disease is a fundamental component in the construction of prediction models. The concept of risk factor has been a part of the public lexicon for several decades, ever since the landmark Framingham Heart Study first reported in the early 1960s that cigarette smoking, elevated blood cholesterol, and high blood pressure were predictors of the likelihood of dying from heart disease. During the course of the study, other risk factors were identified and Framingham investigators started developing predictive models to evaluate the global risk of cardiovascular disease based on the summation of all major risk factors (D'Agostino et al., 2000, 2001; D'Agostino and Nam, 2004). Because longitudinal data were sparse, the initial models predicting risk levels due to cardiovascular risk factors relied heavily on statistical modeling. As newer data from this and other studies were made available, more robust mathematical models were developed and, eventually, simplified point systems were established to facilitate assessment of an individual's global risk of progression to an atherosclerotic cardiovascular event.

Risk assessment and prevention has contributed significantly to reduce mortality from cardiovascular disease. The successful implementation of risk assessment in cardiovascular medicine has stimulated its application to several other areas. Recently, the concept of risk assessment has also been applied to ophthalmology, more specifically, for assessment

*Corresponding author. Tel.: 858-8224592; Fax: 858-5341625;
E-mail: fmedeiros@eyecenter.ucsd.edu

of the risk of development and progression of glaucoma. The purpose of this chapter is to review the motivations and steps involved in the construction and validation of predictive models for the development and progression of glaucoma.

Risk assessment in ocular hypertension and glaucoma

Glaucoma is a neurodegenerative disease of the optic nerve that presents to the practitioner at various stages of a continuum and is characterized by accelerated retinal ganglion cell death, subsequent axonal loss and optic-nerve damage, and eventual visual-field loss (Fig. 1). These initial changes in the retina and optic nerve are often asymptomatic and undetectable with existing diagnostic tests. Also, there is no agreement on the criteria for the diagnosis of early damage that precedes standard achromatic visual-field loss. This suggests that awaiting overt signs of disease involves accepting some irreversible damage and probable progression. As optic-nerve damage progresses, severe visual dysfunction and blindness may ensue in a small group of patients. Since many patients present in the early stages of the disease, the goal of treatment is to arrest or delay the progression of early optic-nerve damage to significant visual impairment.

It is estimated that approximately 8% of adults over the age of 40 years in the United States have ocular hypertension (Tielsch et al., 1991). While ocular hypertension is a common finding, eye-care providers do not know which patients to treat or which patients to monitor without treatment. In 2002, the publication of the results of the Ocular Hypertension Treatment Study (OHTS) stimulated a reassessment of the ways in which to evaluate and manage patients with ocular hypertension (Gordon et al., 2002; Kass et al., 2002). Since the OHTS publication, several strategies for risk assessment in ocular hypertension have been proposed and some have been successfully implemented in clinical practice. Several predictive models (or risk calculators) have been proposed and their use in clinical practice is likely to provide a more objective and evidence-based approach to the management of patients with ocular hypertension.

Risk factors for glaucoma development

The development of predictive models requires a series of complex steps which initially involve the acquisition and analysis of data from one or multiple longitudinal studies that have carefully followed patients over time. A critical step is the identification of the risk factors associated with the

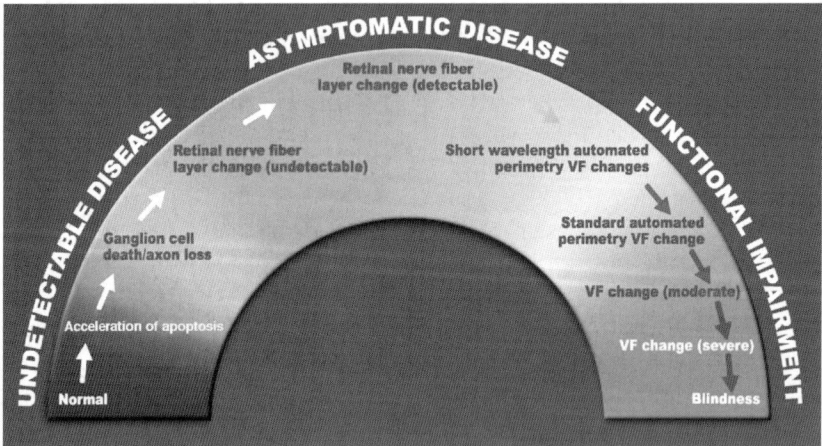

Fig. 1. The glaucoma continuum. Adapted with permission from Weinreb et al. (2004).

outcome one wants to predict. A few large, prospective, longitudinal studies have provided evidence with regard to the risk factors for conversion from ocular hypertension to glaucoma. From these studies, two were randomized clinical trials, the OHTS (Gordon et al., 2002) and the European Glaucoma Prevention Study (EGPS) (Miglior et al., 2007). These two studies have provided the basis for development and validation of the prediction models for glaucoma development available today. Both studies have evaluated a large number of predictive factors for their potential association with the risk of converting to glaucoma. When pooled analyses of the OHTS and EGPS data were conducted, only five baseline factors were identified as significantly associated with the risk of converting to glaucoma: age, intraocular pressure (IOP), central corneal thickness, the measurement of the vertical cup/disc ratio of the optic nerve, and the visual-field index pattern standard deviation (PSD) (Gordon et al., 2007). Table 1 shows relative risks for the baseline predictive factors found to be significantly associated with the risk of developing glaucoma in these two studies. These predictive factors have been incorporated into predictive models to estimate the risk of converting from ocular hypertension to glaucoma.

Several other factors were not found to be statistically significantly related to the risk of conversion to glaucoma in the OHTS/EGPS pooled dataset, such as diabetes mellitus, history of heart disease, and race, among others. It is important to emphasize, however, that even for the OHTS/EGPS combined dataset, the power of the study was probably not enough to detect a significant predictive value for many of the evaluated risk factors. Also, methodological weaknesses precluded a better investigation of the real value of potential risk factors, such as positive family history of glaucoma. As no relatives of the study subjects were examined, investigators had to rely on self-reported family history with its potential inaccuracy. It is likely that this contributed to the lack of association between family history and risk of glaucoma development as reported by these investigations.

Below, we review some of the evidence with regard to the predictive value of risk factors reported to be associated with glaucoma development.

Intraocular pressure

In the OHTS, 1636 ocular hypertensive patients were randomized to either observation or treatment and followed for a median time of 72 months. Ocular hypertension was defined based on the presence of qualifying IOP between 24 and 32 mmHg in one eye and 21 and 32 mmHg in the other eye, gonioscopically open angles, normal visual fields, and normal optic discs (Gordon and Kass, 1999). Participants randomized to medication began treatment to achieve a target IOP of 24 mmHg or less and a minimum of 20% reduction in IOP from the average of the qualifying IOP and IOP at the baseline randomization visit. At baseline, mean IOP was 24.9 ± 2.6 and 24.9 ± 2.7 mmHg in the treated and observation groups, respectively. The average IOP reduction in the treated group was $22.5 \pm 9.9\%$ compared to $4.0 \pm 11.6\%$ in the observation group. At 60 months, the cumulative probability of developing POAG was 4.4% in the medication group compared to 9.5% in the observation group, which

Table 1. OHTS vs. EGPS — risk factors

	OHTS observation $N = 819$		EGPS placebo $N = 522$	
	HR	95% CI	HR	95% CI
Age (per decade)	1.12	0.91–1.39	1.39	1.01–1.91
IOP (per mmHg)	1.22	1.12–1.32	1.10	0.97–1.26
CCT (per 40 μm thinner)	2.03	1.61–2.55	2.12	1.51–2.97
PSD (per 0.2 dB greater)	1.19	0.97–1.45	1.06	0.96–1.17
Vertical C/D ratio (per 0.1 larger)	1.27	1.14–1.43	1.26	1.03–1.53

translates into a 54% relative reduction in the risk of developing POAG with treatment. In the analysis of baseline predictive factors for development of POAG, 1 mmHg higher baseline IOP was associated with a 10% higher risk of developing POAG during follow-up, after adjustment for other predictive factors in a multivariate model (Gordon et al., 2002). For this calculation, baseline IOP was calculated from four to six baseline IOP measurements per eye.

The EGPS (Miglior et al., 2005) was also designed to investigate whether the onset of POAG can be prevented or delayed in ocular hypertensive patients by medical hypotensive therapy. Inclusion criteria for the EGPS were similar to the OHTS, requiring participants to have normal visual fields and normal optic discs at baseline. However, qualifying IOP had to be between 22 and 29 mmHg in at least one eye on two consecutive measurements taken at least 2 h apart. There was no mention with regard to the IOP in the other eye in the study protocol (Miglior et al., 2002). The EGPS randomized 1081 patients to treatment with Dorzolamide or placebo, with a planned follow-up of 5 years. However, only 64% of patients randomized to Dorzolamide and 75% of the patients randomized to placebo completed the study. Mean IOP at baseline was 23.4 and 23.5 mmHg in the Dorzolamide and placebo groups, respectively. Mean IOP reduction at 5 years was 22.1% in the Dorzolamide group and 18.7% in the placebo group. At the completion of the study, there was no statistically significant difference in the cumulative probability of developing POAG between patients randomized to Dorzolamide versus placebo (13.4 vs. 14.1%, respectively; HR = 0.86; 95% CI: 0.58–1.26).

Several reasons have been proposed to explain the conflicting results between the OHTS and EGPS including regression to the mean effects, lack of target IOP, and selective loss to follow-up (Quigley, 2005; Parrish, 2006). However, despite the fact that the EGPS could not find significant differences between Dorzolamide and placebo groups on the rate of POAG development, its results are compatible with higher IOP being a risk factor for POAG incidence. A 1 mmHg higher baseline IOP was associated with 18% higher risk of developing POAG (HR = 1.18; 95% CI: 1.06–1.31; $P = 0.002$) in a multivariable model containing age, presence of cardiovascular disease, CCT, and presence of pseudoexfoliation (Miglior et al., 2007).

In the pooled analysis of the OHTS and EGPS control groups (1319 patients followed without treatment), 1 mmHg higher baseline IOP was associated with 9% higher risk of developing POAG (HR = 1.09; 95% CI: 1.03–1.17), after adjustment for other predictive factors (Gordon et al., 2007). It is important to note that even for this pooled analysis, the 95% confidence interval was still relatively large, ranging from 1.03 to 1.17. That is, each 1 mmHg increased IOP could be associated with 3%–17% increased risk.

Corneal thickness

Corneal thickness is another factor that has been associated with the risk of conversion from OHT to glaucoma. IOP as assessed by applanation tonometry may be overestimated or underestimated in thick or thin corneas, respectively (Ehlers et al., 1975; Whitacre et al., 1993; Herndon et al., 1997; Copt et al., 1999; Doughty and Zaman, 2000; Brandt, 2001). A considerable subset of patients classified as having ocular hypertension may simply have thicker than average corneas that result in an overestimation of what is likely a normal, true IOP. As a consequence, OHT patients with thicker corneas may be at a lower risk for glaucoma development. In fact, the OHTS showed that CCT was a powerful predictor of development of primary open-angle glaucoma among ocular hypertensive eyes (Gordon et al., 2002). Eyes with CCT of 555 μm or less had a threefold greater risk of developing glaucoma compared with participants who had CCT of more than 588 μm. A 40 μm thinner cornea was associated with a 71% increase in the risk of converting to glaucoma among OHTS patients in a multivariate model adjusting for other risk factors. Similar results were found by the EGPS, with a 40 μm thinner cornea being associated with a 32% increase in the risk of converting to glaucoma in the multivariate model. Recent results from the Barbados Eye Study, a population-based cohort study involving over 3000 participants, also found that thinner corneas were associated with a higher incidence of glaucoma over time. A 40 μm thinner cornea was associated with approximately 40% increase in the risk of developing glaucoma.

The relationship between ocular hypertension, corneal thickness, and presence of structural and functional glaucomatous abnormalities has been investigated by Medeiros et al. (2003a, b). The authors showed that OHT patients with thin corneas have a higher prevalence of abnormalities detected by function-specific perimetric tests, such as short-wavelength automated perimetry (SWAP) and frequency doubling technology perimetry (FDT), even though they still have normal results on standard automated perimetry (SAP). Similar results were found when structural analysis was performed by retinal nerve fiber layer assessment using scanning laser polarimetry. OHT patients with thinner corneas had a higher prevalence of abnormalities on this test compared to patients with thicker corneas. This additional evidence of the association between thinner corneas and development of glaucomatous functional and structural damage supports the importance of considering central corneal thickness in the assessment of risk for the development of glaucoma in patients with ocular hypertension.

The mechanism by which CCT influences the risk of developing glaucoma has not been completely established. Although the effect of corneal thickness could potentially be attributed to an artifact of tonometric measurements, it is also possible that CCT could be a marker for biomechanical and structural characteristics of ocular tissues, which may influence the risk of development of glaucomatous neuropathy. Eyes with thinner corneas could have a particular structural susceptibility that would make them more prone to develop glaucomatous damage. Further studies are necessary to evaluate this hypothesis.

Age

There is strong evidence that older age is an independent risk factor for the progression of ocular hypertension and glaucoma. Older age has been reported as a risk factor for the development of glaucoma in patients with ocular hypertension in multiple longitudinal studies. Several population-based studies have also found that the incidence of OAG increases with older age. Both the OHTS and the EGPS found that older patients with ocular hypertension had an increased risk of converting to glaucoma over time.

Cup/disc ratio and pattern standard deviation

The OHTS as well as the EGPS and several other longitudinal studies have found that certain indicators of structural and functional integrity at baseline are predictive factors for development of overt glaucomatous optic neuropathy or visual-field defects in the future. Two indices that have consistently been associated with higher risk of developing glaucoma are the vertical cup/disc ratio and the visual-field PSD, both measured at the baseline visit. These factors cannot be considered strictly as risk factors for the disease, as they are part of its definition. However, if their assessment proves to be helpful in predicting which patients are more likely to develop clinically important stages of disease in the future, their inclusion in predictive models is justified. Both vertical cup/disc ratio and PSD were significantly associated with risk of developing glaucoma in the multivariate model combining OHTS and EGPS datasets. A 0.1 increase in vertical cup/disc ratio was associated with 19% higher chance of developing glaucoma. For PSD, a 0.2 dB increase in the baseline PSD value was associated with a 13% increase in risk.

The need for predictive models

Although the information on individual risk factors may already help clinicians in management decisions, it is often difficult to integrate the information on the several risk factors and provide a global assessment for a particular patient. In that situation, predictive models or risk calculators may help clinicians in providing a more objective assessment of risk. Mansberger and Cioffi (2006) performed a survey of ophthalmologists to estimate their ability to predict the risk of glaucoma development in ocular hypertensive patients.

Ophthalmologists had the benefit of an oral review and written handouts summarizing the OHTS results. They found that ophthalmologists tended to underestimate the risk when compared to the actual risk found by a risk calculator. Ophthalmologists also had a large range of predictions, sometimes differing from the actual risk by 40%, illustrating the need for a more standardized method for risk assessment.

Predictive models for glaucoma development

The development of predictive models (or risk calculators) involves use of statistical methods to develop models for prediction of outcome using one or more explanatory variables. In 2005, we published the results on the development of a risk calculator to assess the risk of an ocular hypertensive patient to develop glaucoma (Mcdeiros et al., 2005). The risk calculator was derived based on the results published by the OHTS (Gordon et al., 2002; Coleman et al., 2004) and it incorporated the variables that were described by that study as being significantly associated with the risk of developing glaucoma over time. The risk calculator was designed to estimate the chance of an ocular hypertensive patient to develop glaucoma if left untreated for 5 years. To simplify the use of the risk calculator, a point system and an electronic version of the calculator were made available for clinicians (Fig. 2).

A predictive model that is derived from a particular dataset is not guaranteed to work on a different group of patients. In fact, the performance of regression models (or risk calculators) used as diagnostic or prediction tools is generally better on the dataset on which the model has been constructed (derivation set) compared to the performance of the same model on new data. Therefore, before risk calculators can be successfully incorporated into clinical practice they need to be validated on different populations. By validation we mean establishing that the risk calculator works satisfactorily for patients other than those from whose data the model was derived. Along with the steps involved in the development of the risk calculator, we also presented the results of its validation on an independent population of 126 patients with ocular hypertension who were followed as part of a prospective longitudinal study conducted at the University of California San Diego (DIGS — Diagnostic Innovations in Glaucoma Study).

Several steps were taken to validate the OHTS-derived model. In the first step, the importance of the prognostic variables that had been previously identified by the OHTS study was evaluated on the new data set (DIGS data set). All the variables had similar performance, except for diabetes mellitus, which was not significantly associated with the risk of developing glaucoma in the DIGS data. Subsequently, the predictive performance of the model was investigated on the new data set. The ability of the OHTS-derived risk calculator to discriminate DIGS subjects who developed glaucoma from those who did not was reasonably good with a c-index of approximately 0.7. The c-index is a measure of the discriminating ability of a model (similar to the area under the receiver operating characteristic [ROC] curve) and a c-index of 0.7 indicates that, in approximately 70% of the cases, the model allocated a higher predicted probability for a subject who actually developed glaucoma than for a subject who did not. The closer the c-index gets to 1, the better is the discriminating ability of the model. The values of c-index found for the OHTS-derived risk calculator when applied to DIGS subjects were similar to those found when risk models such as the Framingham coronary prediction scores are used to predict coronary heart disease events (D'Agostino et al., 2001; Liu et al., 2004). D'Agostino et al. (2001) reported c-indexes ranging from 0.63 to 0.83 when the Framingham functions were applied to six different cohorts of patients.

The OHTS-derived risk calculator also had a good calibration when applied to the DIGS data set (Fig. 3). Checking calibration is another important step in validating a predictive model. A reliable or well-calibrated model will give predicted probabilities that agree numerically with the actual outcomes. For example, let us consider a group of 100 ocular hypertensive patients. If the model assigns an average probability of 12% for conversion to glaucoma for this group of subjects, it is expected that approximately 12 subjects will

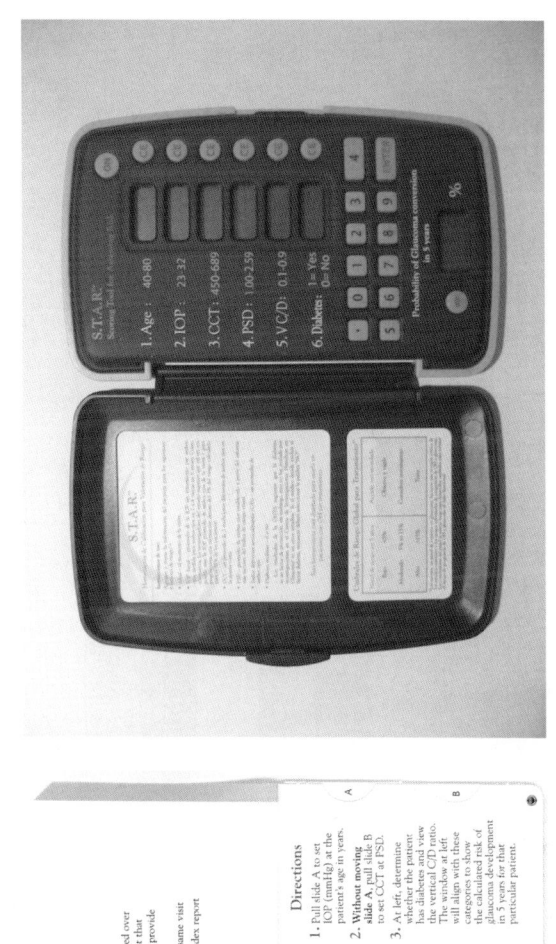
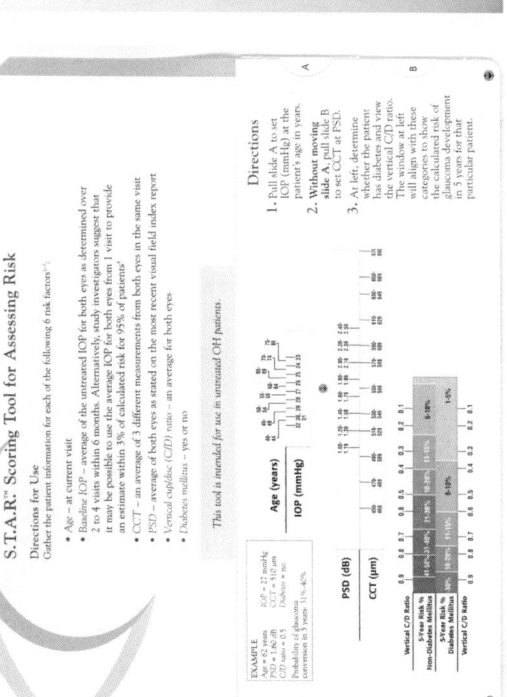

Fig. 2. Point system and electronic version of the risk calculator.

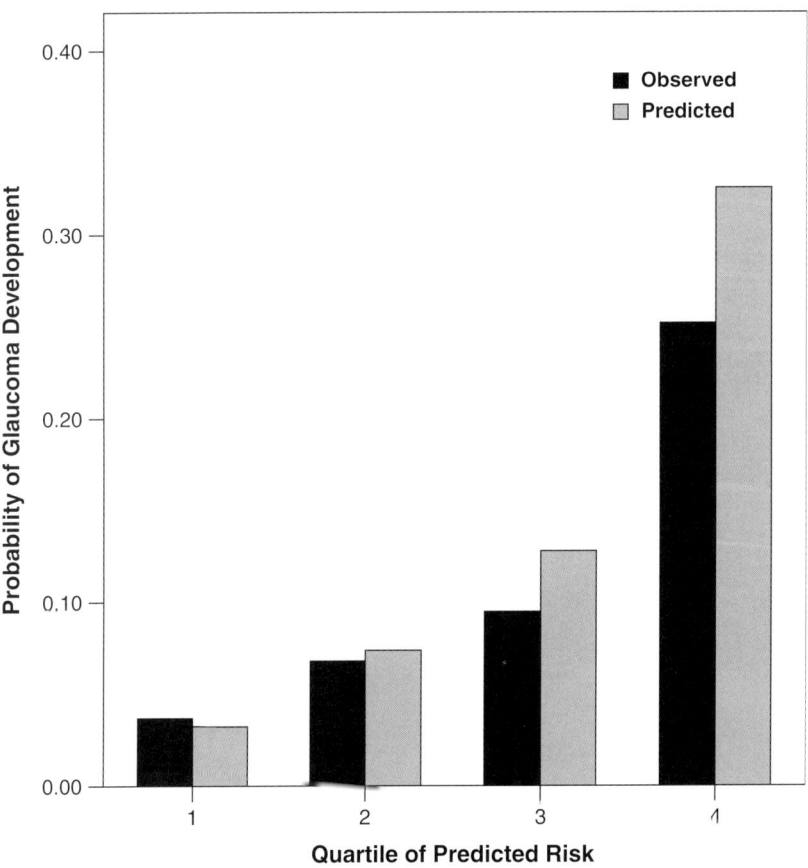

Fig. 3. Comparison of observed and predicted 5-year incidence of glaucoma when the OHTS-derived risk calculator was applied to the diagnostic innovations in glaucoma study (DIGS) data set. Predicted probabilities agreed closely with observed outcomes. Adapted with permission from Medeiros et al. (2005).

convert to glaucoma over time. That is, for a well-calibrated model, the predicted probabilities of conversion to glaucoma will agree closely with the observed probabilities of conversion. Figure 3 shows that the OHTS-derived risk calculator performed well on the DIGS data set. For patients in whom the model predicted a high chance of converting to glaucoma, there was a high observed conversion rate; whereas for patients in whom the model predicted a low conversion rate, there was a low observed conversion rate.

In 2007, OHTS and EGPS investigators published results of the development and validation of a risk calculator for glaucoma based on the analysis of the combined OHTS/EGPS dataset (Gordon et al., 2007). The results were similar to the predictive model published in 2005, and the risk calculator contained the five variables significantly associated with the risk of glaucoma conversion: age, IOP, CCT, PSD, and vertical cup/disc ratio. The risk model from the pooled OHTS/EGPS sample of over 1100 ocular hypertension patients demonstrated excellent fit with a c-statistic of 0.74 and good calibration. The OHTS/EGPS risk calculator is available on the web at http://ohts.wustl.edu/risk

Predictive models for glaucoma progression

Estimation of the risk of patients with existing glaucoma of developing progressive damage over time is at least as important as estimating the risk of unaffected patients developing glaucoma. The development of predictive models for glaucoma progression could use the same principles as those used to develop and validate models for glaucoma development. Initially, longitudinal studies that followed patients with glaucoma would have to be reviewed to identify risk factors associated with progressive disease.

Several studies have investigated the risk factors for progression in patients with established glaucomatous damage. The Early Manifest Glaucoma Trial (EMGT) (Heijl et al., 2002) was designed specifically to evaluate the effect of IOP-lowering treatment on progression of glaucoma. The EMGT enrolled 255 newly diagnosed, previously untreated, open-angle glaucoma patients who had reproducible visual-field defects at baseline. Patients with advanced visual-field loss or IOP greater than 30 mmHg at baseline were excluded. Patients were randomized to 360° trabeculoplasty plus betaxolol versus no treatment. Eyes stayed in their allocation arms unless significant progression occurred. If the IOP in treated eyes exceeded 25 mmHg at two consecutive follow-ups or 35 mmHg in control eyes, latanoprost was added. Patients were followed for a median of 6 years, with excellent retention. Baseline IOP in treated and untreated groups were 20.6 ± 4.1 and 20.9 ± 4.1 mmHg, respectively. Mean IOP reduction was 25% in the treated group, with no changes in the control group. At study closure, the proportion of patients who developed progression was significantly larger in the control versus the treatment group (62% vs. 45%, respectively; HR = 0.60; 95% CI: 0.42–0.84; $P = 0.003$). Differences between treated and untreated patients remained when results were stratified by baseline IOP level (<21 mmHg or ≥ 21 mmHg), degree of visual-field damage, age, or presence of exfoliation.

Besides IOP, several other risk factors were identified by the EMGT as significantly associated with the risk of glaucoma progression: older age, exfoliation, presence of bilateral disease, and worse mean deviation on the baseline visual fields. Recently, the EMGT also published results on the long-term follow-up of the original cohort and concluded that thinner central corneal thickness and decreased ocular perfusion pressure were also associated with higher risk of visual-field progression over time.

A predictive model could theoretically be developed based on the results from the EMGT incorporating all the risk factors found to be significantly associated with progressive disease. Such a model would be helpful in estimating which glaucoma patients are at higher risk for developing progressive loss of visual function. It is important to emphasize that any predictive model developed on the basis of the EMGT or other studies evaluating risk factors for glaucoma progression would have to be validated on an independent population of patients, as described above for the risk calculators in ocular hypertension.

Limitations of predictive models

The use of predictive models in clinical practice has several limitations. Predictive models are based on restricted populations of patients that were selected based on strict inclusion and exclusion criteria and that may not be representative of all patients seen at everyday clinical settings. Use of these models should be restricted to those patients who are similar to the ones included in the studies used to develop and/or validate it. It is also important to emphasize that although predictive models can provide a more objective evaluation of risk, their use does not replace the judgment of a clinician when making management decisions. For example, current risk calculators to estimate risk of glaucoma development do not include important information to guide treatment such as medical health status and life expectancy, patient's willingness to treatment, costs of medications, and overall effect of treatment on quality of life. Also, it is important to emphasize that current risk calculators for glaucoma have been designed to estimate the risk of development of the earliest signs of disease, which do not necessarily have an impact on the quality of vision of the patient. Finally, as more evidence regarding risk factors for disease development and progression accumulates, newer and better refined predictive models will be developed that should replace current existing ones.

References

Brandt, J.D. (2001) The influence of corneal thickness on the diagnosis and management of glaucoma. J. Glaucoma, 10(5 (Suppl 1)): S65–S67.

Coleman, A.L., Gordon, M.O., et al. (2004) Baseline risk factors for the development of primary open-angle glaucoma in the ocular hypertension treatment study. Am. J. Ophthalmol., 138(4): 684–685.

Copt, R.P., Thomas, R., et al. (1999) Corneal thickness in ocular hypertension, primary open-angle glaucoma, and normal tension glaucoma. Arch. Ophthalmol., 117(1): 14–16.

D'Agostino, R.B. and Nam, B.H. (2004). Evaluation of the performance of survival analysis models: discrimination and calibration measures. In: N. Balakrishnan and C.R. Rao (Eds.). Handbook of Statistics, Vol. 23. Elsevier, Amsterdam.

D'Agostino, R.B., Sr., Grundy, S., et al. (2001) Validation of the Framingham coronary heart disease prediction scores: results of a multiple ethnic groups investigation. JAMA, 286(2): 180–187.

D'Agostino, R.B., Russell, M.W., et al. (2000) Primary and subsequent coronary risk appraisal: new results from the Framingham study. Am. Heart J., 139(2 (Pt 1)): 272–281.

Doughty, M.J. and Zaman, M.L. (2000) Human corneal thickness and its impact on intraocular pressure measures: a review and meta-analysis approach. Surv. Ophthalmol., 44(5): 367–408.

Ehlers, N., Bramsen, T., et al. (1975) Applanation tonometry and central corneal thickness. Acta Ophthalmol. (Copenh.), 53(1): 34–43.

Gordon, M.O., Beiser, J.A., et al. (2002) The ocular hypertension treatment study: baseline factors that predict the onset of primary open-angle glaucomaArch. Ophthalmol., 120(6): 714–720; discussion 829–830.

Gordon, M.O. and Kass, M.A. (1999) The ocular hypertension treatment study: design and baseline description of the participants. Arch. Ophthalmol., 117(5): 573–583.

Gordon, M.O., Torri, V., et al. (2007) Validated prediction model for the development of primary open-angle glaucoma in individuals with ocular hypertension. Ophthalmology, 114(1): 10–19.

Heijl, A., Leske, M.C., et al. (2002) Reduction of intraocular pressure and glaucoma progression: results from the early manifest glaucoma trial. Arch. Ophthalmol., 120(10): 1268–1279.

Herndon, L.W., Choudhri, S.A., et al. (1997) Central corneal thickness in normal, glaucomatous, and ocular hypertensive eyes. Arch. Ophthalmol., 115(9): 1137–1141.

Kass, M.A., Heuer, D.K., et al. (2002) The ocular hypertension treatment study: a randomized trial determines that topical ocular hypotensive medication delays or prevents the onset of primary open-angle glaucomaArch. Ophthalmol., 120(6): 701–713; discussion 829–830.

Liu, J., Hong, Y., et al. (2004) Predictive value for the Chinese population of the Framingham CHD risk assessment tool compared with the Chinese multi-provincial cohort study. JAMA, 291(21): 2591–2599.

Mansberger, S.L. and Cioffi, G.A. (2006) The probability of glaucoma from ocular hypertension determined by ophthalmologists in comparison to a risk calculator. J. Glaucoma, 15(5): 426–431.

Medeiros, F.A., Sample, P.A., et al. (2003a) Corneal thickness measurements and frequency doubling technology perimetry abnormalities in ocular hypertensive eyes. Ophthalmology, 110(10): 1903–1908.

Medeiros, F.A., Sample, P.A., et al. (2003b) Corneal thickness measurements and visual function abnormalities in ocular hypertensive patients. Am. J. Ophthalmol., 135(2): 131–137.

Medeiros, F.A., Weinreb, R.N., et al. (2005) Validation of a predictive model to estimate the risk of conversion from ocular hypertension to glaucoma. Arch. Ophthalmol., 123(10): 1351–1360.

Miglior, S., Pfeiffer, N., et al. (2007) Predictive factors for open-angle glaucoma among patients with ocular hypertension in the European glaucoma prevention study. Ophthalmology, 114(1): 3–9.

Miglior, S., Zeyen, T., et al. (2002) The European glaucoma prevention study design and baseline description of the participants. Ophthalmology, 109(9): 1612–1621.

Miglior, S., Zeyen, T., et al. (2005) Results of the European glaucoma prevention study. Ophthalmology, 112(3): 366–375.

Parrish, R.K., 2nd (2006) The European glaucoma prevention study and the ocular hypertension treatment study: why do two studies have different results? Curr. Opin. Ophthalmol., 17(2): 138–141.

Quigley, H.A. (2005) European glaucoma prevention study-Ophthalmology, 112(9): 1642–1643; author reply 1643–1645.

Tielsch, J.M., Katz, J., et al. (1991) A population-based evaluation of glaucoma screening: the Baltimore eye survey. Am. J. Epidemiol., 134(10): 1102–1110.

Weinreb, R.N., Friedman, D.S., Fechtner, R.D., Cioffi, G.A., Coleman, A.L., Girkin, C.A., Liebmann, J.M., Singh, K., Wilson, M.R., Wilson, R. and Kannel, W.B. (2004) Risk assessment in the management of patients with ocular hypertension. Am. J. Ophthalmol., 138(3): 458–467.

Whitacre, M.M., Stein, R.A., et al. (2004) The effect of corneal thickness on applanation tonometry. Am. J. Ophthalmol., 115(5): 592–596.

CHAPTER 3

Intraocular pressure and central corneal thickness

Gianluca Manni[1,2,*], Francesco Oddone[2], Vincenzo Parisi[2], Adriana Tosto[1] and Marco Centofanti[1,2]

[1]*Biopathology Department, University of Tor Vergata, Rome, Italy*
[2]*G.B. Bietti Eye Foundation — IRCCS, Rome, Italy*

Abstract: From the results of the Ocular Hypertension Treatment Study emerged the conclusion that ocular hypertensive subjects with thinner central corneal thickness (CCT) are at increased risk of developing glaucoma. Although possible underlying biases that could have led to this conclusion are still under investigation, there is an increasing interest in the scientific community to understand the potential mechanisms of this increased risk profile. It has been proposed that interindividual differences in CCT might be purely responsible for inaccuracies of the tonometric readings with potential underestimation of the true IOP in subjects with thinner CCT although it is becoming progressively clearer that the true IOP is unpredictable with linear correction formulas for CCT, and it is likely that other material properties of the cornea contribute, together with CCT, to the tonometric artifact. Recently, it has become possible to measure the biomechanical properties of the cornea in vivo and it has been suggested that differences in corneal biomechanics may be the expression of interindividual structural differences of the ocular tissues (including lamina cribrosa), with potential consequences on the interindividual susceptibility to the glaucomatous damage under the same IOP level. A possible underlying biological risk related to thinner CCTs, independent of the influence on tonometric reading, has been proposed and largely studied after the results of the OHTS were published. Besides the understanding of the mechanism underlying the role of CCT as a risk factor for the development of glaucoma, it is important to understand how the information about CCT should be integrated in the clinical management of both ocular hypertension (OHT) and glaucoma and whether other ocular properties should be measured to better understand the individual risk profile.

Keywords: corneal thickness; corneal biomechanics; tonometry; glaucoma; ocular hypertension; risk factor

Main text

Intraocular pressure (IOP) is an important risk factor for the development of glaucoma from OHT (Gordon et al., 2002) as well as for the progression of an already established glaucoma (Leske et al., 1999; Anderson et al., 2003). The results of the Ocular Hypertension Treatment Study (OHTS) published in 2002 brought to the attention of the scientific and clinical communities the importance of central corneal thickness (CCT) in the clinical management of OHT (Gordon et al., 2002).

Indeed, CCT proved to be the most potent predictor of which OHT subjects would develop glaucoma in a multivariate model of baseline characteristics. Specifically, OHT subjects with thinner corneas were found to be at increased risk of developing glaucoma compared to subjects

*Corresponding author. Tel./Fax: +39-(0)6-20902968;
E-mail: glmanni@inwind.it

Only recently, it has become possible to measure the biomechanical properties of the cornea in vivo and the role of corneal biomechanics has been the subject of a recent review by Kotecha (2007). This review pointed out that the importance of corneal biomechanics to the glaucoma clinician primarily rests with its effects on IOP measurement, although it is not possible to completely exclude the fact that corneal biomechanics may give an indication of the structural integrity of the optic nerve head. The Ocular Response Analyzer (ORA; Reichert Corporation; Depew, USA) has been recently developed by Reichert: this instrument is able to measure the corneal response to a rapid jet of air. The jet of air generates a corneal indentation consisting of an initial inward applanation and a second outward applanation when the cornea reverts to its steady shape. The instrument is able to quantify the force required to flatten the cornea during the first and second applanation separately. It has been found that the second applanation occurs at a lower IOP than the first, and the difference between the two pressures is called corneal hysteresis (CH). CH is believed to be a measure of corneal biomechanics and may contribute, together with CCT, to explain the corneal behavior during applanation tonometry. It has been observed that CH is reduced in eyes with keratoconus (Shah et al., 2007), Fuch's endothelial dystrophy (Luce, 2005), and congenital glaucoma (Kirwan et al., 2006), especially if Haab's striae are present. A marked decrease of CH following laser in situ keratomileusis has also been reported (Kirwan and O'Keefe, 2007; Ortiz et al., 2007). Since CH is not independent from CCT, and from the level of true IOP, further studies are required to elucidate the role of this property during applanation tonometry.

Other parameters such as corneal resistance factor (CRF) and a corneal constant factor (CCF) have been developed from the ORA measurement and both are believed to be relatively unaffected by the IOP level despite being positively associated with CCT (Kotecha, 2007). However, further studies are required to clearly understand which biomechanical properties are represented by these parameters and how they may influence applanation tonometry.

It has also been suggested that differences in corneal biomechanics may be the expression of interindividual structural differences of the ocular tissues (including lamina cribrosa), with potential consequences on the interindividual susceptibility to the glaucomatous damage under the same IOP level. A retrospective chart review by Congdon et al. (2006) has recently reported that low values of CH are associated with visual field progression, despite larger and longer studies are required to determine the role of CH in determining the glaucoma susceptibility.

The evidence of the influence of CCT on the GAT reading stimulated the development of new technologies to measure IOP independently from CCT, and among the new tonometers, the dynamic contour tonometer (DCT; Swiss MicrotechnologyAG, Port, Switzerland) has been proposed to reduce the corneal effect and to improve the accuracy of IOP assessment.

DCT is a new digital nonapplanation contact tonometer with a concave surface of the tonometer tip that matches the contour of the cornea, creating an equilibrium between capillary force, rigidity force, appositional force, and force exerted on the cornea by IOP. A piezoelectric sensor integrated into the contoured surface of the tip measures IOP once the corneal contour is perfectly matched. In a clinical observational study on 176 eyes, Kamppeter and Jonas (2005) observed a lower dependence of DCT–IOP on CCT than applanation tonometry, and this result is in agreement with several reports performed on mixed populations of healthy and glaucomatous eyes (Martinez de la Casa et al., 2006; Francis et al., 2007; Medeiros et al., 2007; Ceruti et al., 2008; Herdener et al., 2008). However, a significant correlation between CCT and DCT–IOP was reported by Grieshaber et al. (2007) in POAG patients. The authors hypothesized that in contrast to healthy subjects, patients with POAG have increased IOP, which is independent of CCT and, furthermore, the corneal rigidity in patients with glaucoma may be altered primarily or secondarily to topical drugs, possibly affecting IOP measurements, as some antiglaucomatous drugs may modulate the extracellular matrix (Ito et al., 2006; Brandt, 2007). Therefore, the potential advantage of DCT relative to CCT independence may not hold true for patients with POAG.

The answer to the clinical questions of whether CCT measurement is useful in the clinical practice and how the pachimetric data should influence the clinical decision process in the management of glaucoma suspects, established glaucomas, or OHT is all but straightforward.

The evidence that CCT is a reliable indicator of risk for progression of OHT to glaucoma is consistent, as shown by the OHTS and EGPS results. The decision to treat a patient with OHT depends on an assessment of risk, and CCT is an important and necessary part of that determination while there is little evidence that CCT is useful in predicting progression of glaucoma as shown by the results of EMGT. Besides the hypothesis that CCT might influence the underlying biological risk to develop glaucoma, there is the universally acknowledged influence of CCT on IOP measurements, and the fact that IOP is an important parameter for diagnosing glaucoma and represents the only risk factor modifiable with therapy.

As previously discussed, a correction formula for CCT would be useful to improve the accuracy of GAT readings, but considering the variety and inconsistency of the published correction algorithms, the arbitrary selection of one algorithm carries the risk of introducing further errors rather than removing them.

Moreover, the accuracy in measuring true IOP might not be absolutely necessary in every stage of glaucoma management and the error induced in an individual case is likely to be constant, not impairing monitoring of IOP changes over time.

It is generally accepted and consistent in most reports in the literature that thicker corneas are associated with an overestimation of the true IOP and thinner corneas are associated with an underestimation of the true IOP although it is likely that for the majority of patients the inaccuracy would be small with little clinical impact.

On the basis of the scientific knowledge available so far, it is likely that a reasonable approach is, as proposed by James D. Brandt, to "take care of patients simply by categorizing corneas as thin, average or thick, just as it is important to recognize that optic discs come in small, medium, and large, allowing the clinician to interpret the configurations accordingly" (Brandt, 2007).

References

Aghaian, E., Choe, J.E., Lin, S. and Stamper, R.L. (2004) Central corneal thickness of Caucasians, Chinese, Hispanics, Filipinos, African Americans, and Japanese in a glaucoma clinic. Ophthalmology, 111(12): 2211–2219.

Anderson, D.R.Normal Tension Glaucoma Study. (2003) Collaborative normal tension glaucoma study. Curr. Opin. Ophthalmol., 14(2): 86–90.

Brandt, J.D. (2007) Central corneal thickness, tonometry, and glaucoma risk — a guide for the perplexed. Can. J. Ophthalmol., 42(4): 562–566.

Ceruti, P., Morbio, R., Marraffa, M. and Marchini, G. (2008) Comparison of Goldmann applanation tonometry and dynamic contour tonometry in healthy and glaucomatous eyes. Eye, 25(1):

Congdon, N.G., Broman, A.T., Bandeen-Roche, K., Grover, D. and Quigley, H.A. (2006) Central corneal thickness and corneal hysteresis associated with glaucoma damage. Am. J. Ophthalmol., 141: 868–875.

Ehlers, N., Bramsen, T. and Sperling, S. (1975) Applanation tonometry and central corneal thickness. Acta Ophthalmol. (Copenh.), 53: 34–43.

Foster, P.J., Baasanhu, J., Alsbirk, P.H., Munkhbayar, D., Uranchimeg, D. and Johnson, G.J. (1998) Central corneal thickness and intraocular pressure in a Mongolian population. Ophthalmology, 105: 969–973.

Foster, P.J., Machin, D., Wong, T.Y., Ng, T.P., Kirwan, J.F., Johnson, G.J., Khaw, P.T. and Seah, S.K. (2003) Determinants of intraocular pressure and its association with glaucomatous optic neuropathy in Chinese Singaporeans: the Tanjong Pagar Study. Invest. Ophthalmol. Vis. Sci., 44(3): 885–891.

Francis, B.A., Hsieh, A., Lai, M.Y., Chopra, V., Pena, F., Azen, S. and Varma, R.Los Angeles Latino Eye Study Group. (2007) Effects of corneal thickness, corneal curvature, and intraocular pressure level on Goldmann applanation tonometry and dynamic contour tonometry. Ophthalmology, 114(1): 20–26.

Goldmann, H. and Schmidt, T. (1957) Applanation tonometry. Ophthalmologica, 134(4): 221–242.

Gordon, M.O., Beiser, J.A., Brandt, J.D., Heuer, D.K., Higginbotham, E.J., Johnson, C.A., Keltner, J.L., Miller, J.P., Parrish II, R.K., Wilson, M.R. and Kass, M.A. (2002) The ocular hypertension treatment study: baseline factors that predict the onset of primary open-angle glaucomaArch. Ophthalmol., 120: 714–720. discussion 829–830

Grieshaber, M.C., Schoetzau, A., Zawinka, C., Flammer, J. and Orgul, S. (2007) Effect of central corneal thickness on dynamic contour tonometry and Goldmann applanation tonometry in primary open-angle glaucoma. Arch. Ophthalmol., 125(6): 740–744.

Herdener, S., Hafizovic, D., Pache, M., Lautebach, S. and Funk, J. (2008) Is the PASCAL-tonometer suitable for measuring intraocular pressure in clinical routine? Long- and short-term reproducibility of dynamic contour tonometry. Eur. J. Ophthalmol., 18(1): 39–43.

Ito, T., Ohguro, H., Mamiya, K., Ohguro, I. and Nakazawa, M. (2006) Effects of antiglaucoma drops on MMP and TIMP balance in conjunctival and subconjunctival tissue. Invest. Ophthalmol. Vis. Sci., 47(3): 823–830.

Johnson, M., Kass, M.A., Moses, R.A. and Grodzki, W.J. (1978) Increased corneal thickness simulating elevated intraocular pressure. Arch. Ophthalmol., 96: 664–665.

Kamppeter, B.A. and Jonas, J.B. (2005) Dynamic contour tonometry for intraocular pressure measurement. Am. J. Ophthalmol., 140(2): 318–320.

Kirwan, C. and O'Keefe, M. (2007) Corneal hysteresis using the Reichert ocular response analyser: findings pre- and post-LASIK and LASEK. Acta Ophthalmol. Scand., 21.

Kirwan, C., O'Keefe, M. and Lanigan, B. (2006) Corneal hysteresis and intraocular pressure measurement in children using the Reichert ocular response analyzer. Am. J. Ophthalmol., 142(6): 990–992.

Kotecha, A. (2007) What biomechanical properties of the cornea are relevant for the clinician? Surv. Ophthalmol., 52(Suppl 2): 109–114.

La Rosa, F.A., Gross, R.L. and Orengo, N.S. (2001) Central corneal thickness of Caucasians and African Americans in glaucomatous and nonglaucomatous populations. Arch. Ophthalmol., 119: 23–27.

Leske, M.C., Heijl, A., Hyman, L., Bengtsson, B., Dong, L. and Yang, Z. (1999) Early manifest glaucoma trial: design and baseline data. Ophthalmology, 106: 2144–2153.

Leske, M.C., Heijl, A., Hussein, M., Bengtsson, B., Hyman, L. and Komaroff, E.Early Manifest Glaucoma Trial Group. (2003) Factors for glaucoma progression and the effect of treatment: the early manifest glaucoma trial. Arch. Ophthalmol., 121(1): 48–56.

Liu, J. and Roberts, C.J. (2005) Influence of corneal biomechanical properties on intraocular pressure measurement: quantitative analysis. J. Cataract Refract. Surg., 31: 146–155.

Luce, D.A. (2005) Determining *in vivo* biomechanical properties of the cornea with an ocular response analyzer. J. Cataract Refract. Surg., 31(1): 156–162.

Martinez de la Casa, J.M., Garcia Feijoo, J., Vico, E., Fernandez Vidal, A., Benitez del Castillo, J.M., Wasfi, M. and Garcia Sanchez, J. (2006) Effect of corneal thickness on dynamic contour, rebound, and Goldmann tonometry. Ophthalmology, 113(12): 2156–2162.

Medeiros, F.A., Sample, P.A. and Weinreb, R.N. (2007) Comparison of dynamic contour tonometry and Goldmann applanation tonometry in African American subjects. Ophthalmology, 114(4): 658–665.

Miglior, S., Pfeiffer, N., Torri, V., Zeyen, T., Cunha Vaz, J. and Adamsons, I. (2007) Predictive factors for open-angle glaucoma among patients with ocular hypertension in the European glaucoma prevention study. Ophthalmology, 114(1): 3–9.

Nicolela, M.T., Soares, A.S., Carrillo, M.M., Chauhan, B.C., LeBlanc, R.P. and Artes, P.H. (2006) Effect of moderate intraocular pression changes on topographic measurement with confocal scanning laser tomography in patients with glaucoma. Arch. Ophthalmol., 124(5): 633–640.

Orssengo, G.J. and Pye, D.C. (1999) Determination of the true intraocular pressure and modulus of elasticity of the human cornea in vivo. Bull. Mathematical Biol., 61: 551–572.

Ortiz, D., Piñero, D., Shabayek, M.H., Arnalich Montiel, F. and Alió, J.L. (2007) Corneal biomechanical properties in normal, post-laser in situ keratomilleusis, and keratoconic eyes. J. Cataract Refract. Surg., 33(8): 1371–1375.

Pfeiffer, N., Torri, V., Miglior, S., Zeyen, T., Adamsons, I. and Cunha Vaz, J. (2007) Central corneal thickness in the European glaucoma prevention study. Ophthalmology, 114(3): 454–459.

Shah, S., Laiquzzaman, M., Bhojwani, R., Mantry, S. and Cunliffe, I. (2007) Assessment of the biomechanical properties of the cornea with the ocular response analyzer in normal and keratoconic eyes. Invest. Ophthalmol. Vis. Sci., 48(7): 3026–3031.

CHAPTER 4

Angle-closure: risk factors, diagnosis and treatment

Nishani Amerasinghe and Tin Aung*

Singapore National Eye Centre, 11 Third Hospital Avenue, Singapore, Singapore

Abstract: Introduction: Primary angle-closure glaucoma (PACG) is the leading cause of blindness in East Asia. The disease can be classified into primary angle-closure suspect, primary angle closure (PAC), and PACG. Pupil-block, anterior nonpupil-block (plateau iris and peripheral iris crowding), lens related and retrolenticular mechanisms have been suggested as the four main mechanisms of angle closure. Risk factors: The risk factors for PAC are female gender, increasing age, Inuit or East Asian ethnicity, shallow anterior chamber, shorter axial length, and genetic factors. Diagnosis: The diagnosis of acute PAC is mainly clinical. Diagnosis can be made with careful slit lamp examination, including intraocular pressure (IOP) measurement and gonioscopy. The diagnosis of chronic PAC and chronic PACG also require a careful history to assess risk factors, slit lamp examination including IOP and gonioscopy. Further investigations may also be required including visual fields, ultrasound biomicroscopy, and other imaging methods. Management: In acute PAC, rapid control of the IOP needs to be achieved to limit optic-nerve damage. This can be carried out medically, and/or by laser iridoplasty. Both the affected and fellow eye should undergo laser peripheral iridotomy (PI). The aim of treating chronic PAC is to eliminate the underlying pathophysiological mechanism and to reduce IOP. This can be done by carrying out laser PI, iridoplasty, medical therapy, or surgery (trabeculectomy, lens extraction, combined lens extraction with trabeculectomy and goniosynechialysis). Conclusion: Angle-closure glaucoma is usually an aggressive, visually destructive disease. By assessing the risk factors and diagnosing the mechanism involved in a patient's condition, the management of that patient can be tailored appropriately.

Keywords: primary angle-closure suspect; primary angle closure; primary angle-closure glaucoma; risk factors; diagnosis; management

Introduction

Primary angle-closure glaucoma (PACG) is a leading cause of blindness in East Asia (Foster et al., 1996, 2000; Foster and Johnson, 2001). It has a greater tendency to cause bilateral blindness than primary open angle glaucoma (POAG) (Seah et al., 1997; Wong et al., 2000). In China, PACG is responsible for over 90% of bilateral glaucoma blindness (Foster and Johnson, 2001). It is also responsible for most bilateral glaucoma blindness in Singapore and India (Dandona et al., 2000; Foster et al., 2000).

Due to the problem of variability of nomenclature for angle closure in early papers, recent studies have adopted the following definitions. Primary angle-closure suspect (PACS) is the term for an eye in which contact between the peripheral iris and posterior trabecular meshwork is considered possible, but there are no other abnormalities in the eye

*Corresponding author. Tel.: (65) 6322 4592; Fax: (65) 6322 4598; E-mail: tin11@pacific.net.sg

(Aung et al., 2001; Foster et al., 2002). Primary angle closure (PAC) is present when there are features indicating that trabecular meshwork obstruction by the peripheral iris has occurred with consequences in the eye such as peripheral anterior synechiae (PAS), increased intraocular pressure (IOP), iris whorling, glaucomfleken, lens opacities, or excessive pigment deposition on the trabecular meshwork. At this stage, the optic disc does not have signs of glaucomatous damage. PACG is PAC with evidence of glaucomatous optic neuropathy (GON). Table 1 summarizes the definitions.

A patient with acute primary angle closure (APAC) usually has the following symptoms: ocular or periocular pain, nausea and/or vomiting, a history of intermittent blurring of vision with haloes, IOP > 21 mmHg: and the following signs: conjunctival injection, corneal epithelial edema, mid-dilated unreactive pupil, shallow anterior chamber, and the presence of an occludable angle.

Mechanism

The mechanism responsible for angle closure is also important, especially in planning clinical management. Pupil-block, anterior nonpupil-block (plateau iris and peripheral iris crowding), lens related and retrolenticular mechanisms have been suggested as the four main mechanisms of angle closure, though they may coexist. When assessing a case of angle closure both staging and mechanism should be taken into account (Ritch and Lowe, 1996a).

Ritch et al. described the mechanisms of angle closure resulting in iris blocking aqueous outflow through the trabecular meshwork. This is caused by forces acting at four anatomic levels: the iris, ciliary body, the lens, and vectors posterior to the lens. It should be noted that each level of the block may have a component of the preceding levels and a combination of mechanisms may coexist in the same patient. Therefore, treatment can become more complex for each level of the block as the lower levels of block may also require treatment (Ritch and Lowe, 1996a).

Level 1 — iris and pupil

Pupillary block is the most common mechanism of angle closure (Nolan et al., 2000), the majority of other causes of angle closure will have an element of pupil block. In East Asians, the mechanism is predominantly mixed (He et al., 2006). In pupillary block, there is resistance to aqueous flow through the pupil in the area of iridolenticular contact. This causes a limitation of aqueous flow from the nonpigmented ciliary epithelium (where it is produced) in the posterior chamber to the anterior chamber. This creates an increased pressure gradient between the anterior and posterior chambers causing anterior bowing of the iris, narrowing of the angle, and acute or chronic iridotrabecular contact. Usually the anterior segment structures appear normal; however, occasionally, there may be abnormalities of the iris architecture (thickness, orientation, muscle tone) that may be contributing factors. Laser iridotomy relieves the pressure difference between the anterior and posterior chambers. This reduces the iris convexity, the iris

Table 1. Classification of primary angle closure (PAC)

Acute primary angle closure (APAC)	Symptoms: ocular/periocular pain, nausea and/or vomiting, a history of intermittent blurring of vision with halos, IOP >21 mmHg
	Signs: conjunctival injection, corneal epithelial oedema, mid-dilated unreactive pupil, shallow anterior chamber, and/or occudable angle
Primary angle closure suspect (PACS)	Contact between peripheral iris and posterior trabecular meshwork is considered possible Eye otherwise normal
Primary angle closure (PAC)	Occludable drainage angle with trabecular meshwork obstruction by peripheral iris by PAS, raised IOP, iris whorling, glaucomflecken, iris opacities, excessive pigment deposition on trabecular surface
	No optic disc damage
Primary angle closure glaucoma (PACG)	PAC with evidence of GON

becomes flatter and the iridocorneal angle widens. The area of iridolenticular contact increases, as aqueous flows through the iridotomy rather than the pupillary space (Foster et al., 2006).

Level II — ciliary body

Abnormal ciliary body position leads to anteriorly positioned ciliary processes, these force the peripheral iris into the angle causing angle closure. This is known as plateau iris. Examining the angle with gonioscopy will show the iris root angulated forward and centrally. Laser iridotomy partially opens or fails to open the angle. Therefore laser iridoplasty is the treatment of choice. Plateau iris syndrome occurs when angle closure develops, either spontaneously or after pupillary dilation, in an eye with plateau iris configuration despite a patent iridotomy. Other disorders of the ciliary body may mimic plateau iris configuration and include iridociliary cysts, tumors, or edema (Foster et al., 2006).

Level III — lens-induced glaucoma

A large lens (due to intumescence) may press against the iris and ciliary body, forcing them forward and therefore causing acute or chronic angle-closure glaucoma (phacomorphic glaucoma). This may also occur if there is anterior subluxation of the lens (Foster et al., 2006).

Level IV — malignant glaucoma

In this form of glaucoma, the pressure difference is created between the vitreous and aqueous compartments due to aqueous misdirection into the vitreous. This pushes the lens–iris diaphragm forward. Anterior rotation of the ciliary body with forward rotation of the lens–iris diaphragm and relaxation of the zonular apparatus causes anterior lens displacement causing angle closure by pushing the iris against the trabecular meshwork. A shallow supraciliary detachment maybe present and it is this effusion that is thought to cause the anterior rotation of the ciliary body (Foster et al., 2006).

Other causes of angle closure

These include anterior subluxation of the lens, iris, or ciliary body cysts, ciliary body tumors or inflammation, as well as air or gas bubbles after intraocular surgery. PAS can be caused by iris and angle neovascularization, iridocorneal endothelial syndrome, or anterior uveitis. These disorders need to be identified and treated specifically (Foster et al., 2006).

Risk factors

A list of the common/important risk factors for angle-closure glaucoma are summarized in Table 2.

Age and gender

The prevalence of angle closure increases with age (Seah et al., 1997; Wong et al., 2000; Lai et al., 2001; Vijaya et al., 2007). The prevalence of narrow angles, PAC, and PACG is higher in females compared to males (Teikari et al., 1987; Seah et al., 1997; Wong et al., 2000; Lai et al., 2001; Ivanisevic et al., 2002; Vijaya et al., 2007).

Ethnicity

East Asian populations (Chinese from Singapore and Hong Kong) have the highest incidence rates of acute angle closure. Compared to East Asians, South and South East Asians (Indians, Thais, and Malays) have lower rates of angle-closure. It is noted that only 25–35% of angle closure in Asian people causes symptoms (Foster et al., 1996, 2000; Congdon et al., 1996; Yip and Foster, 2006). The highest rates of angle-closure glaucoma are found in the Inuits of Alaska, Canada, and Greenland (Alsbirk, 1976;

Table 2. Risk factors for angle closure

Increasing age
Female gender
Ethnicity: Inuits and East Asians
Shallow anterior chamber depth
Shorter axial length
Genetic factors

Arkell et al., 1987). Studies from India show that PACG is more common in India than in Europeans and has a tendency to be asymptomatic (Jacob et al., 1998; Dandona et al., 2000). PAC is not a common condition in Europeans; its prevalence rate is 0.1% or less in people over 40 years old (Hollows and Graham, 1966; Coffey et al., 1993; Wensor et al., 1998). In the African population, a study has shown that the rate of PAC was equal among black and white populations in Johannesburg (Luntz, 1973); however, studies looking at people of Bantu ethnicity have shown a prevalence of PACG at 0.5% much lower then the prevalence of other glaucomas (Buhrmann et al., 2000).

Ocular biometry

Anterior chamber depth (ACD) is the most important anatomical risk factor for angle-closure (Yip and Foster, 2006). A shallower ACD leads to an increased risk of angle closure, and this collates to demographic factors. ACD is shallower in females and decreases with increasing age (Alsbirk, 1974; Foster et al., 1997). This age related change is also more pronounced in women than men (Alsbirk, 1976; Foster et al., 1997).

As the position and thickness of the lens determines the ACD, these factors are the ultimate determinants of the risk of angle-closure. A study comparing normals with angle closure in Australians found the ACD to be 1.0 mm shallower in the PAC group. 66% of this difference was due to the more anterior positioning of the lens and 33% was due to the lens being thicker than normal (Lowe, 1969). A later study of Chinese people found lens thickness was the major determinant of a shallow ACD. The lens position only accounted for 4% of the difference between angle closure and normal eyes (Friedman et al., 2003).

Subjects with angle closure have shorter axial lengths (Lowe, 1977). Studies in both Chinese and Indian populations have shown that affected people have shorter axial lengths compared to those classified as normal. Also, eyes suffering from "acute" angle closure have shorter axial lengths than those affected by the "chronic," asymptomatic angle closure (Lin et al., 1997; Sihota et al., 2000; George et al., 2003). Therefore in patients with angle closure, the anterior segment is more crowded as the lens is thicker and more anteriorly located. This is confirmed by ultrasound biomicroscopy (UBM), which demonstrates forward rotation of the ciliary processes (Marchini et al., 1998).

Studies from Mongolia and Singapore clearly show PAS development increases with reduction in ACD (Aung et al., 2005a). Interestingly, in the Singaporean population there was a consistent increase in PAS across the range of ACD but in the Mongolian population there was a clear threshold (2.4 mm) at or above which PAS was very uncommon (Aung et al., 2005a).

Genetics

Relative to POAG, PACG has been poorly researched. A genetic locus for PACG has not been published and candidate-gene-association studies have not been reported. Tornquist first suggested that PACG was transmitted by a single dominant gene in 1953. Ocular characteristics related to angle-closure glaucoma, namely, the anterior positioning of the lens, increased lens thickness, and shallow anterior chambers are more common in close relatives of affected patients than in the general population (Lowe, 1964, 1972; Tomlinson and Leighton, 1973; Alsbirk, 1975).

Recently, a study into the heritability of ACD as an intermediate phenotype of angle closure found additive genetic effects appear to be the major contributor to the variation of ACD and relative ACD (ACD/axial length) in Chinese twins (He et al., 2008). Small studies have reported that PACG subjects may carry a mutation in the myocilin gene (MYOC) (Faucher et al., 2002; Vincent et al., 2002). It is thought that in POAG, mutant MYOC proteins accumulate in the trabecular meshwork, impairing aqueous humor outflow (Jacobson et al., 2001). However, molecular analysis of the MYOC gene in 106 Chinese patients with chronic PACG was negative and did not support the role of MYOC mutation in the pathogenesis of chronic PACG in the Chinese population (Aung et al., 2005b). An association between a single nucleotide polymorphism in matrix metallopeptidase 9 (MMP-9) gene and acute PAC has also been reported (Wang et al., 2006).

Diagnosis

Acute primary angle closure

The diagnosis of APAC is mainly clinical. There is a sudden, usually symptomatic rise in IOP. This is usually unilateral, but bilateral simultaneous attacks can occur. Patients complain of periocular or ocular pain, headache, nausea, and/or vomiting. The patients also complain of blurring of vision with haloes. Slit lamp examination will reveal conjunctival injection, corneal epithelial edema with a mid dilated sluggish/nonreacting pupil. The IOP is likely to be greater than 30 mmHg. The anterior chamber will be shallow and gonioscopy will show occludable or occluded drainage angles in both eyes. In some instances, if there is a severe anterior chamber reaction, with hypopyon, the IOP maybe normal or low due to ciliary body shutdown. It is important that these episodes of APAC are not misdiagnosed as uveitis. Also, certain types of open angle glaucoma can be associated with an acute rise in IOP with ocular pain, conjunctival injection, and corneal edema. These include phacolytic glaucoma, Posner–Schlossman syndrome, pseudoexfoliative glaucoma, and neovascular glaucoma.

Angle assessment in angle closure

In angle closure, the drainage angle is occludable and there are features indicating that trabecular obstruction has occurred like PAS. The gold standard technique to diagnose angle closure is gonioscopy. There are three widely used grading systems. The Scheie scheme is based on the angle structures seen during gonioscopy (Scheie, 1957). The Shaffer system requires the assessment of the angular distance between the iris and cornea (Becker and Shaffer, 1965). The Spaeth scheme allows for more detailed recording of the angle characteristics (geometric angle, iris profile, true and apparent level of insertion) (Spaeth, 1971). Table 3 gives a summary of the grading systems (Scheie,

Table 3. Gonioscopy grading systems

	0	I	II	III	IV
Shaffer	Closed	10°	20°	30°	40°
Modified Shaffer	Schwalbe's line not visible	Schwalbe's line visible	Anterior trabecular meshwork is visible	Scleral spur is visible	Ciliary band is visible
Scheie	Ciliary band is visible	Last roll of iris obscures ciliary body	Nothing posterior to the trabecular meshwork is visible	Posterior portion of trabecular meshwork is hidden	No structures posterior to Schwalbe's line visible

Spaeth system	
(1) Iris insertion	Anterior to Schwalbe's line
	Behind Schwalbe's line
	Centred at scleral spur
	Deep to scleral spur
	Extremely deep/on ciliary band
(2) Angle width	Slit
	10°
	20°
	30°
	40°
(3) Peripheral iris configuration	Queerly concave
	Regular
	Steep
(4) Trabecular meshwork pigment	0 (none) to 4 (maximal)

1957; Becker and Shaffer, 1965; Spaeth, 1971; South East Asia Glaucoma Interest Group, 2008). The Goldmann lens gives a stable, clear view of the important landmarks but indentation of appositionally closed angles using this lens has not been validated, and is difficult as the curvature of the lens is more than the corneal curvature. Therefore, the use of a four-mirror, like the Zeiss four-mirror is necessary. This lens has the same radius of curvature as the cornea so the patient's own tear film functions as a coupling agent. The Goldmann-type lenses require an optical coupling agent.

Gonioscopy technique

Gonioscopy should be carried out in a darkened room. The patient should have adequate topical anasthesia and should be looking in the primary position. The slit lamp beam should be 1 mm high and narrow. The light must be kept away from the pupil, at the lowest illumination that will allow angle visualization. The lens can be moved minimally along the cornea to see over the convexity of the iris, however, care must be taken not to apply pressure and cause indentation. Using high magnification, the termination of the corneal wedge (which marks the anterior edge of the trabecular meshwork) can be identified. Additionally, it is important to locate the scleral spur as the trabecular meshwork is directly anterior to this structure. Assessment of whether the iris is in contact with the trabecular meshwork is done. If it is not, the angle between the trabecular meshwork and adjacent peripheral iris is estimated and the level of the most anterior point of contact between the iris and angle structures is described. This is carried out for all four quadrants, then dynamic gonioscopy can be carried out.

If Goldmann-style lenses are being used, the patient should be instructed to look toward the mirror, the examiner should then press on the rim of the lens overlying the mirror, so as to indent the central cornea. The accuracy of indentation using this method has not been validated. The ideal technique involves using another goniolens with a diameter smaller than the corneal diameter, e.g. a four-mirror Zeiss lens. Pressure should be applied over the cornea, so as to displace aqueous from the centre of the anterior chamber into the periphery, pushing the iris posteriorly, falsely opening the drainage angle. This allows one to assess whether the iridotrabecular contact is appositional or synechial (i.e. permanent). The extent of the synechial closure should be assessed. Once it is determined that the angle is indeed occludable, the slit beam height and illumination and room lights should be turned up ideally prior to indentation gonioscopy to look for PAS. Any pseudo-PAS would open up with bright light besides the pressure applied on the cornea. Iris processes should not be confused with PAS. Iris processes are uveal extensions from the iris on to the trabecular meshwork and occur in normal angles. Figure 1 shows the normal angle anatomy.

Ultrasound biomicroscopy (UBM)

UBM gives good qualitative information about the drainage angle including visualization of the ciliary body. However, highly reproducible quantitative information is dependent on examiner technique and experience. UBM is usually performed with the patient in the supine position. A suitably sized eye cup (around 20 mm) is inserted between the eyelids and the coupling medium (e.g. methylcellulose and/or normal saline) is inserted into it. The probe is then inserted into the medium and real time images are displayed on a video monitor. These can be stored and/or printed out for analysis. It should be noted that room illumination and accommodation must be kept constant. Also the configuration of the anterior segment and the proportions of the structures seen depends on the plane of the section and any degree of tilt in the scanning probe (Liebmann, 2006). Figure 2 shows an UBM scan of narrow angles.

Kumar et al. (2008) have used the UBM to define plateau iris. The features of UBM are defined in each quadrant, and include the presence of an anteriorly directed ciliary body, an absent ciliary sulcus, a steep iris root from its point of insertion followed by a downward angulation from the corneoscleral wall, presence of a central flat iris plane, and irido-angle contact. At least two quadrants have to fulfill the above criteria for plateau iris to be defined (Kumar et al., 2008). Figure 3 shows the features.

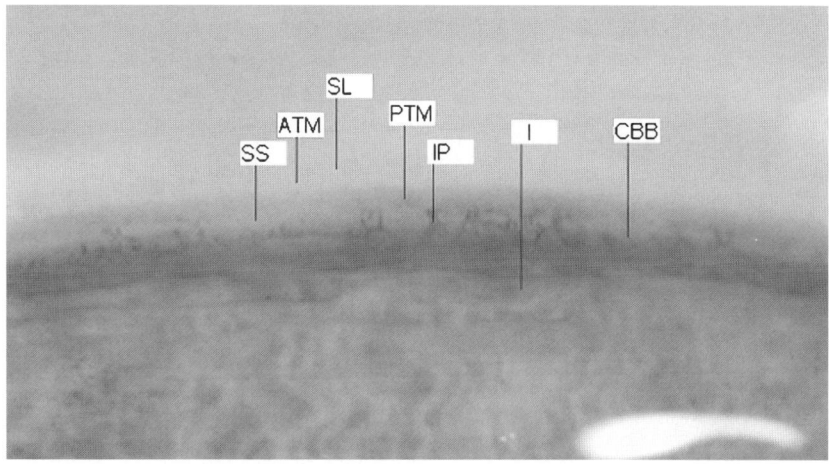

Fig. 1. Gonioscopic view of normal angle anatomy, showing iris (I), ciliary body band (CBB), scleral spur (SS), posterior trabecular meshwork (PTM), anterior trabecular meshwork (ATM), and Schwalbe's line (SL). Iris processes can also be clearly seen (IP). (Courtesy of Lisandro Sakata, MD, PhD, University of Alabama, Birmingham, USA.) (See Color Plate 4.1 in color plate section.)

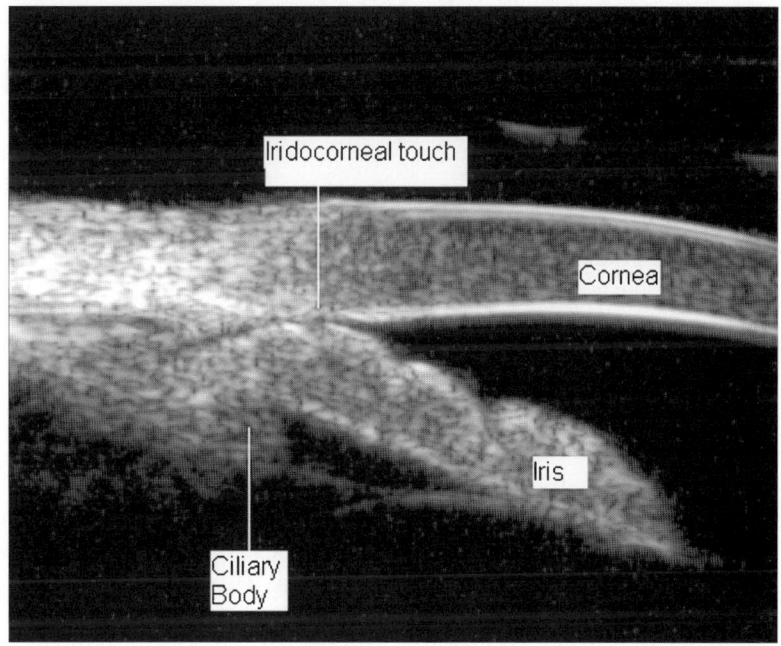

Fig. 2. The figure shows an ultrasound biomicroscopy scan of a closed angle; there is iridocorneal touch, obstructing the trabecular meshwork.

Anterior segment optical coherence tomography (AS-OCT)

The AS-OCT is a noncontact instrument that rapidly obtains high-resolution images of the angle and anterior chamber using infrared light (Baskaran, 2006). Unlike the UBM it cannot image the ciliary body. The image capture scan takes a few seconds and is akin to taking a photograph. The device allows qualitative and quantitative angle

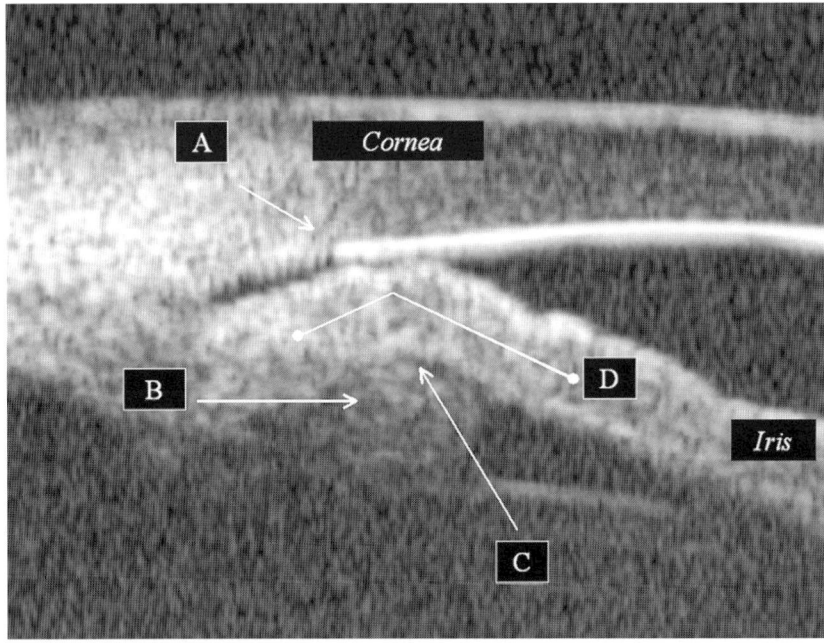

Fig. 3. The figure shows an ultrasound biomicroscopy image of a quadrant showing plateau iris after laser peripheral iridotomy. Features shown: (A) irido-angle touch, (B) anteriorly rotated ciliary process, (C) absent ciliary sulcus, and (D) iris angulation. (Courtsey of Rajesh Kumar, MS, Singapore National Eye Centre, Singapore.)

imaging, which is objective and reproducible. Research comparing UBM, AS-OCT, and gonioscopy shows the AS-OCT is good at identifying narrow angles; however, the device does identify more subjects as having closed angles than gonioscopy (Radhakrishnan et al., 2005, 2007). Figure 4 shows an AS-OCT scan of an eye with narrow angles.

Scanning peripheral anterior chamber depth analyzer (SPAC)

The SPAC does not image the angle per se but takes rapid slit images of the central and peripheral anterior chamber using an optical method and creates an iris anterior surface contour using these measurements. This is then graded and compared to the normative database and the resultant grade gives a risk assessment for the patient (Kashiwagi et al., 2004). The SPAC correlates well with the modified van Herick system in grading peripheral ACD. However, it overestimates the proportion of narrow angles relative to gonioscopy and the modified van Herick grading system (Baskaran et al., 2007).

Visual-field loss

It has been observed that the pattern of visual-field loss in PACG is different from that of POAG. Gazzard et al. showed that subjects with POAG had greater superior hemifield loss than in the inferior hemifield. This difference between the two hemifields was less pronounced in the PACG patients. However, the PACG group exhibited more severe visual-field loss compared to the POAG group. The authors postulated that POAG is thought to be due to a combination of pressure dependent and independent mechanisms whereas PACG is predominantly pressure related. This may be why there is less of a difference between the two hemifields in PACG patients. The reason for the more severe field loss is less clear and maybe due to the tendency of PACG patients to present later (Gazzard et al., 2002).

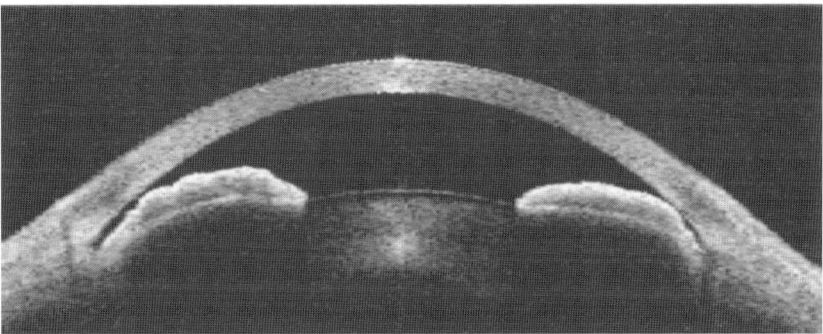

Fig. 4. The figure shows an anterior segment-optical coherence tomography scan of narrow angles. (Courtesy of Lavayana Raghavan, DO, Singapore National Eye Centre, Singapore.)

Management

Acute primary angle closure

The aims of the treatment of APAC are to achieve rapid control of IOP, so as to limit optic-nerve damage and eliminate pupil block. The patient should then be monitored for subsequent IOP elevation and development of PACG. The management should be tailored for the fellow eye as well.

Medical therapy

This aids the rapid reduction in IOP. Intravenous carbonic anhydrase inhibitors (e.g. acetazolamide) usually have a rapid IOP lowering affect. 500 mg of acetazolamide is usually given intravenously, together with an oral dose of 500 mg (if the patient is not vomiting). The side effects of acetazolamide include paraesthesia, drowsiness, confusion, loss of appetite, polydipsia, and polyuria. It can also cause metabolic acidosis and electrolyte disturbance, respiratory failure, and Stevens–Johnson syndrome (Lam et al., 2007). Therefore, it is important to test blood for urea and electrolytes.

The inflammation in the eye is controlled by topical steroids (e.g. dexamethasone 0.1% or prednisolone 1%). Topical IOP lowering medications (e.g. beta blockers and alpha2 adrenergics) should also be given. The patients can be in pain with nausea and vomiting, so analgesics and antiemetics should also be given as supportive measures. Also the patient should be laid supine to prevent further forward movement of the lens (Choong et al., 1999).

Miotics usually open the angle by pulling the peripheral iris away from the angle. However, in some eyes, miotics may increase the axial lens thickness and loosen the zonules, allowing for anterior lens movement, and thereby inducing further angle closure (Kobayashi et al., 1999). Due to this, it is preferable to withhold pilocarpine until the IOP has been reduced. Usually the pressure is rechecked 1 h after the commencement of treatment and pilocarpine is given then (Choong et al., 1999).

After 2 h, if the IOP is still above 35 mmHg, a hyperosmotic agent (e.g. 20% mannitol 1–2 g/kg) should be given intravenously for over 45 min. If there is no vomiting oral hyperosmotic agent (e.g. glycerol 1 g/kg) may be given as an alternative. Due care must be given as acidosis, pulmonary edema, congestive heart failure, dehydration, and acute renal failure are all side effects of hyperosmotics. Glycerol may cause ketoacidosis in diabetics.

Argon laser peripheral iridoplasty (ALPI)

If there are contraindications to systemic medications or the IOP is still elevated after some time, the next stage is to perform an ALPI. It involves the placement of a ring of contraction burns on the peripheral iris to contract the iris stroma near the angle, mechanically pulling open the angle. This allows for the eye to become quiet before the definitive treatment of laser peripheral iridotomy (PI) can be carried out. It has the advantage of being able to be performed in eyes with relatively hazy corneas and shallow angles; also it avoids the risks of complications from systemic therapy especially in APAC patients who are usually elderly with coexisting medical conditions. ALPI also opens the angles in eyes with plateau iris (Ritch et al., 2004).

The laser is performed with either an argon or diode laser. Either an Abraham (+66 dioptre) or Wise lens (+103 dioptre) is used. The burns are placed at the iris periphery as close to the limbus as possible. If bubbles or charring occur, the laser energy should be reduced. Four to six burns are placed per quadrant for a total of two to four quadrants (spot size 200–500 μm, energy 100–400 mW, duration 0.2–0.5 s).

Complications of iridoplasty include corneal endothelial cell damage, iris atrophy, inflammation, and PAS.

In cases where laser iridoplasty is not available, immediate anterior chamber paracentesis has been proposed as an alternative procedure to rapidly lower the IOP in APAC (Lam et al., 2007).

Laser peripheral iridotomy (PI)

Once the IOP is controlled and the cornea has cleared sufficiently, a laser PI can be carried out. This eliminates the pupil block. In Asian eyes, (which usually have thick brown irides) sequential use of the argon and neodyuim: yttrium-aluminum-garnet (Nd:YAG) lasers are used to create a PI, allowing for less total laser energy (Lim et al., 1996). In Caucasian eyes, with thinner irides, Nd:YAG laser may only be used. Burns with a small spot size, short duration, and high energy are placed as far in the iris periphery as possible between 11 and 2 o'clock positions. Care should be taken to avoid corneal burns.

First the iridotomy lens (Wise or Abraham's) is placed on the eye and argon laser is used to place four to six spots of about 600 mW in an overlapping pattern. The centre is then deepened with the energy increasing to achieve patency (spot size 50 μm, energy 600–1400 mW, duration 0.01–0.1 s).

The PI is then enlarged with the Nd:YAG laser (power 1.4–6 mJ).

Complications of PI include corneal endothelial damage, hemorrhage, cataract formation, imperforate PI, glare, retinal burns, IOP spikes, and malignant glaucoma. The post laser spikes can be reduced with perioperative use of alpha-2 agonists like brimonidine or apraclonidine (Chen et al., 2001).

Lens extraction

Removal of the crystalline lens deepens the anterior chamber and widens the drainage angle. In one report, it had a success rate of about 70% even if the eyes have significant PAS. Also, there were fewer additional surgical interventions and sight-threatening complications compared to trabeculectomy (Ming Zhi et al., 2003).

The preliminary results of a randomized control trial comparing phacoemulsification and conventional argon laser PI after APAC has shown an IOP rise of 3.2% in the phacoemulsification group compared to 28.3% in the iridotomy group. Surgery was performed 7 ± 3 days after aborting the APAC. The authors felt that phacoemulsification can be carried out soon after aborting APAC but was not without risk and felt a better option was to perform surgery about 4 weeks after APAC when the eye had settled down adequately and the IOP had not yet risen (Lam et al., 2007).

It is notable that lens removal does not remove the risk of angle closure in eyes with plateau iris syndrome (Tran et al., 2003).

Monitoring for subsequent IOP rise in eyes with APAC

In Caucasians, 70% of APAC is controlled by PI alone. However, in Asians, as many as 58% of eyes develop an increased IOP after resolution of the acute event. 77% of these eyes developed the increase within 6 months of the acute episode and 33% required glaucoma filtering surgery (Aung et al., 2001). These eyes require long-term follow up, especially those with optic-nerve damage, >50% PAS, failure of initial medical therapy to lower initial IOP, and delayed presentation (Nolan et al., 2000).

Fellow eye of APAC

The fellow eye will have the same dimensions as the attack eye and therefore is at high risk of APAC as well. Therefore a prophylactic PI should be carried out for the fellow eye. This will prevent APAC, however 10% may still have a rise in IOP over time and therefore require close monitoring including regular gonioscopy. Figure 5 is a flow diagram for the management of APAC.

Chronic primary angle-closure glaucoma (CACG)

The aim of treatment is to eliminate the underlying pathophysioloical mechanism causing the angle closure (mainly pupil block and peripheral angle crowding/plateau iris).

Laser peripheral iridotomy

All CACG patients should undergo PIs. However, this is an unsatisfactory long-term therapy in the sense that additional treatment is often required, particularly for eyes with GON. The majority of patients require further medication or surgery to control IOP (Alsagoff et al., 2000).

Laser iridoplasty

If the patient has plateau iris or peripheral crowding of the angle, i.e. the angle remains occludable after PI, then peripheral iridoplasty can be considered. It should be carried out early in the disease course as laser iridoplasty might not be effective when PACG is well established or when medical therapy has already failed, or in the presence of extensive PAS.

Medical therapy

The need for further medical therapy after iridotomy is determined by IOP and the extent of glaucomatous damage. This is common in Asian patients and can be difficult to manage (Ritch and Lowe, 1996b). It is known as "residual CACG after iridectomy or iridotomy." Usually, nonpupil block mechanisms including lens factors, plateau iris, and damage to the trabecular meshwork, are the cause (Ritch and Lowe, 1996b).

Medical therapy entails the use beta-adrenergic agonists, alpha2 adrenergic agonists, mitotics, topical carbonic anhydrase inhibitors, and prostaglandins. Of note long-term miotic treatment in the absence of an iridotomy may expedite the development of acute angle closure. Low dose pilocarpine can be used as an alternative to iridoplasty in plateau iris syndrome. Studies have shown that the prostaglandin, latanoprost, is more effective than the beta-blocker, timolol, in lowering IOP in PACG patients. It is thought that latanoprost enhances aqueous humor access to the ciliary body via the still open part of the drainage angle (Aung et al., 2000; Chew et al., 2004).

Trabeculectomy

This is indicated when the IOP cannot be controlled by laser iridotomy or medication, if there is evidence of continuing glaucomatous damage, poor compliance or intolerance to medical treatment, or poorly controlled glaucoma at the time of planned cataract surgery. The use of antiscarring agents is also similar to that of POAG, i.e. eyes at high risk of failure of surgery (e.g. those with previous failed trabeculectomy), eyes with advanced disease (extensive PAS, optic-nerve damage, and visual-field loss), and eyes on multiple medications (Aung, 2006).

Lens extraction

It is postulated that removal of the lens leads to deepening of the anterior chamber, resulting in reduction of angle crowding and the relief of pupil block. A prospective case series carried out in Hong Kong found both the IOP and requirement for glaucoma drugs reduced significantly after cataract extraction (Lai et al., 2006).

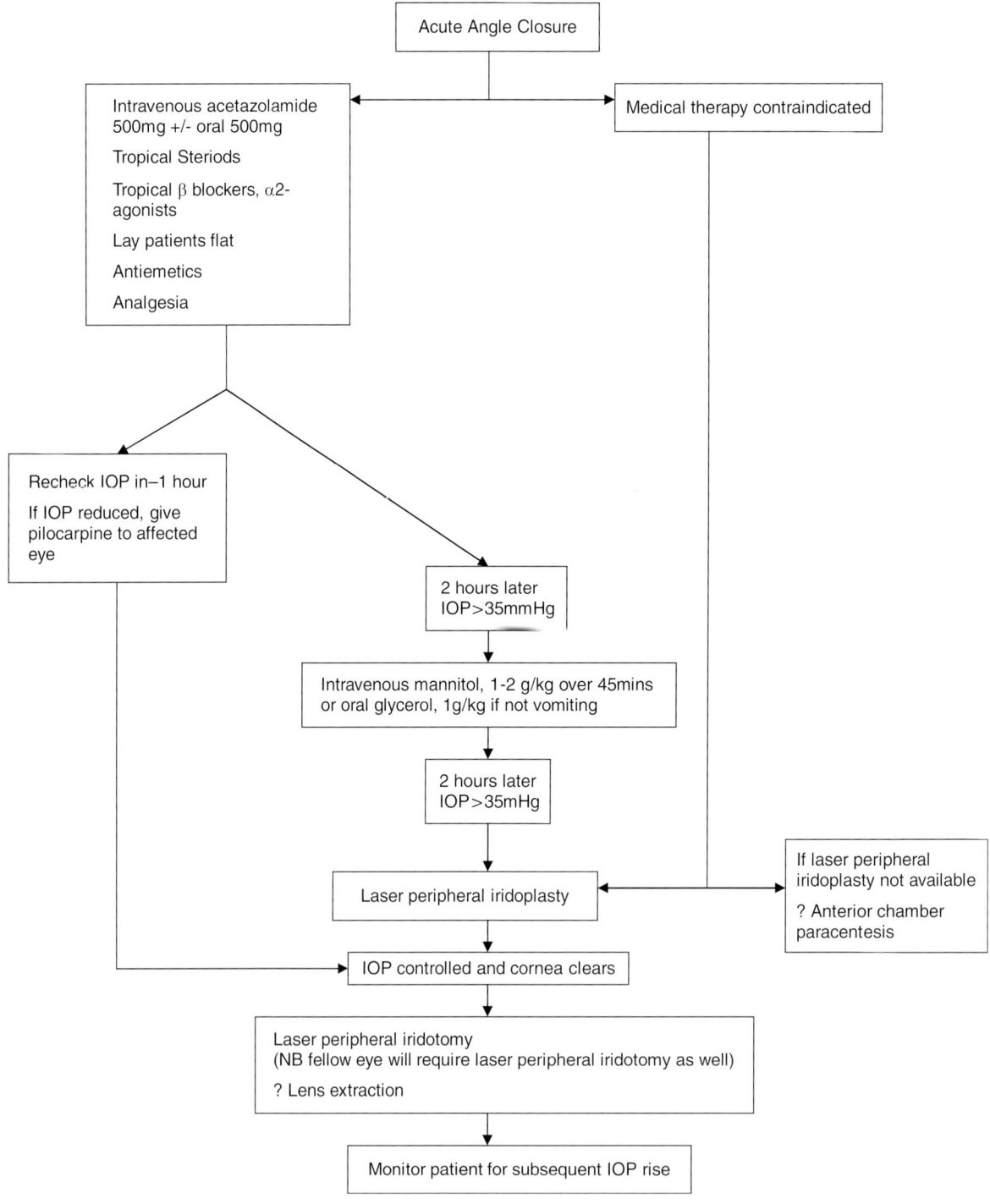

Fig. 5. Management of acute angle closure.

Combined lens extraction and trabeculectomy surgery

Combined lens extraction and filtration surgery also has a role in the management of CACG. This allows for visual rehabilitation after cataract extraction, prevention of IOP spikes in the immediate post-operative period, and widening of the angle after lens removal with improved IOP control. A retrospective study from Singapore showed that combined phacoemulisfication with posterior chamber intraocular lens implantation and trabeculectomy was associated with good IOP control and visual outcome in patients with PACG (Aung, 2006). Another study found the complication rates of phacotrabeculectomy were similar in PACG and POAG (Aung, 2006).

Goniosynechialysis

This procedure involves the stripping of PAS from the angle wall, therefore opening the angle and restoring trabecular function. It is another surgical option for the treatment of CACG and is more effective if the synechiae have been present for less than a year (Aung, 2006).

Summary

In summary, angle-closure glaucoma is usually an aggressive, visually destructive disease. By assessing the risk factors and diagnosing the mechanism involved in a patient's condition, the management of that patient can be tailored appropriately.

List of abbreviations

ACD	anterior chamber depth
ALPI	argon laser peripheral iridoplasty
APAC	acute primary angle closure
AS-OCT	anterior segment optical coherence tomography
CACG	chronic primary angle-closure glaucoma
g	grams
GON	glaucomatous optic neuropathy
IOP	intraocular pressure
kg	kilograms
mg	milligrams
mmHg	millimeters of mercury
MMP-9	matrix metallopeptidase 9
mW	illiwatts
MYOC	myocilin gene
Nd:YAG	neodyuim:yttrium-aluminum-garnet
PAC	primary angle closure
PACG	primary closed angle glaucoma
PACS	primary angle-closure suspect
PAS	peripheral anterior synechiae
PI	laser peripheral iridotomy
POAG	primary open angle glaucoma
s	seconds
SPAC	scanning peripheral anterior chamber depth analyzer
UBM	ultrasound biomicroscopy
%	percent
μm	micrometers

References

Alsagoff, Z., Aung, T., Ang, L.P. and Chew, P.T. (2000) Long-term clinical course of primary angle-closure glaucoma in an Asian population. Ophthalmology, 107: 2300–2304.

Alsbirk, P.H. (1974) Anterior chamber depth in Greenland Eskimos. A population study of variation with age and sex. Acta Ophthalmol. (Copenh.), 52: 551–564.

Alsbirk, P.H. (1975) Anterior chamber depth and primary angle-closure glaucoma. II. A genetic study. Acta Ophthalmol. (Copenh.), 53: 436–449.

Alsbirk, P.H. (1976) Primary angle-closure glaucoma. Oculometry, epidemiology, and genetics in a high-risk population. Acta Ophthalmol. (Copenh.), 54: 5–31.

Arkell, S.M., Lightman, D.A., Sommer, A., et al. (1987) The prevalence of glaucoma among Eskimos of Northwest Alaska. Arch. Ophthalmol., 105: 482–485.

Aung, T. (2006) In: Weinreb R.N. and Friedman D.S. (Eds.), Angle Closure and Angle Closure Glaucoma Consensus Series 3. Kugler Publications.

Aung, T., Ang, L.P., Chan, S.P. and Chew, P.T. (2001) Acute primary angle closure: long-term intraocular pressure outcome in Asian eyes. Am. J. Ophthalmol., 131: 7–12.

Aung, T., Nolan, W.P., Machin, D., et al. (2005a) Anterior chamber depth and the risk of primary angle closure in 2 East Asian populations. Arch. Ophthalmol., 123: 527–532.

Aung, T., Yong, V.H.K., Chew, P.T.K., et al. (2005b) Molecular analysis of the Myocilin gene in Chinese subjects with chronic primary-angle closure glaucoma. Invest. Ophthalmol. Vis. Sci., 46: 1303–1306.

Aung, T., Wong, H.T., Yip, C.C., et al. (2000) Comparison of the intraocular pressure-lowering effect of latanoprost and timolol in patients with chronic angle-closure glaucoma: a preliminary study. Ophthalmology, 107: 1178–1183.

Baskaran, M. (2006) In: Weinreb R.N. and Friedman D.S. (Eds.), Angle closure and angle closure glaucoma Consensus series 3. Kugler Publications Appendix C.

Baskaran, M., Oen, F.T., Chan, Y.H., et al. (2007) Comparison of the scanning peripheral anterior chamber depth analyzer and the modified van Herick grading system in the assessment of angle closure. Ophthalmology, 114: 501–506.

Becker, R. and Shaffer, R.N. (1965) Diagnosis and Therapy of the Glaucomas. CV Mosby, St. Louis, pp. 42–53.

Buhrmann, R.R., Quigley, H.A., Barron, Y., et al. (2000) Prevalence of glaucoma in a rural East African population. Invest Ophthalmol. Vis. Sci., 41: 40–48.

Chen, T.C., Ang, R.T., Grosskreutz, C.L., et al. (2001) Brimonidine 0.2% versus apraclonidine 0.5% for prevention of intraocular pressure elevations after anterior segment laser surgery. Ophthalmology, 108: 1033–1038.

Chew, P.T., Aung, T., Aquino, M.V. and Rojanapongpun, P.EXACT Study Group. (2004) Intraocular pressure-reducing effects and safety of latanoprost versus timolol in patients with chronic angle-closure glaucoma. Ophthalmology, 111: 427–434.

Choong, Y.F., Irfan, S. and Manage, M.J. (1999) Acute angle closure glaucoma: an evaluation of a protocol for acute treatment. Eye, 13: 613–616.

Coffey, M., Reidy, A., Wormald, R., et al. (1993) Prevalence of glaucoma in west of Ireland. Br. J. Ophthalmol., 77: 17–21.

Congdon, N., Quigley, H.A., Hung, P.T., et al. (1996) Screening techniques for angle-closure in rural Taiwan. Acta Ophthalmol. Scand., 74: 113–119.

Dandona, L., Dandona, R., Mandal, P., et al. (2000) Angle closure glaucoma in an urban population in Southern India. The Andhra Pradesh Eye Disease Study. Ophthalmology, 107: 1710–1716.

Faucher, M., Anctil, J.L., Rodrigue, M.A., et al. (2002) Founder TIGR/MYOC mutations for glaucoma in the Quebec population. Hum. Mol. Genet., 11: 2077–2090.

Foster, P.J., Alsbirk, P.H., Baasanhu, J., et al. (1997) Anterior chamber depth in Mongolians. Variation with age, sex and method of measurement. Am. J. Ophthalmol., 124: 40–53.

Foster, P.J., Bassanhu, J., Alsbrik, P.H., et al. (1996) Glaucoma in Mongolia. A population based-survey in Hovsgol Province, northern Mongolia. Arch. Ophthalmol., 114: 1235–1241.

Foster, P.J., Buhrmann, R., Quigley, H.A. and Johnson, G.J. (2002) The definition and classification of glaucoma in prevalence surveys. Br. J. Ophthalmol., 238–242.

Foster, P.J. and Johnson, G.J. (2001) Glaucoma in China. How big is the problem? Br. J. Ophthalmol., 85: 1277–1282.

Foster, P.J., Oen, F.T., Machin, D., et al. (2000) The prevalence of glaucoma in Chinese residents of Singapore: a cross sectional population survey of the Tanjong Pagar district. Arch. Ophthalmol., 118: 1105–1111.

Foster, P., He, M. and Liebmann, J. (2006) In: Weinreb R.N. and Friedman D.S. (Eds.), Angle Closure and Angle Closure Glaucoma Consensus Series 3. Kugler Publications, p. 20.

Friedman, D.S., Gazzard, G., Foster, P.J., et al. (2003) Ultrasonograghic biomicroscopy, Scheimpflug photography, and novel provocative tests in contralateral eyes of Chinese patients initially see with acute angle closure. Arch. Ophthalmol., 121: 633–642.

Gazzard, G., Foster, P.J., Viswanathan, A.C., et al. (2002) The severity and spatial distribution of visual field defects in primary glaucoma: a comparison of primary open-angle glaucoma and primary angle-closure glaucoma. Arch. Ophthalmol., 120: 1636–1643.

George, R., Paul, P.G., Baskaran, M., et al. (2003) Ocular biometry in occludable angles and angle closure glaucoma: a population based survey. Br. J. Ophthalmol., 87: 399–402.

He, M., Foster, P.J., Johnson, G.J. and Khaw, P.T. (2006) Angle-closure glaucoma in East Asian and European people. Different diseases? Eye, 20: 3–12.

He, M., Wang, D., Zheng, Y., et al. (2008) Heritability of anterior chamber depth as an intermediate phenotype of angle-closure in Chinese: The Guangzhou Twin Eye Study. Invest Ophthalmol. Vis. Sci., 49: 81–86.

Hollows, F.C. and Graham, P.A. (1966) Intraocular pressure, glaucoma and glaucoma suspects in a defined population. Br. J. Ophthalmol., 50: 570–586.

Ivanisevic, M., Ereeg, M., Smoljanovic, A. and Trosic, Z. (2002) Incidence and seasonal variations of acute primary angle-closure glaucoma. Coll. Antropol., 26: 41–45.

Jacob, A., Thomas, R., Koshi, S.P., et al. (1998) Prevalence of primary glaucoma in an south Indian population. Ind. J. Ophthalmol., 46–86.

Jacobson, N., Andrews, M., Shepard, A.R., et al. (2001) Non-secretion of mutant proteins of the glaucoma gene, myocilin in cultured trabecular meshwork cells and in aqueous humour. Hum. Mol. Genet., 10: 117–125.

Kashiwagi, K., Abe, K. and Tsukahara, S. (2004) Quantitative evaluation of changes in anterior segment biometry by peripheral laser iridotomy using newly developed scanning peripheral anterior chamber depth analyser. Br. J. Ophthalmol., 88: 1036–1041.

Kobayashi, H., Kobayashi, K., Kiryu, J. and Kondo, T. (1999) Pilocarpine induces an increase in the anterior chamber angular width in eyes with narrow angles. Br. J. Ophthalmol., 83: 553–558.

Kumar, K.S., Baskaran, M., Chew, P.T., et al. (2008) Prevalence of plateau iris in primary angle closure suspects an ultrasound biomicroscopy study. Ophthalmology, 115: 430–434.

Lai, J.S., Liu, D.T., Tham, C.C., et al. (2001) Epidemiology of acute primary angle-closure glaucoma in the Hong Kong Chinese population: prospective study. Hong Kong Med. J., 7: 118–123.

Lai, J.S., Tham, C.C. and Chan, J.C. (2006) The clinical outcomes of cataract extraction by phacoemulisfication in eyes with primary angle-closure glaucoma (PACG) and co-existing cataract: a prospective case series. J. Glaucoma, 15: 47–52.

Lam, D.S., Tham, C.C., Lai, J.S. and Leung, D.Y. (2007) Current approaches to the management of acute primary angle closure. Curr. Opin. Ophthalmol., 18: 146–151.

Liebmann, J. (2006) In: Weinreb R.N. and Friedman D.S. (Eds.), Angle Closure and Angle Closure Glaucoma Consensus Series 3. Kugler Publications Appendix B.

Lim, L., Seah, S.K. and Lim, A.S. (1996) Comparison of argon laser iridotomy and sequential argon laser and Nd:YAG laser iridotomy in dark irides. Ophthalmic Surg. Lasers, 27: 285–288.

Lin, Y.W., Wang, T.H. and Hung, P.T. (1997) Biometric study of acute primary angle-closure glaucoma. J. Formos. Med. Assoc., 96: 908–912.

Lowe, R.F. (1964) Primary angle-closure glaucoma. Family histories and anterior chamber depths. Br. J. Ophthalmol., 48: 191–195.

Lowe, R.F. (1969) Causes of shallow anterior chamber in primary angle closure glaucoma. Ultrasonic biometry of normal and angle-closure eyes. Am. J. Ophthalmol., 67: 87–93.

Lowe, R.F. (1972) Anterior lens curvature. Comparisons between normal eyes and those with primary angle-closure glaucoma. Br. J. Ophthalmol., 56: 409–413.

Lowe, R.F. (1977) Primary angle-closure glaucoma: A review of ocular biometry. Aust. J. Ophthalmol., 5: 9–17.

Luntz, M.H. (1973) Primary angle-closure glaucoma in urbanized South African Caucasoid and Negroid communities. Br. J. Ophthalmol., 57: 445–446.

Marchini, G., Pagliarusco, A., Toscano, A., et al. (1998) Ultrasound biomicroscopic and conventional ultrasonographic study of ocular dimensions in primary angle closure glaucoma. Ophthalmology, 105: 2091–2098.

Ming Zhi, Z., Lim, A.S. and Yin Wong, T. (2003) A pilot study of lens extraction in the management of acute primary angle-closure glaucoma. Am. J. Ophthalmol., 135: 534–536.

Nolan, W.P., Foster, P.J., Devereux, J.G., et al. (2000) YAG laser iridotomy treatment for primary angle closure in East Asian eyes. Br. J. Ophthalmol., 84: 1255–1259.

Radhakrishnan, S., Goldsmith, J., Huang, D., et al. (2005) Comparison of optical coherence tomography and ultrasound biomicroscopy for detection of narrow anterior chamber angles. Arch. Ophthalmol., 123: 1053–1059.

Radhakrishnan, S., See, J., Smith, S.D., et al. (2007) Reproducibility of anterior chamber angle measurements obtained with anterior segment optical coherence tomography. Invest. Ophthalmol. Vis. Sci., 48: 1303–1306.

Ritch, R. and Lowe, R.F. (1996a) In: Ritch R., Shields M.B. and Krupin T. (Eds.), Classifications and Mechanisms of the Glaucomas, The Glaucomas (2nd edn.). Mosby, St. Louis, p 752.

Ritch, R. and Lowe, R.F. (1996b) Angle-closure glaucoma: therapeutic overview. In: Ritch R., Shields M.B. and Krupin T. (Eds.), The Glaucomas (2nd edn.). Mosby, St. Louis, pp 1521–1531.

Ritch, R., Tham, C.C. and Lam, D.S. (2004) Long-term success of argon laser peripheral iridoplasty in the management of plateau iris syndrome. Ophthalmology, 111: 104–108.

Scheie, H.G. (1957) Width and pigmentation of the angle of the anterior chamber. A system of grading by gonioscopy. Arch. Ophthalmol., 58: 510–512.

Seah, S.K., Foster, P.J., Chew, P.T., et al. (1997) Incidence of acute angle closure glaucoma in Singapore. An island-wide survey. Arch. Ophthalmol., 115: 1436–1440.

Sihota, R., Lakshimaiah, N.C., Agrawal, H.C., et al. (2000) Ocular parameters in the subgroups of angle closure glaucoma. Clin. Exp. Ophthalmol., 28: 253–258.

South East Asia Glaucoma Interest Group website. Available at http://www.seagig.org/pdf/APGGuidelinesNMview.pdf. Accessed February 29th, 2008

Spaeth, G.L. (1971) The normal development of the human anterior chamber angle: a new system of descriptive grading. Trans. Ophthalmol. Soc. UK, 91: 709–739.

Teikari, J., Raivio, I. and Nurminen, M. (1987) Incidence of acute glaucoma in Finland from 1973 to 1982. Graefes Arch. Clin. Exp., 225: 357–360.

Tomlinson, A. and Leighton, D.A. (1973) Ocular dimensions in the heredity of angle-closure glaucoma. Br. J. Ophthalmol., 57: 475–486.

Tran, H.V., Liebmann, J.M. and Ritch, R. (2003) Iridociliary apposition in plateau iris syndrome persists after cataract extraction. Am. J. Ophthalmol., 135: 40–43.

Vijaya, L., George, R. and Arvind, H., et al. (2007) Prevalence of primary angle-closure disease in an urban South Indian population and comparison with a rural population The Chennai Glaucoma Study. Ophthalmology (Epub ahead of print).

Vincent, A.L., Billingsley, G., Buys, Y., et al. (2002) Digenic inheritance of early-onset glaucoma: CYP1B1, a potential modifier gene. Am. J. Hum. Genet., 70: 448–460.

Wang, I.J., Chiang, Th., Shih, Y.F., et al. (2006) The association of single nucleotide polymorphisms in the MMP-9 genes with susceptibility to acute primary angle closure glaucoma in Taiwanese patients. Mol. Vis., 12: 1223–1232.

Wensor, M.D., Mc Carty, C.A., Stanislavsky, Y.L., et al. (1998) The prevalence of glaucoma in the Melbourne Visual Impairment Project. Ophthalmology, 105: 733–739.

Wong, T.Y., Foster, P.J., Seah, S.K. and Chew, P.T. (2000) Rates of hospital admissions for primary angle closure glaucoma among Chinese, Malays and Indians in Singapore. Br. J. Ophthalmol., 84: 990–992.

Yip, J.L.Y. and Foster, P.J. (2006) Ethnic differences in primary angle-closure glaucoma. Curr. Opin. Ophthalmol., 17: 175–180.

CHAPTER 5

Early diagnosis in glaucoma

David F. Garway-Heath[1,2,*]

[1]*NIHR Biomedical Research Centre for Ophthalmology, Moorfields Eye Hospital NHS Foundation Trust and UCL Institute of Ophthalmology, London, UK*
[2]*G.B. Bietti Foundation — IRCCS (Istituto di Ricovero e Cura a Carattere Scientifico), Rome, Italy*

Abstract: This chapter reviews the evidence for the clinical application of vision function tests and imaging devices to identify early glaucoma, and sets out a scheme for the appropriate use and interpretation of test results in screening/case-finding and clinic settings. In early glaucoma, signs may be equivocal and the diagnosis is often uncertain. Either structural damage or vision function loss may be the first sign of glaucoma; neither one is consistently apparent before the other. Quantitative tests of visual function and measurements of optic-nerve head and retinal nerve fiber layer anatomy are useful to either raise or lower the probability that glaucoma is present. The posttest probability for glaucoma may be calculated from the pretest probability and the likelihood ratio of the diagnostic criterion, and the output of several diagnostic devices may be combined to achieve a final probability. However, clinicians need to understand how these diagnostic devices make their measurements, so that the validity of each test result can be adequately assessed. Only then should the result be used, together with the patient history and clinical examination, to derive a diagnosis.

Keywords: glaucoma; perimetry; imaging; diagnosis; probability; likelihood ratio

Introduction

The process of diagnosis involves the gathering of information about a patient, through history-taking, clinical examination, and diagnostic tests, to formulate the probability that a disease is either present or not present. At each stage of this process, a piece of information either increases or decreases the probability. This process is called Bayesian inference and clinicians use this in everyday practice, although usually subconsciously and not in a formalized, quantitative manner — the clinician uses his or her experience to put pieces of information together to arrive at a conclusion.

Identification of the glaucomatous neuropathy, when the disease is moderately advanced, is not a difficult task. The typical features of diffuse with superimposed focal rim narrowing and retinal nerve fiber layer (RNFL) loss, and associated characteristic patterns of visual-field loss, are easily recognized. However, diagnosis of the early stages of glaucoma can be challenging, with equivocal features in the optic-nerve head (ONH), RNFL, or visual field. Additional diagnostic tests may aid the clinician in making a diagnosis of early glaucoma and much research effort has been directed to develop more sensitive tests to reliably identify early glaucomatous visual function loss

*Corresponding author. Tel.: +44 20 7566 2059;
Fax: +44 20 7566 2059;
E-mail: david.garway-heath@moorfields.nhs.uk

and imaging devices to identify the earliest signs of structural damage. Yet the role of these tests in the clinical routine is unclear. The main emphasis of this chapter will be on the interpretation and integration of test results in the diagnostic process.

To make appropriate use of diagnostic tests, and the results they produce, one should consider how early in the course of the disease it is necessary to make a definite diagnosis, and how much time and resources should be devoted to this task. The aim of glaucoma management is to prevent symptomatic vision loss during a patient's lifetime. Patients at greatest risk of symptomatic vision loss are those in whom the diagnosis has been made late in the disease process (Chen, 2003; Forsman et al., 2007). As glaucoma is generally a slowly progressive disease, patients with very early or equivocal disease are not at great risk of symptomatic vision loss. Thus, in these cases of diagnostic uncertainty, an initial diagnosis of "suspect glaucoma" is probably sufficient. Such cases can be monitored (at a frequency dictated by the patients risk profile) and the diagnosis made when definite change (progression) is identified or the findings of the clinical examination or conventional tests become unequivocal.

The stage of disease that one aims to identify depends on the clinical situation. When screening, or case-finding, for glaucoma, the false–positive "identification" of cases needs to be limited to avoid over-burdening the health care system and causing unnecessary distress to individuals who are, in fact, healthy. The actual number of false positive cases identified is related to the prevalence of the condition (or prior probability that the disease is present) and the specificity (true negative, or 1 − false positive rate) of the test being applied. The specificity of any test is lower (and the false positive rate is higher) when selecting a diagnostic criterion that is sensitive in earlier stages of disease. Therefore, one has to balance the stage that one aims to identify against the expected false–positive rate. When screening or case-finding in the general population, this may mean not attempting to identify the very earliest signs of glaucoma, but aiming to identify moderate (unequivocal) disease. However, in individuals already referred to secondary care, the probability that glaucoma is present is already much higher than it is in the general population and the clinician can afford to aim to identify earlier stages of glaucoma without generating too many false–positive diagnoses.

History and examination

The purpose of history-taking is to establish the background probability that a particular condition is present. For glaucoma, this includes identifying the family history — glaucoma tends to run in families and a positive family history for glaucoma increases the probability that glaucoma is present. Other aspects of the patient history may raise the probability for glaucoma being present — age, ethnic background, myopic refractive error, previous ocular injury or surgery, previous steroid use, and other factors (Boland and Quigley, 2007). This background (or prior) probability becomes the context for the clinical examination, and the findings of the clinical examination are interpreted in the context of the patient history.

Important aspects of the clinical examination include intraocular pressure (IOP) measurement, gonioscopy, and ONH and RNFL examination.

The level of the IOP has a moderately strong effect on the probability that glaucoma is present, because although glaucoma can be present at all levels of IOP, the prevalence is much greater at higher IOP. This is illustrated by the findings of the Baltimore Eye Survey (Fig. 1). Similarly, the gonioscopy findings will influence the probability that glaucoma is present.

Clinical examination of the ONH and RNFL directly evaluates the eye for signs of glaucomatous damage. The signs, in the context of the history, may be so obvious that further tests are not needed to make the diagnosis. For instance, notching of the rim with adjacent wedge-shaped RNFL loss in an eye with an IOP of 28 mmHg is almost certain to have glaucoma. Additional testing is not required to make the diagnosis (although visual-field testing and imaging are required to document the extent of damage and to monitor for progression). However, lesser degrees of neural rim narrowing and uncertain

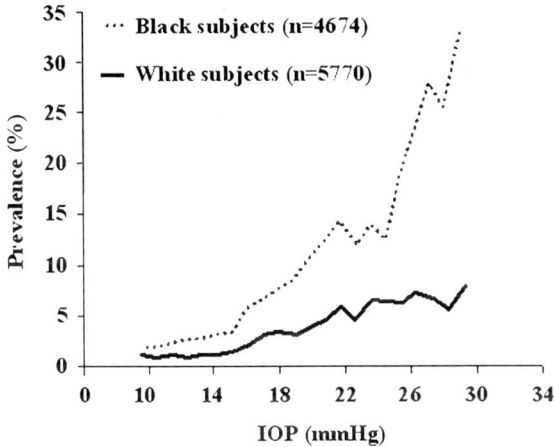

Fig. 1. The prevalence of glaucoma at various levels of intraocular pressure (adapted with permission from Sommer et al., 1991b).

thinning of the RNFL are associated with lower levels of probability and additional tests are helpful to raise or lower the probability.

Quantitative tests and the diagnostic process

Various quantitative tests are available to aid glaucoma diagnosis. These include standard automated perimetry (SAP) and "selective" tests of visual function, such as frequency doubling technology (FDT) perimetry and short-wavelength automated perimetry (SWAP), and imaging, such as confocal scanning laser ophthalmoscopy, scanning laser polarimetry, and optical coherence tomography.

There is a temptation for a busy clinician to read the output from a test (for instance, the Glaucoma Hemifield Test in Humphrey perimetry or the Moorfields Regression Analysis [MRA] in the Heidelberg retina tomograph [HRT]) and take it to be the "diagnosis." However, clinicians need to remember that "devices cannot diagnose our patients' conditions, but the findings they provide frequently alter the *probability* that a subject has a particular condition" (Garway-Heath and Friedman, 2006). Quantitative test results can be formally combined, using Bayesian statistics, to derive a probability for a disease being present. There are several steps of reasoning that the clinician should go through before and after ordering a diagnostic test — deciding what is the probability of glaucoma before the test (and whether the test being ordered will usefully alter that probability), deciding whether the test result is valid, and then deciding how the test result has altered the probability that glaucoma is present.

Pretest probability

The probability of glaucoma before application of the diagnostic test can be estimated in a semiquantitative manner by combining information from the history and clinical examination. An example of a quantitative estimation of glaucoma probability on the basis of IOP measurement can be derived from data reported from the Egna–Neumarkt Glaucoma Study (Bonomi et al., 2001). With a criterion of IOP >21 mmHg, 2.1% had ocular hypertension, 1.4% had hypertensive primary open-angle glaucoma, and 0.6% had normal tension glaucoma. Therefore, 3.5% of the population had an IOP >21 mmHg. The probability of glaucoma in those with an IOP >21 mmHg is $1.4/3.5 = 40\%$. The probability of glaucoma with an IOP <22 mmHg is $0.6/(1-0.035) = 0.62\%$.

This estimation can act as the "pretest probability," from which the "posttest probability" can be calculated, knowing the performance of the diagnostic test (see below).

Test validity

Before using the result of any diagnostic test, the clinician should evaluate the validity of the test. The validity depends on a number of factors: test quality and reproducibility, presence of confounding factors, and the appropriateness of the instrument reference database to the patient (Jaeschke et al., 2001).

Examples of factors affecting test quality include false–positive responses and learning effects (a particular problem for the newly-referred patient) for perimetry or scan quality for the quantitative imaging devices.

Confounding factors include central corneal thickness for IOP measurements, cataract or retinal pathology for perimetry, and image

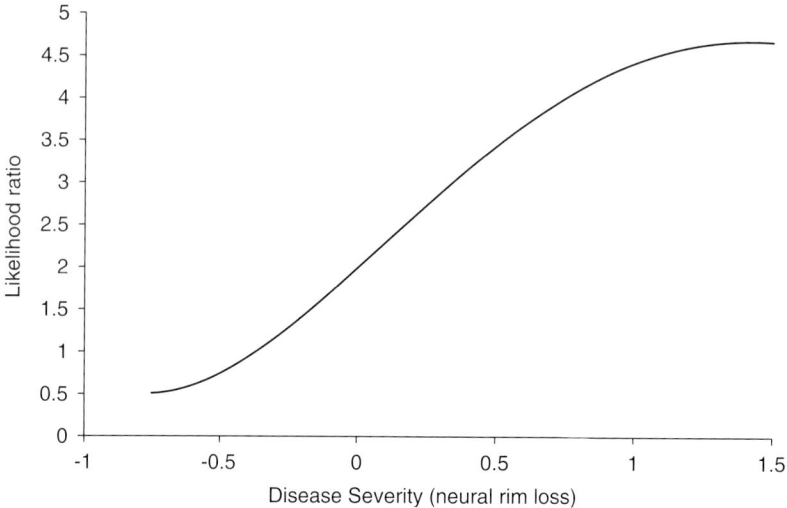

Fig. 4. Likelihood ratio values for a visual field mean deviation criterion at various stages of glaucoma, defined by the extent of neural rim loss at the optic-nerve head (adapted with permission from Stroux et al., 2003).

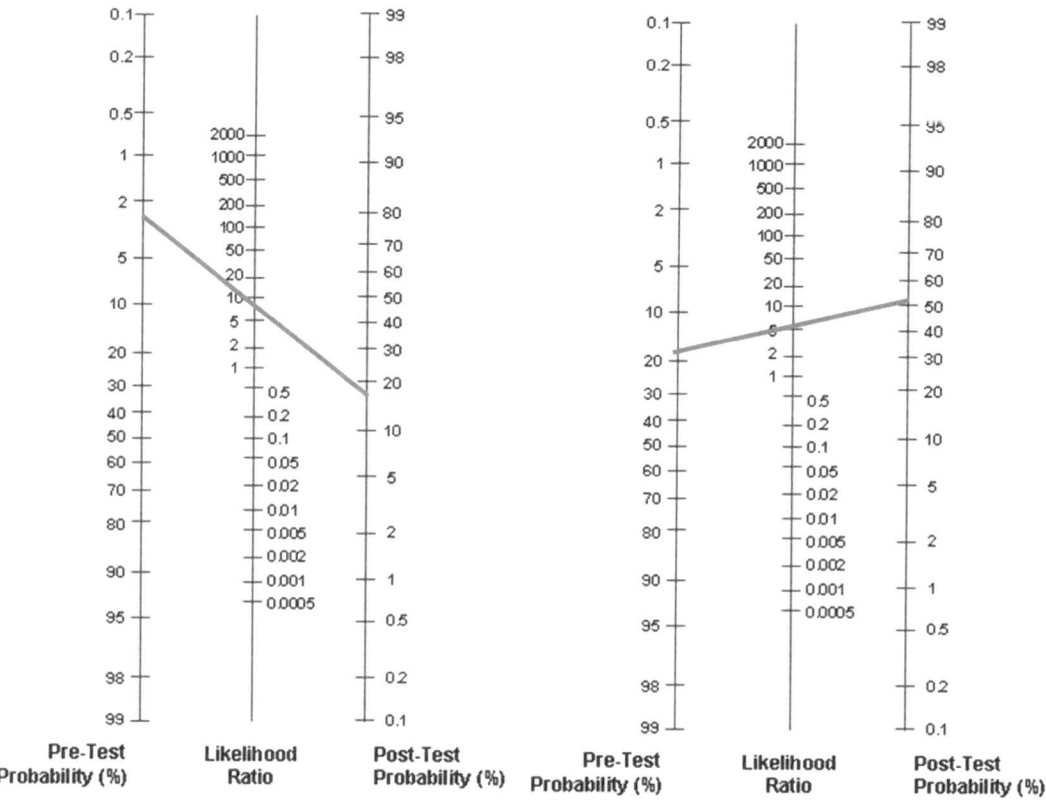

Fig. 5. Combining diagnostic test results. The panels represent the application of two diagnostic tests, in a population with a glaucoma prevalence of 2.5%. The first test has a likelihood ratio of 8 and the second a likelihood ratio of 5. The final probability is 51%.

for progression. In cases where the diagnosis is not certain from the history and clinical examination, the field test provides data that raise or lower the probability for glaucoma. This can be done formally, using the quantitative data reported by the test (such as the mean deviation and pattern standard deviation) and knowing the performance of the test (Stroux et al., 2003). However, additional information, deriving from the distribution of abnormal test points within the visual field, further influences the probability that glaucoma is present. The glaucoma hemifield test makes a quantitative comparison of the differential light sensitivity in regions of the upper and lower hemifields. The experience of the clinician is also valuable in assessing the distribution of abnormal points — the glaucomatous neuropathy is associated with characteristic patterns of visual-field loss, such as the arcuate distribution and nasal step, and artifacts, such superior defects related to lid ptosis, also have a characteristic appearance. Thus, given current data interpretation software, the evaluation of the visual-field test result cannot be entirely automated.

There is a widely held belief that SAP is not a sensitive test in early glaucoma. This stems from reports that a large proportion of retinal ganglion cells may be lost before the visual field becomes statistically abnormal (Quigley et al., 1988; Kerrigan-Baumrind et al., 2000) and that evidence of structural damage (ONH changes and RNFL loss) may be seen in some patients in the presence of a visual-field test "within normal limits" (Sommer et al., 1991a; Mohammadi et al., 2004). This gave rise to the idea of a "functional reserve" of ganglion cells. However, there is a growing body of evidence that there is no functional reserve, but a continuous structure/function relationship, so that the measured function relates directly to the number of retinal ganglion cells (Garway-Heath et al., 2000; Swanson et al., 2004). The implication is that structural and functional damage occurs at the same time, so that when a ganglion cell dies, some function is lost (Garway-Heath et al., 2002; Harwerth et al., 2004; Harwerth and Quigley, 2006). There are several factors that may disturb this one-to-one relationship, such as retinal ganglion cell dysfunction and media opacity, which may result in lower measurements of function than would be expected from the measurement of structure, and architectural changes to the ONH or RNFL structure which may not be directly related to ganglion cell loss.

The early identification of glaucoma, statistically, is limited by between-subject variability, so that 40–50% of retinal ganglion cells need to be lost before the visual function loss exceeds the 95% confidence limits for normality in the population (Harwerth et al., 2004). Similar findings are seen with between-subject variability in structural measurements, with the lower 98% confidence limit for the normal range of ONH neural rim area being about 65% of the average value — suggesting that 35% of the rim area needs to be lost before it becomes smaller than the lower end of the normal range (Garway-Heath and Hitchings, 1998a). This means that there needs to be a substantial amount of neural tissue loss before either structural or functional measurement fall below the statistically defined normal ranges. Thus, depending on the method for measurement and the individual, some eyes will have measurable damage first by structural measurements, whereas with another measurement method or in another individual, functional loss will manifest first.

There are many studies in the literature reporting structural damage evident years before visual-field loss (Sommer et al., 1991a; Mohammadi et al., 2004), but this does not mean that this is the rule. No studies have addressed the question the other way around — in other words, no study has followed a group of patients with visual-field loss and apparently normal structure to see how long it takes for the structural damage to become evident (i.e. looking for evidence of functional loss preceding structural loss). That there are such patients is evident from the many cross-sectional studies evaluating the sensitivity and specificity of imaging devices. Most studies find that, when test specificity is fixed at around 95%, the test sensitivity is around 70% (Medeiros et al., 2004). This means that around 30% of eyes with early visual-field loss have structural measurements within the normal range. Some may argue that imaging devices are not as sensitive to early structural damage as clinicians evaluating the patient or photographs of the ONH or RNFL. However, there is little support for this view in the literature, with studies showing that only the most expert clinicians (or a consensus of experts) perform better than quantitative imaging devices (Wollstein et al., 2000; Deleon-Ortega et al., 2006; Reus et al., 2007).

Selective tests of visual function

An early histological study of glaucoma suggested that there may be preferential loss of larger retinal ganglion cells (Quigley et al., 1987) and this finding spawned the development of "selective" psychophysical tests that stimulate subsets of ganglion cells with larger axons — those subserving the blue/yellow (koniocellular) and the motion and flicker (magnocellular) pathways. However, more recent work suggests that there is no preferential loss of larger fibers, but shrinkage of all fibers as part of the glaucomatous process (Morgan, 2002; Yucel et al., 2003).

Psychophysical tests have been used to probe the parvocellular (acuity), koniocellular, and magnocellular pathways in glaucoma, in an attempt to identify preferential damage in one or other pathway. The results of these studies provide conflicting evidence (Anderson and O'Brien, 1997; Harwerth et al., 1999; Ansari et al., 2002). The literature concerning the sensitivity of selective vision function tests to identify early glaucoma is extensive, yet appropriately designed studies are rare. Most studies suffer from design bias similar to that outlined above for studies assessing the value of structural measurements in early glaucoma — subjects are selected for having normal SAP and then are tested with "selective" visual function tests, such as FDT perimetry and SWAP. Unsurprisingly, glaucoma suspect patients with normal SAP are found with abnormal FDT and SWAP and the erroneous conclusion is drawn that FDT and SWAP are more sensitive than SAP. However, by definition (as a result of the study design), these patients have to have normal SAP, and there is no opportunity to find patients that have abnormal SAP and normal SWAP or FDT. The appropriate study design to compare SAP, SWAP, and FDT, or any other test of visual function, is to use a reference standard for glaucoma that is independent of the tests being evaluated, such as a structural measurement of the ONH or RNFL (Stroux et al., 2003; Sample et al., 2006). Similarly, when quantitative imaging devices are being evaluated, the appropriate reference is the visual field (and not the appearance of the ONH and RNFL) (Garway-Heath and Hitchings, 1998b; Medeiros, 2007).

Early glaucoma diagnosis from quantitative test results

The sensitivity of various imaging devices, available as clinical tools, to identify early glaucoma has been compared against the appropriate reference standard of early SAP loss (Medeiros et al., 2004). The study by Medeiros and colleagues did not use a clinical evaluation of the ONH or RNFL as a selection criterion for glaucomatous subjects in the study, although they did require normal subjects to have a healthy appearance of the ONH and RNFL. This may result in an over-estimation of the diagnostic performance of the devices. The sensitivity of the devices (GDxVCC Nerve Fiber Analyzer, Heidelberg retina tomograph II, and Stratus Optical Coherence Tomograph), at a fixed specificity of 95%, for early glaucoma (average mean deviation −4.9 dB) is given in Table 1.

In a study with a similar design, Sample and colleagues evaluated the sensitivity of various vision function tests (SAP, SWAP, FDT, and high-pass resolution perimetry) to identify early glaucoma with the appropriate reference standard of ONH structure (Sample et al., 2006). Normal eyes had a normal-appearing ONH and glaucoma eyes had either a glaucomatous-appearing ONH or progressive changes in ONH appearance. There was little difference in the performance of the various vision function diagnostic tests in the subjects with a glaucomatous-appearing ONH. In the eyes with progressive changes in ONH appearance, FDT performed better than the other tests. Table 2 gives the sensitivity of the tests in all glaucoma subjects.

Taking the results of these two studies together, it can be seen that neither structure nor function is consistently abnormal before the other in early

Table 1. The sensitivity to identify early glaucoma, defined by visual field loss, of three imaging devices

	Sensitivity at 95% specificity (%)
GDxVCC	61
HRT II	59
Stratus OCT	71

Note: GDxVCC, GDx nerve fiber analyzer with variable corneal compensator; HRT II, Heidelberg retina tomograph II; Stratus OCT, Stratus optical coherence tomograph.

Table 2. The sensitivity to identify early glaucoma, defined by structural damage to the optic nerve head, of four vision function tests

	Sensitivity at 80% specificity (%)
SAP	51
SWAP	50
FDT	66
HPRP	51

Note: SAP, standard automated perimetry; SWAP, short-wavelength automated perimetry; HPRP, high-pass resolution perimetry.

glaucoma, but either one may be abnormal and the other normal. In the clinical environment, this leads to diagnostic uncertainty. Of course, these tests are never interpreted in isolation, and when considered in the context of the history and clinical examination, an otherwise borderline result in a quantitative test may be more useful. However, the results illustrate the difficulty in making a diagnosis early in the disease course.

Progression to make a diagnosis

Glaucoma is a chronic progressive neuropathy. In many eyes with early glaucoma, the diagnosis is uncertain when the patient is first seen. Given the progressive nature of glaucoma, careful follow-up to identify changes in the ONH, RNFL, or visual function will enable a diagnosis to be made. The degree of probability that glaucoma is present before a clinician offers a diagnosis to the patient will vary between clinicians. As the risk of symptomatic vision loss when a patient presents with very early disease is low, the impact of a false diagnosis on a patient is potentially great, and the diagnosis can be made with greater certainty if change (progression) is identified, it may be wise to require a high probability for glaucoma before making the diagnosis. Those with a lower probability (uncertain cases) can be followed for signs of progression. At the same time, it should be remembered that making the diagnosis and making a treatment decision, although linked, are not the same thing. A clinician may decide to treat a patient even when the diagnosis is uncertain if there are significant risk factors for future vision loss, such as a high IOP.

Conclusions

Signs in early glaucoma are frequently equivocal, and either structural damage or vision function loss may be the first sign of glaucoma. Quantitative tests, such as SAP and imaging devices, are useful adjuncts to the clinical evaluation. The results of the tests raise or lower the probability that glaucoma is present, and the results of the tests may be combined mathematically to establish probability levels. The validity of test results from quantitative devices (including data quality and sources of error) should always be considered before the results are used for patient management.

Abbreviations

FDT	frequency doubling technology
GDxVCC	GDx nerve fiber analyzer variable corneal compenator
HRT	Heidelberg retina tomograph
HPRP	high-pass resolution perimetry
IOP	intraocular pressure
OCT	optical coherence tomography
ONH	optic-nerve head
RNFL	retinal nerve fiber layer
SAP	standard automated perimetry
SWAP	short-wavelength automated perimetry

Acknowledgments

The corresponding author has received a proportion of his funding from the Department of

Health's National Institute for Health Research Biomedical Research Centre at Moorfields Eye Hospital and the UCL Institute of Ophthalmology. The views expressed in this publication are those of the authors and not necessarily those of the Department of Health.

References

Anderson, R.S. and O'brien, C. (1997) Psychophysical evidence for a selective loss of M ganglion cells in glaucoma. Vision Res., 37(8): 1079–1083.

Ansari, E.A., Morgan, J.E. and Snowden, R.J. (2002) Psychophysical characterisation of early functional loss in glaucoma and ocular hypertension. Br. J. Ophthalmol., 86(10): 1131–1135.

Boland, M.V. and Quigley, H.A. (2007) Risk factors and open-angle glaucoma: classification and application. J. Glaucoma, 16(4): 406–418.

Bonomi, L., Marchini, G., Marraffa, M. and Morbio, R. (2001) The relationship between intraocular pressure and glaucoma in a defined population. Data from the Egna–Neumarkt Glaucoma Study. Ophthalmologica, 215(1): 34–38.

Centre for Evidence-Based Medicine. Fagan nomogram. http://www.cebm.net/index.aspx?o = 1043. Accessed 5th April, 2008.

Chen, P.P. (2003) Blindness in patients with treated open-angle glaucoma. Ophthalmology, 110(4): 726–733.

Deleon-Ortega, J.E., Arthur, S.N., Mcgwin, G., Jr., Xie, A., Monheit, B.E. and Girkin, C.A. (2006) Discrimination between glaucomatous and nonglaucomatous eyes using quantitative imaging devices and subjective optic nerve head assessment. Invest. Ophthalmol. Vis. Sci., 47(8): 3374–3380.

Fagan, T.J. (1975) Letter: nomogram for Bayes theorem. N. Engl. J. Med., 293(5): p. 257.

Forsman, E., Kivela, T. and Vesti, E. (2007) Lifetime visual disability in open-angle glaucoma and ocular hypertension. J. Glaucoma, 16(3): 313–319.

Garway-Heath, D.F., Caprioli, J., Fitzke, F.W. and Hitchings, R.A. (2000) Scaling the hill of vision: the physiological relationship between ganglion cell numbers and light sensitivity. Invest. Ophthalmol. Vis. Sci., 41(7): 1774–1782.

Garway-Heath, D.F. and Friedman, D.S. (2006) How should results from clinical tests be integrated into the diagnostic process?. Ophthalmology, 113(9): 1479–1480.

Garway-Heath, D.F. and Hitchings, R.A. (1998a) Quantitative evaluation of the optic nerve head in early glaucoma. Br. J. Ophthalmol., 82(4): 352–361.

Garway-Heath, D.F. and Hitchings, R.A. (1998b) Sources of bias in studies of optic disc and retinal nerve fibre layer morphology. Br. J. Ophthalmol., 82(9): p. 986.

Garway-Heath, D.F., Holder, G.E., Fitzke, F.W. and Hitchings, R.A. (2002) Relationship between electrophysiological, psychophysical, and anatomical measurements in glaucoma. Invest. Ophthalmol. Vis. Sci., 43(7): 2213–2220.

Halkin, A., Reichman, J., Schwaber, M., Paltiel, O. and Brezis, M. (1998) Likelihood ratios: getting diagnostic testing into perspective. Q.J. Med., 91(4): 247–258.

Harwerth, R.S., Carter-Dawson, L., Shen, F., Smith, E.L., 3rd and Crawford, ML. (1999) Ganglion cell losses underlying visual field defects from experimental glaucoma. Invest. Ophthalmol. Vis. Sci., 40(10): 2242–2250.

Harwerth, R.S., Carter-Dawson, L., Smith, E.L., 3rd., Barnes, G., Holt, W.F. and Crawford, M.L. (2004) Neural losses correlated with visual losses in clinical perimetry. Invest. Ophthalmol. Vis. Sci., 45(9): 3152–3160.

Harwerth, R.S. and Quigley, H.A. (2006) Visual field defects and retinal ganglion cell losses in patients with glaucoma. Arch. Ophthalmol., 124(6): 853–859.

Jaeschke, R., Guyatt, G.H. and Lijmer, J. (2001) Diagnostic tests. In: Guyatt G.H. and Rennie D. (Eds.), Users' Guides to the Medical Literature: Essentials of Evidence-Based Clinical Practice. American Medical Association, Chicago.

Kerrigan-Baumrind, L.A., Quigley, H.A., Pease, M.E., Kerrigan, D.F. and Mitchell, R.S. (2000) Number of ganglion cells in glaucoma eyes compared with threshold visual field tests in the same persons. Invest. Ophthalmol. Vis. Sci., 41(3): 741–748.

Medeiros, F.A. (2007) How should diagnostic tests be evaluated in glaucoma? Br. J. Ophthalmol., 91(3): 273–274.

Medeiros, F.A., Zangwill, L.M., Bowd, C. and Weinreb, R.N. (2004) Comparison of the GDx VCC scanning laser polarimeter, HRT II confocal scanning laser ophthalmoscope, and stratus OCT optical coherence tomograph for the detection of glaucoma. Arch. Ophthalmol., 122(6): 827–837.

Mitchell, P., Smith, W., Attebo, K. and Healey, P.R. (1996) Prevalence of open-angle glaucoma in Australia. The Blue Mountains Eye Study. Ophthalmology, 103(10): 1661–1669.

Mohammadi, K., Bowd, C., Weinreb, R.N., Medeiros, F.A., Sample, P.A. and Zangwill, L.M. (2004) Retinal nerve fiber layer thickness measurements with scanning laser polarimetry predict glaucomatous visual field loss. Am. J. Ophthalmol., 138(4): 592–601.

Morgan, J.E. (2002) Retinal ganglion cell shrinkage in glaucoma. J. Glaucoma, 11(4): 365–370.

Quigley, H.A., Dunkelberger, G.R. and Green, W.R. (1988) Chronic human glaucoma causing selectively greater loss of large optic nerve fibers. Ophthalmology, 95(3): 357–363.

Quigley, H.A., Sanchez, R.M., Dunkelberger, G.R., Nl, L.H. and Baginski, T.A. (1987) Chronic glaucoma selectively damages large optic nerve fibers. Invest. Ophthalmol. Vis. Sci., 28(6): 913–920.

Reus, N.J., De Graaf, M. and Lemij, H.G. (2007) Accuracy of GDx VCC, HRT I, and clinical assessment of stereoscopic optic nerve head photographs for diagnosing glaucoma. Br. J. Ophthalmol., 91(3): 313–318.

Sample, P.A., Medeiros, F.A., Racette, L., Pascual, J.P., Boden, C., Zangwill, L.M., Bowd, C. and Weinreb, R.N. (2006) Identifying glaucomatous vision loss with visual-function-specific perimetry in the diagnostic innovations in glaucoma study. Invest. Ophthalmol. Vis. Sci., 47(8): 3381–3389.

"Collaborative Normal-Tension Glaucoma Study (CNTGS)," and ocular hypertension — the "Ocular Hypertension Study (OHTS)" and the "European Glaucoma Prevention Study (EGPS)." Early open-angle glaucoma patients were included in the EMGT trial as well as in the "Collaborative Initial Glaucoma Treatment Study (CIGTS)," whereas advanced glaucoma patients were examined in the "Advanced Glaucoma Intervention Study (AGIS)." In all of these studies, SAP represented a prominent diagnostic tool as well as a primary endpoint of the study. (The Advanced Glaucoma Intervention Study (AGIS): 1, 1994; Gordon and Kass, 1999; Musch et al., 1999; European Glaucoma Prevention Study (EGPS) Group, 2002; Heijl et al., 2002; Miglior et al., 2007).

SAP visual field assessment: interpretation and reliability factors

Fixation losses. The HFA periodically checks the patient's fixation by presenting stimuli within their blind spot (Heijl–Krakau technique). If the number of fixation losses is greater than 20%, a symbol will appear next to the fixation losses.

False-negative errors. A brighter stimulus is presented at a test point in the field that was earlier seen at a lesser luminance: If the patient does not respond to the brighter stimulus, a false-negative error is displayed. High false-negative rate might indicate fatigue or inattentive patient, but are frequent in very damaged VFs.

False-positive errors. The patient responds to a stimulus that was not projected by the projector: A false-positive error is displayed.

Short-term fluctuation (SF). SF represents intratest variability in threshold measurements of selected VF locations. Two main reasons rely on an abnormal SF: an inattentive patient or a patient with a damaged visual system. Normal SF values are lesser than 3 dB when the same VF location is tested twice.

On the contrary, long-term fluctuation (LF) is the variability in threshold measurements between two VF sessions. LF represents a physiologic variation in differential light sensitivity over time; it exceeds the quote due to SF (intratest measurement variability) or learning effect (Flammer et al., 1984).

Although it is an actual change in differential light sensitivity, LF is not quantified in clinical perimetry. However, its occurrence must be considered when evaluating multiple VFs over time.

Graytone map. It represents both tested points and nontested VF intermediate points, which have been assigned values interpolated from neighboring VF locations. For this reason, the accuracy of this representation is poor.

Total deviation (TD). TD is the difference at each test point between the patient's measured threshold sensitivity and the median normal value for the patient's age.

Pattern deviation (PD). PD plot shows localized defects that may not be revealed because of either a generalized depression or an elevation of the hill of vision. To obtain the PD plot, an adjustment for the general height of the hill of vision is made on the basis of the 24-2 pattern: The seventh most sensitive (compared with normal) point, that is, the 85th percentile best point of the 24-2 grid, is adjusted to 0 deviation to represent the overall general height of the hill of vision.

It has been pointed out that ophthalmologists should evaluate both total and PD analyses to make informed decisions on VF progression in glaucoma.

Probability of abnormality. The P-value represents the probability whereto measured threshold values have deviated from the expected normal values. The probability statement is based on the hill of vision distribution observed in the normal population. This P-value is computed from the TD and the PD plots. $P<1\%$ means that this deviation happens in less than 1% of the normal population and must be considered highly suspicious.

Global indices

HFA: MD, SF, PSD, CPSD

Mean deviation (MD). The MD is the mean difference in decibels (dB) between the "normal" expected hill of vision and the patient's hill of vision. If the deviation is significantly outside the norms, a *P*-value will be given. $P < 0.5\%$ means that less than 0.5% of the normal population showed an MD value larger than the value found for this test. This index is a measure of overall depression, elevation of the field, or significantly deep losses in one part of the field and not in others.

Pattern standard deviation (PSD). This is a measurement of the degree by which the shape of the patient's measured field or hill of vision departs from the "normal" age-corrected reference field model. The value is expressed in decibels, and any value of 2 dB or greater will have a (*P*) value next to it, indicating the significance of the deviation.

Short-term fluctuation (SF). It represents an index of the consistency of the patient's responses during the field testing. This value is obtained when 10 preselected points are tested twice and the difference, in decibels, of the patient's responses is compared.

Corrected pattern standard deviation (CPSD). This is a calculated measurement in decibels of how much the total shape of the patient's hill of vision deviates from the shape of the "normal" hill of vision for the patient's age, after being corrected for intratest variability. In calculating the CPSD, the STATPAC attempts to determine if the irregularities in the hill of vision are real by removing the SF, which may mask a relative scotoma.

Glaucoma hemifield test

In the glaucoma hemifield test (GHT) (30-2 or 24-2 program; FT or Swedish interactive threshold algorithm [SITA] strategies), five areas in the upper hemifield are compared with five mirroring areas in the lower one. A score is assigned to each zone on the basis of the percentile deviations in the PD plot of the locations in that area. A comparison of each zone is made with the mirroring zone. The difference in scores between the two mirroring areas is then compared with significance limits of a database of normal subjects. GHT produces five possible messages: outside normal limits, borderline, general reduction of sensitivity, abnormal high sensitivity, and within normal limits (Anderson and Patella, 1999).

Octopus indices: MD, SF, CLV

The Octopus perimeter (program G1) provides four equivalent global indices: mean defect, corresponding to HFA MD; loss variance (LV), corresponding to HFA PSD; SF; and corrected loss variance (CLV), corresponding to HFA CPSD. However, they differ in that positive values of MD represent loss of sensitivity, and the indices of localized loss are given as the variance instead of the standard deviation. Nevertheless, they share common significance with HFA indices.

OCTOPUS seven-in-one report (Fig. 2)

The reliability factor (RF) is a percentage value of the positive and negative catch trials. For reliable results, the RF should be under 15%.

Bebie curve. This method allows an easy recognition of diffuse as well as local damage. According to Kaufmann and Flammer (1989), the clinical application of the cumulative defect curve in different diseases allows an easy recognition of diffuse as well as local VF damage. Ranking all defects values from left to right, the Bebie curve shows the actual VF results together with the normal bandwidth. It differentiates overall depression (parallel to the band of normality) from focal defects (steep decline).

Statistical information about uniform (MD) and localized loss (LV) provide an easy assessment of the field. For Octopus perimetry, the MD normal tolerance range is from -2 to $+2$ dB (Paetzold et al., 2007).

The VF indices attempt to summarize the distribution of differential sensitivity within the VF in a few numeric parameters and to give useful information on the functional loss (Flammer, 1986). VF indices can also be used in staging VF defects in glaucoma. In this case, however, it must be considered that VF indices miss to show any spatial distribution of defects; either threshold fluctuations or a variety of artifacts can influence any classification based on VF indices only. Nevertheless, different studies seem to demonstrate that the MD, considered together with the CPSD (or PSD, CLV, or LV) indices, may be useful in the staging of functional damage in

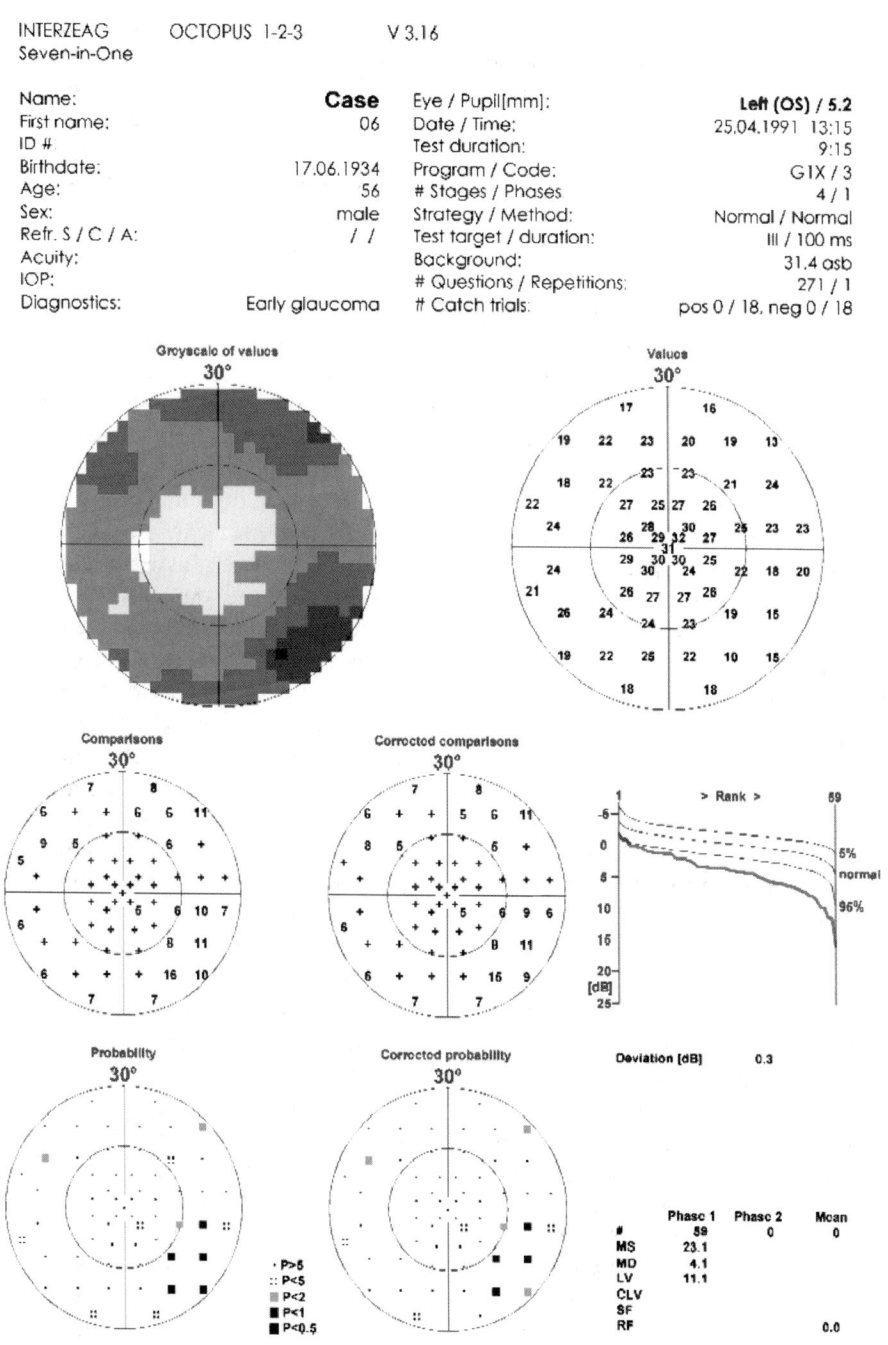

Fig. 2. Octopus seven-in-one report (*Source*: http://www.octopus.ch/products).

glaucoma. Keltner et al. (1993) developed a VF classification system for the Optic Neuritis Treatment Trial that also has been used in the Ocular Hypertension Treatment Study (OHTS).

SAP VF assessment: full-threshold strategy

Deciding which perimetric program to apply requires knowledge of both the pathology concerned and the characteristics of the various programs available. One can nonetheless affirm that the majority of ocular and neuro-ophthalmological diseases affecting the VF can be studied analyzing the central VF with a 30° fixed grid of points set at 6° or at a variable density, using a III white target stimulus (program 30/2 or 24/2 Humphrey, G1/G2 or 30/2 Octopus) (Anderson and Patella, 1999).

The FT is the most suitable strategy to evaluate VF in glaucoma patients, except in certain cases (the elderly, patients who are particularly fatigued, etc.) when fast threshold strategies (Humphrey Fastpac, Octopus tendency-oriented perimetry [TOP] strategies) could be better.

In some cases with very advanced perimetric damage in which only a small area of residual central functionality remains, or in patients suffering from macular pathologies, the examination pattern can be focused on the central 10° field, with specific programs (10/2 and Macula for Humphrey and M1 or M2 for Octopus) and/or using a V target stimulus in the case of low-vision patients.

Before carrying out the examination, it is important to explain to the patient what automated perimetry is, how it is carried out, and its importance in diagnosing and/or treating the pathology with which the patient is affected. Indeed, the patient's informed awareness is fundamental in order to obtain reliable and clinically useful results. It must not be forgotten that SAP is a subjective, psychophysical method in which collaboration between the patient and the person carrying out the examination is essential and the importance of the learning effect should also be taken into consideration (Kulze et al., 1990; Marra and Flammer, 1991) (Fig. 3).

In other words, the patient must "learn" to carry out an automated perimetry exam, and it is therefore always advisable to repeat the test if the results are in doubt or not consistent with the clinical situation. It is also necessary to consider the psychological behavior of the patient undergoing perimetry. In most cases, the patient is suffering from glaucoma and is aware of the chronic nature of the disease and the way it develops. Consequently, the VF examination often causes anxiety in that it represents the moment of verifying effective ocular conditions and progression of the disease. It is therefore up to the person carrying out the exam to reassure the patient and warn him that certain light stimuli may not be seen but that this does not necessarily mean having pathological implications.

Particular care must also be taken in the correct insertion of the data. As is known, the single perimetric exam results are compared with a normative database that takes into account the age of the patient. Therefore, an erroneous date of birth can lead to comparison with the wrong age group and the possibility of diagnosing nonexistent perimetric defects or masking real defects.

The pupil diameter influences the quantity of light reaching the retina and consequently its adaptation. It is therefore advisable to carry out and repeat a perimetric examination under the same pupil diameter conditions. Furthermore, normal perimetric data are gathered from subjects with nondilated pupils. Consequently, it is best to carry out an automated perimetry in physiological conditions, monitoring successive examinations for the constancy of the pupil diameter in order to ensure a correct comparison of the results obtained (Park and Youn, 1994).

Another important aspect is the use of optical corrections. For refractive defects of up to 1 spherical and/or cylindrical diopter, it is advisable not to use corrective lenses because the support of these can lead to artifacts that can compromise a correct clinical and statistical interpretation of the results (Zalta, 1989). In cases in which an additional optical correction is required, it is absolutely necessary to use lenses with metallic frames to avoid false "annular" scotomata due to the lens frame (Fig. 4). Anatomical anomalies such

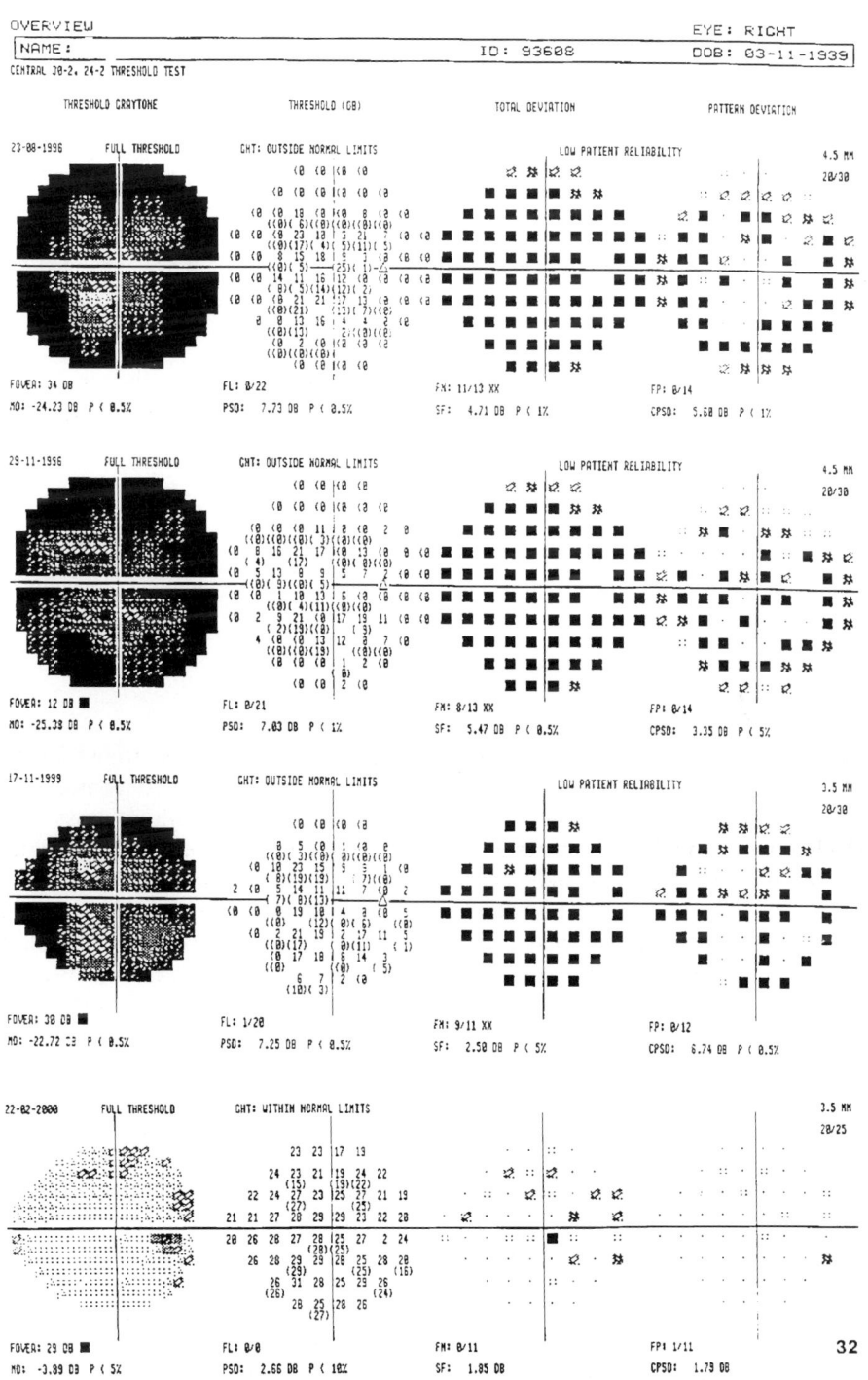

Fig. 3. Learning effect: Note the improvement in differential light sensitivity on the fourth visual field.

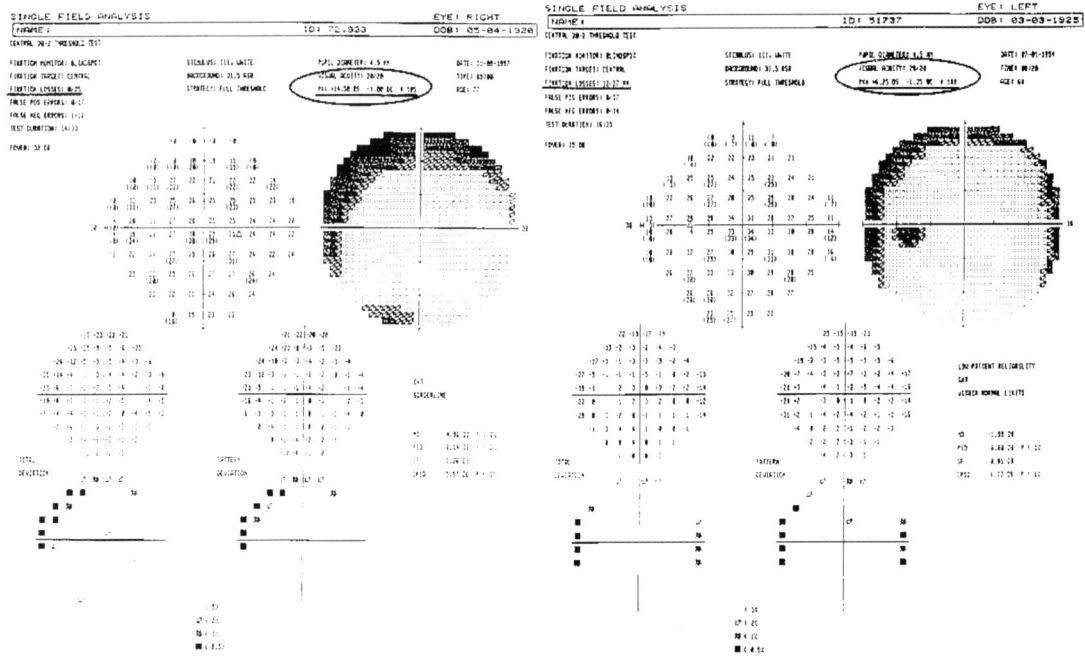

Fig. 4. Artifact due to corrective lens.

as congenital or acquired ptosis, facial asymmetries, and prominent noses should always be noted in the exam report, because they may cause perimetric artifacts.

As mentioned above, SAP results represented a primary outcome in most of the clinical trials and population-based studies on glaucoma and ocular hypertension. Each trial adopted different criteria to define a glaucomatous VF defect (Heijl et al., 2008).

SAP VF defects assessment: OHTS criteria

In the OHTS, a reliable VF was considered to be abnormal if GHT and/or corrected CPSD global index results were outside normal limits ($P<0.01$ and $P<0.05$, respectively).

Borderline GHT results were not included as an abnormal field. These results were an indication that the results were suspicious but not clearly abnormal.

An OHTS VF was classified as reliable if none of the reliability indices (false-positive errors, false-negative errors, or fixation losses) exceeded the 33% limit.

In OHTS, glaucomatous VF damage was defined as follows: two consecutively abnormal SAP test results, with at least one of the two abnormal tests being within the 6-month interval (Gordon and Kass, 1999).

Only few studies suggested the possibility that the VF index MD would be more likely than PSD to capture a generalized depression of sensitivity that may be a prominent feature of early glaucoma cases. The OHTS showed that PSD, but not MD, is a predictor of glaucoma development among ocular hypertensive subjects. PSD may be important for identification of early glaucoma cases. PSD was found to be a good predictor of the development of OAG in the EGPS (Miglior et al., 2007).

However, an average loss of 27.3% of retinal ganglion cells is necessary for the corrected PSD index of standard achromatic perimetry to fall below the 95% normal confidence limits (Kerrigan–Baumrind et al., 2000).

In another study, using SAP-SITA PSD at 95% specificity, the average percentage loss of rim area of the patients with glaucoma identified as abnormal by HRT II analysis was 30% (Haymes et al., 2005).

SAP VF defects assessment: AGIS criteria

The AGIS used the HFA 24-2 TD printout to determine a patient's VF score. In this study, if VF threshold values were depressed by 5–9 dB from the normal values, as measured in decibels, it was considered abnormal, depending on the field location. To qualify as abnormal, a depression had to be greater in the superior field than in the inferior field, and a depression had to be greater in the periphery than centrally.

Scores were assigned on the basis of the decibel depressions found in the various areas. A score of zero indicated no VF loss, while 20 was the maximum possible score. One stage of field loss progression was arbitrarily defined as an increase in score of four (The Advanced Glaucoma Intervention Study [AGIS]: 1, 1994).

A recent study showed a strong association between VF defects in the central 24° field and the risk of motor vehicle collisions (MVCs) among patients with glaucoma. In this study, as the AGIS score increased (VF defect was more severe) for both the better and worse eyes, there was a corresponding increase in the odds of an MVC (Hall and Owsley, 2005).

This finding means that the convention of assessing each eye's field by itself, even just the central 24° field, may be informative with respect to crash risk. Thus, not only the binocular VF assessed by the Esterman grid methods or similar programs but also SAP VF assessment of each eye separately, as in the clinical routine, has to be considered by clinicians.

SAP VF defects assessment: CIGTS

The CIGTS used a scoring methodology similar to that used by the AGIS. Glaucoma staging in this study was also on the basis of the HFA 24-2 program, but the categories of probability values on the TD probability plot rather than decibel deviations as in AGIS were used. Scores were scaled from 0 to 20 (Musch et al., 1999).

Three of major glaucoma trials (AGIS, EMGT, and CIGTS), all using the same Humphrey VF tests, developed different criteria to define either VF damage or progression.

A recent study found that among progressing eyes, 58% were identified using AGIS criteria, 75% were identified with CIGTS criteria, and 96% were identified with EMGT criteria. The specificity for EMGT criteria was 89%, lower ($P<0.05$) than that of AGIS (98%) and CIGTS (99%) criteria (Heijl et al., 2008).

However, the prolonged test time of the original FT tests was tiring for patients, increasing the likelihood of fatigue artifact, which artificially depresses the threshold values of the test. Moreover, FT tests were time consuming and not practical in the clinical setting. These findings led to the development of new test strategies in order to shorten VF examination time. It is noteworthy that in the course of long-term clinical trials, key methods used to assess outcomes may become obsolete.

The authors of CIGTS evaluated the impact of converting from Humphrey 24-2 FT VF testing to SITA-standard (SS) VF testing during the follow-up phase of this clinical trial. They found that the CIGTS VF score and GHT results differed between SS and FT tests. Any change in the protocol used to identify a study's primary VF outcome requires careful evaluation of converting from old test strategies to new ones (Musch et al., 2005).

Fastpac

The Fastpac strategy varies stimulus intensity in 3 dB steps depending on the patient response. Testing stops when the sequence crosses threshold, without inverting direction in smaller steps. In other words, threshold is recorded at the last seen stimulus intensity.

Since the Fastpac strategy measures threshold without the double crossing of threshold, sensitivity values determined by this algorithm show more intratest fluctuation. Nevertheless, shortening of test times achieved with Fastpac (up to 70%) has been judged useful for various clinical purposes.

Swedish interactive threshold algorithm

Many efforts have been made to shorten thresholding strategies for VF assessment without reducing the accuracy of the test. Newer algorithms for measuring VFs have been recently developed.

The Humphrey SS and SITA fast demonstrated, in different studies, excellent sensitivity and specificity for glaucomatous VF loss with remarkable savings in time.

Among a cohort of patients experienced with automated perimetry, changing the order of eye testing using the SS 24-2 did not have a significant effect on the MD or the test reliability. While inter-eye fatigue may not be clinically significant with the SITA, fixation losses represent a problem with the use of this algorithm.

The SITA algorithm showed reduced between-subject variability. Nevertheless, SS produced slightly higher mean sensitivity values compared with that of existing algorithms. Furthermore, HFA II SS underestimates patients' false-positive errors, particularly among normal subjects. High false-positive error frequencies can have adverse effects on MD and PSD, leading clinicians and researchers to an inaccurate determination of the amount and severity of VF loss.

Nevertheless, SS reduced test-taking time from FT by 52% in normal subjects and 47% in glaucoma patients (Wild et al., 1999a, b; Barkana et al., 2006; McGwin et al., 2006).

Hoffmann et al. (2006) investigated the inter-eye correspondence between patterns of VF loss on SAP in patients diagnosed with glaucomatous optic neuropathy. All participants performed two SAP VFs using the 24-2 program and SITA thresholding algorithm of the HFA. They found that more advanced VF defects (e.g., partial arcuate) show higher correspondence rates between the eyes than less advanced defects.

Another fast strategy algorithm, the TOP, has been developed for the Octopus perimeter. The TOP strategy tests each position in the VF once and extrapolates this information to the surrounding points. High correlation was found between the SITA fast (Humphrey) and TOP (Octopus) strategies for measurements of global indices.

The TOP strategy tended to underestimate focal VF loss compared with SITA fast. The TOP strategy resulted faster than SITA fast, while the two algorithms showed similar sensitivity and specificity (King et al., 2002).

SAP VF assessment: the glaucoma staging system

The classification of VF defects identified with SAP has been used to distinguish between healthy and glaucomatous eyes, to establish homogeneous criteria for grading severity of disease, to allow accurate follow-up of disease.

SAP is still the accepted technique for quantifying functional damage in patients suffering from glaucoma.

Several classification systems exist to identify SAP VF defects and patterns of VF loss. None of the different classification methods have been proposed in the past yet have obtained a widespread use.

The Aulhorn and Karmeyer's classification relies on manual perimetry, a testing procedure that is obsolete; nevertheless, this five-stage classification is still considered to be a reference point in glaucoma clinic and research.

Nowadays, one of the most commonly used methods to stage glaucomatous VF loss severity is the classification proposed by Hodapp et al. in 1993.

In this method, the VF defect is classified as early, moderate, and severe defect on the basis of the MD value and the number of defective points in the STATPAC-2 PD probability map; the defect proximity to the fixation point is also considered. This classification requires an accurate analysis of every single VF result, and therefore, it is time consuming. Furthermore, it may be inaccurate for a fine categorization of VF defects.

Recently, the Bascom Palmer (Hodapp–Anderson–Parrish) classification was selected for a retrospective glaucoma staging system (GSS) on the basis of Humphrey VFs. This system, modified by a panel of glaucoma specialists after a review of published GSSs was conducted, underwent additional modifications after pilot testing to cover the full range of disease progression, from pre-glaucoma diagnosis to complete blindness.

This GSS, based on the Humphrey perimetry, consists of six ordered stages. It was validated by reviewing patient charts from 12 US glaucoma centers.

Several studies adopted the AGIS Investigators classification method (1994). The AGIS score is based on the number and depth of adjacent depressed test locations. This score ranges from 0 to 20, and it can be used to classify the defect into five severity categories.

The CIGTS classification method is similar to the AGIS method. Both AGIS and CGITS scoring systems need a discrete training to be applied in daily clinical practice, although they are analytical and accurate.

Brusini pointed out that both the AGIS and the Hodapp–Parrish–Anderson methods are accurate with regard to localized defects but fail to take into consideration slight diffuse sensitivity depressions, which may at times be due to an early glaucomatous damage.

Most of these methods have been specifically designed to be used with the 30-2 and 24-2 programs of the early Humphrey perimeters. Moreover, they often miss information regarding the characteristic of the defect (Brusini and Filacorda, 2006).

The GSS (Brusini), introduced in 1996 (Fig. 5), is based on the two main VF global indices: MD or mean defect and CPSD, or CLV, plotted on an X–Y coordinate diagram.

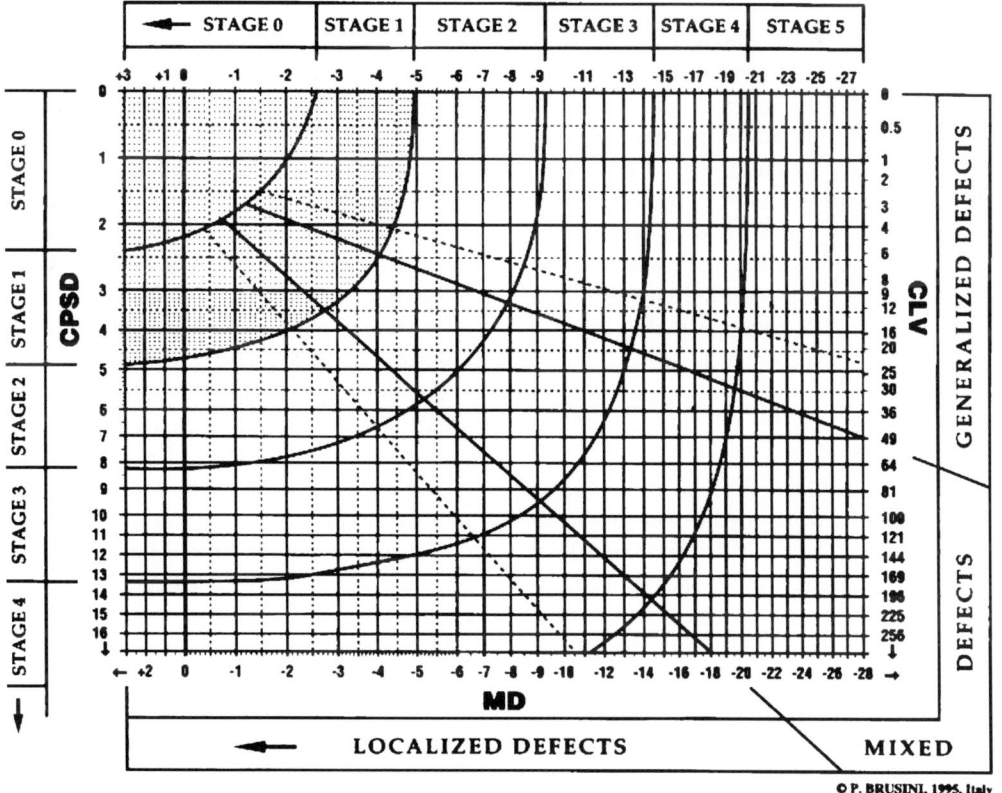

Fig. 5. The glaucoma staging system. Adapted with permission from P. Brusini, Udine, Italy.

The GSS was developed on the basis of 500 automated VFs (332 Humphrey 30-2 FT VFs and 168 Octopus G1/G1X VFs, normal strategy, three phases) from 471 patients with primary open-angle glaucoma (POAG) at various stages of severity.

A new intermediate stage between Stage 0 and Stage 1 has been added in the more recent GSS 2 (Brusini and Filacorda, 2006) to include borderline cases. One hundred and twenty-eight automated VF tests performed with the HFA 30-2 FT test were analyzed to define the two lines that separate this new borderline stage from both Stage 0 and Stage 1.

Brusini's GSS has shown to be useful both in staging the damage severity and in identifying the different components of VF loss (generalized, localized, and mixed) and to monitor defect progression over time.

SAP: interocular asymmetries in OHTS

Binocular VF test point asymmetry can assist in clinical evaluation of eyes at risk of development of POAG. In OHTS, asymmetric VF threshold values proved to be an important predictor of POAG. Eyes from an asymmetric pair that have reduced sensitivity were at 59% greater risk of developing POAG.

The OHTS investigators developed a new index, MP (mean prognosis), to discriminate between eyes that developed POAG from eyes that did not (Levine, 2006).

SAP, VF progression

The aim of computerized perimetry over time is to follow the evolution of a glaucomatous defect. The most commonly noticeable progression in worsening glaucoma is an increase in the depth of VF defects, followed progressively by their widening. In approximately half of the cases, the appearance of a new scotoma in different areas is seen (Mikelberg and Drance, 1984).

In order to correctly assess the evolution of the glaucoma, computerized perimetry must be considered the surest method, as long as important rules are followed. First of all, it is wise to always use the same threshold strategy examinations with the same parameters during the follow-up.

It is necessary to identify the baseline VF, taking this as a first one, while being careful of all possible forms of artifacts: first the learning effect, which is responsible for an improvement of the second test in comparison to the first; the fatigue effect, which can cause a depression of the threshold during the examination; and the demotivation effect, sometimes the reason for a worsening that cannot be otherwise explained, due to the lack of motivation with which a patient faces the examination (Katz et al., 1997; Anderson et al., 2000).

The LF invariably tends to pollute the data so that a suitable number of tests (at least four or five) must be available in order to issue a correct clinical judgment on the evolution of a perimetric defect.

As regards the frequency with which the perimetry should be carried out, it is necessary to consider the extent of the glaucoma, that is to say the perimetric damage at the outset, the intraocular pressure control over time, the age of the patient, his degree of cooperation, and the current therapy (the lesser the care, the more the VF must be controlled). Basically, a frequency of 6 months will be sufficient in most cases, while subjects at risk (e.g., those with a threatened fixation point) should be tested every 3–4 months. Where a test appears clearly different (improved or worsened) in comparison to previous ones, it is always advisable to repeat the test.

At this stage it is essential to establish if the variations observed during the follow-up are important or not. Due to LFs, a not-always-perfect cooperation, altered examination conditions, and many other disturbing factors, it is often difficult to establish if a change should be taken into account or ignored as being without clinical importance. To express such a judgment, one can rely on experience, use standardized criteria, or use special statistical programs (Spry and Johnson, 2002).

The first method offers the advantage of considering the clinical context and of being flexible, but it is subjective and requires considerable knowledge about perimetry so as not to make quite simple mistakes.

The standardized criteria (Hodapp et al., 1993; AMO, AGIS, and CIGTS) are based on the variations in the number of altered points and the depth of the defect: They provide less subjective data, but are too strict for a routine clinical use (Walsh, 1990).

The statistical programs have the advantage of not being influenced by what are only apparently important variations, but deal mathematically with data that is biological and thus subject to physiological fluctuations. Finally, it should be remembered that statistical importance is not always the same as clinical importance.

The most well-known statistical programs are STATPAC, Humphrey, PeriTrend Octopus, and Peridata. The STATPAC's Glaucoma Change Probability relies on repeatability of multiple negative pointwise changes to thresholds on either the TD or the PD plot.

The Glaucoma Progression Analysis (GPA) is based on PD, defined as the deviation from age-corrected values adjusted for the general height of the VF and relies on EMGT progression algorithm.

The Progressor and Peridata softwares are based on pointwise linear regression analysis (PLRA).

The STATPAC Humphrey program makes it possible to collect up to 16 examinations in one printout without losing any information called "Overview" (Fig. 6). The advantages of the Overview option is to have a simple view of several and in-sequence exams in the same page, while the main limit is that only a clinical and subjective evaluation is possible.

With "Change Analysis" option, the main indices (MD, PSD, SF, and CPSD) are graphed, and if at least five examinations are available, the significance of the variations of the MD over time is produced, based on linear regression analysis (Heijl et al., 1991). The new version of the program (STATPAC-2) also makes it possible to assess point by point the significance of the variations in different light sensitivity with respect to the baseline VF, obtained from the average of the first two examinations called "Glaucoma Change Probability" (Fig. 7). In the Glaucoma Change Probability option, changes in the VF over time are expressed by symbols: Δ indicates points where the different light sensitivity is improved and ▲ points where the different light sensitivity is decreased with respect to baseline exams.

With the new Humphrey GPA software, even examinations carried out with SITA strategy can be analyzed. This program also provides a statistical judgment on the possible or probable progression of a perimetrical defect on the basis of the possibility of repeating a worsening that regards the same area in two or more consecutive tests (Fig. 8).

Minimum of three tests are required: two as baseline and other as follow-up exams. Each follow-up exam is compared to averaged thresholds of two baseline exams. Additional follow-up exams are compared both to baseline and to two most recent follow-ups. The results are expressed using the following symbols:

Δ indicates progression at 95% significance level.
▲ indicates progressing point repeated in two consecutive exams.
☐ indicates progressing point repeated in three consecutive exams.

GPA Alert TM signal is based on the presence of symbols: Three ▲ in one exam denotes possible progression, and three ☐ indicates likely progression of the VF defect in time.

The PeriTrend Octopus program uses the analysis of linear regression for estimating variations in the MD and LV indices over time. The different colors of the lines for MD and LV indicate the significant or insignificant changes of these indices over time.

With the Peridata program it is possible to visualize and print a series of VFs in chronological order, choosing between types of display, including a pointwise analysis of linear regression (Weber and Caprioli, 2000).

The Progressor (Fitzke et al., 1996) program uses linear regression for highlighting in different colors any significant variations in sensitivity in each point examined.

The GSS can also be used for graphically following the evolution of a defect over time (Koçak et al., 1997).

Other programs that use more complex systems of analysis seem now to be useful but are still not

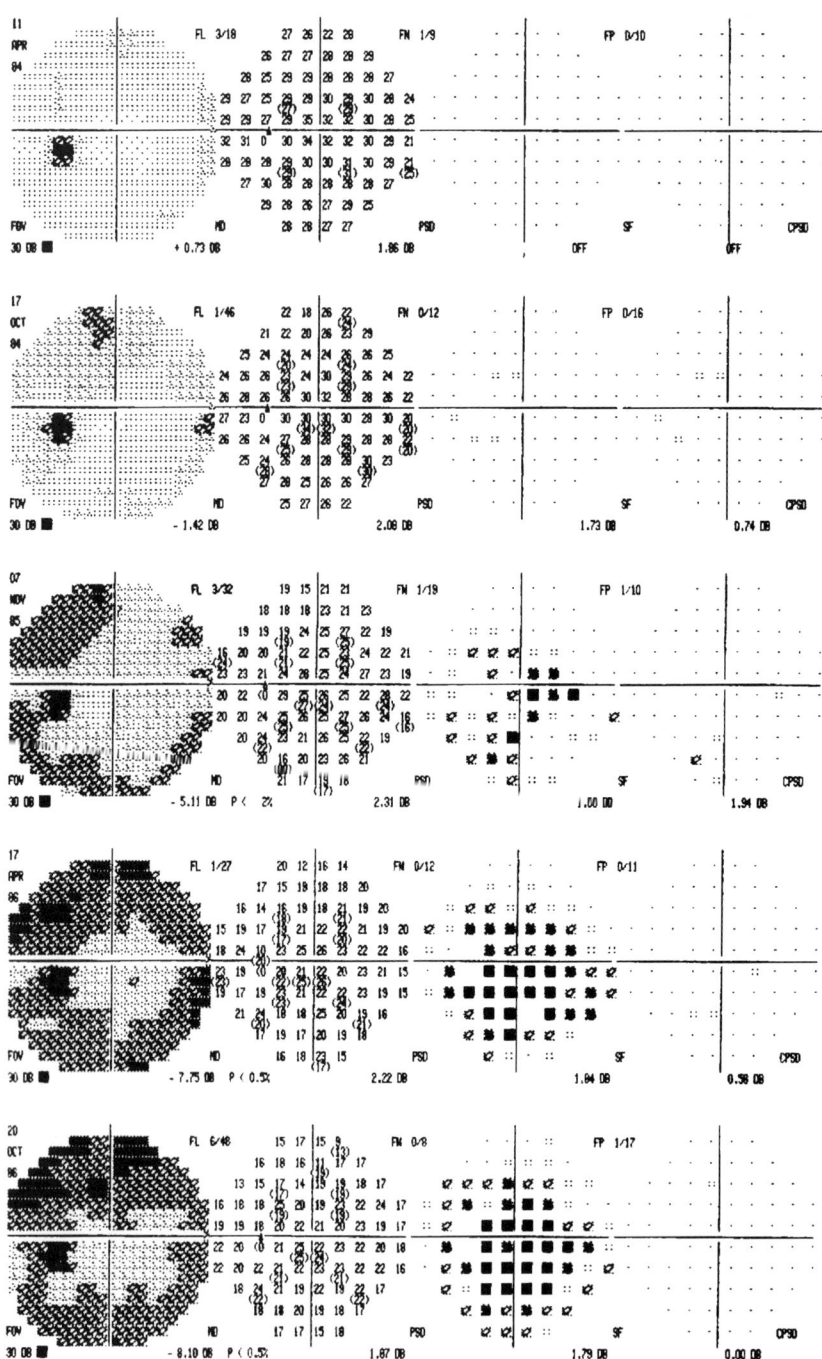

Fig. 6. STATPAC Overview.

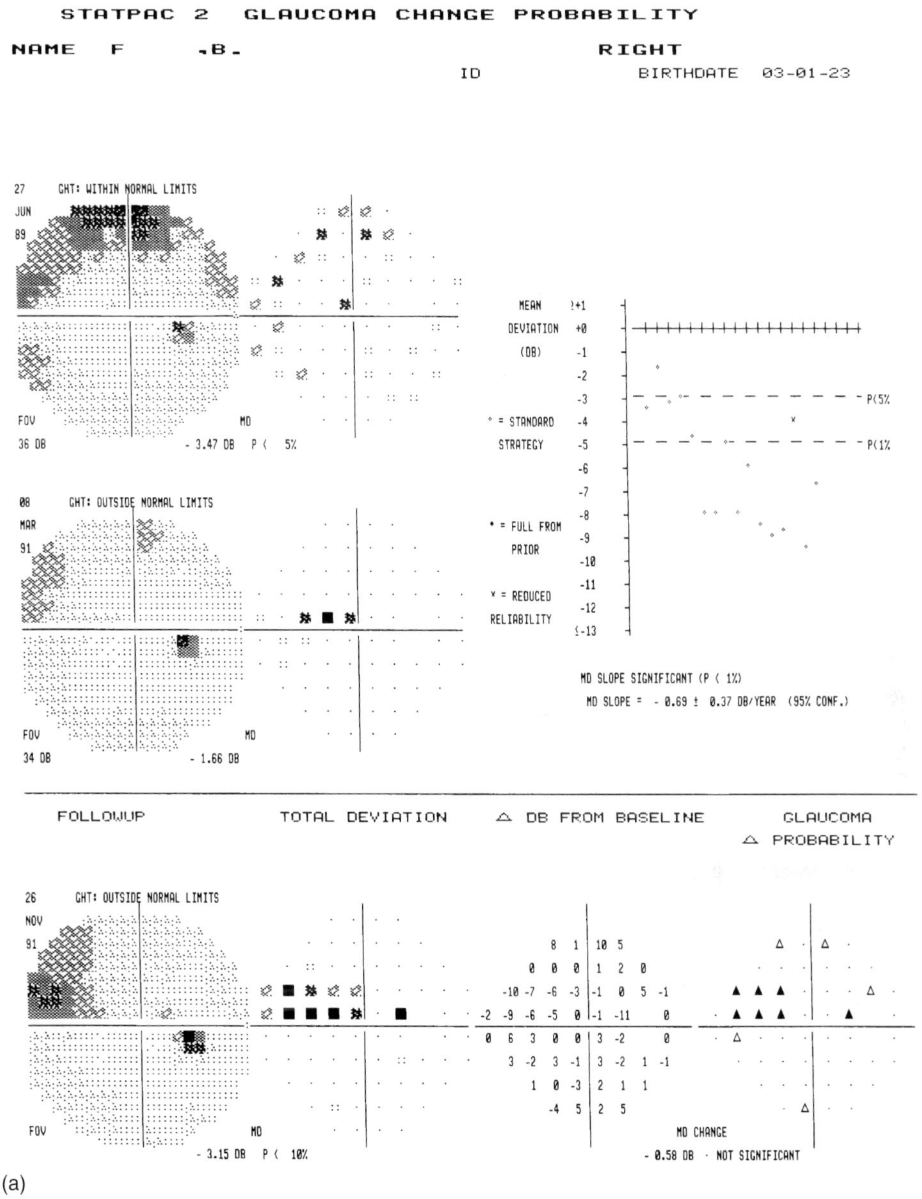

Fig. 7. Glaucoma Change Probability (see the section on "SAP, VF progression" for explanation).

used in clinical practice (Nouri-Mahdavi et al., 1997; Lin et al., 2003; Vesti et al., 2003).

No matter what method is used for assessing the evolution of a perimetric defect, it should however be emphasized that the results of the perimetry must be compared to the other clinical information, inasmuch as a variation in the VF could be due to reasons other than glaucoma.

Finally, it should be remembered that, in order to be of such a clinical importance as to change current therapy, a worsening of the perimetric defect must be statistically important, capable of being reproduced in a second test and relative to the current illness.

SAP offers different advantages: It is probably the test most used in glaucoma over the past two

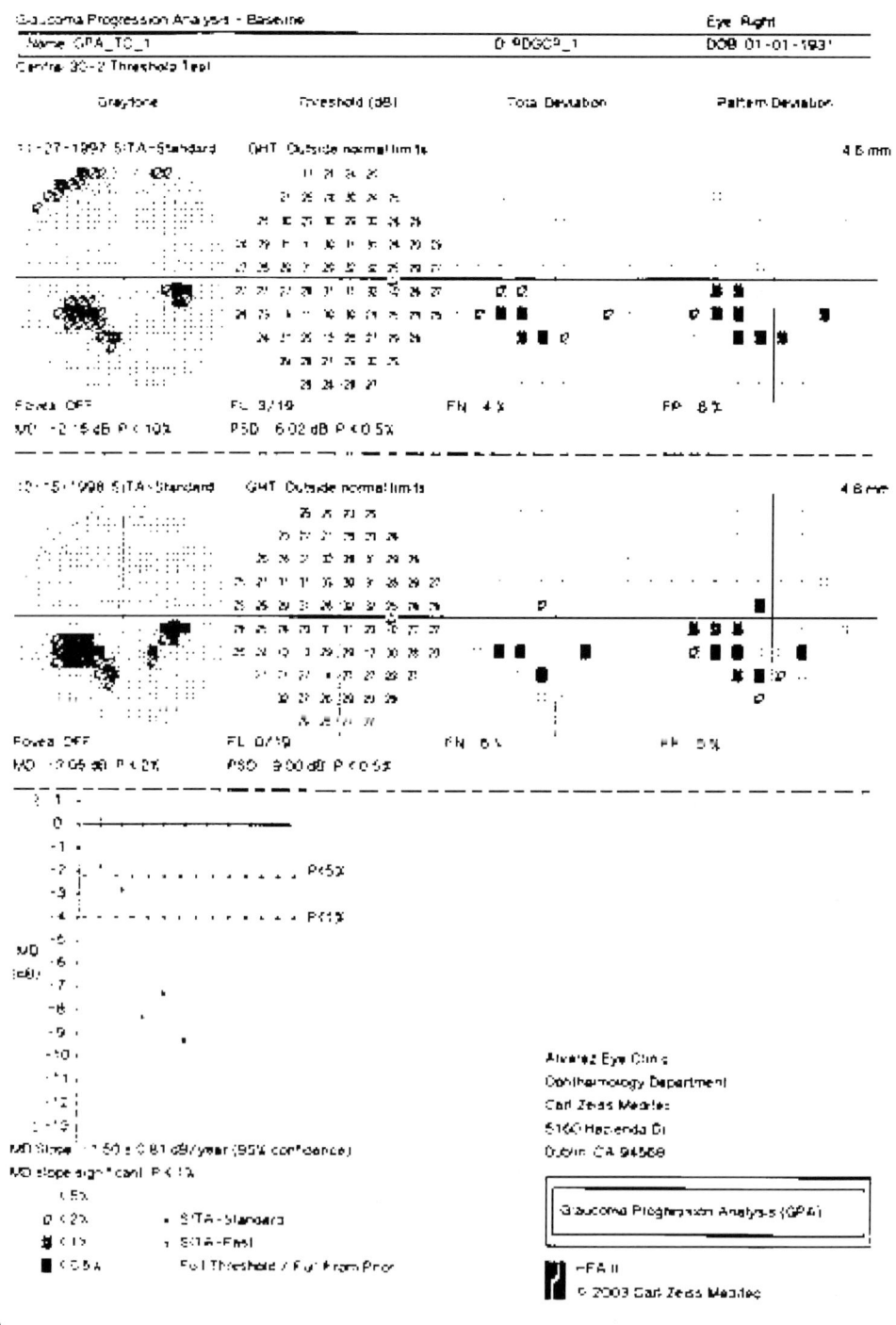

Fig. 7. (*Continued*).

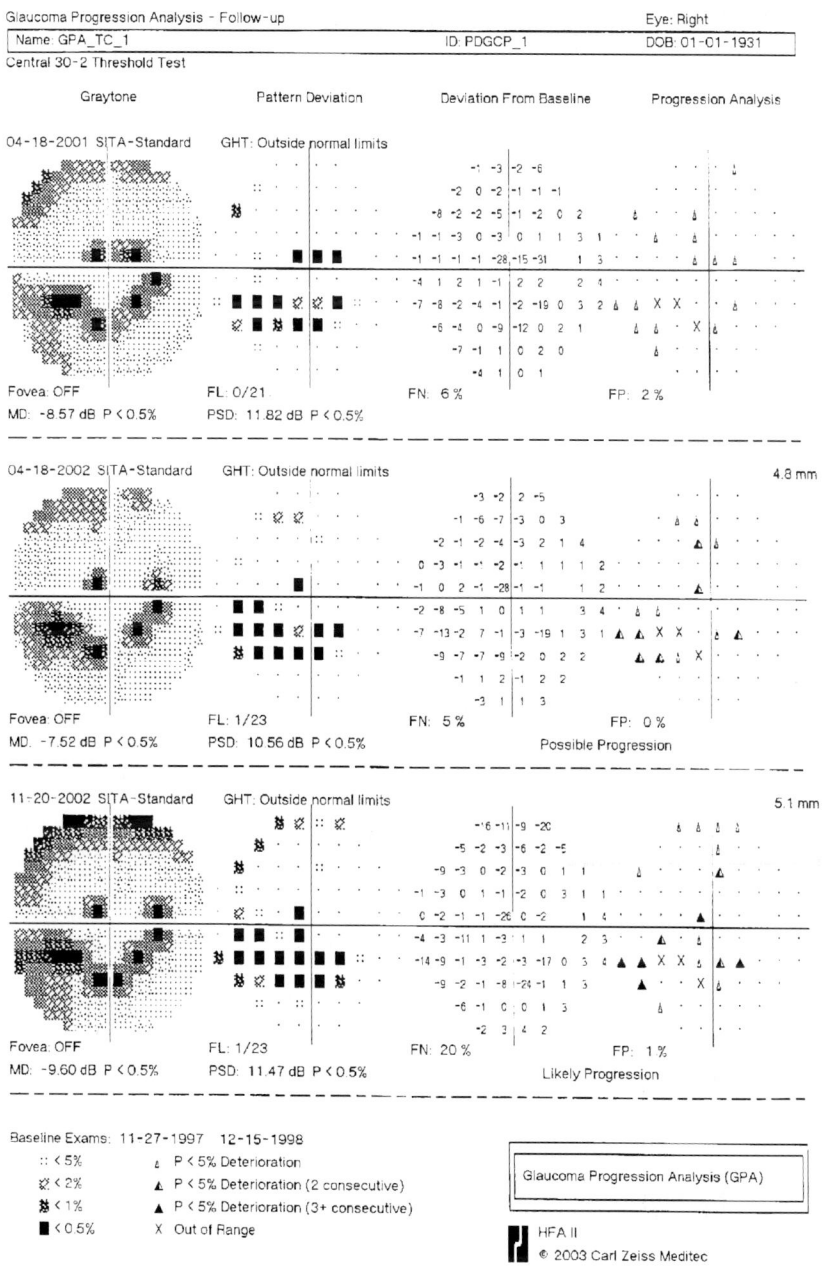

Fig. 8. Glaucoma Progression Analysis (GPA) (see the section on "SAP, VF progression" for explanation).

decades; it was implemented with complex statistic techniques that allowed to shorten test time without loss of accuracy. The SITA algorithm decreased test time (and patient's fatigue) but maintained reproducibility.

Studies on SITA show significant reduction in testing time (50%) without affecting diagnostic accuracy. It is associated with a greater sensitivity and reproducibility and less intertest variability when compared with standard FT testing. Part of this positive effect seems to be related to better performance due to shorter test times. Nevertheless, a new baseline should be established when patients are switched from the standard threshold automated perimetry algorithm to SITA.

Furthermore, correlation between SITA MD and PSD indices and FT MD and PSD, as well as retinal nerve fiber layer (RNFL) thickness, was demonstrated.

On the other hand, inaccuracy of fixation monitoring and lack of sensitivity in detecting small scotomata still affect SITA SAP.

The impact of converting from Humphrey 24-2 FT VF testing to SS VF testing during the follow-up phase of a clinical trial has been evaluated by Musch et al. (1999) in the CIGTS. The CIGTS VF score and GHT differed between SS and FT tests. Changing the approach used to measuring a study's primary VF outcome should be accompanied by a critical evaluation of the change's impact.

Clusters of related locations of the VF have been used to investigate the structure–function relationship and to aid in the detection of glaucomatous progression by reducing the effect of long-term variability.

Machine learning classifiers (MLC) can learn complex patterns and trends in data; the application of MLC to longitudinal SAP data from ocular hypertensive eyes allowed to predict the confirmed abnormality much earlier (on average, 3.92 ± 0.55 years) than traditional STATPAC-like methods.

Future work is needed to investigate the potential for using MLC to optimize VF assessment, in particular if field data are combined with structural imaging data.

SAP: the relationship to other functional and structural diagnostic tests in glaucoma

SAP is not selective for a particular ganglion cell type. Any of the primary ganglion cell subtypes can respond to a static achromatic incremental stimulus presented on an achromatic background. Since there is considerable overlap in the receptive fields of retinal ganglion cells, a test with a nonselective stimulus may not be highly sensitive for the earliest loss of retinal ganglion cells. The reason relies on the considerable redundancy in the coverage of a given location in the retina. Therefore, SAP may not detect VF loss until the optic nerve has already suffered considerable damage, unless the damage is very localized (Delgado et al., 2002).

On the contrary, Short-Wavelength Automated Perimetry (SWAP), Frequency Doubling Perimetry (FDP), Motion Automated Perimetry (MAP), and High-Pass Resolution Perimetry (HPRP) are considered as selective tests.

SAP, FDP-Matrix

FDP and SAP perform similarly in their ability to detect VF defects in early-to-moderate glaucoma. Larger and deeper defects detected with FDP suggest the possibility of earlier detection at high specificity.

These benefits, however, can only be conclusively demonstrated in prospective longitudinal studies.

In comparison with SAP, the sensitivity and specificity of FDP were 92% and 98% with GHT criteria and 98% and 93% with PSD < 5% criteria, respectively.

Similarly, high diagnostic precision was found with MD and PSD (at 95% specificity, MD and PSD sensitivity was 82% and 90%, respectively). However, other authors found the Matrix examination did not detect 36% of abnormal SITA fields. Matrix field defects were smaller and deeper than those appearing in SITA perimetry.

In another study, the FDP-Matrix and SAP-SITA detected abnormal visual function in 51% and 44%, respectively, of GAOD eyes, and both perimetry techniques identified 11% of healthy eyes as abnormal. Agreement between FDP-Matrix and SAP-SITA was moderate (Artes et al., 2005a, b; Haymes et al., 2005; Pierre-Filho et al., 2006; Ferreras et al., 2007; Leeprechanon et al., 2007; Patel et al., 2007).

SAP, SWAP, HPRP, FDT

In a recent study, SAP performance resulted equal to or slightly better than SWAP and not significantly different from FDT and HPRP. This finding differs from other results of previous studies. Furthermore, since in this study none test type was always affected in patients with glaucoma whereas the other test types remained normal, the authors suggest that a combination of test types may be most efficient in identifying early loss and confirming the region of optic pathways affected by glaucoma (Sakata et al., 2007a, b).

These findings seem to confirm a prominent clinical role of the SAP in the diagnosis and follow-up of glaucoma. Although a true gold standard for glaucoma diagnosis is still lacking, none of the newer test types can entirely replace SAP at this time.

SAP: the relationship between function and structure

Accurate measurement of both functional and structural changes is critical in both clinical practice and clinical trials to support the diagnosis of glaucoma, and also to follow up any progression of the disease over time.

The relationship between structure and function, however, is not entirely explained.

The relationship between structure, observed by different imaging techniques, like scanning laser polarimetry (SLP), and functions, assessed by SAP, has been investigated in various studies. With SLP with variable corneal compensation (VCC; commercially available in the GDx Nerve Fiber Analyzer; Carl Zeiss Meditec, Inc., Dublin, CA), the structure–function relationship has been shown to be curvilinear when VF sensitivity is expressed in a decibel scale.

Other studies showed that the relationship between function and structure is curvilinear for the correlations between decibel-DLS and both the number of ganglion cells and neuroretinal rim.

However, if VF differential light sensitivity is expressed in an antilog (1/Lambert) scale, function relates linearly to structure, as has been shown by Garway-Heath (Johnson et al., 2000; Harwerth et al., 2007).

On the other hand, both the OHTS and the EGPS showed that, in many eyes, structural defects develop before functional defects, whereas in a similar number of other eyes, functional defects develop first (with both kinds of defects developing in some eyes simultaneously). It is reasonable that diseased retinal ganglion cells begin to malfunction before dying: In this case, a decrease in differential light sensitivity can be measured without a detectable structural loss (Miglior et al., 2007).

These findings suggest that both the SAP VF and the optic disc must be monitored with equal attention, because either may show the first evidence of damage due to glaucoma. VF changes like an increase in PSD or evidence of a nasal step and/or partial arcuate defect and changes in the ONH based on stereo-photographic observation (i.e., a slight increase in C/D ratio or rim thinning) should be sought through repeat testing and correlation with other clinical results because they suggest an increased risk of developing glaucoma.

SAP, confocal scanning laser ophthalmoscopy, SLP-VCC

Retinal ganglion cell function assessed by SAP relates only slightly to measurements of neuroretinal rim area using confocal scanning laser ophthalmoscopy (CSLO) and of RNFL thickness using SLP-VCC. The curvilinearity of the relationship between function and structure is mainly due to the standard decibel scale in SAP, as previously reported (Racette et al., 2007).

On the other hand, measurements of neuroretinal rim area using CSLO compare well with measurements of RNFL thickness using SLP-VCC.

Nevertheless, a stronger structure–function relationship between RNFL retardation and SAP VF sensitivity has been found in images obtained with the GDx ECC than with the GDx VCC.

In other words, more accurate RNFL imaging may show a stronger correlation of structural imaging with SAP results (Mai et al., 2007).

SAP, optical coherence tomography

The relationship between RNFL thickness, as measured by optical coherence tomography (OCT), and VF sensitivity, as measured by SAP, has been evaluated in normal subjects and in patients with glaucoma. A strong curvilinear regression has been found in POAG eyes with PSD 1.9 dB and RNFL average thickness below 70 µ, whereas no correlation was detectable above these values.

Other studies showed that SAP measures of VF defects and OCT measures of RNFL defects are correlated measures of glaucomatous neuropathy. They conclude that RNFL thickness may be a more sensitive measurement for early stages and perimetry a better measure for moderate-to-advanced stages of glaucoma (Ajtony et al., 2007).

Nevertheless, it has been stated that the fluctuation in thresholding of automated perimetry testing in advanced glaucoma is too large. For this reason, the magnitude of change required to have statistical confidence that the change is real often exceeds the dynamic range of the perimeter.

SAP and functional magnetic resonance imaging

A correlation between functional magnetic resonance imaging (fMRI) responses and measurements of optic disc damage for OCT (RNFL), HRT (mean height contour), and GDx (RNFL) has been recently demonstrated using a novel method for projecting VF scotomas onto the flattened cortical representation.

Cortical activity for viewing through the glaucomatous versus fellow eye was compared by alternately presenting each eye with a contrast-reversing checkerboard pattern. The resultant fMRI response was then compared to interocular differences in RNFL or mean height contour for analogous regions of the VF.

Although indirectly, these findings suggest that the pattern of cortical activity in V1 may be correlated with the pattern of VF loss measured with SAP (Duncan et al., 2007).

Because perimetry is a psychophysical measure, the values of differential sensitivity obtained are not dependent on the functional architecture of the visual system alone but also on a variety of factors. Excessive pupillary constriction or dilatation, improper refraction, lens rim artifact, and blepharoptosis, all affect luminance thresholds. Moreover, cognitive factors including learning effect, subject attention, fatigue, motivation, and response bias can influence the obtained thresholds. In addition, perimetrist's instructions can significantly affect obtained automated perimetry thresholds (Kutzko et al., 2000).

Other authors suggest that, during perimetry, subjects who are distracted or anxious may produce VF alterations that are extremely similar to the typical nerve fiber bundle defects due to glaucoma.

SAP showed significant intertest variability in individual glaucoma patients and patients with high risk factors for developing glaucomatous damage. Therefore, it may be difficult to determine with statistical confidence whether SAP VF glaucomatous damage is present. It may be also hard to differentiate between true progression or fluctuation unless the test is repeated multiple times. Repeating the VF examination is a time-consuming procedure not always accomplished in daily practice, despite the published evidence. A report by the American Academy of Ophthalmology (2002) on automated perimetry stated: "*Although standard automated perimetry has become one of the most useful tools in the detection and management of glaucoma, there is lack of a true gold standard.*"

Despite these important caveats, recent findings validate the clinical role of SAP in glaucoma.

SAP results were correlated to structure of the visual system as measured by morphological techniques (CSLO, SLP, and OCT) and, with obvious limitations, to fMRI results.

Many other efforts are necessary to optimize SAP VF assessment: newer statistical methods, the use of MLC, and combining of SAP field data with structural data seem to be relevant ways to achieve this important result.

In particular, MLC are computational methods that enable machines to learn from experience. They were effective in VF interpretation, identification of patterns of VF loss, diagnosis of glaucoma through structural measures, and detection of progression of glaucoma. According to Bowd et al. (2008), combining OCT and SAP measurements of healthy and early glaucomatous eyes using two types of MLC increased diagnostic performance in glaucoma is obtained.

MLC, newer thresholding algorithms strategies, more sophisticated statistic software, and better fixation monitoring seem to be promising tools in order to improve precision and accuracy of this well-known and widespread diffused method of examination in glaucoma management.

References

Ajtony, C., Balla, Z., Somoskeoy, S. and Kovacs, B. (2007) Relationship between visual field sensitivity and retinal nerve fiber layer thickness as measured by optical coherence tomography. Invest. Ophthalmol. Vis. Sci., 48(1): 258–263.

Anderson, D.R. and Patella, V.M. (1999) Automated Static Perimetry (2nd edn.). Mosby, St. Louis.

Anderson, D.R., Chauhan, B., Johnson, C., Katz, J., Patella, V.M. and Drance, S.M. (2000) Criteria for progression of glaucoma in clinical management and in outcome studies. Am. J. Ophthalmol., 130: 827–829.

Artes, P.H., Hutchison, D.M., Nicolela, M.T., LeBlanc, R.P. and Chauhan, B.C. (2005a) Threshold and variability properties of matrix frequency-doubling technology and standard automated perimetry in glaucoma. Invest. Ophthalmol. Vis. Sci., 46(7): 2451–2457.

Artes, P.H., Nicolela, M.T., LeBlanc, R.P. and Chauhan, B.C. (2005b) Visual field progression in glaucoma: total versus pattern deviation analyses. Invest. Ophthalmol. Vis. Sci., 46: 4600–4606.

Barkana, Y., Gerber, Y., Mora, R., Liebmann, J.M. and Ritch, R. (2006) Effect of eye testing order on automated perimetry results using the Swedish interactive threshold algorithm standard 24-2. Arch. Ophthalmol., 124(6): 781–784.

Bowd, C., Hao, J., Tavares, I.M., Medeiros, F.A., Zangwill, L.M., Lee, T.W., Sample, P.A., Weinreb, R.N. and Goldbaum, M.H. (2008) Bayesian machine learning classifiers for combining structural and functional measurements to classify healthy and glaucomatous eyes. Invest. Ophthalmol. Vis. Sci., 49: 945–953.

Brusini, P. and Filacorda, S. (2006) Enhanced glaucoma staging system (GSS2) for classifying functional damage in glaucoma. J. Glaucoma, 15: 40–46.

Delgado, M.F., Nguyen, N.T., Cox, T.A., Singh, K., Lee, D.A., Dueker, D.K., Fechtner, R.D., Juzych, M.S., Lin, S.C., Netland, P.A., Pastor, S.A., Schuman, J.S. and Samples, J.R.American Academy of Ophthalmology. Ophthalmic Technology Assessment Committee 2001–2002 Glaucoma Panel. (2002) Automated perimetry: a report by the American Academy of Ophthalmology. Ophthalmology, 109(12): 2362–2374.

Duncan, R.O., Sample, P.A., Weinreb, R.N., Bowd, C. and Zangwill, L.M. (2007) Retinotopic organization of primary visual cortex in glaucoma: a method for comparing cortical function with damage to the optic disk. Invest. Ophthalmol. Vis. Sci., 48(2): 733–744.

European Glaucoma Prevention Study (EGPS) Group. (2002) The European Glaucoma Prevention Study: design and baseline description of the participants. Ophthalmology, 109: 1612–1621.

Ferreras, A., Polo, V., Larrosa, J.M., Pablo, L.E., Pajarin, A.B., Pueyo, V. and Honrubia, F.M. (2007) Can frequency-doubling technology and short-wavelength automated perimetries detect visual field defects before standard automated perimetry in patients with preperimetric glaucoma? J. Glaucoma, 16(4): 372–383.

Fitzke, F.W., Hitchings, R.A., Poinoosawmy, D., McNaught, A.I. and Crabb, D.P. (1996) Analysis of visual field progression in glaucoma. Br. J. Ophthalmol., 80: 40–48.

Flammer, J. (1986) The concept of visual field indices. Graefes Arch. Clin. Exp. Ophthalmol., 224(5): 389–392.

Flammer, J., Drance, S.M. and Zulauf, M. (1984) Differential light threshold, short- and long-term fluctuation in patients with glaucoma, normal controls, and patients with suspected glaucoma. Arch. Ophthalmol., 102: 704–706.

Gordon, M.O. and Kass, M.A. (1999) Ocular hypertension treatment study group. The ocular hypertension treatment study: design and baseline description of the participants. Arch. Ophthalmol., 117: 573–583.

Hall, T.A. and Owsley, C. (2005) Visual field defects and the risk of motor vehicle collisions among patients with glaucoma. Invest. Ophthalmol. Vis. Sci., 46: 4437–4441.

Harwerth, R.S., Vilupuru, A.S., Rangaswamy, N.V. and Smith III, E.L. (2007) The relationship between nerve fiber layer and perimetry measurements. Invest. Ophthalmol. Vis. Sci., 48: 763–773.

Haymes, S.A., Hutchison, D.M., McCormick, T.A., Varma, D.K., Nicolela, M.T., LeBlanc, R.P. and Chauhan, B.C. (2005) Glaucomatous visual field progression with frequency-doubling technology and standard automated perimetry in a longitudinal prospective study. Invest. Ophthalmol. Vis. Sci., 46(2): 547–554.

Heijl, A., Lindgren, G. and Lindgren, A. (1991) Extended empirical statistical package for evaluation of single and multiple fields: Statpac 2. In: Mills R.P. and Heijl A. (Eds.), Perimetry Update 1990/1991. Kugler and Ghedini, New York, pp. 303–315.

Heijl, A., Leske, M.C. and Bengtsson, B. (2002) Reduction of intraocular pressure and glaucoma progression: results from the Early Manifest Glaucoma Trial. Arch. Ophthalmol., 120: 1268–1279.

Heijl, A., Bengtsson, B., Chauhan, B.C., Lieberman, M.F., Cunliffe, I., Hyman, L., Leske, M.C. (2008) A comparison of visual field progression criteria of 3 major glaucoma trials in early manifest glaucoma trial patients. *Ophthalmology*, March, Epub ahead of print.

Hodapp, E., Parrish II, R.K. and Anderson, D.R. (1993) Clinical Decisions in Glaucoma. Mosby Company, St. Louis, pp. 98–114.

Hoffmann, E.M., Boden, C., Zangwill, L.M., Bourne, R.R., Weinreb, R.N. and Sample, P.A. (2006) Inter-eye comparison of patterns of visual field loss in patients with glaucomatous optic neuropathy. Am. J Ophthalmol., 141(4): 703–708.

Johnson, C.A., Cioffi, G.A., Liebmann, J.R., Sample, P.A., Zangwill, L.M. and Weinreb, R.N. (2000) The relationship between structural and functional alterations in glaucoma: a review. Semin. Ophthalmol., 15(4): 221–233.

Katz, J., Gilbert, D., Quigley, H.A. and Sommer, A. (1997) Estimating progression of visual field loss in glaucoma. Ophthalmology, 104: 1017–1025.

Kaufmann, H. and Flammer, J. (1989) The Bebié curve (cumulative defect curve) for differentiating local and diffuse visual field defects. Fortschr Ophthalmol., 86(6): 687–691.

Keltner, J.L., Johnson, C.A., Spurr, J.O. and Beck, R.W. (1993) Baseline visual field profile of optic neuritis. The experience of the optic neuritis treatment trial. Optic Neurities Study Group. Arch Ophthalmol., 111(2): 231–234.

Kerrigan–Baumrind, L.A., Quigley, H.A., Pease, M.E., Kerrigan, D.F. and Mitchell, R.S. (2000) Number of ganglion cells in glaucoma eyes compared with threshold visual field tests in the same persons. Invest. Ophthalmol. Vis. Sci., 41: 741–748.

King, A.J.W., Taguri, A., Wadood, A.C. and Azuara-Blanco, A. (2002) Comparison of two fast strategies, SITA Fast and TOP, for the assessment of visual fields in glaucoma patients. Graefes Arch. Clin. Exp. Ophthalmol., 240: 481–487.

Koçak, I., Zulauf, M., Hendrickson, P. and Stumpfig, D. (1997) Evaluation of the Brusini Staging System for follow-up in glaucoma. Eur. J. Ophthalmol., 7: 345–350.

Kulze, J.C., Stewart, W.C. and Sutherland, S.E. (1990) Factors associated with a learning effect in glaucoma patients using automated perimetry. Acta Ophthalmol., 68: 681–686.

Kutzko, K.E., Brito, C.F. and Wall, M. (2000) Effect of instructions on conventional automated perimetry. Invest. Ophthalmol. Vis. Sci., 41: 2006–2013.

Leeprechanon, N., Giangiacomo, A., Fontana, H., Hoffman, D. and Caprioli, J. (2007) Frequency-doubling perimetry: comparison with standard automated perimetry to detect glaucoma. Am. J. Ophthalmol., 143(2): 263–271.

Levine, R. (2006) Asymmetries and visual field summaries as predictors of glaucoma in the ocular hypertension treatment study. Invest. Ophthalmol. Vis. Sci., 47: 3896–3903.

Lin, A.L., Hoffman, D., Gaasterland, D.E. and Caprioli, J. (2003) Neural networks to identify glaucomatous visual field progression. Am. J. Ophthalmol., 135: 49–54.

Mai, T.A., Reus, N.J. and Lemij, H.G. (2007) Structure–function relationship is stronger with enhanced corneal compensation than with variable corneal compensation in scanning laser polarimetry. Invest. Ophthalmol. Vis. Sci., 48(4): 1651–1658.

Marra, G. and Flammer, J. (1991) The learning and fatigue effect in automated perimetry. Graefes Arch. Clin. Exp. Ophthalmol., 229: 501–504.

McGwin, G., Jr., Xie, A., Mays, A., Joiner, W., DeCarlo, K.D., Newkirk, M.R., Gardiner, S.K., Demirel, S. and Johnson, C.A. (2006) Assessment of false positives with the Humphrey Field Analyzer II Perimeter with the SITA algorithm. Invest. Ophthalmol. Vis. Sci., 47: 4632–4637.

Miglior, S., Torri, V., Zeyen, T., Pfeiffer, N., Vaz, J.C. and Adamsons, I.EGPS Group. (2007) Intercurrent factors associated with the development of open-angle glaucoma in the European glaucoma prevention study. Am. J. Ophthalmol., 144(2): 266–275.

Mikelberg, F.S. and Drance, S.M. (1984) The mode of progression of visual field defects in glaucoma. Am. J. Ophthalmol., 98: 443–445.

Musch, D.C., Lishter, P.R., Guire, K.E. and Standardi, C.L.CIGTS Study Group. (1999) The collaborative initial glaucoma treatment study. Ophthalmology, 106: 653–662.

Musch, D.C., Gillespie, B.W., Motyka, B.M., Niziol, L.M., Mills, R.P. and Lichter, P.R.on behalf of the CIGTS Study Group. (2005) Converting to SITA-standard from full-threshold visual field testing in the follow-up phase of a clinical trial. Invest. Ophthalmol. Vis. Sci., 46: 2755–2759.

Nouri-Mahdavi, K., Brigatti, L., Weitzman, M. and Caprioli, J. (1997) Comparison of methods to detect visual field progression in glaucoma. Ophthalmology, 104: 1228–1236.

Park, H.J. and Youn, D.H. (1994) Quantitative analysis of changes of automated perimetric thresholds after pupillary dilation and induced myopia in normal subjects. Korean J. Ophthalmol., 8: 53–60.

Paetzold, H.A., Vonthein, R., Krapp, E., Rauscher, S. and Schiefer, U. (2007) Age-dependent normative values for differential luminance sensitivity in automated static perimetry using the Octopus 101. Acta Ophthalmol. Scand., 7.

Patel, A., Wollstein, G., Ishikawa, H. and Schuman, J.S. (2007) Comparison of visual field defects using matrix perimetry and standard achromatic perimetry. Ophthalmology, 114(3): 480–487.

Pierre-Filho, P.T., Schimiti, R.B., de Vasconcellos, J.P. and Costa, V.P. (2006) Sensitivity and specificity of frequency-doubling technology, tendency-oriented perimetry, SITA Standard and SITA Fast perimetry in perimetrically inexperienced individual. Acta Ophthalmol. Scand., 84(3): 345–350.

Racette, L., Medeiros, F.A., Bowd, C., Zangwill, L.M., Weinreb, R.N. and Sample, P.A. (2007) The impact of the perimetric measurement scale, sample composition, and statistical method on the structure-function relationship in glaucoma. J. Glaucoma, 16(8): 676–684.

Sakata, LM., Deleon-Ortega, J., Arthur, S.N., Monheit, B.E. and Girkin, C.A. (2007a) Detecting visual function abnormalities using the Swedish interactive threshold algorithm and matrix perimetry in eyes with glaucomatous appearance of the optic disc. Arch. Ophthalmol., 125(3): 340–345.

Sakata, L.M., DeLeón-Ortega, J. and Girkin, C.A. (2007b) Selective perimetry in glaucoma diagnosisCurr. Opin. Ophthalmol., 18(2): 115–121., Review.

Spry, P.G.D. and Johnson, C.A. (2002) Identification of progressive glaucomatous visual field loss. Surv. Ophthalmol., 47: 158–173.

The Advanced Glaucoma Intervention Study (AGIS): 1. (1994) Study design and methods and baseline characteristics of study patients. Control Clin. Trials, 15(4): 299–325.

Vesti, E., Johnson, C.A. and Chauhan, B.C. (2003) Comparison of different methods for detecting glaucomatous visual field progression. Invest. Ophthalmol. Vis. Sci., 44: 3873–3879.

Walsh, T.J. (1990) Visual Fields. Examination and Interpretation. American Academy of Ophthalmology, San Francisco.

Weber, J. and Caprioli, J. (2000) Atlas of Computerized Perimetry. WB Saunders Comp, Philadelphia, pp. 215–231.

Wild, J.M., Pacey, I.E., Hancock, S.A. and Cunliffe, I.A. (1999a) Between-algorithm, between-individual differences in normal perimetric sensitivity: full threshold, FASTPAC, and SITA Swedish Interactive Threshold algorithm. Invest. Ophthalmol. Vis. Sci., 40: 1152–1161.

Wild, J.M., Pacey, I.E., O'Neill, E.C. and Cunliffe, I.A. (1999b) The SITA perimetric threshold algorithms in glaucoma. Invest. Ophthalmol. Vis. Sci., 40: 1998–2009.

Zalta, A.H. (1989) Lens rim artifact in automated threshold perimetry. Ophthalmology, 96: 1302–1311.

CHAPTER 8

Short-wavelength automated perimetry and frequency-doubling technology perimetry in glaucoma

Paolo Fogagnolo[1,*], Luca Rossetti[2], Stefano Ranno[2], Antonio Ferreras[3] and Nicola Orzalesi[2]

[1]*G.B. Bietti Foundation – IRCCS (Istituto di Ricovero e Cura a Carattere Scientifico), Rome, Italy*
[2]*Eye Clinic, Department of Medicine, Surgery and Odontoiatry, San Paolo Hospital, University of Milan, Milan, Italy*
[3]*Department of Ophthalmology, Miguel Servet University Hospital, Zaragoza, Spain*

Abstract: Standard automated perimetry (SAP) is today still the clinical standard for the management of glaucoma and its progression, though it has been shown that it may detect the disease only after the death of a high number of retinal ganglion cells (RGCs). A number of "unconventional" perimetries have recently been evaluated by several clinical studies which showed their ability to identify the earliest glaucoma changes; the most promising of these techniques are short-wavelength automated perimetry (SWAP) and frequency-doubling technology perimetry (FDT). The applicability of these techniques is still limited by a number of factors: the limited economic resources allocated to perimetry; the paucity of well-conducted, prospective longitudinal studies showing the superiority of SWAP and FDT over SAP; and the lack of a consensus on the criteria to define test abnormality with these techniques. The aim of this article is to review the rationale, the limits, and the potentiality of SWAP and FDT for glaucoma management and to summarize the tasks required to improve the clinical usefulness of these two instruments in the future.

Keywords: glaucoma; standard automated perimetry; short-wavelength automated perimetry (SWAP); frequency-doubling technology perimetry (FDT); matrix; retinal ganglion cells

Introduction

Perimetry still remains the "gold standard" for diagnosis of glaucoma and of its progression, as confirmed by the recent glaucoma trials (AGIS, 1994; CNTSG, 1998; Gordon and Kass, 1999; Leske et al., 1999; Musch et al., 1999; CGS, 2006),

*Corresponding author. Tel.: +390281844301; Fax: +390250323750; E-mail: fogagnolopaolo@googlemail.com

on which visual field changes were considered the primary endpoint for progression. Nevertheless, it is often quoted that glaucomatous defects involving about 30% to 50% of retinal ganglion cells (RGCs) may be undetected by automated achromatic static perimetry (standard automated perimetry, SAP) (Harwerth et al., 1999).

The logical extension of this finding is that by the time SAP defects are detected, the disease is already at a relatively advanced stage, questioning the justification of investments to perform SAP tests. Hence, over the last two decades, research has also focused on the development of new, "unconventional" perimetric techniques for diagnosing and monitoring the earliest glaucomatous changes. Although a number of techniques are now available (short-wavelength automated perimetry, SWAP, or blue-on-yellow perimetry; frequency-doubling technology perimetry, FDT; motion-automated perimetry, MAP; high-pass resolution perimetry, HPRP), the most largely used in clinical settings are SWAP and FDT. The aim of this article is to review the rationale, the most relevant clinical data available in literature, and the tasks required to improve the clinical usefulness of these two instruments.

Retinal ganglion cells: anatomy and function

In the human visual system, RGCs project to relay cells in the layers of the dorsal lateral geniculate nucleus (LGN), which project to the primary visual cortex. All retinal and optic nerve head (ONH) diseases determine the death of the first axon of this pathway (in glaucoma through apoptosis). The primary damage causes a secondary atrophy of the LGN (Weinreb et al., 1994) and, ultimately, a loss of information projected to the visual cortex (Yucel et al., 2003; Gupta and Yucel, 2007). A full comprehension of the anatomy and the physiology of RGC is therefore required to understand the mechanisms by which the disease may induce changes in the visual function.

Recent electrophysiology studies on primates provided evidence that the retina is endowed with three primary pathways: parvocellular (P-cells), magnocellular (M-cells), and koniocellular (K-cells) (Kaplan, 2004; Callaway, 2005). This subdivision into different pathways is maintained through the LGN (Kaplan, 2004; Callaway, 2005), while the projections to the cortex are both anatomically and functionally much less segregated (Dobkins and Albright, 2004; Kaplan, 2004). In addition to these pathways, a number of other ganglion cell types connect to the LGN (Polyak, 1941) whose functional properties and postsynaptic targets still remain unidentified (Callaway, 2005).

Table 1 summarizes the main features of the subgroups of RGCs. Although this classification is almost universally adopted, it must be remembered that the fiber diameter is influenced not only by the ganglion cell type but also by eccentricity: fibers within the central retina are thinner than those projecting from the peripheral retina. As a consequence, larger diameter fibers are not exclusively magnocellular: some eccentric parvocellular RGC axons may be even larger than more central magnocellular ones.

P-cells represent nearly 80% of RGC. They are generally small in size and located in the whole retina, although they have a very high concentration in the macular region; their receptive fields are much smaller than M- and K-pathways and they have substantial overlapping. Conducting velocity of P-cells is slower than K- and M-cells. P-cells are responsible for detecting, encoding, and transmitting information about colored, high-contrast, low temporal frequency (i.e., static) stimuli, although M- and K-cells can detect the same stimuli, albeit with a lower sensitivity (Solomon et al., 2002). An example of a selective stimulation of P-neurons is represented by the projection to the retina of the smallest letters of a standard Snellen chart used to test visual acuity.

P-cells can be classified as central and peripheral. Central P-cells are sensitive to color vision, and they can be subdivided in red-ON, red-OFF, green-ON, green-OFF, and, probably, blue-OFF. Peripheral P-cells comprise two groups (luminance-ON and -OFF) and they are sensitive to luminance (Callaway, 2005). P-cells can be also classified as type I and II on the basis of their receptive organization (type I has a centre–surround organization; type II is less diffuse and has coextensive ON and OFF regions) (Callaway, 2005).

Within the same retina region (see above), M-cells are larger than the other RGCs; they represent about 10% of all RGCs and they are endowed with low redundancy (Sample et al., 2000a, b). M-cells are located in the peripheral retina and they have large receptive fields with very limited overlapping. Within RGCs, M-cells have the fastest conduction of the stimulus. They are sensitive to low-contrast, high-temporal frequency (i.e., motion) stimuli (Solomon et al., 2005); for example, a black car rapidly passing by a driver's side window at night would selectively stimulate M-pathway neurons.

Table 1. Subtypes and features of RGCs

	Cell pathway		
	K	M	P
Percentage	9%	10%	80%
Receive input from	Mainly bistratified (blue-ON) retinal ganglion cells	Parasol retinal ganglion cells	Midget retinal ganglion cells
Location in LGN	Within and between principal layers (interlaminar)	Most ventral (layers 1 and 2)	Most dorsal (layers 3 to 6)
Sensitive to	Shorter wavelengths, moderate spatial resolution	Higher temporal frequencies (movement)	Higher spatial frequencies (detail), colors, luminance
Retinal location	Diffuse	Increasing with eccentricity	Decreasing with eccentricity
Redundancy	Low	Low	High
Conducting velocity	Intermediate	Fast	Slow
Receptive field size	Very large	Large	Small
Segregation	High	High	Low
Isolation	15 dB	Unknown	Unknown

It has been speculated that a portion of the M-cell population, the M_y cells, serves as the primary basis for the frequency-doubling phenomenon (Maddess and Henry, 1992). The response of this subgroup of cells is supposed to be independent by the wavelength of the stimulus, a feature that differentiates them from all the other RGCs, which show a biphasic response at the variation of stimulus wavelength (Solomon et al., 2005).

The third subgroup of RGCs is represented by koniocells. Literally, "koniocells" means "cells as small as dust." This term was used because, due to their small size, it was very difficult to detect them in the context of the peripheral retina, where they are located. Studies on animal models recently improved our knowledge of K-cells: They represent a small subgroup (about 9%) of RGCs of small size (though they are larger than P-cells) and little redundancy (DeMonasterio, 1979); they are sparse in location and their receptive field is very large with no surrounds (Callaway, 2005); they are heterogeneous both in structure and function, and different subtypes have been identified. An important reduction of the subtype of koniocells expressing CaMKII-α has been shown in a model of glaucoma in monkeys compared to controls ($10,456 \pm 8770$ neurons in glaucoma vs. $73,303 \pm 15,776$ in controls) (Yucel et al., 2003).

The main function of K-cells is to process blue–yellow color vision (Dacey and Lee, 1994). Information on the blue–yellow axis is captured by blue-ON receptors, projected by koniocells within and between the principal layers of LGN; the stimulus is conducted with intermediate velocity. K-cells also respond to stimuli with moderate spatial resolution (i.e., moderate contrast).

Is glaucoma damage selective for any subgroup of RGCs?

Over the last decades, a strong debate arose on the RGC-selectivity of glaucoma damage. Scientists dealing with unconventional perimetries supported the hypothesis that glaucoma damage was selective for the subgroups of RGCs isolated by those visual field techniques (Maddess and Henry, 1992). Yet the hypothesis of the selectivity of glaucoma damage was brilliantly confuted by Harwerth et al. (1999). In their experimental study in primate animal models of glaucoma, evaluated with psychophysics, electrophysiology, anatomy, and histochemistry, the authors showed that glaucomatous atrophy causes a nonselective reduction of metabolism of magnocellular and parvocellular neurons in the afferent visual pathway. Such findings were confirmed by Yucel et al. (2003), who showed the absence of selective cell loss within the LGN in experimental glaucoma models. Although few scientists still argue that not all glaucoma cases behave in the same way (with individuals showing damage first to K-cells, others to M-cells, and others to P-cells), after these findings the nonselectivity of glaucoma damage seems demonstrated: as M-, K-, and P-cells are equally affected by the disease, a hypothetical 10,000-fiber loss would determine the loss of 8000 P-cells, 1000 M-cells, 900 K-cells, and 100 non-P, -M, -K-cells.

As a corollary to the assumption of nonselectivity of glaucoma damage, one would expect that perimetric techniques which use stimuli that are detected by all ganglion subsets (such as SAP) would have the same diagnostic power than those which selectively test a single pathway (such as SWAP and FDT), a fact that has been refuted by many clinical studies (see below). How can this discrepancy be possible? In order to answer this question, some functional aspects of the visual system must be considered in more detail: they are segregation, isolation, receptive field organization, and redundancy.

Segregation

We previously stated that visual pathways have a well-defined anatomical segregation up to the LGN. Functional segregation is present as well, but probably only for M- and K-cells. In the presence of damage to these cells, the visual system has a reduced ability to use other subsets of RGCs to compensate their information (Kaplan, 2004; Callaway, 2005). On the opposite end, both M- and K-cells are also sensitive to the colored, high-contrast, static stimuli, and they can therefore substitute P-cells on the transmission of information to the visual cortex in the case of damage to the P-pathway.

This only partial functional segregation may explain by itself how, though all subtypes of RGCs are damaged in glaucoma, tests that favor detection of a stimulus by one visual pathway (for example, FDT for M-cells, and SWAP for K-cells) reduce the ability of the visual system to use other pathways to compensate for the damaged RGC type (Kaplan, 2004; Callaway, 2005). When visual function is tested by SAP, such a compensation would occur.

Isolation

In a visual pathway, isolation defines the amount of sensitivity, which has to be lost before another cell type could assist in responding to the stimulus.

Up to now, the amount of isolation is unknown for retinal pathways, except for K-cells stimulated by blue-on-yellow targets, which provide approximately 15 dB of isolation. This means that the blue–yellow ganglion cell system would have to lose 15 dB of sensitivity before another cell type could assist in responding to the SWAP stimulus (Sample et al., 1996). Together with segregation, isolation reduces the likelihood of other visual pathways to compensate for initial damage (at least for K-cells), thus confirming the potential diagnostic superiority of techniques which selectively test one visual pathway.

Receptive field structure and redundancy

The receptive field is the area of the visual space where presentation or withdrawal of light causes changes in the action potential firing of the visual responsive unit (Hartline, 1940; Polyak, 1941; Kuffler, 1953; Hubel and Wiesel, 1959; Solomon et al., 2002). Visual responsive units therefore represent the functional units of the peripheral visual system; each unit is composed of a variable number of retinal receptors, intraretinal cells, and a single RGC, which brings information to the brain. Intraretinal cells down- and up-regulate the excitability of the whole functional unit and of the neighboring ones, and they create a variable overlapping between contiguous receptive fields; they comprise bipolar cells (which connect receptors to RGCs), horizontal cells (which interconnect receptors inside and outside the receptive field), and amacrine cells (which create a network of connections between contiguous RGCs).

Receptive fields are circular and their dimensions are proportional to the number of retinal receptors, which are connected to a single RGC. The size of the receptive fields increases with eccentricity. In the fovea, the ratio between receptors and RGCs is about 16:1, while in the peripheral retina, nearly 1500 receptors project to a single RGC. As a consequence, central vision has a very high spatial discrimination, whereas peripheral vision is much less defined. In order to preserve the foveal highly detailed vision, inter receptors connections are absent; this means that receptive fields have no overlapping. On the other hand, in the peripheral retina there is a high interconnection among receptors, intraretinal cells, and ganglion cells, leading to a partial overlapping of the contiguous receptive fields.

The concept of overlapping receptive fields is also known as redundancy. Due to redundancy, a single stimulus may simultaneously stimulate two or more RGCs; in the case of retinal damage, this guarantees a higher probability of the stimulus to be transmitted to the brain. As a consequence, the higher the redundancy, the more minor the probability of detection of early damages of the RGCs. It has been demonstrated that, apart from the foveal area, P-pathway is endowed with high redundancy (Johnson, 1994). On the other hand, it is likely that the receptive fields of a sparse subset of RGCs (such as M- and K-cells) have a lower overlap, though low redundancy has until now been demonstrated only for M-cells (Haymes et al., 2005).

FDT: rationale and perimetric techniques

FDT is based on a phenomenon described about 40 years ago by Kelly, who observed that when an achromatic sinusoidal grating of low spatial frequency undergoes counterphase flickering at a high-temporal frequency, the apparent spatial frequency of the grating appears to be doubled (Kelly, 1966). He reported that this "frequency-doubling" phenomenon occurred for sinusoidal gratings having a spatial frequency less than approximately 3 cyc/degree undergoing a counterphase flickering at a temporal frequency greater than 7 Hz.

The FDT stimulus predominantly stimulates the M-cell pathway, which is primarily involved in motion and flicker detection and represents about 10% of all RGCs. Some believe that the frequency-doubling illusion in humans is mediated by the subgroup of M_y cells (Maddess and Henry, 1992). However, the existence of M-cells that exhibit nonlinear response properties, M_y cells, is not universally agreed. White et al. (2002) reported that there is no evidence of a separate nonlinear M-cell class in the primate visual system. They suggested that a cortical loss of temporal phase discrimination is the principal cause of the illusion and proposed that the mechanisms underlying the illusion resemble those underlying the detection of full-field flicker, which appears to be accomplished through the M-cell pathway. Thus, FDT (and particularly Matrix FDT) is most likely a probe of contrast sensitivity: it does not depend on whether the stimulus is perceived as doubled but simply measures detection thresholds of the M-cell pathway.

FDT perimeter (Welch Allyn, Skaneateles Falls, NY; and Humphrey Instruments, San Leandro, CA) uses a vertical sine wave grating of low spatial frequency (0.25–0.50 cyc/degree) that undergoes counterphase flickering at a high-temporal frequency (12–25 Hz). The contrast of the stimulus is modified in each location to calculate threshold sensitivity.

FDT comprises first- and second-generation tests. The major difference between the two groups is the number of tested locations, the dimension, and the characteristics of the stimuli (Table 2); Fig. 1 shows the FDT tests most commonly used in glaucoma practice.

In the first-generation FDT, two programs are available: C-20 and N-30. C-20 consists of a grid of 16 square locations of 10×10 degrees each, projecting on the central 20 degrees of the retina, and a foveal circular grid location of 5 degrees. The N-30 program also tests two additional locations between 20 and 30 degrees of the nasal retina immediately above and below the horizontal line, for a total of 19 locations.

Two strategies are available: screening (suprathreshold) and full-threshold. Both can be performed using the C-20 or the N-30 programs. The average test duration is about 1–1.5 min for the screening test and 4–5 min for the full-threshold mode, and it is directly related to the presence of visual field defects. The greater the visual field loss, the longer the test duration.

The screening strategy compares the point-to-point results with an age-corrected normative database and assigns one of the following values: within normal limits ($P \geqslant 1\%$), mild relative loss

Table 2. Features of the programs of the first-generation FDT (above) and the novel programs of Matrix FDT (bottom)

Program	C-20 screening	C-20 threshold	N-30 threshold	N-30 screening
Spatial frequency (cyc/degree)	0.25	0.25	0.25	0.25
Temporal frequency (Hz)	25	25	25	25
Stimulus size	10°[a]	10°[a]	10°[a]	10°[a]
Tested area	20°	20°	20° (30° nasally)	20° (30° nasally)
Number of tested locations	17	17	19	19
Test time (in normal subjects)	45 s	4 min	5 min	90 s
Program	24-2 threshold	30-2 threshold	10-2 threshold	Macula screening
Spatial frequency (cyc/degree)	0.5	0.5	0.5	0.5
Temporal frequency (Hz)	18	18	12	12
Stimulus size	5°	5°	2°	2°
Tested area	24° (30° nasally)	30°	10°	5°
Number of tested locations	55	69	44	16
Test time (in normal subjects)	5 min	6 min	4 min	90 s

[a]The size of the foveal stimulus is 5°.

($P<1\%$), moderate relative loss ($P<0.5\%$), and severe loss (point not seen; failure to respond to maximum contrast level). In the full-threshold strategy, point-to-point threshold is calculated using a Method of Binary Search (MOBS) procedure (Tyrell and Owens, 1998) that is a different threshold approach compared to the staircase procedure used in Humphrey-SAP. Then the threshold values are compared to the normative database; each location is deemed as normal ($P>5\%$) or abnormal at different levels of probability ($P<5\%$, $<2\%$, $<1\%$, $<0.5\%$) in two maps: total deviation map and pattern deviation map. Point-to-point results are depicted on a colored scale corresponding to the different probability levels.

Reliability indices (fixation errors, false-positive and false-negative) are provided in both strategies. Similarly to the SAP programs with short duration and absence of short-term fluctuation (such as SITA-standard and SITA-FAST), full-threshold FDT also gives two global indices: mean defect (MD) and pattern standard deviation (PSD).

The FDT database contains data from more than 700 eyes of 450 normal subjects with ages ranging from 18 to 85. As it has been shown that FDT sensitivity is lower for the second tested eye versus the first (probably due to cortical adaptation to the FDT stimulus), the machine automatically adjusts the values of the second eye.

The first FDT version has few manufacturing limitations: a system to monitor fixation throughout testing is unavailable (the operator cannot check whether the patient is fixating properly or pause the test to adjust patient alignment during the examination); in order to examine the two nasal points in the N-30 strategy, the fixation target is moved temporally, which may be confusing or cause improper fixation in some patients. FDT includes a small printer which can provide a brief printout; an external computer using specific software, Windows ViewFinder, is needed to obtain more detailed printouts and store data.

The second-generation FDT perimeters include a new program called Matrix which is very similar to the previous version except for the use of a smaller size of the stimulus (5 degrees for the 24-2 and the 30-2 programs and 2 degrees for the 10-2 program), a higher number of tested points, a more efficient method to calculate threshold, and a slightly longer duration (about 6 min). The programs of this FDT version are closer and more comparable to the same programs of SAP; other advantages are the presence of a video eye monitor to check patient alignment and cooperation during the test, a bigger screen to examine the nasal

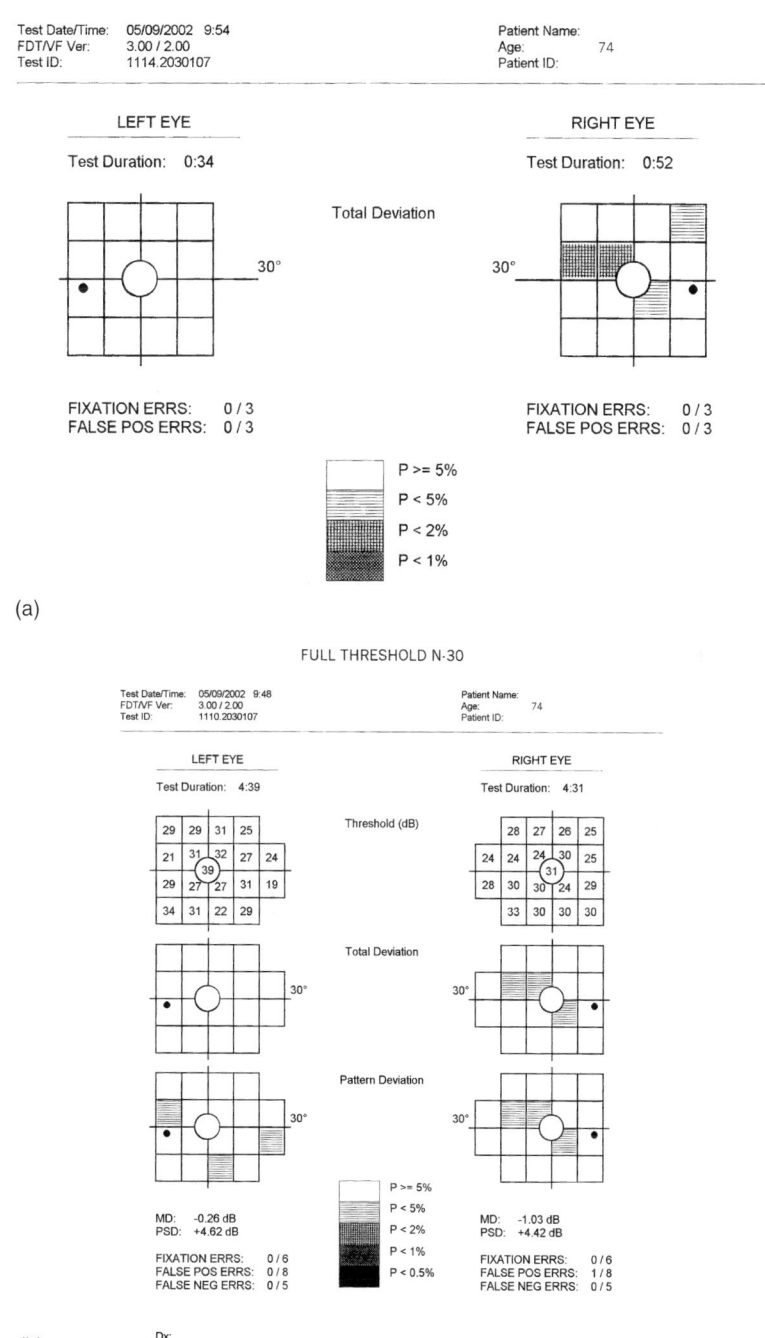

Fig. 1. Printouts of a C-20 screening test (a), an N-30 full-threshold test (b), and a 24-2 Matrix FDT (c).

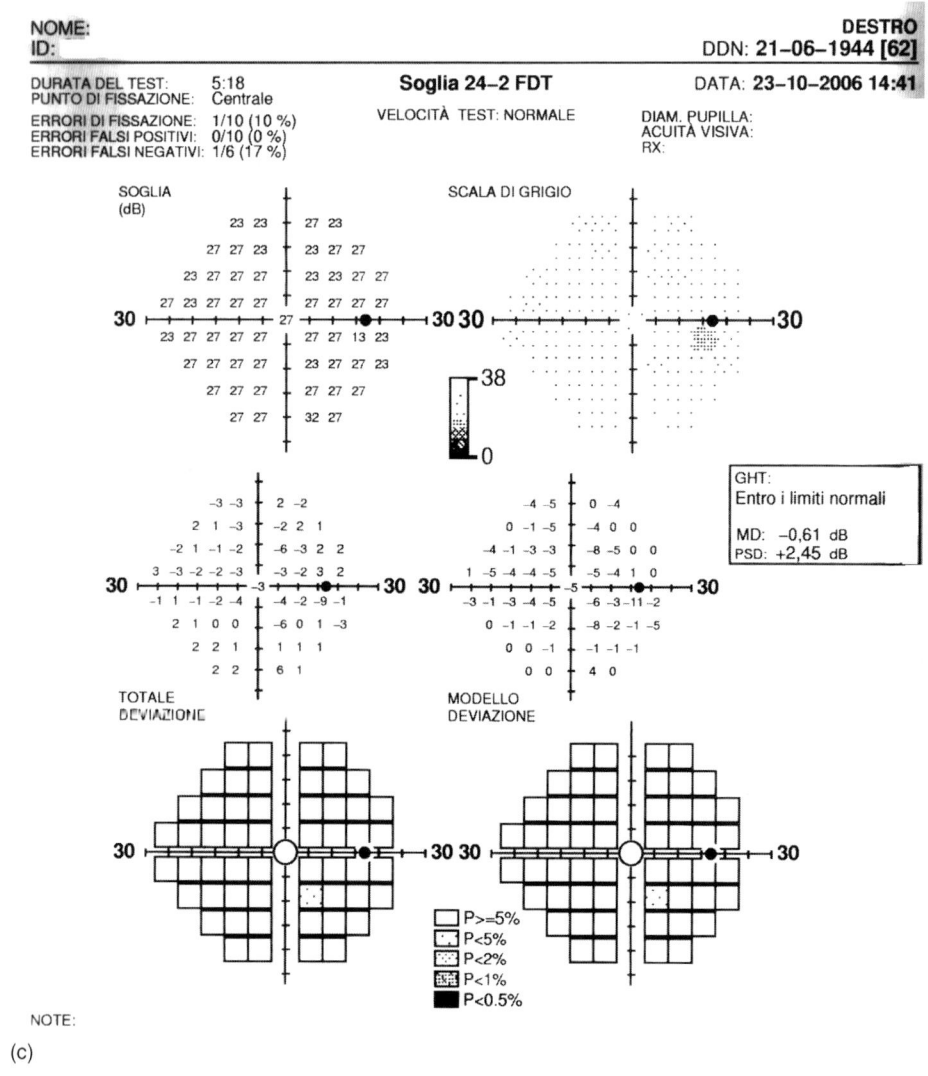

Fig. 1. (Continued).

targets within the central 30° without a moving fixation target, the possibility to store data on the machine, and an enhanced software for result interpretation. Matrix FDT interpretation is based on a database of 270 subjects with age ranging from 18 to 85.

SWAP: rationale and perimetric techniques

The evidence of a deficiency in color vision, particularly in the blue–yellow axis, has been clearly demonstrated in patients affected by glaucoma more than 25 years ago (Drance et al., 1981). Information on the blue–yellow axis is projected to LGN by koniocells, which are a small subgroup of RGCs endowed with little redundancy, a fact which is supposed to allow early glaucoma detection.

The commercially available SWAP version is a modification of the SAP test obtained by Humphrey Field Analyzer (Humphrey-Zeiss, Dublin, CA) using a V stimulus (1.8 degree) of blue light at 440 nm projected to a background illuminated

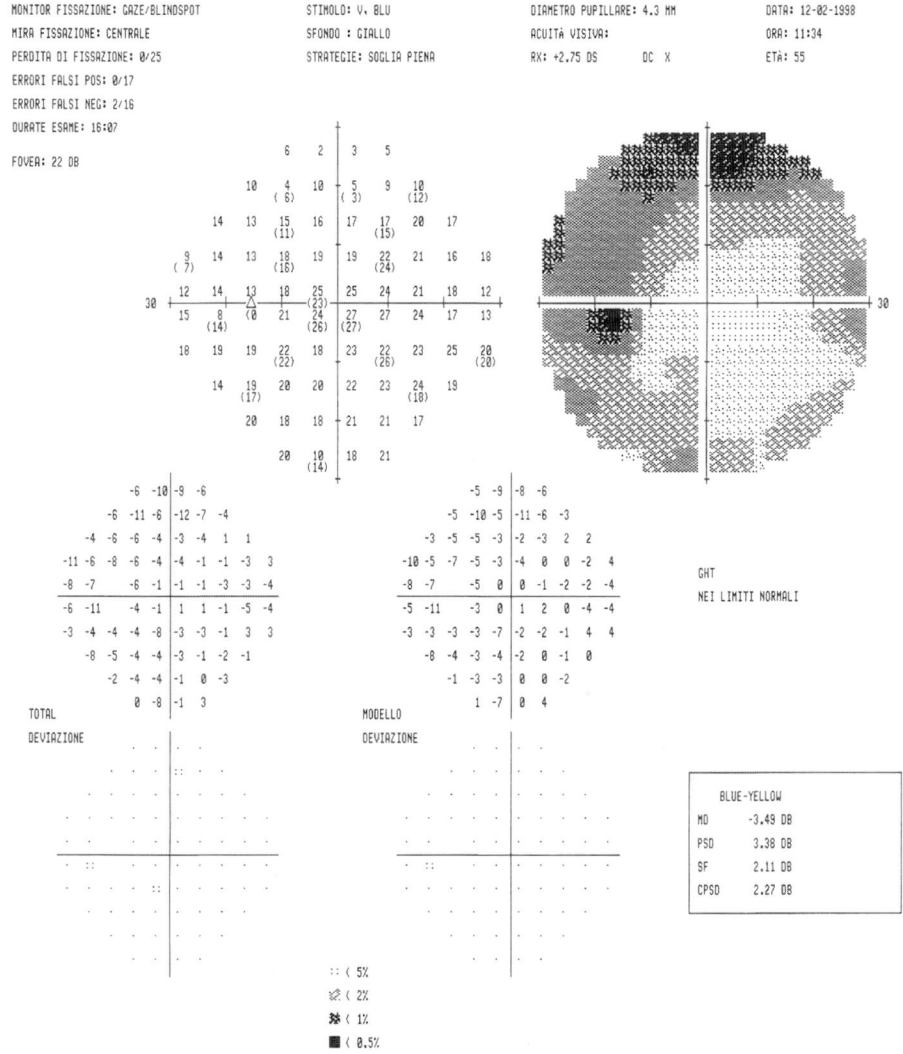

Fig. 2. Printout of a full-threshold SWAP.

with yellow light (570–590 nm) at a luminance of 200 cd/m². The test can be performed over both the 30 and 24 central degrees of the retina; respectively, 76 and 54 locations are tested.

As evident in Fig. 2, SWAP tests are performed and analyzed similarly to SAP: reliability parameters (false-positive, false-negative, and fixation loss) and perimetric indices (MD; PSD; short fluctuation, SF; corrected PSD, CPSD) are calculated, gaze tracking is available and test duration is recorded; the point-to-point sensitivity is reported in decibel and greytone maps (the latter of which should be ignored; it is usually misleading because it is calibrated on SAP sensitivities, which are higher than SWAP). Point-to-point sensitivities are matched to those of normal subjects of the same age in order to identify diffuse (total deviation map) and localized (pattern deviation map) defects. A glaucoma hemifield test (GHT) is finally performed by comparison of symmetric areas of the superior and inferior hemifields.

The full-threshold strategy is the most commonly used but it has a long duration (up to 20 min per eye) and, as a consequence, a high intra-test variability compared to SAP and FDT (Hutchings et al., 2001), which is the result of a number of factors: difficulty in detecting the stimulus, intra-test fatigue effect, higher range of sensitivities within the tested areas (the hill of vision has a steeper shape than SAP and FDT, as described above) (Landers et al., 2006). Overall, these features enlarge the confidence intervals for normality, thus reducing the diagnostic power of this perimetry.

A novel software, SITA SWAP, has been included in the latest version of the Humphrey Field Analyzer (HFA II-i); although clinical data are still missing, it may hopefully overcome the problems of the full-threshold strategy and improve the clinical use of this perimetry for glaucoma detection. A preliminary work confirmed that this procedure could reduce 50% of test duration with encouraging diagnostic results; the confidence intervals for the point-to-point sensitivity are reduced compared to full-threshold SWAP, and this program may identify the same number of visual field losses as the previous version (Bengtsson and Heijl, 2006).

FDT: clinical data

Maddess and Henry were the first to suggest that the frequency-doubling illusion could be useful in detecting glaucomatous field loss. In their study, a group of ocular hypertensive (OH) patients with initial SAP defects were tested with FDT. Compared to a control group, these patients detected stimuli only when an abnormally high luminance was applied, thus suggesting that the measurement of the contrast sensitivity of a frequency-doubled grating may represent a good indicator of neural damage from elevated intraocular pressure (Maddess and Henry, 1992).

Thereafter, several clinical studies have been conducted to determine the accuracy of glaucoma detection by FDT. Johnson and Samuels reported a sensitivity of 93% and a specificity of 100% when testing 15 normal subjects and 15 age-matched patients with early or moderate glaucoma damage with full-threshold FDT (Johnson and Samuels, 1997). Trible found a specificity of 91% and a sensitivity of 35, 88, and 100% for early, moderate, and severe glaucoma, respectively (Trible et al., 2000). These ancillary findings on the comparison between performances at SAP and FDT were confirmed by a number of studies; overall, when conventional perimetry was used as a "gold standard," FDT obtained good specificity, with sensitivities ranging from moderate to excellent depending on the stage of the disease. A review by the American Academy of Ophthalmology stated that FDT "showed sensitivity and specificity greater than 97% for detecting moderate and advanced glaucoma, and sensitivity of 85% and specificity of 90% for early glaucoma" (Delgado et al., 2002). Very similar sensitivities and specificities were obtained by the second-generation FDT using the 24-2 program (Brusini et al., 2006a, b; Spry et al., 2007).

One of the most interesting applications of FDT is to detect early glaucomatous defects. Abnormal FDT results were obtained in 20–54% of patients with retinal nerve fiber layer (RNFL) defects but normal SAP (the so-called "pre-perimetric" glaucomas) (Brusini et al., 2006a, b; Ferreras et al., 2007; Kim et al., 2007; Lee et al. 2007). These findings are in contrast with two studies which, using a morphological "gold standard," obtained similar performances for SAP and FDT (Spry et al., 2005; Burgansky-Eliash et al., 2007). The major limitation of these studies is the cross-sectional design, which does not allow any inspection on the true potentiality of FDT in diagnosing early-stage patients who will develop SAP abnormalities only after years of follow-up.

Longitudinal, comparative data on FDT and SAP have recently been provided (Haymes et al., 2005). In their study of 65 glaucoma patients, Haymes and coworkers performed FDT and SAP every 6 months for a mean follow-up of 3.5 years. They showed that FDT and SAP detected progression in the same number of patients (49%), although the proportion of patients who showed progression with both FDT and SAP was small (25%), possibly indicating that the two techniques identified patients with different patterns of the disease.

Several diagnostic parameters (presence of at least 1 or 2 location(s) with $P<5\%$ or $<1\%$ in the total or pattern deviation map) and scoring systems have been proposed for first-generation FDT; being Matrix FDT similar to SAP 24-2, it is reasonable to adopt the criteria for abnormality used for conventional perimetry (Hodapp et al., 1993). Until now, none of these FDT criteria has been clearly validated over the others. As a general rule, a higher specificity is obtained using a cut-off for abnormality of $P<1\%$; on the other hand, sensitivity may increase when using the full-threshold N-30 test at a strategy of $P<5\%$ and selecting looser diagnostic criteria. In any case, the diagnostic power of FDT seems to be only marginally affected by the generation of the perimeter, the criteria adopted to define abnormality, the program (C-20 vs. N-30 vs. 24-2), or the strategy (full-threshold vs. screening) (Delgado et al., 2002; Fogagnolo et al., 2005).

Regardless of the criteria used to define abnormality, in the case of apparently abnormal results retest is recommended, since this would improve specificity with a negligible loss in sensitivity (Gardiner et al., 2006). This is particularly convenient for screening procedures (which are low time-consuming) and it is strongly suggested when a test with low MD is obtained from subjects unexperienced to perimetry, as they may be prone to learning effect (Brush and Chen, 2004; Contestabile et al., 2007).

Criteria for progression are also lacking (Sample et al., 2000a, b), although the staging systems proposed by Brusini et al. may represent a useful tool to stage the severity of the functional damage and to correctly distinguish among generalized, localized, and mixed defects (Brusini and Tosoni, 2003; Brusini, 2006).

Compared to the other perimetric techniques, FDT has a number of advantages. Intra- and inter-test variability for FDT is comparable to SAP in healthy subjects. In glaucomatous patients, SAP variability seems to increase as defect severity increases while it remains stable and low for FDT (Artes et al., 2005). Moreover, the shape of the "hill of vision" for FDT is significantly flatter than for SAP and SWAP (Landers et al., 2006). This topography is probably the result of the retinotopic distribution of M-RGCs (increasing cell density with increasing eccentricity) and it allows lower point-to-point confidence intervals and, hence, a more precise discrimination between normal and abnormal responses. The learning effect (which is a well-known phenomenon occurring in many psychophysical examinations, defined as the improvement of performances over test repetitions) is absent in a large number of tested subjects; for the small percentage of subjects showing improvement of performance upon retest, only the first examination seems to be affected (Brush and Chen, 2004; Contestabile et al., 2007). As for other perimetries, MD was the parameter most sensitive to learning. FDT has been successfully used in clinical practice also to test children (Blumenthal et al., 2004). Finally, FDT is supposed to be unaffected by defocus (Anderson and Johnson, 2003), although a previous study was in contrast with this finding (Artes et al., 2003).

SWAP: clinical data

The first two studies showing the clinical efficacy of SWAP in glaucoma were published by Johnson and coworkers in 1993. In the first study, they tested 76 OH and 124 normal eyes with SAP and SWAP at baseline every 12 months for 5 years (Johnson et al., 1993a). At baseline, SAP was normal in all cases, whereas nine OH patients had abnormal SWAP tests. At the end of the study, five out of nine of these patients developed SAP glaucomatous defects, thus showing that SWAP can predict the development of glaucomatous defects from 3 to 5 years before SAP. The second study aimed at validating SWAP as a tool to identify early glaucoma progression (Johnson et al., 1993b). Thirty-two eyes of 16 glaucoma patients underwent SWAP and SAP tests once a year for 5 years and were deemed stable or progressing on the basis of the eventual changes at SAP during the study period. The authors showed that, at baseline, SWAP defects were larger than SAP in 80% of cases. Whereas SWAP defects were twice as large as SAP in the group of stable patients, they were three to four times larger in the group with progressive field loss. Therefore, the presence of large SWAP defects may predict glaucoma progression at SAP.

The results of these ancillary studies were confirmed by a series of cross-sectional data (Girkin et al., 2000; Polo et al., 2002; Ferreras et al., 2007; Leeprechanon et al., 2007). In particular, a good association between RNFL defects and SWAP abnormalities has been shown (Polo et al., 1998; Mok et al., 2003; Sánchez-Galeana et al., 2004).

A prospective study was conducted on 47 glaucoma patients tested every 6 months with both SAP and SWAP over a mean follow-up of 4 years. Stability or progression of the disease was defined independently from SAP (it was in fact defined on the basis of ONH stereophotography). At the end of the study, progression was found in 22/47 patients and SWAP obtained a better area under the curve compared to SAP, thus confirming that SWAP may improve the detection of progressive glaucoma compared to SAP (Girkin et al., 2000).

Other studies reporting data on sensitivity and specificity for SWAP in comparison to FDT are discussed in the next session. In their review of published literature, Delgado and coworkers confirmed the clinical usefulness of SWAP, reporting a mean sensitivity and specificity of 88 and 92%, respectively (Delgado et al., 2002).

Also for SWAP, a consensus on criteria for abnormality is still missing. One study reported that the optimum criterion to define glaucomatous abnormalities is the presence of a cluster of four points lower than $P<5\%$ or a cluster of three points lower than $P<1\%$ (Polo et al., 2001). Another study suggested that GHT is the most sensitive parameter to identify the disease and its progression (Johnson et al., 2002). Contrary to FDT, a high variability in results is generally obtained if different criteria are used to define abnormality (Reus et al., 2005).

Unfortunately, SWAP applicability in clinical settings is still limited by a number of factors. SWAP is a demanding test, since stimuli are more difficult to detect than those of SAP and FDT; the full-threshold strategy has a high duration (about 15–18 min); hence, patients are prone to a "fatigue effect" during examination.

SWAP must be conducted only on patients with clear media, since the presence of opacities, in particular cataract, affects the results (abnormally lower sensitivities were also found in the case of modest opacities, Sample et al., 1996).

A correction for light absorption can be performed, but this procedure is time-consuming (about 35 min). Patients suffering from migraine (McKendrick et al., 2002), epilepsy (Hosking and Hilton, 2002), ocular conditions such as diabetic maculopathy (Remky et al., 2000), and optic neuropathies (Keltner and Johnson, 1995) and those using drugs interfering with neurotransmitters (Paczka et al., 2001) may frequently obtain false-positive results at SWAP.

One of the main limitations of this perimetry is the presence of learning effect (Rossetti et al., 2006; Wild et al., 2006). We conducted a study on 30 patients at risk for glaucoma and already experienced with SAP (which represented a group of subjects who could highly benefit from early detection of the disease by SWAP); they performed a battery of 5 SWAP within 1 month. Eighty five percent of patients showed a significant learning effect: MD improved 0.6 dB per repetition and it was supposed to reach a plateau only at the sixth repetition; mean duration decreased 17 s per examination without reaching a plateau at the end of the study. The analysis of demographic, clinical, and perimetric data excluded the possibility of identifying a subgroup of patients more prone to learning effect at SWAP before carrying out the battery of tests. We concluded that at least three repetitions are required to rule out the presence of a learning effect, although a subgroup of patients could need up to five repetitions before providing clinically useful results. In Fig. 3, the first and the last SWAP of a patient in this study are shown, and substantial learning effect is evident for perimetric indices, duration, and number of abnormal points.

Finally, SWAP is also limited by statistical biases. In normal subjects, the inter-individual threshold variability is higher for SWAP than

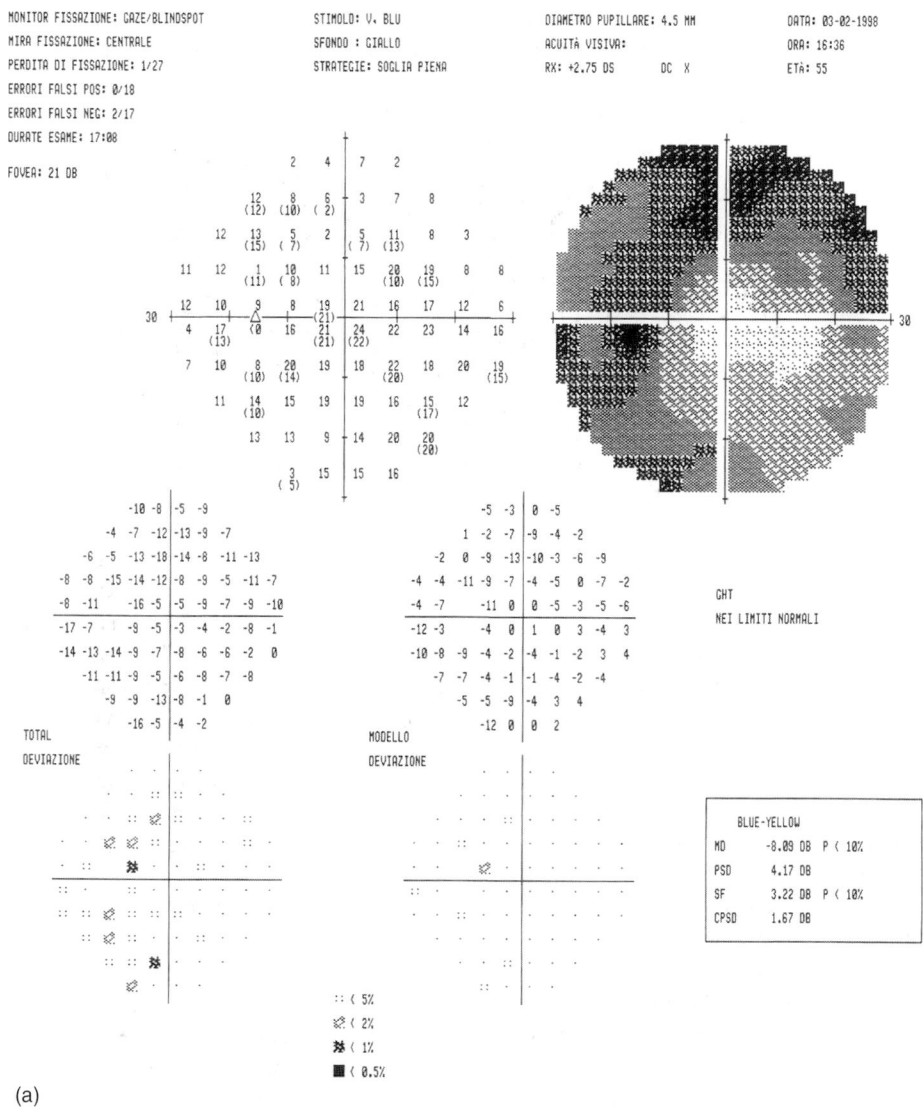

(a)

Fig. 3. SWAP printouts showing a significant learning effect. The first SWAP (a) had abnormal MD (−8.09 dB) and SF (3.22 dB), with presence of clusters of abnormal points in the total deviation map and a duration of 17.08 min. The fifth repetition (b) was performed 1 month later it was perfectly normal, with improvement of all these parameters: MD was −1.82 dB, SF 2.08 dB, no abnormal clusters were found in both maps, and duration reduced to 16.04 min.

SAP (Wild et al., 1998; Blumenthal et al., 2003); therefore, confidence intervals are wider, thus negatively affecting the ability to discriminate between normal and glaucoma cases. SWAP intertest variability in suspect and manifest glaucoma is also augmented at about 0.5–0.7 dB compared to SAP (Wild et al., 1998; Hutchings et al., 2001; Blumenthal et al., 2003), which makes it difficult to assess progression accurately. Several factors could respond to this high variability: the difficulty of stimulus detection, the long test duration, the high sensitivity of SWAP to pupil

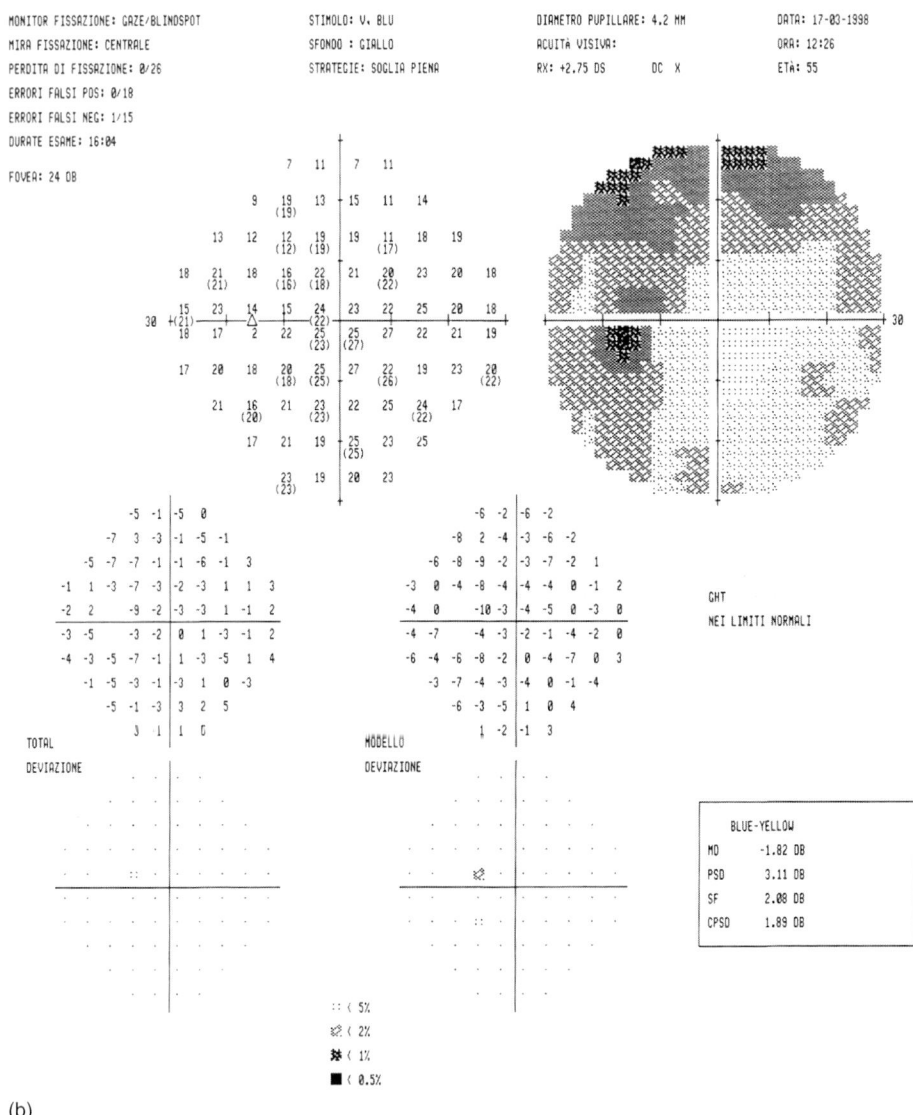

Fig. 3. (Continued).

size (Eisner et al., 2006), and a flaw-normative database that did not rule out, at least, the cases with a major learning effect. SITA SWAP, a novel program included in the latest version of the HFA II-i, has a shorter duration (between 3 and 6 min) and may at least partially reduce the impact of such limitations (Bengtsson and Heijl, 2003, 2006) and improve the performances of this perimetry.

Clinical data comparing FDT and SWAP

The studies comparing the diagnostic efficacy of FDT and SWAP are the most pertinent to the purpose of this review, but, to the best of our knowledge, only 10 have been published. Ideally, each study should answer to two questions: "Is it worth testing patients with glaucoma or glaucoma suspect with unconventional perimetries?" and

"Which one, among FDT and SWAP, is preferable?" The first question implies a longitudinal design to verify whether unconventional perimetries are more effective than SAP in detecting conversion to the disease or progression. As funding allocation is common for all perimetries, in the case of their validation, unconventional perimetries would be performed at the expense of SAP. This hypothetical superiority over SAP should therefore be demonstrated using an independent "gold standard" for diagnosis (ONH or RNFL appearance). Unfortunately, as shown in Table 3, no studies fulfilled all these features, and therefore any conclusion on the efficacy of unconventional perimetries over SAP must be considered with great caution.

Landers conducted a study over 62 OH patients (normal ONH and SAP) to verify whether FDT and SWAP could detect underlying earlier visual field loss (Landers et al., 2003). Patients underwent SAP, SWAP, and FDT every year for a 3-year follow-up. At the end of the study, nine subjects had abnormal SWAP and 10 abnormal FDT. Field loss at SAP developed in five subjects, all of whom had pre-existing abnormal SWAP and FDT results; no SAP defects developed in patients with normal SWAP or FDT. This study suggested that both FDT and SWAP are useful tools to predict the development of SAP loss in OH patients.

A number of cross-sectional studies obtained similar results, thus suggesting that FDT and SWAP were equally effective in diagnosing pre-perimetric glaucoma cases (Bayer and Erb, 2002; Bayer et al., 2002; Leeprechanon et al., 2007), with a sensitivity ranging from 20 (Ferreras et al., 2007) to 72% (Leeprechanon et al, 2007) for FDT, and from 20 (Ferreras et al., 2007) to 54% (Leeprechanon et al., 2007) for SWAP; specificity was similar for the two techniques (53% for FDT vs. 44% for SWAP) (Leeprechanon et al., 2007).

Other studies suggested that FDT may achieve better diagnostic performances than SWAP (Bowd et al., 2001; Soliman et al., 2002). An interesting study compared the diagnostic ability of several morphological and functional tests in diagnosing early glaucoma (Bowd et al., 2001). Two different definitions were adopted, based on ONH appearance and SAP. The area under the curve (AUC) was calculated for each test; overall, FDT had better diagnostic power than SWAP (AUCs of 0.87 and 0.88 compared to 0.76 and 0.78 for SWAP).

Another relevant piece of information from this study was that, when specificity was set at 90%, the two techniques obtained a poor diagnostic agreement. This means that, at the initial stage of the disease, no single perimetric test was always affected, whereas the other remained normal, a fact that has been recently confirmed (Sample et al., 2006).

On the other hand, performing a battery of SWAP and FDT is a good strategy for identifying the largest number of early glaucoma cases as possible (Ferreras et al., 2007; Horn et al., 2007).

Detection of early glaucoma cases can be further maximized in both screening (Tóth et al., 2007)

Table 3. Characteristics of the design of the studies comparing FDT and SWAP available in literature

	Cross-sectional	Longitudinal (follow-up)	Gold standard			Which is better?
			SAP only	SAP+ONH	ONH only	
Bowd et al., 2001	X		X		X	FDT
Bayer et al., 2002	X			X		FDT = SWAP
Soliman et al., 2002	X			X		FDT
Landers et al., 2003		X (36 months)		X		FDT = SWAP
Bagga et al., 2006	X			X		FDT = SWAP
Sample et al., 2006	X				X	FDT
Shah, 2007	X		X		X	FDT
Ferreras et al., 2007	X			X		FDT = SWAP
Horn et al., 2007	X			X		FDT = SWAP
Leeprechanon et al., 2007	X			X		FDT = SWAP

and specialistic (Horn et al., 2003; Bagga et al., 2006) settings if a combination of functional and morphological examinations is obtained. It has been shown that adding FDT perimetry to each of the best structural parameters led to a significant increase in sensitivity without a significant change in specificity compared with structural parameters alone, whereas adding SWAP to each of the best structural parameters led to a significant increase in sensitivity and a significant decrease in specificity compared with each structural parameter alone (Shah et al., 2007).

Table 4. Items that are required to improve the clinical applicability of unconventional perimetries

1. Provide high-quality scientific evidence of the importance of increasing the frequency and the regularity of visual test repetitions (with both conventional and, eventually, unconventional techniques) in order to obtain a more generous resource allocation for perimetry in glaucoma management
2. Provide high-quality scientific evidence of the superiority, if any, of unconventional perimetries in the early diagnosis of glaucoma:
   - Prospective, longitudinal studies
   - Possibly multicentric studies
   - Morphological "gold standard"
   - Comparative data between conventional and unconventional techniques
   - Large sample size to provide a reasonably high number of progressing cases
3. Unify and validate the diagnostic criteria for FDT and SWAP
4. As for SAP, generate software to evaluate FDT and SWAP progression
5. Validate the new SITA SWAP program

Conclusions

FDT and SWAP are two relatively novel perimetric techniques that can provide additional or confirmatory evidence in glaucoma patients. Furthermore, other perimetric techniques (MAP and HPRP) may have a role in the management of the disease, but due to their poor diffusion their use is still limited to experimental settings.

Some functional features of the RGCs, such as segregation (Kaplan, 2004; Callaway, 2005), isolation (Sample et al., 1996), and redundancy (Johnson, 1994; Haymes et al., 2005), have been only recently clarified; these aspects provided confirmation that both SWAP and FDT, thanks

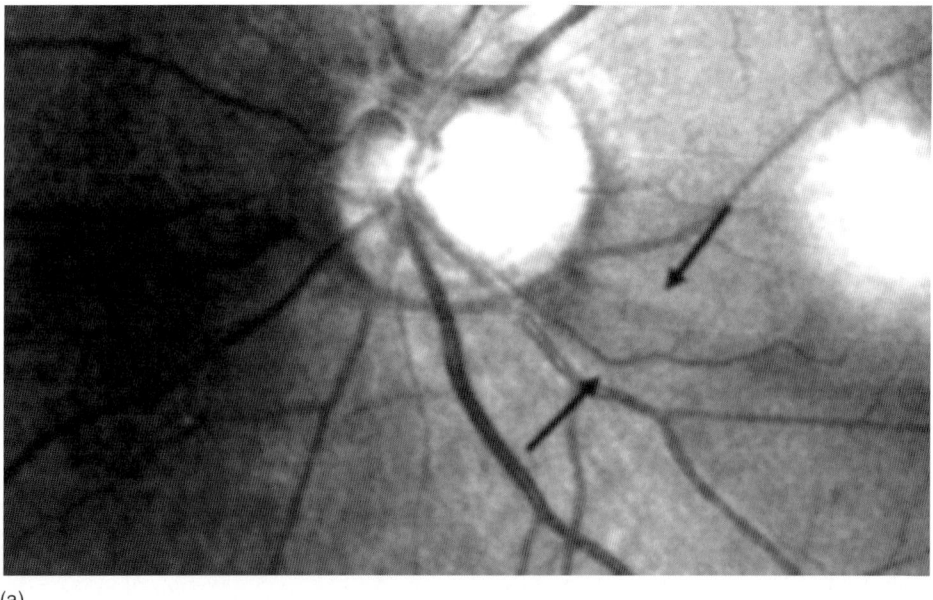

(a)

Fig. 4. The case of a 71-year-old man showing RNFL defects (a), normal SAP (b), and FDT (c) and full-threshold SWAP (d) abnormalities at baseline who developed glaucomatous SAP defects after 7 years (e).

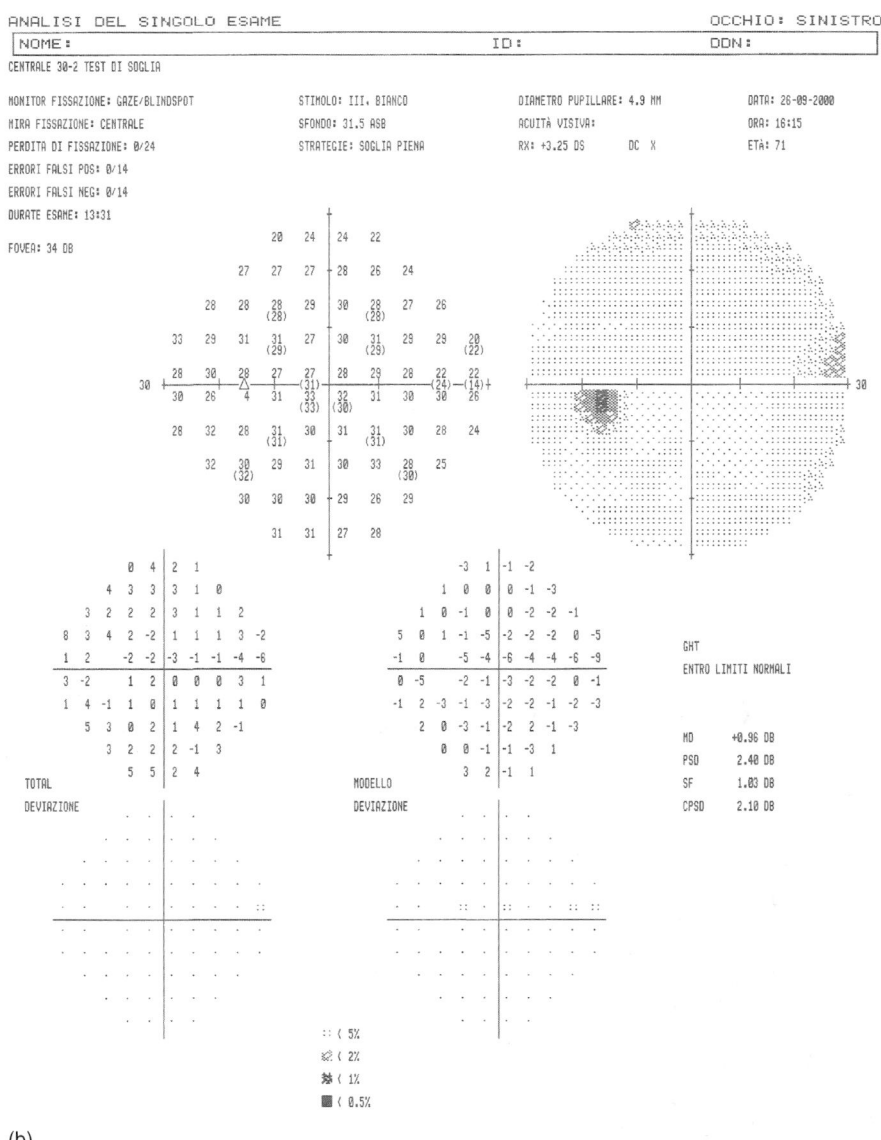

(b)

Fig. 4. (Continued).

to their selective evaluation of subgroups of RGCs with low number and redundancy, actually investigate functions that are impaired in the early stages of the disease (Johnson et al., 1993a, b; Trible et al., 2000; Landers et al., 2003).

Clinical studies on SWAP and FDT are growing in number, and they seem to confirm the ability of both techniques in detecting abnormalities in early glaucoma cases otherwise judged as normal on the basis of SAP results (the so-called "pre-perimetric" glaucomas) (Johnson et al., 1993a, b; Johnson and Samuels, 1997). Overall, FDT shows a slightly better diagnostic power, a more solid database (with lower intra- and inter-individual variability), and more reliable results than SWAP (Soliman et al., 2002; Zangwil et al., 2006).

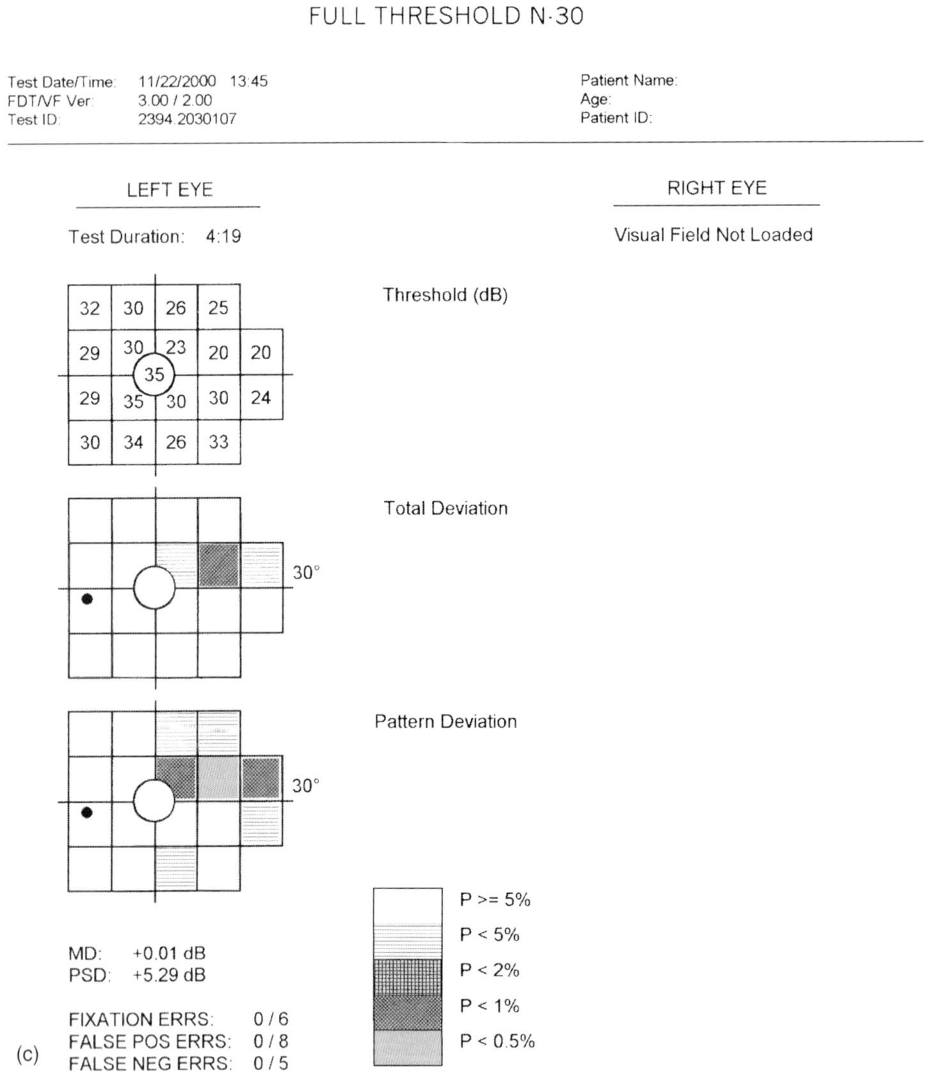

FULL THRESHOLD N-30

Test Date/Time: 11/22/2000 13:45
FDT/VF Ver: 3.00 / 2.00
Test ID: 2394 2030107

Patient Name:
Age:
Patient ID:

LEFT EYE

Test Duration: 4:19

RIGHT EYE

Visual Field Not Loaded

Threshold (dB)

32	30	26	25	
29	30	23	20	20
	35			
29	35	30	30	24
30	34	26	33	

Total Deviation

Pattern Deviation

P >= 5%
P < 5%
P < 2%
P < 1%
P < 0.5%

MD: +0.01 dB
PSD: +5.29 dB

FIXATION ERRS: 0 / 6
FALSE POS ERRS: 0 / 8
FALSE NEG ERRS: 0 / 5

(c)

Fig. 4. (Continued).

On the other side, the clinical applicability of both procedures is still limited by a number of factors. Serious doubts on the validity of the full-threshold SWAP database have been raised (high intra- and intertest variability; presence of a learning effect also in patients already experienced with SAP) (Bengtsson and Heijl, 2003; Rossetti et al., 2006), whereas no data derived from clinical settings are currently available for the novel SITA SWAP program.

The testing strategy for SWAP is very similar to SAP, and therefore the same diagnostic criteria could be used to detect glaucoma changes (although some authors raised skepticism on the validity of SAP criteria for SWAP) (Johnson et al., 2002). On the other side, several criteria to define FDT abnormality have been proposed (Delgado et al., 2002), but no criterion has been clearly validated and accepted. As for all physical measurements, also for first-generation FDT, it has been shown that the loosest criterion (i.e., abnormality defined in the presence of at least one location with $P < 5\%$) obtains the highest sensitivity but it

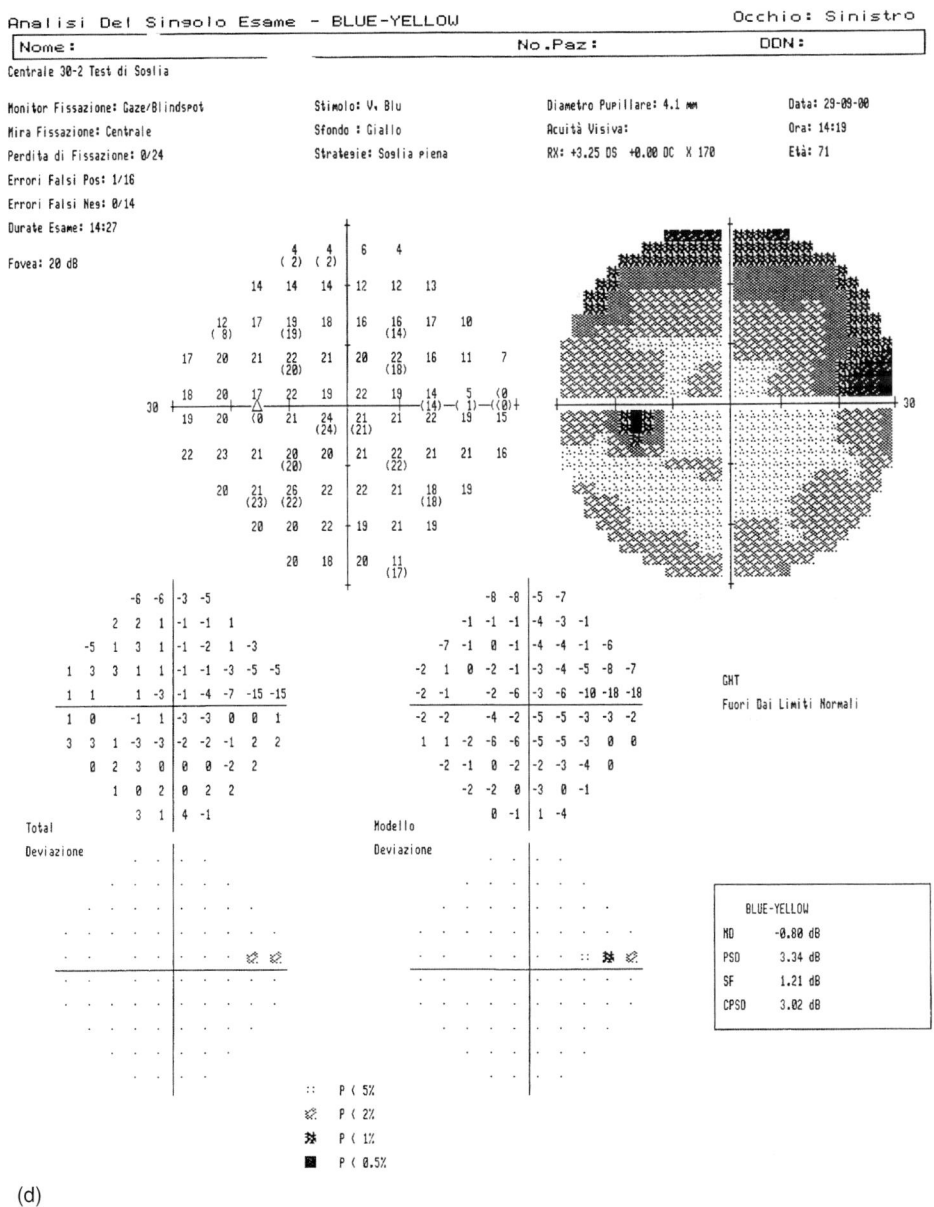

Fig. 4. (Continued).

invariably causes an increase of the false-positive rate. The novel Matrix FDT uses a testing program that is more similar to SAP; this could also probably allow the introduction of more uniform criteria for abnormality for this technique.

The large part of the literature on SWAP and FDT is composed of cross-sectional studies; very few prospective data are available, and they were conducted on small groups of patients. Many studies had a selection bias, since only patients with normal SAP were recruited; as a consequence, the diagnostic power of SAP may be underestimated, a fact that has been confirmed by the recent studies considering the appearance of ONH

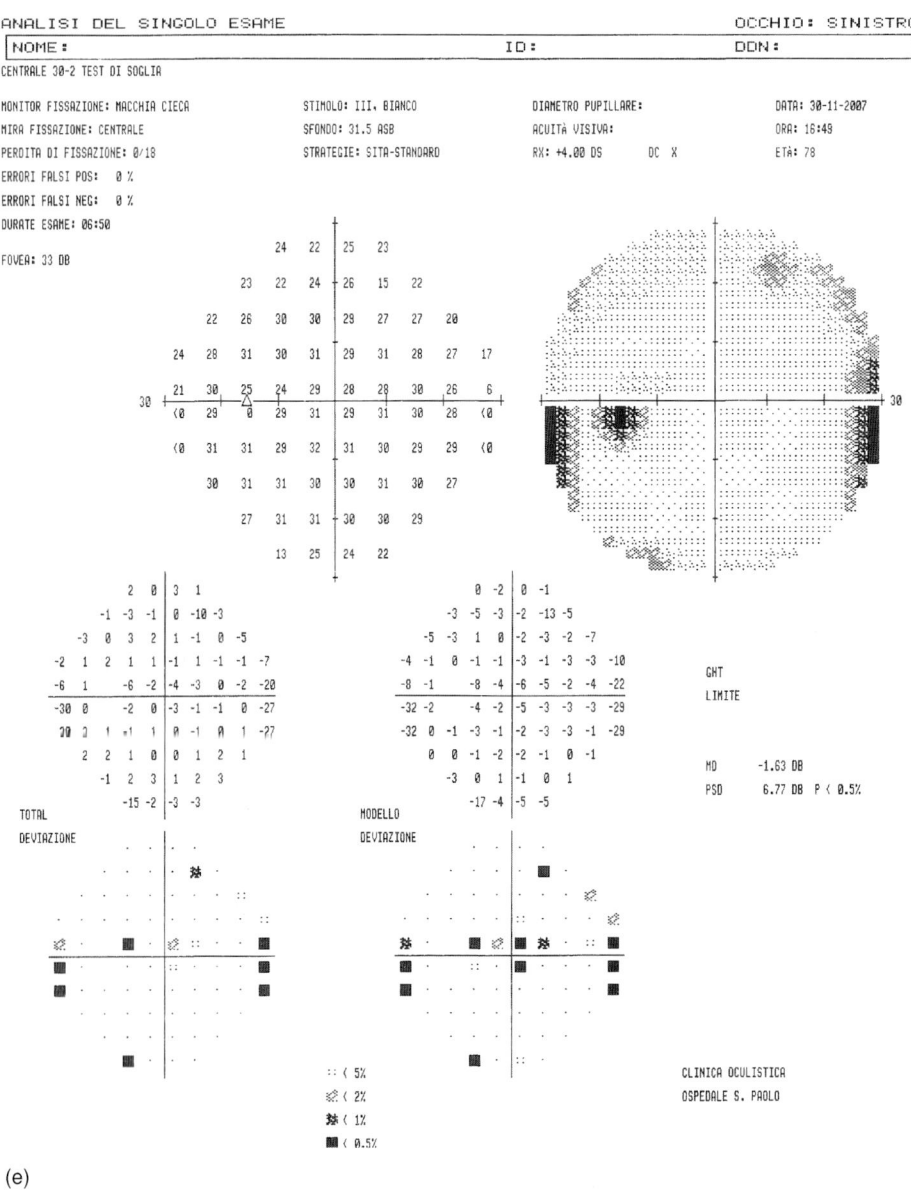

(e)

Fig. 4. (Continued).

as the "gold standard" for diagnosis (Bagga et al., 2006; Sample et al., 2006). Finally, very scanty data are available on the efficacy of SWAP and FDT in detecting the progression of glaucoma.

Being resource allocation common to all perimetries, in routine clinical practice the use of unconventional perimetries is strictly related to a reduction in the number of SAP tests. Unfortunately, the number of SAP examinations already falls substantially below published recommendations for maintaining minimum practice standards in the majority of clinical settings (Friedman et al., 2005). Based on the actual knowledge of unconventional perimetry, a further reduction seems inappropriate, above all in consideration that SAP provides the only parameters having a direct relevance to quality-of-life measures (Gutierrez et al., 1997; Parrish et al., 1997; Sherwood et al.,

1998; Jampel et al., 2002; Nelson et al., 2003) and that our knowledge on visual field progression derives almost exclusively from SAP data.

Based on these considerations, it is still unrealistic to consider FDT and SWAP as diagnostic tools to be routinely used in clinical practice. A number of items, summarized in Table 4, need to be clarified in order to reconsider the role of unconventional perimetry in clinical practice. Nevertheless, the usefulness of both FDT and SWAP, at least in the subgroup of patients more at risk for developing the disease (i.e., patients with high intraocular pressure values, strong familiarity for glaucoma, severe vascular diseases), cannot be denied: their ability in anticipating by years the development of SAP abnormalities has been demonstrated (Ferreras et al., 2007; Leeprechanon et al., 2007) and is commonly ascertained by clinicians using these techniques. An example of this diagnostic ability is reported in Fig. 4, which shows the case of a patient with RNFL defects, normal SAP, and FDT and SWAP abnormalities at baseline, who developed glaucomatous defects at SAP 7 years later.

Considering that glaucoma still nowadays causes an unacceptably high number of visually impaired or even blind patients (Hattenhauer et al., 1998; Kwon et al., 2001; Oliver et al., 2002; Chen, 2003; Eid et al., 2003; Zahari et al., 2006; Forsman et al., 2007) and that motor vehicle accidents are significantly related to the level of SAP loss (McGwin et al., 1998; Szlyk et al., 2005; Haymes et al., 2007) in spite of patients being well within the legal visual requirements to hold a driving license and having only early field damage in the *worse* eye (Haymes et al., 2007), the possibility of adopting strategies to detect the earliest glaucoma defects and to prevent their evolution must be considered with extreme interest due to their possible positive effect on the socio-economic impact of the disease.

References

Anderson, A.J. and Johnson, C.A. (2003) Frequency-doubling technology perimetry and optical defocus. Invest. Ophthalmol. Vis. Sci., 44: 4147–4152.

Artes, P.H., Hutchison, D.M., Nicolela, M.T., LeBlanc, R.P. and Chauhan, B.C. (2005) Threshold and variability properties of matrix frequency-doubling technology and standard automated perimetry in glaucoma. Invest. Ophthalmol. Vis. Sci., 46: 2451–2457.

Artes, P.H., Nicolela, M.T., McCormick, T.A., LeBlanc, R.P. and Chauhan, B.C. (2003) Effects of blur and repeated testing on sensitivity estimates with frequency doubling perimetry. Invest. Ophthalmol. Vis. Sci., 44: 646–652.

Bagga, H., Feuer, W.J. and Greenfield, D.S. (2006) Detection of psychophysical and structural injury in eyes with glaucomatous optic neuropathy and normal standard automated perimetry. Arch. Ophthalmol., 124: 169–176.

Bayer, A.U. and Erb, C. (2002) Short wavelength automated perimetry, frequency doubling technology perimetry, and pattern electroretinography for prediction of progressive glaucomatous standard visual field defects. Ophthalmology, 109: 1009–1017.

Bayer, A.U., Maag, K.P. and Erb, C. (2002) Detection of optic neuropathy in glaucomatous eyes with normal standard visual fields using a test battery of short-wavelength automated perimetry and pattern electroretinography. Ophthalmology, 109: 1350–1361.

Bengtsson, B. and Heijl, A. (2003) Normal intersubject threshold variability and normal limits of the SITA SWAP and full threshold SWAP perimetric programs. Invest. Ophthalmol. Vis. Sci., 44: 5029–5034.

Bengtsson, B. and Heijl, A. (2006) Diagnostic sensitivity of fast blue-yellow and standard automated perimetry in early glaucoma: a comparison between different test programs. Ophthalmology, 13(7): 1092–1097.

Blumenthal, E.Z., Haddad, A., Horani, A. and Anteby, I. (2004) The reliability of frequency-doubling perimetry in young children. Ophthalmology, 111: 435–439.

Blumenthal, E.Z., Sample, P.A., Berry, C.C., Lee, A.C., Girkin, C.A., Zangwill, L., Caprioli, J. and Weinreb, R.N. (2003) Evaluating several sources of variability for standard and SWAP visual fields in glaucoma patients, suspects, and normals. Ophthalmology, 110: 1895–1902.

Bowd, C., Zangwill, L.M., Berry, C.C., Blumenthal, E.Z., Vasile, C., Sanchez-Galeana, C., Bosworth, C.F., Sample, P.A. and Weinreb, R.N. (2001) Detecting early glaucoma by assessment of retinal nerve fiber layer thickness and visual function. Invest. Ophthalmol. Vis. Sci., 42(9): 1993–2003.

Brush, M.B. and Chen, P.P. (2004) Learning effect among perimetric novices with screening C-20-1 frequency doubling technology perimetry. Am. J. Ophthalmol., 137: 551–552.

Brusini, P. (2006) Frequency doubling technology staging system 2. J. Glaucoma, 15: 315–320.

Brusini, P. and Tosoni, C. (2003) Staging of functional damage in glaucoma using frequency doubling technology. J. Glaucoma, 12(5): 417–426.

Brusini, P., Salvetat, M.L., Zeppieri, M. and Parisi, L. (2006a) Frequency doubling technology perimetry with the Humphrey Matrix 30-2 test. J. Glaucoma, 15: 77–83.

Brusini, P., Salvetat, M.L., Zeppieri, M., Tosoni, C., Parisi, L. and Felletti, M. (2006b) Visual field testing with the new Humphrey Matrix: a comparison between the FDT N-30 and Matrix N-30-F tests. Acta Ophthalmol. Scand., 84: 351–356.

Burgansky-Eliash, Z., Wollstein, G., Patel, A., Bilonick, R.A., Ishikawa, H., Kagemann, L., Dilworth, W.D. and Schuman,

J.S. (2007) Glaucoma detection with matrix and standard achromatic perimetry. Br. J. Ophthalmol., 91: 933–938.

Callaway, E.M. (2005) Structure and function of parallel pathways in the primate early visual system. J. Physiol., 566(Pt 1): 13–19.

Canadian Glaucoma Study: 1. (2006) Study design, baseline characteristics, and preliminary analyses. The Canadian Glaucoma Study Group. Can. J. Ophthalmol., 41: 566–575.

Chen, P.P. (2003) Blindness in patients with treated open-angle glaucoma. Ophthalmology, 110: 726–733.

Collaborative Normal-Tension Glaucoma Study Group. (1998) Comparison of glaucomatous progression between untreated patients with normal-tension glaucoma and patients with therapeutically reduced intraocular pressures. Am. J. Ophthalmol., 126: 487–497.

Contestabile, M.T., Perdicchi, A., Amodeo, S., Recupero, V. and Recupero, S.M. (2007) The influence of learning effect on frequency doubling technology perimetry (Matrix). J. Glaucoma, 16: 297–301.

Dacey, D.M. and Lee, B.B. (1994) The "blue-on" opponent pathway in primate retina originates from a distinct bistratified ganglion cell type. Nature, 367: 731–735.

Delgado, M.F., Nguyen, N.T., Cox, T.A., Singh, K., Lee, D.A., Dueker, D.K., Fechtner, R.D., Juzych, M.S., Lin, S.C., Netland, P.A., Pastor, S.A., Schuman, J.S. and Samples, J.R. (2002) Automated perimetry: a report by the American Academy of Ophthalmology. Ophthalmology, 109: 2362–2374.

DeMonasterio, F.M. (1979) Asymmetry of on- and off-pathways of blue-sensitive cones of the retina of macaque. Brain Res., 166: 39–48.

Dobkins, K.R. and Albright, T.D. (2004) Merging processing streams: colour clues for motion detection and interpretation. In: Chalupa L.M. and Werner J.S. (Eds.), The Visual Neurosciences. The MIT Press, Cambridge, MA, pp. 1217–1228.

Drance, S.M., Lakowski, R., Schulzer, M. and Douglas, G.R. (1981) Acquired colour vision changes in glaucoma: use of 100-Hue Test and Pickford anomaloscope as predictors of glaucomatous field change. Arch. Ophthalmol., 99: 829–831.

Eid, T.M., Spaeth, G.L., Bitterman, A. and Steinmann, W.C. (2003) Rate and amount of visual loss in 102 patients with open-angle glaucoma followed up for at least 15 years. Ophthalmology, 110: 900–907.

Eisner, A., Toomey, M.D., Incognito, L.J., O'Malley, J.P. and Samples, J.R. (2006) Contrasting blue-on-yellow with white-on-white visual fields: roles of visual adaptation for healthy peri- or postmenopausal women younger than 70 years of age. Invest. Ophthalmol. Vis. Sci., 47: 5605–5614.

Ferreras, A., Polo, V., Larrosa, J.M., Pablo, L.E., Pajarin, A.B., Pueyo, V. and Honrubia, F.M. (2007) Can frequency-doubling technology and short-wavelength automated perimetries detect visual field defects before standard automated perimetry in patients with preperimetric glaucoma? J. Glaucoma, 16: 372–383.

Fogagnolo, P., Mazzolani, F., Rossetti, L. and Orzalesi, N. (2005) Detecting glaucoma with frequency-doubling technology perimetry a comparison between N-30 and C-20 screening programs. J. Glaucoma, 14: 485–491.

Forsman, E., Kivela, T. and Vesti, E. (2007) Lifetime visual disability in open-angle glaucoma and ocular hypertension. J. Glaucoma, 16: 313–319.

Friedman, D.S., Nordstrom, B., Mozaffari, E. and Quigley, H.A. (2005) Glaucoma management among individuals enrolled in a single comprehensive insurance plan. Ophthalmology, 112: 1500–1504.

Gardiner, S.K., Anderson, D.R., Fingeret, M., McSoley, J.J. and Johnson, C.A. (2006) Evaluation of decision rules for frequency-doubling technology screening tests. Optom. Vis. Sci., 83: 432–437.

Girkin, C.A., Emdadi, A., Sample, P.A., Blumenthal, E.Z., Lee, A.C., Zangwill, L.M. and Weinreb, R.N. (2000) Short-wavelength automated perimetry and standard perimetry in the detection of progressive optic disc cupping. Arch. Ophthalmol., 118: 1231–1236.

Gordon, M.O. and Kass, M.A. (1999) The ocular hypertension treatment study: design and baseline description of the participants. Arch. Ophthalmol., 117: 573–583.

Gupta, N. and Yucel, Y.H. (2007) What changes can we expect in the brain of glaucoma patients?. Surv. Ophthalmol., 52(suppl 2): S122–S126.

Gutierrez, P., Wilson, M.R., Johnson, C., Gordon, M., Cioffi, G.A. and Ritch, R. (1997) Influence of glaucomatous visual field loss on health-related quality of life. Arch. Ophthalmol., 115: 777–784.

Hartline, H.K. (1940) The receptive fields of optic nerve fibers. Am. J. Physiol., 130: 700–711.

Harwerth, R.S., Carter-Dawson, L. and Shen, F. (1999) Ganglion cell losses underlying visual field defects from experimental glaucoma. Invest. Ophthalmol. Vis. Sci., 40: 2242–2250.

Hattenhauer, M.G., Johnson, D.H., Ing, H.H., Herman, D.C., Hodge, D.O. and Yawn, B.P. (1998) The probability of blindness from open-angle glaucoma. Ophthalmology, 105: 2099–2104.

Haymes, D., Hutchison, D.M., McCormick, T.A., Varma, D.K., Nicolela, M.T., LeBlanc, R.P. and Chauhan, B.C. (2005) Glaucomatous visual field progression with frequency-doubling technology and standard automated perimetry in a longitudinal prospective study. Invest. Ophthalmol. Vis. Sci., 46: 547–554.

Haymes, S.A., Leblanc, R.P., Nicolela, M.T., Chiasson, L.A. and Chauhan, B.C. (2007) Risk of falls and motor vehicle collisions in glaucoma. Invest. Ophthalmol. Vis. Sci., 48: 1149–1155.

Hodapp, E., Parrish II, R. and Anderson, D. (1993) Clinical Decisions in Glaucoma. CV Mosby, St Louis, pp. 11–63.

Horn, F.K., Brenning, A., Jünemann, A.G. and Lausen, B. (2007) Glaucoma detection with frequency doubling perimetry and short-wavelength perimetry. J. Glaucoma, 16: 363–371.

Horn, F.K., Nguyen, N.X., Mardin, C.Y. and Jünemann, A.G. (2003) Combined use of frequency doubling perimetry and polarimetric measurements of retinal nerve fiber layer in glaucoma detection. Am. J. Ophthalmol., 135: 160–168.

Hosking, S.L. and Hilton, E.J. (2002) Neurotoxic effects of GABA-transaminase inhibitors in the treatment of epilepsy: ocular perfusion and visual performance. Ophthalmic. Physiol. Opt., 22(5): 440–447.

Hubel, D.H. and Wiesel, T.N. (1959) Receptive fields of single neurons in the cat's striate cortex. J. Physiol., 148: 574–591.

Hutchings, N., Hosking, S.L., John, M. and Flanagan, J.G. (2001) Long-term fluctuation in short wavelength automated perimetry in glaucoma suspects and glaucoma patients. Invest. Ophthalmol. Vis. Sci., 42: 2332–2337.

Jampel, H.D., Friedman, D.S., Quigley, H. and Miller, R. (2002) Correlation of the binocular visual field with patient assessment of vision. Invest. Ophthalmol. Vis. Sci., 43: 1059–1067.

Johnson, C.A. (1994) Selective versus nonselective losses in glaucoma. J. Glaucoma, 3(suppl 1): 32–44.

Johnson, C.A. and Samuels, S.J. (1997) Screening for glaucomatous visual field loss with frequency-doubling perimetry. Invest. Ophthalmol. Vis. Sci., 38: 413–425.

Johnson, C.A., Adams, A.J., Casson, E.J. and Brandt, J.D. (1993a) Blue-on-yellow perimetry can predict the development of glaucomatous visual field loss. Arch. Ophthalmol., 111: 645–650.

Johnson, C.A., Adams, A.J., Casson, E.J. and Brandt, J.D. (1993b) Progression of early glaucomatous visual field loss for blue-on-yellow and standard white-on-white automated perimetry. Arch. Ophthalmol., 111: 651–656.

Johnson, C.A., Sample, P.A., Cioffi, G.A., Liebmann, J.R. and Weinreb, R.N. (2002) Structure and function evaluation (SAFE): I. criteria for glaucomatous visual field loss using standard automated perimetry (SAP) and short wavelength automated perimetry (SWAP). Am. J. Ophthalmol., 134: 177–185.

Kaplan, E. (2004) The M, P, K pathways of the primate visual system. In: Chalupa L.M. and Werner J.S. (Eds.), The Visual Neurosciences. The MIT Press, Cambridge, MA, pp. 481–493.

Kelly, D.H. (1966) Frequency doubling in visual responses. J. Opt. Soc. Am., 56: 1628–1633.

Keltner, J.L. and Johnson, C.A. (1995) Short-wavelength automated perimetry in neuro-ophthalmologic disorders. Arch. Ophthalmol., 113(4): 475–481.

Kim, T.W., Zangwill, L.M., Bowd, C., Sample, P.A., Shah, N. and Weinreb, R.N. (2007) Retinal nerve fiber layer damage as assessed by optical coherence tomography in eyes with a visual field defect detected by frequency doubling technology perimetry but not by standard automated perimetry. Ophthalmology, 114: 1053–1057.

Kuffler, S.W. (1953) Discharge patterns and functional organization of mammalian retina. J. Neurophysiol., 16: 37–68.

Kwon, Y.H., Kim, C.S., Zimmerman, M.B., Alward, W.L. and Hayreh, S.S. (2001) Rate of visual field loss and long-term visual outcome in primary open-angle glaucoma. Am. J. Ophthalmol., 132: 47–56.

Landers, J.A., Goldberg, I.F. and Graham, S.L. (2003) Detection of early visual field loss in glaucoma using frequency-doubling perimetry and short-wavelength automated perimetry. Arch. Ophthalmol., 121: 1705–1710.

Landers, J., Sharma, A., Goldberg, I. and Graham, S. (2006) Topography of the frequency doubling perimetry visual field compared with that of short wavelength and achromatic automated perimetry visual fields. Br. J. Ophthalmol., 90: 70–74.

Lee, M.J., Kim, D.M., Jeoung, J.W., Hwang, S.S., Kim, T.W. and Park, K.H. (2007) Localized retinal nerve fiber layer defects and visual field abnormalities by Humphrey Matrix frequency doubling technology perimetry. Am. J. Ophthalmol., 143: 1056–1058.

Leeprechanon, N., Giaconi, J.A., Manassakorn, A., Hoffman, D. and Caprioli, J. (2007) Frequency doubling perimetry and short-wavelength automated perimetry to detect early glaucoma. Ophthalmology, 114: 931–937.

Leske, M.C., Heijl, A., Hyman, L. and Bengtsson, B. (1999) Early manifest glaucoma trial: design and baseline data. Ophthalmology, 106: 2144–2153.

Maddess, T. and Henry, G.H. (1992) Performance of nonlinear visual units in ocular hypertension and glaucoma. Clin. Vis. Sci., 7: 371–383.

McGwin, G., Jr., Owsley, C. and Ball, K. (1998) Identifying crash involvement among older drivers: agreement between self-report and state records. Accid. Anal. Prev., 30: 781–791.

McKendrick, A.M., Cioffi, G.A. and Johnson, C.A. (2002) Short-wavelength sensitivity deficits in patients with migraine. Arch. Ophthalmol., 120(2): 154–161.

Mok, K.H., Lee, V.W. and So, K.F. (2003) Retinal nerve fiber layer measurement by optical coherence tomography in glaucoma suspects with short-wavelength perimetry abnormalities. J. Glaucoma, 12: 45–49.

Musch, D.C., Lichter, P.R., Guire, K.E. and Standardi, C.L. (1999) The collaborative initial glaucoma treatment study: study design, methods, and baseline characteristics of enrolled patients. Ophthalmology, 106: 653–662.

Nelson, P., Aspinall, P., Papasouliotis, O., Worton, B. and O'Brien, C. (2003) Quality of life in glaucoma and its relationship with visual function. J. Glaucoma, 12: 139–150.

Oliver, J.E., Hattenhauer, M.G., Herman, D., Hodge, D.O., Kennedy, R. and Fang-Yen, M. (2002) Blindness and glaucoma: a comparison of patients progressing to blindness from glaucoma with patients maintaining vision. Am. J. Ophthalmol., 133: 764–772.

Paczka, J.A., Friedman, D.S., Quigley, H.A., Barron, Y. and Vitale, S. (2001) Diagnostic capabilities of frequency-doubling tecnology, scanning laser polarimetry and nerve fiber layer photograph to distinguish glaucomatous damage. Am. J. Ophthalmol., 131: 188–197.

Parrish, R.K., 2nd, Gedde, S.J., Scott, I.U., Feuer, W.J., Schiffman, J.C. and Mangione, C.M. (1997) Visual function and quality of life among patients with glaucoma. Arch. Ophthalmol., 115: 1447–1455.

Polo, V., Abecia, E., Pablo, L.E., Pinilla, I., Larrosa, J.M. and Honrubia, F.M. (1998) Short-wavelength automated perimetry and retinal nerve fiber layer evaluation in suspected cases of glaucoma. Arch. Ophthalmol., 116: 1295–1298.

Polo, V., Larrosa, J.M., Pinilla, I., Pablo, L. and Honrubia, F.M. (2001) Optimum criteria for short-wavelength automated perimetry. Ophthalmology, 108: 285–289.

Polo, V., Larrosa, J.M., Pinilla, I., Perez, S., Gonzalvo, F. and Honrubia, F.M. (2002) Predictive value of short-wavelength automated perimetry: a 3-year follow-up study. Ophthalmology, 109: 761–765.

Polyak, S.L. (1941) The Retina. University of Chicago Press, Chicago.

Remky, A., Arend, O. and Hendricks, S. (2000) Short-wavelength automated perimetry and capillary density in early diabetic maculopathy. Invest. Ophthalmol. Vis. Sci., 41(1): 274–281.

Reus, N.J., Colen, T.P. and Lemij, H.G. (2005) The prevalence of glaucomatous defects with short-wavelength automated perimetry in patients with elevated intraocular pressures. J. Glaucoma, 14: 26–29.

Rossetti, L., Fogagnolo, P., Miglior, S., Centofanti, M., Vetrugno, M. and Orzatesi, N. (2006) Learning effect of short-wavelength automated perimetry in patients with ocular hypertension. J. Glaucoma, 15(5): 399–404.

Sample, P.A., Bosworth, C.F., Blumenthal, E.Z., Girkin, C. and Weinreb, R.N. (2000a) Visual function specific perimetry for indirect comparison of different ganglion cell populations in glaucoma. Invest. Ophthalmol. Vis. Sci., 41: 1783–1790.

Sample, P.A., Bosworth, C.F. and Weinreb, R.N. (2000b) The loss of visual function in glaucoma. Semin. Ophthalmol., 15: 182–193.

Sample, P.A., Johnson, C.A., Haegerstrom-Portnoy, G. and Adams, A.J. (1996) Optimum parameters for short-wavelength automated perimetry. J. Glaucoma, 5: 375–383.

Sample, P.A., Medeiros, F.A., Racette, L., Pascual, J.P., Boden, C., Zangwill, L.M., Bowd, C. and Weinreb, R.N. (2006) Identifying glaucomatous vision loss with visual-function–specific perimetry in the diagnostic innovations in glaucoma study. Invest. Ophthalmol. Vis. Sci., 47: 3381–3389.

Sánchez-Galeana, C.A., Bowd, C., Zangwill, L.M., Sample, P.A. and Weinreb, R.N. (2004) Short-wavelength automated perimetry results are correlated with optical coherence tomography retinal nerve fiber layer thickness measurements in glaucomatous eyes. Ophthalmology, 111: 1866–1872.

Shah, N.N., Bowd, C., Medeiros, F.A., Weinreb, R.N., Sample, P.A., Hoffmann, E.M. and Zangwill, L.M. (2007) Combining structural and functional testing for detection of glaucoma. Ophthalmology, 114: p. 1414.

Sherwood, M.B., Garcia-Siekavizza, A., Meltzer, M.I., Hebert, A., Burns, A.F. and McGorray, S. (1998) Glaucoma's impact on quality of life and its relation to clinical indicators. A pilot study. Ophthalmology, 105: 561–566.

Soliman, M.A., De Jong, L.A., Ismaeil, A.A., Van den Berg, T.J. and De Smet, M.D. (2002) Standard achromatic perimetry, short wavelength automated perimetry, and frequency doubling technology for detection of glaucoma damage. Ophthalmology, 109: 444–454.

Solomon, S.G., Barry, B.L., Andrew, J.R., White, L.R. and Martin, P.R. (2005) Chromatic organization of ganglion cell receptive fields in the peripheral retina. J. Neurosci., 25: 4527–4539.

Solomon, S.G., White, A.J.R. and Martin, P.R. (2002) Extraclassical receptive fields in primate LGNJ. Neurosci., 22(1): 338–349

Spry, P.G., Hussin, H.M. and Sparrow, J.M. (2005) Glaucoma detection with matrix and standard achromatic perimetry. Br. J. Ophthalmol., 91: 933–938.

Spry, P.G., Hussin, H.M. and Sparrow, J.M. (2007) Performance of the 24-2-5 frequency doubling technology screening test: a prospective case study. Br. J. Ophthalmol., 91: 1345–1349.

Szlyk, J.P., Mahler, C.L., Seiple, W., Edward, D.P. and Wilensky, J.T. (2005) Driving performance of glaucoma patients correlates with peripheral visual field loss. J. Glaucoma, 14: 145–150.

The Advanced Glaucoma Intervention Study (AGIS) 1. (1994) Study design and methods and baseline characteristics of study patients. Control Clin. Trials, 15: 299–325.

Tóth, M., Kóthy, P., Vargha, P. and Holló, G. (2007) Accuracy of combined GDx-VCC and matrix FDT in a glaucoma screening trial. J. Glaucoma, 16: 462–470.

Trible, J.R., Schultz, R.O., Robinson, J.C. and Rothe, T.L. (2000) Accuracy of glaucoma detection with frequency-doubling perimetry. Am. J. Ophthalmol., 129: 740–745.

Tyrell, R.A. and Owens, D.A. (1998) A rapid technique to assess the resting states of the eyes and other threshold phenomena: The Modified Binary Search (MOBS). Behav. Res. Methods Instr. Computers, 20: 137–141.

Weinreb, R.N., Lindsey, J.D. and Sample, P.A. (1994) Lateral geniculate nucleus in glaucoma. Am. J. Ophthalmol., 118: 126–129.

White, A.J., Sun, H., Swanson, W.H. and Lee, B.B. (2002) An examination of physiological mechanisms underlying the frequency-doubling illusion. Invest. Ophthalmol. Vis. Sci., 43: 3590–3599.

Wild, J.M., Cubbidge, R.P., Pacey, I.E. and Robinson, R. (1998) Statistical aspects of the normal visual field in short-wavelength automated perimetry. Invest. Ophthalmol. Vis. Sci., 39(1): 54–63.

Wild, J.M., Kim, L.S., Pacey, I.E. and Cunliffe, I.A. (2006) Evidence for a learning effect in short-wavelength automated perimetry. Ophthalmology, 113(2): 206–215.

Yucel, Y.H., Zhang, Q., Weinreb, R.N., Kaufman, P.L. and Gupta, N. (2003) Effects of retinal ganglion cell loss on magno-, parvo-, koniocellular pathways in the lateral geniculate nucleus and visual cortex in glaucoma. Prog. Retin. Eye Res., 22(4): 465–481.

Zahari, M., Mukesh, B.N., Rait, J.L., Taylor, H.R. and McCarty, C.A. (2006) Progression of visual field loss in open angle glaucoma in the Melbourne Visual Impairment Project. Clin. Experiment Ophthalmol., 34: 20–26.

Zangwil, L.M.L., Bowd, C. and Weinreb, R.N. (2006) Identifying glaucomatous vision loss with visual-function–specific perimetry in the diagnostic innovations in glaucoma study. Invest. Ophthalmol. Vis. Sci., 47: 3381–3389.

CHAPTER 9

Scanning laser polarimetry and confocal scanning laser ophthalmoscopy: technical notes on their use in glaucoma

Felicia Ferreri*, Pasquale Aragona and Giuseppe Ferreri

Section of Ophthalmology, Department of Surgical Specialties, Azienda Ospedaliera Universitaria Policlinico G. Martino, Messina, Italy

Abstract: The mere intraocular pressure (IOP) measurement and the visual field (VF) examination do not allow early primary open-angle glaucoma (POAG) diagnosis. At present, the morphological and morphometric analysis of the optic disk is considered very important for an early diagnosis and follow-up of the disease. The recent introduction of laser systems equipped with new polarimetry techniques (GDx) and confocal tomography (HRT) allows an objective, quantitative, and reproducible evaluation of the morphometry and morphology of the optic disk and retinal nerve fiber layer (RNFL). The GDx, scanning laser polarimetry, studies the RNFL. The HRT, confocal scanning laser tomography, examines several optic disk and peripapillar area parameters. These devices allow obtaining objective and quantitative data concerning RNFL and optic nerve head. They represent complementary and important examinations in case of uncertain POAG diagnosis. The correct evaluation of the parameters studied by these techniques and the knowledge of the instruments' limits are needed for an adequate interpretation of the results obtained.

Keywords: GDx; HRT; primary open-angle glaucoma; laser polarimetry; confocal scanning laser polarimetry

Primary open-angle glaucoma (POAG) is a progressive chronic optic neuropathy characterized by optic nerve head and retinal nerve fiber layer (RNFL) morphological alterations, induced by ganglion cells death with consequent visual field (VF) loss.

POAG, the most frequent type of glaucoma (Klein et al., 1992), has controversies about pathogenetic, diagnostic, and therapeutic issues. Classically, the POAG diagnosis is based on the triad: ocular hypertension, VF damage, and optic nerve atrophy. Clinical evidence show that hypertension on its own is not sufficient to determine the glaucomatous damage, but other factors of different nature may be involved in causing the diseases. Therefore, the ocular pressure can be considered as a risk factor for developing POAG (Quigley et al., 1999). VF damage, studied by standard automated perimetry (SAP), is the sign of the ganglion cells damage, but it becomes evident when 30% of the RNFL is damaged (Quigley et al., 1982).

The mere intraocular pressure (IOP) measure and the VF examination do not allow achieving an

*Corresponding author. Tel.: 0902213958; Fax: 0902212400; E-mail: fferreri@unime.it

early POAG diagnosis. At present, the optic head morphological and morphometric analysis is considered to give important clues to the early diagnosis of the disease and the follow-up of its progression. In fact, the ganglion cells death, which is responsible for the functional damage, directly affects both the optical head morphology and the RNFL thickness, thus anticipating the clinical and VF damage.

The examination of the optic nerve head and the papillary area should be simple, quick, objective, accurate, and reproducible. Recently, the new lasers based on polarimetric technique (GDx) and on confocal tomography (Heidelberg retinal tomograph, HRT) have allowed the objective and reproducible morphological and morphometric evaluation of both optic nerve head and RNFL.

The GDx scanning laser polarimeter

The GDx (Laser Diagnostic Technologies, Inc., San Diego, CA), performing a scanning laser polarimetry study of the retina, is a noninvasive diagnostic technique used to estimate the RNFL thickness using a polarized laser beam with a wavelength of 780 nm (Hollo et al., 1997; Weinreb et al., 1998; Yamada et al., 2000).

The linear birefringence property of RNFL is determined by the presence of the microtubules with a parallel disposition within the nervous fiber, inducing a variation of polarization of the light beam, which passes through them. These polarization changes, called retardation, are linearly related to the histological characteristic of the analyzed structure and can be registered by a polarization detector, giving an evaluation of the RNFL thickness.

The polarization detector measures the retardation of the light coming from the analyzed retinal point. The thickness of RNFL, obtained through an algorithm, is shown by means of color representations, in an image of 256 × 256 pixels that is acquired and stored in the computer memory.

The image acquisition is possible also with a miotic pupil, unless the pupillary diameter is not lower than 2 mm, although a regular pupillary diameter is essential to perform the examination correctly (Weinreb et al., 1995).

Since other birefringent ocular tissues, such as the cornea, may interfere with the results of the test, a corneal compensator was added to allow the evaluation of the variation in birefringence due to the corneal interference, which is peculiar for each patient (Greenfield et al., 2000; Weinreb et al., 2002a). This latest version of the RNFL analyzed is named GDx-VCC (variable corneal compensator), which compensates the individual corneal birefringence hence allowing a good correction of the corneal polarization effect. In this way, a more reliable evaluation of the RNFL thickness can be achieved (Weinreb et al., 2002b; Tannenbaum et al., 2004).

Before performing the test, it is necessary to insert patients' personal and clinical data, such as date of birth, gender, ethnicity, associated systemic disease, and spherical equivalent. Each eye will be examined on its own. The test progression is as follows.

A first scansion, called corneal compensation, is carried out on the macular region where the birefringence is assent. This measure is used to obtain the compensation of the corneal birefringence; the examiner must check the exact positioning of the ellipse corresponding to the macular region, which will appear homogeneously colored in blue (Fig. 1).

Once the corneal birefringence compensation has been carried out, the "acquisition" phase will take place. In this phase, it will be possible to get

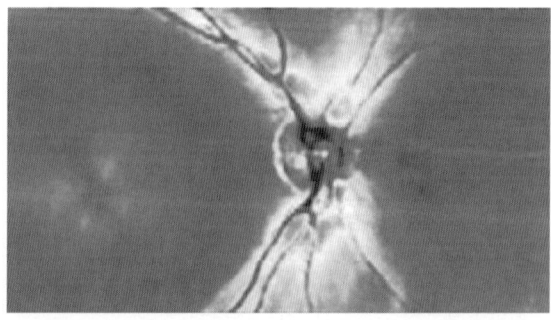

Fig. 1. Image of the macular region evenly stained in blue due to the absence of the birefringence.

compensated images of the RNFL. The compensation check is carried out only once, at the first test, and the value obtained is stored in a database to be used in future examinations. It is possible that a new measurement of corneal compensation is required in case of overcoming cataract or refractive surgery. In this phase, an ellipse delimiting the optic nerve head will be shown. This ellipse can be modified in order to compensate the presence of optic nerve anomalies, such as peripapillary atrophy and scleral crescent.

During the acquisition phase, two concentric circles will be shown around the papilla. The area of calculation is included between the two circles. Three images will be automatically acquired. The final image is obtained from a mean of these three images. This image is divided into four segments centered on the optic nerve head: superior and inferior 120° wide, nasal 70° wide, and temporal 50° wide. In this way are identified the areas from where the data are obtained and showed in the graphic representation called TSINT (temporal, superior, nasal, inferior, temporal). The measures obtained in the different areas will contribute to obtain the so-called nerve fiber index (NFI), which is supposed to reflect the probability that the patient has a glaucomatous damage on a scale from 0 to 100, where values above 40 are considered abnormal.

The area where the calculation is performed is automatically determined and is set on a minimal dimension. It extends only for 35 pixels with an inner diameter of 27 pixels and 8 pixels width (Fig. 2).

In the emmetropic eye, it is possible to have the dimensions expressed in millimeters, as they are measured on the retina: outer ray 1.628 mm and inner ray 1.256 mm. The values expressed in pixels are worked out so that 256 values, uniformly distributed along the circular area, are obtained.

Three areas of different dimensions are available for the calculation of the parameters: small, medium, and large. In Table 1 are summarized the dimensions of each variable.

In general, it is advisable to use a small calculation area because it allows better quality and more reliable results.

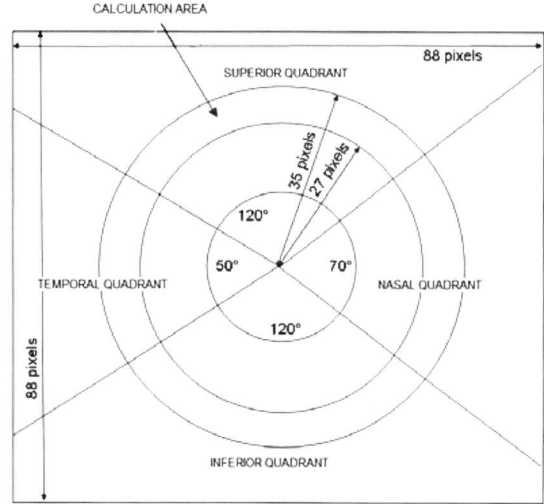

Fig. 2. Calculation area.

Table 1. Parameters of three area used for calculation

Areas	Inner ray	Outer ray	Calculation rectangle
Small	27 pixels (1.256 mm)	35 pixels (1.628 mm)	88 × 88 pixels
Medium	35 pixels (1.628 mm)	43 pixels (2.0 mm)	88 × 88 pixels
Large	43 pixels (2.0 mm)	51 pixels (2.372 mm)	104 × 104 pixels

In case of patients with peripapillary atrophy, myopic crescent, or other optic nerve head morphological anomalies, it is better to use larger areas because with the small one the data collected could be unreliable.

Performing the measurements, the computer calculates the light retardation from all the areas considered. The data obtained from each patient are compared with the database obtained in normal subjects of the same age. In this way are evaluated 13 indexes, which include several measurements obtained in the peripapillary region.

The results are represented in a printout (Fig. 3), where data and images, obtained for each eye, are presented separately according to two types of analysis, namely, nerve fiber analysis (NFA) and serial analysis.

128

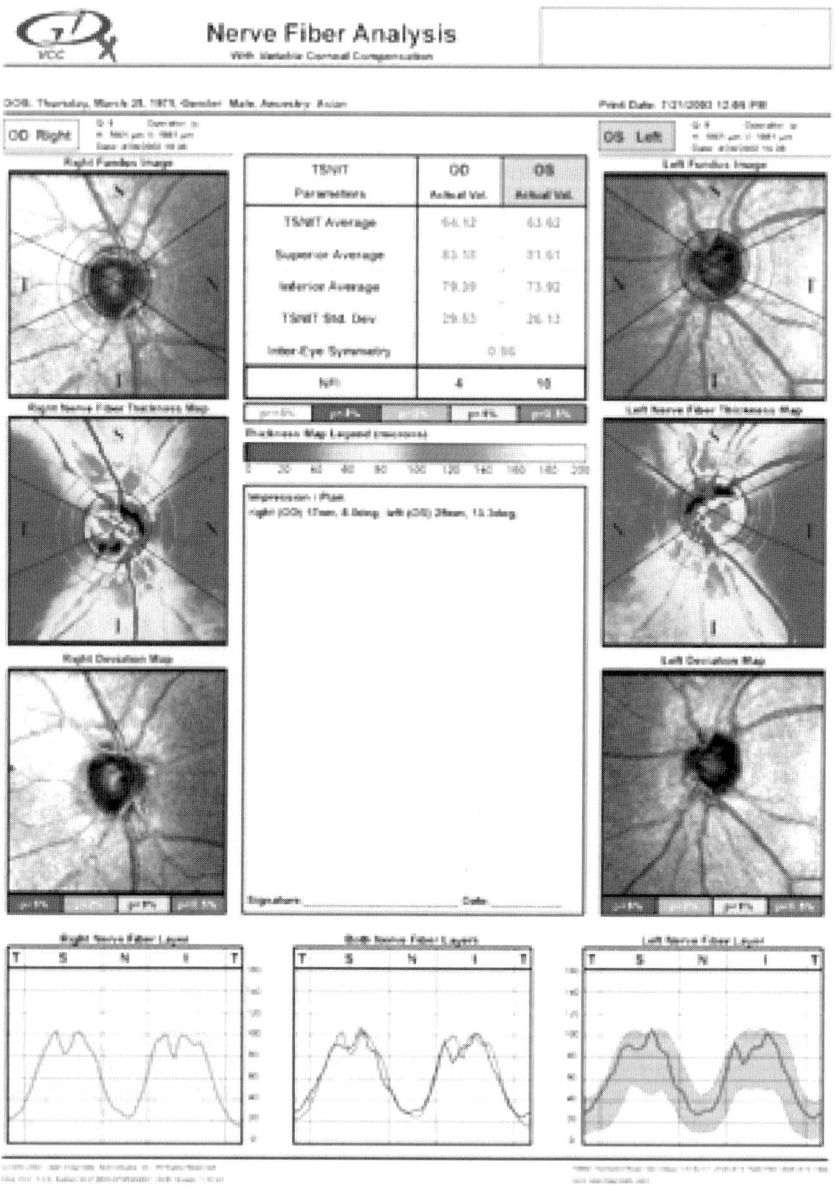

Fig. 3. Printout of a GDx examination.

Nerve fiber analysis

The NFA includes fundus images, RFNL thickness map, TSINT parameters, NFI, TSINT graphic, and deviation from reference map.

The central TSNIT graphic shows data from both eyes in order to facilitate the evaluation of the interocular symmetry, which is only shown in this graph.

- Color image: Color has been artificially added to this image of the optic nerve head and the peripapillary retina in order to assist their observation and to allow the evaluation of the quality of the scan performed (Fig. 4).

- Thickness (polarization) map: This map gives a color-coded image of the measured points to indicate RNFL thickness. Bright colors (red and yellow) are associated with thicker areas, signifying healthy RNFL. Dark colors are associated with thinner areas, indicating less healthy RNFL. In the scheme from normal subjects, light yellow and red colors are allocated in the superior and inferior sectors, while green and blue colors are located in nasal and temporal sectors, respectively (Fig. 5).
- Standard deviation map: This is a superpixel map showing different colors with respect to the probability to deviation from normal values of reference (Fig. 6).
- TSNIT (double hump) graph: In TSNIT graph are shown the normal values (shaded area) and patient values (dark line) relative to the RNFL thickness on the data obtained in the calculation area. Looking at the TSNIT graph, from left to right, are shown the thickness values of the temporal, superior, nasal, inferior, and temporal again. In normal condition, the RNFL profile shows a double hump aspect with higher thickness in the superior and inferior sectors and lower values in nasal and temporal sectors (Fig. 7).
- TSNIT symmetry graph: This is a confrontation of the TSINT graphs from both eyes of the patient. In this graph, it is possible to evaluate if there are differences in the RNFL

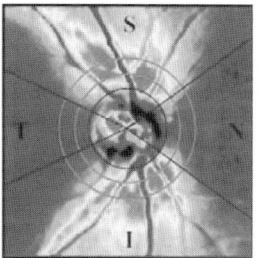

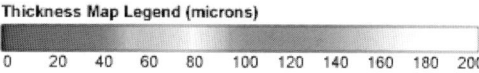

Fig. 5. Thickness polarization map.

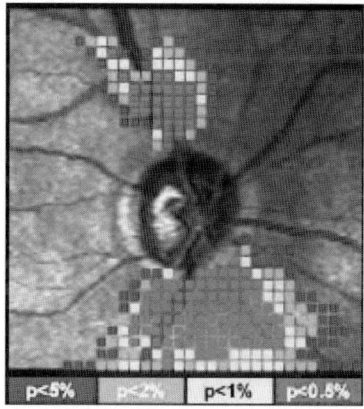

Fig. 6. Standard deviation map.

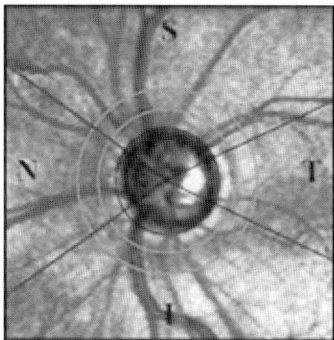

Fig. 4. Color image of the optic nerve head and the peripapillary area.

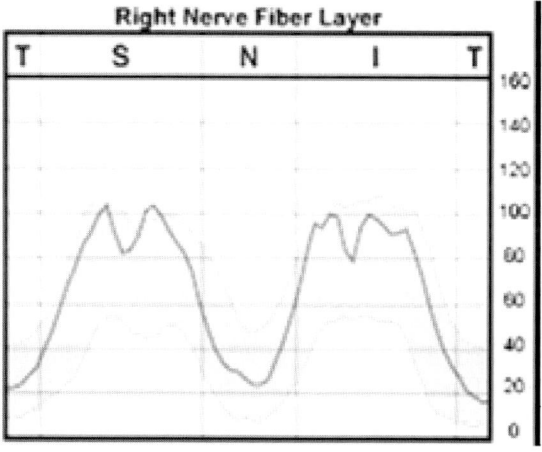

Fig. 7. TSNIT double hump graph.

thickness and TSNIT form and position between the two eyes.
- Deviation from normal map: In this map, the RNFL thickness of the patients is compared with normative database. Small colored squares indicate the percentage of deviation from normal value in a certain evaluated point. The area from where the squares are evaluated is shown on black and white image of the ocular fundus for reference. A color legend defines the statistical significance of the deviation from normal values with a significance comprised between $p > 5\%$ and $p < 0.5\%$.
- TSNIT parameters: The data from the calculation area of the patients are matched with those from a normative database. The parameters are encoded according to different colors to indicate the deviation from normal values and can help in distinguishing between glaucomatous and normal subjects. However, these parameters should be considered together with other clinical data of the patients.

  In the TSNIT table are considered the following parameters:
  1. TSNIT average: It indicates the mean thickness values in the calculation area.
  2. Superior average: It indicates the mean thickness of the pixels in the superior 120°.
  3. Inferior average: It indicates the mean thickness of the pixels in the inferior 120°.
  4. TSNIT standard deviation.
  5. Inter-eye symmetry: It shows the correlation of the TSNIT data for corresponding points of both eyes. If the ratio is close to 1, the RNFL will be symmetric in both eyes.
- Nerve fiber indicator: It is an indicator of the probable presence of the POAG. The GDx-VCC system has an algorithm aimed to optimize the confrontation between normal and abnormal RNFL; to obtain this algorithm, images from glaucomatous and normal eyes were used.

The NFI is shown in a numeric format from 0 to 100. The higher the number, the higher is the probability that the patient is affected by glaucoma. This index is not related to the gravity or the progression of POAG.

Although there may be some exceptions, it is possible to use the following scale as a guideline for NFI evaluation:

< 30, low probability of POAG
30–50, suspect of POAG
> 50, high probability of POAG

The NFI depends on the good positioning of the circle, which must be centered on the optic nerve head; a modification of the circle position can interfere with the NFI value obtained.

- Symmetry: This is the mean ratio between the mean of the 210 thickest measurements from the superior and inferior sectors. The closer the ratio to 1, the higher the RNFL symmetry in these sectors.
- Superior ratio: This is the mean ratio between the mean of the 210 thickest measures from the superior and temporal sectors.
- Inferior ratio: This is the mean ratio between the mean of the 210 thickest measures from the inferior and temporal sectors.
- Superior/nasal: This is the mean ratio between the mean of the 210 thickest measures from the superior and nasal sectors.
- Max modulation: It gives an indication of the differences existing between the thickest and the thinnest areas of the RNFL. The higher this number, the higher the difference between thick and thin RNFL areas. In normal eyes, where the inferior and superior RNFL thickness is higher than that of nasal and temporal, the number obtained is generally higher than 1.
- Superior maximum: This is the mean of the 210 thickest measurements of the superior sector.
- Inferior maximum: This is the mean of the 210 thickest measurements of the inferior sector.
- Ellipse modulation: Similar to the max modulation parameters, it indicates the difference existing between the thickest and thinnest areas of RNFL. These parameters used only the points along the ellipse circumscribing the optic nerve.

- Normalized superior area: This parameter examines the data obtained exclusively in the superior part of the ellipse; higher value represents a physiological condition, while a lower value indicates an RNFL loss.
- Normalized inferior area: This parameter examines the data obtained exclusively in the inferior part of the ellipse; higher value represents a physiological condition, while a lower value indicates an RNFL loss.
- Ellipse standard deviation: It indicates the standard deviation of the values obtained in the calculation area.

Serial analysis

In the serial analysis are included up to four scansions of the same eye reported chronologically, in order to allow an evaluation of RNFL variations with time:

- TSNIT confrontation graph: It compares, by means of a superimposition, the TSNIT graphs of two scansion of the same eye, taken during two consecutive visits. With this graph, it is possible to evaluate the variations of the RNFL thickness with time.
- TSNIT serial analysis graph: It compares, by means of a superimposition, the TSNIT graphs of two, three, or four subsequent scansions of the same eye, taken during different visits. With this graph, it is possible to evaluate the variations of the RNFL thickness with time.
- Deviation from reference map: It shows the variation of the RNFL thickness occurring during different visits. The colored areas and dots indicate the possible significant clinical variation. The colored legend defined variations with 20 μm increase.

Limits

The GDx, in the latest version called GDx-VCC, shows a high reproducibility with a good discrimination within normal and glaucomatous subjects (Greenfield et al., 2002; Weinreb et al., 2003; Brusini et al., 2005), and a close correlation between VF defects and RNFL damage (Bowd et al., 2003; Reus and Lemij, 2005).

However, the examination shows some limitation due to the ellipse localization that is operator dependent; furthermore, the interpretation of the results needs to be integrated with the clinical examination to achieve a precise diagnosis. There are still some limitations of the use of this instrument, for example, corneal refractive surgery (Choplin et al., 2005; Zangwill et al., 2005), lens surgery, presence of chorioretinal atrophy or scars, presence of myelinic fibers; all these conditions may interfere with birefringence. In addition, corneal opacity, pupil diameter lower than 2 mm, and significant vitreal opacities do not allow to perform the test.

The NFI can be considered positive when the value obtained is equal to or higher than 40 and this index has a sensitivity of 76.8% and a reproducibility of 89.1% (Colen et al., 2004).

Also, the progression of the damage should be considered with caution, since it was not demonstrated that variations from two consecutive examinations are certainly due to the progression of the disease rather than due to a physiological variation (Boehm et al., 2003).

The Heidelberg retinal tomograph

The HRT (Heidelberg Engineering, Heidelberg, Germany) is specifically designed to analyze the optic nerve head and gives an indirect evaluation of the RNFL. It is a confocal laser that uses a red diode laser of 670 nm wavelength. It performs a three-dimensional evaluation of the characteristics of the optic nerve head and peripapillary retina, with no need for mydriasis. It captures, in a time interval between 1.2 and 4.5 s, 32 optical pictures parallel to the retinal plane, analyzing the optic disk up to 15°.

The scansions are obtained by a periodic deflection of the laser beam, by means of swinging mirrors, and using the confocal characteristics of the instrument, so that only the light coming from a determined focal plane is captured by the detector. The light coming from contiguous plans is discarded through two inner diaphragms, one of which is in front of the laser source and the other in front of the detector. In this way, the light from each retinal point is reflected toward the detector and is represented as a pixel on the screen. Each pixel height is automatically calculated with respect to the reference plane, located 50 µm behind the papillomacular bundle (Vihanninjoki et al., 2002).

A topographical image is acquired and, subsequently, a three-dimensional image of the optic nerve is obtained by computer analysis.

Before the image acquisition, it is important to upload patients' personal and clinical data, corneal ray of curvature, and refraction: myopia and hyperopia of up to 11 diopters can be corrected, while high astigmatisms can be corrected by the use of adjunctive lenses. In patients with higher refractive errors, it is not possible to perform the examination.

At the start of the examination, the instrument makes an automatic scansion on 32 different planes on the papillary area (from a prepapillary to a retrolaminar plane) so that 32 bidimensional images are obtained.

The examination is repeated thrice so that three different images are obtained. From these images, the computer elaborates a mean topographic image 384 × 384 pixels wide, with three-dimensional reconstruction of the optic disk.

The instrument is able to evaluate the image quality by means of two quality parameters: interscan standard deviation (i.e., the mean test–retest variability: SD) and the mean confidence interval (CI) of the highest of the three images (good >20 µm, sufficient >50 µm). Furthermore, the instrument is able to give suggestions about the acquisition and to correct the scan depth and/or the refractive defects.

The mean image is shown with two maps: topographic map and reflectivity map. In the topographic map, the depth value is expressed with several colors (blue green is the deepest area; in the reflectivity map the reflectivity of each pixel is shown). On the topographic image, the operator can delimit the optic nerve head (contour line) along the inner border of the Elschnig's scleral ring.

Fig. 8. HRT three-dimensional image of optic nerve head and peripapillary area.

Following this demarcation, the instrument automatically chooses a standard reference plane localized 50 µm under the mean peripapillary retinal thickness, along the contour line in the temporal sector between 350° and 356°. This reference plane utilizes most part of the considered parameters circumscribed within the two areas of the optic disk: above the neuroretinal ring (green color) and below the excavation zone (red color).

The three-dimensional analysis allows obtaining planimetric and volumetric parameters (23 global and 13 partial) related to the optic nerve head and the RNFL measurement on the optical nerve head external border (Fig. 8).

The principal analyses obtained are the following:

- Cross-section analysis: It allows to evaluate the three-dimensional aspect of the optic disk along one of the three Cartesian axes by means of a cursor.
- Topographic map: It gives, in the absolute value or in mean ± standard deviation, the height of each pixel analyzed by the system.
- RNFL thickness diagram: It analyzes the thickness variation with double hump image from the temporal to the inferior sectors (TSNIT graph).
- Stereometric parameters of the optic nerve head:
  1. Dependent on the reference plane: cup area, cup/disk area ratio, rim area, cup volume, rim volume, RNFL cross-sectional area, and mean RNFL thickness.
  2. Independent of the reference plane: disk area, height variation contour, maximum

contour elevation and depression, CLM temporal superior and temporal inferior, mean cup depth, maximum cup depth, and cup shape measurement (morphological index of the cup or CSM) (Table 2).

- Moorfields regression analysis (MRA): It compares the volumes of two stereometric parameters (rim and cup) in six papillary sectors with values obtained from normal subjects and early glaucoma patients, with optic disk diameter between 1.2 and 2.8 mm^2. Data obtained from this classification give a good specificity and sensibility (Wollstein et al., 1998; Miglior et al., 2003) and can be shown in graphic and numerical details in comparison with the predictive values per age and optic disk diameter in the 95.0, 99.0, and 99.9% of the normative database. The graphs give quick visualization of the site and entity of the damage, evaluated according to a score in normal, borderline, and outline (Fig. 9).

- Interactive measurement: It gives an interactive horizontal and vertical profile of the optic nerve (Fig. 10).

Among the instrument functions, it is worth mentioning the possibility to study the glaucoma progression with time, which depends on the reproducibility of the several examinations performed.

The possibility to automatically display the previous contour line increases the reproducibility of the test (Verdonck et al., 2002; Tan et al., 2003) and, therefore, gives to the instrument the possibility to objectively analyze the papillary damage progression, by two different measurements:

1. Stereometric progression chart: Two sequential examinations are necessary; it also evaluates the variation of single stereometric parameters with time. The stability of the disease is indicated by the average normalized parameters value of 0, while the progression of the disease gives a value of −0.05. The

Table 2. HRT stereometric parameters of the optic nerve head

Parameters	Predefined segments						
	Global	Temporal	tmp/sup	tmp/inf	Nasal	nsl/sup	nsl/inf
Disk area (mm^2)	1.490	0.344	0.209	0.208	0.355	0.185	0.189
Cup area (mm^2)	0.317	0.199	0.062	0.035	0.007	0.008	0.006
Rim area (mm^2)	1.173	0.146	0.147	0.173	0.348	0.176	0.183
Cup/disk area ratio	0.213	0.577	0.297	0.171	0.019	0.044	0.033
Rim/disk area ratio	0.787	0.423	0.703	0.829	0.981	0.956	0.967
Cup volume (mm^3)	0.065	0.036	0.017	0.007	0.002	0.002	0.001
Rim volume (mm^3)	0.326	0.008	0.028	0.042	0.125	0.069	0.053
Mean Cup depth (mm)	0.198	0.235	0.250	0.166	0.129	0.192	0.070
Maximum cup depth (mm)	0.701	0.692	0.724	0.613	0.675	0.710	0.490
Height variation contour (mm)	0.378	0.108	0.231	0.180	0.089	0.070	0.025
Cup shape measure (mm)	−0.298	−0.178	−0.168	−0.341	−0.519	−0.403	−0.482
Mean RNFL thickness (mm)	0.254	0.065	0.242	0.264	0.349	0.389	0.312
RNFL cross-sectional area (mm^2)	1.099	0.068	0.138	0.148	0.365	0.214	0.173
Horizontal cup/disk ratio	0.486	–	–	–	–	–	–
Vertical cup/disk ratio	0.301	–	–	–	–	–	–
Maximum contour elevation (mm)	−0.085	–	–	–	–	–	–
Maximum contour depression (mm)	0.293	–	–	–	–	–	–
CLM temporal–superior (mm)	0.177	–	–	–	–	–	–
CLM temporal–inferior (mm)	0.198	–	–	–	–	–	–
Average variability (SD) (mm)	0.013	–	–	–	–	–	–
Reference height (mm)	0.330	–	–	–	–	–	–
FSM discriminant function value	2.076	–	–	–	–	–	–
RB discriminant function value	1.539	–	–	–	–	–	–

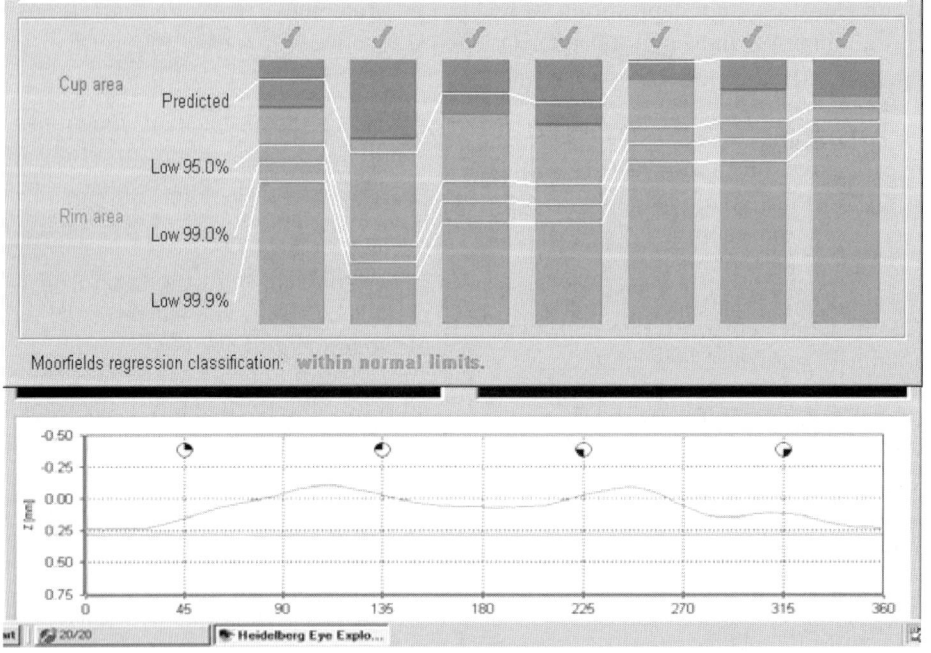

Fig. 9. HRT Moorfields regression analysis graphs.

normalized parameters can be analyzed singularly and globally or with three different sector combinations: superotemporal sector (from 45° to 90°), inferotemporal sector (from −90° to −45°), superior sector (from 22.5° to 112.5°), inferior sector (from −112.5° to 22.5°), superior hemisphere (from 0° to 180°), and inferior hemisphere (from −0° to −180°). Studying the variation of single parameters, the instrument can evaluate a difference of values between baseline and follow-up (Fig. 11).

2. Progression analysis: Three sequential examinations are necessary. It is independent of the contour line and evaluates the modification of values from each pixel allowing the formation of the three-dimensional image (Chauhan et al., 2000). By studying the local variability, the test indicates if a change can be due to a modification of the parameter (change probability), which indicates a significant variation if lower than 0.05 compared with the basal examination.

In the progression analysis, the modified areas, with respect to previous evaluations, are shown with red pixels in the refractive map; a variation is considered significant if an area of at least 20 adjacent pixels is involved. Three examinations must be of excellent quality and perfectly aligned to perform the correct analyses, since the worst is the quality, the higher the variability.

It remains difficult to bring into evidence the damage progression because the criteria to state it are still not precise (Fig. 12). Recent studies have brought into evidence a long-term fluctuation of HRT parameters similar to that occurring with SAP (Chauhan et al., 2001; Funk and Mueller, 2003). The present version of the HRT, with the software 3.0 (HRT III), offers an option for alternative analysis that does not require placement of a contour line that also may introduce interoperator variability (Garway-Heath et al., 1999; Iester et al., 2001; Miglior et al., 2002).

Although the normative database in HRT II included 349 subject for the stereoscopic

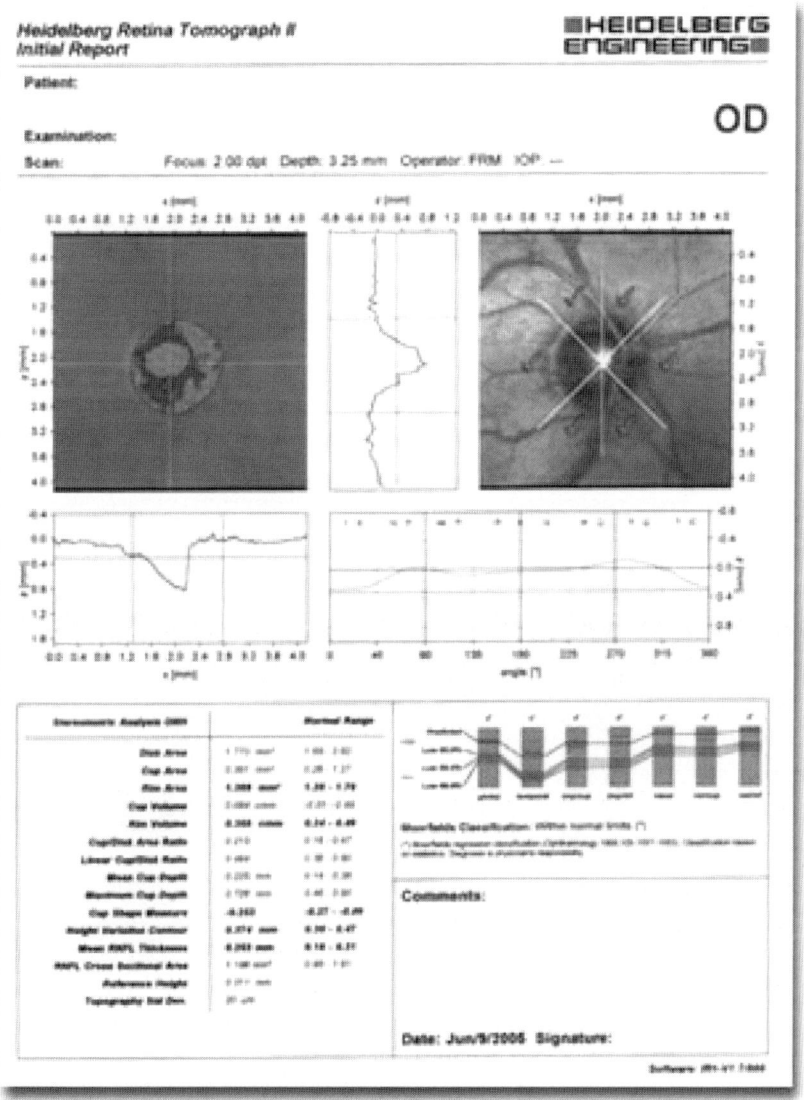

Fig. 10. HRT interactive measurement graph.

parameters and 110 subjects for the MRA, the HRT III normative database included 733 healthy Caucasian eyes and 215 healthy African eyes (Burgansky-Eliash et al., 2007). Based on the enlarged database, the equations of the MRA were modified between HRT II and HRT III.

The technique provides stereometric data by applying an automatic model of the optic nerve head shape, and the resultant morphological parameters are analyzed by a machine-learning classifier (relevance vector machine) resulting in a glaucoma probability score (GPS).

The GPS analysis provides a disease probability value based on the three-dimensional shape of the optic nerve and RNFL, and this classification represents the likelihood of glaucoma and not

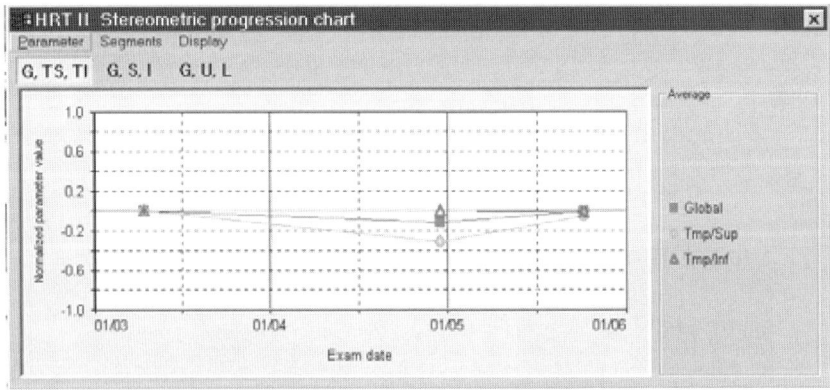

Fig. 11. HRT stereometric progression chart.

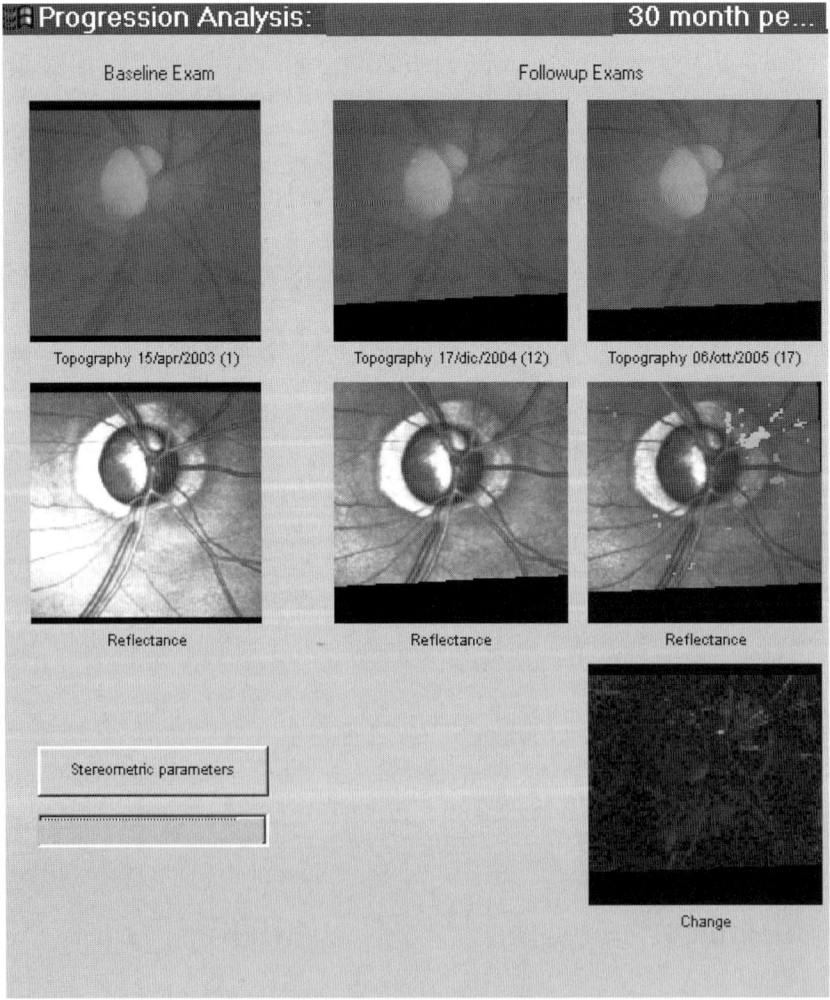

Fig. 12. HRT progression analysis chart.

the level of damage; thus, higher GPS values do not necessarily indicate more advanced disease (Ferreras et al., 2008).

Limits

The limits are represented by the presence of opacity of dioptric means, miotic pupils, optic nerve head abnormalities (colobomata and drusen), previous refractive surgery, and filtering surgery in the previous 6 months (Bresson-Dumont et al., 2003).

Conclusions

The GDx-VCC and the HRT both allowed obtaining objective and quantitative measurements of the RNFL and optic nerve head and representing important and complementary examinations in those cases with uncertain diagnosis: ocular hypertension with normal VF, ocular normotension with normal VF, and suspect papillary aspects (disk cupping, asymmetry, papillary bird hemorrhages, papillary pallor, and so on).

The constant improvement of these instruments and their analysis programs gives a hope to obtain, in the next future, an even more precise evaluation of the structural damage in course of glaucoma. The correct evaluation of the analyzed parameters and the knowledge of the instruments' limits are fundamental to achieve an adequate interpretation of the result obtained.

References

Boehm, M.D., Nedrud, C., Greenfield, D.S. and Chen, P.P. (2003) Scanning laser polarimetry and detection of progression after optic disc hemorrhage in patients with glaucoma. Arch. Ophthalmol., 121(2): 189–194.

Bowd, C., Zangwill, L.M. and Weinreb, R.N. (2003) Association between scanning laser polarimetry measurements using variable corneal polarization compensation and visual field sensitivity in glaucomatous eyes. Arch. Ophthalmol., 121(7): 961–966.

Bresson-Dumont, H., Lamy, C., Hinard, L. and Santiago, P.Y. (2003) HRT and glaucoma surgery follow-up. J. Fr. Ophtalmol., 26(spec no. 2): S7–S9.

Brusini, P., Salvetat, M.L., Parisi, L., Zeppieri, M. and Tosoni, C. (2005) Discrimination between normal and early glaucomatous eyes with scanning laser polarimeter with fixed and variable corneal compensator settings. Eur. J. Ophthalmol., 15: 468–476.

Burgansky-Eliash, Z., Wollstein, G., Patel, A., Bilonick, R.A., Ishikawa, H., Kagemann, L., Dilworth, W.D. and Schuman, J.S. (2007) Glaucoma detection with matrix and standard achromatic perimetry. Br. J. Ophthalmol., 91(7): 933–938.

Chauhan, B.C., Blanchard, J.W., Hamilton, D.C. and LeBlanc, R.P. (2000) Technique for detecting serial topographic changes in the optic disc and peripapillary retina using scanning laser tomography. Invest. Ophthalmol. Vis. Sci., 41(3): 775–782.

Chauhan, B.C., McCormick, T.A., Nicolela, M.T. and LeBlanc, R.P. (2001) Optic disc and visual field changes in a prospective longitudinal study of patients with glaucoma: comparison of scanning laser tomography with conventional perimetry and optic disc photography. Arch. Ophthalmol., 119(10): 1492–1499.

Choplin, N.T., Schallhorn, S.C., Sinai, M., Tanzer, D., Tidwell, J.L. and Zhou, Q. (2005) Retinal nerve fiber layer measurements do not change after LASIK for high myopia as measured by scanning laser polarimetry with custom compensation. Ophthalmology, 112(1): 92–97.

Colen, T.P., Tang, N.E., Mulder, P.G. and Lemij, H.G. (2004) Sensitivity and specificity of new GDx parameters. J. Glaucoma, 13: 28–33.

Ferreras, A., Pablo, L.E., Pajarín, A.B., Larrosa, J.M., Polo, V. and Pueyo, V. (2008) Diagnostic ability of the Heidelberg Retina Tomograph 3 for glaucoma. Am. J. Ophthalmol., 145(2): 354–359.

Funk, J. and Mueller, H. (2003) Comparison of long-term fluctuations: laser scanning tomography versus automated perimetry. Graefes Arch. Clin. Exp. Ophthalmol., 241(9): 721–724.

Garway-Heath, D.F., Poinoosawmy, D., Wollstein, G., Viswanathan, A., Kamal, D., Fontana, L. and Hitchings, R.A. (1999) Inter- and intraobserver variation in the analysis of optic disc images: comparison of the Heidelberg Retina Tomograph and computer assisted planimetry. Br. J. Ophthalmol., 83(6): 664–669.

Greenfield, D.S., Knighton, R.W., Feuer, W.J., Schiffman, J.C., Zangwill, L. and Weinreb, R.N. (2002) Correction for corneal polarization axis improves the discriminating power of scanning laser polarimetry. Am. J. Ophthalmol., 134: 27–33.

Greenfield, D.S., Knighton, R.W. and Huang, X.R. (2000) Effect of corneal polarization axis on assessment of retinal nerve fiber layer thickness by scanning laser polarimetry. Am. J. Ophthalmol., 129(6): 715–722.

Hollo, G., Suveges, I., Nagymihaly, A. and Vargha, P. (1997) Scanning laser polarimetry of the retinal nerve fibre layer in primary open angle and capsular glaucoma. Br. J. Ophthalmol., 81: 857–861.

Iester, M., Mikelberg, F.S., Courtright, P., Burk, R.O., Caprioli, J., Jonas, J.B., Weinreb, R.N. and Zangwill, L.

(2001). Interobserver variability of optic disk variables measured by confocal scanning laser tomography. Am. J. Ophthalmol., 132(7): 57–62.

Klein, B.E., Klein, R., Sponsel, W.E., Franke, T., Cantor, L.B., Martone, J. and Menage, M.J. (1992) Prevalence of glaucoma: the Beaver Dam Eye Study. Ophthalmology, 99(10): 1499–1504.

Miglior, S., Albé, E., Guareschi, M., Rossetti, L. and Orzalesi, N. (2002) Intraobserver and interobserver reproducibility in the evaluation of optic disc stereometric parameters by Heidelberg Retina Tomograph. Ophthalmology, 109(6): 1072–1077.

Miglior, S., Guareschi, M., Albé, E., Gomarasca, S., Vavassori, M. and Orzalesi, N. (2003) Detection of glaucomatous visual field changes using the Moorfields regression analysis of the Heidelberg retina tomograph. Am. J. Ophthalmol., 136(1): 26–33.

Quigley, H.A., Addicks, E.M. and Green, W.R. (1982) Optic nerve damage in human glaucoma. III. Quantitative correlation of nerve fiber loss and visual field defect in glaucoma, ischemic neuropathy, papilledema, and toxic neuropathy. Arch. Ophthalmol., 100(1): 135–146.

Quigley, H.A., Varma, R., Tielsch, J.M., Katz, J., Sommer, A. and Gilbert, D.L. (1999) The relationship between optic disk area and open angle glaucoma: the Baltimora Eye Survey. J. Glaucoma, 8(6): 347–352.

Reus, N.J. and Lemij, H.G. (2005) Relationships between standard automated perimetry, HRT confocal scanning laser ophthalmoscopy, and GDx VCC scanning laser polarimetry. Invest. Ophthalmol. Vis. Sci., 46(11): 4182–4188.

Tan, J.C., Garway-Heath, D.F. and Hitchings, R.A. (2003) Variability across the optic nerve head in scanning laser tomography. Br. J. Ophthalmol., 87(5): 557–559.

Tannenbaum, D.P., Hoffman, D., Lemij, H.G., Garway-Heath, D.F,, Greenfield, D.S. and Caprioli, J. (2004) Variable corneal compensation improves discrimination between normal and glaucomatous eyes with the scanning laser polarimeter. Ophthalmology, 111(2): 259–264.

Verdonck, N., Zeyen, T., Van Malderen, L. and Spileers, W. (2002) Short-term intra-individual variability in Heidelberg Retina Tomograph II. Bull. Soc. Belge Ophtalmol., 28(6): 51–57.

Vihanninjoki, K., Burk, R.O., Teesalu, P., Tuulonen, A. and Airaksinen, P.J. (2002) Optic disc biomorphometry with the Heidelberg Retina Tomograph at different reference levels. Acta Ophthalmol. Scand., 80(1): 47–53.

Weinreb, R.N., Bowd, C., Greenfield, D.S. and Zangwill, L.M. (2002a) Measurement of the magnitude and axis of corneal polarization with scanning laser polarimetry. Arch. Ophthalmol., 120(7): 901–906.

Weinreb, R.N., Bowd, C. and Zangwill, L.M. (2002b) Assessment of the retinal nerve fiber layer of the normal and glaucomatous monkey with scanning laser polarimetry. Trans. Am. Ophthalmol. Soc., 100: 161–166.

Weinreb, R.N., Bowd, C. and Zangwill, L.M. (2003) Glaucoma detection using scanning laser polarimetry with variable corneal polarization compensation. Arch. Ophthalmol., 121: 218–224.

Weinreb, R.N., Shakiba, S. and Zangwill, L. (1995) Scanning laser polarimetry to measure the nerve fiber layer of normal and glaucomatous eyes. Am. J. Ophthalmol., 119(5): 627–636.

Weinreb, R.N., Zangwill, L., Berry, C.C., Bathija, R. and Sample, P.A. (1998) Detection of glaucoma with scanning laser polarimetry. Arch. Ophthalmol., 116: 1583–1589.

Wollstein, G., Garway-Heath, D.F. and Hitchings, R.A. (1998) Identification of early glaucoma cases with the scanning laser ophthalmoscope. Ophthalmology, 105(8): 1557–1563.

Yamada, N., Chen, P.P., Mills, R.P., Leen, M.M., Stamper, R.L. and Lieberman, M.F. (2000) Glaucoma screening using the scanning laser polarimeter. J. Glaucoma, 9: 254–261.

Zangwill, L.M., Abunto, T., Bowd, C., Angeles, R., Schanzlin, D.J. and Weinreb, R.N. (2005) Scanning laser polarimetry retinal nerve fiber layer thickness measurements after LASIK. Ophthalmology, 112(2): 200–207.

CHAPTER 10

The role of OCT in glaucoma management

Monica M. Pagliara, Domenico Lepore[*] and Emilio Balestrazzi

Department of Ophthalmology, A. Gemelli University Hospital, Catholic University of the Sacred Heart, Rome, Italy

Abstract: Clinical examination of the optic nerve and achromatic automated perimetry is the gold standard for the management of glaucoma. However, there is an increasing need for an objective evaluation of the optic nerve structure, particularly for preperimetric glaucoma. Optical coherence tomography (OCT) is a noninvasive tool that measures retinal nerve fiber layer (RNFL) thickness based on its optical properties. Computer image processing algorithms estimate NFL thickness from circumpapillary OCT images that are acquired in cylindrical sections surrounding the optic disc. Average values of NFL thickness can be calculated in the four quadrants or the 12 o'clock position sectors around the optic disc. The mean NFL thickness around the entire disc can also be calculated. NFL thickness values may be compared to a normative database. Although this technique offers objectivity, rapidity, and reproducibility, it is largely influenced by the variability of optic disc size and the number of nerve fibers among individuals. At present, OCT is a good instrument to diagnose early glaucoma, but cannot be used to exclude it. New technologies, like spectral domain and ultra-high resolution, which are already available, will overcome the limitations of OCT.

Keywords: OCT; imaging; retinal fiber layer thickness; early glaucoma

Introduction

With glaucoma, there is progressive and irreversible loss of both the retinal nerve fiber layer (RNFL) and ganglion cell layer, with corresponding optic disc and visual field changes.

The structural changes that are most clinically recognized include generalized or localized thinning of the neuroretinal rim and deepening of the optic cup. Several techniques are currently available to detect, document, and quantify optic disc changes, such as clinical examination, photography, and modern imaging devices. The wider availability of the latter, with automated image acquisition and objective analysis, represents an important step in the clinical management of glaucoma by enhancing the detection of structural changes.

Devices allowing early detection and prevention of glaucomatous damage are therefore of vital importance in patient management. The detection of glaucomatous degeneration of the RNFL is likely to be valuable in the early diagnosis of glaucoma because it may precede the clinical findings of optic disc cupping and visual field loss.

The optical properties of the RNFL have allowed recent advances in ocular imaging technology to obtain thickness measurements (Hougaard et al., 2007). Optical coherence tomography (OCT) is a noncontact, noninvasive, diagnostic tool to measure RNFL thickness. It is able to precisely distinguish (8 to 10 μm resolution) the interface between the vitreous cavity and the retinal nerve fiber surface

---
[*]Corresponding author. Tel./Fax: +390630154853;
E-mail: dlepore@rm.unicatt.it

anteriorly and the retinal nerve fibers and retinal ganglion cells posteriorly. OCT obtains measurements of RNFL thickness with a fast image acquisition rate (400 A-scans per second) and good resolution.

How OCT works

OCT imaging is similar to ultrasound B-mode imaging; however, it uses light waves instead of acoustic waves. OCT performs cross-sectional imaging by measuring the echo time delay and intensity of light that is scattered or reflected back from tissue source in question. OCT images are two- or three-dimensional datasets that represent variations in optical backscattering of reflection in a cross-sectional plane or volume of tissue (Fujimoto et al., 2004).

As opposed to sound waves, light is highly scattered or absorbed in most biological tissue, and therefore optical imaging is restricted to tissues that are optically accessible. OCT uses near-infrared interferometry, and is therefore not affected by refractive status, axial length, or by changes in nuclear sclerotic cataract density. However, posterior subcapsular and cortical cataracts do impose limitations to the performance of OCT. Another advantage to OCT is that it can be performed without physically coming into contact with the patient's eye, thus minimizing discomfort.

Furthermore, light provides a significantly higher spatial resolution than ultrasound waves. Standard OCT images have an axial resolution of 10 μm, which is 10 to 20 times finer than that of standard ultrasound B-mode imaging. Research OCT systems for ultra-high resolution imaging can achieve an even finer resolution of 3 μm.

To perform cross-sectional optical imaging, OCT first measures the distance between tissues, similar to ultrasound A mode. This is accomplished by directing the light into the eye and measuring the echo time delay and intensity of backscattering or back reflection from different eye structures. Because light travels much faster than sound, a direct measurement of optical echoes is not possible. OCT uses a correlation technique known as low-coherence interferometry. Due to extremely weak backscattering or back reflection, OCT requires a high-sensitivity detection known as "optical heterodyne detection," which was originally developed for optical communications systems. The quality of OCT imaging is largely influenced by image resolution, pixel density, and speed of image acquisition.

Longitudinal (axial) image resolutions and transverse image resolutions are determined by completely different mechanisms when using OCT. The axial resolution depends upon the coherence length of the light source. This determines the accuracy with which distance can be measured and is characterized by its frequency or wavelength bandwidth. In contrast to conventional microscopy, OCT has a good axial resolution and is independent of focusing conditions and depth of field. This is achieved with a small focusing angle (or small numerical aperture focusing) and a large depth of focus using low-coherence interferometry. With the use of short pulse laser light sources that have broader bandwidths (100–200 nm) and shorter coherence lengths, axial resolutions of 3 μm can be achieved.

The transverse resolution is determined by the same principles as the transverse resolution in conventional optical microscopy, particularly the spot size of the focused OCT imaging beam, the diffraction properties of light, and the focusing parameters used. To achieve an extremely small spot, it is necessary to have a large beam diameter and a lens with a short focal length. Similar to optical microscopy, higher magnifications have a very short depth of field, and while smaller spot sizes improve the transverse resolution, they have a very shallow depth of focus (the depth of field decreases with the square of the focused spot size). The smallest spot size that can be achieved on the retina is limited by optical aberrations in the eye. Typical ophthalmic OCT systems have a 20 μm spot size on the retina.

Pixel density influences the quality of OCT imaging. It is analogous to pixel density in digital photography. The image must have sufficient pixel density in order to be able to visualize small features with a given resolution. Because OCT images are generated by acquiring successive axial measurements (axial scans) of back reflection or backscattering versus depth at different transverse positions, the number of pixels in the transverse direction is equal to the number of axial scans. A typical OCT retinal image is 6 mm wide in transverse direction and has 512 transverse pixels, so the pixels are 6 mm/512 = 11.7 μm wide. In contrast, the transverse resolution for a typical retinal image is determined by the focused spots size and is typically 20 μm. The pixel density in the axial direction is determined by the speed at which the computer can record the electronic signal from the axial scan of backscattering or back reflection versus depth. The speed can be very rapid. The typical OCT retinal image is 2 mm deep in axial direction and acquires 1024 pixels, therefore the pixel size is 2/1024 = 1.9 μm in axial depth. Finally, in order to utilize the full instrumental resolution, the size of the pixel must be smaller than the instrument resolution.

The speed of image acquisition is another important factor. Rapid acquisition not only minimizes image distortion and artifact, but also improves patient comfort. The speed is directly related to the sensitivity of the measurement because imaging more rapidly reduces signal-to-noise performance. This performance could be improved using higher incident optical power, but has to be limited due to safety standards. In nonophthalmic imaging applications, higher incident optical powers can be used to dramatically increase the acquisition speed. Image acquisition time also increases in proportion to the number of transverse pixels in an image. If a higher transverse pixel resolution is desired, then more axial scans are required and the acquisition time increases proportionally. Conversely, if only low-transverse resolution imaging is necessary, then the number of transverse pixels may be reduced and the image acquisition time will be proportionally decreased. The trade-off between pixel density and acquisition speed is important for imaging protocols. Commercially available OCT has an axial scan repletion rate of 400 axial scans per second. The high transverse pixel density image has 512 transverse pixels and is 6 mm wide, with an acquisition speed of 512/400 = 1.28 s. The same OCT having a low-transverse pixel density with 128 transverse pixels has a speed of 128/400 = 0.32 s.

In conclusion, a larger number of transverse pixels are desirable to improve the visualization of retinal features; however, this high-density image requires a longer time and therefore artifacts from eye motion can increase. A more precise registration is possible with faster image acquisition, but the image will appear grainier because of a reduced transverse pixel density.

How OCT is performed

Pupillary dilatation is recommended before glaucoma assessment using the Stratus OCT. Studies found that the dilated scans were more reproducible and of higher quality than the undilated scans (Smith et al., 2007).

The operator must choose among different scanning protocols to perform RNFL or optic nerve head (ONH) OCT. Commercially available OCT machines have various scan protocols. They are composed of two basic types of scans: line and circle. Each protocol is composed of line or circle scans with different parameters (number, angle, length, and diameter). Once the protocol is selected, the patient is positioned behind the instrument and the examiner is provided with a video camera to view the scanning probe beam on the fundus. At the same time, a computer monitor shows the OCT image acquired in real time. It is possible to center the circular scan on the optic nerve head while the subject fixates with the eye being studied (internal fixation technique). The centering technique depends on the examiner's ability to perform fine positioning of the circular scan. After the image is acquired, various analysis protocols are available to represent RNFL or ONH OCT images.

Evaluation of RNFL thickness

Evaluation of the RNFL thickness is essential for diagnosing and managing glaucoma. The RNFL comprises ganglion cell axons, neuroglia, and astrocytes (Radius and Anderson, 1979; Varma et al., 1996). The ophthalmoscopic appearance of the RNFL was first described by Vogt in 1913 (Vogt, 1913). In the early seventies, numerous studies (Hoyt and Newmann, 1972; Hoyt et al., 1973) emphasized the importance of examining the

RNFL in studying glaucoma. These studies suggested that, in some cases of glaucoma, RNFL atrophy is primarily related to a degeneration of the ganglion cell axons, followed by a thinning of the NFL. Thus, when attempting to detect early glaucomatous optic nerve damage, it would be useful to obtain objective quantitative measurements of the thickness of the RNFL in addition to qualitative observation of the defects in the RNFL.

RNFL can be assessed subjectively with a slit lamp and a handheld lens, but this technique requires experience and offers only qualitative data that are difficult to compare over time. Several instruments such as scanning laser ophthalmoscopy (SLO), scanning laser polarimetry (SLP), and OCT allow objective and quantitative evaluation of RNFL.

Nerve fiber layer thickness is assumed to be correlated with the extent of the red, highly reflective layer at the vitreoretinal interface (Fig. 1). Boundaries are located by searching for the first points on each scan where the reflectivity exceeds a certain threshold. The posterior margin of the nerve fiber layer is located by starting within the photoreceptor layer and searching upward in the image (Schuman et al., 1996).

Using the most commercially available OCT (Stratus OCT 3, Carl Zeiss Meditec Inc., Dublin, California), two protocols are available for managing glaucoma. The *RNFL thickness (3.4)* protocol is designed to acquire three circle scans of diameter of 3.4 mm around the optic disc. The *fast RNFL* protocol compresses the three RNFL thickness (3.4) circle scans into one scan to simplify the process and shorten the acquisition time.

OCT acquires images in cylindrical sections surrounding the optic disc. Computer image processing allows for an estimation of RNFL thickness from circumpapillary scans: the RNFL thickness is defined as the number of pixels between the anterior and the posterior edges on the RNFL identified by an edge detection algorithm.

The protocol used most to study RNFL thickness is the *RNFL thickness average analysis* protocol (Fig. 2). The NFL thicknesses are reported as averages over either each quadrant or clock hour and as a graph of NFL thickness along the scan line (Schuman et al., 1995, 1996).

The RNFL thickness average analysis protocol is extremely useful in comparing the measurements of the NFL between the eyes. The *RNFL thickness map analysis* protocol is used to obtain a map of NFL thickness of the peripapillary area. To assess changes in NFL thickness between examinations, "*RNFL thickness change*" and "*RNFL thickness serial*" analysis protocols are useful.

Several studies evaluated the optimal circle diameter for OCT NFL scans based on reproducibility of measurements and disc size. Fixation technique (internal or external) and its effect on the reproducibility of OCT measurements have been evaluated. The standard circle diameter selected is 3.4 mm, which is ideal for several reasons. It is large enough to avoid overlap with the optic nerve head

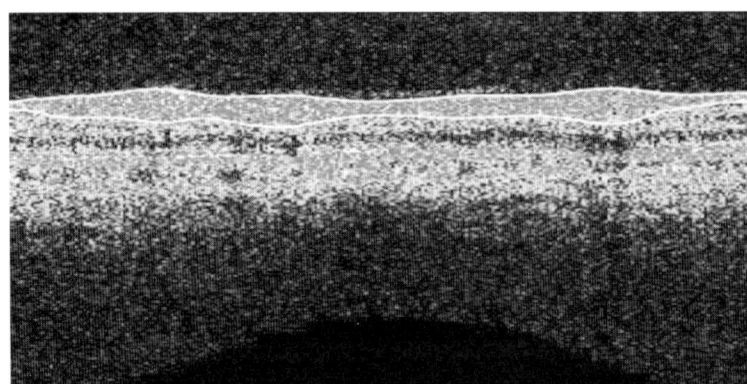

Fig. 1. The nerve fiber layer appears in the OCT images as a highly backscattering layer in the superficial retina. NFL thickness may be assessed at individual points on a cylindrical or linear tomogram in the peripapillary region.

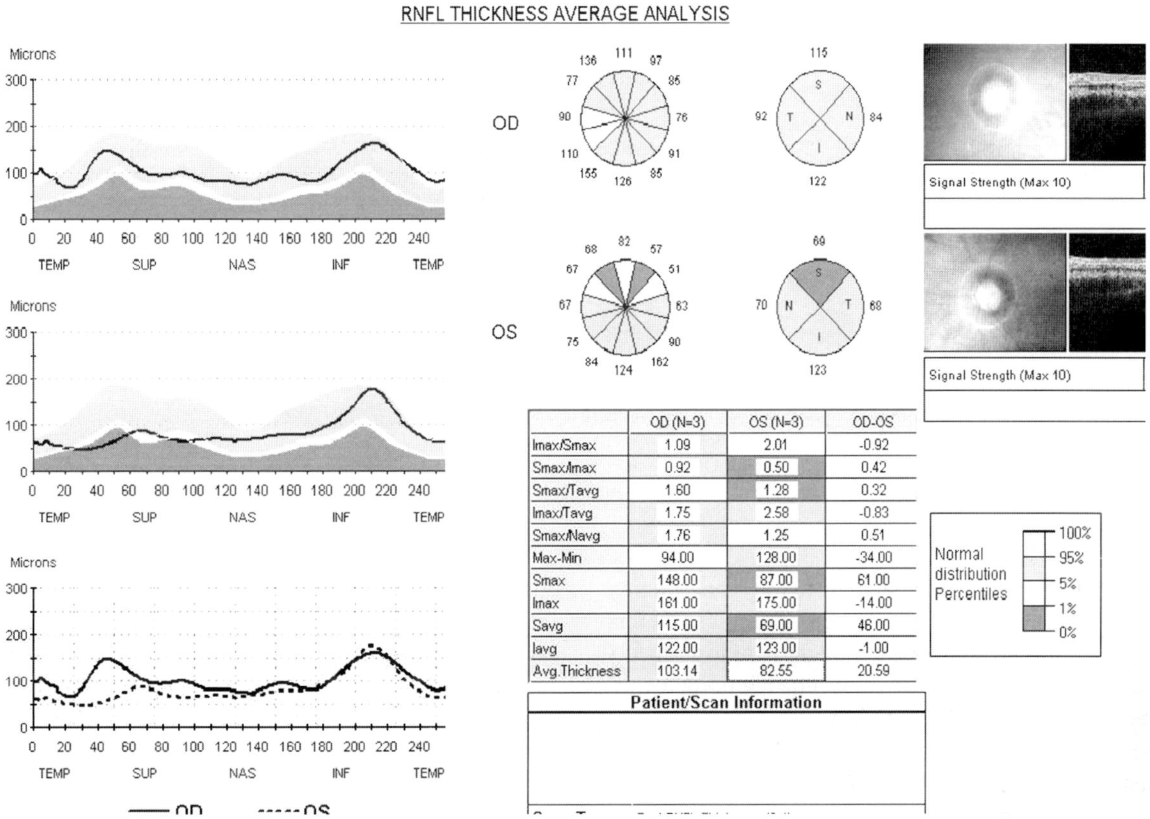

Fig. 2. RNFL thickness average analysis protocol: NFL thickness is reported individually for each A-scan as averages over each quadrant (superior, inferior, temporal, nasal), as averages for each clock hour, or as averages over the entire cylindrical section. Smax and Imax represent the maximum thickness of NFL in superior and inferior quadrant, respectively. Similarly, Tavg, Navg, Savg, and Iavg represent the average thickness of NFL in each quadrant. The graphs of a person with normal eyes show the typical double-hump pattern of normal NFL thickness that is thicker superiorly and inferiorly. This pattern actually contains two peaks superiorly and a peak and a shoulder inferiorly, but by convention this pattern is referred to as the "double hump" pattern. Several spikes may be seen, and they are typically blood vessels.

in nearly all eyes and the NFL to be measured in a thicker area, thus permitting a higher sensitivity to subtle NFL defects (Schuman et al., 1996).

It is important to emphasize that Fast RNFL scan is the only peripapillary RNFL scan type available in the Stratus OCT software that has a normative database analysis. The values obtained can be compared against a normative database of age-matched controls to derive percentile values. The four percentile values included in the OCT software are the top 5th percentile, top 95th percentile, bottom 5th percentile, and bottom 1st percentile. The normative data used in the software were collected by studying approximately 350 normal individuals equally distributed into decades between the ages of 20 and 80 years. In the Fast scan program, subject values are compared with values obtained in the normative database. There is no correction for other demographic factors, such as ethnicity or gender, because these factors have not been demonstrated to affect RNFL thickness to date (Budenz et al., 2005).

Evaluation of optic disc

Changes in optic nerve head are a well-established marker for glaucoma. Determining whether

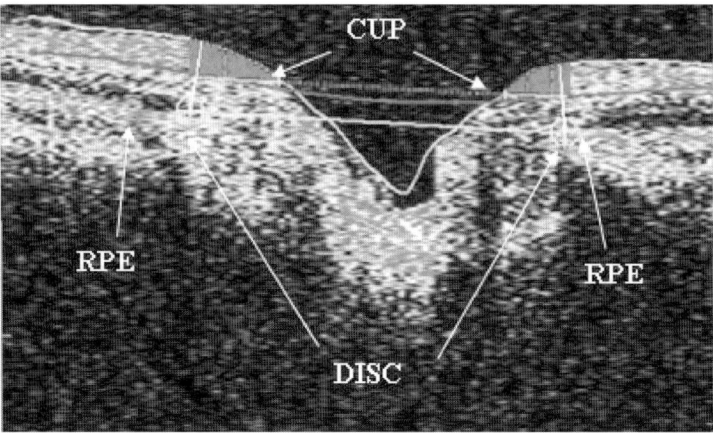

Fig. 3. Measurements from the multiple radial OCT images at varying angular orientations can be used to construct a two-dimensional map of optic nerve head. The disc area, cup area, neuroretinal rim area, as well as various cup-to-disc ratios, can be calculated.

changes have occurred in the optic disc remains one of the most important and challenging aspects of glaucoma management. Structural changes that are most clinically recognized include generalized or localized thinning of the neuroretinal rim (Airaksinen and Drance, 1985) and deepening of the optic cup. These changes reflect a loss or alteration of retinal ganglion cell axons and/or the structures that support them (Quigley, 1999). Optic disc size can be estimated roughly by disc photographs, with correction for ocular parameters such as corneal curvature and axial length (Balazsi et al., 1984). Modern imaging devices are able to document optic disc morphology and quantify reduction of retinal fiber thickness. Stratus OCT can readily provide accurate estimates of optic disc size. Cup and rim area estimates are also valuable in determining the degree and rate of change. Expert algorithms can also be used to perform OCT image analysis in order to assess the optic nerve head and measure cup and disc parameters.

The Stratus OCT 3 uses two protocols. The *optical disc* protocol consists of a series of 6–24 equally spaced line scans through a common center. The default pattern has six lines that are 4 mm in length. The scans created with this protocol are used with the optic nerve head analysis protocol. The *fast optical disc* protocol compresses the six optical disc scans into one scan. This protocol consists of six 4 mm radial line scans. The resolution of fast protocols is lower, but the chance of error from patient movement is less.

As shown in Fig. 3, the boundary of the disc can be determined from each OCT image by the point at which the photoreceptor layer, RPE, and choriocapillaris terminate at the laminar cribrosa. This point can be located automatically by expert image processing algorithms and then viewed and confirmed by the operator. The disc diameter can be determined by measuring the distance between the disc boundaries on opposite sides of the disc. The cup diameter can be measured by constructing a line parallel to and offset anteriorly by a standard amount to the line that defines the disc diameter.

Optic disc topography measured by OCT has shown to be in agreement with other disc-measuring instruments (Bowd et al., 2001; Zangwill et al., 2001; Bowd et al., 2002; Williams et al., 2002; Aydin et al., 2003; Guedes et al., 2003; Schuman et al., 2003).

OCT in glaucoma management

Visual field examination can be considered the gold standard in glaucomatous patient management, and any new technique, such as OCT imaging, must be compared to it.

Typical visual field abnormalities reliably establish glaucomatous nerve atrophy. Visual fields can be preceded by significant loss of retinal ganglion cells. In the early seventies, Hoyt (Hoyt et al., 1973) correlated slit-like defects in the appearance of the nerve fiber layer with early clinically detectable manifestations of glaucomatous damage. Since Hoyt's initial report, numerous experimental and observational studies suggested that in glaucomatous eyes, the atrophy of the retinal fiber layer is primarily related to the degeneration of ganglion cell axons, followed by NFL thinning (Quigley et al., 1977; Sommer et al., 1984; Airaksinen et al., 1985).

Subsequent reports correlate abnormalities in the RNFL to an arcuate loss of visual field, and in doing so, these reports emphasize the importance of examining the RNFL in glaucoma. Several authors affirm that visual field defects are anticipated by loss of RNFL, and furthermore, early glaucoma cannot be excluded without excluding the presence of RNFL defects (Quigley et al., 1982; Caprioli and Miller, 1990; Mikelberg et al., 1995). Of note, Sommer (Sommer et al., 1991) observed that in 60% of eyes, NFL layer loss appears approximately 6 years before any detectable visual field defects.

The RNFL assessment with a slit lamp and a handheld lens requires experience and offers only subjective and qualitative data that is difficult to compare over time. When attempting to detect early glaucomatous optic nerve damage, it would be useful to supplement qualitative observations of RNFL defects with objective quantitative measurements of the NFL thickness.

According to Soliman (Soliman et al., 2002), there is a nonlinear relationship between RNFL loss (measured by OCT RNFL) and visual field damage, and it can be better approximated using an exponential model represented by the curved line showing relationship between OCT RNFL and standard achromatic perimetry (SAP) pattern standard deviation. The graphic shows that a considerable amount of RNFL is lost before the development of observable visual field damage. In early glaucoma, RNFL loss can occur without visual field changes and can be easily detected with OCT. However, in late stages, the progress of glaucomatous damage can be better detected with visual fields, as the variation in RNFL thickness is too small to be detected, while the variation in visual field is larger and easily detectable.

Measurements of retinal NFL with the OCT have demonstrated a reproducible difference between normal eyes and eyes with open-angle glaucoma, as defined by abnormal achromatic visual fields (Pieroth et al., 1998).

Schumann (Schumann et al., 1995) showed that the diagnosis of glaucoma was associated with a significantly thinner RNFL, especially in the inferior quadrant, as compared with measurements in normal eyes. In 2007, Johnson (Johnson et al., 2007) clearly stated that subjective biomicroscopic examination of the fundus is the current gold standard for detecting glaucomatous structural damage, but it relies on the examiner's experience. OCT, one of the several imaging technologies introduced to measure RNFL thickness, has been proposed as a diagnostic tool for the detection of glaucoma because of its ability to provide quantitative, reproducible, and objective data. Patients with early glaucoma provide a diagnostic challenge, and it is this group for whom it is hoped that OCT can provide useful information. Using quality assessment for the diagnostic accuracy of OCT in glaucoma, Johnson concluded that reporting on this subject is "suboptimal." Keeping this "warning" in mind, the following representative data from "milestone" papers describe the use of OCT in the diagnosis of glaucoma.

Nouri-Madhavi in 2004 (Nouri-Madhavi et al., 2004) found that OCT effectively differentiates early perimetric glaucoma from normal eyes, while its discriminating power in glaucoma suspects (eyes with suspicious optic disc cupping and normal achromatic visual fields) is less adequate. This study confirmed earlier reports by the same group: Greaney (Greaney et al., 2002) compared various quantitative optic nerve imaging methods [OCT, confocal scanning laser ophthalmoscopy (CSLO), SLP] with the qualitative disc assessment by experienced observers (ONHP: optic nerve head photographs). None of these quantitative techniques, when used alone, was sufficiently able to discriminate between normal eyes and early to moderate glaucoma. Only after combining parameters from four different optic nerve imaging techniques, the authors were able to improve this diagnostic ability.

Recently, Parikh et al. (2007) compared their data with data from international literature and affirmed that OCT has moderate sensitivity with high specificity. Data from inferior hemi-meridian showed the best combination between sensitivity and specificity compared to superior hemi-meridian.

Although it is clear that further scientific evidence is needed to clarify the role of OCT in glaucoma, we can state that OCT at the present time is a good instrument to diagnose early glaucoma but cannot be used to exclude it.

New perspective

Improvements in OCT technology have recently been introduced.

Podoleanu and Jackson in 1997 (Podoleanu et al., 1997) introduced a transverse scanning technique (parallel to the retina surface) to capture real-time coronal planes (C-scans) or sagittal planes (B-scans) simultaneous with confocal SLO images (OCT/SLO). The distance between scans is four times less in transverse scan than in A-scan. Transverse retinal scanning follows retinal layers, allowing better imaging of the outer retinal region. Coronal OCT scans visualize details that are often lost in B-scan with a more complex interpretation. Moreover, SLO channel maintains multifunctional capabilities of angiography, microperimetry and mfERG.

Ultra-high resolution (UHR) OCT introduced by Drexel and Fujimoto in 2001 (Drexel et al., 2001) provided a better axial resolution (up to 2 µm) using an improved super luminescent diode. The enhanced anatomical details were limited by slow time acquisition (4–5 s) and alignment dependence on macular fixation.

Wojtkoswki in 2002 (Wojtkoswki et al., 2002) first used spectrometers with high-speed cameras to capture sets of axial scans to allow multiple signal (up to 200) returns. The use of this spectral domain technique accelerated data collection from 400 A-scans/s of one conventional OCT to 27 A-scans/s. With a faster acquisition, there is a more stable image that is not affected by patient motion. A stack of one hundred or more cross-sectional scans (high-density scan) can be acquired in the same time that it takes for six cross-sectional scans (low-density scan) with conventional time-domain OCT, thus allowing for a three-dimensional representation of images.

The use of the improved super luminescent diode, coupled with the spectral domain OCT (SDOCT) technique, allows us to acquire, in vivo, cross-sectional retinal images with an axial resolution up to five times higher and an imaging speed that is 60 times faster than conventional OCT.

This increase in resolution and scanning speed permits high-density faster scanning of retinal tissue while minimizing eye motion artifacts.

By combining high resolution OCT, SDOCT, and SLO, it is finally possible to obtain a device with the ability to generate high-speed (200 frame per second) B-scan OCT/SLO, and to reveal internal three-dimensional anatomic details along with surface features with a 35 degree image field. All of these properties are reached without losing the multifunctionality of SLO.

Using this new device, it is possible to detect and segment the RNFL in each faster OCT image and use these data to construct a detailed RNFL thickness map. In particular, UHR/SDOCT/SLO has shown that RNFL thickness is generally inversely related to the distance from the ONH center in the peripapillary region of healthy subjects. Aside from the nasal segment, all areas show an initial increase in RNFL, followed by a peak and gradual linear decrease.

Like any other technology that has not yet completed the ascending phase of its cycle of use, at the moment UHR/SDOCT/SLO and, in general, OCT have many drawbacks. First of all, normative data are required for the new system. Moreover, large datasets need a plan for storage and backup, faster image processing, and integration with conventional images. Because a large number of new SDOCT systems (Table 1) are now commercially available, it is difficult to integrate data across different clinics, thus creating a challenge for new clinical trials.

Table 1. Spectral domain OCT systems commercially availables

Company	Name of device	Acquisition speed (A-scans/s)	Axial resolution (μm)
Bioptigen	3D SDOCT	20,000	~6
Carl Zeiss Meditec	Cirrus	27,000	~5
Heidelberg	Spectralis	40,000	~7
OTI	Spectral OCT/SLO	27,000	5–6
Optovue	RTVue-100	26,000	~5
Optopol	Copernicus	25,000	~6
Topcon	3D OCT 1000	20,000	5

Abbreviations

CSLO	confocal scanning laser ophthalmoscopy
OCT	optical coherence tomography
ONH	optic nerve head
ONHP	optic nerve head photographs
RNFL	retinal nerve fibers layer
SDOCT	spectral domain OCT
SLP	scanning laser polarimetry
UHR	ultra-high resolution

References

Airaksinen, P.J. and Drance, S.M. (1985) Neuroretinal rim area and retinal nerve fiber layer in glaucoma. Arch. Ophthalmol., 103: 203–204.

Airaksinen, P.J., Drance, S.M., Douglas, G.R., Mawson, D.K. and Nieminen, H. (1985) Diffuse and localized nerve fiber loss in glaucoma. Am. J. Ophthalmol., 98(5): 566–571.

Aydin, A., Wollstein, G., Price, L.L., Fujimoto, J.G. and Schuman, J.S. (2003) Optical coherence tomography assessment of retinal nerve fiber layer thickness changes after glaucoma surgery. Ophthalmology, 110(8): 1506–1511.

Balazsi, A.G., Drance, S.M., Schulzer, M. and Douglas, G.R. (1984) Neuroretinal rim area in suspected glaucoma and early chronic open angle glaucoma. Arch. Ophthalmol., 102(7): 1011–1014.

Bowd, C., Zangwill, L.M., Berry, C.C., Blumenthal, E.Z., Vasile, C., Sanchez-Galeana, C.A., Bosworth, C.F., Sample, P.A. and Weinreb, E.N. (2001) Detecting early glaucoma by assessment of retinal nerve fiber layer thickness and visual function. Invest. Ophthalmol. Vis. Sci., 42(9): 1993–2003.

Bowd, C., Zangwill, L.M., Blumenthal, E.Z., Vasile, C., Boehm, A.G., Gokhale, P.A., Mohammadi, K., Amini, P., Sankary, T.M. and Weinreb, R.N. (2002) Imaging of the optic disc and retinal nerve fiber layer: the effects of age, optic disc area, refractive error, and gender. J. Opt. Soc. Am. A Opt. Image Sci. Vis., 19(1): 197–207.

Budenz, D.L., Michael, A., Chang, R.T., Chang, R.T., McSoley, J. and Katz, J. (2005) Sensitivity and specificity of the stratus OCT for perimetric glaucoma. Ophthalmology, 112: 3–9.

Caprioli, J. and Miller, J.M. (1990) Measurement of relative nerve fiber layer surface height in glaucoma. Ophthalmology, 97: 358–365.

Drexel, W., Morgner, U., Ghanta, R.K., Schuman, J.S., Kartbner, F.X. and Fujimoto, J.G. (2001) Ultrahigh resolution ophthalmological optical coherence tomography. Nat. Med., 7: 502–507.

Fujimoto, J.G., Huang, D., Hee, M.R., Ko, T., Swanson, E., Puliafito, C.A. and Schuman, J.S. (2004) Physical principles of optical coherence tomography. In: Schuman, J.S., Puliafito, C.A. and Fujimoto J.G. (Eds.), Optical coherence tomography of ocular diseases, 2nd ed., Appendix A. Slack Inc., Thorofare, NJ, pp. 677–688.

Greaney, M.J., Hoffman, D.C., Garway-Heath, D.F., Nakla, M., Coleman, A.L. and Caprioli, J. (2002) Comparison of optic nerve imaging methods to distinguish normal eyes from those with glaucoma. Invest. Ophthalmol. Vis. Sci., 43: 140–145.

Guedes, V., Schuman, J.S., Hertzmark, E., Wollstein, G., Correnti, A., Mancini, R., Lederer, D., Voskanian, S., Velasquez, L., Pakter, H.M., Pedut-Kloizman, T., Fujimoto, J.G. and Mattox, C. (2003) Optical coherence tomography measurement of macular and nerve fiber layer thickness in normal and glaucomatous human eyes. Ophthalmology, 110(1): 177–189.

Hougaard, J.L., Heijl, A. and Bengtsson, B. (2007) Glaucoma detection by Stratus OCT. J. Glaucoma, 16(3): 302–306.

Hoyt, W.F., Frisen, L. and Newman, N.M. (1973) Funduscopy of nerve fiber layer defects in glaucoma. Invest. Ophthalmol. Vis. Sci., 12: 814–829.

Hoyt, W.F. and Newmann, N.M. (1972) The earliest observable defects in glaucoma. Lancet, 1: 692–693.

Johnson, Z.K., Rehman Siddiqui, M.A. and Azuara-Blanco, A. (2007) The quality of reporting of diagnostic accuracy studies of optical coherence tomography in glaucoma. Ophthalmology, 114: 1607–1612.

Mikelberg, F.S., Yidegiligne, H.M. and Shulzer, M. (1995) Optic nerve axon count and axon diameter in patient with ocular hypertension and normal visual field. Ophthalmology, 102: 342–348.

Nouri-Madhavi, K., Hoffman, D.C., Tannenbaum, D.P., Law, S.K. and Caprioli, J. (2004) Identifying early glaucoma with optical coherence tomography. Am. J. Ophthalmol., 37: 228–235.

Parikh, R.S., Parikh, S., Sekhar, G.C., Kumar, R.S., Prabakaran, S., Babu, J.G. and Thomas, R. (2007) Diagnostic capability of optical coherence tomography (Stratus OCT 3) in early glaucoma. Ophthalmology, 114: 2238–2243.

Pieroth, L., Schuman, J.S., Hertzmark, E., Hee, M.R., Wilkins, J.R., Coker, J., Mattox, C., Pedut-Kloizman, T., Puliafito, C.A., Fujimoto, J.G. and Swanso, E. (1998) Evaluation of focal defects of the nerve fiber layer using optical coherence tomography. Ophthalmology, 106: 570–579.

Podoleanu, A.G., Dobre, G.M., Webb, D.J. and Jackson, D.A. (1997) Simultaneous en-face imaging for two layers in the human retina by low coherence interferometry. Opt. Lett., 22(13): 1039–1041.

Quigley, H.A. (1999) Neuronal death in glaucoma. Prog. Retin. Eye Res., 18: 39–57.

Quigley, H.A., Adicks, E.M. and Green, R. (1982) Optic nerve damage in human glaucoma. III: Quantitative correlation of nerve fiber loss and visual field defect in glaucoma, ischemic neuropathy, papilledema and toxic neuropathy. Arch. Ophthalmol., 100: 135–146.

Quigley, H.A., Davis, E.B. and Anderson, E.R. (1977) Descending optic nerve degeneration in primates. Invest. Ophthalmol. Vis. Sci., 16: 841–849.

Radius, R.L. and Anderson, D.R. (1979) The histology of the retinal nerve fiber layer bundles and bundle defects. Arch. Ophthalmol., 97: 948–951.

Schuman, J.S., Hee, M.R., Puliafito, C.A., Wong, C., Pedut-Kloizman, T., Lin, C.P., Hertzmark, E., Izatt, J.A., Swanson, E.A. and Fujimoto, J.G. (1995) Quantification of nerve fiber layer thickness in normal and glaucomatous eyes using optical coherence tomography. Arch. Ophthalmol., 113: 586–596.

Schuman, J.S., Pedut-Kloizman, T., Hertzmark, E., Hee, M.R., Wilkins, J.R., Coker, J.G., Puliafito, C.A., Fujimoto, J.G. and Swanson, E.A. (1996) Reproducibility of nerve fiber layer thickness measurements using optical coherence tomography. Ophthalmology, 103: 1889–1898.

Schuman, J.S., Wollstein, G., Farra, T., Hertzmark, E., Aydin, A., Fujimoto, J.G. and Paunescu, L.A. (2003) Comparison of optic nerve head measurements obtained by optical coherence tomography and confocal scanning laser ophthalmoscopy. Am. J. Ophthalmol., 135(4): 504–512.

Smith, M., Frost, A., Graham, C.M. and Shaw, S. (2007) Effect of pupillary dilatation on glaucoma assessments using optical coherence tomography. Br. J. Ophthalmol., 91: 1686–1690.

Soliman, M.A.E., Van den Berg, T.J.T.P., Ismaeil, A.A., De Jong, L.A.M.S. and De Smet, M.D. (2002) Retinal nerve fiber layer analysis: relationship between optical coherence tomography and red-free photography. Am. J. Ophthalmol., 133: 187–195.

Sommer, A., Katz, J., Quigley, H.A., Miller, N.R., Robin, A.L., Richter, R.C. and Witt, K.A. (1991) Clinically detectable nerve fiber atrophy precedes the onset of glaucomatous field loss. Arch. Ophthalmol., 109: 77–83.

Sommer, A., Quigley, H.A., Robin, A.L., Miller, N.R., Katz, J. and Arkell, S. (1984) Evaluation of nerve fiber layer assessment. Arch. Ophthalmol., 102(12): 1766–1771.

Varma, R., Skaf, M. and Barron, E. (1996) Retinal nerve fiber layer thickness in normal human eyes. Ophthalmology, 103: 2114–2119.

Vogt, A. (1913) Herstellung eines gelbblauen Lichtfiltrates, in welchem die macula centralis in vivo in gelber Farbung erscheint, die nervenfasern der netzhaut und andere feine einzelhelten derselben sichtbar warden und der grad der gelbfarbung der lines ophthalmoskopisch nachweisbar ist. Arch. Ophthalmol., 84: 293–311.

Williams, Z.Y., Schuman, J.S., Gamell, L., Nemi, A., Hertzmark, E., Fujimoto, J.G., Mattox, C., Simpson, J. and Wollstein, G. (2002) Optical coherence tomography measurement of nerve fiber layer thickness and the likelihood of a visual field defect. Am. J. Ophthalmol., 134(4): 538–546.

Wojtkoswki, M., Leitgeb, R., Kowalczyk, A., Bajraszewski, T. and Fercher, A.F. (2002) In vivo human retinal imaging by Fourier domain optical coherence tomography. J. Biomed. Opt., 7(3): 457–463.

Zangwill, L.M., Bowd, C., Berry, C.C., Williams, J., Blumenthal, E.Z., Sanchez-Galeana, C.A., Vasile, C. and Weinreb, E.N. (2001) Discriminating between normal and glaucomatous eyes using the Heidelberg retina tomograph, GDx nerve fiber analyzer, and optical coherence tomography. Arch. Ophthalmol., 119(7): 985–993.

CHAPTER 11

Functional laser Doppler flowmetry of the optic nerve: physiological aspects and clinical applications

Charles E. Riva[1] and Benedetto Falsini[2],*

[1]Dipartimento di Discipline Chirurgiche, Rianomatorie e dei Trapianti "Antonio Valsalva", Universitá di Bologna, Bologna, Italy
[2]Istituto di Oftalmologia, Universita' Cattolica del S. Cuore, Roma, Italy

Abstract: The present paper reviews the methodology and clinical results of recording, by laser Doppler flowmetry, the hemodynamic response of the optic nerve head elicited by visual stimulation. The basic mechanism underlying this novel technique (which is called here functional laser Doppler flowmetry (FLDF)) is the coupling between visually evoked neural activity and vascular activity within the neural tissue of the optic nerve (neurovascular coupling). The blood flow responses elicited by various visual stimuli (luminance and chromatic flicker, focal and pattern stimulation) have been characterized in humans by FLDF. These responses are similar to those assessed by electrophysiological methods (flicker and pattern electroretinography) evoked by the same stimuli. In addition, a significant correlation has been demonstrated between the hemodynamic responses and the neural activity induced electrical signals arising from the inner retina, providing evidence in support of the presence of a neurovascular coupling in humans. The application of FLDF in patients with ocular hypertension and early glaucoma demonstrates that the visually evoked hyperemic responses are significantly depressed even when neural retinal activity may be still relatively preserved, suggesting that abnormal optic nerve head autoregulation in response to visual stimuli may be altered early in the disease process. FLDF may open new avenues of investigation in the field of glaucoma and other neuro-ophthalmic disorders, providing new pathophysiological data and outcome measures for potential neuro-protective treatments.

Keywords: functional hyperemia; glaucoma; laser Doppler flowmetry; neural activity; optic nerve; ERG; neurovascular coupling

Introduction

Recent experimental data in both animals and humans indicate that visual stimulation by uniform field flicker or pattern contrast reversal induces an increase in optic nerve head blood flow (F_{onh}) coupled with an increase of retinal neural activity (Riva et al., 1991, 1992, 2001, 2005). These flow and activity responses are symbolized here as RF_{onh} and RA. In cats, the coupling between RF_{onh} and RA is particularly tight when F_{onh} is measured by laser Doppler flowmetry (LDF) and RA is derived from the electrical signal generated by the ganglion cells (Riva and Buerk, 1998) or from the K^+ ion concentration (Buerk et al., 1995), both measured at the surface of the optic nerve.

*Corresponding author. Tel.: +39 06 30156344;
Fax: +39 06 3051274; Email: bfalsini@rm.unicatt.it

This relationship between RA and RF_{onh}, the so-called neurovascular coupling as originally postulated for the brain (Roy and Sherrington, 1890), has led to a large number of studies in the brain aimed at exploring the physiological parameters, as well as the clinical conditions which, in humans, might influence this coupling. Similar investigations in the optic nerve, a tissue consisting of axons and glia (white matter) but deprived of neuron somata (Ransom and Orkand, 1996), are more recent, although this tissue, which can be accessed noninvasively, is involved in the pathogenesis of a number of severe ocular diseases.

The aim of the present review is to report the current status of knowledge on the RA-induced RF_{onh}, the latter measured by LDF. This functional LDF (FLDF) has been employed to investigate the dependence of RF_{onh} upon the characteristics of the visual stimuli, to quantitatively measure the coupling between RF_{onh} and RA, and to determine whether, and to what extent, pathologic conditions such as glaucoma may affect this coupling. The findings obtained so far indicate that (i) visual stimulation is a powerful modulator of F_{onh}; (ii) a coupling can be quantified in the optic nerve by simultaneously recording RF_{onh} and electrical activity changes in the neural tissue of the retina; (iii) RF_{onh} may be altered in certain stress conditions and in an ocular pathological condition involving the optic nerve head microcirculation, i.e. ocular hypertension and chronic glaucoma.

Technology

Laser Doppler flowmetry of the optic nerve

This technique is based on the Doppler effect, which describes the change in frequency (Δf) of an electromagnetic wave of frequency f emitted from a source moving toward or away from an observer. When a laser beam impinges on red blood cells (RBCs) moving in a small region of the tissue of the optic nerve, the multiplicity of directions and magnitudes of the velocity of the RBCs, as well as the scattering of the laser light in many directions by the nonmoving tissue structures give rise to a distribution of Doppler shift frequencies, the so-called Doppler shifts power spectrum (DSPS). As discussed elsewhere (Riva et al., 2000), from which the following text has been borrowed with permission from the publisher, it can be said that (i) since blood occupies only 1–5% of the sampled tissue volume, only a small fraction of the scattered light is Doppler frequency-shifted through interaction with the RBCs. The rest of the scattered light has been predominantly scattered by nonmoving structural components of the tissue; (ii) the DSPS is generated by the heterodyne mixing of the nonshifted light with the shifted scattered light at the surface of the photodetector. In accordance with Bonner and Nossal's (1990) theory of LDF, the following parameters are extracted from the DSPS:

$$\text{Vel}_{onh} = \frac{\int_{30 \text{ Hz}}^{3000 \text{ Hz}} \Delta f \cdot P(\Delta f) \, d\Delta f}{\int_{30 \text{ Hz}}^{3000 \text{ Hz}} P(\Delta f) \, d\Delta f}$$

$$\text{Vol}_{onh} = \frac{1}{A_{dc}^2} \int_{30 \text{ Hz}}^{3000 \text{ Hz}} P(\Delta f) \, d\Delta f$$

$$F_{onh} = \frac{1}{A_{dc}^2} \int_{30 \text{ Hz}}^{3000 \text{ Hz}} \Delta f \cdot P(\Delta f) \, d\Delta f$$

Vel_{onh} is the mean Doppler shift frequency (unit Hz), which is proportional to the mean velocity of the RBCs moving in the sampled volume. Vol_{onh} is proportional to the number of moving RBCs. $F_{onh} = \text{Vel}_{onh} \times \text{Vol}_{onh}$. Both F_{onh} and Vol_{onh} are expressed in relative units and Δf in Hz. $P(\Delta f)$ is the power of the DSPS at each Δf and A_{dc} is the amplitude of the direct current produced by the light incident on the detector. Doppler shifts below 30 Hz are filtered out to avoid slow tissue motion artifacts. In the optic nerve, the Doppler shifts are below 3000 Hz.

F_{onh} varies linearly with the actual flux of the RBCs, which is identical to blood flow if the hematocrit does not vary during the experiment (Riva and Petrig, 1997). This condition has been assumed to be fulfilled in all experiments to be discussed in this review.

The LDF data presented in this review has been obtained with a near-infrared LDF system mounted on a fundus camera (Fig. 1) (Riva et al., 2000;

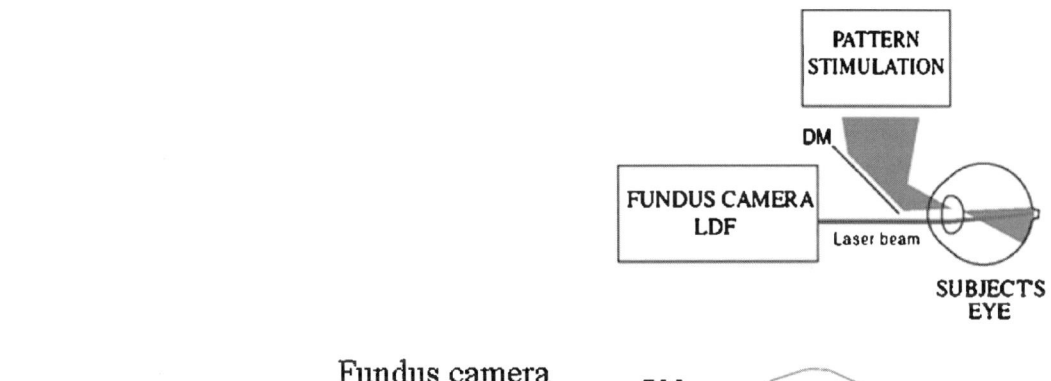
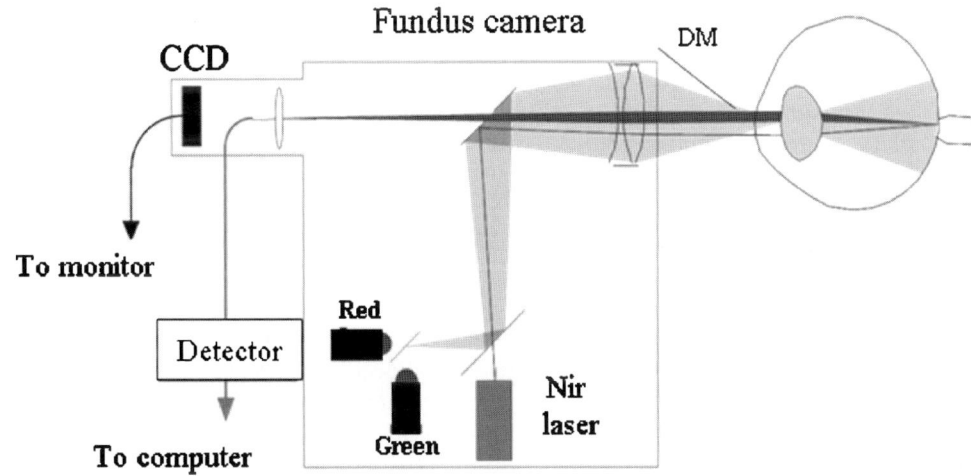

Fig. 1. Schematic representation of fundus camera-based near-infrared LDF system. The red/green emitting diodes allow chromatic flicker stimulation. They can be replaced by a white light flicker illumination via an optical fiber guide. Not shown is the fundus illumination system in the near infrared. Upper right: Arrangement to deliver contrast reversal checkerboard patterns through the dichroic mirror DM.

Logean et al., 2005). Briefly, with this system, the laser beam (wavelength between 750 and 810 nm) followed the optical path of the fundus illumination system and was aimed at a site on the rim of the optic disc, avoiding the main retinal vessels. The light scattered by the RBCs and tissue structures was picked up by an optical fiber whose aperture at the disc is $\approx 160\,\mu m$ and guided to an avalanche photodetector. The fundus was illuminated in near-infrared light (826 nm) and viewed on the screen of a CCD camera, together with the aperture of the fiber collecting the scattered light. The aperture of an optical fiber (50 μm diameter) mounted on an x-y-z micromanipulator placed in one of the retinal planes of the fundus camera was illuminated by a red light to provided a fixation target. By moving this target, the operator, guided by photographs of the disc and the pitch of the Doppler sound, aimed the laser at the desired site of measurement.

The output signal of the detector was analyzed using software implemented on a NeXT computer and the LDF parameters were displayed on a monitor at a rate of 21 values per second (Petrig and Riva, 1988). Figure 2 shows recordings of Vel_{onh}, Vol_{onh}, and F_{onh} obtained from the optic disc of a human volunteer.

Visual stimulation

Two types of stimulation were used in humans: (A) constant luminance (red/black or green/black flicker) and chromatic red/green flicker. These

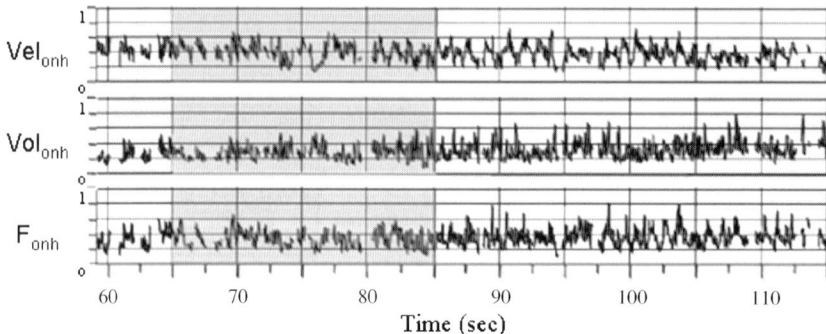

Fig. 2. Recording of the LDF flow parameters from the optic nerve head of a human volunteer using the system shown in Fig. 1.

stimuli were generated by light-emitting diodes located in an image plane of the eye pupil. These diodes could uniformly illuminate a field as large as 50° at the fundus with a maximum luminance of 40 lux for each color; (B) focal contrast reversal checkerboards. These were displayed on the monitor of a pattern generating system (VGS series 5, Cambridge Research System, England) and focused at the subject's retina through a dichroic mirror placed between the fundus camera objective and the eye of the subject and a spectacle lens with a dioptric power specific for the tested eye, as shown in Fig. 1 (Logean et al., 2005).

Reproducibility and habituation of RF_{onh}

Defining RF_{onh} as $RF_{onh} = 100 \times ((F_{onh,st}-F_{onh,bl})/F_{onh,bl})$, where $F_{onh,bl}$ is the average flow during baseline (no stimulus) and $F_{onh,st}$ is the mean flow during the last 20 s of a 1 min stimulation, test–retest coefficient of variation of RF_{onh} data determined in five subjects under various conditions of stimulation ranged from 10 to 48% (average, 26%) (Falsini et al., 2002).

Habituation, an effect that may lead to a reduction of RF_{onh} over time, was investigated in a volunteer using a 15 Hz luminance flicker applied for 1 min and repeated several times over periods of 12 and 20 min. Linear regressions across both series of data did not reveal any significant trend (Falsini et al., 2002), suggesting the absence of this phenomenon at the level of the optic nerve. Furthermore, 8 min recordings of F_{onh} from a temporal site of the disc during luminance flicker showed no significant trend when a linear regression was determined for the period of flicker between 30 s and 8 min (Falsini et al., 2002).

Retinal neural activity as assessed from the electroretinogram (ERG)

The visually evoked neural retinal activity changes can be induced by diffuse flicker and pattern contrast reversal stimuli and assessed by means of the flicker and pattern ERG (F-ERG and P-ERG, respectively), which are now briefly reviewed.

F-ERG

The F-ERG has two major components: a linear component at the same temporal frequency as the stimulus (fundamental harmonic, 1F) and a non-linear component at twice the stimulation frequency (second harmonic, 2F). The 2F may be dominant at low temporal frequencies (Baker and Hess, 1984). Many lines of evidence indicate that the 2F, unlike the 1F, component (which is probably of receptoral origin) reflects the activity of proximal retinal layers (Porciatti and Falsini, 1993). Current source density analysis in primate retinas disclosed several sources for the 2F, with at least one in the inner retina (Baker et al., 1988).

P-ERG

A steady-state P-ERG can be recorded in response to black–white pattern reversal modulation of checkerboards. Experimental evidence in cats and monkeys (Maffei and Fiorentini, 1986) indicates that the P-ERG is associated with the activity of inner retinal neurons (perhaps the ganglion cells themselves). In some clinical situations, the F-ERG and the P-ERG second harmonic dissociate and there are differences between generators for the second harmonic of the ERG evoked by modulation of luminance versus spatial contrast (Porciatti et al., 1989; Falsini et al., 1991).

P-ERG can be recorded in both primates and humans by red–green equiluminant patterns, alternated in counterphase (Porciatti et al., 1989; Falsini et al., 1991; Morrone et al., 1994a). The response displays characteristics comparable to those of retinal P-ganglion cells (see below), such as low-pass response function as the spatial and temporal frequency of the stimulus are varied, and a significant latency increase compared to responses evoked by luminance contrast. There is also evidence that the P-ERG to chromatic contrast is altered in glaucoma (Korth et al., 1994; Porciatti et al., 1997). When compared to the P-ERG elicited by luminance contrast, the chromatic P-ERG in glaucoma can be altered even to a greater extent, provided that the optimal spatiotemporal parameters are used. P-ERG abnormalities involve both amplitude and latency of the response (Porciatti et al., 1997).

The Parvo (P)- and Magno (M)-cellular pathways

Psychophysical studies have shown that luminance and chromatic modulations may allow differential assessment of the activity of the M- and P-cellular pathways, the two main cellular systems of the primate visual pathway. The P-system consists of tonic, wavelength-opponent retinal ganglion cells projecting to the P-cellular layers of the lateral geniculate nucleus (Lee, 1990). Most neurons belonging to this system receive antagonistic input from medium-wavelength-sensitive and long-wavelength-sensitive cones. The M-system is made up of phasic, nonopponent ganglion cells, which project to the M-cellular layers of the lateral geniculate nucleus. M-cellular ganglion cells receive combined input from medium- and long-wavelength-sensitive cones at both center and surround. Uniform field stimuli, whose luminance is varied periodically over time (i.e. the luminance flicker modulation) at relatively high temporal frequencies, are known to be optimal, though not specific, for evoking the response of the M-neurons. On the other hand, low temporal frequency, counterphase modulation of red and green fields, whose luminance has been matched by heterochromatic flicker photometry (i.e. the equiluminant chromatic flicker modulation), represents a strong stimulus for the P-cellular neurons, eliciting predominantly their activity. At both retinal and postretinal levels, sensitivity of the human visual system to luminance modulation is maximal at intermediate temporal frequencies (10–20 Hz) and falls-off at both low and high frequencies, whereas sensitivity to equiluminant chromatic modulation is maximal at low temporal frequencies (<5 Hz) and falls-off at frequencies greater than 5–10 Hz (Lee, 1990; Korth et al., 1994; Morrone et al., 1994a, b; Fiorentini et al., 1996; Porciatti et al., 1997).

Physiology

Magnitude and time course of RF_{onh} in humans

Figure 3 shows the group average time course of $RVel_{onh}$, $RVol_{onh}$, and RF_{onh} to photopic diffuse luminance flicker obtained from the temporal rim of the optic disk in a group of normal subjects (Riva et al., 2004a). Time constants of the increases (τ_i) and decreases (τ_d) of each LDF parameter after onset and offset of the stimulation were obtained using a two-parameter exponential function. For the $RVel_{onh}$, $RVol_{onh}$, and RF_{onh}, τ_i were 3.4, 12.7, and 9.1 s, respectively, and τ_d were 3.8, 9.8, and 7.5 s, respectively (Riva et al., 2005). RF_{onh} at onset of stimulation and after stimulation stop closely resembles that of cerebral blood flow in response to motor and visual stimulation (Krüger et al., 1999; Miller et al., 2001). Although the local vasodilatation reflected in $RVol_{onh}$ could be due to capillary recruitment, capillary dilatation, or venous dilatation, various considerations

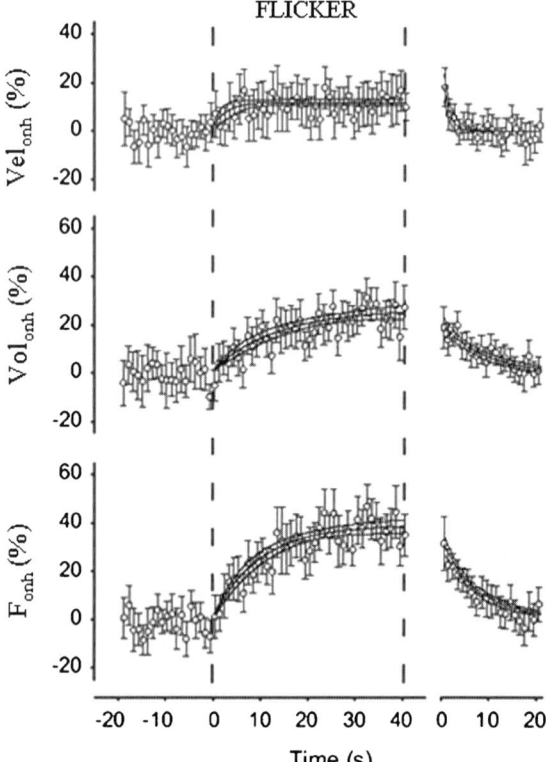

Fig. 3. Group average time course of $RVel_{onh}$, $RVol_{onh}$, and RF_{onh} to photopic diffuse luminance flicker from the temporal rim of the optic disk in a group of 15 normal volunteers. The time constants of the increases at onset and decreases at cessation of flicker are given in the text. Adapted with permission from Riva et al. (2004a).

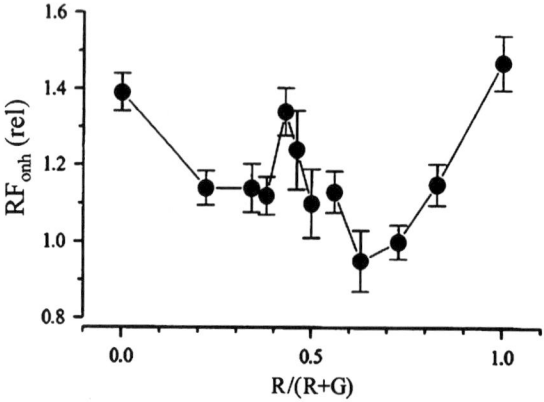

Fig. 4. RF_{onh} plotted as a function of stimulus color ratio (see text) for a 15 Hz flicker. Adapted with permission from Riva et al. (2001).

suggest a mechanism involving predominantly capillary dilatation (Riva et al., 2005).

Varying the parameters of the stimulus on RF_{onh}

Luminance versus chromatic modulation

RF_{onh} as a function of the color ratio $r\,[=R/(R+G)]$ for a 15 Hz flicker was largest at pure luminance ($r=0$ and 1), declined at mixed luminance and chromatic modulations, and reached a secondary maximum at $r=0.45$, the value of psychophysical equiluminance (Fig. 4) (Riva et al., 2001). Thus, in humans, an RF_{onh} can be evoked by heterochromatic flicker, modulated either in luminance or in chromatic equiluminance. Presumably, RF_{onh} is specific for luminance and chromatic modulation, similar to neural responses dominated by the M- and P-cellular activities, respectively.

Frequency

RF_{onh} versus flicker frequency (2–40 Hz) displayed the characteristics of a band-pass function, with a maximum at intermediate frequencies, for the luminance flicker stimulus and of a low-pass function for the equiluminant flicker stimulus (Fig. 5) (Riva et al., 2001).

Effect of stimulus area and site of measurement of RF_{onh}

RF_{onh} increases linearly with the area of an 8 Hz diffuse luminance field centered at the disk (Fig. 6) (Riva et al., 2005). The stimulation consisted of a circular field centered at the optic disk. A small response was detectable when the stimulation just covered the optic disk. The data suggest that in humans, RF_{onh} is generated mainly from the retinal area directly stimulated and therefore shows spatial summation properties, similarly to that observed for the pattern ERG (Hess and Baker, 1984).

Effect of pattern stimulation

Recordings of F_{onh} in response to checkerboard pattern reversed in counterphase at 8 Hz (check

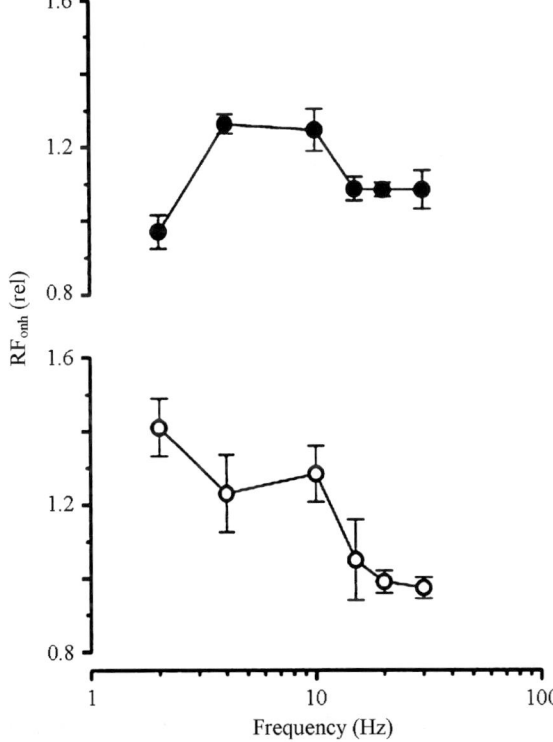

Fig. 5. RF_{onh} versus flicker frequency for luminance red–black flicker (top) and chromatic red–green flicker (bottom). Adapted with permission from Riva et al. (2001).

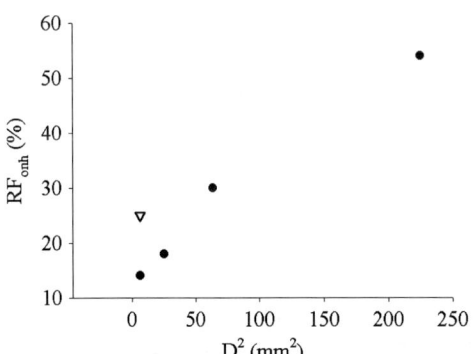

Fig. 6. Effect of increasing the area of the flicker stimulation on RF_{onh}. (●) The flicker field was centered at the ONH. (▽) The flicker field of the size of the optic disk was moved from the disk to the fovea. Adapted with permission from Riva et al. (2005).

size 32 min arc, $15° \times 15°$ field, approximately centered on the disk) were performed using the setup shown in Fig. 1. After a baseline measurement of F_{onh} at constant, uniform luminance, the black–white checkerboard pattern was presented for 1 min and then switched off to the constant uniform luminance (same mean luminance as the pattern). The experiment was repeated four times (Fig. 7). ERG was measured simultaneously with F_{onh} and the 2P component was determined. Average RF_{onh} ($\pm SD$) (%) was 38 ± 6%. After the stimulus was turned off, F_{onh} returned back to the baseline in about 10 s (Riva et al., 2005). The P-ERG data recorded simultaneously with F_{onh}, show a clear and reliable second harmonic component, with a temporal period of 60 ms (corresponding to the contrast reversal period) and a peak-to-peak amplitude of 1.9 μV (SD error 0.49 μV).

Neurovascular coupling in humans

Measurements in five normal observers (age between 28 and 61 years), in which F_{onh} was recorded simultaneously with the ERG in response to counterphased red (R) and green (G) flickering lights (30° field at the retina), revealed a significant correlation between RF_{onh} and the changes of the ERG components, particularly with the 2F harmonic, further supporting the presence of a coupling between RF_{onh} and inner retinal activity (Fig. 8) (Falsini et al., 2002). RF_{onh} and the ERG changes were documented as a function of (a) the frequency (2.3–64 Hz) of luminance and chromatic modulations, (b) the mean illuminance of 10 Hz luminance modulation (1–13 lux), and (c) the color ratio r of a 15 Hz chromatic modulation. The first harmonic 1F- and 2F-ERG amplitudes were determined. In both observers, RF_{onh} and 2F amplitude displayed similar frequency-dependent changes, for either luminance or chromatic modulations. Both RF_{onh} and 2F amplitude increased and then saturated with increasing mean illuminance of 10 Hz luminance modulation, and decreased with a similar functional shape as the 15 Hz chromatic modulation approached the R-G equiluminance value. In each observer, under specific experimental conditions (frequency response for the equiluminant red–green flicker, effect of mean illuminance for the luminance green flicker), RF_{onh} was positively correlated with corresponding 1F or 2F amplitudes.

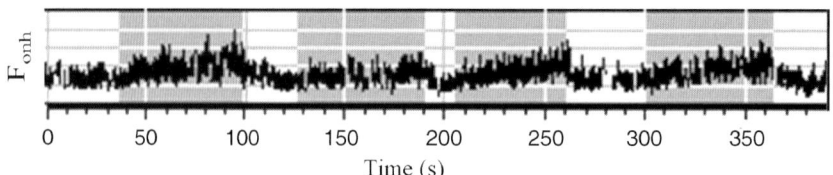

Fig. 7. F_{onh} in response to a checkerboard pattern. Check size: 36 min arc, field: 15° × 15°, frequency: 8 Hz. Grey zones indicate period with pattern on. Average ± SD (%) for RF_{onh} was 38 ± 6%. Adapted with permission from Riva et al. (2005).

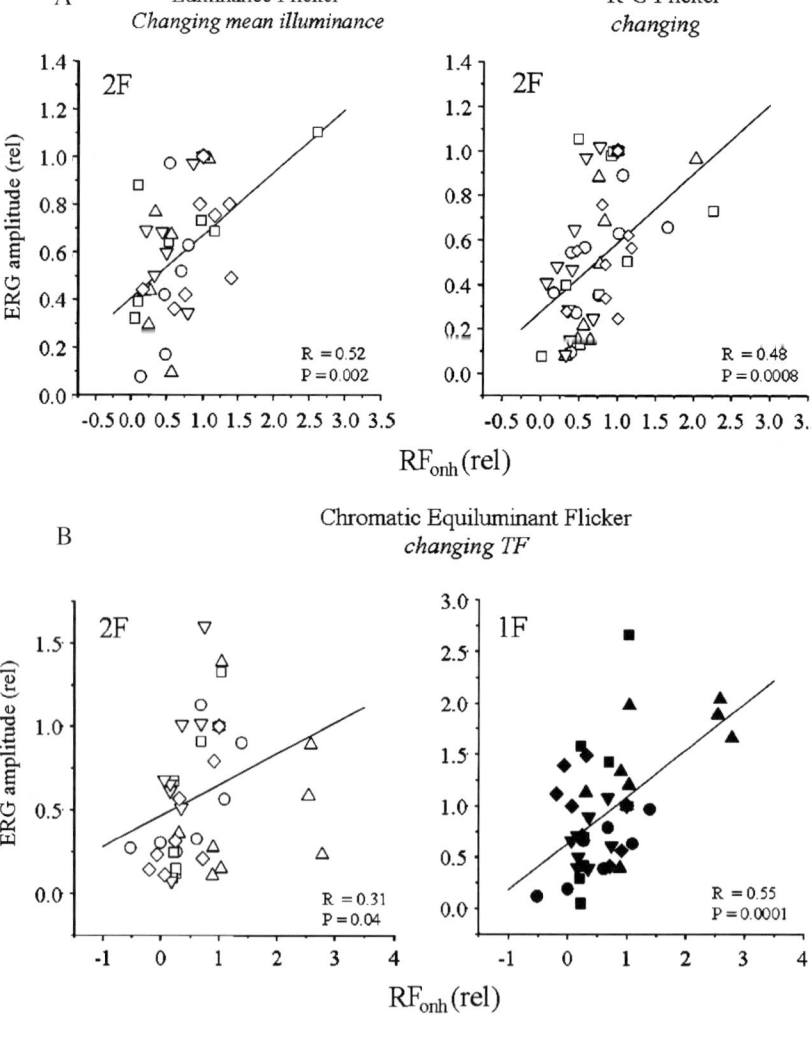

Fig. 8. Correlations between RF_{onh} and flicker ERG 1F or 2F components obtained from the normal subjects in whom RF_{onh} and flicker ERGs were recorded simultaneous. Adapted with permission from Falsini et al. (2002).

Clinical application

ERG response to heterochromatic flicker and pattern contrast reversal stimulation

Results in humans (Porciatti et al., 1987, 1989; Ghirlanda et al., 1991; Falsini et al., 1991, 1995; Porciatti and Falsini, 1993) showed that the 2F component is selectively altered (i.e. with normal 1F component) in diseases affecting primarily the inner retina, such as glaucoma, optic neuritis, or optic nerve compression (Porciatti and Falsini, 1993), and appeared to be highly sensitive to retinal vascular disorders, such as diabetic retinopathy (Ghirlanda et al., 1991) and branch occlusion of the central retinal artery or vein (Porciatti et al., 1989; Falsini et al., 1995). Different clinical studies on cohorts of patients with hypertension (OHT) or glaucoma (Porciatti et al., 1987; Falsini et al., 1991; Colotto et al., 1995) have demonstrated that the P-ERG, a retinal potential requiring, for its generation, the functional integrity of the inner retina, is altered in a substantial proportion of OHT eyes and in the great majority of eyes with early glaucoma. Diagnostic accuracy of this test in detecting glaucomatous damage to the optic nerve is relatively higher than that observed for VEPs and psychophysical tests, such as conventional automated perimetry, blue-on-yellow perimetry, and contrast sensitivity. One possible reason is that the P-ERG, unlike other electrophysiological and psychophysical tests, can provide a direct and objective probe of inner retinal function, and does not reflect possible compensatory processes of peripheral neural damage occurring at higher levels along the visual pathways. It has been also demonstrated that the P-ERG can be of value in the clinical follow-up of glaucoma (Falsini et al., 1992). In a 24 months follow-up study on OHT and glaucoma patients under topical beta-blocker treatment (Falsini et al., 1992), progressive P-ERG amplitude losses were associated with poor intraocular pressure (IOP) control. By contrast, a recovery or stabilization of the response amplitude was found in patients with satisfactory IOP control. No visual field changes were observed in patients during the follow-up. These data suggest the possible use of the P-ERG to estimate the "target" IOP in an individual patient undergoing medical treatment. In agreement with this suggestion, a predictive value of the P-ERG response for the development of early visual field defects has been found in a 24-month follow-up of treated OHT patients (Colotto et al., 1995), confirming previous results obtained by others (Pfeiffer et al., 1993). More recently, subclinical functional damage in OHT patients, as determined by the P-ERG, was compared with early anatomical abnormalities of the optic disc, found in the same patients by using confocal scanning laser ophthalmoscopy (Salgarello et al., 1999). All patients had normal conventional white-on-white perimetry and optic disc appearance at biomicroscopic examination. P-ERG amplitude was found to be significantly correlated with the shape of the optic disc cup, with increasing losses as the cup shape approached abnormal values. These data provided evidence that, already in the OHT stage, a parallel involvement of both function and morphology may occur, and stressed the value of combined structural and functional analysis in detection of early glaucomatous damage.

RF_{onh} in OHT and glaucoma patients

In a previous work, from which some of the following text is borrowed with permission from the publisher, RF_{onh} to luminance (l-fl) and chromatic (c-fl) flicker was evaluated in OHT and early glaucoma (EOAG) patients (Riva et al., 2004b). These stimuli generate neural activity dominated by M- and P-cellular system, respectively. OHT, POAG patients, and normal controls were selected based on their ability to maintain adequate head stability, excellent target fixation, a pupil dilatation >6 mm or more and no cataract. Fifteen early EOAG, 27 OHT, and 12 age-matched control subjects were examined. RF_{onh} was determined in response to a 15 Hz green luminance flicker (30° field). Two to three responses were collected at different temporal sites of the optic disk. The highest RF_{onh} value was used.

As compared to controls, OHT and EOAG patients showed a decrease ($p<0.01$) in both mean baseline F_{onh} and RF_{onh} ($p<0.001$). In individual patients, both F_{onh} and RF_{onh} were positively

($r = 0.44$) correlated ($p<0.01$) with the neuroretinal rim area ($r = 0.44$). RF_{onh} showed a weak, borderline significant ($p<0.10$) correlation ($r = 0.30$) with pattern ERG amplitudes and perimetric mean deviation. From this study, it was concluded that RF_{onh} is abnormally reduced in OHT and EOAG patients, indicating an impairment of luminance (M-cellular)-mediated vasoactivity. Flicker-evoked F_{onh} changes appear to be only weakly correlated with functional indicators of neural damage, suggesting that a loss of neurovascular coupling of the optic nerve head may occur independently of neural activity loss early in the disease process.

Discussion

FLDF and neurovascular coupling in humans

Previous studies in animals have demonstrated a tight correlation between local RF_{onh} and neuronal activity response, the latter obtained from the flicker-induced changes in the electrical signal (Riva and Buerk, 1998) and K^+ ion concentration (Buerk et al., 1995), both measured at the surface of the optic disk. In humans, luminance and heterochromatic flicker stimulation generate a RF_{onh} that displays characteristics comparable to those of ganglion cell spiking activity evoked by the same stimuli, when RF_{onh} is plotted as a function of frequency, modulation depth, and relative chromaticity (Riva et al., 2001). Although these findings had suggested the existence of a coupling between vaso- and neuronal activity, the exact relationship between both physiological activities remained to be established. With this goal in mind, and since noninvasive methods to measure local activity at the optic nerve rim are not yet available, the 1F and 2F components of the flicker ERG and the simultaneously recorded RF_{onh} were correlated. Although the pattern ERG, which is generated mainly by ganglion cell activity, would appear to be the most appropriate stimulus to correlate activity and RF_{onh} in humans, the components of the F-ERG were chosen because of the substantial amount of data available from humans. In addition, the hypothesis of an association between RF_{onh} and the 1F and 2F harmonics of the flicker ERG was justified based on the origin of these components of retinal neural activity.

The results of investigations in the normal human eye (Falsini et al., 2002) indicated that, under specific experimental conditions, the stimulus-evoked changes in RF_{onh} are similar and significantly correlated to the changes in the flicker ERG harmonic components. In particular, the associations were the most significant (a) when (Fig. 8) the temporal frequency of the stimulus was changed in the condition of heterochromatic red–green equiluminant flicker; (b) when the mean illuminance was varied for a green illuminance flicker; and (c) when the color ratio, r, was varied for the heterochromatic R-G flicker. In case (a), RF_{onh} was correlated with both 1F and 2F amplitudes. In the others, RF_{onh} was correlated only with the 2F amplitude.

The significant correlation found between RF_{onh} and each of 1F and 2F amplitudes for the case of R-G equiluminant flicker responses recorded as a function of temporal frequency, as well as between RF_{onh} and 2F amplitudes for R-G flicker responses as a function of r, suggests a coupling arising from the pooled activity of the middle- and long-wavelength sensitive cones. Given the retinal origin of the flicker ERG 2F, the present findings provide support to the hypothesis that blood flow changes recorded at the ONH in response to flicker stimulation parallel the corresponding changes in the neural function of the inner retina. Under some experimental conditions (chromatic flicker modulated at different temporal frequencies), however, RF_{onh} changes may also be similar to changes in the 1F amplitude, implying common properties of optic nerve vaso- and neural activity changes in the middle retina (i.e. ON and OFF bipolar cells).

In the case of pure luminance flicker experiments, when changing temporal frequency, no significant association was found between RF_{onh} and both components' amplitudes. A possible physiological basis for this finding is that the generators underlying RF_{onh} and ERG response to luminance flicker may differ depending on flicker frequency. Multiple generators for both responses, whose relative contribution changes according to stimulus frequency, may obscure the correlation when measurements are obtained over a range of frequencies. On the other hand, it is possible that the blood flow and neural responses to a specific stimulus frequency, i.e. the luminance flicker at 10 Hz, share common generators in the inner retina, thus revealing the association between RF_{onh} and ERG 2F amplitude. In contrast, the R-G heterochromatic flicker reveals a significant coupling between RF_{onh} and both 1F and 2F amplitudes, suggesting a contribution from both retinal layers to the coupling between the neural and vascular responses.

Although several previous studies have suggested a neurovascular coupling in the human retina (Scheiner et al., 1994; Riva et al., 1995), direct evidence of it has not been provided in the past. The observed correlations may not directly prove an effect of retinal neural activity on blood flow measured at the optic disk. It is however reasonable to suggest that the pooled response of neural generators underlying the 1F and 2F components may induce a vaso-active mechanism resulting in a corresponding blood flow change. Putative mediators underlying this neurovascular response have been discussed in detail in previous reports (Buerk et al., 1995, 1996; Kondo et al., 1997; Riva and Buerk, 1998), the most prominent being nitric oxide (NO). The present findings support the hypothesis that the strength of neurovascular coupling may be dependent upon the type of flicker stimulus. From the data shown in Fig. 8 it can be concluded that both luminance and chromatic flicker may be appropriate to study the coupling between RF_{onh} and 2F amplitude, whereas the chromatic equiluminant stimulus appears to be more suitable for investigating the association between RF_{onh} and 1F.

Comments on clinical application of FLDF in glaucoma

The data reviewed here demonstrated that RF_{onh} is diminished in both OHT and EOAG patients compared to normal controls. The group-averaged time course of the flicker-induced increase in F_{onh} in EOAG was not significantly different from that found in the normal subjects. Presumably, in the EOAG patients the flicker-induced vasodilatation in the apparently healthy rim tissue of the anterior optic nerve has a normal temporal evolution.

Our results strongly suggest that the decrease in RF_{onh} reflects an altered ganglion cell activity response to the flicker. Nevertheless, other possible explanations of this decrease need to be addressed to. One possibility could be that resting blood flow in the rim tissue of the OHT and EOAG patients could be too low in comparison with the amount of tissue it supplies to satisfy the metabolic needs. Consequently, it would be unable to fully respond to the additional stress evoked by the flicker stimulation. However, the lack of correlation between RF_{onh} and F_{onh} (Riva et al., 2004b) goes against this hypothesis. Another possibility is that in the EOAG group, because the nerve fiber layer is thinner than in the normal control and OHT groups, the choroid would contribute more to the LDF signal. This is unlikely since the markedly higher flow velocities in the choriocapillaris did not result in a significantly greater $RVel_{onh}$ at rest in the EOAG compared to the control group. Furthermore, RF_{onh} was also found to be reduced in the OHT group, although the nerve fiber layer thickness in this group was not significantly different from the control group.

Studies in the human eye using a range of flicker frequencies and modulations suggest that the RF_{onh} loss observed in our patients when using a 15 Hz luminance flicker occurs predominantly at the level of the ganglion cell M-cellular pathway (Riva et al., 2001; Falsini et al., 2002). This hypothesis agrees with the large body of anatomical evidence indicating that, in early glaucoma, large ganglion cells, subserving primarily the M-cellular pathway, are selectively or predominantly damaged (Quigley et al., 1989; Glovinski et al., 1991; Kerrigan-Baumrind et al., 2000).

Patients' RF_{onh}, in contrast, were poorly correlated with functional indicators of early damage, such as pattern ERG and Humphrey perimetric indices. Thus, the data obtained from the OHT patients show that, in some of these patients, the pattern ERG was normal whereas RF_{onh} was substantially diminished. In other OHT patients, the opposite was found. This poor correlation may be related to the inherent variability of the techniques employed in this study, or, alternatively, may reflect an altered neurovascular coupling in early stages of diseases. Alteration of the neurovascular coupling could occur at various points of the chain of events coupling the activity to the local vasodilatation (Zonta et al., 2003), such as the flicker-induced release of NO, K^+, or other substances (Buerk et al., 1996; Kondo et al., 1997; Zonta et al., 2003), the glutamate release from axonal terminals, the activation of astrocytes and subsequent release of vasodilating products and possibly others. Clearly, future work should be aimed at elucidating this important question.

RF_{onh} was positively correlated with neuro-retinal rim area when the two groups of patients were pooled. In EOAG patients, although this correlation did not reach significance, possibly because of the low number of patients, both neuro-retinal rim area and RF_{onh} were smaller in the average than the corresponding values in OHT and normal controls. It has been previously shown that a reduction in neuro-retinal rim area, together with an increase in cup/disk area ratio and cup shape measure, may be highly accurate in detecting early glaucomatous damage (Uchida et al., 1996; Blumenthal and Weinreb, 2001). These results indicate that a reduction in ONH vaso-activity assessed by flicker stimulation is associated with early loss of nerve fiber layer in eyes with EOAG.

A number of studies have compared F_{onh} measurements obtained in normal subjects and in glaucoma patients. Although the results vary considerably between studies (Piltz-Seymour et al., 2001), F_{onh} has been found to be reduced in primary open-angle glaucoma patients compared to normal controls (Michelson et al., 1996; Nicolela et al., 1996; Hafez et al., 2003; Riva et al., 2004b). Furthermore, F_{onh} also tended to be reduced in the OHT patients (Riva et al., 2004b).

The decrease of F_{onh} in glaucoma has been interpreted by a number of investigators as evidence of an actual reduction in blood flow, although some caution about the validity of this interpretation has been expressed in view of the limitations of the LDF technique. As discussed elsewhere (Riva and Petrig, 2003), comparisons between F_{onh} values in terms of actual blood flow are strictly valid only if the scattering properties of the tissue from which F_{onh} is measured are identical. This is due to the fact that the DSPS depends not only upon the number and velocity of RBCs in the sampled volume, but also on the optical characteristics (i.e. the absorption and scattering) of the tissue (nonmoving scatterers) sampled. In general, increased tissue scattering broadens the power spectrum, causing an artificial increase in the measured flow. To our present knowledge, it is not clear yet how the scattering of light by the glaucomatous rim tissue differs from that of the healthy eye and therefore how it may affect F_{onh}.

In contrast to F_{onh}, RF_{onh} is not affected by the scattering properties of the tissue. Changes in light scattering may be expected during increased neural activity, but these changes are too fast and probably too small to have a notable effect on the LDF spectrum and F_{onh} (Gratton and Fabiani, 2001). Furthermore, the changes in F_{onh} are proportional to the actual flow changes as they are within the range of linearity of the LDF technique (Riva et al., 2000). Both of these aspects make the LDF technique most appropriate to investigate the regulation of F_{onh} in response to various physiological stimuli.

Using flicker to investigate this regulation offers additional advantages over previously used physiological stimuli such as decreases in ocular perfusion pressure (OPP) achieved by increasing the IOP with a suction cup (Pillunat et al., 1997; Riva et al., 1997), increases in OPP by means of isometric exercises (Movaffaghy et al., 1998), and the breathing of various gas mixtures (Harris et al., 1996). Flicker is not invasive and a more physiological stimulus since modulation of light exposure is the most natural stimulus for the visual system, leaving also the systemic circulation unperturbed.

The LDF measurements were obtained by directing the probing laser beam at the temporal site of the neuro-retinal rim of the optic disk. In humans, as in monkeys (Petrig et al., 1999), LDF is probably predominantly sensitive to blood flow changes occurring only in the most superficial layers (supplied by the retinal circulation) of the optic nerve head. However, regardless of which layer of the optic nerve head circulation was in fact sampled by the LDF technique, our data may be considered as reflecting the blood flow changes (elicited by a visual stimulus) in a limited area of the microcirculatory district belonging to the neuro-retinal rim, a key anatomical component of the optic nerve (Jonas and Budde, 2000). Although we cannot speculate about the possible implications of the present data for the pathophysiology of glaucomatous optic neuropathy, the finding of an abnormal flicker-evoked flow regulation in the superficial layers of the optic nerve recorded in both OHT and EOAG eyes clearly merits further investigation.

In conclusion, this LDF study demonstrates a decreased blood flow response to flicker stimulation measured at the neuro-retinal rim of the optic disk in both OHT and EOAG patients. This decrease, which in EOAG is associated with early loss of nerve fiber layer, could be due to a decrease in the activity of the M-cellular pathway in response to flicker. Furthermore, the relation of blood flow changes with visual function data suggests that neural activity and vascular response can be independently altered early in the disease process.

Conclusions and futures directions

Evidence from experimental and clinical studies has been accumulated so far indicating that retinal and optic nerve head microcirculation is modulated by visually evoked neural activity. It has been shown, indeed, that functional hyperemia measured at the neuro-retinal rim of the optic nerve is tightly coupled with retinal activity, mostly generated in inner retinal layers. The human optic nerve may represent an open "window" to the brain where neurovascular and neurometabolic coupling can be noninvasively investigated by optical techniques such as LDF and activity-induced changes in intrinsic optic signals. As recently discussed (Girouard and Iadecola, 2006), the functional hyperemia involves the coordinated interaction of neurons, glia, and vascular cells. Recent evidence in the central nervous system suggests that various vaso-active agents are implicated in functional hyperemia. These have been reviewed in details in a recent paper (Iadecola and Nedergaard, 2007). They include adenosine, nitric oxide, and arachidonic acid metabolites.

OHT and glaucoma may disrupt the neurovascular coupling, indicating that novel clinical protocols aimed at evaluating both visually induced functional hyperemia and activity changes are potentially valuable either for early detection of neuro-ophthalmic disorders or for better understanding of their pathophysiology.

Clearly, much remains to be investigated in this area. The mediators and modulators of the neurovascular and neurometabolic coupling need to be fully elucidated. In normal subjects, the full set of visual stimuli that, being parametrically modulated in their characteristics (i.e. space, time, chromaticity, adaptation state), are able to reveal the neurovascular coupling at the optic nerve needs to be established. Finally, more systematic clinical studies in different ophthalmic disorders will reveal the role of the current proposed approaches in the diagnosis and management of optic nerve diseases.

Acknowledgements

This study was supported by research grant from Fondazione Cassa di Risparmio, in Bologna (to C.E.R. and B.F.) and Fondi di Ateneo, ex 60% (to B.F.).

References

Baker, C.L. and Hess, R.F. (1984) Linear and nonlinear components of human electroretinogram. J. Neurophysiol., 51(5): 952–967.

Baker, C.L., Jr., Hess, R.R., Olsen, B.T. and Zrenner, E. (1988) Current source density analysis of linear and non-linear components of the primate electroretinogram. J. Physiol., 407: 155–176.

Blumenthal, E.Z. and Weinreb, R.N. (2001) Assessment of the retinal nerve fiber layer in clinical trials of glaucoma neuroprotection. Surv. Ophthalmol., 45(Suppl 3): 305–310.

Bonner, R.F. and Nossal, R. (1990) Principles of laser Doppler flowmetry. In: Shepherd A.P. and Oberg P.A. (Eds.), Laser-Doppler Blood Flowmetry. Kluwer Academic Publishers, Boston, pp. 57–72.

Buerk, D.G., Riva, C.E. and Cranstoun, S.D. (1995) Frequency and luminance-dependent blood flow and K^+ ion changes during flicker stimuli in cat optic nerve head. Invest. Ophthalmol. Vis. Sci., 36: 2216–2227.

Buerk, D.G., Riva, C.E. and Cranstoun, S.D. (1996) Nitric oxide has a vasodilatory role in cat optic nerve head during flicker stimuli. Microvasc. Res., 52: 13–26.

Colotto, A., Salgarello, T., Giudiceandrea, A., De Luca, L.A., Coppè, A., Buzzonetti, L. and Falsini, B. (1995) Pattern electroretinogram in treated ocular hypertension: a cross-sectional study after timolol maleate therapy. Ophthalmic Res., 27: 168–177.

Falsini, B., Colotto, A. and Porciatti, V. (1991) Macular flicker- and pattern-ERGs are differently affected in ocular hypertension and glaucoma. Clin. Vision Sci., 6: p. 429.

Falsini, B., Colotto, A., Porciatti, V. and Porrello, G. (1992) Follow-up study with pattern erg in ocular hypertension and glaucoma patients under timolol maleate treatment. Clin. Vision Sci., 7: 341–347.

Falsini, B., Porciatti, V., Fadda, A., Merendino, E., Iarossi, G. and Cermola, S. (1995) The first and second harmonics of the macular flicker electroretinogram: differential effects of retinal diseases. Doc. Ophthalmol., 90: 157–167.

Falsini, B., Riva, C.E. and Logean, E. (2002) Flicker-evoked changes in human optic nerve blood flow: relationship with retinal neural activity. Invest. Ophthalmol. Vis. Sci., 43: 2721–2726.

Fiorentini, A., Porciatti, V., Morrone, M.C. and Burr, D.C. (1996) Visual ageing: unspecific decline of the responses to luminance and colour. Vision Res., 36.

Ghirlanda, G., Di Leo, M.A., Falsini, B., Porciatti, V., Marietti, G. and Greco, A.V. (1991) Detection of inner retina dysfunction by steady-state focal electroretinogram pattern and flicker in early IDDM. Diabetes, 40: 1122–1127.

Girouard, H. and Iadecola, C. (2006) Neurovascular coupling in the normal brain and in hypertension, stroke, and Alzheimer disease. J. Appl. Physiol., 100: 328–335.

Glovinski, Y., Quigley, H.A. and Dunkelberger, G.R. (1991) Retinal ganglion cells loss is size dependent in experimental glaucoma. Invest. Ophthalmol. Vis. Sci., 32: 484–491.

Gratton, G. and Fabiani, M. (2001) The event-related optical signal: a new tool for studying brain function. Int. J. Psychophysiol., 42: 109–121.

Hafez, A.S., Bizzarro, R.L. and Lesk, M.R. (2003) Evaluation of optic nerve head and peripapillary retinal blood flow in glaucoma patients, ocular hypertension, and normal subjects. Am. J. Ophthalmol., 136: 1022–1031.

Harris, A., Anderson, D.R., Pillunat, L., Joos, K., Knighton, R.W., Kagemann, L. and Martin, B.J. (1996) Laser Doppler flowmetry measurement of changes in human optic nerve head blood flow in response to blood gas perturbations. J. Glaucoma, 5: 258–265.

Hess, R.F. and Baker, C.L. (1984) Human pattern-evoked electroretinogram. J. Neurophysiol., 51(5): 939–951.

Iadecola, C. and Nedergaard, M. (2007) Glial regulation of the cerebral microvasculature. Nat. Neurosci., 10: 1369–1376.

Jonas, J.B. and Budde, W.M. (2000) Diagnosis and pathogenesis of glaucomatous optic neuropathy: morphological aspects. Prog. Retin. Eye Res., 19: 1–40.

Kerrigan-Baumrind, L.A., Quigley, H.A., Pease, M.E., Kerrigan, D.F. and Mitchell, R.S. (2000) Number of ganglion cells in glaucoma eyes compared with threshold visual field tests in the same persons. Invest. Ophthalmol. Vis. Sci., 41: 741–748.

Kondo, M., Wang, L. and Bill, A. (1997) The role of nitric oxide in hyperaemic response to flicker in the retina and optic nerve in cats. Acta Ophthalmol. Scand., 75: 232–235.

Korth, M., Nguyen, N.X., Horn, F. and Martus, P. (1994) Scotopic threshold response and scotopic PII in glaucoma. Invest. Ophthalmol. Vis. Sci., 35: 619–625.

Krüger, G., Kastrup, A., Atsushi, T. and Glover, G. (1999) Simultaneous monitoring of dynamic changes in cerebral blood flow and oxygenation during sustained activation of the human visual cortex. Neuroreport, 10: 2939–2943.

Lee, B.B. (1990) Luminance and chromatic modulation sensitivity of macaque ganglion cells and human observers. J. Opt. Soc. Am. A, 7: 2223–2236.

Logean, E., Geiser, M.H. and Riva, C.E. (2005) Laser Doppler instrument to investigate retinal neural activity-induced changes in optic nerve head blood flow. Opt. Lasers Eng., 43: 591–602.

Maffei L. and Fiorentini A. (Eds.), (1986). Generators Sources of the Pattern ERG in Man and Animals. Liss, New York.

Michelson, G., Langhans, M.J. and Groh, M.J.M. (1996) Perfusion of the juxtapapillary retina and the neuroretinal rim area in primary open angle glaucoma. J. Glaucoma, 5: 91–98.

Miller, K.L., Luh, W.M., Liu, T.T., Martinez, A., Obata, T., Wong, E.C., Frank, L.R. and Buxton, R.B. (2001) Nonlinear temporal dynamics of the cerebral blood flow response. Hum. Brain Mapp., 13: 1–12.

Morrone, C., Fiorentini, A. and Bisti, S. (1994a) Pattern-reversal electroretinogram in response to chromatic stimuli: II. Monkey. Vis. Neurosci., 11: 873–884.

Morrone, C., Porciatti, V., Fiorentini, A. and Burr, D.C. (1994b) Pattern-reversal electroretinogram in response to chromatic stimuli: I. Humans. Vis. Neurosci., 11: 861–871.

Movaffaghy, A., Chamot, S.R., Petrig, B.L. and Riva, C.E. (1998) Blood flow in the human optic nerve head during isometric exercise. Exp. Eye Res., 67: 561–568.

Nicolela, M.T., Hnik, P. and Drance, S.M. (1996) Scanning laser Doppler flowmeter study of retinal and optic disk blood flow in glaucomatous patients. Am. J. Ophthalmol., 122: 775–783.

Petrig, B.L. and Riva, C.E. (1988) Retinal laser-Doppler velocimetry: towards its computer assisted clinical use. Appl. Opt., 27: 1126–1134.

Petrig, B.L., Riva, C.E. and Hayreh, S.S. (1999) Laser Doppler flowmetry and optic nerve head blood flow. Am. J. Ophthalmol., 127: 413–425.

Pfeiffer, N., Tillmon, B. and Bach, M. (1993) Predictive value of the pattern electroretinogram in high-risk ocular hypertension. Invest. Ophthalmol. Vis. Sci., 34: 1710–1715.

Pillunat, L.E., Anderson, D.R., Knighton, R.W., Joos, K.M. and Feuer, W.J. (1997) Autoregulation of human optic nerve head circulation in response to increased intraocular pressure. Exp. Eye Res., 64: 737–744.

Piltz-Seymour, J.R., Grunwald, J.E., Hariprasad, S.M. and Dupont, J. (2001) Optic nerve blood flow is diminished in eyes of primary open-angle glaucoma suspects. Am. J. Ophthalmol., 132: 63–69.

Porciatti, V., di Bartolo, E., Nardi, M.M. and Fiorentini, A. (1997) Responses to chromatic and luminance contrast in glaucoma: a psychophysical and electrophysiological study. Vision Res., 37(14): 1975–1987.

Porciatti, V. and Falsini, B. (1993) Inner retina contribution to the flicker electroretinogram: a comparison with the pattern electroretinogram. Clin. Vision Sci., 8: 435–447.

Porciatti, V., Falsini, B., Brunori, S., Colotto, A. and Morreti, G. (1987) Pattern electroretinogram as a function of spatial frequency in ocular hypertension and early glaucoma. Doc. Ophthalmol., 65: 349–355.

Porciatti, V., Falsini, B., Fadda, A. and Bolzani, R. (1989) Steady-state analysis of the focal ERG to pattern and flicker: relationship between ERG components and retinal pathology. Clin. Vision Sci., 4: 323–332.

Quigley, H.A., Dunkelberger, G.R. and Green, W.R. (1989) Retinal ganglion cell atrophy correlated with automated perimetry in human eyes with glaucoma. Am. J. Ophthalmol., 15: 453–464.

Ransom, B.R. and Orkand, R.K. (1996) Glial–neuronal interactions in non-synaptic areas of the brain: studies in the optic nerve. Trends Neurosci., 19: 352–358.

Riva, C.E. and Buerk, D.G. (1998) Dynamic coupling of blood flow to function and metabolism in the optic nerve head. Neuro-ophthalmology, 20: 45–54.

Riva, C.E., Falsini, B. and Logean, E. (2001) Flicker-evoked responses of human optic nerve head blood flow: luminance versus chromatic modulation. Invest. Ophthalmol. Vis. Sci., 42: 756–762.

Riva, C.E., Harino, S., Petrig, B.L. and Shonat, R.D. (1992) Laser Doppler flowmetry in the optic nerve. Exp. Eye Res., 55: 499–506.

Riva, C.E., Harino, S., Shonat, R.D. and Petrig, B.L. (1991) Flicker evoked increase in optic nerve head blood flow in anesthetized cats. Neurosci. Lett., 128: 291–296.

Riva, C.E., Hero, M., Titzé, P. and Petrig, B.L. (1997) Autoregulation of human optic nerve head blood flow in response to acute changes in ocular perfusion pressure. Graefe's Arch. Clin. Exp. Ophthalmol., 235: 618–626.

Riva, C.E., Logean, E. and Falsini, B. (2004a) Temporal dynamics and magnitude of the blood flow response at the optic disk in normal subjects during functional retinal flicker-stimulation. Neurosci. Lett., 356: 75–78.

Riva, C.E., Logean, E. and Falsini, B. (2005) Visually evoked hemodynamical response and assessment of neurovascular coupling in the optic nerve and retina. Prog. Retin. Eye Res., 24: 183–215.

Riva, C.E., Mendel, M.J. and Petrig, B.L. (1995) Flicker-induced optic nerve blood flow change. In: Ocular Blood Flow. New Insights into the Pathogenesis of Ocular Diseases. Karger, Basel, pp. 128–137.

Riva, C.E. and Petrig, B.L. (1997) Laser Doppler flowmetry in the optic nerve head. In: Drance S.M. (Ed.), Vascular Risk Factors and Neuroprotection in Glaucoma — Update 1996. Kugler Publications, Amsterdam/New York, pp 43–55.

Riva, C.E. and Petrig, B.L. (2003) Laser Doppler techniques in ophthamology — principles and applications. In: Fankhauser F. and Kwasniewska S. (Eds.), Lasers in Ophthalmology — Basic, Diagnostic and Surgical Aspects. Kugler Publications, The Hague, The Netherlands, pp. 51–59.

Riva, C.E., Petrig, B.L., Falsini, B. and Logean, E. (2000) Neuro-vascular coupling at the optic nerve head studied by laser Doppler flowmetry. OSA TOPS Vision Sci. Appl., 35: 64–76.

Riva, C.E., Salgarello, T., Logean, E., Colotto, A., Galan, E. and Falsini, B. (2004b) Flicker-evoked response measured at the optic disk rim is reduced in ocular hypertension and early glaucoma. Invest. Ophthalmol. Vis. Sci., 45: 3662–3668.

Roy, C.S. and Sherrington, C.S. (1890) On the regulation of the blood-supply of the brain. J. Physiol., 11: 85–108.

Salgarello, T., Colotto, A., Falsini, B., Buzzonetti, L., Cesari, L., Iarossi, G. and Scullica, L. (1999) Correlation of pattern electroretinogram with optic disc cup shape in ocular hypertension. Invest. Ophthalmol. Vis. Sci., 40: 1989–1997.

Scheiner, A.J., Riva, C.E., Kazahaya, K. and Petrig, B.L. (1994) Effect of flicker on macular blood flow assessed by the blue field simulation technique. Invest. Ophthalmol. Vis. Sci., 35: 3436–3441.

Uchida, H., Brigatti, L. and Caprioli, J. (1996) Detection of structural damage from glaucoma with confocal laser image analysis. Invest. Ophthalmol. Vis. Sci., 37: 2393–2401.

Zonta, M., Angulo, M.C., Gobbo, S., Rosengarten, B., Hossmann, K.A., Pozzan, T. and Carmignoto, G. (2003) Neuron-to-astrocyte signaling is central to the dynamic control of brain microcirculation. Nat. Neurosci., 6: 43–50.

CHAPTER 12

Advances in neuroimaging of the visual pathways and their use in glaucoma

Francesco Giuseppe Garaci[1,2,*], Valeria Cozzolino[1], Carlo Nucci[3], Fabrizio Gaudiello[1], Andrea Ludovici[1], Tommaso Lupattelli[4], Roberto Floris[1] and Giovanni Simonetti[1]

[1]*Department of Diagnostic Imaging and Interventional Radiology, University of Rome "Tor Vergata," V.le Oxford 81, Rome, Italy*
[2]*IRCCS San Raffaele Pisana, Via della Pisana 235, Rome, Italy*
[3]*Physiopathological Optics, Department of Biopathology, University of Rome Tor Vergata, Rome, Italy*
[4]*Interventional Radiology Department, IRCCS Multimedica, Sesto San Giovanni, Milan, Italy*

Abstract: Recently developed neuroimaging techniques such as diffusion tensor (DT) magnetic resonance (MR) imaging, functional MR imaging (fMRI), and MR spectroscopy can be used to evaluate the microstructural integrity of white-matter fibers and the functional activity of gray matter. They have been widely employed to investigate various diseases of the central nervous system, and they can be useful tools for assessing the integrity and functional connections of the visual pathways and areas that play key roles in glaucoma. In vivo degeneration of the optic nerves can be noninvasively demonstrated by DT MR imaging. DT fiber tractography provides valuable information on the axonal density of postgeniculate fibers (optic radiation), and fMRI studies of patients with primary open-angle glaucoma (POAG) have demonstrated alterations involving the human visual cortex that are consistent with clinically documented losses of visual function. This article reviews some of the more recent data supporting the use of MR imaging techniques as reliable, noninvasive tools for monitoring the progression of human glaucoma.

Keywords: MRI; diffusion tensor imaging; functional MR imaging; tractography; glaucoma

Introduction

Recently developed neuroimaging techniques such as diffusion tensor (DT) magnetic resonance (MR) imaging and functional MR imaging (fMRI) can be used to evaluate the microstructural integrity of white-matter fibers and the functional activity of gray matter. Their introduction has allowed a reexploration of the normal anatomy of white-matter tracts in the living human brain and the elaboration of connectional models of brain function. During the last 10 years, these techniques have also been used for the in vivo study of a variety of brain pathologies. First used in the investigation of ischemic stroke, they are now becoming increasingly important in the evaluation of intracranial neoplasms, inflammatory disorders, developmental anomalies of the central nervous system (CNS), and neurodegenerative diseases.

Both techniques can also be utilized to investigate the integrity and functional connections of the visual pathways, including areas that play key

*Corresponding author. Tel.: +39 06 2090 2401;
Fax: +39 06 2090 2404; E-mail: garaci@gmail.com

roles in glaucoma: pre- and postgeniculate white-matter fibers, the lateral geniculate nuclei (LGNs), and various areas of the visual cortex. This article provides an overview of some of the newer MRI techniques being used in neuro-ophthalmology and their specific applications in the study of patients with glaucoma.

Conventional MR imaging and the visual pathways

The macroscopic anatomy of the visual system can be demonstrated and assessed in vivo with conventional MR imaging. The optic nerve is a white-matter tract that is sheathed by the leptomeninges and dura mater, and its subarachnoid space is continuous with the intracranial subarachnoid space. On oblique coronal MR images, the intraorbital segment of the optic nerve ranges in thickness from 3.1 (anteriorly) to 2.5 mm (posteriorly), whereas the mean dural diameter measures between 5.1 (anteriorly) and 3.8 mm (posteriorly). The width of the subarachnoid space between the pial and dural sheaths ranges from 0.4 to 0.6 mm.

The optic nerve itself can be divided into intraocular, intraorbital, intracanalicular, and intracranial segments. The intraorbital tract, which is approximately 3 cm long, follows a sinuous course in both the horizontal and vertical planes. The intracanalicular portion is about 8 mm long. It passes through the optic canal (the channel at the apex of the orbital cavity) and the anterior end of the optic groove, which lies just below the lesser wing of the sphenoid bone. The ophthalmic artery runs through the same canal. The two optic nerves emerge from the optic foramina, ascend at an angle of approximately 45°, and join to form the optic chiasm, which lies beneath the floor of the third ventricle, approximately 10 mm above the diaphragma sella.

The orbit is lined with adipose tissue, which is well organized and divided by fibrovascular septa. The contrast that makes the orbital structures stand out on an MR image is produced by this fat, which is associated with high signal intensity on T1-weighted images and high signal intensity on conventional T2-weighted images.

The anatomy of the optic nerve can be studied on an inversion recovery T1-weighted volume set, which has a high signal-to-noise ratio, provides excellent contrast resolution, and can be reformatted in several planes. Use of this sequence facilitates differentiation of the optic nerve from the adipose tissue that surrounds it. The macroscopic anatomy can be better visualized on axial, coronal, and oblique–sagittal (along the long axis of the optic nerve) reconstructions with thin thickness and no gaps between the sections (Fig. 1). T2-weighted images can also be used to study the normal anatomy of the optic nerve (Fig. 2), but they are more useful for excluding the presence of lesions and/or atrophy. In the presence of optic nerve atrophy, the volume of the cerebrospinal fluid surrounding the nerve (which produces high-intensity signals on T2-weighted images) is increased. Again, T1-weighted images are helpful for visualizing the anatomy of the chiasm and for assessment of post-contrast enhancement on fat-suppressed images (Fig. 3).

The optic tracts are visualized using the same method. A three-dimensional T2-weighted sequence can also be used to image the structures in and around the optic chiasm, thanks to the natural contrast furnished by the abundant cerebrospinal fluid in the chiasmatic cistern (Fig. 4). The optic tract segments close to the LGN are difficult to visualize on morphological sequences owing to the partial volume effect produced by the intraaxial brain structures. T1-weighted inversion recovery images are also used to identify the LGNs based on the contrast between the MR signals generated by gray and white matter. Postgeniculate fibers are not visualized with conventional morphological imaging because these sequences can only distinguish brain structures characterized by "natural contrast" (e.g., white vs. gray matter, nerves vs. CSF) (Jacobs and Galetta, 2007).

For distinguishing the white-matter fibers and mapping the visual gray matter areas (LGN and cortex), the functional neuroimaging modalities discussed below (diffusion-weighted MR and fMRI) are valuable tools.

Diffusion MR imaging

Diffusion is a random process that results from the thermal translational motion of molecules. In

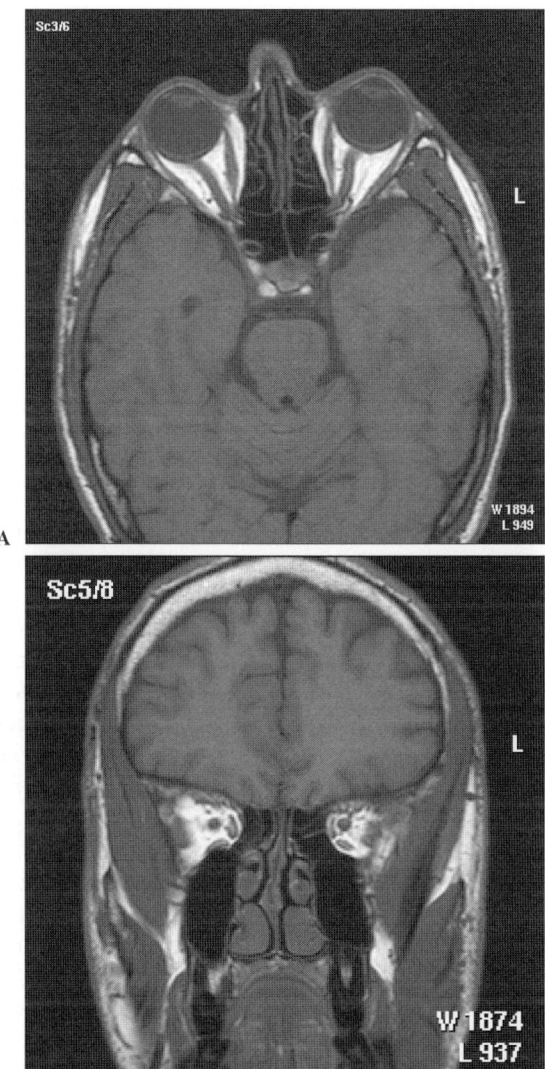

Fig. 1. Conventional axial (A) and coronal (B) T1-weighted images demonstrating the macroscopic anatomy of the intraorbital portion of the optic nerve. Note the hyperintense signal of intraorbital adipose tissue lining the nerves and the hypointense signal from the extraocular muscles.

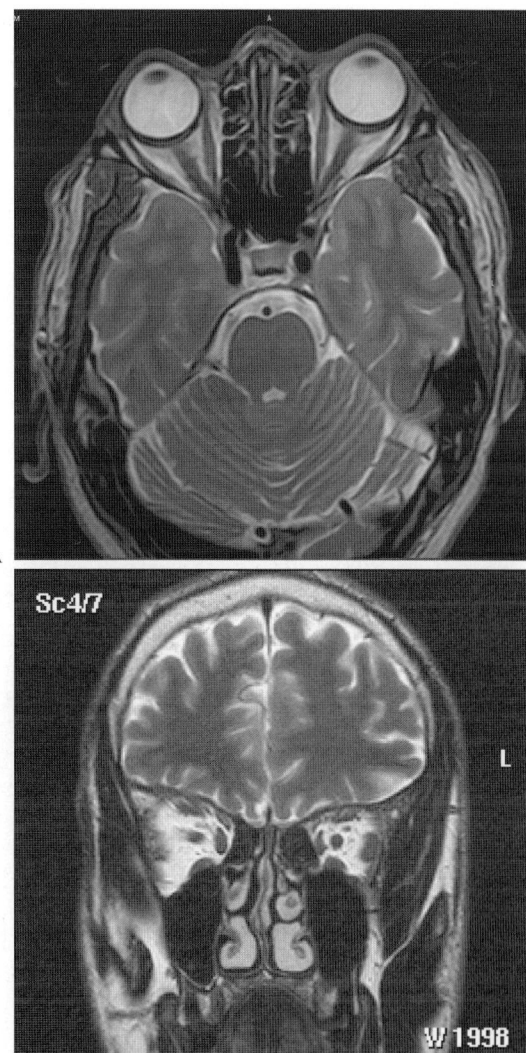

Fig. 2. Conventional morphological axial (A) and coronal (B) T2-weighted images obtained by using a 3 T magnet showing the intraorbital segment of optic nerves.

biological tissues, water diffuses inside, outside, around, and through cellular structures. Cellular membranes hinder these movements and force the water to take more tortuous paths. Diffusion is restricted by the presence of cellular swelling or increased cellularity, whereas necrosis, which involves the breakdown of cellular membranes, decreases diffusion-path tortuosity and increases apparent diffusivity (Le Bihan, 1995). Studies of water diffusion can thus provide information on cellular integrity and pathology. In conventional MR imaging, the effect of diffusion contributes very little to the overall signal intensity. As the diffusing protons move through intrinsic and extrinsic magnetic field gradients, they experience phase shifts that result in a loss of transverse

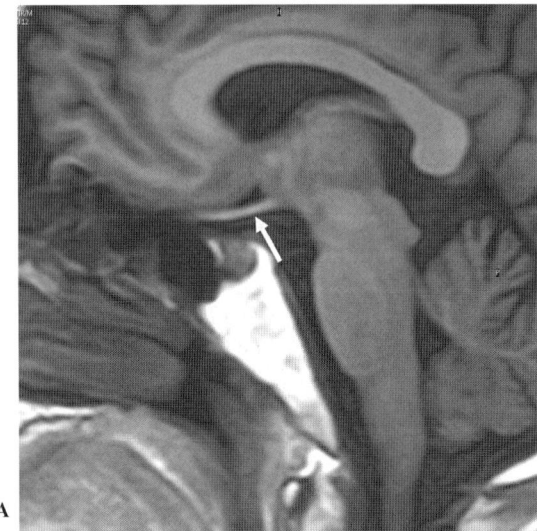

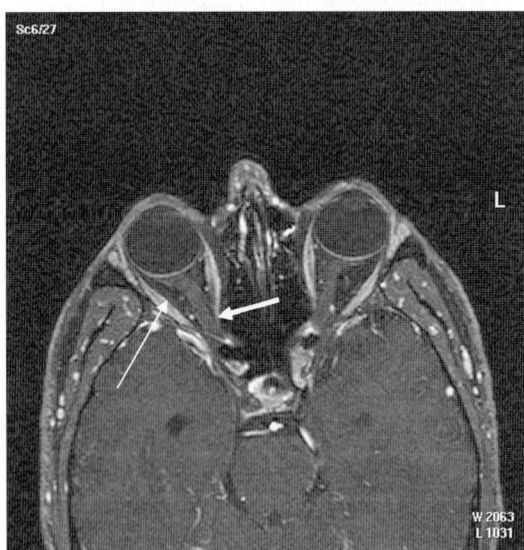

Fig. 3. Sagittal T1-weighted image (A) showing the chiasmatic region and the optic tract (arrow); (B) Post-contrast axial T1-weighted image with fat-suppression technique, note the normal enhancement of the extraocular muscles (long arrow) and the absence of enhancement of the optic nerve (thick arrow).

with respect to the tracts being examined (Hajnal et al., 1991). Fiber tracts lying parallel to an applied gradient have the highest ADC, and the lowest coefficients are observed with tracts that lie transverse or oblique to a sensitizing gradient. This observation highlighted the importance of directionality in the assessment of diffusing molecules. The diffusion of water molecules (especially in white matter) does not proceed equally in all directions: it is *anisotropic*, i.e., characterized by the predominance of movement in one direction (Basser, 1995).

Diffusion-weighted MR imaging has found numerous applications in the field of neuroscience, including the evaluation of stroke, brain development, tumors, and demyelinating disorders. The most common approach is to obtain an echoplanar sequence with supplementary diffusion-sensitive gradients in at least three main axes in space (xx, yy, zz). Two strong gradients of opposite polarity are applied with a short interval in between. All stationary spins are equally dephased and rephased by the action of these gradients. Spins moving between the two gradients are subjected to the effects of a different field at the time of the second application. Their phase recovery is thus incomplete, and the result is signal attenuation. The sensitivity of the sequence to microscopic motion such as the diffusion of water is related to the strength and the duration of the applied gradients, which are collectively expressed by the b-value. In general, a b-value of about $800-1000\,\text{s/mm}^2$ is used. Acquisition of a diffusion-weighted image is always accompanied by the acquisition of a reference image with a b-value of 0 (no applied diffusion gradients). The latter consists of an echoplanar image weighted in $T2^*$ (Fig. 5A).

The final signal produced by the water molecules depends on their ability to diffuse within a tissue: increased diffusion is reflected by a reduction in the intensity of the tissue signal. This phenomenon can be exploited to quantify the diffusivity of water molecules in different directions. The prevalence of diffusion in one direction, for example, along the course of white-matter fibers — diffusion anisotropy — can be quantified by calculation of a second-order symmetric tensor of six elements (or diffusivities) (Moseley et al., 1990; Conturo et al.,

magnetization. This phenomenon is exploited to create diffusion maps, which represent the spatial distribution of diffusion coefficients for the imaged tissue. An early study on the diffusivity of white-matter fibers in normal subjects showed that the apparent diffusion coefficients (ADC) are dependent on the orientation of the diffusion gradients

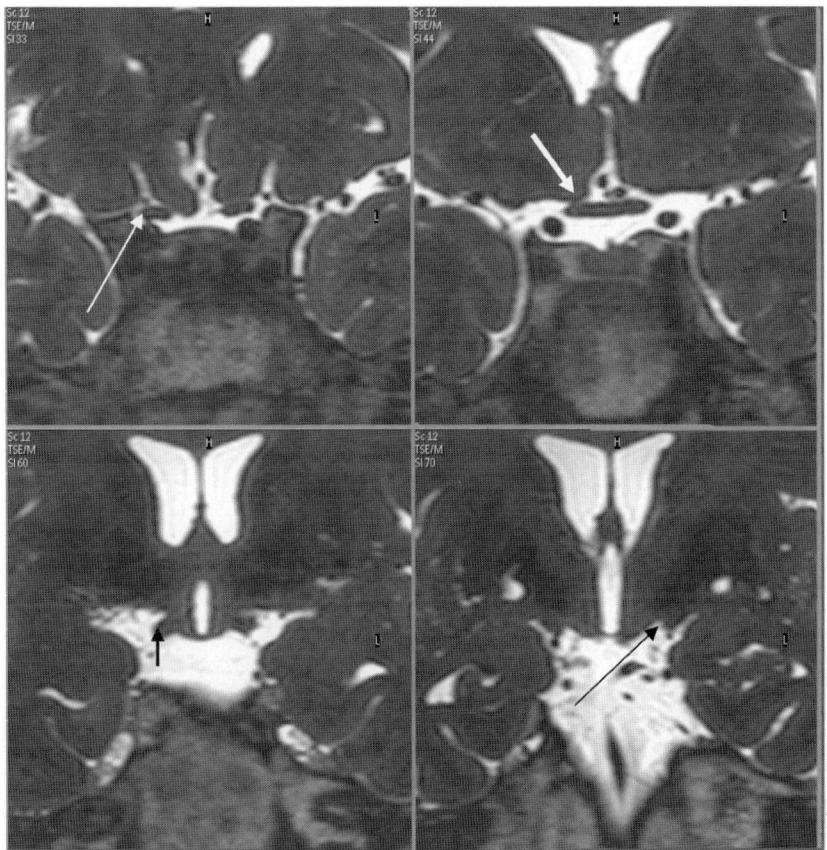

Fig. 4. Set of coronal T2-weighted images from a three-dimensional sequence showing the prechiasmatic segment of the optic nerve (long white arrow), chiasm (thick white arrow), optic tracts (short black arrow), and pregeniculate segment of the optic tract (long black arrow).

1999). The information derived from this DT can be used to elaborate quantitative maps (Fig. 5B–D) showing mean diffusivity (MD) and fractional anisotropy (FA), which are useful for comparing individuals or populations. Anisotropy data can also be used to deduce the pathways of major nerve-fiber tracts, a technique known as tractography.

Diffusion tensor imaging (DTI) has been used to assess axonal degeneration within the visual pathways. Trip et al. (2006) found that MD is significantly increased and FA is significantly reduced in optic nerves affected by optic neuritis (compared with unaffected contralateral nerves and with those of healthy controls). These findings, which appeared compatible with the presence of axonal disruption, were similar to those reported in studies of chronic brain lesions in multiple sclerosis (Werring et al., 1999; Filippi and Inglese, 2001; Dong et al., 2004). Correlation between clinical and electrophysiological parameters and diffusion-weighted MR data has been observed in optic nerves with chronic post-inflammatory lesions (Hickman et al., 2005).

Postmortem histological studies have demonstrated that a lesion located between the retina and lateral geniculate body can cause full-length degeneration of the retinal ganglion cell (RGC) axons that develops in both the anterograde and retrograde directions (Hoyt and Luis, 1962). Ueki et al. (2006) used a PROPELLER (periodically rotated overlapping parallel lines with enhanced reconstruction) sequence (Pipe et al., 2002) to

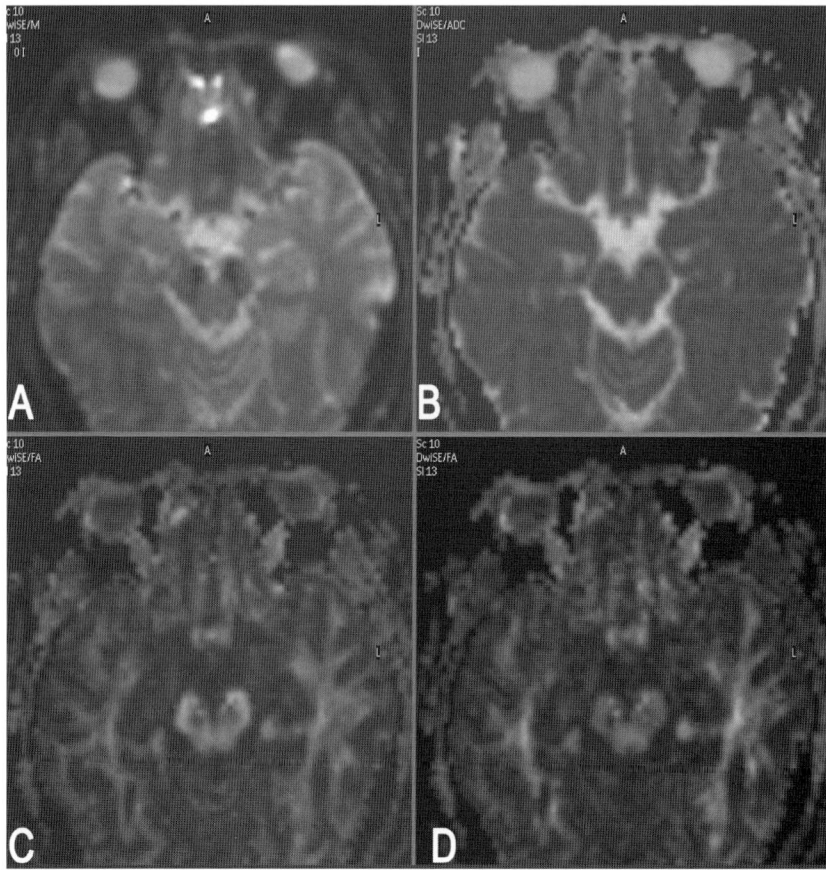

Fig. 5. Set of diffusion images, axial plane. (A) Non-weighted image b = 0; (B) mean diffusivity (MD) map; (C) fractional anisotropy (FA); (D) color map showing the direction of the fibers: green, fronto-occipital; blue, cranio-caudal; red, lateral.

obtain diffusion-weighted images of the visual pathway and found that increased diffusivity was the main alteration associated with RGC degeneration. Recently, Wu et al. (2007) demonstrated that diffusion changes are closely correlated with the total axolemmal cross-sectional area of the prechiasmatic segment of the murine optic nerve. These quantitative histological data confirm that optic nerve diffusivity is a reliable quantitative index of ultrastrucural axonal degeneration.

Irreversible degeneration of the optic nerve axons has been demonstrated in glaucoma (Nickells, 1996; Artes and Chauhan, 2005). DTI of the optic nerve in rats has revealed that FA decreases and MD increases with time after the experimental induction of glaucoma (Hui et al., 2007). The changes observed on DTI in this study were associated with histological evidence of reductions in the number of RGC axons in the glaucomatous optic nerve. (The loss of these axons results in an expansion of the extracellular space, which enhances diffusion.) In a study being conducted by our group, DTI changes are also being found in the optic nerves of patients with different degrees of glaucoma, and these changes seem to be significantly correlated with those of the Humphrey field analysis. These data suggest that MR imaging could be a reliable, noninvasive tool for monitoring the progression of human glaucoma based on quantitative assessments of axonal loss.

DT tractography is a more recently developed technique, which can disclose the complex arrangement of fiber tracts and provide fundamental information on cerebral connectivity (Fig. 6). It requires calculation of various parameters of the DT ellipsoid, a graphical representation of the DT within each voxel that highlights the three-dimensional character of diffusion directionality. One of the most important is the direction of the largest eigenvector ($\lambda 1$), i.e., the direction of greatest diffusivity, which is generally assumed to coincide with that of the fiber bundles (Mori et al., 1999; Masutani et al., 2003). In DT ellipsoid-based fiber tractography, the path of a reconstructed fiber is determined by the $\lambda 1$ in each voxel. Tractography usually models the water diffusion characteristics within each voxel using the DT and assumes that the tensor field within and across voxels reflects the underlying axonal architecture (Basser and Pierpaoli, 1996). In this manner, white matter can be parcellated into fiber structures containing similarly oriented axonal fascicles (Mori and van Zijl, 2002).

DT fiber tractography can be used to visualize the longitudinal direction of the optic nerve fibers (Figs. 6 and 7), to depict the optic radiation (also known as the geniculocalcarine tract) (Yamamoto et al., 2007), and to assess the axonal density of postgeniculate fibers (Figs. 8 and 9).

The optic radiation consists of a broad, thin layer of fiber tracts that extend from the LGN to the primary visual cortex (which corresponds to Brodmann area 17) and are organized into anterior, central, and posterior bundles. The anterior bundle runs in an anterolateral direction and then sweeps forward and courses along the inferior horn of the lateral ventricle in the temporal lobe, thus forming the Meyer loop. It then runs posteriorly, along the lateral wall of the

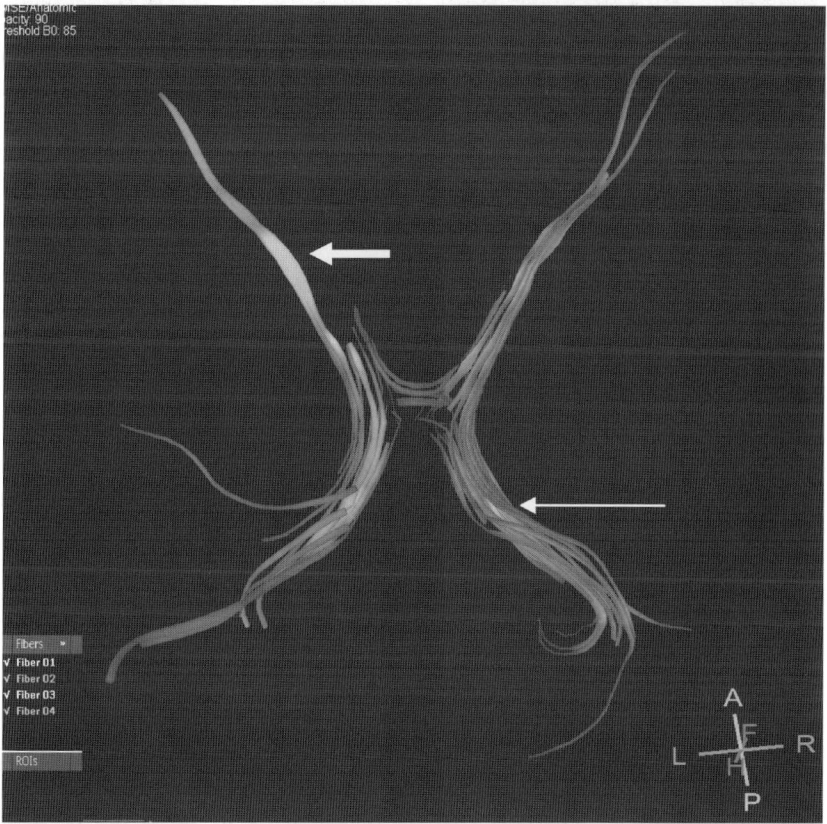

Fig. 6. Reconstructed diffusion tensor fiber tractography of the optic nerves (short arrow), chiasm and the optic tracts (long arrow). (See Color Plate 12.6 in color plate section.)

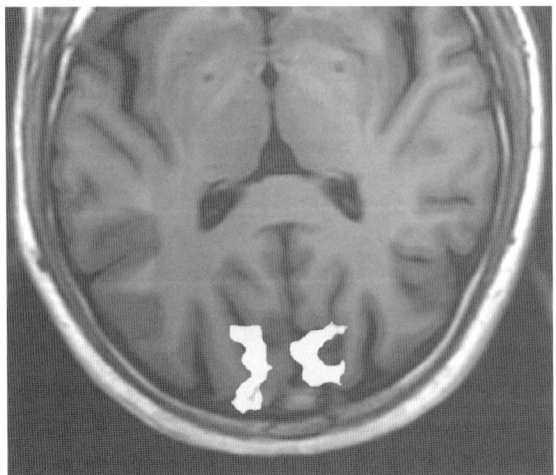

Fig. 10. Functional magnetic resonance imaging (fMRI) results of a single subject following visual task performance.

monkeys exposed to visual stimulation. In accordance with the results of previous studies (Bandettini et al., 1992; Kollias et al., 2000), their findings confirmed that localized increases in BOLD contrast reflect stimulus-evoked increases in neural activity. The response to a visual stimulus begins within a few hundred milliseconds of neuronal stimulation. The decrease in signal intensity observed 0.5–2.0 s after stimulus onset has been attributed to an early focal deoxygenation phase, which precedes increases in oxygenated hemoglobin caused by local enhancement of blood flow (Ernst and Hennig, 1994).

Direct relations between stimulus intensity and occipital responses have also been reported. Alvarez-Linera Prado et al. (2007) demonstrated correlation between the BOLD response and visual-stimulus intensity and found that cortical reactivity to these stimuli (especially those within the low-to-medium-intensity range) is higher in patients with photophobia than normal controls.

fMRI was first employed to establish the extent of striate and extrastriate cortical areas and to map the retinotopic borders of the visual cortex (Courtney and Ungerleider, 1997; Engel et al., 1997). Precise retinotopic mapping was impossible prior to the mid 1990s, when Engel et al. (1997) developed the technique of phase-encoded retinal stimulation. They used a contrast-reversing checkerboard (contrast reversal rate, 8 Hz) presented at the center of gaze to create a strong neural response within area V1 of the human visual cortex. Subjects were exposed to expanding-ring and contracting-ring stimuli, which generated a traveling wave of neural activity within visual cortex. With this new approach, the investigators were able to distinguish the cortical representation of the fovea in the posterior cortex from that of the peripheral retina.

Many efforts were made to correlate neurological deficits with damage to specific brain areas. The first attempts to identify the anatomic sources of specific cerebral functions were based on the study of correlations between loss of function and brain lesion location. The early studies of visual impairments produced by focal lesions suggested that the human visual cortex is organized into two anatomically and functionally distinct units, the ventral and the dorsal pathways. Symptoms like visual object agnosia, prosopagnosia, and achromatopsia are produced by occipitotemporal lesions involving the ventral pathway, whereas optic ataxia, visual neglect, constructional apraxia, gaze apraxia, akinotopsia, and disorders of spatial cognition are the result of occipitoparietal lesions within the dorsal pathway (Boller and Grafman, 1989; Laskowitz et al., 1998).

Anatomical neuroimaging may not be sufficient to determine the extent of brain damage caused by cerebrovascular lesions, neoplasms, inflammatory states, or congenital disease. Many studies have demonstrated good agreement between fMRI findings and the results of traditional visual examinations. Patients with congruous homonymous hemianopia caused by retrochiasmatic lesions (Miki et al., 1996a) and those with visual field defects caused by prechiasmatic and chiasmatic lesions (Miki et al., 1996b) showed abnormal cortical activation that was concordant with the visual defect. fMRI thus appears to be a potentially valuable tool for assessing local brain function in patients with visual deficits.

In patients with space-occupying lesions involving the optic radiation, fMRI examinations revealed activation patterns in the visual cortex that were consistent with the patients' visual field deficits and with traditional retinotopic representations.

Optic neuritis is a common manifestation of multiple sclerosis, a disease characterized by the development of multifocal, inflammatory, demyelinating lesions of the white matter. In patients with unilateral optic neuritis, fMRI has revealed that the response to visual stimuli of the affected eye is characterized by reductions in the area of activation in the primary visual cortex and by significant increases in the latency of the major positive component of the visual-evoked potential (VEP). fMRI can thus be considered a reliable method for obtaining detailed topographic information related to functional deficits in multiple sclerosis (Gareau et al., 1999).

fMRI can also be used to identify epileptogenic foci in patients with epilepsy based on the demonstration of abnormal activation patterns that are concordant with seizure onset and interictal epileptiform discharges; these studies showed agreement between fMRI data and EEG (Lengler et al., 2007).

Dyslexia is a developmental disorder characterized by low reading achievement in individuals whose cognitive abilities, motivation, and education are adequate for accurate, fluent reading. fMRI studies of subjects with dyslexia frequently exhibit hypoactivation in the left parietotemporal cortex together with hyperactivation in the left inferior frontal cortex. Hoeft et al. (2007) recently reported that the areas of hyperactivation reflected processes related to the subject's current level of reading ability, independent of the dyslexia, whereas the areas of hypoactivation seen in the left parietotemporal and occipitotemporal lobes represent functional atypicalities related to the dyslexia itself.

In schizophrenics who experience visual hallucinations, fMRI has revealed increased cerebral activity in the ventral extrastriate visual cortex and in the hippocampus (Oertel et al., 2007).

Miki et al. (1996b) performed fMRI during monocular visual stimulation in patients with visual field losses caused by lesions of the optic nerve and chiasm, and the results showed agreement with the pattern of visual field defects. In patients with unilateral optic neuropathy, like that frequently seen in glaucoma, stimulation of the affected eye failed to activate the portion of the primary visual cortex corresponding to the central visual field defects and reduced activity in the associated visual cortex. Patients with chiasmal compression, monocular stimulation caused markedly asymmetrical activation of the primary visual cortex, which corresponded to the visual field abnormality.

Recently, Duncan et al. (2007) demonstrated correlation between functional organization of the visual cortex (V1) and visual field deficits in patients with primary open-angle glaucoma (POAG). A retinotopic map of visual space was obtained for visual areas in occipital cortex. Templates were used to project regions within the visual field onto a flattened representation of cortex. The resulting BOLD fMRI response was compared to interocular differences in thresholds for corresponding regions of the visual field. The spatial pattern of activity observed in the flattened representation was consistent with the pattern of visual field loss. In patients withPOAG, the BOLD signal in human visual cortex is altered in a manner that is consistent with the loss of visual function, suggesting that fMRI can be used to quantify functional changes in glaucoma.

Proton MR spectroscopy

MR spectroscopy is a noninvasive tool for investigating the chemical environment of the brain. The proton MR spectrum is characterized by at least three peaks, which represent (1) the compounds creatine and phosphocreatine (Cr), which are generally associated with cellular energy metabolism; (2) choline (Cho), which is associated with cell membrane synthesis; and (3) N-acetyl aspartate (NAA), a marker of neuronal integrity. Several studies have demonstrated the usefulness of MR spectroscopy in clinical settings, for the study of seizure foci and neoplasms and for distinguishing recurrent tumor from radiation necrosis, metabolic diseases, and white-matter disease (Brandão and Domingues, 2004).

In a recent study, Boucard et al. (2007) performed proton MR spectroscopy with a single-voxel technique in patients with glaucoma. They found that absolute levels of NAA in the occipital brain of subjects with progressive visual field defects caused by age-related macular degeneration or glaucoma were not significantly different from those found in a group of control subjects. Visual field degeneration in both these diseases progresses slowly, and the authors hypothesized the rate of progression might not be high enough to provoke a decrease in NAA concentration. An alternative explanation is that the cortical area corresponding to the affected retinal region is too small to provoke changes in the NAA concentration that are detectable with proton MR spectroscopy (Boucard et al., 2007).

References

Alvarez-Linera Prado, J., Ríos-Lago, M., Martín-Alvarez, H., Hernández-Tamames, J.A., Escribano-Vera, J. and Sánchez-delRío, M. (2007) Functional magnetic resonance imaging of the visual cortex: relation between stimulus intensity and bold response. Rev. Neurol., 45(3): 147–151.

Artes, P.H. and Chauhan, B.C. (2005) Longitudinal changes in the visual field and optic disc in glaucoma. Prog. Retin. Eye Res., 24: 333–354.

Bandettini, P.A., Wong, E.C., Hinks, R.S., Tikofsky, R.S. and Hyde, J.S. (1992) Time course EPI of human brain function during task activation. Magn. Reson. Med., 25(2): 390–397.

Basser, P.J. (1995) Inferring microstructural features and the physiological state of tissues from diffusion-weighted images. NMR Biomed., 8(7–8): 333–344.

Basser, P.J. and Pierpaoli, C. (1996) Microstructural and physiological features of tissues elucidated by quantitative-diffusion-tensor. J. Magn. Reson. B, 111(13): 209–219.

Boller, F. and Grafman, J. (1989) Disorders of visual behavior. In: Goodglass H. and Damasio A.R. (Eds.), Handbook of Neuropsychology, Vol. 2. Elsevier, Amsterdam, pp 333–356.

Boucard, C.C., Hoogduin, J.M., van der Grond, J. and Cornelissen, F.W. (2007) Occipital proton magnetic resonance spectroscopy (1H-MRS) reveals normal metabolite concentrations in retinal visual field defects. PLoS ONE, 2(2): p. e222.

Brandão, L.A. and Domingues, R.C. (2004) MR Spectroscopy of the Brain. Lippincott Williams & Wilkins, Philadelphia.

Conturo, T.E., Lori, N.F., Cull, T.S., Akbudak, E., Snyder, A.Z., Shimony, J.S., McKinstry, R.C., Burton, H. and Raichle, M.E. (1999) Tracking neuronal fiber pathways in the living human brain. Proc. Natl. Acad. Sci. U.S.A., 96(18): 10422–10427.

Courtney, S.M. and Ungerleider, L.G. (1997) What fMRI has taught us about human vision. Curr. Opin. Neurobiol., 7(4): 554–561.

Dong, Q., Welsh, R.C., Chenevert, T.L., Carlos, R.C., Maly-Sundgren, P., Gomez-Hassan, D.M. and Mukherji, S.K. (2004) Clinical applications of diffusion tensor imaging. J. Magn. Reson. Imaging, 19(1): 6–18.

Duncan, R.O., Sample, P.A., Weinreb, R.N., Bowd, C. and Zangwill, L.M. (2007) Retinotopic organization of primary visual cortex in glaucoma: comparing fMRI measurements of cortical function with visual field loss. Prog. Retin. Eye Res., 26(1): 38–56.

Ebeling, U. and Reulen, H-J. (1988) Neurosurgical topography of the optic radiation in the temporal lobe. Acta Neurochir. (Wien.), 92: 29–36.

Engel, S.A., Glover, G.H. and Wandell, B.A. (1997) Retinotopic organization in human visual cortex and the spatial precision of functional MRI. Cereb. Cortex, 7(2): 181–192.

Ernst, T. and Hennig, J. (1994) Observation of a fast response in functional MR. Magn. Reson. Med., 32(1): 146–149.

Filippi, M. and Inglese, M. (2001) Overview of diffusion-weighted magnetic resonance studies in multiple sclerosis. J. Neurol. Sci., 186(1): S37–S43.

Gareau, P.J., Gati, J.S., Menon, R.S., Lee, D., Rice, G., Mitchell, J.R., Mandelfino, P. and Karlik, S.J. (1999) Reduced visual evoked responses in multiple sclerosis patients with optic neuritis: comparison of functional magnetic resonance imaging and visual evoked potentials. Mult. Scler., 5(3): 161–164.

Hajnal, J.V., Doran, M., Hall, A.S., Collins, A.G., Oatridge, A., Pennock, J.M., Young, I.R. and Bydder, G.M. (1991) MR imaging of anisotropically restricted diffusion of water in the nervous system: technical, anatomic, and pathologic considerations. J. Comput. Assist. Tomogr., 15(1): 1–8.

Hickman, S.J., Wheeler-Kingshott, C.A.M., Jones, S.J., Miszkiel, K.A., Barker, G.J., Plant, G.T. and Miller, D.H. (2005) Optic nerve diffusion measurement from diffusion-weighted imaging in optic neuritis. AJNR Am. J. Neuroradiol., 26(4): 951–956.

Hoeft, F., Meyler, A., Hernandez, A., Juel, C., Taylor-Hill, H., Martindale, J.L., McMillon, G., Kolchugina, G., Black, J.M., Faizi, A., Deutsch, G.K., Siok, W.T., Reiss, A.L., Whitfield-Gabrieli, S. and Gabrieli, J.D. (2007) Functional and morphometric brain dissociation between dyslexia and reading ability. Proc. Natl. Acad. Sci. U.S.A., 104(10): 4234–4239.

Hoyt, W.F. and Luis, O. (1962) Visual fiber anatomy in the infrageniculate pathway of the primate. Arch. Ophthalmol., 68: 94–106.

Hui, E.S., Fu, Q.L., So, K.F. and Wu, E.X. (2007) Diffusion tensor MR study of optic nerve degeneration in glaucoma. Conf. Proc. IEEE Eng. Med. Biol. Soc., 2007: 4312–4315.

Jacobs, D.A. and Galetta, S.L. (2007) Neuro-ophthalmology for neuroradiologists. AJNR Am. J. Neuroradiol., 28(1): 3–8.

Kollias, S.S. (2004) Investigations of the human visual system using functional magnetic resonance imaging (FMRI). Eur. J. Radiol., 49(1): 64–75.

Kollias, S.S., Golay, X., Boesiger, P. and Valavanis, A. (2000) Dynamic characteristics of oxygenation-sensitive MR signal during application of varying temporal protocols for imaging of human brain activity. Neuroradiology, 42(8): 591–601.

Laskowitz, D., Liu, G.T. and Galetta, S.L. (1998) Acute visual loss and other disorders of the eyes. Neurol. Clin., 16(2): 323–353.

Le Bihan, D. (1995) Molecular diffusion, tissue microdynamics and microstructure. NMR Biomed., 8: 375–386.

Lengler, U., Kafadar, I., Neubauer, B.A. and Krakow, K. (2007) fMRI correlates of interictal epileptic activity in patients with idiopathic benign focal epilepsy of childhood. A simultaneous EEG-functional MRI study. Epilepsy Res., 75(1): 29–38.

Logothetis, N.K., Pauls, J., Augath, M., Trinath, T. and Oeltermann, A. (2001) Neurophysiological investigation of the basis of the fMRI signal. Nature, 412(6843): 150–157.

Masutani, Y., Aoki, S., Abe, O., Hayashi, N. and Otomo, K. (2003) MR diffusion tensor imaging: recent advance and new techniques for diffusion tensor visualization. Eur. J. Radiol., 46(1): 53–66.

Miki, A., Nakajima, T., Fujita, M., Takagi, M. and Abe, H. (1996a) Functional magnetic resonance imaging in homonymous hemianopsia. Am. J. Ophthalmol., 121(3): 258–266.

Miki, A., Nakajima, T., Takagi, M., Shirakashi, M. and Abe, H. (1996b) Detection of visual dysfunction in optic atrophy by functional magnetic resonance imaging during monocular visual stimulation. Am. J. Ophthalmol., 122(3): 404–415.

Mori, S., Crain, B.J., Chacko, V.P. and van Zijl, P.C. (1999) Three-dimensional tracking of axonal projections in the brain by magnetic resonance imaging. Ann. Neurol., 45(2): 265–269.

Mori, S. and van Zijl, P.C. (2002) Fiber tracking: principles and strategies — a technical review. NMR Biomed., 15(7–8): 468–480.

Moseley, M.E., Cohen, Y., Kucharczyk, J., Mintorovitch, J., Asgari, H.S., Wendland, M.F., Tsuruda, J. and Norman, D. (1990) Diffusion-weighted MR imaging of anisotropic water diffusion in cat central nervous system. Radiology, 176(2): 439–445.

Nickells, R.W. (1996) Retinal ganglion cell death in glaucoma: the how, the why, and the maybe. J. Glaucoma, 5(5): 345–356.

Oertel, V., Rotarska-Jagiela, A., vandeVen, V.G., Haenschel, C., Maurer, K. and Linden, D.E. (2007) Visual hallucinations in schizophrenia investigated with functional magnetic resonance imaging. Psychiatry Res., 15(3): 269–273.

Ogawa, S., Lee, T.M., Kay, A.R. and Tank, D.W. (1990) Brain magnetic resonance imaging with contrast dependent on blood oxygenation. Proc. Natl. Acad. Sci. U.S.A., 87: 9868–9872.

Pipe, J.G., Farthing, V.G. and Forbes, K.P. (2002) Multishot diffusion weighted FSE using PROPELLER MRI. Magn. Reson. Med., 47(1): 42–52.

Trip, S.A., Wheeler-Kingshott, C., Jones, S.J., Li, W.Y., Barker, G.J., Thompson, A.J., Plant, G.T. and Miller, D.H. (2006) Optic nerve diffusion tensor imaging in optic neuritis. Neuroimage, 30(2): 498–505.

Ueki, S., Fujii, Y., Matsuzawa, H., Takagi, M., Abe, H., Kwee, I.L. and Nakada, T. (2006) Assessment of axonal degeneration along the human visual pathway using diffusion trace analysis. Am. J. Ophthalmol., 142(4): 591–596.

Werring, D.J., Clark, C.A., Barker, G.J., Thompson, A.J. and Miller, D.H. (1999) Diffusion tensor imaging of lesions and normal-appearing white matter in multiple sclerosis. Neurology, 52(8): 1626–1632.

Wu, Q., Butzkueven, H., Gresle, M., Kirchhoff, F., Friedhuber, A., Yang, Q., Wang, H., Fang, K., Lei, H., Egan, G.F. and Kilpatrick, T.J. (2007) MR diffusion changes correlate with ultra-structurally defined axonal degeneration in murine optic nerve. Neuroimage, 37(4): 1138–1147.

Yamamoto, A., Miki, Y., Urayama, S., Fushimi, Y., Okada, T., Hanakawa, T., Fukuyama, H. and Togashi, K. (2007) Diffusion tensor fiber tractography of the optic radiation: analysis with 6-, 12-, 40-, and 81-directional motion-probing gradients, a preliminary study. AJNR Am. J. Neuroradiol., 28(1): 92–96.

Yamamoto, T., Yamada, K., Nishimura, T. and Kinoshita, S. (2005) Tractography to depict three layers of visual field trajectories to the calcarine gyri. Am. J. Ophthalmol., 140(5): 781–785.

SECTION III

Current Therapy

CHAPTER 13

Primary open angle glaucoma: an overview on medical therapy

Michele Vetrugno*, Francesco Cantatore, Giuseppe Ruggeri, Paolo Ferreri, Alfonso Montepara, Alessandra Quinto and Carlo Sborgia

Glaucoma Center, Department of Ophthalmology, University of Bari, Bari, Italy

Abstract: The purpose of this review is to discuss the topics relevant to the use of intraocular pressure-lowering strategies, which remains the first line in the management of glaucoma. Estimates of blindness from glaucoma and identification of risk factors remain of interest for all ophthalmologists. New functional tests offer promise for better detection and more accurate diagnosis of glaucoma. We finally discuss the impact of various glaucoma therapies, the principles of monotherapy and fixed combinations, which offer benefits of convenience, cost, and safety.

Keywords: glaucoma; medical therapy; prostaglandins; risk factors

Introduction

Glaucoma is nowadays defined as a progressive optic neuropathy (Gupta and Weinreb, 1997) with a typical associated visual field loss, and it is one of the leading causes of preventable blindness in developed countries.

It is estimated that glaucoma approximately causes the 10% of all blindness (Quigley, 1996).

Since life expectancy is increasing, all the efforts need to be focused on maintaining the quality of patient's life, and alleviating the social and economic burden of glaucoma.

Glaucoma treatment has been available for more than a century. Nevertheless, due to the unproven efficacy of glaucoma therapy and also to the additional treatment modalities, which have expanded the available options, there is a considerable controversy within the glaucoma researchers community concerning how the open angle glaucoma should be treated, and, particularly, which weapons should be employed.

Glaucoma prevention consists in identifying the risk factors associated with the optic neuropathy and attempting to treat those factors for which a therapy exists. For decades intraocular pressure (IOP) has been considered the only risk factor associated with glaucoma and for this reason the goal of many therapeutic options is to treat it. Having the researchers recognized the existence of other treatable risk factors, new therapeutic options should include blood flow, neuroprotection, and genetically based agents.

This work aims to provide an overview on the medical treatment, especially referring to three matters:

- When to treat
- Whom to treat
- How to treat

*Corresponding author. Tel.: +39805592525; Fax: +39805478918; E-mail: m.vetrugno@oftalmo.uniba.it

When to treat

This is the first step in glaucoma treatment, and this item is directly linked to the current opinions about the early glaucoma diagnosis or high-risk ocular hypertension.

Affirmed that, since glaucoma is a slowly progressing disease, someone believes that an early diagnosis and a consequently early treatment may not be essential.

A definition of early glaucoma is needed, in order to guide physicians in their diagnostic and therapeutical decisions. Early glaucoma is a silent condition, without symptoms, as many patients do not even know they are suffering from this disease until it has further progressed. Thus, all the efforts should be dedicated in the population screening, in order to identify affected people.

Current findings suggest that glaucoma may remain undetected in approximately 50% of the population, until some loss of vision has occurred; moreover, even nowadays, a delay in diagnosis is possible for a great proportion of patients (Caprioli and Garway-Heath, 2007).

Early detection, together with an appropriate treatment, can actually improve patients' outcomes.

Referring to the impact of an early treatment on its outcome, first it has to be clarified if the intended outcome is the field loss or the disability prevention: the former refers to the visual function, whereas the latter is related to the quality of life.

The existence of a tolerated IOP range without optic nerve damage, and, on the other side, of IOP levels commonly considered as normal characterized instead by a glaucoma rate of progression, is to correlate with an individual optic nerve damage sensibility.

It is still up for discussion whether a treatment should start as soon as possible, since there are findings showing that considerable retinal ganglion cells (RGCs) death may go undetected.

Schwarzt and Budenz, early treatment supporters, believe that prolonged elevation of the IOP triggers a series of events that results in the ganglion cells progressive loss, even after the IOP is adequately controlled (Schwarzt and Budenz, 2004). This hypothesis may explain why some patients continue to progress despite an adequate control of the IOP. It has been generally agreed that an early detection does not automatically imply an early treatment: the early detection and the early treatment should be considered separately.

If the early treatment loses importance, then an alternative strategy in the ocular hypertension management is waiting for signs of manifest glaucoma (optic nerve changes or visual field abnormalities).

Although a decision of an early treatment may not be made, other appropriate measures, such as close monitoring, may be considered (Caprioli and Garway-Heath, 2007).

Early detection may benefit the physician–patient relationship. It is important that the physician educates and counsels the patient about his condition to ensure that he becomes engaged in the disease management.

The patient surveillance should match the risk level. For example, a patient who is highly suspected for glaucoma may be asked to come back for follow-up in 6 months, instead of 1–5 years check-up for patients with a lower risk of disease.

Although the ultimate impact of delaying the treatment is currently unknown, there is evidence showing that early treatment can prevent or at least delay progression to glaucoma.

The Ocular Hypertension Treatment Study (OHTS) first findings, showed that lowering the IOP with topical hypotensive medications, can prevent or delay progression to glaucoma in OHT patients without definite evidences of glaucomatous damage (Kass et al., 2002).

In this study, at 5 years follow-up, the cumulative probability of progression to glaucoma in the treated group was less than a half, comparing with the untreated group.

Early IOP-lowering intervention was also found to reduce the rate of conversion from OHT to glaucoma in the European Glaucoma Prevention Study (EGPS, 2005). Although findings suggest that the rate of conversion to glaucoma from OHT is relatively low, it is essential that those in whom the disease progresses receive appropriate care.

It was suggested that follow-up study to the Early Manifest Glaucoma Trial (EMGT) should be conducted to investigate progression rates of patients who received early treatment, compared with those who received later treatment.

In the 6-year EMGT, early intervention delayed disease progression in the treatment group with early glaucoma compared with untreated patients (Heijil et al., 2002).

In other words, the matter is to determine if early treatment is beneficial for the long-term visual outcome or whether it is acceptable to observe progression behavior and then treat the patient on the basis of observed progression behavior.

For social and economic reasons, the glaucoma screening is a useful and necessary task, with possible benefits for individuals and the health care system, thanks to the early diagnosis and early therapy of patients suffering from glaucoma. Knowledge from literature shows that an early treatment of patients with glaucoma decreases blindness risk and lowers the direct and indirect costs for patients with glaucoma.

Whom to treat

In prescribing initial medical therapy for glaucoma or ocular hypertension, a number of factors have to be considered. It was thought that treatment should begin if it is deemed necessary to preserve the quality of life, but that initiation should be considered on an individual base. According to this, discovering and treating people at risk of visual function's loss with an individualized management is preferable.

As recommended in the OHTS (Caprioli and Garway-Heath, 2007), not all hypertensive patients should be treated, but several parameters should be taken into account, among which

- social-economic impact of a long-term treatment
- likelihoods that the treatment is useful for the patient
- patient's health status
- life expectancy
- patient's relative risk of developing glaucoma.

Considering more specifically the medical therapy, the following parameters have to be considered:

- efficacy
- side effects
- costs
- convenience of dosing
- diurnal fluctuation.

For these reasons, glaucoma therapy can vary according to the single patient but its principle aim is to preserve visual function at the lowest risks, costs, and side effects for the patient.

As for the chance of beginning the medical treatment, some risk factors have to be considered: the presence of glaucomatous damage in the fellow eye, a family history in the first degree relatives, a pre-treatment tonometry readings, the age, and the race.

Genetics

Much research has been focused on the gene expression, protein processing, and mutations of MYOC/TIGR, which is associated with both juvenile- and adult-onset primary open angle glaucoma (POAG). Investigations of other glaucoma related genes such as PITX2, FOXC1, and CYP1B1, are enabling a better understanding of anterior segment development and its relation to glaucoma (WuDunn, 2002).

Age

Glaucoma is age-related, and its prevalence increases up to 0.2% per year in patients aged from 50 to 54; up to 2% in an elder population aged above 70 years and its incidence is estimated to be 0.11% per year in people between 55 and 74. Perhaps the increased prevalence in advanced ages indicates a longer exposure to higher pressure levels or a major optic nerve susceptibility (EGS, 2003). The burden of a visual impairment is not uniformly distributed throughout the world; the least developed regions support the largest share. Visual impairment is also unequally distributed across age groups, being largely confined to adults aged 50 and over (Caprioli and Garway-Heath, 2007).

Race

The Baltimore Eye Survey indicates that POAG prevalence is four times higher in Afro-Americans than in other races (Tielsh et al., 1991). Moreover Afro-Americans glaucomatous blindness probability is four to eight times higher than in Caucasian-Americans. The role of race in glaucoma rate of progression has not been cleared up but basically a hereditary mechanism occurs.

Ocular and systemic abnormalities

Other possible but controversial risk factors are myopia >4 diopters, pseudoexfoliation, thin cornea, and vascular diseases like systemic pressure, vasospasm (Raynaud syndrome), migraine, prinzmetal angina, and diabetes (EGS, 2003).

IOP fluctuations

Unless the IOP is so high to be danger for patient visual function, a medical treatment should not be started at the first visit. The IOP should be measured more than once and preferably more times a day, in order to quantify the diurnal fluctuation. The IOP is characterized by diurnal fluctuations in a large proportion of healthy people. Its variations are generally less than 4 mmHg, with higher levels in the morning. Differential IOP suggests the efficacy of the drug used, above all when it is not affected by diurnal fluctuation. A recent study (Schwarzt and Budenz, 2004) suggests that high IOP diurnal fluctuation, even in treated patients, can result in more progression, compared with patients who do not show high diurnal fluctuations. The range of the initial daily IOP variation was more predictive of the risk of visual field loss, progression than was the mean or the peak IOP.

This concept that a greater range of diurnal IOP variation is more damaging to the eye has suggested a different approach to glaucoma therapy, namely that the treatment should be aimed at trying to minimize the diurnal fluctuation in the IOP as well as to eliminate pressure peaks. The duration of many antiglaucoma drugs action is fairly short, so that some of the drug's effect is already wearing off before the next dose is administered (Wilensky, 2004).

Tonometry and pachymetry

Glaucoma screening usually uses only one parameter (IOP) to detect and to discriminate glaucoma patients but glaucoma actually requires a variety of diagnoses, therefore a single test would not be sensible and specific enough to detect glaucoma. With the use of a single device such as tonometry, there is a high probability of a false-positive. The diagnosis should be confirmed by the disc damage assessment and by the trends in visual field.

Based on this rationale, the rate of progression then determines how treatment's targets should be set. Corneal thickness can influence IOP assessment, as well as its curvature and hydration. Central corneal thickness (CCT) evaluation by pachymetry is intended as an aid to correct the IOP measurements, above all when it is necessary to cast doubts about starting glaucoma treatment.

Optic nerve condition and visual field stage

The presence of risk factors is important in order to establish treatment guidelines in preventing optic nerve rate of decay. The most important indications about the relative glaucoma damage risk are the current damage and the rate of progression (EGS, 2003).

Another screening parameter should be the optic nerve head morphology, since the diagnosis of glaucoma is very closely associated with a morphologic change in the optic nerve head. Damage stage can be evaluated by optic nerve and visual field assessment.

Optic nerve stereoscopic evaluation is based on glaucoma damage signs such as neuroretinal rim thinning (at superior and inferior poles), notching, splinter optic disc hemorrhages, cup asymmetry, parapapillary atrophy, bared circumlinear vessels.

- Neuroretinal rim thinness can affect all disc sectors, but generally it is often remarkable at inferior and superior poles, so that inferotemporal edge is not characteristically the thickest.
- Disc hemorrhages represent a sign of local vascular damage and their presence is likely to be pathological.
- A parapapillary atrophy can be present in no glaucomatous eyes, so it is intended as a clue.
- An early sign of acquired rim thinning is a bared circumlinear vessel at the edge of the disc.
- In early glaucoma stages slit-like, grove-like, and spindle shaped defects are more evident, coexisting a generalized thinning of nerve fiber layer (NFL).

An initial alteration is generally characterized by both diffuse thinning and one or more localized defects. Since those NFL defects are present up to 3% in no glaucomatous eyes, they are likely to be pathological. An optic nerve or visual field damage rate of progression is typical of glaucoma, but it is very difficult to evidence it during the first patient assessment. Rate of progression can be determined by a continuous follow-up.

An advanced optic nerve damage, clinically evident, rapidly progressing, affected by risk factors, requires an aggressive treatment by lowering the IOP. Although the IOP lowering by medical therapy has been shown to be beneficial in delaying or preventing the glaucoma onset in ocular hypertensive and delaying or preventing visual field loss in people with glaucoma, there is a potential therapy downside it (Schwarzt and Budenz, 2004).

If the medical treatment is not effective in obtaining stated IOP target or in preventing the decrease in clinical data, other therapeutical options, such as laser therapy or surgery, can be considered on the individual needs.

According to Schwarzt and Budenz (2004) in a 90-year-old ocular hypertensive patient with no visual field loss, for example, observation might be a better strategy than lowering the IOP by 20%, especially if the therapy introduces the risk of ocular or systemic side effects or high medication costs.

At the other end of the spectrum, the authors consider a 60-year-old patient with severe, progressive glaucoma who has IOP in the mid-20s on maximal medical therapy and has already received laser trabeculoplasty. The risk of permanent disability is high without IOP lowering, and the benefits of trabeculectomy are high. In this case, it is probably worth to accept the small complications risk from trabeculectomy surgery.

How to treat

A medical treatment is considered effective when the mean effect produced by that drug is similar to published average effect on general population and this effect should be higher than the ones found by tonometry, that are affected by errors and variations (EGS, 2003).

IOP lowering is the most effective therapeutical approach to avoid function loss, because a high IOP is the main risk factor for glaucomatous damage onset. Normal IOP level is a statistical outcome, based on population measurements, so it cannot be applied to all patients arbitrarily.

Although the disease progression is usually slow, it may be faster in individuals whose optic nerve is more susceptible to IOP-related damage. There is substantial evidence which confirms that lowering IOP is effective in reducing glaucoma rate of progression in some or in the majority of the patients. Before setting medical treatment, baseline IOP levels should be measured and compared with the ones found during the follow-up. Moreover, the relative risk of developing optic nerve damage depends on the mean IOP, the maximum IOP, and the IOP fluctuations.

Generally, the more advanced is the damage, the lower is the target IOP: the goal of glaucoma therapy in ocular hypertension is lowering IOP by at least a 20% in patients with moderate to high risk of progression. In patients with perimetry-proven glaucoma, IOP should be lowered by at least a 30% in early to moderate glaucoma, and perhaps a 40 to 50% in severe glaucoma (Schwarzt and Budenz, 2004).

Target IOP can vary during glaucoma natural history, and this is the reason for a continuous re-evaluation of treatment efficacy. Several trials have been carried out comparing drugs safety and efficacy in lowering the IOP, but they are not very reliable, due to the diversity of population's samples studied.

According to EGS guidelines (EGS, 2003) topical treatment should be started in one eye at a time.

It is preferable to start with monotherapy and in the last few years there has been a gradual change in medical treatment choice.

If the first drug used is not effective in decreasing the IOP or is not well tolerated, it is preferable to change agent category.

Moreover, if the topical agent used in monotherapy does not produce side effects but it is not sufficient in decreasing the IOP, another topical agent can be added.

The antiglaucoma topical agents can also be associated to achieve the target pressure.

Drugs having the same action cannot be combined with each other (for example, two different beta-blockers or two prostaglandin analogs).

It is also recommended to use no more than two combined drugs where the second one should be added only if it is useful to achieve the target IOP.

An increase in the required drug dosage does not produce further therapeutical effects but only side effects.

When evaluating any class of medications, we must consider the therapeutic index (Robin, 1997). This is a measure of the relative potency, considering both the efficacy and the safety of a medication. The therapeutic index is a ratio of the toxic dose, typically at the 50% of the response level. The greater is the therapeutic index, the greater is the difference between the amount of medication that causes a beneficial effect and that dosage that commonly induces life-threatening side effects. The smaller is the therapeutic index, the smaller is the difference between a therapeutically desirable effect and a potentially serious side effect.

The medications now available fall into five classes:

- beta-blockers
- prostaglandin analogs
- alpha-agonists
- carbonic anhydrase inhibitors (CAIs)
- myotics.

All drugs work by lowering IOP, either by improving the aqueous humor outflow or reducing its production (Schwarzt and Budenz, 2004); they have been shown to be effective in lowering IOP and in preventing glaucoma progression, so the decision of which class of drug should be preferred, is never really based on efficacy only.

Ocular and systemic tolerability, dosing regimen, and cost must be considered as well.

If the starting IOP is higher, then the lowering percentage may be more than if the starting IOP is lower. Also these approximations only apply if the medicine is used at the frequency recommended by the drug leaflet.

As already mentioned, the duration of action of many antiglaucoma drugs is so short that some of the drug effect is already wearing off before the next dose is administered. This is clearly the case with pilocarpine, topical CAIs and alpha adrenergics (Talluto et al., 1997), while the prostaglandin analogs such as latanoprost, travoprost, and bimatoprost have a much longer duration of action.

Beta-blockers

Beta-blockers have been introduced for glaucoma treatment in 1979 and they have been the first-line therapy until recently.

Although beta-blockers have proven to be very effective and safe when used as eye drops, there are several long-term side effects.

Side effects are generally associated more with non-selective beta-blockers, such as timolol and levobunolol; thus, on the other side, a beta1-selective beta-blocker, like betaxolol, is not as effective as a non-selective beta-blocker (Caprioli and Garway-Heath, 2007) or as the ones with intrinsic sympathomimetic activity (ISA) such as carteolol and pindolol.

All the beta-blockers are less effective in dark colored eyes (Soltau and Zimmerman, 2002).

They all decrease IOP by the reduction of the aqueous humor production (Hayreh et al., 1999) and their peak effect occurs in 2 h (EGS, 2003).

At the starting time, the dose regimen should be at the lowest concentration and the administrations should be once or twice a day, but if the clinical response is not adequate, the dosage may be increased to higher concentrations.

Dosing more than twice daily will not give any further pressure-lowering effect. The washout time needed for beta-blockers is 2–5 weeks.

No dose–response curves for the different beta-blockers treatment have been established (EGS, 2003).

Non-selective beta-blockers are usually well tolerated, but may cause an exacerbation of respiratory symptoms in patients with reactive airway disease (such as asthma) and bradycardia (Schwarzt and Budenz, 2004).

It may be worthwhile to avoid beta-blockers in smoking patients and in those with a history of bronchospastic disorders. They should be used with caution in diabetics because they may mask the symptoms of hypoglycemia. These agents should be used with caution in any patient with heart disease, heart block, or cardiac failure.

Recently, reports (Hayreh et al., 1999) suggest that beta-blockers are associated with nocturnal hypotension, which may be a risk factor in glaucomatous progression.

After a prolonged use, depression, mood alterations, memory loss, hallucinations, decreased libido, impotence, and decreased exercise tolerance have also been reported with beta-blockers.

An easy and effective way to reduce systemic side effects is to perform a nasolacrimal occlusion after the topical application, reducing plasma levels by up to 70% (EGS, 2003).

Uncommon ocular side effects are epithelial keratopathy and a slight reduction in corneal sensitivity. Caution should be used in the co-administration of beta-adrenergic blocking agents with oral and intravenous calcium antagonists, digitalis, and catecholamine-depleting drugs.

Prostaglandins

This group of medications has had the biggest impact in the last 10 years of glaucoma treatment. For a long time it has been a skepticism on using these drugs, because of the inflammation they cause. After years of dedicated researches, it was found that by slightly modifying the prostaglandin molecule, it is possible to achieve an IOP lowering and reducing the inflammation at the same time.

Prostaglandins lower the IOP by increasing the aqueous outflow through an alternative pathway.

The first drug of this class has been latanoprost, followed by travoprost and bimatoprost. These agents have been approved for both first-line and adjunctive therapy, depending on the country (Hylton and Robin, 2003).

All the above once-daily prostaglandin analogs are at least as effective, if not even more, in lowering the IOP than timolol maleate. Regarding the efficacy within prostaglandin derivatives' class, the only conclusive study (Hylton and Robin, 2003) shows that latanoprost, bimatoprost, and travoprost appear to have a similar efficacy in reducing IOP of a 20–35% from baseline.

The reduction of the IOP starts approximately 2–4 h after the administration, with a peak effect reached approximately at 8–12 h. This intraocular-lowering effect persists, but is less evident at 48 h after the administration (EGS, 2003).

The maximum IOP lowering is often achieved at 3–5 weeks from the beginning of the treatment.

The most attractive feature of the prostaglandin analogs is their ability to significantly reduce the IOP with only once-daily administration in patients with glaucoma or ocular hypertension, preferably in the evening. In fact, their IOP-lowering ability is decreased if used more than once daily.

This once-daily usage, together with the favorable local and systemic side effect profile has increased rates of compliance and reduced rates of discontinuation of therapy. The relatively lower incidence of adverse systemic effects seen with topical prostaglandins compared with more traditional glaucoma therapies has promoted their usage as the ophthalmic market sales leader.

Prostaglandin analogs are similar to each other, with regard to their overall favorable safety profile. Some non-specific associations of adverse systemic effects, including upper respiratory tract infections, headache, flu-like syndrome, and musculoskeletal pain have been reported with all the three agents.

As topical side effects have been reported: redness in the treated eye for the first weeks of use, burning, tearing, recurrent erythema, itching, hyperemia, hypertrichosis, and increased iris and periocular skin pigmentation. Conjunctival hyperemia and eyelash growth are other side effects shared by all the three compounds. In a longer term we could find a change in the iris color to brown; the change is permanent but it is not intended as pathological. However, the highest incidence of hyperemia is seen in patients on bimatoprost therapy.

A treatment with prostaglandin analogs in pseudophakic and aphakic patients has reported to be associated with cystoid macular edema (CME). The incidence of CME in patients on prostaglandin therapy appears to be higher in those patients with a compromised blood–aqueous barrier. Pseudophakic or aphakic patients who have had complicated cataract surgery or vitrectomy are at a higher risk of developing CME, even without the addition of prostaglandin agents. It has been suggested that prostaglandins can accelerate the disruption of the blood–aqueous barrier after cataract surgery (Hylton and Robin, 2003).

Alpha-agonists

This new class of drugs is vaguely related to an older drug called dipivefrin that fell from favor (Robin, 1997). There are selective and non-selective alpha-agonists. The non-selective alpha-agonists are dipivefrin and epinephrine, which can be deleterious for occludable angles and aphakic patients because of the macular edema that can occur.

Clonidine was the first relatively selective alpha2-agonist identified for clinical application.

The concern over the topical clonidine's systemic side effects, such as lowering systemic blood pressure and decreasing blood flow to the optic nerve, may have led to the clinical development of other selective alpha2-agonists, including apraclonidine and brimonidine. None of the three is purely alpha1 or alpha2-selective.

Apraclonidine causes no systemic hypotension, and brimonidine causes less frequent systemic hypotension than other agents. They primarily lower the IOP by suppressing the aqueous humor production, but also by altering the ocular blood flow to the ciliar body and by decreasing the episcleral venous pressure. They can be used twice a day in combination with other drugs, or sometimes three times a day if used alone.

Brimonidine tartrate, in contrast to apraclonidine, is a lipophilic drug and its primary route is the cornea. It is administered at a twice-daily dosing regimen. The effect seems to markedly diminish at 6 h after dosing, till a maximum of 12 h; the maximum IOP's decrease range is between 20 and 30% (EGS, 2003).

The washed-out time needed for these compounds to completely lose their action is 1–3 weeks.

Brimonidine 0.2% is the most effective dose because it is not only at the top of the dose–response curve, but it has also showed the fewest systemic and local side effects (Robin, 1997).

The most frequent side effects reported with brimonidine are dry mouth, conjunctival blanching, systemic hypotension, fatigue, and drowsiness, especially in children.

There is a tendency for a number of patients to develop allergy to the drops after several months: this propensity does not mean that the drugs should not be used, but merely that if allergy develops they should be stopped and changed to an alternative treatment.

Its selectivity for alpha2 vs. alpha1 receptors results in no mydriasis and in the absence of vasoconstriction.

Brimonidine has been associated with respiratory and cardiac depression in infants and is controindicated under the age of 2; caution is actually indicated in all the pediatric patients and the nursing mothers.

There is some hopeful research suggesting that brimonidine may have a "direct neuroprotective" effect and may prevent the retinal nerve cells degeneration (Robin, 1997).

Neuroprotection is intended as a preservation from an early ganglion cells loss, caused by toxins and ischemia. A neuroprotective treatment should be aimed at recovering dysfunctioning RGCs subsequent to glaucoma related damage. It is likely that these cells do not die immediately when

Table 4. Indications to perform neuroimaging evaluation in normal-tension glaucoma

General
- Young age (less than 50 years)
- New onset or increased severity of headaches
- Localizing neurologic symptoms other than migraine
- Neurologic visual abnormalities

Ocular
- Color vision abnormalities
- Pallor of the remaining neuroretinal rim
- Highly asymmetric cupping
- Unilateral or highly asymmetric abnormalities
- Lack of disc and visual field correlation
- Visual field defect respecting vertical midline

evaluation is not performed in NTG diagnosis. Studies have shown that anterior visual pathway compression is a rare finding in neuroimaging of patients with a presumptive diagnosis of NTG (Greenfield et al., 1998). Relative indications to perform neuroimaging include younger age, decreased visual acuity, vertically aligned visual field defects that respect the midline, significant asymmetry, and neuroretinal rim pallor that exceeds the amount of cupping (Table 4).

Therapy

IOP reduction

The mainstay of treatment for glaucoma remains IOP reduction. The CNTGS recommendation is to reduce IOP by at least 30% to reduce the incidence of visual field progression. The modalities for IOP reduction include glaucoma medications, laser trabeculopasty, and glaucoma surgery. Glaucoma medications are the usual initial therapy for treatment. Glaucoma filtering surgery with antifibrotic agents (5-fluorouracil or mitomycin C) is the preferred surgery for NTG as it can achieve a very low postoperative IOP. The rate of cataract formation is higher after glaucoma surgery as shown in the CNTGS and other studies (AGIS Investigators, 2001). In fact, the protective effect of IOP reduction on visual field was masked by the progression of cataract in the treated group. Only by removing the data affected by cataract, the protective effect of IOP lowering on preservation of visual field was made evident (CNTGS, 1998a, b). It is important to point out that cataracts are reversible through a highly successful surgical procedure whereas visual field loss from glaucoma is not. Consequently, both the stage of glaucoma and the longevity of the patient need to be considered before embarking on surgical treatment that may accelerate cataract formation.

Systemic medications

New glaucoma therapies target IOP-independent mechanisms. Calcium channel blockers may protect visual field by increasing the capillary perfusion of the optic nerve head by relieving the effect of vasospasm in susceptible individuals, although results are conflicting and nonpersuasive. Various studies have suggested some benefits of nifedipine, verapamil, and nimodipine in protecting against glaucoma progression (Kitazawa et al., 1989; Bose et al., 1995; Netland et al., 1995), while other studies have shown no beneficial effect. Experts in the field of glaucoma do not commonly use calcium channel blockers because of a lack of clear evidence of efficacy and potential harmful side effects such as postural hypotension. If ophthalmologists wish to use calcium channel blocker therapy, they should coordinate treatment with a primary care physician because of potentially dangerous side effects such as systemic hypotension (Lumme et al., 1991). Another new but still experimental treatment is the use of angiotensin-converting enzyme (ACE) inhibitors. One small retrospective study found ACE inhibitors may have a favorable effect on visual field in NTG (Hirooka et al., 2006). The clinical significance of this study remains uncertain. A randomized controlled study is needed to evaluate the effect of ACE inhibitors in the prevention of visual field progression in NTG. Existing cardiovascular abnormalities (i.e., anemia, congestive heart failure, transient ischemic attack, arrhythmias) should be treated to ensure maximum perfusion of the optic nerve head.

Neuroprotection

IOP-independent mechanisms of glaucomatous optic nerve damage may play an important role in

NTG as a component of damage, or in a subset of susceptible patients. A goal of current glaucoma research is to develop neuroprotective treatment strategies to prevent retinal ganglion cell death (Kuehn et al., 2005). Memantine is a promising new drug that is currently being investigated for treatment of POAG. Memantine is thought to protect the optic nerve from the toxic glutamate levels that may lead to apoptosis of retinal ganglion cells in glaucoma. Any new treatment regime should have a rational scientific basis, be delivered safely to the site of damage, and show both efficacy and safety through randomized prospective clinical trials. Since glaucoma is a slowly progressive disease, it can take many years to detect a significant benefit of new treatments, particularly when added to robust IOP-lowering treatments.

Noncompliance

Noncompliance may be an important reason that some glaucoma suspects or glaucoma patients under "treatment" go on to develop severe deterioration of vision. Many of these individuals may have retained meaningful vision if appropriate therapy had been effectively applied. Several studies have evaluated factors that predispose glaucoma patients to noncompliance. Low socioeconomic status, language barrier, and aspects of treatment regime (i.e., dose frequency, number of medications, number of clinic visits) have all been linked to noncompliance. Few studies have been conducted to look at factors for noncompliance in the NTG population specifically. One study showed that approximately 50% of patients classified as NTG suspects lacked appropriate follow-up care. The lack of health insurance was a significant barrier for these patients in this study (Ngan et al., 2007). Another study showed that even within a single comprehensive insurance plan patients thought to require treatment for glaucoma were not being monitored at recommended intervals set by medical guidelines (Friedman et al., 2005). The Glaucoma Adherence and Persistency Study has shown that patient adherence to glaucoma medications is poor and comparable to other chronic diseases (Friedman et al., 2007).

More research is needed to better identify individuals at greatest risk for noncompliance. The Patient Care Improvement Project conducted by the American Glaucoma Society is attempting to identify both the barriers to compliance and tools to overcome these obstacles (American Glaucoma Society). This project has suggested that tools such as memory aids and tracking tools for medications, appointment reminders, and a new bottle design that would alert patients that it is time to refill medications may prove to be helpful. Also, social programs that empower the patient like support groups and accessible patient education classes would be helpful. As the glaucoma population grows, it will become very important that we address the problem of noncompliance. This significant issue will become an even larger burden on society as it takes people out of the work force and makes them dependent on social programs.

Genetics of NTG

A great deal of ongoing research is dedicated to identifying a genetic basis for NTG. An OPA1 gene polymorphism (OPA1 IVS $8+32$ T/C) has been associated with NTG, and one study showed that it may be used as a marker for this disease association (Mabuchi et al., 2007). Little is known, however, about the function of the OPA1 protein and how this polymorphism may cause glaucomatous neuropathy. An optineurin sequence variation, Glu50Lys OPTN, has been associated with familial NTG (Alward et al., 2003). This change, however, is responsible for less than 0.1% of open-angle glaucomas. It is unclear exactly what the function of this novel gene is. It is thought to protect the optic nerve from TNF-α-mediated apoptosis, and consequently a loss of function of this protein may decrease the threshold for ganglion cell apoptosis. Also, studies of lymphocytes in NTG have shown altered expression of the p53 gene, which is a known regulator of apoptosis (Wiggs, 2005). These results indicate that abnormal regulation of retinal ganglion cell apoptosis may be one of the IOP-independent mechanisms of optic nerve damage in glaucoma. It is unlikely that a single gene or even a small set of genes will be accountable for the clinical disease. It is likely that downstream effects, including but not limited to proteonomics, play a significant role in mechanisms of damage.

These require further intensive investigation to unravel and are likely to be quite complex.

Abbreviations

ACE	angiotensin-converting enzyme
CCT	central corneal thickness
CNTGS	Collaborative Normal-Tension Glaucoma Study
IOP	intraocular pressure
NTG	normal-tension glaucoma
POAG	primary open-angle glaucoma

References

AGIS (Advanced Glaucoma Intervention Study) Investigators. (2001) The advanced glaucoma intervention study: 8. Risk of cataract formation after trabeculectomy. Arch. Ophthalmol., 119(12): 1771–1779.

Allingham, R.R., Damji, K., Freedman, S., Moroi, S., Shafranov, G. and Shields, M.B. (2005) Chronic open-angle glaucoma and normal-tension glaucoma. In: Pine J. and Murphy J. (Eds.), Shields' Textbook of Glaucoma (5th edition). Lippincott Williams & Wilkins, Philadelphia, PA, pp 197–207. Chapter 11.

Alward, W., Kwon, Y., Kawase, K., Craig, J., Hayreh, S., Johnson, A., Khanna, C., Yamamoto, T., Mackey, D., Roos, B., Affatigato, L., Sheffield, V. and Stone, E. (2003) Evaluation of optineurin sequence variations in 1,048 patients with open-angle glaucoma. Am. J. Ophthalmol., 136: 904–910.

American Glaucoma Society. Patient Care Improvent Project: Pearls to Improve Patient Compliance. http://www.glaucomaweb.org/associations/5224/files/02Booklet_Pearls.F.pdf

Bose, S., Piltz, J.R. and Brenton, M.E. (1995) Nimodipine, a centrally active calcium antagonist, exerts a beneficial effect on contrast sensitivity in patients with NTG and in control subjects. Ophthalmology, 102(8): 1236–1241.

Butt, Z., McKillop, G., O'Brien, C., Allan, P. and Aspinall, P. (1995) Measurement of ocular blood flow velocity using color Doppler imaging in low tension glaucoma. Eye, 9(Pt 1): 29–33.

Caprioli, J. (1998) The treatment of normal-tension glaucoma. Am. J. Ophthalmol., 126(4): 578–581.

Caprioli, J. and Spaeth, G.L. (1984) Comparison of visual field defects in the low-tension glaucomas with those in the high-tension glaucomas. Am. J. Ophthalmol., 97: 730–737.

Caprioli, J. and Spaeth, G.L. (1985) Comparison of the optic nerve head in high- and low-tension glaucoma. Arch. Ophthalmol., 130: 1145–1149.

Cartwright, N.J. and Anderson, D.R. (1988) Correlation of asymmetric damage with asymmetric intraocular pressure in NTG (low tension glaucoma). Arch. Ophthalmol., 106: 898–900.

Cartwright, M.J., Grajewski, A.L., Friedberg, M.L., Anderson, D.R. and Richards, D.W. (1992) Immune-related disease and normal-tension glaucoma: a case–control study. Arch. Ophthalmol., 110(4): 500–502.

Crichton, A., Drance, S.M., Douglas, G.R. and Schulzer, M. (1989) Unequal intraocular pressure and its relation to asymmetric visual field defects in low-tension glaucoma. Ophthalmology, 96: 1312–1314.

Cursiefen, C., Wisse, M., Cursiefen, S., Junemann, A., Martus, P. and Korth, M. (2000) Migraine and tension headache in high-pressure and normal-pressure glaucoma. Am. J. Ophthalmol., 129: 102–104.

Drance, S., Anderson, D.R. and Schulzer, M. (2001) Risk factors for progression of visual field abnormalities in normal-tension glaucoma. Am. J. Ophthalmol., 131: 699–708.

Friedman, D.S., Nordstrom, B., Mozaffari, E. and Quigley, H.A. (2005) Glaucoma management among individuals enrolled in a single comprehensive insurance plan. Ophthalmology, 112: 1500–1504.

Friedman, D.S., Quigley, H.A., Gelb, L., Tan, J., Margolis, J., Shah, S.N., Kim, E.E., Zimmerman, T. and Hahn, S.R. (2007) Using pharmacy claims data to study adherence to glaucoma medications: methodology and findings of the Glaucoma Adherence and Persistency Study (GAPS). Invest. Ophthalmol. Vis. Sci., 48: 5052–5057.

Gaasterland, D.E., Allingham, R.R., Gross, R.L., Jampel, H.D., Kwon, Y.H., Prum, B.E. and Gordon, M.O. (2005) In: Garratt S. (Ed.), American Academy of Ophthalmology. Primary Open-Angle Glaucoma, Preferred Practice Pattern. San Francisco, American Academy of Ophthalmology, pp. 1–23. Available at www.aao.org/ppp

Geijssen, H.C. (1991) Studies on Normal-Pressure Glaucoma. Kugler, Amstelveen, The Netherlands.

Graefe, A.V. (1857) Uber die Iridectomie bei Glaucoma und über den Glaucomatösen Prezess. Albrecht von Graefes Archiv. Klin. Exp. Ophthalmol., 13: 456–650.

Greenfield, D.S., Liebmann J.M., Ritch, R., Krupin, T. and Low-Pressure Glaucoma Study Group. (2007) Visual field and intraocular pressure asymmetry in the Low-Pressure Glaucoma Treatment Study. Ophthalmology, 114(3): 460–465.

Greenfield, D.S., Siatkowski, R.M., Glaser, J.S., Schatz, N.J. and Parrish, R.K. (1998) The cupped disc: who needs neuroimaging? Ophthalmology, 105: 1866–1874.

Harrington, D.E. (1969) The pathogenesis of the glaucoma field: clinical evidence that circulatory insufficiency in the optic nerve is the primary cause of visual field in glaucoma. Am. J. Ophthalmol., 47(2): 177–185.

Harris, A., Rechtman, E., Siesky, B., Jonescu-Cuypers, C., McCranor, L. and Garzozi, H.J. (2005) The role of optic nerve blood flow in the pathogenesis of glaucoma. Ophthalmol. Clin. North Am., 18(3): 345–353.

Hayreh, S.S., Podhajsky, P. and Zimmerman, M.B. (1999) Beta-blocker eye drops and nocturnal arterial hypertension. Am. J. Ophthalmol., 128(3): 301–309.

Hayreh, S.S., Zimmerman, M.B., Podhajsky, P. and Alward, W.L. (1994) Nocturnal arterial hypotension and its role in optic nerve head and ocular ischemic disorders. Am. J. Ophthalmol., 117(5): 603–624.

Hirooka, K., Baba, T., Fujimura, T. and Shiraga, F. (2006) Prevention of visual field defect progression with

XFS leads not only to severe, chronic open-angle glaucoma, but also to lens subluxation, angle-closure, blood–aqueous barrier impairment, and serious complications at the time of cataract extraction, such as zonular dialysis, capsular rupture, and vitreous loss. There is increasing evidence for an etiological association of XFS with cataract formation and with retinal vein occlusion. Systemic associations, primarily ischemic vascular disease, are being increasingly reported. Plasma, aqueous humor, and tear fluid homocysteine levels are elevated in XFS with and without glaucoma (Leibovitch et al., 2003; Vessani et al., 2003; Bleich et al., 2004; Puustjärvi et al., 2004; Altintas et al., 2005; Roedl et al., 2007a, b). The recent discovery of two polymorphisms in the lysyl oxidase-like 1 (LOXL1) gene that confer susceptibility to exfoliative glaucoma, primarily through XFS (Thorliefsson et al., 2007) will hopefully open the door for much more research on this understudied disease and lead to new approaches to treatment.

Epidemiology

The reported prevalence of XFS both with and without glaucoma has varied widely. This reflects true differences due to racial, ethnic, or other as-yet-unknown factors; the age and sex distribution of the patient cohort or population group examined; the clinical criteria used to diagnose XFS; the ability of the examiner to detect early stages and/or more subtle signs; the method and thoroughness of the examination; and the awareness of the observer (Aasved, 1969). Although long and erroneously thought peculiar to Scandinavia, XFS comprises over half the open-angle glaucoma in such diverse countries as Norway, Ireland, Greece, Saudi Arabia, and India. Previously thought rare in Africa, recent reports suggest that it is common in Ethiopia, where 25% of open-angle glaucoma patients had XFS (Bedri and Alemu, 1999). It is also found in Navajo Indians and in Australian Aborigines and in South African Zulus. In the United States, it is much more common in Caucasians than in persons of African ancestry, comprising about 12% of glaucoma populations (Gradle and Sugar, 1947; Roth and Epstein, 1980;

Horns et al., 1983). There are geographic and ethnic variations, being less in the southern United States and in African-Americans. Although common in Japan and Mongolia, it is rare in southern China.

The prevalence of XFS increases steadily with age in all populations. About two-thirds of patients have clinically unilateral involvement, but XFS can be diagnosed prior to the clinically visible appearance of classic exfoliation material (XFM) on the lens surface by conjunctival biopsy, showing that the disease is present microscopically before it is detected clinically in the fellow eye (Prince et al., 1987). The reasons for this presentation remain unknown, but it may be analogous to that of uveitis, which is also often clinically unilateral or markedly asymmetric.

Ethnic and local variations also exist. In New Mexico, Spanish-American men develop XFS six times as often as Anglo-Americans (Jones et al., 1992). In France, XFS is much more common in Brittany, the population of which has Celtic origins, than in southeastern France (Colin et al., 1988). It accounts for about 60% of the open-angle glaucoma in Ireland and in the Isle of Man, but only 10% in neighboring England. In central Norway, the prevalence in two adjacent towns (20%) was twice that in a third, adjacent town. Interestingly, the prevalence of XFS in both members of 343 married couples (3.2%) was significantly higher ($P = 0.022$) than would be expected, suggesting the possibility of an infectious origin (Ringvold et al., 1988). In Nepal, XFS was found in 12% of members of one ethnic group, the Gurung, and only 0.24% of non-Gurung of similar ages (Shakya et al., 2004). Other examples exist and why this is so has yet to be explained.

The prevalence of XFS in glaucoma cohorts is significantly higher than in age-matched nonglaucomatous populations. In 100 consecutive patients with XFS, Kozart and Yanoff (1982) found glaucomatous optic nerve or visual field damage in 7% and ocular hypertension in 15%. This is approximately six times the chance of finding elevated IOP in eyes without XFS. Similar figures for elevated IOP with or without glaucomatous damage have been reported in XFS patients in Europe (Moreno Montañés et al., 1989; Kozobolis et al., 1997), Australia (Mitchell et al., 1999), and Japan (Shimizu et al., 1988).

The prognosis of exfoliative glaucoma (XFG) is more severe than that of primary open-angle glaucoma (POAG). Patients with XFS are twice as likely to convert from ocular hypertension to glaucoma and when glaucoma is present, to progress (Leske et al., 2003; Bengtsson and Heijl, 2005). The mean IOP is greater in normotensive patients with XFS than in the general population and greater in XFG patients at presentation than in POAG patients. At any specific IOP level, eyes with XFS are more likely to have glaucomatous damage than are eyes without XFS. There is greater 24-h IOP fluctuation, greater visual field loss, and optic disc damage at the time of detection, poorer response to medications, more rapid progression, greater need for surgical intervention, and greater proportion of blindness.

Clinical findings

XFS is diagnosed by the presence of typical XFM on the anterior lens surface or pupillary border (Fig. 1). All anterior segment structures are involved. The classic pattern of deposition on the lens consists of three distinct zones that become visible when the pupil is fully dilated, a central disc, intermediate clear zone created by the iris rubbing XFM from the lens surface during its physiologic excursions, and a granular peripheral zone. XFM is often found at the pupillary border.

Just as the iris scrapes XFM from the lens surface, the material on the lens causes rupture of iris pigment epithelial cells with concomitant pigment dispersion into the anterior chamber, leading to iris transillumination defects, loss of the ruff, and trabecular hyperpigmentation (Figs. 2 and 3). Pigment dispersion after pupillary dilation may be profuse. Marked intraocular (IOP) rises can occur and IOP should be measured routinely in all patients after dilation.

The diagnosis should be suspected in the absence of XFM in the presence of these signs, which define patients as "exfoliation suspects"

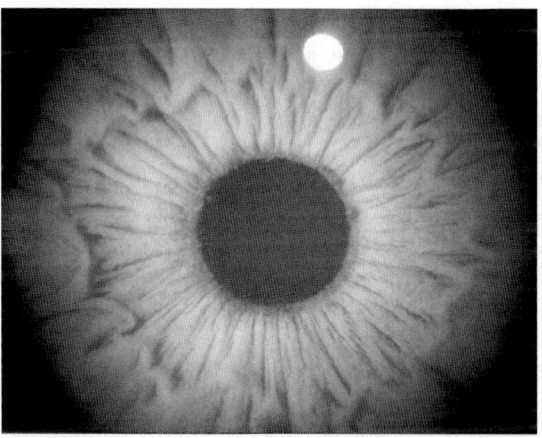

Fig. 2. Loss of iris pigment ruff in XFS. Exfoliation material is present on the pupillary border. (See Color Plate 15.2 in color plate section.)

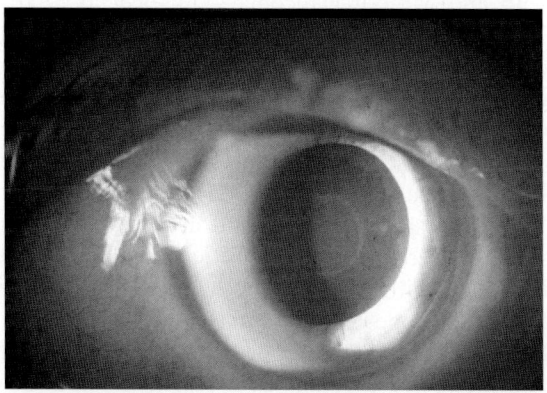

Fig. 1. Classic appearance of XFM on the anterior lens surface consists of a central disc, intermediate clear zone, and granular peripheral zone. (See Color Plate 15.1 in color plate section.)

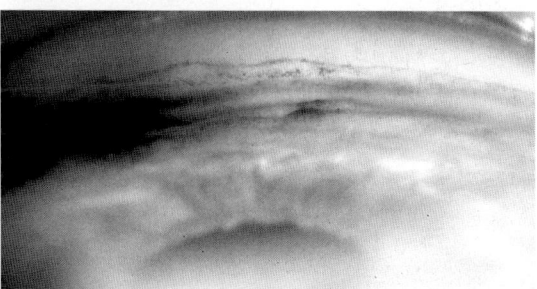

Fig. 3. Hyperpigmentation of the trabecular meshwork. Pigment is also found on Schwalbe's line and on the peripheral corneal shelf (Sampaolesi line). (See Color Plate 15.3 in color plate section.)

(Prince and Ritch, 1986; Prince et al., 1987). Transmission electron microscopy of conjunctival biopsy specimens from such patients, who had previously been diagnosed to have POAG, revealed XFM in 8/23 suspect eyes. These pigment-related signs also correlated with the presence of XFM in eyelid skin (Schlötzer-Schrehardt et al., 1993).

The zonules are often frayed and broken (Fig 4). Phacodonesis is common, and spontaneous lens subluxation can occur. Complications of cataract surgery are more common. Zonular fragility, poor pupillary dilation, and posterior synechiae increase the risk of lens dislocation, zonular dialysis, and vitreous loss.

Ocular and systemic associations

Ocular associations

Cataract and XFS are etiologically associated. Aqeous humor ascorbic acid concentrations are reduced in XFS (Koliakos et al., 2002), and malondialdehyde concentrations are much higher (Yilmaz et al., 2005), suggesting a faulty antioxidant defense system. Increased 8-*iso*-Prostaglandin F_{2a} suggests a role for free radical-induced oxidative damage (Koliakos et al., 2003).

Ocular ischemia is the rule and iris abnormalities can be detected both on angiography (Laatikainen, 1971) and histopathologically (Hammer et al., 2001), not only in affected eyes, but in unaffected fellow eyes of patients with clinically unilateral involvement. Vessel lumens are often narrowed and may become obliterated, with marked alteration of the iris vasculature in advanced cases. Vessel dropout with collateral formation and iris hypoperfusion lead to anterior chamber hypoxia and patchy iris microneovascularization (Ritch and Schlötzer-Schrehardt, 2001; Schlötzer-Schrehardt and Naumann, 2006). There is chronic breakdown of the blood–aqueous barrier and increased protein in the anterior chamber.

Retinal vein occlusion has been associated with XFS at variable frequencies and based upon retrospective studies, which employed either slit-lamp examination or histopathology (Saatci et al., 1999; Cursiefen et al., 2001). Elevated IOP and glaucoma have been suggested as a cause for the association of XFS with CRVO.

Systemic associations

An emerging clinical spectrum of associations with cardiovascular and cerebrovascular diseases elevates XFS to a condition of general medical importance. Deposits of XFM have been identified in the walls of posterior ciliary arteries, vortex veins, and central retinal vessels as they exit the optic nerve. Ocular, retrobulbar, and cerebral blood flow are reduced in patients with XFS both with and without glaucoma (Yüksel et al., 2001; Yüksel et al., 2006). Blood flow of the lamina cribrosa and neural rim decreases with increasing glaucomatous damage (Harju and Vesti, 2001). In clinically unilateral cases, ipsilateral pulsatile ocular blood flow and carotid blood flow are reduced (Scullica et al., 1993; Sibour et al., 1997). Recently, pathological carotid artery function as well as altered parasympathetic vascular control increasing with both age and higher homocysteine levels was reported (Visontai et al., 2006). In a large study, XFS was reported to be an important risk factor for coronary artery disease (Andrikopoulos et al., 2007). Patients with exfoliative glaucoma had lower baseline fingertip cutaneous capillary perfusion than those with POAG or

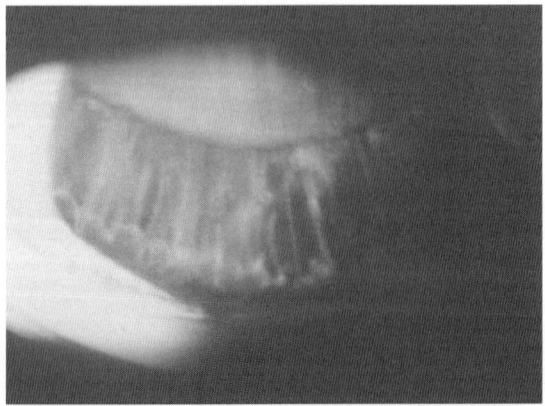

Fig. 4. Fragmented zonules in XFS. (See Color Plate 15.4 in color plate section.)

controls, longer time to maximal cold-induced flow reduction, and longer recovery time (Holló et al., 1998). These findings have been summarized in a recent editorial (Irkec, 2006).

Electron microscopy reveals aggregates of XFM in heart, lung, liver, kidney, gall bladder, and meninges in patients with ocular XFS (Schlötzer-Schrehardt et al., 1992; Streeten et al., 1992). In addition to the vascular abnormalities described above, an increasing number of associations with specific systemic disorders is being reported, including transient ischemic attacks, hypertension, angina, myocardial infarction, stroke, asymptomatic myocardial dysfunction, Alzheimer's disease, and hearing loss (Hagadus et al., 1989; Repo et al., 1995; Mitchell et al., 1997; Linnér et al., 2001; Cahill et al., 2002; Shaban and Asfour, 2004; Aydogan Ozkan et al., 2006). Some of these associations have been disputed and there is yet no clear evidence of increased mortality in patients with XFS, which one might expect with these associations, nor has there been shown a clear-cut association of XFS with a systemic disease with conclusive evidence of a functional deficit caused by the presence of XFS.

Pathogenesis of exfoliation syndrome

The pathologic process is characterized by the chronic accumulation of an abnormal fibrillar matrix product, which is either the result of an excessive production or insufficient breakdown or both, and which is regarded as pathognomonic for the disease, based on its unique light microscopic and ultrastructural criteria (Naumann et al., 1998).

Immunohistochemical, biochemical, and molecular biologic data give strong support to the elastic microfibril theory of pathogenesis, first proposed by Streeten et al. (1986) on the basis of histochemical similarities between XFM and zonular fibers and which explains XFS as a type of elastosis affecting elastic microfibrils. The characteristic fibrils, composed of microfibrillar subunits surrounded by an amorphous matrix comprising various glycoconjugates, contain predominantly epitopes of elastic fibers, such as elastin, tropoelastin, amyloid P, vitronectin, and components of elastic microfibrils, such as fibrillin-1, microfibril-associated glycoprotein (MAGP-1), and latent TGF-β binding proteins (LTBP-1 and LTBP-2) by immunohistochemistry (Ritch and Schlötzer-Schrehardt, 2001). Recently, a direct analytical approach by using liquid chromatography coupled with tandem mass spectrometry (LC-MS/MS) has been accomplished and showed XFM to consist of the elastic microfibril components fibrillin-1, fibulin-2, and vitronectin, the proteoglycans syndecan and versican, the extracellular chaperone clusterin, the cross-linking enzyme lysyl oxidase, and other proteins, confirming many of the previously reported immunohistochemical data (Ovodenko et al., 2007). The currently proposed pathogenetic concept of XFS describes the condition as a specific type of stress-induced elastosis, an elastic microfibrillopathy, associated with the excessive production of elastic microfibrils and their aggregation into typical mature fibrils by a variety of potentially elastogenic cells (Schlötzer-Schrehardt and Naumann, 2006).

A set of genes primarily involved in extracellular matrix metabolism and in cellular stress was found to be differentially expressed in anterior segment tissues of XFS eyes [Zenkel, 2005 # 14541] (Zenkel et al., 2006), suggesting that the underlying pathophysiology of XFS is associated with excess production of elastic microfibril components, enzymatic cross-linking processes, overexpression of the transforming growth factor (TGF-β1), a proteolytic imbalance between matrix metalloproteinases (MMPs) and their tissue inhibitors (TIMPs), increased cellular and oxidative stress, and an impaired cellular stress response (Schlötzer-Schrehardt et al., 2003; Ho et al., 2005; Rönkkö et al., 2007). Growth factors, particularly TGF-β1, increased cellular and oxidative stress, an impaired cellular protection system, and the stable aggregation of misfolded stressed proteins appear to be involved in this fibrotic process (Koliakos et al., 2001). Due to an imbalance between MMPs and TIMPs and extensive cross-linking processes involved in fiber formation, the pathologic material is not properly degraded but progressively accumulates within the tissues over time.

Proof of the elastotic concept was supplied by a recent milestone study showing that two common single nucleotide polymorphisms in the coding region of the *lysyl oxidase-like 1* (*LOXL1*) gene located on chromosome 15 were specifically associated with XFS and XFG in two Scandinavian populations accounting for virtually all XFS cases (Thorliefsson et al., 2007). These findings have been substantiated in other populations (Fingert et al., 2007; Hewitt et al., 2007; Hayashi et al., 2008; Aragon-Martin et al., 2008).

LOXL1 is a member of the lysyl oxidase family of enzymes, which are essential for the formation, stabilization, maintenance, and remodeling of elastic fibers and prevent age-related loss of elasticity of tissues (Oleggini et al., 2007). Lysyl oxidase serves both as a cross-linking enzyme and a scaffolding element which ensures spatially defined deposition of elastin (Liu et al., 2004). It also regulates the promoter of elastin (Oleggini et al., 2007). Genetic variation in LOXL1 has been suggested to be a factor in spontaneous cervical artery dissection, a common cause of stroke in relatively young patients (Kuhlenbaumer et al., 2007). Overactivity of lysyl oxidase, with localization of the enzyme in blood vessel walls and in plaque-like structures, has been found in Alzheimer's disease (Gilad et al., 2005). LOXL1 protein is a major component of exfoliation deposits and appears to play a role in its accumulation and in concomitant elastotic processes in intra and extraocular tissues of XFS patients.

A number of nongenetic factors, including dietary factors, autoimmunity, infectious agents, and trauma, have also been hypothesized to be involved in the pathogenesis of XFS. Altogether, it appears that XFS represents a complex, multifactorial, late-onset disease, involving both genetic and nongenetic factors in its pathogenesis.

Mechanisms of glaucoma development

Friction between the iris and the XFM covering the lens surface leads to disruption of the iris pigment epithelium at the sphincter and pigment liberation. Just as the iris scrapes XFM from the lens surface, the material on the lens causes rupture of iris pigment epithelial cells at the ruff and sphincter region with concomitant dispersion of pigment into the anterior chamber. Blockage of aqueous outflow by a combination of pigment and XFM deposited in the intertrabecular spaces, and XFM in the juxtacanalicular meshwork, and beneath the endothelium of Schlemm's canal is believed to be the major cause of elevated IOP.

Although XFG is characteristically a high-pressure disease, pressure-independent risk factors, such as an impaired ocular and retrobulbar perfusion and abnormalities of elastic tissue of the lamina cribrosa, may be present and further increase the individual risk for glaucomatous damage. In a prospective study, Puska et al. (1999) found that in normotensive XFS patients with clinically unilateral involvement, in whom IOP was equal throughout the follow-up period, disc changes took place only in the involved eye, suggesting that the exfoliation process itself may be a risk factor for optic disc changes.

Management

Medical therapy

Initial medical treatment of XFG generally consists of topical prostaglandin analogs and aqueous suppressants. In one crossover trial, bimatoprost provided a statistically greater IOP reduction for all time points and for the mean diurnal curve after 3 months of therapy (35% reduction with bimatoprost vs. 31% reduction for latanoprost; Konstas et al., 2007a, b). Another 24-h study showed that both latanoprost and travoprost reduced IOP at each time point and for the 24-h curve from untreated baseline (Konstas et al., 2007a, b). However, travoprost provided a slightly greater hypotensive effect for the 24-h curve (mean difference between groups was 0.5 mmHg). Konstas et al. (2004) demonstrated the benefit of IOP reduction and suggested a target IOP of ≤17 mmHg and lower to prevent or slow progressive damage. Latanoprost provided a narrower range of diurnal IOP fluctuation compared to timolol. In addition to lowering IOP, prostaglandin analogs may interfere with the disease process. Latanoprost treatment had a marked effect on the aqueous concentration of TGF-β1, MMP-2, and TIMP-2 in XFG patients (Konstas et al., 2006a, b).

Aqueous suppressants do not interfere with the mechanism of the cause of progression of trabecular damage, i.e. iridolenticular friction and disruption of the iris pigment epithelial cells. Cholinergic agents have multiple beneficial actions in eyes with XFS. Not only do they lower IOP, but by increasing aqueous outflow, they should enable the trabecular meshwork to clear more rapidly, and by limiting pupillary movement, should slow the progression of the disease. Aqueous suppressants, on the other hand, by decreasing aqueous secretion, result in decreased aqueous flow through the trabecular meshwork. Becker, 1995 has presented suggestive evidence that treatment with aqueous suppressants leads to worsening of trabecular function.

Pilocarpine has multiple beneficial actions in eyes with XFS. Not only does it lower IOP, but by increasing aqueous outflow, it should enable the trabecular meshwork to clear more rapidly, and by limiting pupillary movement, should slow the progression of the disease. Theoretically, miotics should be the first line of treatment. The use of miotics, unfortunately, has almost disappeared from use in glaucoma on the basis that they are considered a q.i.d. drug and because many patients have nuclear sclerosis and miotics may reduce visual acuity or dim vision sufficiently to create difficulty. Pilocarpine has also been shown to blunt an early morning IOP spike which often occurs in the supine position after a night's sleep (Barkana et al., 2006). The long-term use of miotics may lead to the development of posterior synechiae in patients with XFS. However, we have found that 2% pilocarpine q.h.s. can provide sufficient limitation of pupillary mobility without causing these side effects. An international, multiinstitutional prospective trial (International Collaborative Exfoliation Syndrome Treatment Study) comparing latanoprost and 2% pilocarpine q.h.s versus timolol/Cosopt for patients with XFS and ocular hypertension or glaucoma has been completed and analysis is in progress.

Laser surgery

Argon laser trabeculoplasty (ALT) is particularly effective in eyes with XFS. The baseline IOP is usually higher in XFG than in eyes with POAG undergoing ALT and thus the initial drop in IOP is greater in XFG. Primary ALT can delay the use of medical therapy for up to 8 years in a significant proportion of these patients (Odberg and Sandvik, 1999). There is a gradual reduction in success over time, with long-term rates dropping to approximately 35–55% at 3–6 years.

Approximately 20% of patients develop sudden, late rises of IOP within the first 2 years after treatment (Ritch and Podos, 1983). Continued pigment liberation may overwhelm the restored functional capacity of the trabecular meshwork, and maintenance miotic therapy to minimize papillary movement after ALT might counteract this. We have found empirically that 2% pilocarpine q.h.s. is sufficient to provide this protection.

Selective laser trabeculoplasty (SLT) needs to be further evaluated as an effective and safe alternative to ALT in the treatment of XFG. In a prospective trial, Gracner (2002) found that the success rate after 12 months (Kaplan–Meier survival analysis) was 97%, after 24 months 88%, after 36 months 76%, and after 48 months 71%. Further studies are necessary. It should be noted that SLT in eyes with heavily pigmented trabecular meshworks may cause marked IOP elevations, even necessitating filtration surgery (Harasymowicz et al., 2005).

Operative surgery

The results of trabeculectomy are comparable to those in POAG. In a recent study, trabeculectomy with MMC obtained better 24-h IOP control than successful maximal medical therapy in patients with advanced XFG or POAG (Konstas et al., 2006a, b). However, surgical complications are more common in patients with XFG. In some cases high preoperative IOP may predispose to choroidal hemorrhage, or effusion. Weakened zonular support may allow intraoperative lens movement, or even in extreme cases subluxation.

This could lead to inadvertent lens damage during iridectomy, vitreous loss, or late incarceration of vitreous into the internal ostium. Hyphema from the surgical iridectomy could be the result of undetected iris fine neovascularization, which occurs in XFG. After trabeculectomy, there is an increased possibility of cataract progression in patients with XFG. The more advanced the disease and the longer the duration of it the higher the likelihood for complications to occur.

Trabeculotomy, performed with the rationale that it may bypass mechanical blockage of the trabecular meshwork, has been reported successful (Gillies, 1977; Tanihara et al., 1993; Honjo et al., 1998). Along similar lines of reasoning, Jacobi et al. (1998) described a procedure termed trabecular aspiration, designed to improve outflow facility.

Deep sclerectomy and similar procedures including a deroofing of Schlemm's canal are becoming popular choices in some centers owing to the reduced risk profile of nonpenetrating surgery. Patients with XFG had significantly better success than those with POAG following deep sclerectomy with either an absorbable collagen implant or a nonabsorbable hydrophylic acrylic implant (Drolsum, 2003, 2006). Another study showed equal effectiveness (Rekonen et al., 2006). Phacoemulsification combined with penetrating and nonpenetrating procedures does not seem to adversely influence success rate.

Future treatment of exfoliation syndrome and exfoliative glaucoma

For most of the twentieth century, glaucoma was equated with elevated IOP, and all therapy has been guided at lowering IOP. Directed therapy simply means devising specific treatments for specific diseases. There was little incentive to attempt to distinguish between various open-angle glaucomas if the treatments were essentially the same. However, this view also prevented the application of directed therapy in those instances in which such was available and applicable.

Aqueous outflow

Miotics, which improve trabecular outflow, as mentioned above, have nearly passed from the armamentarium for treatment of glaucoma. Drugs which affect the integrity of the cytoskeleton of the trabecular meshwork, may be a boon to the treatment of XFS and XFG, although they have not yet been examined for this purpose. Over 30 years ago, Kaufman and Bárány (1977) showed that intracameral cytochalasin B caused an increase in outflow facility in the eyes of cynomolgus monkeys. Bill et al. (1980) described cell separation in the trabecular meshwork after intracameral injection of sodium-EDTA. There was also distention of the juxtacanalicular meshwork, washout of extracellular material, and disintegration of the denuded trabecular cores. Use of these drugs never came to fruition in patients because of potential toxicity, particularly to the cornea, but it can be seen that a washout of extracellular material from the trabecular meshwork might be a breakthrough approach to the treatment of XFS if the blockage to the trabecular meshwork could be removed and have a long-lasting effect. The serine–threonine protein kinase inhibitor H-7 was shown to have a similar effect on the trabecular meshwork by a different mechanism (Tian et al., 1999) probably by inhibiting cell contractility, cytoskeletal support, and cell–cell adhesions in the trabecular meshwork (Tian et al., 1998).

More recently, latrunculin B administered topically has been shown to increase aqueous outflow facility by a similar mechanism (Peterson et al., 2000) without having any adverse effect on the retinal vascular permeability or electrophysiology (Kiland et al., 2006) or affecting the cornea (Okka et al., 2004; Sabanay et al., 2006). Latrunculins bind to the free actin in the cell, preventing it from polymerizing into microfilaments. The existing actin cytoskeleton gradually degenerates, leading to a large increase in outflow facility (Ethier et al., 2006). Depending on its effect on the XFM and pigment blocked in the intertrabecular spaces, an effect as yet undetermined, latrunculin may turn out to be a drop which might be given much less than once a day, perhaps even once a week or once a month. Early trials in human patients are now in progress.

Treatment directed at exfoliation material

Reduction of IOP should not be the sole focus of therapy in XFG. As yet, there are no medications proven in controlled, prospective trials to provide either neuroprotection or vasoprotection in glaucoma. Understanding the mechanisms leading to elevated IOP in XFS could allow us to develop new more logical approaches to therapy. The eventual goal is to prevent the development of XFM, thus effectively curing this disease. A treatment which would eliminate the formation of XFM or depolymerize it when once formed should be a prime goal. Possible approaches include finding a means to prevent it from aggregating initially, prevent it from cross-linking, disaggregating the fibrils, and depolymerizing the microfibrils.

Homocysteine. Homocysteine is a highly cytotoxic amino acid derived from methionine metabolism, and elevated serum levels result from disturbed methionine metabolism. Mild hyperhomocysteinemia, a recognized cardiovascular risk factor, is common and may result from a variety of causes affecting homocysteine metabolic pathways (Selhub, 2006). Hyperhomocysteinemia in animals is associated with disruption of the elastic fiber component of the extracellular matrix, with resulting vascular complications (Starcher and Hill, 2005). Elevated homocysteine levels are present in blood, aqueous humor, and tear film in patients with XFS (Leibovitch et al., 2003; Vessani et al., 2003; Bleich et al., 2004; Puustjärvi et al., 2004; Altintas et al., 2005; Roedl et al., 2007a, b). The systemic abnormalities associated with XFS are also associated with hyperhomocysteinemia, which thus appears to be a common thread extending through both XFS and the systemic disorders associated with it.

A large study of 24,968 healthy women found homocysteine levels to be inversely associated with intake of folate and vitamins B2, B6, and B12 (Zee et al., 2007). Treatment with folic acid and vitamins B6 and B12 reduce homocysteine concentrations in patients with coronary artery disease (Lobo et al., 1999). Homocysteine might be a modifiable risk factor for XFS. A decreased serum concentration of vitamins B6 and B12 and folate has recently been reported in patients with XFS (Roedl et al., 2007a, b). Because of the strong association with elevated homocysteine levels, one must also consider the possibility that these patients may benefit from lowering of plasma homocysteine levels by supplemental vitamins B6, B12, and folic acid (Vessani et al., 2003).

Folate deficiency leads to altered expression of genes involved in cell signaling, the cytoskeleton, and the extracellular matrix (Katula et al., 2007). Actin disrupting agents, such as latrunculin B, reversibly increase the proportion of receptors on the cell surface and increase (Holven et al., 2006) the rate of 5-methyltetrahydrofolate delivery (Lewis et al., 1998).

Vitamin B_6 is an essential micronutrient involved in a variety of critical metabolic reactions including carbohydrate metabolism, sphingolipid biosynthesis and degradation, amino acid metabolism (including that of homocysteine), and neurotransmitter metabolism (Merrill and Henderson, 1987). Therefore, deficiency of this essential micronutrient in humans leads to a variety of adverse conditions and to disturbances in normal cellular metabolism. Pyridoxine also plays a role in the integrity of the extracellular matrix (Massé et al., 1995).

Vitamin B12 is important for the normal functioning of the brain and nervous system and for the formation of blood. It is normally involved in the metabolism of every cell of the body, especially affecting the DNA synthesis and regulation but also fatty acid synthesis and energy production. Most symptoms of B12 deficiency are actually folate deficient symptoms.

Inflammation. C-reactive protein is a marker of inflammation and a predictor of cardiovascular disease, while interleukin-6, a regulator of C-reactive protein plays a key role in the initiation of inflammation (Holven et al., 2006). Patients with hyperhomocysteinemia have elevated levels of these compounds (Holven et al., 2006). Evidence that XFS is accompanied by low-grade inflammation has been presented (Ovodenko et al., 2007).

Lysyl oxidase. Lysyl oxidase in vascular endothelia is inhibited by high concentrations of homocysteine. This downregulation impairs the endothelial barrier function and could be involved in homocysteine-induced endothelial dysfunction (Raposo et al., 2004). Endothelial dysfunction induced by tumor necrosis factor-alpha (TNF-α) is also associated with a decrease of lysyl oxidase expression or activity (Rodríguez et al., 2007). Hyperhomocysteinemia in animals is associated with disruption of the elastic fiber component of the extracellular matrix, with resulting vascular complications (Starcher and Hill, 2005).

Transforming growth factor-β1 (TGF-β1) also interacts with lysyl oxidase to influence the formation of elastic tissue (Oleggini et al., 2007), and levels of TGF-β1 are significantly elevated in the aqueous humor of eyes with XFS and are believed to be both responsible for overproduction of extracellular matrix and an important causative factor for the production of XFM (Schlötzer-Schrehardt et al., 1999). TGFβ-1 and β-2 contribute to conjunctival scarring after filtering surgery (Kottler et al., 2005). Modification of TGF-β1 activity may improve both the disease itself and the surgical treatment necessitated by it.

Our eventual goal is to prevent the development of XFM, thus effectively curing this disease. A treatment which would eliminate the formation of XFM or depolymerize it once formed should be a prime goal. Possible approaches include finding a means to prevent it from aggregating initially, prevent it from cross-linking, disaggregating the fibrils, and depolymerizing the microfibrils.

Interactions between lysyl oxidase, homocysteine, TGF-β1, and their effect upon elastic tissue in XFS require clarification. How do the different polymorphisms of the *LOXL1* gene affect these interactions and alter elastogenesis? Will mutations in *LOXL1* be discovered? Different mutations may lead to greater or lesser disease severity. In the future, will we be able to modulate *LOXL1* gene activity to alter the course of XFS? These and other questions are now open to further exploration in an attempt to manage this very common and severe glaucoma.

References

Aasved, H. (1969) The geographical distribution of fibrillopathia epitheliocapsularis. Acta Ophthalmol., 47: 792–810.

Altintas, O., Maral, H., Yuksel, N., Karabas, V.L., Dillioglugil, M.O. and Caglar, Y. (2005) Homocysteine and nitric oxide levels in plasma of patients with pseudoexfoliation syndrome, pseudoexfoliation glaucoma, and primary open-angle glaucoma. Graefes Arch. Clin. Exp. Ophthalmol., 243: 677–683.

Andrikopoulos, G.K., Mela, E.K., Georgakopoulos, C.D., et al. (2007) Pseudoexfoliation syndrome prevalence in Greek patients with cataract and its association to glaucoma and coronary artery disease. Eye, Epub Oct 12.

Aragon-Martin, J.A., Ritch, R., Liebmann, J., O'Brien, C., Blaaow, K., Mercieca, F., Spiteri, A., Cobb, C., Damji, K., Tarkkanen, A., Rezaie, T., Child, A. and Sarfarazi, M. (2008) Evaluation of LOXL1 gene polymorphisms in exfoliation syndrome and exfoliation glaucoma. Mol. Vis., 14: 533–541.

Aydogan Ozkan, D., Yuksel, N., Kookin, G., et al. (2006) Homocysteine levels in plasma and sensorineural hearing loss in patients with pseudoexfoliation syndrome. Eur. J. Ophthalmol., 16: 542–547.

Barkana, Y., Anis, S., Liebmann, J.M., Tello, C. and Ritch, R. (2006) Intraocular pressure monitoring outside of normal office hours in patients with glaucoma is clinically useful. Arch. Ophthalmol., 124: 793–799.

Becker, B. (1995) Does hyposecretion of aqueous humor damage the trabecular meshwork? (editorial). J. Glaucoma, 4: 303–305.

Bedri, A. and Alemu, B. (1999) Pseudoexfoliation syndrome in Ethiopian glaucoma patients. East Afr. Med. J., 76: 278–280.

Bengtsson, B. and Heijl, A. (2005) A long-term prospective study of risk factors for glaucomatous visual field loss in patients with ocular hypertension. J. Glaucoma, 14: 135–138.

Bill, A., Lütjen-Drecoll, E. and Svedbergh, B. (1980) Effects of intracameral Na_2 EDTA and EGTA on aqueous outflow routes in the monkey eye. Invest. Ophthalmol. Vis. Sci., 19: 492–504.

Bleich, S., Roedl, J., Von Ahsen, N., Schlötzer-Schrehardt, U., Reulbach, U., Beck, G., Kruse, F.E., Naumann, G.O.H., Kornhuber, J. and Junemann, A.G.M. (2004) Elevated homocysteine levels in aqueous humor of patients with pseudoexfoliation glaucoma. Am. J. Ophthalmol., 138: 162–164.

Cahill, M., Early, A., Stack, S., Blayney, A.W. and Eustace, P. (2002) Pseudoexfoliation and sensorineural hearing loss. Eye, 16: 261–266.

Colin, J., Le Gall, G., Le Jeune, B., et al. (1988) The prevalence of exfoliation syndrome in different areas of France. Acta Ophthalmol., 66(Suppl 184): 86–89.

Cursiefen, C., Hammer, T., Küchle, M., Naumann, G.O.H. and Schlötzer-Schrehardt, U. (2001) Pseudoexfoliation syndrome in eyes with ischemic central retinal vein occlusion. A histopathologic and electron microscopic study. Acta Ophthalmol. Scand., 79: 476–478.

Drolsum, L. (2003) Deep sclerectomy in patients with capsular glaucoma. Acta Ophthalmol. Scand., 81: 567–572.

Drolsum, L. (2006) Longterm follow-up after deep sclerectomy in patients with pseudoexfoliative glaucoma. Arch. Ophthalmol. Scand., 84: 502–506.

Ethier, C.R., Read, A.T. and Chan, D.W. (2006) Effects of latrunculin-B on outflow facility and trabecular meshwork structure in human eyes. Invest. Ophthalmol. Vis. Sci., 47: 1991–1998.

Fingert, J.H., Alward, W.L., Kwon, Y.H., Wang, K., Streb, L.M., Sheffield, V.C. and Stone, E.M. (2007). LOXL1 mutations are associated with exfoliation syndrome in patients from the Midwestern United States. Am. J. Ophthalmol., 144: 974–975.

Gilad, G.M., Kagan, H.M. and Gilad, V.H. (2005) Evidence for increased lysyl oxidase, the extracellular matrix-forming enzyme, in Alzheimer's disease brain. Neurosci. Lett., 376: 210–214.

Gillies, W.E. (1977) Trabeculotomy in pseudoexfoliation of the lens capsule. Br. J. Ophthalmol., 61: 297–305.

Gracner, T. (2002) Intraocular pressure response of capsular glaucoma and primary open angle glaucoma to selective Nd:YAG laser trabeculoplasty: a prospective, comparative clinical trial. Eur. J. Ophthalmol., 12: 287–292.

Gradle, H.S. and Sugar, H.S. (1947) Glaucoma capsulare. Am. J. Ophthalmol., 30: 12–19.

Hagadus, R.J., Wandel, T., Ritch, R., et al. (1989) Pseudoexfoliation in patients with Alzheimer's disease. Invest. Ophthalmol. Vis. Sci., 30(Suppl): p. 27.

Hammer, T., Schlötzer-Schrehardt, U. and Naumann, G.O.H. (2001) Unilateral or asymmetric pseudoexfoliation syndrome? An ultrastructural study. Arch. Ophthalmol., 119: 1023–1031.

Harasymowicz, P.J., Papamatheakis, D.G., Latina, M., de Leon, M., Lesk, M.R. and Damji, K.F. (2005) Selective laser trabeculoplasty (SLT) complicated by intraocular pressure elevation in eyes with heavily pigmented trabecular meshworks. Am. J. Ophthalmol., 139(6): 1110–1113.

Harju, M. and Vesti, E. (2001) Blood flow of the optic nerve head and peripapillary retina in exfoliation syndrome with unilateral glaucoma or ocular hypertension. Graefes Arch. Clin. Exp. Ophthalmol., 239: 271–277.

Hayashi, H., Gotoh, N. and Ueda, Y. et al., (2008) Lysyl oxidase-like 1 polymorphisms and exfoliation syndrome in the Japanese population. Am. J. Ophthalmol., Jan 15 Epub ahead of print.

Hewitt, A.W. et al., (2007) Ancestral LOXL1 variants are associated with pseudoexfoliation in Caucasian Australians but with markedly lower penetrance than in Nordic people. Hum. Mol. Genet., Nov. 23 Epub ahead of print.

Ho, S.L., Dogar, G.F., Wang, J., et al. (2005) Elevated aqueous humour tissue inhibitor of matrix metalloproteinase-1 and connective tissue growth factor in pseudoexfoliation syndrome. Br. J. Ophthalmol., 89: 169–173.

Holló, G., Lakatos, P. and Farkas, K. (1998) Cold pressor test and plasma endothelin-1 concentration in primary open-angle glaucoma and capsular glaucoma. J. Glaucoma, 7: 105–110.

Holven, K.B., Halvorsen, B., Schulz, H., et al. (2006) Increased levels of C-reactive protein and interleukin-6 in hyperhomocysteinemic subjects. Scand. J. Clin. Lab. Invest., 66: 45–54.

Honjo, M., Tanihara, H., Inatani, M., et al. (1998) Phacoemulsification, intraocular lens implantation, and trabeculotomy to treat pseudoexfoliation syndrome. J. Cataract Refract. Surg., 24: 781–786.

Horns, D.J.,, et al. (1983) Argon laser trabeculoplasty for open angle glaucoma: a retrospective study of 380 eyes. Trans. Ophthalmol. Soc. UK, 103: 288–294.

Irkec, M. (2006) Exfoliation and carotid stiffness (editorial). Br. J. Ophthalmol., 90: 529–530.

Jacobi, P.C., Dietlein, T.S. and Krieglstein, G.K. (1998) Bimanual trabecular aspiration in pseudoexfoliation glaucoma — an alternative in nonfiltering glaucoma surgery. Ophthalmology, 105: 886–894.

Jones, W., White, R.E. and Magnus, D.E. (1992) Increased occurrence of exfoliation in the male, Spanish American population of New Mexico. J. Am. Optom. Assoc., 63: 643–648.

Katula, K.S., Heinloth, A.N. and Paules, R.S. (2007) Folate deficiency in normal human fibroblasts leads to altered expression of genes primarily linked to cell signaling, the cytoskeleton and extracellular matrix. J. Nutr. Biochem., 18: 541–552.

Kaufman, P.L. and Bárány, E.H. (1977) Cytochalasin B reversibly increases outflow facility in the eye of the cynomolgus monkey. Invest. Ophthalmol. Vis. Sci., 16: 47–53.

Kiland, J.A., Miller, C.L., Kim, C.B., et al. (2006) Effect of H-7 and Lat-B on retinal physiology. Curr. Eye Res., 31: 441–455.

Koliakos, G.G., Konstas, A.G.P., Schlötzer-Schrehardt, U., et al. (2002) Ascorbic acid concentration is reduced in the aqueous humor of patients with exfoliation syndrome. Am. J. Ophthalmol., 134: 879–883.

Koliakos, G.G., Konstas, A.G.P., Schlötzer-Schrehardt, U., et al. (2003) 8-Isoprostaglandin F2a and ascorbic acid concentration in the aqueous humour of patients with exfoliation syndrome. Br. J. Ophthalmol., 87: 353–356.

Koliakos, G.G., Schlötzer-Schrehardt, U., Konstas, A.G., Bufidis, T., Georgiadis, N. and Dimitriadou, A. (2001) Transforming and insulin like growth factor in the aqueous humor of patients with exfoliation syndrome. Graefes Arch. Clin. Exp. Ophthalmol., 239: 482–487.

Konstas, A.G.P., Holló, G., Astakhov, Y.S., et al. (2004) Factors associated with long-term progression or stability in exfoliation glaucoma. Arch. Ophthalmol., 122: 29–33.

Konstas, A.G.P., Holló, G., Irkec, M., Tsironi, S., Durukan, I., Goldenfeld, M. and Melamed, S. (2007b) Diurnal IOP control with bimatoprost vs latanoprost in exfoliative glaucoma: a crossover observer-masked 3-center study. Br. J. Ophthalmol., 91: 757–760.

Konstas, A.G.P., Koliakos, G.G., Karabatsas, C.H., et al. (2006a) Latanoprost therapy reduces the levels of TGF beta 1 and gelatinases in the aqueous humour of patients with exfoliative glaucoma. Exp. Eye Res., 82: 319–322.

Konstas, A.G.P., Kozobolis, V.P., Katsimpris, I.E., et al. (2007a) Efficacy and safety of latanoprost versus travoprost in exfoliative glaucoma patients. Ophthalmology, 114: 653–657.

Konstas, A.G.P., Topouzis, F., Leliopoulou, O., et al. (2006b) 24-hour intraocular pressure control with maximum medical therapy compared with surgery in patients with advanced open-angle glaucoma. Ophthalmology, 113: 761–765.

Kottler, U.B., Jünemann, A.G., Aigner, T., Zenkel, M., Rummelt, C. and Schlötzer-Schrehardt, U. (2005) Comparative effects of TGF-beta1 and TGF-beta2 on extracellular matrix production, proliferation, migration, and collagen contraction of human Tenon's capsule fibroblasts in pseudoexfoliation and primary open-angle glaucoma. Exp. Eye Res., 80: 121–134.

Kozart, D.M. and Yanoff, M. (1982) Intraocular pressure status in 100 consecutive patients with exfoliation syndrome. Ophthalmology, 89: 214–218.

Kozobolis, V.P., Papatzanaki, M., Vlachonikolis, I.G., Pallikaris, I.G. and Tsambarlakis, I.G. (1997) Epidemiology of pseudoexfoliation in the island of Crete (Greece). Acta Ophthalmol., 75: 726–729.

Kuhlenbaumer, G., Friedrichs, F., Kis, B., et al. (2007) Association between single nucleotide polymorphisms in the lysyl oxidase-like 1 gene and spontaneous cervical artery dissection. Cerebrovasc. Dis., 24: 343–348.

Laatikainen, L. (1971) Fluorescein angiographic studies of the peripapillary and perilimbal regions in simple, capsular and low-tension glaucoma. Acta Ophthalmol., III(Suppl.): 3–83.

Leibovitch, I., Kurtz, S., Shemesh, G., et al. (2003) Hyperhomocystinemia in pseudoexfoliation glaucoma. J. Glaucoma, 12: 36–39.

Leske, M.C., Heijl, A., Hussein, M., et al. (2003) Factors for glaucoma progression and the effect of treatment. The Early Manifest Glaucoma Trial. Arch. Ophthalmol., 121: 48–56.

Lewis, C.M., Smith, A.K. and Kamen, B.A. (1998) Receptor-mediated folate uptake is positively regulated by disruption of the actin cytoskeleton. Cancer Res., 15: 2952–2956.

Linnér, E., Popovic, V., Gottfries, C.G., et al. (2001) The exfoliation syndrome in cognitive impairment of cerebrovascular or Alzheimer's type. Acta Ophthalmol. Scand., 79: 283–285.

Liu, X., Zhao, Y., Gao, J., et al. (2004) Elastic fiber homeostasis requires lysyl oxidase-like 1 protein. Nat. Genet., 36: 178–182.

Lobo, A., Naso, A., Arheart, K., et al. (1999) Reduction of homocysteine levels in coronary artery disease by low-dose folic acid combined with vitamins B6 and B12. Am. J. Cardiol., 83: 821–825.

Massé, P.G., Yamauchi, M., Mahuren, J.D., et al. (1995) Connective tissue integrity is lost in vitamin B-6-deficient chicks. J. Nutr., 125: 26–34.

Merrill, A.H. and Henderson, M. (1987) Diseases associated with defects in vitamin B6 metabolism or utilization. Annu. Rev. Nutr., 7: 137–156.

Mitchell, P., Wang, J.J. and Hourihan, F. (1999) The relationship between glaucoma and pseudoexfoliation: the Blue Mountains Eye Study. Arch. Ophthalmol., 117: 1319–1328.

Mitchell, P., Wang, J.J. and Smith, W. (1997) Association of pseudoexfoliation with increased vascular risk. Am. J. Ophthalmol., 124: 685–687.

Moreno Montañés, J., Alcolea Paredes, A. and Campos García, S. (1989) Prevalence of pseudoexfoliation syndrome in the northwest of Spain. Acta Ophthalmol., 67: 383–385.

Naumann, G.O.H., Schlötzer-Schrehardt, U. and Küchle, M. (1998) Pseudoexfoliation syndrome for the comprehensive ophthalmologist: intraocular and systemic manifestations. Ophthalmology, 105: 951–968.

Odberg, T. and Sandvik, L. (1999) The medium and long-term efficacy of primary argon laser trabeculoplasty in avoiding topical medication in open-angle glaucoma. Acta Ophthalmol., 77: 176–181.

Okka, M., Tian, B. and Kaufman, P.L. (2004) Effect of low-dose latrunculin B on anterior segment physiologic features in the monkey. Arch. Ophthalmol., 122: 1482–1488.

Oleggini, R., Gastaldo, N. and Di Donato, A. (2007) Regulation of elastin promoter by lysyl oxidase and growth factors: cross control of lysyl oxidase on TGF-beta1 effects. Matrix Biol., 26: 494–505.

Ovodenko, B., Rostagno, A., Neubert, T.A., et al. (2007) Proteomic analysis of lenticular exfoliation deposits. Invest. Ophthalmol. Vis. Sci., 48: 1447–1457.

Peterson, J.A., Tian, B., Geiger, B. and Kaufman, P.L. (2000) Effect of latrunculin-B on outflow facility in monkeys. Exp. Eye Res., 70: 307–314.

Prince, A.M. and Ritch, R. (1986) Clinical signs of the pseudoexfoliation syndrome. Ophthalmology, 93: 803–807.

Prince, A.M., Streeten, B.W., Ritch, R., Dark, A.J. and Sperling, M. (1987) Preclinical diagnosis of pseudoexfoliation syndrome. Arch. Ophthalmol., 105: 1076–1082.

Puska, P., Vesti, E., Tomita, G., Ishida, K. and Raitta, C. (1999) Optic disc changes in normotensive persons with unilateral exfoliation syndrome: a 3-year follow-up study. Graefes Arch. Clin. Exp. Ophthalmol., 237: 457–462.

Puustjärvi, T., Blomster, H., Kontkanen, M., et al. (2004) Plasma and aqueous humour levels of homocysteine in exfoliation syndrome. Graefe's Arch. Clin. Exp. Ophthalmol., 242: 749–754.

Raposo, B., Rodriguez, C., Martínez-Gonzáles, J. and Badimon, L. (2004) High levels of homocysteine inhibit lysyl oxidase (LOX) and downregulate LOX expression in vascular endothelial cells. Atherosclerosis, 177: 1–8.

Rekonen, P., Kannisto, T., Puustjärvi, T., et al. (2006) Deep sclerectomy for the treatment of exfoliation and primary open-angle glaucoma. Arch. Ophthalmol. Scand., 84: 507–511.

Repo, L.P., Suhonen, M.T., Teräsvirta, M.E. and Koivisto, J.K. (1995) Color Doppler imaging of the ophthalmic artery blood flow spectra of patients who have had a transient ischemic attack. Correlations with generalized iris transluminance and pseudoexfoliation syndrome. Ophthalmology, 102: 1199–1205.

Ringvold, A., Blika, S., Elsås, T., et al. (1988) The middle-Norway eye-screening study. I. Epidemiology of the pseudoexfoliation syndrome. Acta Ophthalmol., 66: 652–657.

Ritch, R. (1994) Exfoliation syndrome: the most common identifiable cause of open-angle glaucoma. J. Glaucoma, 3: 176–178.

Ritch, R. and Podos, S.M. (1983) Laser trabeculoplasty in exfoliation syndrome. Bull. N.Y. Acad. Med., 59: 339–344.

Ritch, R. and Schlötzer-Schrehardt, U. (2001) Exfoliation syndrome. Surv. Ophthalmol., 45: 265–315.

Ritch, R., Schlötzer-Schrehardt, U. and Konstas, A.G.P. (2003) Why is glaucoma associated with exfoliation syndrome? Prog. Retinal Eye Res., 22: 253–275.

Rodríguez, C., Alcudia, J.F. and Martínez-González, J. et al., 2007. Lysyl oxidase (LOX) down-regulation by TNFalpha: a new mechanism underlying TNFalpha-induced endothelial dysfunction. Atherosclerosis, Epub ahead of print.

Roedl, J.B., Bleich, S., Reulbach, U., et al. (2007a) Homocysteine in tear fluid of patients with pseudoexfoliation glaucoma. J. Glaucoma, 16: 234–239.

Roedl, J.B., Bleich, S., Reulbach, U., et al. (2007b) Vitamin deficiency and hyperhomocysteinemia in pseudoexfoliation glaucoma. J. Neural. Transm., 114: 571–575.

Rönkkö, S., Rekonen, P., Kaarniranta, K., et al. (2007) Matrix metalloproteinases and their inhibitors in the chamber angle of normal eyes and patients with primary open-angle glaucoma and exfoliation glaucoma. Graefe's Arch. Clin. Exp. Ophthalmol., 245: 697–704.

Roth, M. and Epstein, D.L. (1980) Exfoliation syndrome. Am. J. Ophthalmol., 89: 477–486.

Saatci, O.A., Ferliel, S.T., Ferliel, M., et al. (1999) Pseudoexfoliation and glaucoma in eyes with retinal vein occlusion. Int. Ophthalmol., 23: 75–78.

Sabanay, I., Tian, B., Gabelt, B.A.T., et al. (2006) Latrunculin B effects on trabecular meshwork and corneal endothelial morphology in monkeys. Exp. Eye Res., 82: 236–246.

Schlötzer-Schrehardt, U., Koca, M.R., Naumann, G.O.H. and Volkholz, H. (1992) Pseudoexfoliation syndrome. Ocular manifestation of a systemic disorder? Arch. Ophthalmol., 110: 1752–1756.

Schlötzer-Schrehardt, U., Küchle, M., Dorfler, S. and Naumann, G.O.H. (1993) Pseudoexfoliative material in the eyelid skin of pseudoexfoliation-suspect patients: a clinicohistopathological correlation. German J. Ophthalmol., 2: 51–60.

Schlötzer-Schrehardt, U., Küchle, M., Rummelt, C. and Naumann, G.O.H. (1999) Role of transforming growth factor-b and its latent form binding protein in pseudoexfoliation syndrome. Invest. Ophthalmol. Vis. Sci., 40: p. S278.

Schlötzer-Schrehardt, U., Lommatzsch, J., Küchle, M., Konstas, A.G.P. and Naumann, G.O.H. (2003) Matrix metalloproteinases and their inhibitors in aqueous humor of patients with pseudoexfoliation syndrome/glaucoma and primary open-angle glaucoma. Invest. Ophthalmol. Vis. Sci., 44: 1117–1125.

Schlötzer-Schrehardt, U. and Naumann, G.O.H. (2006) Ocular and systemic pseudoexfoliation syndrome. Am. J. Ophthalmol., 141: 921–937.

Scullica, L., Buceti, R., Castagna, I., Ferreri, G. and Trombetta, C.J. (1993) Functional aspects of pseudoexfoliation: physiopathological features. New Trends Ophthalmol., 8: 163–168.

Selhub, J. (2006) The many facets of hyperhomocysteinemia: studies from the Framingham cohorts. J. Nutr., 136(6 Suppl.): 1726S–1730S.

Shaban, R.I. and Asfour, W.M. (2004) Ocular pseudoexfoliation associated with hearing loss. Saudi Med. J., 25: 1254–1257.

Shakya, S., Koirala, S. and Karmacharya, P.C.D. (2004) Pseudoexfoliation syndrome in Nepal: a hospital-based retrospective study. Asia-Pacific J. Ophthalmol., 16: 13–16.

Shimizu, K., Kimura, Y. and Aoki, K. (1988) Prevalence of exfoliation syndrome in the Japanese. Acta Ophthalmol., 66(Suppl 184): 112–115.

Sibour, G., Finazzo, C. and Boles Carenini, A. (1997) Monolateral pseudoexfoliatio capsulae: a study of choroidal blood flow. Acta Ophthalmol. Scand., 75(Suppl 224): 13–14.

Starcher, B. and Hill, C.H. (2005) Elastin defects in the lungs of avian and murine models of homocysteinemia. Exp. Lung Res., 31: 873–885.

Streeten, B.W., Gibson, S.A. and Dark, A.J. (1986) Pseudoexfoliative material contains an elastic microfibrillar-associated glycoprotein. Trans. Am. Ophthalmol. Soc., 84: 304–320.

Streeten, B.W., Li, Z.Y., Wallace, R.N., Eagle, R.C.J. and Keshgegian, A.A. (1992) Pseudoexfoliative fibrillopathy in visceral organs of a patient with pseudoexfoliation syndrome. Arch. Ophthalmol., 110: 1757–1762.

Tanihara, H., Negi, A., Akimoto, M., Terauchi, H., Okudaira, A., Kozaki, J., Takeuchi, A. and Nagata, M. (1993) Surgical effect of trabeculotomy ab externo on adult eyes with primary open angle glaucoma and pseudoexfoliation syndrome. Arch. Ophthalmol., 111: 1653–1661.

Thorliefsson, G., Magnusson, K.P., Sulem, P., et al. (2007) Common sequence variants in the LOXL1 gene confer susceptibility to exfoliation glaucoma. Science, 317: 1397–1400.

Tian, B., Kaufman, P.L., Volberg, T., Gabelt, B.T. and Geiger, B. (1998) H-7 disrupts the actin cytoskeleton and increases outflow facility. Arch. Ophthalmol., 116: 633–643.

Tian, B., Gabelt, B.T., Geiger, B. and Kaufman, P.L. (1999) Combined effects of H-7 and cytochalasin B on outflow facility in monkeys. Exp. Eye Res., 68: 649–655.

Vessani, R.M., Liebmann, J.M., Jofe, M. and Ritch, R. (2003) Plasma homocysteine is elevated in patients with exfoliation syndrome. Am. J. Ophthalmol., 136: 41–46.

Visontai, Z., Merisch, B.M., Kollai, M. and Hollo, G. (2006) Increase of carotid artery stiffness and decrease of baroreflex sensitivity in exfoliation syndrome and glaucoma. Br. J. Ophthalmol., 360–367: p. 90.

Yilmaz, A., Adiguzel, U., Tamer, L., et al. (2005) Serum oxidant/antioxidant balance in exfoliation syndrome. Clin. Experiment. Ophthalmol., 33: 63–66.

Yüksel, N., Anik, Y., Kilic, A., Karabas, V., Demirci, A. and Caglar, Y. (2006) Cerebrovascular blood flow velocities in pseudoexfoliation. Graefes Arch. Clin. Exp. Ophthalmol., 244: 316–321.

Yüksel, N., Karabas, V.L., Arslan, A., et al. (2001) Ocular hemodynamics in pseudoexfoliation syndrome and pseudoexfoliation glaucoma. Ophthalmology, 108: 1043–1049.

Zee, R.Y., Mora, S., et al. (2007) Homocysteine, 5,10-methylenetetrahydrofolate reductase 677C>T polymorphism, nutrient intake, and incident cardiovascular disease in 24,968 initially healthy women. Clin. Chem., 53: 845–851.

Zenkel, M., Kruse, F.E., Jünemann, A.G., Naumann, G.O.H. and Schlötzer-Schrehardt, U. (2006) Clusterin deficiency in eyes with pseudoexfoliation syndrome may be implicated in the aggregation and deposition of pseudoexfoliative material. Invest. Ophthalmol. Vis. Sci., 47: 1982–1990.

Zenkel, M., Pöschl, E., von der Mark, K., Carmen Hofmann-Rummelt, C., Naumann, G.O.H., Kruse, F.E. and Schlötzer-Schrehardt, U. (2005) Differential gene expression in pseudoexfoliation syndrome. Invest. Ophthalmol. Vis. Sci., 46: 3742–3752.

CHAPTER 16

Laser therapies for glaucoma: new frontiers

G.L. Scuderi* and N. Pasquale

University of Rome, "La Sapienza" 2nd Faculty of Medicine, Rome, Italy

Abstract: Glaucoma is a long-term ocular neuropathy defined by optic disc or retinal nerve fiber structural abnormalities and visual-field defects. Treatment for glaucoma consists of reducing intraocular pressure (IOP) to an acceptable target range to prevent further optic-nerve damage. Currently available treatments include topical drug (single then multidrug combinations) followed, for those patients on maximal tolerated medical therapy who still need additional IOP reduction, by laser treatments. These included laser iridotomy, laser trabeculoplasty, laser iridoplasty, laser cyclophotocoagulation. Although the various types of laser enjoyed great success as glaucoma therapy for many years, recently the excimer laser trabeculotomy is a promising IOP-lowering technique.

Keywords: glaucoma; iridectomy; iridotomy; pupillary block; trabecular ring; ciliary body

Background

For more than half a century, many types of lasers (xenon, krypton, argon, neodymium:YAG (Nd:YAG, diode, excimer) have been used to treat glaucoma in its different forms. The aims and outcomes of these treatments have varied. Some have been abandoned, along with the lasers used to perform them; others continue to play roles of primary importance in the parasurgical treatment of glaucoma.

Laser iridotomy

Von Graefe (1857) proposed surgical iridectomy for the treatment of glaucomatous disease. This approach was widely used for over a century, especially in cases of angle-closure glaucoma and in the treatment of acute attacks. Meyer-Schwickerath (1956) demonstrated the efficacy in this setting of xenon-arc laser iridotomy, but this technique was later abandoned because of its high rate of complications.

By the 1980s, incisional iridectomy for angle-closure glaucoma associated with pupillary block had been largely replaced by laser iridotomy using argon and later Nd:YAG lasers. Iridotomy has become the procedure of choice for preventing acute attacks of glaucoma, and its use has reduced the total number of these pathologic events (Scuderi et al., 1995).

Indications

The indications proposed by the European Glaucoma Society include all clinically relevant cases of pupillary block, regardless of degree. More specifically, the following conditions are considered:

- Acute angle-closure glaucoma: Treatment must be administered during the initial phases

---

*Corresponding author. Tel.: 0633775035; Fax: 0633776628; E-mail: gianluca.scuderi@uniroma1.it

of the attack or after drug therapy aimed at reducing the intraocular pressure (IOP) and attenuating inflammation of the iris and corneal edema.
- Narrow-angle glaucoma with positive results in provocative tests: treatment is indicated in young, high-risk patients with positive family histories.
- Pigment dispersion syndrome and the initial phases of pigmentary glaucoma: the aim here is to eliminate iris concavity and iris-zonular contact. The long-term efficacy of laser therapy in this syndrome has not been demonstrated although its use is supported by the results of numerous studies (Gandolfi et al., 1996; Lagreze et al., 1996; Scuderi et al., 1997; Carassa et al., 1998).
- Preparation for argon laser trabeculoplasty (ALT).

Contraindications

The main contraindications are corneal opacity, neovascularization of the iris, and athalamia.

Patient preparation

After the patient's informed consent has been obtained, miotic drops (e.g., dapiprazole 0.5% or pilocarpine 2% or 4%) are usually administered to constrict the pupil and reduce the thickness of the iris so that it can be perforated more easily. Pilocarpine is a more effective miotic than dapiprazole, but it also causes greater congestion of the iris vessels, which increases the risk of bleeding. Systemic administration of acetazolamide or topical application of 1% apraclonidine 1 h before and immediately after the treatment prevents IOP increases and reduces the bleeding of the iris. Laser treatment is performed under topical anesthesia with 4% benoxinate or 4% oxybuprocaine.

Technique

Peripheral iridectomy must be performed in the upper quadrants, approximately 2/3 of the distance between the margin of the pupil and the limbus. The iridotomy is then covered by the upper eyelid to prevent monocular diplopia. This is particularly important for the large iridotomies created with an argon laser; the location is less important for small YAG iridotomies. The thinner areas of the iris, such as the crypts, are the preferred treatment sites. Visible blood vessels should be avoided.

Peripheral laser iridectomies can be done with an argon laser or with an Nd:YAG laser, with the aid of an Abraham or Wise contact lens. Both lasers are effective, but they have different features. The predominant effect of the argon laser is photocoagulation with energy absorption by the iris pigment; the YAG laser produces photodestruction through a chromophore-independent mechanism. Many authors prefer the Nd:YAG treatment because it is simpler to perform, uses less energy, and is associated with a lower rate of iridotomy closure than argon laser treatment. However, in eyes with visible iris vessels and for patients on anticoagulants, the iridotomy site should ideally be pretreated with argon.

Nd:YAG laser iridectomy

The Nd:YAG laser is the laser of choice for iridectomies.

Power consumption ranges from 1 to 6 mJ. One or more impacts are needed to perforate the iris. Depending on the model, applications consisting of pulse trains or a single pulse may be used; the spot diameter ranges from 50 to 70 μm. Use of a converging contact lens equipped with a magnification area improves laser-spot focusing and reduces energy consumption.

Perforation of the iris is generally accompanied by movement of the aqueous humor, together with particles of iris pigment, from the posterior to the anterior chamber (Fig. 1). Perforation can be verified by the use of transillumination, which allows visualization of the choroidal reflex through the gap in the stroma of the iris.

Mild, posttraumatic hemorrhage is a common finding. It can be controlled by exerting light pressure on the cornea for 10–20 s with the contact lens. If this is not sufficient, argon treatment with long-duration pulses can be used to coagulate the edges of the treatment area.

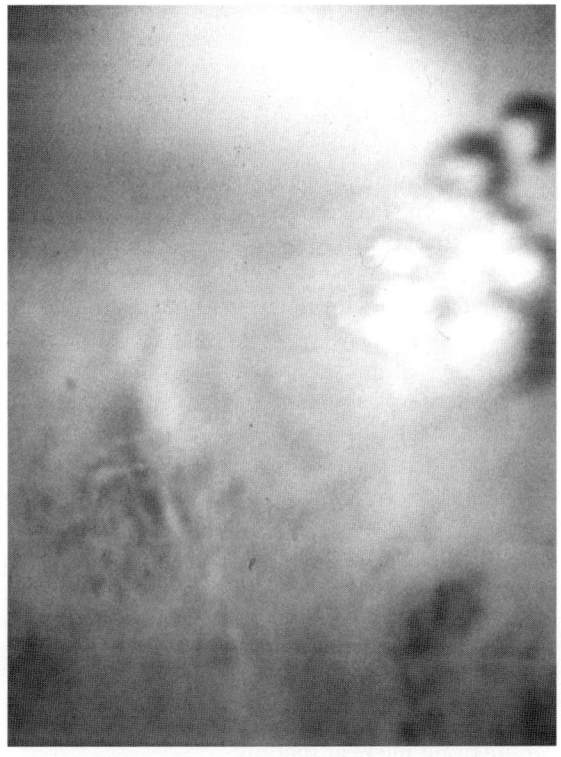

Fig. 1. Iridotomy.

Argon laser iridectomy

Laser parameters vary depending on the type of treatment being administered and the color of the iris. For lightly pigmented irides, energy levels of 800–1000 mW are used with a spot size of 50 μm and an exposure time of 0.2 s. For thick, dark-brown irides, 1500–2500 mW of power is used with an exposure time of 0.02–0.05 s and a spot diameter of 50 μm.

Complications

When the iridotomy is done correctly, both intraoperative and long-term complications are very rare. There are reports in the literature of isolated cases with complications, including some that are serious: retinal detachment, cataract, lens luxation, and rupture of Descemet's membrane (Mastropasqua et al., 1984; Zadok and Chayet, 1999; Liu et al., 2002).

The most common complications are less severe. They include:

- Transient increases in pressure, which occur in approximately one third of treated eyes. They are caused by treatment debris, which impedes outflow of the aqueous humor (Hsiao and Hsu, 2003). This problem can be prevented by the administration of topical timolol and oral acetazolamide (Fleck and Wright, 1997).
- Temporary "fogging."
- Laser-induced iritis; posterior synechiae (rare); the risk of these events can be reduced with the use of topical steroids.
- Diplopia — this occurs when the iridotomy is positioned on the horizontal meridians and not covered by the upper eyelid.
- Bleeding.
- Damage to the corneal endothelium (Wu and Jeng, 2000).
- Incomplete (partial-thickness) iridotomy.

Efficacy

Laser iridotomy is now a valid, effective alternative to surgical procedures for the treatment of narrow-angle glaucoma (Rivera et al., 1985; Fleck and Wright, 1997). It is easy to perform, relatively safe, and rarely associated with complications. The efficacy of this approach has been widely demonstrated over the past 20 years. In 70–85% of the patients treated (Playfair and Watson, 1979; Jiang, 1991; Fleck and Wright, 1997) significant reductions in IOP have been achieved without IOP-reducing drug therapy in follow-ups ranging from 1 to 3 years.

The iridocorneal angle remains open in 70–80% of treated patients, and only a small percentage require retreatment (Thomas and Arun, 1999; Ritch et al., 2004).

LASER trabeculoplasty

The first experimental laser treatments of the iridocorneal angle were the "goniopuncture" procedures performed by Krasnov (1973) to improve aqueous-humor outflow. The method known as ALT was developed laterby Wise and Witter (1979). In the 23 years that have passed since its introduction, ALT has not undergone any substantial modifications. It involves a series of photocoagulations that produce nonperforating thermal lesions uniformly distributed over the trabecular ring.

Treatment technique

Laser trabeculoplasty is frequently used to treat glaucoma, and it produces good long-term results. It is usually done with an argon laser (use of diode lasers is less common) and a gonioscopic contact lens. The most widely used gonioscopy lenses are the Goldmann three-mirror lens and the Ritch gonioprism, which allows better visualization of the trabecular mesh and uses less energy.

The standard settings proposed by Wise and Witter include a spot diameter of 50 µm and an exposure time of 0.1 s. The power varies depending on the response of the tissue and the type of lens (600–1200 mW with the Goldmann lens, 600–800 mW with the Ritch). The spots (20–25 per quadrant) must be uniformly distributed along the entire circumference of the trabecular meshwork. The recommended site of application is the junction of the pigmented and nonpigmented trabecular meshworks. This approach spares part of the filtering trabecula and the posterior trabecular meshwork, thus reducing the incidence of pressure spikes and posttreatment peripheral anterior synechiae (PAS). ALT can be performed in one or two 50-to-60-spot sessions to reduce the risk of laser-induced pressure spikes. The laser produces whitening of the treated zone or the formation of a small bleb in the burst zones.

When an argon laser is used, green light is just as effective as blue-green, and it reduces the risk of damage to the photoreceptors of the operator (Berringer et al., 1989). The operator can also use a contact lens with a metal halide coating to reduce the risk of macular damage caused by reflected light.

Mechanism of action

Various mechanisms of action have been hypothesized; according to the theory proposed by Wise, the thermal effect produced by the laser spots stretches and widens the trabecular meshes and/or the Schlemm's canal (Wise and Witter, 1979). Other authors maintain that the laser triggers physiological changes in the activities of the endothelial cells with increased phagocytotic activity (Bylsma et al., 1988) or increased cell replication within the trabeculate that promote the outflow of aqueous humor (Wilensky and Weinreb, 1983).

Indications for treatment

The European Glaucoma Society's guidelines recommend trabeculoplasty in the following cases:

- Primary open-angle glaucoma that is refractory to medical therapy.
- Patients who are noncompliant or unable to tolerate medical therapy.
- Pseudoexfoliation glaucoma.
- Pigmentary glaucoma.

Contraindications to treatment

Treatment is contraindicated in patients whose angular structures cannot be explored and/or those with uveal inflammation or congenital malformations involving the trabecular structures, as in the following cases:

- Narrow or closed angle.
- Corneal opacity that precludes gonioscopy.
- Absence of trabecular pigmentation.
- Inflammatory or postuveitic glaucoma.
- Juvenile glaucoma.
- Absence of effects in the contralateral eye.

Patient preparation and postoperative follow-up

Written consent should be obtained from the patient after the objectives of the treatment and its potential risks have been presented and discussed.

Before and after the laser treatment apraclonidine 0.5% or brimonidine 0.2% should be added to the patient's normal antiglaucoma regimen to prevent or attenuate iop spikes (Chen and Ang, 2001). Topical anesthesia with 4% benoxinate or 4% oxybuprocaine is generally sufficient.

During the immediate postoperative period (4–7 days), local steroid treatment or topical nonsteroidal anti-inflammatory drugs (NSAIDs) should be applied 3–4 times a day. NSAIDs can also be administered 24 h before the treatment to reduce the release of prostaglandins that are responsible for the inflammatory response (Hotchkiss et al., 1984). The patient should be reexamined 24 h after treatment and at 4–6 weeks to evaluate the effects of the treatment and make sure there are no complications. The European Glaucoma Society recommends closer follow-up (1 and 3 h after treatment) for patients with serious visual-field defects, markedly elevated pretreatment IOPs, exfoliation syndrome, one-eyed patients, and those who have had previous laser trabeculoplasties.

Complications of the treatment

The most common complication of laser trabeculoplasty is the acute, transient elevation of IOP, which is evident immediately after treatment. Thirty to fifty percent of all patients experience IOP increases of 10 mmHg or more 1–7 h after treatment (Krupin et al., 1987; Tuulonen et al., 1989). The incidence and the magnitude of these pressure spikes seem to be reduced when ALT is performed in two 50-spot sessions. The literature contains rare reports of ocular hypertension with onset several weeks after trabeculoplasty, including some that were associated with posttreatment uveitis.

PAS are found in 12–47% of treated eyes, depending on the various statistics; the mean frequency reported in the Glaucoma Laser Trial was 33% (Glaucoma Laser Trial Research Group, 1995). According to this study, synechiae formation is associated with intense pigmentation of the trabecular meshwork; other studies suggest that this complication is dependent on the site of treatment (posterior versus the root of the iris).

Iritis, hemorrhages, transient reductions in visual acuity caused by the gonioscopy fluid (used to enhance contact), and corneal lesions are rare complications.

Efficacy

The effects of the laser treatment are not immediate; 4–6 weeks must pass before the results can be properly evaluated. In 84% of all patients who undergo trabeculoplasty, significant reductions in IOP are observed within 12 months of treatment. However, the rate decreases as the duration of follow-up increases, reaching 55% 5 years after treatment (Fellman et al., 1984; Lunde, 1993). In two other long-term studies, the effect of ALT persisted for 1 year in 67–80% of treated eyes, for 5 years in 35–50%, and for 10 years in 5–30%, with an annual loss of efficacy in the range of 6–10% (Shingleton and Richter, 1987; Spaeth and Baez, 1992).

Patients who have had ALT thus have to be followed constantly, with ongoing monitoring of clinical examinations. It is difficult to predict if or when increases in IOP will occur: they may appear after months or even years of follow-up. Retreatment carries a considerably reduced likelihood of success compared with the initial ALT. In these cases, surgical treatment should probably be considered.

Selective laser trabeculoplasty

Selective laser trabeculoplasty (SLT) is a recently introduced technique that was developed by Latina and Park (1995). In March of 2001, it was approved by the Food and Drug Administration for treatment of open-angle glaucoma. In terms of both efficacy and safety, SLT has been judged to be clinically equivalent to ALT, which was previously used for patients with open-angle glaucoma. It is based on the use of a Q-switched frequency-doubled 532 nm Nd:YAG laser with a spot diameter of 400 µm, a power range of 0.2–2.0 mJ, and a pulse duration of 3–10 ns; it can be used to treat 180° of the trabecular meshwork (50 spots) or the full circumference (100 spots).

Latina and Park showed that a Q-switched laser emitting at 532 nm with a pulse duration in the nanosecond range can selectively lyse the ophthalmologists with more extensive experience with cyclophotocoagulation have extended its indications, noting that the complication rates are lower than once believed (Scuderi et al., 1993; Gayton et al., 1999; Pastor et al., 2001). Cycloablation is still contraindicated, however, when safer options are available and in patients whose visual capacity is high.

Patient preparation

As with all parasurgical procedures, the patient must provide informed consent after the objectives and potential risks of the procedure have been presented and discussed. Cyclophotocoagulation procedures can be performed in an outpatient setting; local anesthesia can be achieved with a peri- or retrobulbar injection of 2% lidocaine + 0.75% bupivacaine with hyaluronidase (50:50). During the preparation phase, systemically administered acetazolamide and/or mannitol can be prescribed to decompress the eye and reduce the risk of choroidal hemorrhage. During the preoperative phase, topical apraclonidine can be given to decongest the conjunctiva and reduce the risk of posttreatment increases in IOP (Chen and Ang, 2001).

Transpupillary cyclophotocoagulation

Introduced by Lee (1971), transpupillary photocoagulation allows destruction of the ciliary processes through the pupil or a wide surgical iridectomy. The laser spot can be focused on the ciliary processes directly or indirectly with the aid of a gonioscopy lens. This treatment is obviously reserved for eyes in which at least one third of the circumference of the ciliary processes can be clearly visualized, i.e., patients with maximum mydriasis, aniridia, wide surgical iridectomy, or retraction of the iris, which may be present in advanced forms of neovascular glaucoma.

Transpupillary cyclophotocoagulation is usually done with an argon laser; less frequently, diode laser is used. Argon lasers are used with a spot diameter of 100–200 µm, exposure time of 0.1–0.2 s, and 700–1000 mW of power; the effect is seen as crater-shaped whitening of the ciliary process.

Transpupillary photocoagulation is not associated with an excessively high rate of complications; however, since its use is limited to rare cases, the transscleral approach is usually preferred.

Endoscopic cyclophotocoagulation

Treatments involving the ciliary processes have been further improved in recent years thanks to the introduction of small endoscopes (21-gauge) that use fiber optics for illumination, observation, and laser treatment. This approach, known as endoscopic cyclophotocoagulation (ecp) is considered a surgical method (Shields et al., 1985) because it requires that an instrument be introduced into the ocular bulb through the pars plana or the limbus. Its advantages include more precise visualization of the anatomic structure being treated, lower energy consumption (normally around 0.3 W per 1 s with around 60 applications) and reduced involvement of contiguous structures. ECP can be performed in association with phacoemulsification procedures (Uram, 1995).

Transscleral cyclophotocoagulation

Transscleral cyclophotocoagulation was introduced by Vucicevic et al. (1961). It involves the transmission of energy across the conjunctiva and the sclera without direct visualization of the ciliary bodies. It can be carried out with or without contact using either an Nd:YAG (continuous-emission or pulsed) or diode laser. Pulsed lasers (Microrupter series) deliver short, high-energy pulses that provoke considerable tissue destruction; continuous-emission lasers (Surgical Laser Technologies, Oaks, PA, Lasag Microruptor III) require longer exposure times and produce coagulative effects. Diode units (Oculigt SLX and DC-3000) are contact-delivery systems; they requires longer exposure times than pulsed high-energy Nd:YAG units and produce tissue effects that are mainly coagulative.

Transscleral noncontact cyclophotocoagulation

Transscleral noncontact photocoagulation is done with an Nd:YAG laser and a slit-lamp; air serves as the energy transmission medium. The laser spot is positioned 1.0–1.5 mm posterior to the sclerocorneal limbus; the distance is calculated with a compass or by positioning the sight at the center of a luminous slit 3 mm in length. The laser can be used in the free-running mode with pulses of 4–8 J, 20-s duration, and the defocus set to 9. With these parameters, one can treat 360° (total: 30–40 spots, 8–9 per quadrant); the 3 and 9 o'clock positions are excluded from treatment to avoid damage to the long posterior ciliary arteries. Treatment is facilitated by the use of the Shields transscleral lens, which improves focusing and facilitates the passage of the laser energy, thus reducing the risk of conjunctival burns.

Postoperative treatment is based on the administration of antibiotics and steroids for approximately 2 weeks; pressure-reducing therapy is continued (with possible withdrawal of miotics), and the patient is checked 1 h, 1 day, and 1 week after treatment.

Transscleral contact cyclophotocoagulation

Transscleral contact cyclophotocoagulation is the most widely used cyclophotocoagulation technique. Contact treatment offers the advantage of energy transmission through the conjunctiva and sclera by means of an optic fiber probe placed directly on the bulb. Compared with the noncontact approach, this technique uses less energy.

It can be carried out with a Nd:YAG laser (1064 nm) in the continuous-emission mode by means of an optical fiber or with a diode laser (810 nm). In the Nd:YAG procedure, the probe is positioned at a right angle to the conjunctiva, 0.5–1 mm behind the sclerocorneal junction; the full circumference (360°) is treated (total: 16–40 spots, 4–6 J) with the exception of the 3 and 9 o'clock positions. The exposure time is 0.5–0.7 s, and scleral indentation facilitates energy transmission. Better results are obtained with a relatively light initial treatment followed by one or more additional treatments as needed.

With diode systems, the energy is delivered by means of the G-probe, a bundle of optical fibers whose contour adapts to the curvature of the sclerocorneal junction. The application site is 0.5–2.0 mm behind the limbus; transillumination is used to visualize the ciliary processes; 8–15 spots (1500–2500 mW lasting 1–2 s) are made from 90° to 270°. The upper temporal zone is generally spared so that trabeculectomy or other surgical procedures can be performed if necessary. The diode laser (DLCP) is a semiconductor laser with a wavelength of 810 nm. Its advantages include good penetration and selective absorption by the pigmented tissue of the ciliary body; transscleral DLCP is thus a selective cycloablative technique that is more conservative than the others.

Complications

In all cases subjected to cyclophotocoagulation treatment, mild reactive iritis develops after treatment. Conjunctival edema and pain during the treatment are common, especially when contact application is used. The most severe complications are seen is eyes treated with noncontact methods, which use high-energy levels and are often associated with arbitrary focusing of the laser beam. In the literature, phthisis bulbi is reported in 10% of all cases, permanent hypotonus in 26%, and anterior-chamber hemorrhage in 10–30%; less frequent complications include detachment of the choroid, hemovitreous, and sympathetic ophthalmia.

Complications are rare in eyes treated with a diode laser:

phthisis bulbi 1.6% (Kramp et al., 2002)
pain during treatment 25%

Pupil ovalization due to the use of high-energy lasers (2000 mW) (very rare) (Pucci and Tappainer, 2003).

Recent studies have assessed the efficacy and complication rates of diode laser treatment with spots of different energy levels in patients with refractory glaucoma: there were no significant differences in the reductions in IOP. However, mitomycin-C (MMC) inhibit fibroblast function and survival when applied locally. However, as these agents have significant side effects, safer and more effective drugs are required. In this chapter, we review current and future agents in development to modulate healing and scarring in glaucoma surgery.

Using surgical and anatomical principles to modify therapy

By revisiting the basic principles of surgery and fluid flow through the bleb, we have shown that simple changes in surgical and antimetabolite application technique can dramatically lower side effects even when the same concentrations of antimetabolites are used (Wells et al., 2003, 2004) (Figs. 1–3). This stresses the importance of combining new therapies with appropriate surgical techniques to maximize the benefit of any therapy. For instance, techniques that minimize tissue damage are clearly important.

The principle of using spacers is well established with the use of tube implants, with plate spacers to keep the subconjunctival space open. Physical spacers using tissue such as amniotic membrane may increase the success of GFS. Human amniotic membrane appears to have antiangiogenic, anti-inflammatory, and antifibrotic characteristics. The effects of amniotic membrane have been tested both in animal models of GFS (Barton et al., 2001; Demir et al., 2002) and in humans (Fujishima et al., 1998; Bruno et al., 2006) with encouraging results. An interesting approach to the application of amniotic membrane in GFS was reported in trabeculectomies of high-risk patients, who had undergone at least two or more trabeculectomies with MMC. Two pieces of amniotic membrane soaked in MMC were sutured, one under the scleral flap and one into the subconjunctival space. After 6–18 months, the IOP was found to be significantly reduced (Drolsum et al., 2006). A bioequivalent gel may in future perform the function of amniotic membranes. Furthermore, amniotic membrane transplantation can be used instead of conjunctival advancement to repair late-onset bleb leakage (Rauscher et al., 2007). Gas, such as perfluoropropane or sodium hyaluronate 2.3%, may improve subconjunctival drainage

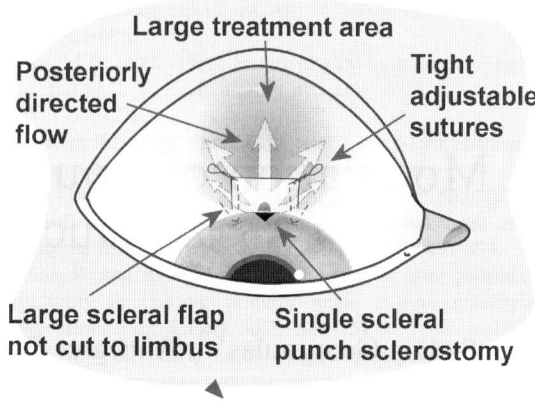

Fig. 1. Simple changes in surgical technique, which have markedly reduced bleb-related complications.

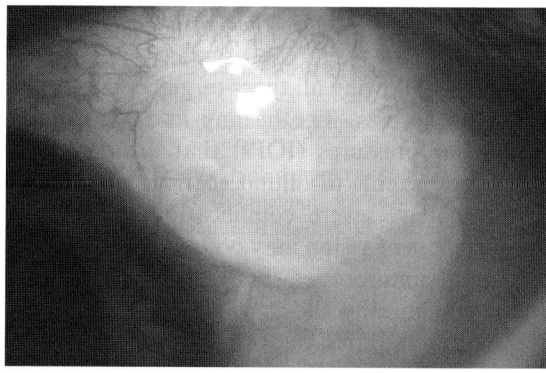

Fig. 2. Focal cystic bleb after exposure to mitomycin C before the changes in surgical technique.

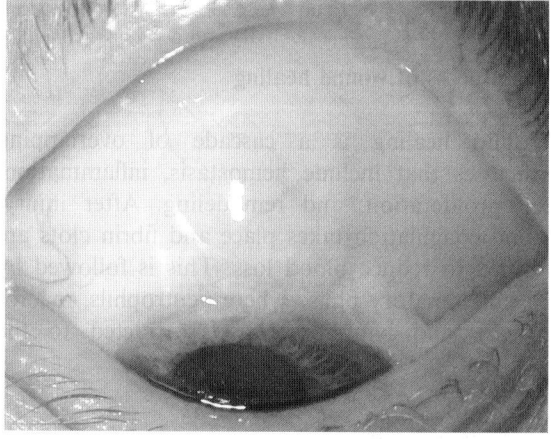

Fig. 3. Improvement in bleb morphology using the new technique and same dose of mitomycin C.

spaces with resultant creation of more diffuse blebs (Wong et al., 1999; Lopes et al., 2006).

In a different area of surgery in the sub-Tenon's space, strabismus surgery, polyurethane sheets, with or without a dexamethasone sustained drug delivery system, significantly reduce adhesions (Kim et al., 2004). Polytetrafluoroethylene (PTFE), Seprafilm (a biodegradable membrane made of sodium hyaluronate and carboxymethylcellulose), ADCON-L (a polyglycan ester), Interceed (a cellulose matrix), sodium hyaluronate and polyglactin mesh, and future variations may also be used in GFS in the future. Devices made of relatively inert materials already used in other spaces such as the suprachoroidal space may also facilitate aqueous outflow by keeping the surgical field free of scar tissue. However, in many cases the continuous inflammatory reaction due to the presence of the biomaterial leads to excessive scarring and poor postoperative results.

Blood clotting and fibrin formation

Based on the principle that fibrin is a critical part of healing, agents that break down fibrin, such as tissue plasminogen activator, are effective in lysing blood clots after surgery (WuDunn, 1997). Although, in the short term, fibrinolytic agents may lower intraocular pressure (IOP), one of the side effects that has inhibited further use of such agents is the increased risk of prolonged bleeding. Furthermore, fibrin breakdown molecules may have a longer-term stimulatory effect on the induction of scarring (Gray et al., 1993).

Inflammatory cells and mediators

Inflammatory cells and mediators released during and after surgery stimulate scarring. The grading system we developed in our long-term Medical Research Council (MRC) trial showed a good correlation between inflammation and the long-term outcome (www.blebs.net; Fig. 4). There is good evidence that topical steroids, used as part of routine postoperative management, are effective in reducing inflammation (Kent et al., 1998). Intrableb triamcinolone acetonide injection at the conclusion of GFS has been reported to be clinically beneficial with regards to IOP reduction and a relatively safe method for steroid administration (Tham et al., 2006). Topical nonsteroidal anti-inflammatory drugs (NSAIDs) may also be effective (Kent et al., 1998), but their use is still controversial.

Other agents reducing inflammation include cyclosporine and cyclooxygenase-2 inhibitors, although the effect of intraoperative or postoperative application of cyclosporine in an animal model was not conclusive (Lattanzio et al., 2005). A novel approach to the inhibition of inflammatory cytokines is the development of dendrimers — hyperbranched nanomolecules that can be chemically synthesized to have precise structural characteristics. In our in vivo model of GFS, water-soluble conjugates of D(+)-glucosamine and D(+)-glucosamine-6-sulfate, with immunomodulatory and antiangiogenic properties applied together, enhanced the long-term success from 30 to 80% (Shaunak et al., 2004). This experimental result is very encouraging and far more effective than that seen with intensive topical steroid drops (Fig. 5).

Growth factors

The tissues in a wound and specifically in GFS as well as the aqueous flowing through the bleb contain a large number of growth factors or cytokines (Chang et al., 2000). Transforming growth factor beta (TGF-β) has been shown to be more stimulatory than other growth factors and cytokines found in the aqueous (Khaw et al., 1994). TGF-β may even reverse the effect of MMC in vivo (Khaw et al., 1994). Therapeutic strategies that modulate the activity of growth factors including TGF-β may be useful inhibitors of fibrosis. Tranilast ((N-(3′,4′)-dimethoxycinnamoyl) anthranilic acid)) is effective against TGF-β activity and has antiscarring potential when used in GFS. TGF-β activity is also inhibited by genistein and suramin. Suramin reduces postoperative scarring in an experimental model of GFS (Mietz et al., 1998), and an early pilot study has been encouraging (Mietz and Krieglstein, 2001). Interferon alpha (IFN-α), a cytokine with antifibrotic action, also reduces scarring activity of

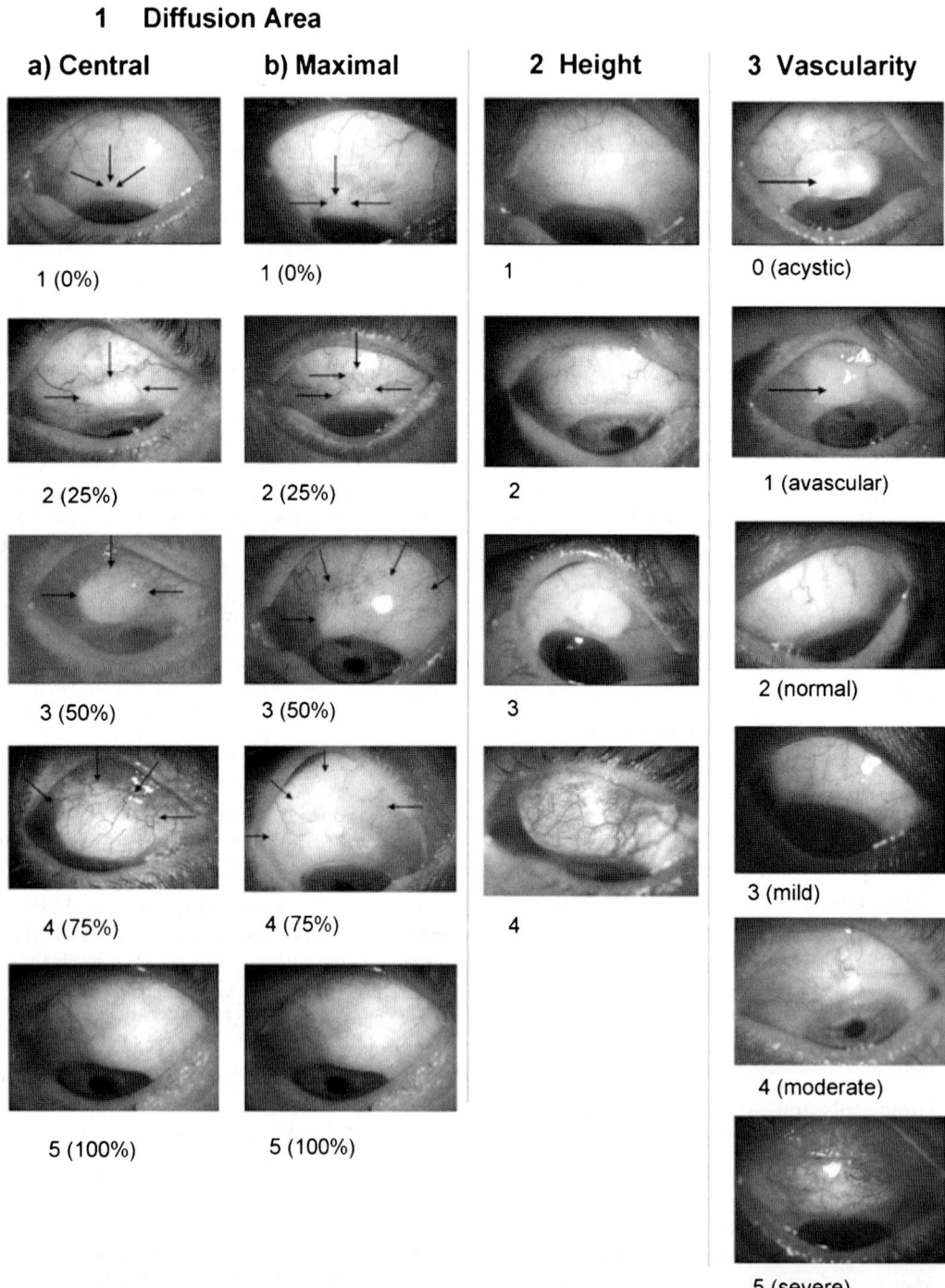

Fig. 4. Moorfields bleb grading system. (See Color Plate 17.4 in color plate section)

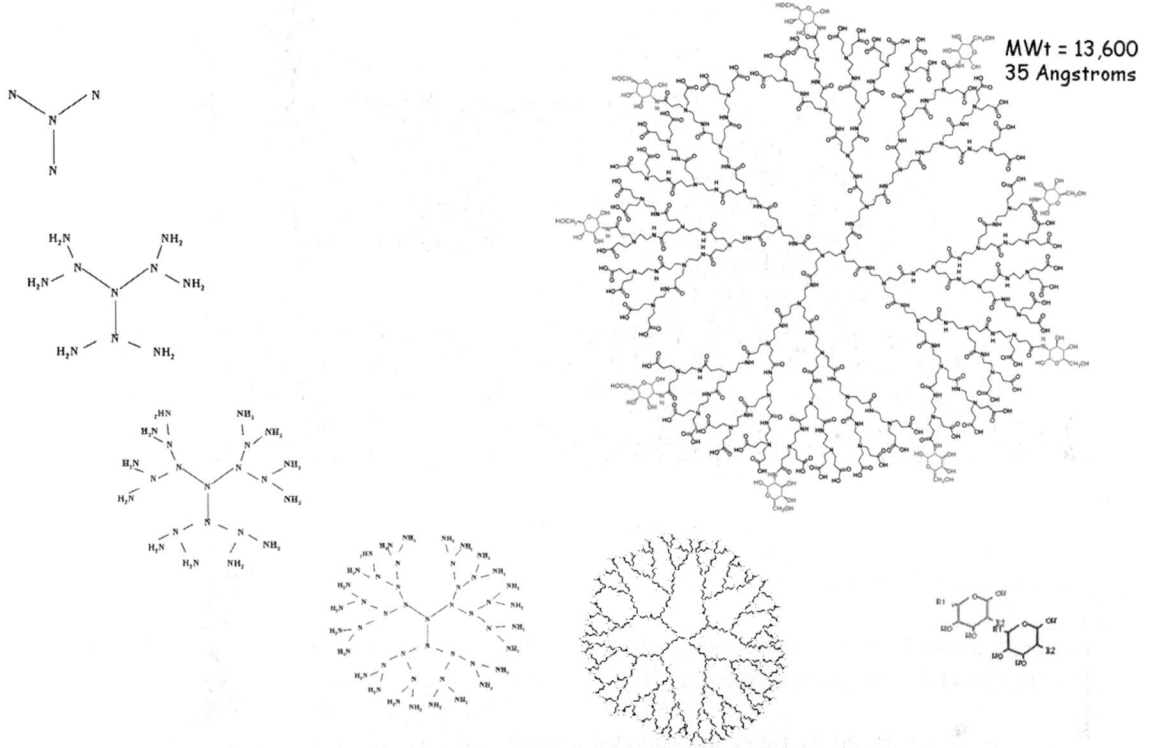

Fig. 5. Dendrimers.

fibroblasts, although a clinical trial did not show it to be significantly better than antimetabolites (Gillies et al., 1999).

As, in the eye, TGF-β seems more important than other growth factors (Khaw et al., 1994), we have used a variety of biological mechanisms to block TGF-β activity, including antisense oligonucleotides (Cordeiro et al., 2003) and a human monoclonal antibody against the active form of human TGF-β2, the predominant isoform in the aqueous (lerdelimumab, Trabio®, Cambridge Antibody Technology, Cambridge, UK). Unlike antimetabolites, one of the theoretical advantages of the monoclonal antibody is its target specificity: it only acts if there is TGF-β2 in the wound, minimizing the risk of adverse events such as hypotony. In an in vivo model of conjunctival scarring, it significantly improved the outcome of GFS (Cordeiro et al., 1999a) and appeared much less destructive to local tissue than MMC. A pilot clinical study of this antibody in GFS demonstrated the absence of significant side effects, inflammatory reaction, and cystic bleb formation, often observed after use of antimetabolites. However, two larger randomized controlled trials have not shown a significant effect on the outcome of GFS (Khaw et al., 2007). Based on the data obtained from an earlier study (Cordeiro et al., 1999a), we believe that the dose used was not sufficient. Subsequent experiments from our laboratory have shown a significantly enhanced effect with a prolonged dosing regimen (Mead et al., 2003), and the data also suggested an enhanced effect in the GFS outcome when the antibody

is combined with intraoperative 5-fluorouracil (5-FU). Due to the encouraging results, further studies are planned with different dosing regimens.

The importance of blocking TGF-β2 in order to control scarring after GFS was shown in a recent study in which tissue transglutaminase (tTgase) and its end product ε-(γ-glutamyl)-lysine were detected in scarred tissue of failed trabeculectomy blebs. tTgases are calcium-dependent enzymes that cross-link proteins using ε-(γ-glutamyl)-lysine bonds. Since vertebrates lack enzymes capable of hydrolyzing these bonds, the protein cross-linking created by tTgases seems to be unbreakable (Priglinger et al., 2006). tTgase cross-links fibronectin and collagen 3; these proteins are produced by human Tenon's layer fibroblasts (HTFs) in vitro and have been detected in the scar tissue deposited in the bleb area after GFS. TGF-β2, known for enhancing conjunctival scarring after GFS, was shown in the same study to trigger the expression of tTgase and subsequently the cross-linking of fibronectin in vitro (Priglinger et al., 2006). In vivo, this pathway might lead to enhanced cross-linking of the newly formed scar tissue and subsequent failure of the GFS, leading again to the conclusion that inhibition of TGF-β2 could extend the success of the surgery.

Alternative inhibitors of TGF-β include decorin, a small proteoglycan, which is a natural TGF-β-inhibitor. In an in vivo experimental model of GFS, the outcome of preoperative and postoperative application of decorin was encouraging as it decreased fibrosis and delayed the increase of IOP (Grisanti et al., 2005). In vitro, silencing RNA (siRNA) against TGF-β receptor II–mRNA reduced the production of TGF-β receptor II as well as the production and deposition of fibronectin and the migration of human corneal fibroblasts. The application of the same siRNA molecule reduced inflammation and deposition of extracellular matrix (ECM) in an in vivo model of subconjunctival scarring after GFS (Nakamura et al., 2004). Furthermore, encapsulated siRNA molecules in poly (D,L-lactide-co-glycolide) microspheres, targeting TGF-β2, applied as a single injection after trabeculectomy led to 100% survival of the bleb for more than a month (Gomes dos Santos et al., 2006).

Targeting intracellular signaling downstream of the TGF-β receptor could be another effective strategy. Within the cell, the main TGF-β signaling pathway runs through proteins that activate transcription of genes that encode the Smad proteins. Of particular relevance is Smad3, which is essential for TGF-β-induced production of ECM proteins (Chen et al., 1999; Massague, 1999). Inhibiting Smad3 in immediate postoperative applications might prove beneficial (Leask and Abraham, 2004). Smad7, acting differently from Smad3, is another potential therapeutic target. As TGF-β can suppress its action through the induction of Smad7 (negative feedback loop), gene transfer of the Smad7 gene has been shown in animal models to have a protective effect against the development of lung, liver, and renal fibrosis (Schiller et al., 2004).

Connective tissue growth factor (CTGF) is a factor that influences ECM production and subsequent scar formation and fibrosis. TGF-β1 triggers the expression of CTGF, which is also necessary for TGF-β stimulation of myofibroblast differentiation and collagen contraction (Garrett et al., 2004). Inhibition of this factor could be a possible future therapeutic target (Wu et al., 2006). Myofibroblasts express a platelet-activating factor (PAF) nuclear receptor and TNF-β receptors; PAF and TNF-β cause time-dependent myofibroblast apoptosis. Future therapeutic approaches may take advantage of the expression of these receptors (He and Bazan, 2006).

TGF-β was recently shown to increase cell tension in HTF cultures and contraction in HTF collagen I gels by triggering the activation of GTPase Rho, which regulates actin cytoskeleton remodeling and cell contractility. Rho activates the serine–threonine kinase, Rho-associated kinase (ROCK), that enhances cytoskeletal tension and results in actomyosin-mediated contraction. ROCK inhibitors reduced cell tension and inhibited the TGF-β-mediated p38 activation, alpha smooth muscle actin (alpha-SMA) expression, and HTF development of enhanced contractile abilities characteristic of the so-called myofibroblast phenotype, not differentiation (Meyer-ter-Vehn et al., 2006). The effectiveness of ROCK inhibitor Y-27632 has been tested in vitro and in vivo (Honjo et al., 2007). In vitro, Y-27632 inhibited contraction of HTF-seeded collagen I gels and alpha-SMA expression by HTFs. In vivo, application of Y-27632 after GFS in rabbits resulted in significant increase in the survival of the blebs compared to controls. Additionally, histological analysis revealed reduction of collagen I deposition and scarring in the treated blebs compared to controls (Honjo et al., 2007). Application of ROCK inhibitors may block TGF-β-induced scarring by downregulating pathways generating mechanical tension and thereby improve the success of GFS.

Another growth factor that activates certain ocular fibroblasts is platelet-derived growth factor (PDGF). Some ocular fibroblasts express high levels of PDGF-β. When stimulated, these fibroblasts proliferate, migrate, and produce ECM molecules, causing contraction, fibrosis, and development of fibrotic membranes such as those observed in proliferative vitreoretinopathy (PVR; Nagineni et al., 2005). Aptamers, nucleic acid–based macromolecules with similar functions as monoclonal antibodies (high affinity and specificity for target proteins), are capable of recognizing, binding, and blocking PDGF-β. Two aptamers, ARC126 and ARC127, have been tested in animal models of PVR (Akiyama et al., 2006), and may also be useful in glaucoma surgery.

The potential involvement of p38 mitogen-activated protein kinase (p38MAPK) in the induction of fibrogenic reaction has also been investigated. p38MAPK is believed to trigger the transcription of Smad2/3, facilitate the phosphorylation and activation of Smad3, and subsequently the formation of Smad3/4 complex, which is important for the development of the fibrotic reaction. The Smad3 signaling pathway is important in retinal fibrosis. Inhibition of Smad3 is associated with a reduction of a cellular fibrotic reaction (Saika et al., 2004). Adenoviral transfer by intravitreal application of a dominant negative p38MAPK gene demonstrated reduced fibrotic reaction of RPE cells after retinal detachment (Saika et al., 2005). In later studies, adenoviral gene transfer of the same dominant negative p38MAPK gene was applied to in vitro cultured HTFs (Meyer-ter-Vehn et al., 2006) and to an in vivo conjunctival scarring model in mice (Yamanaka et al., 2007). Reduction of the differentiation of fibroblasts to myofibroblasts and decline of the CTGF and of the monocyte chemoattractant protein (MCP-1) expression were the main in vitro findings. Inhibition of conjunctival scarring was observed in vivo. As MCP-1 is a chemoattractant for macrophages, one of the main sources of TGF-β, the reduction in levels of MCP-1 through inhibition of p38MAPK seems to reduce the levels of TGF-β; hence its favorable role in conjunctival scarring (Cordeiro et al., 1999b; Dewald et al., 2005; Saika, 2006; Yamanaka et al., 2007). Therefore, the inhibition of p38MAPK has been proposed as a possible future therapeutic option.

Bone morphogenic proteins (BMPs) are growth factors with an important role not only in embryonic development and morphogenesis but also in postnatal life. They are vital for cell proliferation, differentiation, apoptosis, angiogenesis, and other biological functions (Reddi, 1997a, b; Chen et al., 2004). Activins are dimeric proteins that participate in the mechanisms of cell differentiation and proliferation, apoptosis, inflammation and neurogenesis (Tsuchida, 2004). Investigation of the expression of many members of BMPs and activins in normal and scarred conjunctival tissue revealed enhanced mRNA and protein levels of BMP-6 and activin A in scarred compared to normal conjunctiva (Andreev et al., 2006). The use of follistatin, an inhibitor of activin bioactivity, was shown to be effective in decreasing liver and lung fibrosis (Aoki et al., 2005; Patella et al., 2006). Based on the above findings, the inhibition of BMP-6 and activin A could be a future therapeutic option for the control of scarring after GFS (Andreev et al., 2006).

Cellular proliferation and vascularization

The main agents used to inhibit scarring and subsequent IOP increase after trabeculectomy are the anticancer agents 5-FU and MMC. Recently, an interesting approach of a slow release mechanism of MMC was published. MMC-loaded hydrogels were shown to inhibit cell proliferation in an in vitro model. In the future, these gels, placed in the bleb, could find an application in GFS in humans (Blake et al., 2006). It has been reported recently that repeated exposure of HTFs in vitro to MMC increases the expression of P-glycoprotein, a protein that creates multidrug resistance by lowering intracellular concentration of various drugs. This finding may explain failure of repeated trabeculectomies despite the application of MMC.

5-FU seems to be a better option for repeated trabeculectomies as it does not increase the expression of P-glycoprotein (Hueber et al., 2007). Beta-irradiation, which we have shown to have similar effects on controlling the proliferation of fibroblasts (Khaw et al., 1991), significantly improved the success of GFS in a large African trial using a dose based on our laboratory studies (Kirwan et al., 2003, 2006) and may also be useful in pediatric GFS (Akimoto et al., 1998).

Additional future interesting experimental strategies to modulate proliferation include over-expressing genes that inhibit fibroblast proliferation, such as p21 WAF-1/CIP-1 introduced via an adenovirus system (Perkins et al., 2002), antagonizing integrins and their receptors (Paikal et al., 2000), or altering intracellular gene transcription (Akimoto et al., 1998). Paclitaxel, an antineoplasmatic agent that prevents mitosis by blocking the depolymerization of intracellular microtubules, has been reported to inhibit HTF proliferation and prolong the success of GFS. But as paclitaxel is hydrophobic, its administration in the subconjunctival area is not easy. Polyanhydride disks (Jampel et al., 1993) and Carbopol 980 hydrogel (Koz et al., 2007) containing paclitaxel have been tested with encouraging results comparable to the antiscarring effects of MMC, although the development of severe inflammation, especially in the case of polyanhydride disks, has been reported (Jampel et al., 1993).

Photodynamic therapy with a diffuse blue light coupled with a photosensitizing agent to kill fibroblasts may also be another way to control healing in the area of treatment (Grisanti et al., 2000). Over the past decade, many studies have demonstrated that photodynamic therapy reduces neovascularization and subsequent maturation of scar tissue after glaucoma filtration. Intraoperative photodynamic therapy using preoperative subconjunctival application of ethyl etiopurin (Hill et al., 1997) and 2′,7′-bis-(2-carboxyethyl)-5-(and-6)-carboxyfluorescein acetoxymethyl ester (BCECF-AM) as photosensitizers (Grisanti et al., 1999) has proven to be safe and effective against scarring after GFS in vivo. Pilot clinical trials of intraoperative photodynamic therapy with locally administered BCECF-AM have shown promising results (Diestelhorst and Grisanti, 2002; Jordan et al., 2003). Furthermore, postoperative photodynamic therapy has been reported to be effective against neovascularization and scarring as well. Intravenous administration of the photosensitizer verteporfin in the early postoperative period to occlude the newly formed capillaries in the bleb area had favorable results in the modulation of wound healing in rabbits after GFS (Stasi et al., 2006). More recently the VEGF antagonist Bevacizumab has been used clinically in GFS to reduce vascularity, which we have found greatly increases the risk of GFS failure using a masked scoring system.

Cell motility, matrix contraction and synthesis

Bleb analysis using carefully graded masked digital photographs reveals that tissue contraction is a critical component of filtration surgery failure. We have developed a more detailed understanding of the processes that occur during tissue contraction and have recently been able to image cells and matrix simultaneously during the process of contraction. The matrix metalloproteinases (MMPs) are enzymes that degrade ECM. Based on laboratory observations, we discovered that cell-mediated collagen contraction could be inhibited using MMP inhibitors (Daniels et al., 2003). In an experimental model of glaucoma surgery, the use of an MMP inhibitor led to a dramatic reduction of scarring, with retention of normal tissue morphology. This action is equivalent to MMC, but without the deleterious side effects (Wong et al., 2003, 2005). A number of agents can also affect the cytoskeleton of the cell and hence inhibit migration. Taxol and etoposide (microtubule-stabilizing agents) have been used in models of filtration surgery, and they prolong bleb survival (Jampel and Moon, 1998). β-Aminopropionitrile and D-penicillamine interfere with molecular cross-linking of collagen, and there is experimental and clinical evidence that they may work in filtration surgery (Jampel et al., 1998). MMP inhibition also surprisingly results in a reduction of collagen synthesis in vitro (Daniels et al., 2003), which may help to explain the dramatic reduction in scar tissue formation in vivo (Wong et al., 2003).

Drug delivery

At present, postoperative GFS management is vital for the prevention of failure due to prolonged scarring. Therefore, more effective drug delivery methods are needed. The long-term requirement of treatment and the narrow therapeutic window of drugs in current use still limit the control of scarring. Sustained release delivery systems, including liposome encapsulation (Shinohara et al., 2003), microspheres (Herrero-Vanrell and Refojo, 2001), scleral plugs (Yasukawa et al., 2001), and biodegradable implant polymers, may have a very significant future role. With our ability to combine agents will come better efficacy. An example is the combination of 5-FU and heparin we used to prevent PVR (Asaria et al., 2001), which was the first clinical trial to show a statistically significant reduction in PVR.

Future directions: total scarring control and tissue regeneration

Significant advances have been made in developing new treatments and refining existing treatments for the prevention of scarring after disease, trauma, or surgical intervention. In addition to traditional chemical drugs, the advent of new technologies such as dendrimers, antibodies, aptamers, ribozymes, gene therapy with viral vectors, and RNA interference opens the door to a whole new generation of therapies to prevent fibrosis after glaucoma surgery. The ability to fully control fibrotic processes in the eye offers the tantalizing prospect of a near 100% success of glaucoma surgery, with pressure around 10 mmHg associated with minimal progression over a decade as found in our long-term MRC glaucoma surgery study. Finally, most exciting is the prospect that neutralizing the fibrotic response to disease and injury will allow us to revert to the "fetal" mode when regeneration is the "normal" process, such as shown in a recent report which demonstrated that induction of bcl-2 gene expression together with downregulation of gliosis results in axonal regeneration in mice (Cho et al., 2005). Modifying matrix and cell conditions allows intrinsic stem cells to differentiate into different cells of the retina like lower species that can regenerate a severely damaged complex retina. Ultimately, it is likely that our ability to fully modulate the scarring processes will lead toward a much more regenerative reparative process after injury and disease.

Acknowledgments

The authors acknowledge the support of Guide Dogs for the Blind, the Wellcome Trust, Medical Research Council, Moorfields Trustees, the Haymans Trust, the Ron and Liora Moskovitz Foundation, the Michael and Ilse Katz Foundation, A G Leventis Foundation, the Helen Hamlyn Trust in memory of Paul Hamlyn, and Fight for Sight (UK). This research has received a proportion of its funding from the UK Department of Health's National Institute for Health Research Biomedical Research Centre at Moorfields Eye Hospital NHS Foundation Trust and UCL Institute of Ophthalmology. The views expressed in this chapter are those of the authors and not necessarily of the Department of Health.

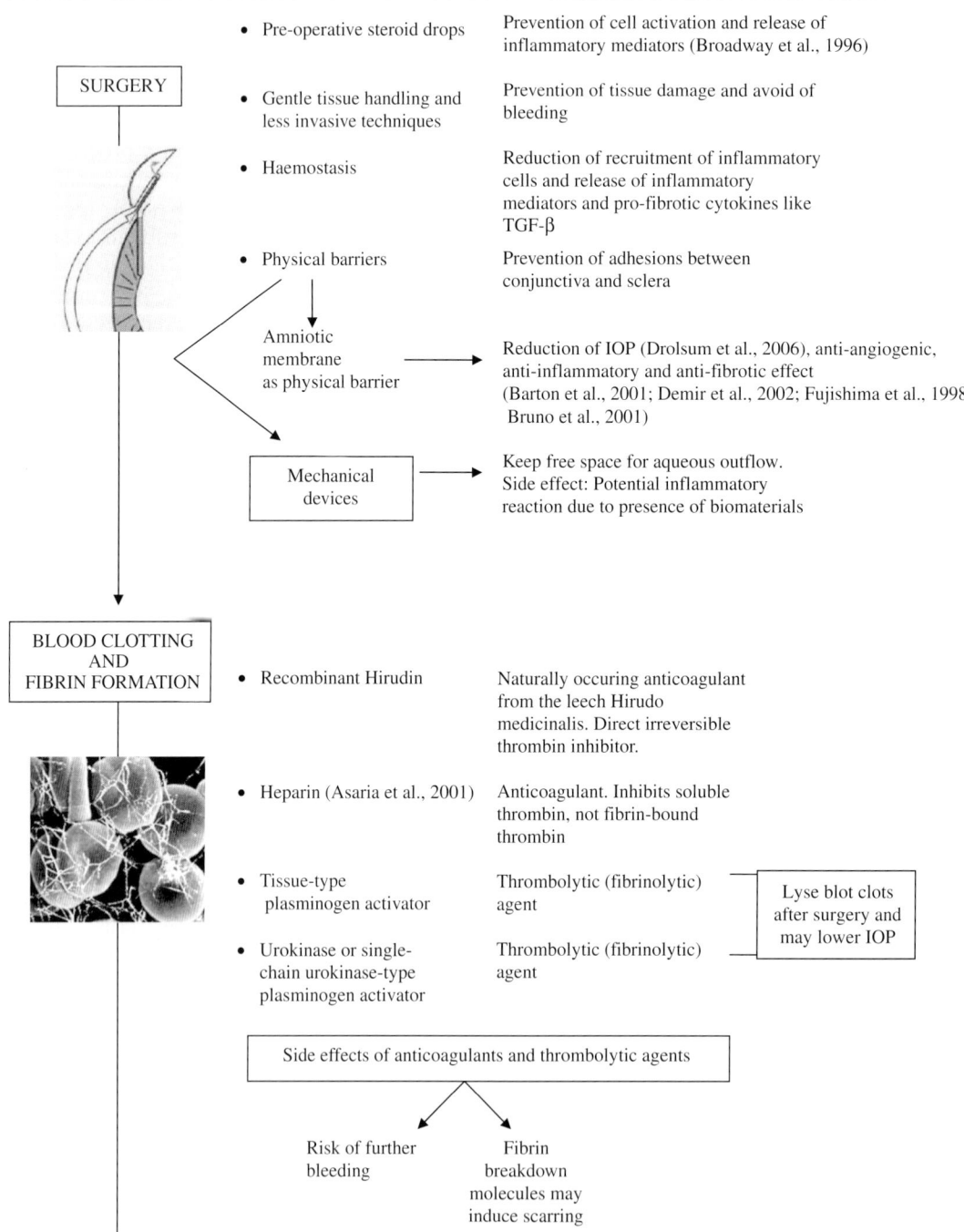

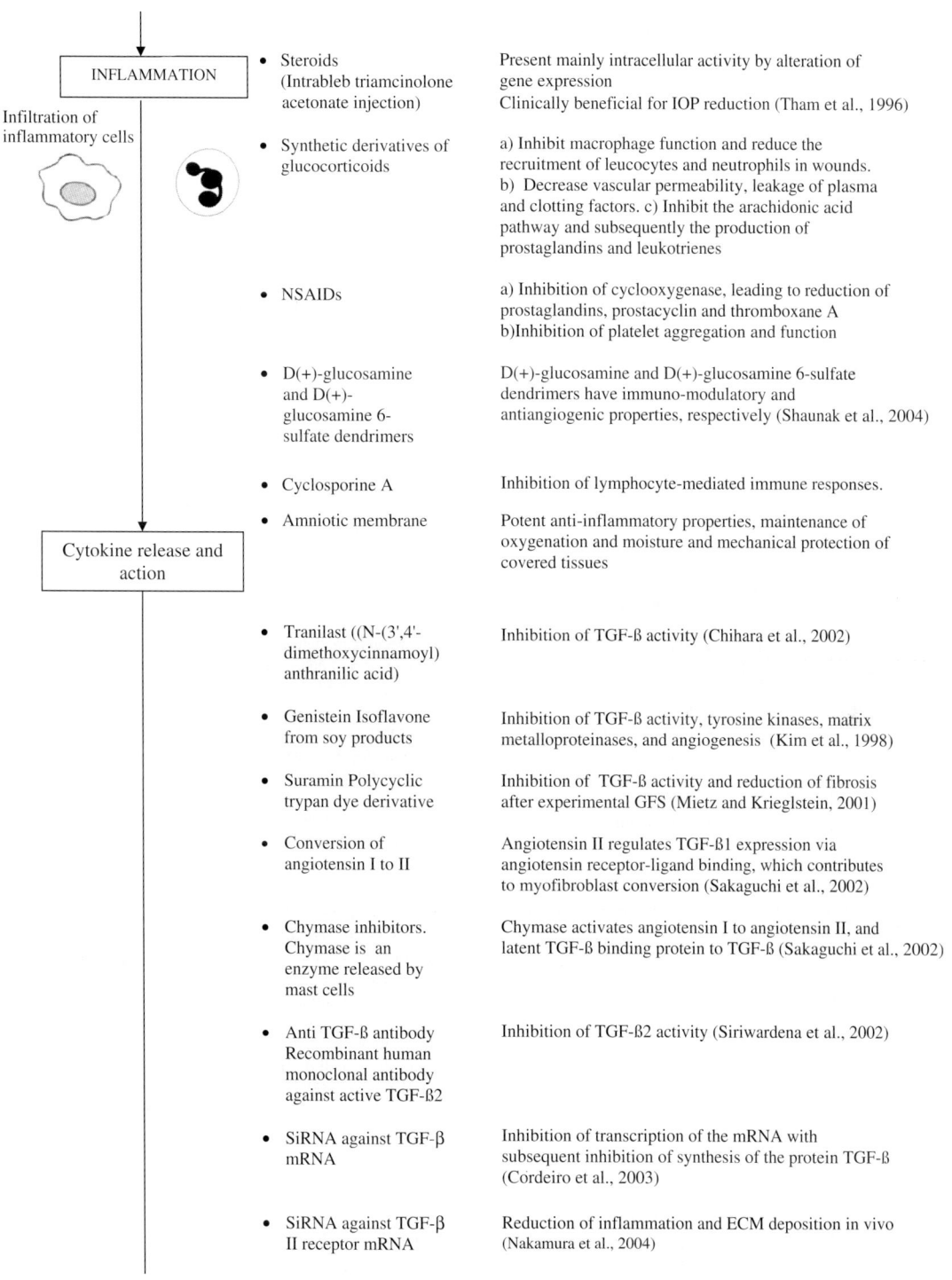

INFLAMMATION	• Steroids (Intrableb triamcinolone acetonate injection)	Present mainly intracellular activity by alteration of gene expression. Clinically beneficial for IOP reduction (Tham et al., 1996)
Infiltration of inflammatory cells	• Synthetic derivatives of glucocorticoids	a) Inhibit macrophage function and reduce the recruitment of leucocytes and neutrophils in wounds. b) Decrease vascular permeability, leakage of plasma and clotting factors. c) Inhibit the arachidonic acid pathway and subsequently the production of prostaglandins and leukotrienes
	• NSAIDs	a) Inhibition of cyclooxygenase, leading to reduction of prostaglandins, prostacyclin and thromboxane A b) Inhibition of platelet aggregation and function
	• D(+)-glucosamine and D(+)-glucosamine 6-sulfate dendrimers	D(+)-glucosamine and D(+)-glucosamine 6-sulfate dendrimers have immuno-modulatory and antiangiogenic properties, respectively (Shaunak et al., 2004)
	• Cyclosporine A	Inhibition of lymphocyte-mediated immune responses.
	• Amniotic membrane	Potent anti-inflammatory properties, maintenance of oxygenation and moisture and mechanical protection of covered tissues
Cytokine release and action	• Tranilast ((N-(3',4'-dimethoxycinnamoyl) anthranilic acid)	Inhibition of TGF-ß activity (Chihara et al., 2002)
	• Genistein Isoflavone from soy products	Inhibition of TGF-ß activity, tyrosine kinases, matrix metalloproteinases, and angiogenesis (Kim et al., 1998)
	• Suramin Polycyclic trypan dye derivative	Inhibition of TGF-ß activity and reduction of fibrosis after experimental GFS (Mietz and Krieglstein, 2001)
	• Conversion of angiotensin I to II	Angiotensin II regulates TGF-ß1 expression via angiotensin receptor-ligand binding, which contributes to myofibroblast conversion (Sakaguchi et al., 2002)
	• Chymase inhibitors. Chymase is an enzyme released by mast cells	Chymase activates angiotensin I to angiotensin II, and latent TGF-ß binding protein to TGF-ß (Sakaguchi et al., 2002)
	• Anti TGF-ß antibody Recombinant human monoclonal antibody against active TGF-ß2	Inhibition of TGF-ß2 activity (Siriwardena et al., 2002)
	• SiRNA against TGF-β mRNA	Inhibition of transcription of the mRNA with subsequent inhibition of synthesis of the protein TGF-ß (Cordeiro et al., 2003)
	• SiRNA against TGF-β II receptor mRNA	Reduction of inflammation and ECM deposition in vivo (Nakamura et al., 2004)

	• Encapsulated SiRNA in poly (D,L-lactide-co-glycolide) microspheres against TGF-β2	Increased bleb survival after experimental GFS (Gomes dos Santos et al., 2006)
	• Smad-7 gene transfer	Suppression of TGF-β action and protection against the development of lung, liver and renal fibrosis (Schiller et al., 2004)
	• ROCK inhibitors (mainly Y-27632)	Control of GTPase Rho activation, which is triggered by TGF-β. Inhibition of HTF collagen I contraction in vitro and increase of bleb survival in vivo (Honjo et al., 2007)
	• Aptamers ARC126 and ARC127	PDGF activates ocular fibroblasts. ARC126 and ARC127 bind and block PDGF-B (Akiyama et al., 2006) and may improve the success of GFS
	• Adenoviral transfer of a dominant negative p38MAPK gene	Inhibition of Smad3 activation. Reduction of the differentiation of fibroblasts to myofibroblasts in vitro (Meyer-ter-Vehn et al., 2006) and reduction of conjunctival scarring in vivo (Yamanaka et al., 2007)
	• Ribozymes. RNA molecules which can cleave specific bonds in other RNA molecules	Cleavage of TGF-ß-mRNA with subsequent inhibition of synthesis of the protein TGF-ß
	• Decorin	Small proteoglycan, natural inhibitor of TGF-β. Delay of IOP increase and decrease of fibrosis after GFS (Grisanti et al., 2005)
	• Simvastatin, Inhibitor of the enzyme, HMG-CoA reductase.	Inhibition of the connective tissue growth factor gene and protein expression, a downstream mediator of TGF-ß (Watts et al., 2005)
	• Follistatin	Inhibitor of activin A. Reduction of liver and lung fibrosis (Aoki et al., 2005; Patella et al., 2006). Potential application in GFS
Cell Proliferation, differentiation and collagen contraction	• Photodynamic therapy	2',7'-bis(2-carboxyethyl)-5(6)-carboxyfluorescein acetoxymethyl ester (BCECF-AM) is a fluorescent probe and is an intracellularly acting photosensitizer. It is applied locally in its inactive form, diffuses into adjacent cells, and is then cleaved and rendered fluorescent by intracellular esterases. After illumination (activation) with blue light, it exerts a photo-oxidative effect that is only cell destructive within the targeted cells (Jordan et al., 2003). Alternatively, ethyl etiopurin (Hill et al., 1997) verteporfin can be used as photosensitizers (Grisanti et al., 1999)
Proliferation of HTFs	• Daunorubicin Anthracycline antibiotic isolated from the fermentation broths of Streptomyces	Inhibits cellular proliferation by several mechanisms, including DNA binding, free radical formation, membrane binding, and metal ion chelation (Zimmermann et al., 1997; Dadeya et al., 2002; Rabowsky et al., 1996)

HTF differentiation to myofibroblasts

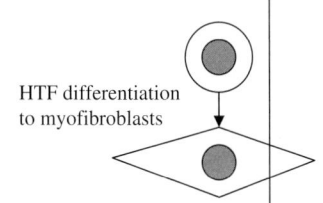

Neovascularisation

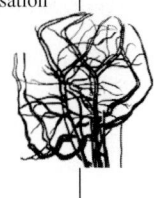

- Paclitaxel (Taxol) (antineoplasmatic agent) First isolated from the bark of the Pacific yew tree, taxus breviofolia, Taxaceae — Inhibits depolymerization of intracellular microtubules and subsequently prevents mitosis (Jampel et al., 1993; Koz et al., 2007; Jampel & Moon, 1998)

- MMC loaded hydrogels — Inhibit cell proliferation in vitro

- Bleomycin Group of related glycopeptide antibiotics isolated from Streptomyces verticillus. — Inhibits cell replication and survival through DNA binding. Creates free radicals, which cause single- and double-strand breaks which lead to inhibition of DNA synthesis (Oshima et al., 1999)

- Thiotepa. Synthetic antimitotic agent similar to nitrogen mustards used in chemical warfare — Polyfunctional alkylating agent

- Retinoic acid and its derivatives Vitamin A derivative — Retinoic acid regulates gene expression by binding to nuclear transcription factors

- Interferon alpha (IFN-alpha). Recombinant protein mimicking the effects of natural IFN-alpha — Interferon alpha regulates cell proliferation and differentiation by affecting several cellular communication and signal transduction pathways (Gillies et al., 1999)

- Lectins (phytoagglutinins). Proteins that agglutinate erythrocytes and other cells — The mushroom lectin from Agaricus bisporus binds to galactosyl-ß-1,3-N-acetyl-galactosamine-alpha (Gal-Gal-NAc) and has a strong antiproliferative effect. The exact mechanism of action is unknown (Batterbury et al., 2002)

- Saporin. Derived from the plant Saponaria officinalis — Ribosome inactivating protein -> cell proliferation inhibitor

- Antiproliferative gene insertion Antiproliferative gene p21(WAF-1/Cip-1) — p21(WAF-1/Cip-1) is a transcription factor that mediates cell cycle arrest in response to cellular stress. Transfection using an adenoviral system resulted in ihibition of scarring (Perkins et al., 2002)

- Bevacizumab (Avastin) intrableb injection(s) post-trabeculectomy — Inhibition of vascularization and reduction of IOP (Kapetansky et al., 2007)

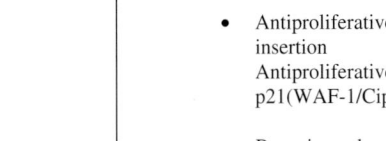

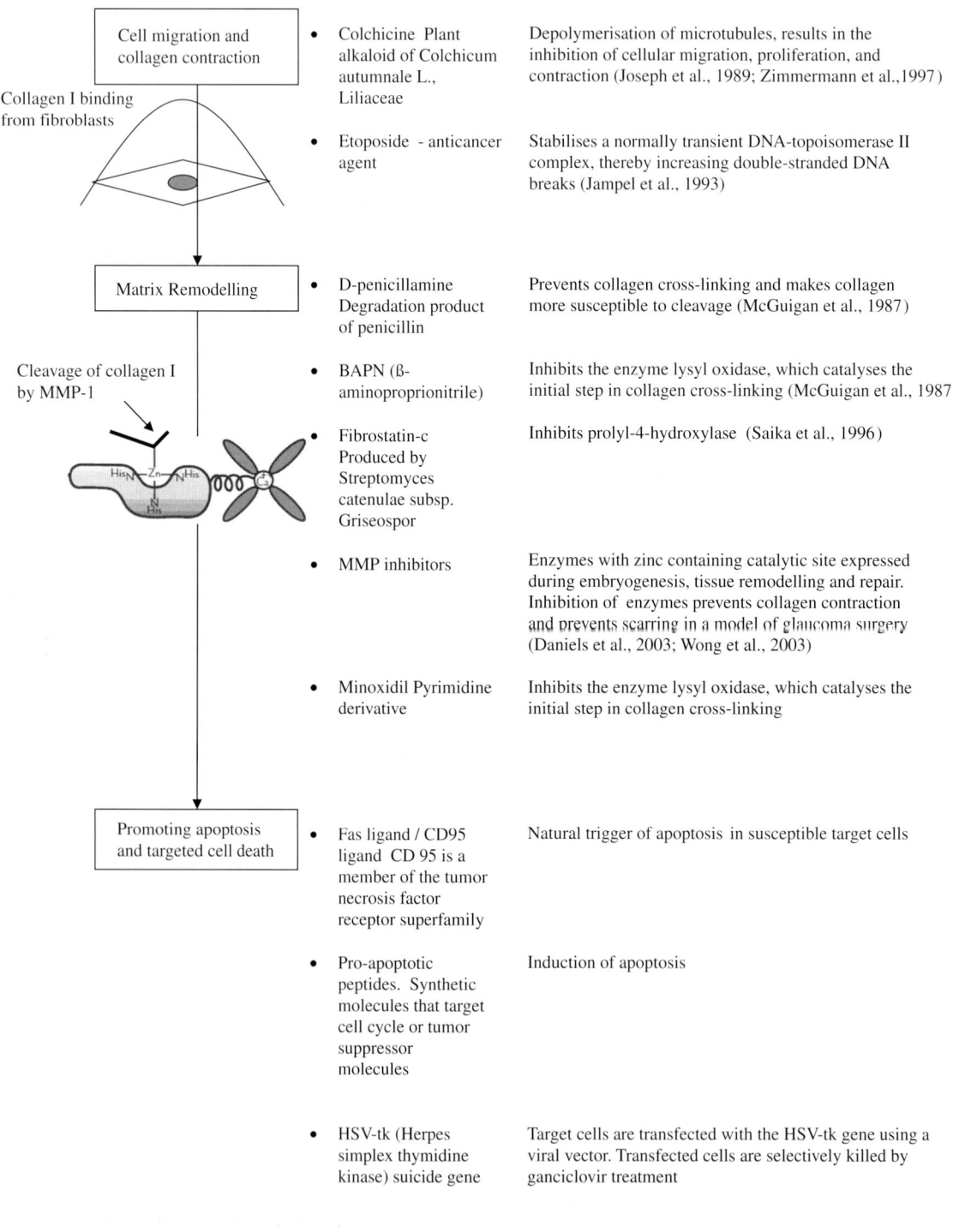

Process	Agent	Mechanism
Cell migration and collagen contraction	Colchicine Plant alkaloid of Colchicum autumnale L., Liliaceae	Depolymerisation of microtubules, results in the inhibition of cellular migration, proliferation, and contraction (Joseph et al., 1989; Zimmermann et al.,1997)
	Etoposide - anticancer agent	Stabilises a normally transient DNA-topoisomerase II complex, thereby increasing double-stranded DNA breaks (Jampel et al., 1993)
Matrix Remodelling	D-penicillamine Degradation product of penicillin	Prevents collagen cross-linking and makes collagen more susceptible to cleavage (McGuigan et al., 1987)
	BAPN (ß-aminoproprionitrile)	Inhibits the enzyme lysyl oxidase, which catalyses the initial step in collagen cross-linking (McGuigan et al., 1987)
	Fibrostatin-c Produced by Streptomyces catenulae subsp. Griseospor	Inhibits prolyl-4-hydroxylase (Saika et al., 1996)
	MMP inhibitors	Enzymes with zinc containing catalytic site expressed during embryogenesis, tissue remodelling and repair. Inhibition of enzymes prevents collagen contraction and prevents scarring in a model of glaucoma surgery (Daniels et al., 2003; Wong et al., 2003)
	Minoxidil Pyrimidine derivative	Inhibits the enzyme lysyl oxidase, which catalyses the initial step in collagen cross-linking
Promoting apoptosis and targeted cell death	Fas ligand / CD95 ligand CD 95 is a member of the tumor necrosis factor receptor superfamily	Natural trigger of apoptosis in susceptible target cells
	Pro-apoptotic peptides. Synthetic molecules that target cell cycle or tumor suppressor molecules	Induction of apoptosis
	HSV-tk (Herpes simplex thymidine kinase) suicide gene	Target cells are transfected with the HSV-tk gene using a viral vector. Transfected cells are selectively killed by ganciclovir treatment

References

Akimoto, M., Hangai, M., Okazak, K., Kogishi, J., Honda, Y. and Kaneda, Y. (1998) Growth inhibition of cultured human Tenon's fibroblastic cells by targeting the E2F transcription factor. Exp. Eye Res., 67: 395–401.

Akiyama, H., Kachi, S., Silva, R.L., Umeda, N., Hackett, S.F., McCauley, D., McCauley, T., Zoltoski, A., Epstein, D.M. and Campochiaro, P.A. (2006) Intraocular injection of an aptamer that binds PDGF-B: a potential treatment for proliferative retinopathies. J. Cell. Physiol., 207(2): 407–412.

Andreev, K., Zenkel, M., Kruse, F., Junemann, A. and Schlotzer-Schrehardt, U. (2006) Expression of bone morphogenetic proteins (BMPs), their receptors, and activins in normal and scarred conjunctiva: role of BMP-6 and activin-A in conjunctival scarring? Exp. Eye Res., 83(5): 1162–1170.

Aoki, F., Kurabayashi, M., Hasegawa, Y. and Kojima, I. (2005) Attenuation of bleomycin-induced pulmonary fibrosis by follistatin. Am. J. Respir. Crit. Care Med., 172(6): 713–720.

Asaria, R.H., Kon, C.H., Bunce, C., Charteris, D.G., Wong, D., Khaw, P.T. and Aylward, G.W. (2001) Adjuvant 5-fluorouracil and heparin prevents proliferative vitreoretinopathy: results from a randomized, double-blind, controlled clinical trial. Ophthalmology, 108(7): 1179–1183.

Barton, K., Budenz, D.L., Khaw, P.T. and Tseng, S.C. (2001) Glaucoma filtration surgery using amniotic membrane transplantation. Invest. Ophthalmol. Vis. Sci., 42(8): 1762–1768.

Batterbury, M., Tebbs, C.A., Rhodes, J.M. and Grierson, I. (2002) Agaricus bisporus (edible mushroom lectin) inhibits ocular fibroblast proliferation and collagen lattice contraction. Exp. Eye Res., 74(3): 361–370.

Blake, D.A., Sahiner, N., John, V.T., Clinton, A.D., Galler, K.E., Walsh, M., Arosemena, A., Johnson, P.Y. and Ayyala, R.S. (2006) Inhibition of cell proliferation by mitomycin C incorporated into P(HEMA) hydrogels. J. Glaucoma, 15(4): 291–298.

Broadway, D.C., Grierson, I., Sturmer, J. and Hitchings, R.A. (1996) Reversal of topical antiglaucoma medication effects on the conjunctiva. Arch. Ophthalmol., 114(3): 262–267.

Bruno, C.A., Eisengart, J.A., Radenbaugh, P.A. and Moroi, S.E. (2006) Subconjunctival placement of human amniotic membrane during high risk glaucoma filtration surgery. Ophthalmic Surg. Lasers Imaging, 37(3): 190–197.

Chang, L., Crowston, J.G., Cordeiro, M.F., Akbar, A.N. and Khaw, P.T. (2000) The role of the immune system in conjunctival wound healing after glaucoma surgery. Surv. Ophthalmol., 45(1): 49–68.

Chen, D., Zhao, M. and Mundy, G.R. (2004) Bone morphogenetic proteins. Growth Factors, 22(4): 233–241.

Chen, S.J., Yuan, W., Mori, Y., Levenson, A., Trojanowska, M. and Varga, J. (1999) Stimulation of type I collagen transcription in human skin fibroblasts by TGF-beta: involvement of Smad 3. J. Invest. Dermatol., 112(1): 49–57.

Chihara, E., Dong, J., Ochiai, H. and Hamada, S. (2002) Effects of tranilast on filtering blebs: a pilot study. J. Glaucoma, 11(2): 127–133.

Cho, K.S., Yang, L., Lu, B., Feng, M.H., Huang, X., Pekny, M. and Chen, D.F. (2005) Re-establishing the regenerative potential of central nervous system axons in postnatal mice. J. Cell Sci., 118(Pt 5): 863–872.

Cordeiro, M.F., Gay, J.A. and Khaw, P.T. (1999a) Human anti-transforming growth factor-beta2 antibody: a new glaucoma anti-scarring agent. Invest. Ophthalmol. Vis. Sci., 40(10): 2225–2234.

Cordeiro, M.F., Mead, A., Ali, R.R., Alexander, R.A., Murray, S., Chen, C., York-Defalco, C., Dean, N.M., Schultz, G.S. and Khaw, P.T. (2003) Novel antisense oligonucleotides targeting TGF-beta inhibit in vivo scarring and improve surgical outcome. Gene Ther., 10(1): 59–71.

Cordeiro, M.F., Schultz, G.S., Ali, R.R., Bhattacharya, S.S. and Khaw, P.T. (1999b) Molecular therapy in ocular wound healing. Br. J. Ophthalmol., 83(11): 1219–1224.

Dadeya, S., Kamlesh, K.C. and Fatima, S. (2002) Intraoperative daunorubicin versus conjunctival autograft in primary pterygium surgery. Cornea, 21(8): 766–769.

Daniels, J.T., Cambrey, A.D., Occleston, N.L., Garrett, Q., Tarnuzzer, R.W., Schultz, G.S. and Khaw, P.T. (2003) Matrix metalloproteinase inhibition modulates fibroblast-mediated matrix contraction and collagen production in vitro. Invest. Ophthalmol. Vis. Sci., 44(3): 1104–1110.

Demir, T., Turgut, B., Akyol, N., Ozercan, I., Ulas, F. and Celiker, U. (2002) Effects of amniotic membrane transplantation and mitomycin C on wound healing in experimental glaucoma surgery. Ophthalmologica, 216(6): 438–442.

Dewald, O., Zymek, P., Winkelmann, K., Koerting, A., Ren, G., bou-Khamis, T., Michael, L.H., Rollins, B.J., Entman, M.L. and Frangogiannis, N.G. (2005) CCL2/Monocyte Chemoattractant Protein-1 regulates inflammatory responses critical to healing myocardial infarcts. Circ. Res., 96(8): 881–889.

Diestelhorst, M. and Grisanti, S. (2002) Photodynamic therapy to control fibrosis in human glaucomatous eyes after trabeculectomy: a clinical pilot study. Arch. Ophthalmol., 120(2): 130–134.

Drolsum, L., Willoch, C. and Nicolaissen, B. (2006) Use of amniotic membrane as an adjuvant in refractory glaucoma. Acta Ophthalmol. Scand., 84(6): 786–789.

Fujishima, H., Shimazaki, J., Shinozaki, N. and Tsubota, K. (1998) Trabeculectomy with the use of amniotic membrane for uncontrollable glaucoma. Ophthalmic. Surg. Lasers, 29(5): 428–431.

Garrett, Q., Khaw, P.T., Blalock, T.D., Schultz, G.S., Grotendorst, G.R. and Daniels, J.T. (2004) Involvement of CTGF in TGF-beta1-stimulation of myofibroblast differentiation and collagen matrix contraction in the presence of mechanical stress. Invest. Ophthalmol. Vis. Sci., 45(4): 1109–1116.

Gillies, M.C., Brooks, A.M.V., Young, S., Gillies, B., Simpson, J.M. and Goldberg, I. (1999) A randomized phase II trial of interferon-alpha2b versus 5-fluorouracil after trabeculectomy. Aust. N. Z. J. Ophthalmol., 27(1): 37–44.

Gomes dos Santos, A.L., Bochot, A., Doyle, A., Tsapis, N., Siepmann, J., Siepmann, F., Schmaler, J., Besnard, M., Behar-Cohen, F. and Fattal, E. (2006) Sustained release of nanosized complexes of polyethylenimine and anti-TGF-beta 2 oligonucleotide improves the outcome of glaucoma surgery. J. Control. Release, 112(3): 369–381.

Gray, A.J., Bishop, J.E., Reeves, J.T. and Laurent, G.J. (1993) A alpha and B beta chains of fibrinogen stimulate proliferation of human fibroblasts. Issue Series Title: J. Cell Sci., 104(Pt 2): 409–413.

Grisanti, S., Diestelhorst, M., Heimann, K. and Krieglstein, G. (1999) Cellular photoablation to control postoperative fibrosis in a rabbit model of filtration surgery. Br. J. Ophthalmol., 83(12): 1353–1359.

Grisanti, S., Gralla, A., Maurer, P., Diestelhorst, M., Krieglstein, G. and Heimann, K. (2000) Cellular photoablation to control postoperative fibrosis in filtration surgery: in vitro studies. Exp. Eye Res., 70(2): 145–152.

Grisanti, S., Szurman, P., Warga, M., Kaczmarek, R., Ziemssen, F., Tatar, O. and Bartz-Schmidt, K.U. (2005) Decorin modulates wound healing in experimental glaucoma filtration surgery: a pilot study. Invest. Ophthalmol. Vis. Sci., 46(1): 191–196.

He, J. and Bazan, H.E. (2006) Synergistic effect of platelet-activating factor and tumor necrosis factor-alpha on corneal myofibroblast apoptosis. Invest. Ophthalmol. Vis. Sci., 47(3): 883–891.

Herrero-Vanrell, R. and Refojo, M.F. (2001) Biodegradable microspheres for vitreoretinal drug delivery. Adv. Drug Deliv. Rev., 52(1): 5–16.

Hill, R.A., Crean, D.H., Doiron, D.R., Ghosheh, F., Ryan, J.A., Kelly, H., Booth, M., Liaw, L.H., Newman, L. and Berns, M.W. (1997) Photodynamic therapy of the ciliary body with tin ethyl etiopurpurin and tin octaethyl benzochlorin in pigmented rabbits. Ophthalmic Surg. Lasers, 28(11): 948–953.

Honjo, M., Tanihara, H., Kameda, T., Kawaji, T., Yoshimura, N. and Araie, M. (2007) Potential role of Rho-associated protein kinase inhibitor Y-27632 in glaucoma filtration surgery. Invest. Ophthalmol. Vis. Sci., 48(12): 5549–5557.

Hueber, A., Esser, J.M., Kociok, N., Welsandt, G., Luke, C., Roters, S. and Esser, P.J. (2007) Mitomycin C induces multidrug resistance in glaucoma surgery. Graefes Arch. Clin. Exp. Ophthalmol., 246: 297–304.

Jampel, H.D. and Moon, J. (1998) The effect of paclitaxel powder on glaucoma filtration surgery in rabbits. J. Glaucoma, 7: 170–177.

Jampel, H.D., Moon, J.H., Quigley, H.A., Barron, Y. and Lam, K.-W. (1998) Aqueous humor uric acid and ascorbic acid concentrations and outcome of trabeculectomy. Arch. Ophthalmol., 116: 281–285.

Jampel, H.D., Thibault, D., Leong, K.W., Uppal, P. and Quigley, H.A. (1993) Glaucoma filtration surgery in nonhuman primates using taxol and etoposide in polyanhydride carriers. Invest. Ophthalmol. Vis. Sci., 34(11): 3076–3083.

Jordan, J.F., Diestelhorst, M. and Grisanti, S. (2003) Photodynamic modulation of wound healing in glaucoma filtration surgery. Br. J. Ophthalmol., 87: 870–875.

Joseph, J.P., Grierson, I. and Hitchings, R.A. (1989) Taxol, cytochalasin B, and colchicine effects on fibroblast migration and contraction: a role in glaucoma filtration surgery? Cur. Eye Res., 8(2): 203–215.

Kapetansky, F.M., Pappa, K.S., Krasnow, M.A., Baker, N.D. and Francis, C.D. (2008). Subconjunctival injection(s) of Bevacizumab for failing filtering blebs. Invest. Ophthalmol. Vis. Sci. ARVO 49 [E-Abstract 4149].

Kent, A.R., Dubiner, H.B., Whitaker, R., Mundorf, T.K., Stewart, J.A., Cate, E.A. and Stewart, W.C. (1998) The efficacy and safety of diclofenac 0.1% versus prednisolone acetate 1% following trabeculectomy with adjunctive mitomycin-C. Ophthalmic Surg. Lasers, 29(7): 562–569.

Khaw, P., Grehn, F., Hollo, G., Overton, B., Wilson, R., Vogel, R. and Smith, Z. (2007) A phase III study of subconjunctival human anti-transforming growth factor beta(2) monoclonal antibody (CAT-152) to prevent scarring after first-time trabeculectomy. Ophthalmology, 114(10): 1822–1830.

Khaw, P.T., Occleston, N.L., Schultz, G., Grierson, I., Sherwood, M.B. and Larkin, G. (1994) Activation and suppression of fibroblast function. Eye, 8(Pt 2): 188–195.

Khaw, P.T., Ward, S., Grierson, I. and Rice, N.S. (1991) Effect of beta radiation on proliferating human Tenon's capsule fibroblasts. Br. J. Ophthalmol., 75(10): 580–583.

Kim, E.K., Peterson, T.G. and Barnes, S. (1998) Mechanisms of action of the soy isoflavone genistein: emerging role for its effects via transforming growth factor beta signaling pathways. Am. J. Clin. Nutr., 68(6 Suppl.): S1418–S1425.

Kim, J.H., Jeong, S.Y., Jung, M.H. and Hwang, J.M. (2004) Use of polyurethane with sustained release dexamethasone in delayed adjustable strabismus surgery. Br. J. Ophthalmol., 88(11): 1450–1454.

Kirwan, J.F., Constable, P.H., Murdoch, I.E. and Khaw, P.T. (2003) Beta irradiation: new uses for an old treatment: a review. Eye, 17(2): 207–215.

Kirwan, J.F., Cousens, S., Venter, L., Cook, C., Stulting, A., Roux, P. and Murdoch, I. (2006) Effect of beta radiation on success of glaucoma drainage surgery in South Africa: randomised controlled trial. BMJ, 333(7575): p. 942.

Koz, O.G., Ozhuy, S., Tezel, G.G., Karaman, N., Unlu, N., Yarangumeli, A. and Kural, G. (2007) The effect of paclitaxel on conjunctival wound healing: a pilot study. J. Glaucoma, 16(7): 610–615.

Lattanzio, F.A., Jr., Crouch, E.R., Jr., Mitrev, P.V., Williams, P.B. and Allen, R.C. (2005) Cyclosporin as an adjunct to glaucoma filtration surgery. J. Glaucoma, 14(6): 441–447.

Leask, A. and Abraham, D.J. (2004) TGF-beta signaling and the fibrotic response. FASEB J., 18(7): 816–827.

Li, Y., Li, D., Khaw, P.T. and Raisman, G. (2008) Transplanted olfactory ensheathing cells incorporated into the optic nerve head ensheathe retinal ganglion cell axons: possible relevance to glaucoma. Neurosci. Lett., 440(3): 251–254 [Epub 2008 May 28].

Lobmann, R., Ambrosch, A., Seewald, M., Dietlein, M., Zink, K., Kullmann, K.H. and Lehnert, H. (2004) Antibiotic therapy for diabetic foot infections: comparison of cephalosporines with chinolones. Diabetes Nutr. Metab., 17(3): 156–162.

Lopes, J.F., Moster, M.R., Wilson, R.P., Altangerel, U., Alvim, H.S., Tong, M.G., Fontanarosa, J. and Steinmann, W.C. (2006) Subconjunctival sodium hyaluronate 2.3% in trabeculectomy: a prospective randomized clinical trial. Ophthalmology, 113(5): 756–760.

Massague, J. (1999) Wounding Smad. Nat. Cell Biol., 1(5): E117–E119.

McGuigan, L.J., Mason, R.P., Sanchez, R. and Quigley, H.A. (1987b) D-pencillamine and betaminopropionitrile effects on experimental filtering surgery. Invest. Ophthalmol. Vis. Sci., 28(10): 1625–1629.

Mead, A.L., Wong, T.T.L., Cordiero, M.F., Anderson, I.K. and Khaw, P.T. (2003) Anti-transforming growth factor-b2 antibody: a new post operative anti-scarring agent in glaucoma surgery. Invest. Ophthalmol. Vis. Sci., 44: 3394–3401.

Meyer-ter-Vehn, T., Gebhardt, S., Sebald, W., Buttmann, M., Grehn, F., Schlunck, G. and Knaus, P. (2006) p38 inhibitors prevent TGF-beta-induced myofibroblast transdifferentiation in human tenon fibroblasts. Invest. Ophthalmol. Vis. Sci., 47(4): 1500–1509.

Mietz, H., Chevez-Barrios, P., Feldman, R.M. and Lieberman, M.W. (1998) Suramin inhibits wound healing following filtering procedures for glaucoma. Br. J. Ophthalmol., 82: 816–820.

Mietz, H. and Krieglstein, G.K. (2001) Suramin to enhance glaucoma filtering procedures: a clinical comparison with mitomycin. Ophthalmic Surg. Lasers, 32(5): 358–369.

Naginen, C.N., Kutty, V., Detrick, B. and Hooks, J.J. (2005) Expression of PDGF and their receptors in human retinal pigment epithelial cells and fibroblasts: regulation by TGF-beta. J. Cell. Physiol., 203(1): 35–43.

Nakamura, H., Siddiqui, S.S., Shen, X., Malik, A.B., Pulido, J.S., Kumar, N.M. and Yue, B.Y. (2004) RNA interference targeting transforming growth factor-beta type II receptor suppresses ocular inflammation and fibrosis. Mol. Vis., 10: 703–711.

Oshima, Y., Sakamoto, T., Nakamura, T., Tahara, Y., Goto, Y., Ishibashi, T. and Inomata, H. (1999) The comparative benefits of glaucoma filtering surgery with an electric-pulse targeted drug delivery system demonstrated in an animal model. 106(6): 1140–1146.

Paikal, D., Zhang, G., Cheng, Q. and Lee, D.A. (2000) The effect of integrin antibodies on the attachment and proliferation of human Tenon's capsule fibroblasts. Exp. Eye Res., 70(4): 393–400.

Patella, S., Phillips, D.J., Tchongue, J., de Kretser, D.M. and Sievert, W. (2006) Follistatin attenuates early liver fibrosis: effects on hepatic stellate cell activation and hepatocyte apoptosis. Am. J. Physiol. Gastrointest. Liver Physiol., 290(1): G137–G144.

Perkins, T.W., Faha, B., Ni, M., Kiland, J.A., Poulsen, G.L., Antelman, D., Atencio, I., Shinoda, J., Sinha, D., Brumback, L., Maneval, D., Kaufman, P.L. and Nickells, R.W. (2002) Adenovirus-mediated gene therapy using human p21WAF-1/Cip-1 to prevent wound healing in a rabbit model of glaucoma filtration surgery. Arch. Ophthalmol., 120(7): 941–949.

Priglinger, S.G., Alge, C.S., Kook, D., Thiel, M., Schumann, R., Eibl, K., Yu, A., Neubauer, A.S., Kampik, A. and Welge-Lussen, U. (2006) Potential role of tissue transglutaminase in glaucoma filtering surgery. Invest. Ophthalmol. Vis. Sci., 47(9): 3835–3845.

Rabowsky, J.H., Dukes, A.J., Lee, D.A. and Leong, K.W. (1996) The use of bioerodible polymers and daunorubicin in glaucoma filtration surgery. Ophthalmology, 103: 800–807.

Rauscher, F.M., Barton, K., Budenz, D.L., Feuer, W.J. and Tseng, S.C. (2007) Long-term outcomes of amniotic membrane transplantation for repair of leaking glaucoma filtering blebs. Am. J. Ophthalmol., 143(6): 1052–1054.

Reddi, A.H. (1997a) BMPs: actions in flesh and bone. Nat. Med., 3(8): 837–839.

Reddi, A.H. (1997b) Bone morphogenetic proteins: an unconventional approach to isolation of first mammalian morphogens. Cytokine Growth Factor Rev., 8(1): 11–20.

Saika, S. (2006) TGFbeta pathobiology in the eye. Lab. Invest., 86(2): 106–115.

Saika, S., Kono-Saika, S., Tanaka, T., Yamanaka, O., Ohnishi, Y., Sato, M., Muragaki, Y., Ooshima, A., Yoo, J., Flanders, K.C. and Roberts, A.B. (2004) Smad3 is required for dedifferentiation of retinal pigment epithelium following retinal detachment in mice. Lab. Invest., 84(10): 1245–1258.

Saika, S., Ooshima, A., Yamanaka, O., Okada, Y., Tanaka, S. and Ohnishi, Y. (1996) Effect of fibrostatin C, an inhibitor of prolyl 4-hydroxylase, on collagen secretion by human Tenon's capsule fibroblasts in vitro. Graefes Arch. Clin. Exp. Ophthalmol., 234(Suppl. 1): S214–S222.

Saika, S., Yamanaka, O., Ikeda, K., Kim-Mitsuyama, S., Flanders, K.C., Yoo, J., Roberts, A.B., Nishikawa-Ishida, I., Ohnishi, Y., Muragaki, Y. and Ooshima, A. (2005) Inhibition of p38MAP kinase suppresses fibrotic reaction of retinal pigment epithelial cells. Lab. Invest., 85(7): 838–850.

Sakaguchi, H., Takai, S., Sakaguchi, M., Sugiyama, T., Ishihara, T., Yao, Y., Miyazaki, M. and Ikeda, T. (2002b) Chymase and angiotensin converting enzyme activities in a hamster model of glaucoma filtering surgery. Curr. Eye Res., 24(5): 325–331.

Schiller, M., Javelaud, D. and Mauviel, A. (2004) TGF-beta-induced SMAD signaling and gene regulation: consequences for extracellular matrix remodeling and wound healing. J. Dermatol. Sci., 35(2): 83–92.

Shaunak, S., Thomas, S., Gianasi, E., Godwin, A., Jones, E., Teo, I., Mireskandari, K., Luthert, P., Duncan, R., Patterson, S., Khaw, P. and Brocchini, S. (2004) Polyvalent dendrimer glucosamine conjugates prevent scar tissue formation. Nat. Biotechnol., 22(8): 977–984.

Shinohara, K., Tanaka, M., Sakuma, T. and Kobayashi, Y. (2003) Efficacy of daunorubicin encapsulated in liposome for the treatment of proliferative vitreoretinopathy. Ophthalmic Surg. Lasers Imaging, 34(4): 299–305.

Singhal, S., Lawrence, J.M., Bhatia, B., Ellis, J.S., Kwan, A.S., Macneil, A., Luthert, P.J., Fawcett, J.W., Perez, M.T., Khaw, P.T. and Limb, G.A. (2008) Chondroitin sulphate proteoglycans and microglia prevent migration and integration of grafted Muller stem cells into degenerating retina. Stem Cells, 26(4): 1074–1082 [Epub 2008 Jan 24].

Siriwardena, D., Khaw, P.T., King A.J., Donaldson, M.L., Overton, B.M., Migdal, C. and Cordeiro, M.F. (2002). Human antitransforming growth factor beta(2) monoclonal antibody–a new modulator of wound healing in trabeculectomy: a randomized placebo controlled clinical study. 109(3): 427–431.

Stasi, K., Paccione, J., Bianchi, G., Friedman, A. and Danias, J. (2006) Photodynamic treatment in a rabbit model of glaucoma surgery. Acta Ophthalmol. Scand., 84(5): 661–666.

Tham, C.C., Li, F.C., Leung, D.Y., Kwong, Y.Y., Yick, D.W., Chi, C.C. and Lam, D.S. (2006) Intrableb triamcinolone acetonide injection after bleb-forming filtration surgery (trabeculectomy, phacotrabeculectomy, and trabeculectomy revision by needling): a pilot study. Eye, 20(12): 1484–1486.

Tsuchida, K. (2004) Activins, myostatin and related TGF-beta family members as novel therapeutic targets for endocrine, metabolic and immune disorders. Curr. Drug Targets. Immune. Endocr. Metabol. Disord., 4(2): 157–166.

Watts, K.L., Sampson, E.M. and Schultz, G.S. (2005) Simvastatin inhibits growth factor expression and modulates profibrogenic markers in lung fibroblasts. Am. J. Respir. Cell Mol. Biol., 32: 290–300.

Wells, A.P., Bunce, C. and Khaw, P.T. (2004) Flap and suture manipulation after trabeculectomy with adjustable sutures: Titration of flow intraocular pressure in guarded filtration surgery. J. Glaucoma., 13(5): 400–406.

Wells, A.P., Cordeiro, M.F., Bunce, C. and Khaw, P.T. (2003) Cystic bleb formation and related complications in limbus-versus fornix-based conjunctival flaps in pediatric and young adult trabeculectomy with mitomycin C. Ophthalmology, 110(11): 2192–2197.

Wong, T.T., Mead, A.L. and Khaw, P.T. (2003) Matrix metalloproteinase inhibition modulates postoperative scarring after experimental glaucoma filtration surgery. Invest. Ophthalmol. Vis. Sci., 44(3): 1097–1103.

Wong, T.T., Mead, A.L. and Khaw, P.T. (2005) Prolonged antiscarring effects of ilomastat and MMC after experimental glaucoma filtration surgery. Invest. Ophthalmol. Vis. Sci., 46(6): 2018–2022.

Wong, H.T., Seah, S.K. and Tym, W.H. (1999) Augmentation of filtering blebs with perfluoropropane gas bubble: an experimental and pilot clinical study. Ophthalmology, 106(3): 545–549.

Wu, X.Y., Yang, Y.M., Guo, H. and Chang, Y. (2006) The role of connective tissue growth factor, transforming growth factor beta1 and Smad signaling pathway in cornea wound healing. Chin. Med. J. (Engl.), 119(1): 57–62.

WuDunn, D. (1997) Intracameral urokinase for dissolution of fibrin or blood clots after glaucoma surgery. Am. J. Ophthalmol., 124(5): 693–695.

Yamanaka, O., Saika, S., Ohnishi, Y., Kim-Mitsuyama, S., Kamaraju, A.K. and Ikeda, K. (2007) Inhibition of p38MAP kinase suppresses fibrogenic reaction in conjunctiva in mice. Mol. Vis., 13: 1730–1739.

Yasukawa, T., Kimura, H., Tabata, Y. and Ogura, Y. (2001) Biodegradable scleral plugs for vitreoretinal drug delivery. Adv. Drug Deliv. Rev., 52(1): 25–36.

Zimmermann, T.J., Kooner, K.S., Sharir, M. and Fechner, R.D. (1997) Textbook of ocular pharmacology. Lippincott Raven, Philadelphia.

CHAPTER 18

Surgical alternative to trabeculectomy

Roberto G. Carassa*

Department of Ophthalmology, University Hospital S. Raffaele, Milano, Italy

Abstract: Non-penetrating glaucoma surgery, represented by deep sclerectomy and viscocanalostomy, is an effective method to lower intraocular pressure (IOP) in glaucomatous patients. Both procedures reduce IOP by allowing aqueous humor drainage without opening the anterior chamber. Deep sclerectomy, similar to trabeculectomy, provides aqueous external filtration in the subconjunctival space. This technique, with the adjunctive use of implants, antimetabolites, and goniopuncture, may provide final IOP comparable to those obtained with trabeculectomy, but with less complications. Viscocanalostomy is less dependent on external filtration since it increases trabecular aqueous outflow facility by micro-disrupting Schlemm's canal walls and juxtacanalicular trabecular meshwork. This technique is very safe, but it provides higher final IOPs compared to trabeculectomy.

Non-penetrating surgery should be therefore considered a surgical alternative to trabeculectomy in specific clinical cases.

Keywords: glaucoma surgery; non-penetrating surgery; deep sclerectomy; viscocanalostomy

Introduction

During the last 20 years, many alternatives to trabeculectomy were proposed, but only "non-penetrating glaucoma surgery" succeeded and was therefore included in guidelines as a surgical option for glaucoma (European Glaucoma Society, 2003; Carassa and Goldberg, 2005). Non-penetrating glaucoma surgery is represented by "*deep sclerectomy*" and by "*viscocanalostomy*" (which was introduced by R. Stegmann in the early 1990s), and is based on the original studies by Krasnov (1972) and by Zimmerman et al. (1984) on "non-penetrating trabeculectomy." Similarly, both procedures are aimed at lowering intraocular pressure (IOP) by allowing drainage of the aqueous humor from the anterior chamber not through a patent scleral opening, but by slow percolation through the inner trabecular meshwork and/or Descemet's membrane ("sclerodescemetic membrane"). This avoids sudden IOP drops, hypotonies, and flat anterior chambers. The absence of anterior chamber opening and iridectomy limits the risk of cataract and infection. Compared to deep sclerectomy, viscocanalostomy is a step forward. In fact, this procedure is aimed not only at taking the advantages of being non-penetrating, as deep sclerectomy, but also, most important, in restoring the physiological outflow pathway, thus avoiding any external filtration. This would make the success of the procedure independent of conjunctival or episcleral scarring, the leading causes of failure in trabeculectomy, with fewer indications for wound healing modulation.

*Corresponding author. Tel.: +39 02 26433591; Fax: +39 02 76311438; E-mail: carassa@tin.it

Moreover, the reduced incidence of the filtering bleb avoids related ocular discomfort, and the procedure can be carried out in any quadrant.

Deep sclerectomy

Deep sclerectomy is aimed at reducing IOP by allowing external filtration of the aqueous humor. Differently from trabeculectomy, aqueous exits the eye not through a patent hole but by slow passage through the sclerodescemetic membrane formed by the internal portion of the posterior and anterior trabecular meshwork and by the adjacent Descemet's membrane. The membrane is created by the removal of the inner wall of Schlemm's canal and juxtacanalicular trabeculum (sites of the increased outflow resistance in glaucoma) and by the exposure of the anterior trabecular meshwork and Descemet's membrane. After exiting the anterior chamber, aqueous fills an intrascleral space called "intrascleral lake" or "decompression chamber," and it is finally drained into the subconjunctival space or is partially reabsorbed into the suprachoroidal space (Chiou et al., 1996, 1998b; Marchini et al., 2001; Sarodia et al., 2007).

The surgical technique varies among surgeons. Different methods of anesthesia are used as topical, local infiltration, peribulbar, or general anesthesia based on surgeon preference. Nevertheless, non-penetrating surgery is a long and difficult procedure; thus, deep and long-lasting anesthesia and akinesia, as the peribulbar block, are to be preferred.

After performing a wide limbus-based conjunctival flap, a 5 × 5 mm rectangular superficial flap, approximately one-third of scleral thickness, is dissected. A limbus-based triangle of deep sclera is then dissected as deep as to leave a thin layer of sclera over the choroid and the ciliary body. The dissection is carried anteriorly until Schlemm's canal is deroofed and 1–2 mm of Descemet's membrane is exposed. At this stage of the procedure, aqueous humor should be seen percolating through the "trabeculodescemetic membrane." In order to increase percolation, some surgeons suggest the removal of the inner wall of Schlemm's canal ("external trabeculectomy"). The inner flap is then removed, and, in order to maintain the space (the "intrascleral lake" or "decompression chamber") and avoid postoperative scarring, different implants are often used. Absorbable porcine collagen implant (Aquaflow™, Staar surgical AG, Nidau, Switzerland), reticulate hyaluronic acid implant (Skgel™, Corneal Laboratoires, Paris, France), non-absorbable implant (T Flux™, Ioltech Laboratoires, La Rochelle, France), or PMMA implant (Homdec SA, Belmont, Switzerland) can be sutured or positioned in the intrascleral space. The superficial scleral flap is then repositioned and sutured with two 10-0 nylon sutures. Finally, the conjunctiva is tightly closed.

Some authors, before dissecting the internal flap or just after the opening of the Schlemm's canal, in all or in selected cases, apply a sponge soaked with Mitomicin C (0.1–0.3 mg/ml) over the sclera for 1–3 min in order to avoid excessive scarring, thus increasing the success rate of the procedure. In the postoperative time, up to 60% of the eyes need to be treated with a Nd:YAG laser goniopuncture of the "trabeculodescemetic" membrane. With a gonioscopy contact lens, the aiming beam of the laser is focused on the semi-transparent trabecular-Descemet's membrane, which often has a concave appearance. In the free-running Q-switched mode with a power of 4–8 mJ, 4–15 shots are applied. This procedure, by creating openings in the membrane itself, increases the outflow of aqueous, thus reducing the IOP.

Results vary between studies due to different follow-up and technique used. The mean final IOP without adjunctive therapy is in the mid- to high-teens ranging from 11 to 20.9 mmHg, while the achievement of an IOP below 21 mmHg is obtained in 57–92.6% at 12 months, in 40–69% at 24 months, in 44–77% at 36 months, and in 34–63% at 48 months (Demailly et al., 1997; Sanchez et al., 1997; Bas and Goethals, 1999; Bechetoille, 1999; Hamard et al., 1999; Karlen et al., 1999; Massy et al., 1999; Sourdille et al., 1999; Dahan and Drusedau, 2000; Mermoud, 2000; Mermoud and Schnyder, 2000; Marchini et al., 2001; Shaarawy et al., 2001; Kozobolis et al., 2002; Auer et al., 2004; Lachkar et al., 2004; Neudorfer et al., 2004; Shaarawy et al., 2004; Anand and Atherley, 2005; Shaarawy and Mermoud, 2005; Khairy et al., 2006; Mansouri et al., 2006; Mielke et al., 2006). In a meta-analysis of 29 articles conducted in 2004 (and thus not including the most updated articles), the success rate of deep sclerectomy (as an IOP < 21 mmHg without medications) was 69.7% without implant, 59.4% with collagen implant, and 71.1% with reticulated hyaluronic acid implant. No significant difference was found among the three techniques (Cheng and Wei, 2004). When compared with trabeculectomy, the success in achieving an IOP below 21 mmHg is comparable even though final IOPs were found lower with trabeculectomy, in some studies (Chiou et al., 1998a; Mermoud et al., 1999; El-Sayyad et al., 2000; Chiselita, 2001; Ambresin et al., 2002; Cillino et al., 2004; Schwenn et al., 2004). The adjunctive use of implants, of antimetabolites, and of goniopuncture allows a greater success rate and lower IOPs, comparable to those obtained with MMC trabeculectomy (Neudorfer et al., 2004; Schwenn et al., 2004; Anand and Atherley, 2005).

Complications are minor and fewer than those reported after trabeculectomy. A reduced induced corneal refractive change was also showed in one study (Egrilmez et al., 2004).

Deep sclerectomy has specific indications and contraindications, based on its intrinsic characteristics and on clinical results. The procedure is not indicated in angle-closure glaucomas, in neovascular glaucoma, and in eyes with wide-angle synechia or diffuse scarring of the conjunctiva in the surgical quadrant. As suggested by the *Terminology and Guidelines for Glaucoma* of the European Glaucoma Society (2003), deep sclerectomy is indicated in primary open-angle glaucoma when target IOP is not very low. The advantages of being non-penetrating make the procedure particularly useful in aphakic eyes with vitreous in the anterior chamber, or in cases where a sudden drop in IOP should be avoided, such as eyes with uncontrolled high pressure or high myopia. The procedure was found effective in uveitic glaucoma (Auer et al., 2004).

Viscocanalostomy

Viscocanalostomy is aimed at reducing IOP by attempting to restore the physiological outflow pathway and not by allowing external filtration of the aqueous humor. The technique is similar to the one used for deep sclerectomy, but for the injection of high molecular weight sodium hyaluronate in Schlemm's canal and the tight suture of the external scleral flap (aimed at making the "intrascleral lake" or "decompression chamber" watertight). The rationale of the technique is the evidence that the site of greater aqueous outflow resistance in open-angle glaucoma is the trabecular meshwork. Viscocanalostomy was aimed at creating a bypass by which aqueous humor could reach Schlemm's canal and leave the eye through the physiological pathway, without passing through the trabecular meshwork. This is made by producing a space inside the sclera (the "intrascleral lake" or "decompression chamber") directly communicating both with Schlemm's canal and with the anterior chamber through the "sclerodescemetic membrane." The aqueous will enter the "intrascleral lake" by percolating through the membrane and will then leave by entering the Schlemm's canal. In reality, viscocanalostomy lowers IOP by increasing the aqueous outflow through different pathways. Injection of viscoelastic into the canal not only dilates the canal and associated collectors, but also disrupts the internal and external walls of Schlemm's canal and adjacent trabecular layers, thus increasing trabecular outflow facility and making the procedure acting as a trabeculotomy (Tamm et al., 2004). Aqueous outflow facility is also increased by damage to the inner wall of Schlemm's canal and adjacent trabeculum at the site of surgery, thus enhancing aqueous outflow into the scleral lake. From here aqueous can leave the eye via three different paths: through the cut ends and previously nonfunctional sectors of Schlemm's canal to collector channels, by external filtration into the subconjunctival space, or by reabsorption into the subchoroidal space. External filtration and filtering blebs are uncommon in viscocanalostomy, while a supraciliary hypoechoic area suggesting aqueous drainage into the subchoroidal space has been shown by the use of ultrasound biomicroscopy (Carassa et al., 1998; Negri-Aranguren et al., 2002).

The surgical technique of viscocanalostomy is similar to the one described for deep sclerectomy.

Nevertheless, some critical differences need to be emphasized.

During the conjunctival fornix-based flap dissection, cautery should be minimized in order to prevent damage to Schlemm's canal and collector channels. In dissecting the internal scleral flap, Schlemm's canal needs to be fully opened and deroofed, leaving two patent and clean openings on the lateral edges of the cut. A paracentesis should always be made in order to decrease IOP, to make incannulation of Schlemm's canal easier, and to reduce bulging of Descemet's membrane during its cleavage from the corneal stroma, which is at high risk of tear formation. To avoid external pressure on the eye, the traction on the bridle suture should also be removed. Using the specific 165-μm cannula, high molecular weight sodium hyaluronate is slowly injected into Schlemm's canal by cannulating the two ostia at the lateral edges of the inner flap. To avoid damage to the canal endothelium, the insertion of the cannula should not exceed 1–1.5 mm from the ostia. The slow injection should be repeated six to seven times on each side to avoid tears and ruptures of the canal. The injection of viscoelastic substance allows progressive dilation of Schlemm's canal over its circumference, disrupting its internal and external walls and adjacent trabecular layers. Moreover, sodium hyaluronate hemostatic properties avoid bleeding and fibrin clot formation, thus limiting healing processes and scarring of Schlemm's canal openings.

Differently from deep sclerectomy, there is little evidence that the use of an implant as the absorbable porcine collagen implant (Aquaflow™, Staar surgical AG, Nidau, Switzerland), the reticulate hyaluronic acid implant (Skgel™, Corneal Laboratoires, Paris, France), and the non-absorbable implant (T Flux™, Ioltech Laboratoires, La Rochelle, France) could be beneficial for the outcome of viscocanalostomy. This is probably related to the differences in the mechanism of action between the two procedures. Nevertheless, as described in the next paragraph, in order to maintain patency of the "intrascleral lake" during the days after surgery, high-weight sodium hyaluronate is used.

In order to seal the "intrascleral lake," the outer scleral flap should be tightly sutured by placing six or seven 10-0 nylon stitches. The step created by the different size of the two flaps allows a better and tight apposition of the external flap. Finally, in order to minimize bleeding and prevent collapsing and scarring of the intrascleral chamber, high molecular weight sodium hyaluronate is injected underneath the flap.

Viscocanalostomy has specific indications and contraindications. It cannot be effective when the angle is closed or neovascularized, or when Schlemm's canal is likely to be damaged. This is the case of previously operated eyes where an extensive cautery of the perilimbar area was made. Due to its final results, the procedure is indicated in primary open-angle glaucoma when target IOP is not very low (as indicated by the *Terminology and Guidelines for Glaucoma* of the European Glaucoma Society (2003) and by the *Consensus Series* book by the Association of International Glaucoma Societies (Carassa and Goldberg, 2005)). The advantage of the absence (or very reduced) external filtration makes the technique safe and particularly indicated in eyes with chronic blepharitis, in lens contact wearer, or when the surgery has to be performed in the lateral or inferior quadrants. Viscocanalostomy was shown effective also in uveitic glaucomas with well-controlled inflammation (Miserocchi et al., 2004), in juvenile glaucomas (Stangos et al., 2005), and in congenital glaucomas (Noureddin et al., 2006).

Viscocanalostomy is an effective procedure in lowering IOP (Carassa et al., 1998; Stegmann et al., 1999; Sunaric-Mégevand and Leuenberger, 2001; Luke et al., 2003; Shaarawy et al., 2003; Yarangumeli et al., 2005). The mean final IOP without adjunctive therapy is in the mid- to high teens ranging from 11.9 to 18.3 mmHg, while the achievement of an IOP below 21 mmHg is obtained in 30–86% at 12 months, in 21–85% at 24 months, and in 35.3–92% at 36 months. In a meta-analysis of eight articles conducted in 2004, the success rate of viscocanalostomy (as an IOP < 21 mmHg without medications) was 72.0% with no significant difference when compared to deep sclerectomy (Cheng and Wei, 2004).

When compared to trabeculectomy, viscocanalostomy may provide higher final mean IOPs and a lower success rate in achieving an IOP below

16 mmHg even if, in most studies, the differences did not achieve statistical significance (Jonescu-Cuypers et al., 2001; Luke et al., 2002; O'Brart et al., 2002; Carassa et al., 2003; O'Brart et al., 2004; Yalvac et al., 2004; Yarangumeli et al., 2004).

Viscocanalostomy is a very safe procedure with few postoperative side effects, mainly related to intraoperative technical complications. It has an easy postoperative management and induces significantly less refractive change and eye discomfort than trabeculectomy, as could be expected considering the absence of the filtering bleb in the majority of the cases (Carassa et al., 2003; Egrilmez et al., 2004).

Conclusions

Non-penetrating glaucoma surgery provides an effective method for lowering IOP in glaucomatous eyes. It has several advantages over trabeculectomy, such as lower complication rate, less incidence of cataract, less risk of infection-related side effects, and less induced refractive change. Deep sclerectomy, with the adjunctive use of antimetabolites, implants and postoperative goniopuncture provides final IOPs in the range of trabeculectomy, but with a safer profile. Viscocanalostomy compared to deep sclerectomy provides higher final IOPs, less postoperative complications, less eye discomfort, and an easy postoperative management. Non-penetrating glaucoma surgery is nevertheless technically demanding; it requires a long learning curve, and it is a longer procedure when compared to trabeculectomy.

References

Ambresin, A., Shaarawy, T. and Mermoud, A. (2002) Deep sclerectomy with collagen implant in one eye compared with trabeculectomy in the other eye of the same patient. J. Glaucoma, 11: 214–220.
Anand, N. and Atherley, C. (2005) Deep sclerectomy augmented with mitomycin C. Eye, 19: 442–450.
Auer, C., Mermoud, A. and Herbort, C.P. (2004) Deep sclerectomy for the management of uncontrolled uveitic glaucoma: preliminary data. Klin. Monatsbl. Augenheilkd., 221: 339–342.
Bas, J.M. and Goethals, M.J.H. (1999) Non penetrating deep sclerectomy preliminary results. Bull. Soc. Belge Ophthalmol., 272: 55–59.
Bechetoille, A. (1999) External trabeculectomy with aspiration: surgical technique. J. Fr. Ophthalmol., 22: 743–748.
Carassa, R.G. and Goldberg, I. (2005) Non penetrating glaucoma drainage surgery. In: Weinreb R.N. and Crowston J.G. (Eds.), Glaucoma Surgery. Open Angle Glaucoma. Kugler, The Hague, pp. 91–106.
Carassa, R.G., Bettin, P., Fiori, M. and Brancato, R. (1998) Viscocanalostomy: a pilot study. Eur. J. Ophthalmol., 8: 57–61.
Carassa, R.G., Bettin, P., Fiori, M. and Brancato, R. (2003) Viscocanalostomy versus trabeculectomy in white adults affected by open-angle glaucoma: a 2-year randomized, controlled trial. Ophthalmology, 110: 882–887.
Cheng, J.W. and Wei, R.L. (2004) Efficacy of non-penetrating trabecular surgery for open angle glaucoma: a meta-analysis. Chin. Med. J., 117: 1006–1010.
Chiou, A.G.Y., Mermoud, A. and Hediguer, A. (1996) Ultrasound biomicroscopy of eyes undergoing deep sclerectomy with collagen implant using an ultrasound biomicroscope. Br. J. Ophthalmol., 80: 541–544.
Chiou, A.G.Y., Mermoud, A. and Jewelewicz, D.A. (1998a) Postoperative inflammation following deep sclerectomy with collagen implant versus standard trabeculectomy. Graefes Arch. Clin. Exp. Ophthalmol., 236: 593–596.
Chiou, A.G.Y., Mermoud, A. and Underdahl, J.P. (1998b) An ultrasound biomicroscopic study of eyes after deep sclerectomy with collagen implant. Ophthalmology, 105: 746–750.
Chiselita, D. (2001) Non penetrating deep sclerectomy versus trabeculectomy in primary open angle glaucoma surgery. Eye, 15: 197–201.
Cillino, S., Di Pace, F. and Casuccio, A. (2004) Deep sclerectomy versus punch trabeculectomy with and without phacoemulsification: a randomised clinical trial. J. Glaucoma, 13: 500–506.
Dahan, E. and Drusedau, M. (2000) Non penetrating filtration surgery for glaucoma: control by surgery only. J. Cataract Refract. Surg., 26: 695–701.
Demailly, P., Lavat, P. and Kretz, G. (1997) Nonpenetrating deep sclerectomy (NPDS) with or without collagen device (CD) in primary open-angle glaucoma: middle-term retrospective study. Int. Ophthalmol., 20: 131–140.
Egrilmez, S., Ates, H., Nalcaci, S., Andac, K. and Yagci, A. (2004) Surgically induced corneal refractive change following glaucoma surgery: nonpenetrating trabecular surgeries versus trabeculectomy. J. Cataract Refract. Surg., 30: 1232–1239.
El-Sayyad, F., Helal, M. and El-Kholify, H. (2000) Nonpenetrating deep sclerectomy versus trabeculectomy in bilateral primary open-angle glaucoma. Ophthalmology, 107: 1671–1674.
European Glaucoma Society. (2003) Incisional surgery. In: Terminology and Guidelines for Glaucoma, 2nd edn., Dogma, Savona, p. 35.

Hamard, P., Plaza, L. and Kopel, H. (1999) Sclerectomie profonde non perforante (SPNP) et glaucome à angle ouvert. Résultats à moyen terme des premiers patients opérés. J. Fr. Ophtalmol., 22: 25–31.

Jonescu-Cuypers, C., Jacobi, P., Konen, W. and Krieglstein, G.K. (2001) Primary viscocanalostomy versus trabeculectomy in white patients with open-angle glaucoma: a randomized clinical trial. Ophthalmology, 108: 254–258.

Karlen, M.E., Sanchez, E. and Schnyder, C.C. (1999) Deep sclerectomy with collagen implant: medium term results. Br. J. Ophthalmol., 83: 6–11.

Khairy, H.A., Green, F.D., Nassar, M.K. and Azuara-Blanco, A. (2006) Control of intraocular pressure after deep sclerectomy. Eye, 20: 336–340.

Kozobolis, V.P., Christodoulakis, E.V., Tzanakis, N., Zacharopoulos, I. and Pallikaris, I.G. (2002) Primary deep sclerectomy versus primary deep sclerectomy with the use of mitomycin C in primary open-angle glaucoma. J. Glaucoma, 11: 287–293.

Krasnov, M.M. (1972) Sinusotomy: foundations, results, prospects. Trans. Am. Ophthalmol. Otolaryngol., 76: 369–374.

Lachkar, Y., Neverauskiene, J., Jeanteur-Lunel, M.N., Gracies, H., Berkani, M., Ecoffet, M., Kopel, J., Kretz, G., Lavat, P., Lehrer, M., Valtot, F. and Demailly, P. (2004) Nonpenetrating deep sclerectomy: a 6-year retrospective study. Eur. J. Ophthalmol., 14: 26–36.

Luke, C., Dietlein, T.S., Jacobi, P.C., Konen, W. and Krieglstein, G.K. (2002) A prospective randomized trial of viscocanalostomy versus trabeculectomy in open angle glaucoma: a 1-year follow-up study. J. Glaucoma, 11: 294–299.

Luke, C., Dietlein, T.S., Jacobi, P.C., Konen, W. and Krieglstein, G.K. (2003) A prospective randomised trial of viscocanalostomy with and without implantation of a reticulated hyaluronic acid implant (SKGEL) in open angle glaucoma. Br. J. Ophthalmol., 87: 599–603.

Mansouri, K., Shaarawy, T., Wedrich, A. and Mermoud, A. (2006) Comparing polymethylmethacrylate implant with collagen implant in deep sclerectomy: a randomised controlled trial. J. Glaucoma, 15: 264–270.

Marchini, G., Marraffa, M. and Brunelli, C. (2001) Ultrasound biomicroscopy and intraocular-pressure-lowering mechanisms of deep sclerectomy with reticulated hyaluronic acid implant. J. Cataract Refract. Surg., 27: 507–517.

Massy, J., Gruber, D., Muraine, G. and Brasseur, G. (1999) Non penetrating deep sclerectomy: mid term results. J. Fr. Ophtalmol., 22: 292–298.

Mermoud, A. (2000) Sinusotomy and deep sclerectomy. Eye, 14: 531–535.

Mermoud, A. and Schnyder, C.C. (2000) Nonpenetrating filtering surgery in glaucoma. Curr. Opin. Ophthalmol., 11: 151–157.

Mermoud, A., Schnyder, C., Sickenberg, M., Chiou, A., Hédiguer, S. and Faggioni, R. (1999) Comparison of deep sclerectomy with collagen implant and trabeculectomy in open angle glaucoma. J. Cataract Refract. Surg., 25: 323–331.

Mielke, C., Dawda, V.K. and Anand, N. (2006) Deep sclerectomy and low dose mitomycin C: a randomised prospective trial in West Africa. Br. J. Ophthalmol., 90: 310–313.

Miserocchi, E., Carassa, R.G., Bettin, P. and Brancato, R. (2004) Viscocanalostomy in patients with uveitis: a preliminary report. J. Cataract and Refract. Surg., 30: 566–570.

Negri-Aranguren, I., Croxatto, O. and Grigera, D.E. (2002) Midterm ultrasound biomicroscopy findings in eyes with successful viscocanalostomy. J. Cataract Refract. Surg., 28: 752–757.

Neudorfer, M., Sadetzki, S., Anisimova, S. and Geyer, O. (2004) Nonpenetrating deep sclerectomy with the use of adjunctive mitomycin C. Ophthalmic Surg. Lasers Imaging, 35: 6–12.

Noureddin, B.N., El-Haibi, C.P., Cheikha, A. and Bashhsur, Z.F. (2006) Viscocanalostomy versus trabeculotomy ab externo in primary congenital glaucoma: 1-year follow-up of a prospective controlled pilot study. Br. J. Ophthalmol., 90: 1281–1285.

O'Brart, D.S.P., Rowlands, E., Islam, N. and Noury, A.M.S. (2002) A randomised, prospective study comparing trabeculectomy augmented with antimetabolites with a viscocanalostomy technique for the management of open angle glaucoma uncontrolled by medical therapy. Br. J. Ophthalmol., 86: 748–754.

O'Brart, D.P., Shiew, M. and Edmunds, B. (2004) A randomised, prospective study comparing trabeculectomy with viscocanalostomy with adjunctive antimetabolite usage for the management of open angle glaucoma uncontrolled by medical therapy. Br. J. Ophthalmol., 88: 1012–1017.

Sanchez, E., Schnyder, C.C. and Sickenberg, M. (1997) Deep sclerectomy: results with and without collagen implant. Int. Ophthalmol., 20: 157–162.

Sarodia, U., Shaarawy, T. and Barton, K. (2007) Nonpenetrating glaucoma surgery: a critical evaluation. Curr. Opin. Ophthalmol., 18: 152–158.

Schwenn, O., Springer, C., Troost, A., Yun, S.H. and Pfeiffer, N. (2004) Deep sclerectomy using hyaluronate implant versus trabeculectomy. A comparison of two glaucoma operations using mitomycin C.. Ophthalmologe, 101: 696–704.

Shaarawy, T., Karlen, M. and Schnyder, C. (2001) Five-year results of deep sclerectomy with collagen implant. J. Cataract Refract. Surg., 27: 1770–1778.

Shaarawy, T., Mansouri, K., Schnyder, C., Ravinet, E., Achache, F. and Mermoud, A. (2004) Long-term results of deep-sclerectomy with collagen implant. J. Cataract Refract. Surg., 30: 1225–1231.

Shaarawy, T. and Mermoud, A. (2005) Deep sclerectomy in one eye vs deep sclerectomy with collagen implant in the contralateral eye of the same patient: long-term follow-up. Eye, 19: 298–302.

Shaarawy, T., Nguyen, C., Schnyder, C. and Mermoud, A. (2003) Five year results of viscocanalostomy. Br. J. Ophthalmol., 87: 441–445.

Sourdille, P., Santiago, P.Y. and Villain, F. (1999) Reticulated hyaluronic acid implant in nonperforating trabecular surgery. J. Cataract Refract. Surg., 25: 332–339.

Stangos, A.N., Whatham, A.R. and Sunaric-Megevand, G. (2005) Primary viscocanalostomy for juvenile open-angle glaucoma. Am. J. Ophthalmol., 140: 490–496.

Stegmann, R., Pienaar, A. and Miller, D. (1999) Viscocanalostomy for open-angle glaucoma in black African patients. J. Cataract Refract. Surg., 25: 316–322.

Sunaric-Mégevand, G. and Leuenberger, P. (2001) Results of viscocanalostomy for primary open angle glaucoma. Am. J. Ophthalmol., 132: 221–228.

Tamm, E.R., Carassa, R.G., Albert, D.M., Gabelt, B.T., Patel, S., Rasmussen, C.A. and Kaufman, P.L. (2004) Viscocanalostomy in rhesus monkeys. Arch. Ophthalmol., 122: 1826–1838.

Yalvac, I.S., Sahin, M., Eksioglu, U., Midillioglu, I.K., Aslan, B.S. and Duman, S. (2004) Primary viscocanalostomy versus trabeculectomy for primary open-angle glaucoma: three-year prospective randomized clinical trial. J. Cataract Refract. Surg., 30: 2050–2057.

Yarangumeli, A., Gurboz Koz, O. and Alp, M.N. (2005) Viscocanalostomy with mitomycin-C: a preliminary study. Eur. J. Ophthalmol., 15: 202–208.

Yarangumeli, A., Gureser, S., Koz, O.G., Elhan, A.H. and Kural, G. (2004) Viscocanalostomy versus trabeculectomy in patients with bilaterl high-tension glaucoma. Int. Ophthalmol., 25: 207–213.

Zimmerman, T.J., Kooner, K.S., Ford, V.J., Olander, K.W., Mandlekorn, R.M., Rawlings, E.F., Leader, B.J. and Koskan, A.J. (1984) Trabeculectomy vs non-penetrating trabeculectomy: a retrospective study of two procedures in phakic patients with glaucoma. Ophthalmic Surg., 15: 734–740.

CHAPTER 19

Modern aqueous shunt implantation: future challenges

Keith Barton[1,2,*] and Dale K. Heuer[3,4]

[1] *Moorfields Eye Hospital, 162 City Road, London EC1V 2PD, UK*
[2] *Department of Epidemiology, Institute of Ophthalmology, University College London, London, UK*
[3] *Department of Ophthalmology, Medical College of Wisconsin Milwaukee, WI 53226, USA*
[4] *Froedtert & Medical College of Wisconsin, Eye Institute Milwaukee, WI 53226, USA*

Abstract: The aqueous shunts that are currently available are based on the principles of the Molteno implant, i.e., a permanent sclerostomy, routing of aqueous to the equatorial subconjunctival space, and an end plate to prevent obstruction, and also to determine the surface area for absorption. While the Ahmed Glaucoma Valve appears to have improved the predictability of early intraocular pressure (IOP) control, the Baerveldt Glaucoma Implant has a tendency towards a lower rate of long-term excessive encapsulation. As a result of improvements in predictability, shunts are used more widely. Because of these positive factors, and ongoing concerns regarding the bleb-related problems associated with mitomycin C trabeculectomy, there is an increasing interest in the use of shunts as primary surgical management for primary glaucoma. At present, the main barrier to wider use of shunts in less-complicated glaucomas will probably be the unknown long-term effect on corneal endothelium, an issue that has not yet been properly addressed.

Keywords: glaucoma; aqueous shunts; surgery; secondary glaucoma; filtering surgery

Background

Although the first attempts at shunting aqueous date as far back as the early part of the 20th century (Rollett and Moreau, 1907; Zorab, 1912), the modern age of shunts essentially began with the Molteno implant in the late 1960s (Molteno, 1969a). The construction of the Molteno implant from a segment of silicone tube approximately 600 μm in external diameter, with a luminal diameter of 300 μm, was intended not only to provide a permanent sclerostomy, but also to divert aqueous away from the traditional trabeculectomy drainage area at the superior limbus to the equatorial subconjunctival space. An end plate was attached to the distal end of the tube through the end plate ridge, preventing fibrous ingrowth from obstructing the tube orifice. This basic design has formed the defining model for subsequent shunt construction. Shunts, such as the Schocket or anterior chamber tube shunt to encircling band (ACTSEB) in which a tube without a fixed end plate was inserted under an encircling band, were found to be less effective in achieving intraocular pressure (IOP) control than fixed one-piece implants (Fig. 1; Lavin et al., 1992; Smith et al., 1992; Wilson et al., 1992).

The shunts currently available, including the Baerveldt Glaucoma Implant (Advanced Medical

*Corresponding author. Tel.: +44 20 7566 2256;
Fax: +44 20 7566 2972;
E-mail: keith.barton@moorfields.nhs.uk

DOI: 10.1016/S0079-6123(08)01119-9

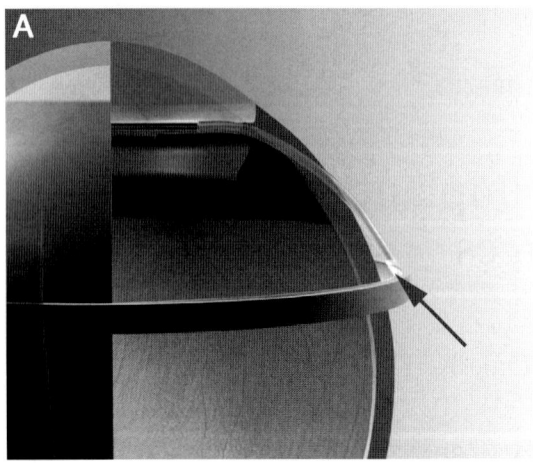 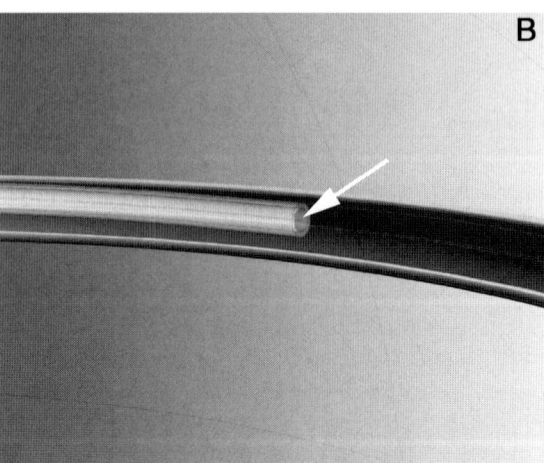

Fig. 1. The anterior chamber tube shunt to encircling band (ACTSEB) or Schocket procedure utilized the capsule of a scleral buckle to act as the bleb capsule for aqueous drainage. The tube portion shunted aqueous from the anterior chamber into the space under an encircling band, or portion thereof (modified Schocket) (A) (Dark grey arrow). A disadvantage of this technique was that the tube opening under the encircling band (B) (white arrow) was prone to occlusion by fibrous ingrowth. Adapted with permission from Shaarawy et al. (2008). Courtesy of Moorfields Eye Hospital.

Optics, Irvine, CA), the Ahmed Glaucoma Valve (New World Medical, Rancho Cucamonga, CA), and more recently the Molteno 3 (Molteno Ophthalmics Limited, Dunedin, New Zealand), are all still based on these concepts (Fig. 2).

Extensive experience with the Molteno implant also highlighted two impediments to achieving safe, predictable, physiological, long-term IOP levels. The first is difficulty in producing a physiological IOP in the early postoperative period, and the second is controlling long-term encapsulation.

Of the two more popular shunts at the time of writing, the Baerveldt Glaucoma Implant and the Ahmed Glaucoma Valve, arguably the Ahmed has substantially overcome the former problem and the Baerveldt, the latter.

Current shunts and factors affecting their function

All of the shunts mentioned above, which are currently available, follow three basic principles, i.e., they provide a permanent sclerostomy, they divert the aqueous to the equatorial rather than limbal subconjunctival space, and the end plate determines the surface area for aqueous absorption (Fig. 3).

They also have a similar luminal diameter and length after implantation and hence a similarly low natural resistance to aqueous flow. However, unlike the Molteno Implant and the Baerveldt Glaucoma Implant, the Ahmed Glaucoma Valve (see below) contains an additional flow resistor, designed to reduce the incidence of early hypotony.

The main determinant of longer-term shunt function is the degree to which the plate encapsulates. All shunt end plates develop a surrounding capsule to some degree (Molteno, 1969a, b; Lloyd et al., 1996). In non-valved shunts, this is the main point of resistance to aqueous flow and therefore the major determinant of IOP in the longer term.

The factors influencing the degree of encapsulation are not well defined but include plate surface area, material, surface profile, flexibility, and the presence or absence of a flow resistor.

Shunt-related factors

Surface area

Although the surface area of the external plate is only one variable that might influence

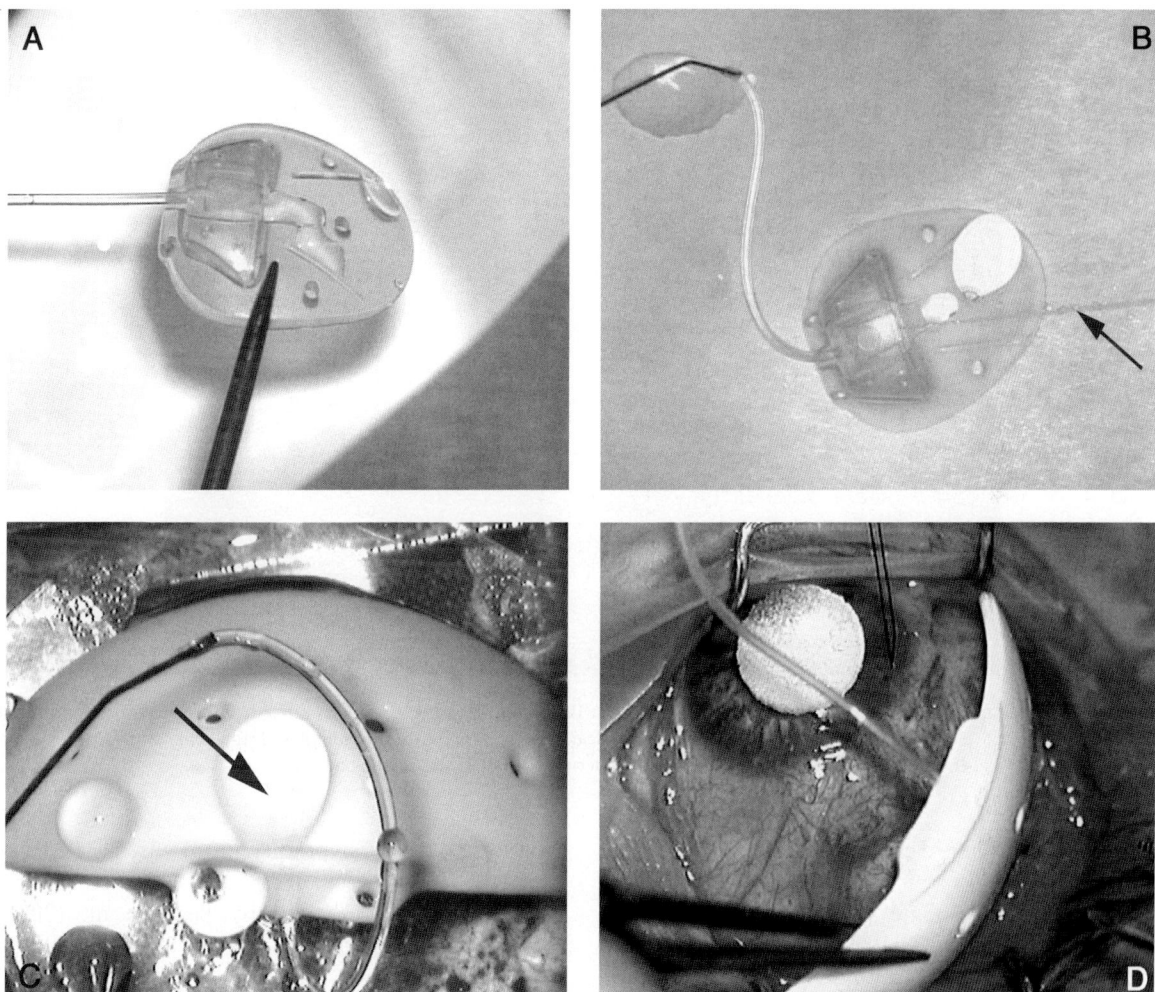

Fig. 2. The Ahmed Glaucoma Valve (A) contains a valve mechanism on the upper surface, which must be primed (B) prior to insertion. The arrow shows balanced salt solution (BSS) emerging from the valve mechanism on priming. (C) The arrow shows infused BSS flowing across the upper surface of the Baerveldt 350 Glaucoma Implant and (D) demonstrates the large plate surface area in comparison with the globe, as well as the curved thin profile on implantation. Adapted with permission from Shaarawy et al. (2008). Courtesy of Moorfields Eye Hospital. (See Color Plate 19.2 in color plate section.)

encapsulation, and hence the major determinant of long-term IOP control, the importance of plate surface area has been well demonstrated in two randomized controlled trials (Heuer et al., 1992; Britt et al., 1999).

Heuer et al. (1992) randomized 132 aphakic or pseudophakic eyes to either a single-plate or double-plate implant. Neovascular glaucomas were excluded. The reported success rate, in terms of IOP control between 6 and 21 mmHg (inclusive), was better at 2 years in the double-plate group (71% vs. 46%). The mean percentage IOP reduction was $46 \pm 33\%$ for the double-plate implant versus $25 \pm 43\%$ for the single-plate implant and there was less hypertensive phase in the former. Both groups required glaucoma medications at 2 years $(1.2 \pm 0.9\%)$ versus $(1.6 \pm 0.9\%)$ for the single plate. However, the rate of complications such as choroidal hemorrhage, flat anterior chamber, corneal decompensation, and phthisis due

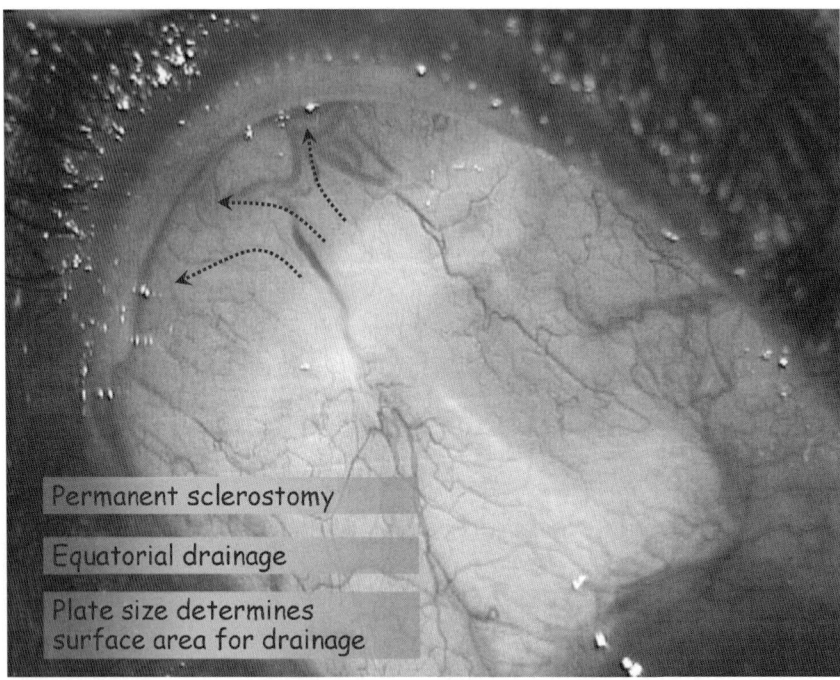

Fig. 3. This illustration shows a diffuse drainage bleb overlying a Baerveldt 350 implant. Shunts differ from trabeculectomies in that there is a permanent sclerostomy (the tube portion), equatorial drainage (arrows), rather than limbal drainage, avoiding limbal bleb-related problems, and the surface area for absorption can be determined by the size of the end plate.

to hypotony was more common in the double-plate group.

The influence of plate size was further investigated by Britt et al. (1999), who compared a 350-mm² Baerveldt Glaucoma Implant (Advanced Medical Optics, Inc.) with a 500-mm² plate in a randomized controlled trial and found that overall the success rate was actually lower with the larger plate size. The success rate at 5 years was 79% in the 350-mm² group compared with 66% in the large-plate group, suggesting that there might be an optimal plate size above which no further increase in size is beneficial. There was no apparent difference in visual acuity, complications, and average IOPs at 5 years, although there was a trend toward more sequelae from hypotony in the 500-mm² group.

In a smaller, nonrandomized clinical series, Molteno found similar results when comparing one-, two-, and four-plate implants. The IOP control with two plates was significantly better than that with one plate. The IOP control with four plates was marginally better again, but at the cost of early hypotony in all cases (Molteno, 1981).

A retrospective study by Seah et al. (2003) comparing 70 Baerveldt 350-mm² implants with 54 Baerveldt 250-mm² implants in Asian eyes found very little difference in IOP reduction between the two groups after a mean follow-up of 33 months.

It seems that shunt size represents a trade-off between smaller plate size and higher long-term pressures, or large plate size and better long-term IOP control, but a higher risk of sequelae from hypotony.

Although plate surface area is only one of a number of implant-related factors that may influence long-term IOP control, it is one of the easiest to modify, given that there are implants of several different sizes on the market. Although there will be individual variation in the response, it seems that while 250–350 mm² appears to afford good pressure control, with the Baerveldt implant, plate sizes greater than 350 mm² are excessive and afford no extra benefit.

Plate material

There is some evidence that the material from which the shunt end plate is manufactured may influence the degree of reaction around the implant and hence the degree of encapsulation. Two studies (Ayyala et al., 1999, 2000) compared the influence of polypropylene with silicone end plates implanted subconjunctivally in rabbits and reported more inflammation with polypropylene than with silicone and more inflammation with rigid than flexible end plates. However, as these plates differ in other factors such as shape, profile, surface texture, contact area with adjacent tissues, flexibility, and micro-motion, all of which might influence the degree of encapsulation, the observed effect may not be exclusively due to the type of plate material or the surface area alone (Lim et al., 1998).

Valved versus non-valved

One of the most important features of an aqueous shunt from the clinician's perspective is the presence or absence of a fixed flow-restrictor, i.e., *valved* or *non-valved*. Although, strictly speaking, the flow-restrictors in the former group have not been shown to act as *valves*, the term has nevertheless entered common parlance (Prata et al., 1995; Lee, 1998).

Valved devices have been defined as those that allow only unidirectional flow with a minimum opening pressure, whereas *non-valved* devices are passive, incapable of influencing either anterograde or retrograde flow (modified from Lieberman and Ewing, 1990). The Ahmed Glaucoma Valve (New World Medical) is an example of the former, whereas the Molteno (Molteno Ophthalmics Limited) and Baerveldt shunts are examples of the latter. These implants have a similar lumen diameter (approximately 300 µm). The flow potential of the non-valved Molteno and Baerveldt shunts is sufficient to far exceed the rate of aqueous production, or, in other words, drain the anterior chamber completely of aqueous in a very short period of time.

In valved shunts, this does not happen in the early postoperative period because of the presence of the flow-restrictor, which prevents hypotony in most, but not all, cases (Syed et al., 2004; Law et al., 2005). With the Ahmed Glaucoma Valve, the implant must be primed with a fluid such as balanced salt solution in order to separate and wet the valve leaflets.

In non-valved shunts, the surgeon must physically stent or ligate the shunt tube at the time of implantation to avoid unrestricted flow and the severe sequelae of hypotony that may result.

A number of techniques have been described to prevent early hypotony with *non-valved* aqueous shunts. The most commonly used technique at the time of writing is external ligation with an absorbable ligature such as 7/0 Vicryl (Ethicon, Johnson & Johnson International, Brussels, Belgium). It is impossible to adjust flow to a clinically safe level with a ligature and therefore ligation needs to completely occlude the tube portion of the implant. Failure to completely occlude the implant will result in severe hypotony. However, to counteract the high IOP that often results from successful ligation, many surgeons will additionally fenestrate the tube proximal to the ligature (*Sherwood slit*). A further disadvantage of external ligation is sudden decompression, usually 5–6 weeks after surgery when the ligature absorbs (Fig. 4). Even if sufficient encapsulation has developed, the initial precipitous drop in pressure in eyes with larger implants, such as the Baerveldt 350, may be sufficient to cause a choroidal hemorrhage in a predisposed individual.

Alternative stenting techniques have been described in conjunction with a ligature so that the rapidity of the drop in pressure is blunted. A stent used regularly by one of the authors (K.B.) is the 3/0 braided nylon stent (Supramid, S. Jackson Inc., Alexandria, VA) as described by Sherwood and Smith (1993).

Commercially available devices

The current most widely available shunts include the Ahmed, Baerveldt, and Molteno.

1. The Ahmed Glaucoma Valve is manufactured with a flexible silicone plate (FP7) or a rigid polypropylene plate (S2) of similar surface

lower part of the normal range, ideally with less dependence on ocular hypotensive medications. The limited available evidence would suggest that the Baerveldt 350 is most likely to achieve this.

Nevertheless, if the eye must sustain a high IOP for several weeks followed by a sudden decompression at an unpredictable time, then there is a high potential cost in achieving a slightly better long-term IOP.

Most of the problems associated with early pressure control can be mitigated to some degree by adjustment of surgical technique. Early hypotony with the Ahmed can be avoided by leaving thicker viscoelastics, such as Healon GV or Healon 5 (Advanced Medical Optics, Irvine, CA), in the anterior chamber at the time of surgery. Early pressure control with the Baerveldt is more difficult to achieve. However, there are methods that can be used to avoid the scenario described above. When complete ligation is used, the tube can also be stented with a 3/0 nylon suture (Supramid, S. Jackson Inc.) in order to reduce the severity of the sudden pressure drop that occurs at the time the ligature opens (Sherwood and Smith, 1993). The early postoperative high pressure can be avoided to some degree with a fenestration (Sherwood and Smith, 1993) or simultaneous orphan trabeculectomy (Budenz et al., 2002). An approach favored by some is to ligate the tube within the anterior chamber with a polypropylene ligature (Budenz et al., 2002). This has the advantage that the ligature can be easily visualized and lasered to reduce the pressure at a preplanned point in time, in theory, more predictable than external ligation with an absorbable ligature. However, a fenestration will be ineffective in this situation and a simultaneous orphan trabeculectomy probably offers a better method of preventing high IOP in the early postoperative period when this type of ligation is used (Budenz et al., 2002).

In eyes that are likely to be especially vulnerable to a sudden drop in IOP, a traditional two-stage implantation technique can be used. Two-stage implantation involves an operation to position and secure the end plate of the implant without implanting the tube portion in the anterior chamber. A second procedure is then performed around 6 weeks later to place the tube portion inside the eye. At that point in time, the end plate will have encapsulated so that implantation of the tube will permit aqueous flow sufficient to control the IOP, but not enough to cause severe hypotony.

A technique that one of the authors (K.B.) has used successfully is to introduce Sherwood's 3/0 nylon Supramid, the entire length of the tube. When the tube plus stent are introduced into the anterior chamber via an entry site in the sclera that is tighter than normal (25 gauge rather than 23), the tube is squeezed where it passes through the sclera. Squeezing the tube around the outside of the stent suture increases the resistance to aqueous flow through the tube to a sufficient degree to prevent early hypotony in most cases without the need to ligate the tube.

When performing this technique, it is essential to test aqueous flow by looking for fluorescein dilution over the end plate portion of the shunt after the tube portion is in the eye. The reason for testing aqueous flow is that manufacturing variability in the suture diameter (200–250 µm) influences resistance to flow. In some cases, aqueous flow will be observed despite this stenting technique, and in such cases the tube must also be ligated. In other cases, resistance will be high and there will be no need to introduce the stent along the entire length of the tube. It is therefore worth examining aqueous flow with fluorescein where the tube exits the end plate in order to ascertain if there is aqueous flow with the stent in place. Ideally, the stent should be introduced into the tube just far enough to prevent any aqueous flow (Fig. 5).

Tolerance of topical ocular hypotensive medications

Clearly a patient who has difficulty with topical medications may have difficulty with any shunt as there is a significant chance that topical medications will still be required. Nevertheless, the Baerveldt implant seems to result in a lower need for topical medications than the Ahmed valve in the longer term.

Aqueous hyposecretion

Aqueous hyposecretion is difficult to quantify and is therefore often deduced from clinical

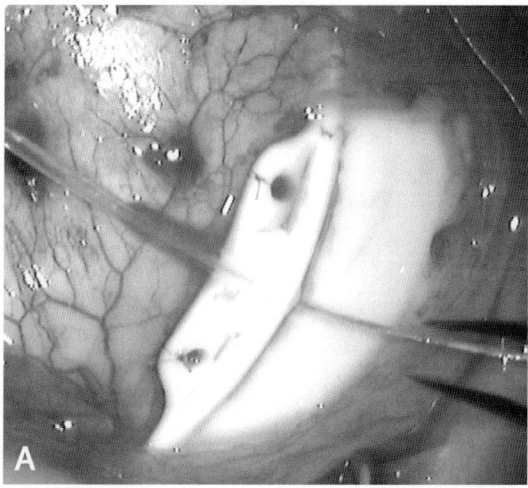
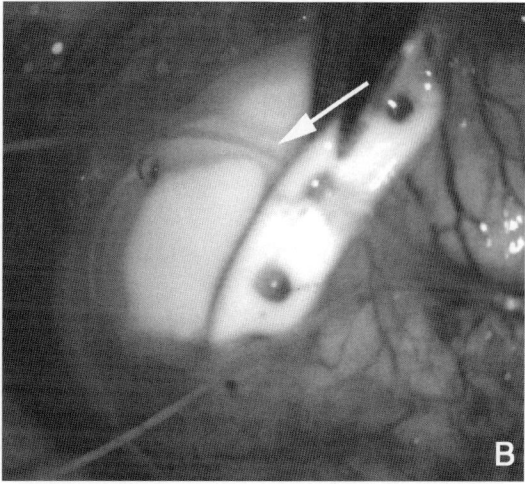

Fig. 5. A 3/0 nylon stenting suture (Supramid, S. Jackson Inc., Alexandria, VA) may be used as an alternative to total ligation with the Baerveldt Glaucoma Implant (A). However, it is important when using this technique to ensure that there is no flow (B, arrow) at the distal end of the tube after it has been inserted into the anterior chamber. Adapted with permission from Shaarawy et al. (2008). Courtesy of Moorfields Eye Hospital.

observations. Eyes with extensive ischemia, such as proliferative diabetic retinopathy or ocular ischemic syndrome; eyes that have had extensive cyclodestruction, e.g., two to three treatment episodes of transscleral diode laser cyclophotocoagulation; or eyes with severe chronic uveitis, especially those associated with juvenile idiopathic arthritis, are at greatest risk of hypotony, due to a relative overdrainage from a normally functioning implant.

Previous ocular surgery

Eyes that have undergone previous conjunctival surgery, especially scleral buckling or vitrectomy with silicone oil, and also strabismus surgery and prior trabeculectomy have a higher risk of scarring and plate encapsulation than other eyes and also may be more difficult to implant.

In general, it is better to implant superotemporal where feasible. Erosion rates are higher in the inferior quadrants and there is a higher risk of diplopia when implanting in the superonasal quadrant. While it may be possible to implant low-profile implants such as the Baerveldt 250 superonasally without induced strabismus, the Baerveldt 350 is more likely to interfere with muscle function, and the Ahmed produces a high-profile bleb that might induce diplopia by displacing the globe slightly. An earlier model of the Baerveldt 350 caused a high rate of diplopia (Smith et al., 1993), but this has been reported less frequently since the introduction of fenestrations through the end plate, permitting fibrous tissue ingrowth through the plate and thereby lower blebs.

In eyes with extensive previous surgery, it may not be possible to implant the wings of the Baerveldt 350 under the muscles. However, if the end plate can be secured tightly to sclera via the fixation holes, it is not essential to implant under the muscles (Fig. 6).

Scleral thinning

Eyes with significant scleral thinning are particularly unsuitable for trabeculectomy surgery, but usually can be implanted safely with a shunt, assuming certain precautions are taken. There are no aspects of implanting an eye with scleral thinning that might alter the authors' choice of shunt. However, it is important, if possible, to secure the plate at a point where there is adequate sclera for the fixation sutures and to enter the

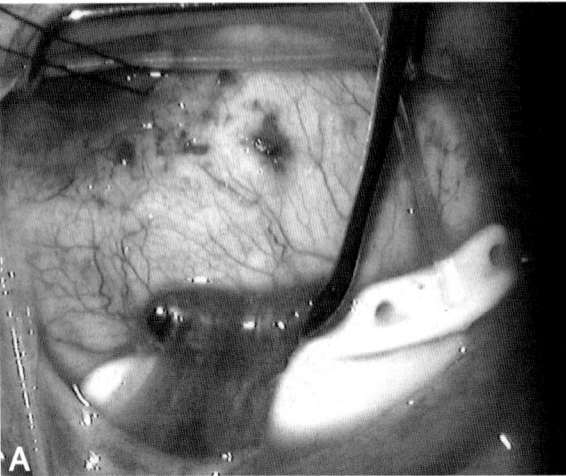 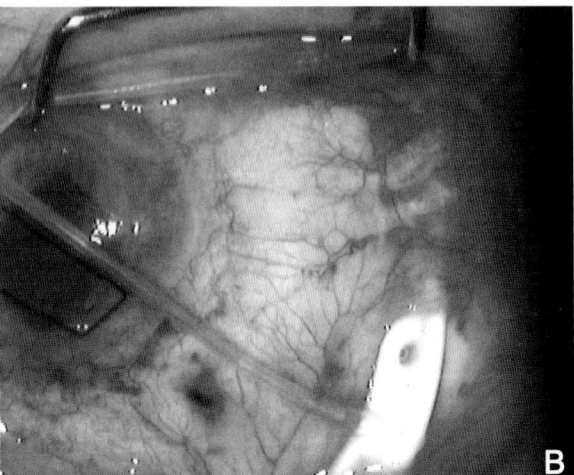

Fig. 6. The wings of the larger 350-mm^2 Baerveldt Glaucoma Implant are usually placed under the superior (A) and lateral (B) rectus muscles. Ideally the implant is placed about 1 mm behind the muscle insertion to avoid restricting muscle action. In eyes with scleral buckles, it may be impossible to place the plate under the muscles and in such circumstances they may be safely placed on top if the plate is securely sutured to sclera.

anterior chamber in an area where there is not excessive scleral thinning.

Patient cooperation for and tolerance of potential slit-lamp interventions

There are other factors that might also influence the choice of implant. In an eye where a patient is unlikely to tolerate postoperative slit-lamp interventions, the individual surgeon will have personal preferences that will influence the choice. This might be to use an Ahmed shunt, leaving viscoelastic in the eye at the time of surgery, or to use a Baerveldt, ligated but partially stented, to prevent a sudden pressure drop at the time of ligature absorption. Although it is difficult to generalize on the course of action required in these situations, it is essential that the surgeon considers the possibility of adapting his or her technique in order to be prepared should the situation arise.

Future challenges

Until recently, aqueous shunt have been reserved for complicated glaucomas that are unresponsive to both medical and other forms of surgical therapy. In such cases, the potential benefits in terms of IOP control have outweighed the risks in terms of unpredictability of early IOP control. However, improving surgical outcomes from aqueous shunts, as demonstrated in the tube versus trabeculectomy (TVT) study (Gedde et al., 2005, 2007a, b), in combination with persistent concern over the long-term risks of mitomycin trabeculectomy, have resulted in renewed interest in the use of aqueous shunts as the primary surgical intervention in primary open-angle glaucoma.

While there is some evidence of comparability in terms of safety and efficacy with trabeculectomy (Gedde et al., 2005, 2007a, b), and also some evidence that shunts may even be relatively safe when used as primary surgery for primary glaucoma (Wilson et al., 2003), there are some additional concerns that should be addressed if aqueous shunts are to be used as a primary alternative to trabeculectomy for primary glaucomas.

These include predictability of early IOP control, cataract formation, and the long-term effect on the cornea.

Predictability

Despite improvements in technique, a remaining concern with aqueous shunt implantation is the predictability in early IOP control, and that shunts do not achieve low IOPs in the longer term. While early predictability is less of an issue with the Ahmed Glaucoma Valve, randomized controlled trials do suggest that IOP control in the lower teens can be achieved with certain implants, specifically the Baerveldt 101-350, albeit that most patients will still require some degree of glaucoma medication. Future improvements in shunt predictability will necessitate the development of implants with some features of both of the above, i.e., a valved implant that also possesses a thin, low-profile, flexible plate of large surface area and good biocompatibility.

Cataract formation

Acceleration of cataract formation (Lichter et al., 2001) and reduced bleb function after cataract extraction (Chen et al., 1998) are well-known consequences of trabeculectomy. As aqueous shunt implantation does not involve a peripheral iridectomy, and may be performed without alteration of anterior chamber depth at the time of surgery, it is possible that aqueous shunt implantation may result in a lower rate of cataract development than trabeculectomy. As most comparative trials, such as the TVT study (Gedde et al., 2005, 2007a, b), involved patients that were largely pseudophakic, it has not been possible to study this adequately. Similarly, the influence of cataract surgery on aqueous shunt function has not been documented.

The long-term effect on the cornea

The long-term influence of aqueous shunts on the cornea (Topouzis et al., 1999) and penetrating keratoplasties (Sherwood et al., 1993; Ayyala, 2000; Arroyave et al., 2001; Kwon et al., 2001; Al Torbak, 2003; Alvarenga et al., 2004) has long been the subject of debate. Promising results reported after 1-year follow-up in the TVT study (Gedde et al., 2007a, b), and following a prior study of primary shunt implantation in primary glaucomas (Wilson et al., 2003), have encouraged some investigators to consider implantation of shunts as the primary surgical treatment for primary glaucomas. The cited benefits of shunt implantation over trabeculectomy with mitomycin C relate to the absence of limbal drainage and hence very low risks of infection or dysaesthesia.

The concept of primary shunt implantation for primary glaucomas is becoming more feasible, especially as the predictability of IOP control will probably improve as shunts continue to evolve and as implantation techniques become more finely tuned. The likely remaining barrier to shunt implantation, as a primary surgical procedure for the treatment of primary glaucomas, is likely to be concern about the long-term effect on the corneal endothelium.

There has been greater focus in the literature on the effect of aqueous shunts on penetrating keratoplasties than on their effect on the normal cornea. However, the best of these are retrospective and employ case–control methodology. This is partly because any prospective study would have a low likely rate of recruitment, necessitating a prolonged recruitment period, but also because of the difficulties in devising a valid control group.

From the available literature, the overall rate of successful IOP control in post-keratoplasty glaucoma appears to be better in the published literature than after mitomycin C trabeculectomy (Ayyala, 2000). However, while the rate of graft failure may be higher in eyes with aqueous shunts, it can be seen in the same literature review by Ayyala (2000) that the average follow-up in aqueous shunt studies has been longer than in mitomycin C trabeculectomy studies.

There are a number of possible reasons why the rate of corneal graft failure might be elevated in eyes with aqueous shunts. Clearly those requiring aqueous shunts might have worse IOP control to start with. As glaucoma itself is a major risk factor for graft failure (The Australian Corneal Graft Registry, 1993), case selection alone might account for the observed risk of graft failure. Mechanical damage to corneal endothelium from the tube portion of the shunt is a likely cause of corneal endothelial damage. Although Kirkness hypothesized a further potential cause, i.e., aqueous shunts might elevate the risk of corneal graft failure because they permit a tidal flow of inflammatory cells from the bleb in and out of the eye, thereby compromising immune privilege (Kirkness et al.,

1988), it seems that neither the influence of mechanical factors has been studied sufficiently to be able to exclude these as a cause nor the influence of case selection fully accounted for.

While the presence of tube-corneal touch is often documented (Topouzis et al., 1999), and in one study, by Al Torbak (2003), physical contact between shunt and endothelium was documented as a significant cause of graft failure (Al Torbak, 2003), studies reporting endothelial cell counts after aqueous shunts are rare (McDermott et al., 1993).

The answer to the question as to whether aqueous shunt implantation in an eye with healthy corneal endothelium, or a healthy penetrating keratoplasty, or, ideally, the shunt implanted before keratoplasty — in order to achieve stable IOP control at the time of keratoplasty, and positioned so that there is no possibility of endothelial contact — does or does not compromise the long-term prognosis for graft clarity has not been adequately investigated.

On available evidence, it is quite possible that shunts positioned behind Schwalbe's line so that the shunt entering the eye is not in contact with corneal endothelium, and sufficiently short so that there is no possibility of tube movement, e.g., on eye-rubbing, causing corneal endothelial-tube contact, may not compromise corneal endothelial function.

There are many situations, such as eyes with extensive peripheral anterior synechiae or small eyes with shallow anterior chambers, where it may not be possible to reliably implant a shunt in the anterior chamber far enough from corneal endothelium to prevent long-term damage.

It therefore remains to be proven whether aqueous shunts can reliably be placed in the anterior chamber in a position that can be guaranteed not to compromise corneal endothelium, e.g., behind Schwalbe's line. Traditionally, it has been difficult to visualize the position of the shunt anterior chamber entry site in eyes with corneal edema. The intraocular course of a shunt can now be visualized in such cases with anterior segment optical coherence tomography (AS-OCT) (Fig. 7; Sarodia et al., 2007). It is likely with better

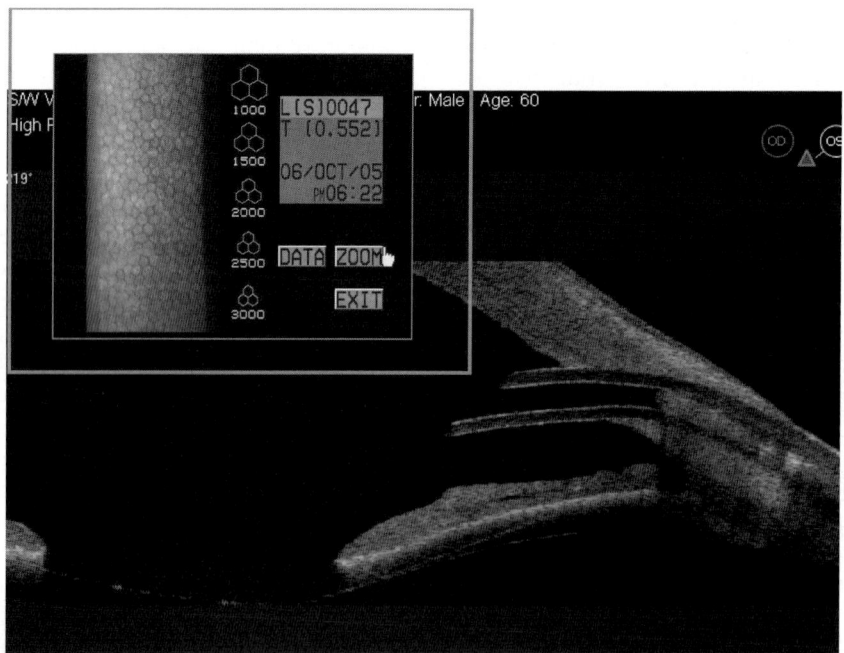

Fig. 7. The long-term influence of aqueous shunts on the corneal endothelium has not been adequately studied. With endothelial specular microscopy (inset) and careful localization of the shunt within the anterior chamber using anterior segment optical coherence tomography, it should now be possible to determine the relationship between the two.

imaging equipment, such as AS-OCT, and non-contact specular microscopy that the mechanical contribution to the influence of shunts on corneal endothelium will eventually be resolved.

References

Al Torbak, A. (2003) Graft survival and glaucoma outcome after simultaneous penetrating keratoplasty and Ahmed glaucoma valve implant. Cornea, 22(3): 194–197.

Alvarenga, L.S., Mannis, M.J., Brandt, J.D., Lee, W.B., Schwab, I.R. and Lim, M.C. (2004) The long-term results of keratoplasty in eyes with a glaucoma drainage device. Am. J. Ophthalmol., 138(2): 200–205.

Arroyave, C.P., Scott, I.U., Fantes, F.E., Feuer, W.J. and Murray, T.G. (2001) Corneal graft survival and intraocular pressure control after penetrating keratoplasty and glaucoma drainage device implantation. Ophthalmology, 108(11): 1978–1985.

Ayyala, R.S. (2000) Penetrating keratoplasty and glaucoma. Surv. Ophthalmol., 45(2): 91–105.

Ayyala, R.S., Harman, L.E., Michelini-Norris, B., Ondrovic, L.E., Haller, E., Margo, C.E. and Stevens, S.X. (1999) Comparison of different biomaterials for glaucoma drainage devices. Arch. Ophthalmol., 117(2): 233–236.

Ayyala, R.S., Michelini-Norris, B., Flores, A., Haller, E. and Margo, C.E. (2000) Comparison of different biomaterials for glaucoma drainage devices: part 2. Arch. Ophthalmol., 118(8): 1081–1084.

Britt, M.T., LaBree, L.D., Lloyd, M.A., Minckler, D.S., Heuer, D.K., Baerveldt, G. and Varma, R. (1999) Randomized clinical trial of the 350-mm^2 versus the 500-mm^2 Baerveldt implant: longer term results: is bigger better? Ophthalmology, 106(12): 2312–2318.

Budenz, D.L., Scott, I.U., Nguyen, Q.H., Feuer, W., Singh, K., Nicolela, M.T., Bueche, M. and Palmberg, P.F. (2002) Combined Baerveldt glaucoma drainage implant and trabeculectomy with mitomycin C for refractory glaucoma. J. Glaucoma, 11(5): 439–445.

Chen, P.P., Weaver, Y.K., Budenz, D.L., Feuer, W.J. and Parrish, R.K. (1998) Trabeculectomy function after cataract extraction. Ophthalmology, 105(10): 1928–1935.

Gedde, S.J., Herndon, L.W., Brandt, J.D., Budenz, D.L., Feuer, W.J. and Schiffman, J.C. (2007a) Surgical complications in the tube versus trabeculectomy study during the first year of follow-up. Am. J. Ophthalmol., 143(1): 23–31.

Gedde, S.J., Schiffman, J.C., Feuer, W.J., Herndon, L.W., Brandt, J.D. and Budenz, D.L. (2007b) Treatment outcomes in the tube versus trabeculectomy study after one year of follow-up. Am. J. Ophthalmol., 143(1): 9–22.

Gedde, S.J., Schiffman, J.C., Feuer, W.J., Parrish, R.K., Heuer, D.K. and Brandt, J.D. (2005) The tube versus trabeculectomy study: design and baseline characteristics of study patients. Am. J. Ophthalmol., 140(2): p. 275.

Heuer, D.K., Lloyd, M.A., Abrams, D.A., Baerveldt, G., Minckler, D.S., Lee, M.B. and Martone, J.F. (1992) Which is better? One or two? A randomized clinical trial of single-plate versus double-plate Molteno implantation for glaucomas in aphakia and pseudophakia. Ophthalmology, 99(10): 1512–1519.

Kirkness, C.M., Ling, Y. and Rice, N.S.C. (1988) The use of silicone drainage tubing to control post-keratoplasty glaucoma. Eye, 2: 583–590.

Kwon, Y.H., Taylor, J.M., Hong, S., Honkanen, R.A., Zimmerman, M.B., Alward, W.L. and Sutphin, J.E. (2001) Long-term results of eyes with penetrating keratoplasty and glaucoma drainage tube implant. Ophthalmology, 108(2): 272–278.

Lavin, M.J., Franks, W.A., Wormald, R.P.L. and Hitchings, R.A. (1992) Clinical risk factors for failure in glaucoma tube surgery. Arch. Ophthal., 110(4): 480–485.

Law, S.K., Nguyen, A., Coleman, A.L. and Caprioli, J. (2005) Comparison of safety and efficacy between silicone and polypropylene Ahmed glaucoma valves in refractory glaucoma. Ophthalmology, 112(9): 1514–1520.

Lee, V.W. (1998) Glaucoma "valves" — truth versus myth. Ophthalmology, 105: 567–568.

Lichter, P.R., Musch, D.C., Gillespie, B.W., Guire, K.E., Janz, N.K., Wren, P.A. and Mills, R.P. (2001) Interim clinical outcomes in the Collaborative Initial Glaucoma Treatment Study comparing initial treatment randomized to medications or surgery. Ophthalmology, 108(11): 1943–1953.

Lieberman, M.F. and Ewing, R.H. (1990) Drainage implant surgery for refractory glaucoma. Int. Ophthalmol. Clin., 30(3): 198–208.

Lim, K.S., Allan, B.D.S., Lloyd, A.W., Muir, A. and Khaw, P.T. (1998) Glaucoma drainage devices; past, present, and future. Br. J. Ophthalmol., 82: 1083–1089.

Lloyd, M.A., Baerveldt, G., Nguyen, Q.H. and Minckler, D.S. (1996) Long-term histologic studies of the Baerveldt implant in a rabbit model. J. Glaucoma, 5(5): 334–339.

McDermott, M.L., Swendris, R.P., Shin, D.H., Juzych, M.S. and Cowden, J.W. (1993) Corneal endothelial cell counts after Molteno implantation. Am. J. Ophthalmol., 115(1): 93–96.

Molteno, A.C. (1969a) New implant for drainage in glaucoma. Br. J. Ophthalmol., 53: 606–615.

Molteno, A.C. (1969b) New implant for drainage in glaucoma. Animal trial. Br. J. Ophthalmol., 53(3): 161–168.

Molteno, A.C. (1981) The optimal design of drainage implants for glaucoma. Trans. Ophthalmol. Soc. N.Z., 33: 39–41.

Prata, J.A., Jr., Mermoud, A., LaBree, L.D. and Minckler, D.S. (1995) In vitro and in vivo flow characteristics of glaucoma drainage implants. Ophthalmology, 102: 894–904.

Rollett, M. and Moreau, M. (1907) Le drainage au crin de la chambre anterieure contre l'hypertonie et la douleur. Rev. Gen. Ophtalmol., 26: 289–292.

Sarodia, U., Sharkawi, E., Hau, S. and Barton, K. (2007) Visualization of aqueous shunt position and patency using anterior segment optical coherence tomography. Am. J. Ophthalmol., 143(6): 1054–1056.

Seah, S.K., Gazzard, G. and Aung, T. (2003) Intermediate-term outcome of Baerveldt Glaucoma Implants in Asian eyes. Ophthalmology, 110(5): 888–894.

Shaarawy, T., Sherwood, M.B., Hitchings, R. and Crowston, J.G. (Eds.) (2008) Glaucoma, Saunders.

Sherwood, M.B. and Smith, M.F. (1993) Prevention of early hypotony associated with Molteno implants by a new occluding stent technique. Ophthalmology, 100(1): 85–90.

Sherwood, M.B., Smith, M.F., Driebe, W.T., Stern, G.A., Beneke, J.A. and Zam, Z.S. (1993) Drainage tube implants in the treatment of glaucoma following penetrating keratoplasty. Ophthalmic Surg., 24(3): 185–189.

Smith, M.F., Sherwood, M.B. and McGorray, S.P. (1992) Comparison of the double-plate Molteno drainage implant with the Schocket procedure. Arch. Ophthal., 110(9): 1246–1250.

Smith, S.L., Starita, R.J., Fellman, R.L. and Lynn, J.R. (1993) Early clinical experience with the Baerveldt 350-mm^2 glaucoma implant and associated extraocular muscle imbalance. Ophthalmology, 100: 914–918.

Syed, H.M., Law, S.K., Nam, S.H., Li, G., Caprioli, J. and Coleman, A. (2004) Baerveldt-350 implant versus Ahmed valve for refractory glaucoma: a case-controlled comparison. J. Glaucoma, 13(1): 38–45.

(1993) The Australian Corneal Graft Registry: 1990 to 1992 Report. Aust. N. Z. J. Ophthalmol., 21(2 Suppl): 1–48.

Topouzis, F., Coleman, A.L., Choplin, N., Bethlem, M.M., Hill, R., Yu, F., Panek, W.C. and Wilson, M.R. (1999) Follow-up of the original cohort with the Ahmed Glaucoma Valve implant. Am. J. Ophthalmol., 128(2): 198–204.

Tsai, J.C., Johnson, C.C., Kammer, J.A. and Dietrich, M.S. (2006) The Ahmed shunt versus the Baerveldt shunt for refractory glaucoma II: longer-term outcomes from a single surgeon. Ophthalmology, 113(6): 913–917.

Wang, J.C., See, J.L. and Chew, P.T. (2004) Experience with the use of Baerveldt and Ahmed glaucoma drainage implants in an Asian population. Ophthalmology, 111(7): 1383–1388.

Wilson, M.R., Mendis, U., Paliwal, A. and Haynatzka, V. (2003) Long-term follow-up of primary glaucoma surgery with Ahmed Glaucoma Valve implant versus trabeculectomy. Am. J. Ophthalmol., 136(3): 464–470.

Wilson, R.P., Cantor, L., Katz, L.J., Schmidt, C.M., Steinmann, W.C. and Allee, S. (1992) Aqueous shunts. Molteno versus Schocket. Ophthalmology, 99(5): 672–676.

Zorab, A. (1912) The reduction in tension in chronic glaucoma. The Ophthalmoscope, 10: 258–261.

SECTION IV

Experimental Approaches to Model Disease

CHAPTER 20

Model systems for experimental studies: retinal ganglion cells in culture

Emilie Goodyear[1] and Leonard A. Levin[1,2,*]

[1]Department of Ophthalmology, University of Montreal, Montreal, Canada
[2]Department of Ophthalmology and Visual Sciences, University of Wisconsin, Madison, USA

Abstract: Glaucomatous optic neuropathy is the most common optic nerve disease. The mechanisms by which retinal ganglion cells (RGCs) die in glaucoma are becoming better understood, but are still poorly defined. Elucidating the pathways that connect risk factors for glaucoma (e.g., elevated intraocular pressure) and RGC death is difficult in patients because of ethical and practical constraints. Even in experimental animals, single-cell observations and cell–cell interactions can be tricky to tease apart. For these reasons, it is helpful to use cell and tissue culture models for studying RGCs and other cellular constituents of the optic nerve. This chapter describes the advantages and disadvantages of several commonly used methods for preparing and studying these cultures, including those most relevant to glaucomatous optic neuropathy.

Keywords: retinal ganglion cells; optic neuropathy; glaucoma; cell culture

The human retina is composed of nine different layers. The innermost layers contain the retinal ganglion cells (RGCs) and their axons in the ganglion cell layer and the nerve fiber layer, respectively. RGCs receive afferents from bipolar and amacrine cells and transmit efferents via action potentials to the brain, specifically the lateral geniculate nucleus, the superior colliculus, the pretectal nuclei, and the suprachiasmatic nucleus.

Optic neuropathies are diseases of the RGC and its axon. The most common optic neuropathy is glaucoma. Optic nerve diseases can be studied in vitro and in vivo using experimental models that range greatly in their applicable to glaucomatous optic neuropathy. There is an approximate hierarchy of models (Fig. 1), where the models higher in hierarchy present an increased similarity to human optic neuropathies and the models lower in the pyramid allow better determination of the mechanisms responsible for the disease. This chapter discusses in vitro methods for studying RGCs (Levin, 2005). There are several types of culture models used for studying the pathophysiology of RGCs: (1) dissociated retinal cells, where the RGCs are either mixed with other cells or identified by labeling; (2) RGCs purified either by immunoaffinity techniques or differential centrifugation; (3) retinal explants; (4) glial and other supporting cell cultures; and (5) RGC-like cell lines.

Mixed RGCs in culture

For mixed primary retinal cultures, neonatal or adult retinas can be dissociated enzymatically and maintained in culture for up to several weeks.

*Corresponding author. Tel.: +1 514 252 3400; Fax: +1 514 251 7094

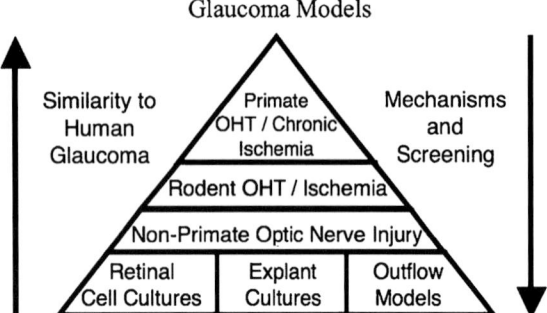

Fig. 1. In vitro and in vivo models of glaucoma can be viewed in a hierarchical pyramid, with the models at the top of the pyramid being more similar to the human disease, but less helpful in studying pathophysiology because of their complexity and experimental limitations. Models toward the bottom of the pyramid do not share clinical features of human glaucoma, but are more useful for studying mechanisms and screening potential therapies (Levin, 2005).

RGCs can then be identified by different techniques. The most commonly used is retrograde labeling with a fluorescent dye. RGC can also be identified by immunolabeling against Thy-1 or Brn-3 (Leifer et al., 1984; Garcia et al., 2002; Leahy et al., 2004) or by using an RGC-specific promoter to drive a reporter gene if cultured from a transgenic animal (Feng et al., 2000). This type of culture allows the observation of interactions among a variety of cell types and the RGC because the diversity is maintained. However, the anatomic arrangement of the retina is not maintained. Axotomy is also inherent to this model because the optic nerve is cut when the eye is taken from the animal, and then the proximal RGC axon is cut when the retina is enzymatically dissociated. Survival is typically short as growth factors are typically not used besides those contained in serum.

Purified ganglion cells

In purified RGC cultures, enzymatic dissociation is performed and then the RGCs are isolated from other retinal neurons via immunoselection. Macrophages are typically removed first with antibodies followed by antibodies to the cell surface marker Thy-1. This methodology was first used by Barres using panning techniques (Barres et al., 1988), and it is the most commonly used method for studying isolated RGCs in vitro. Subsequently, other methods for purification have been developed, including the use of magnetic beads (Tezel and Wax, 2000).

Purified RGC cultures offer the advantage of studying the RGC in isolation without the effects of interactions with other cell types in the retina. RGC purity can be 95% or greater. However, the purification process requires exposure of RGC to antibodies to Thy-1, which might have biologic effects. Growth factors are typically used to maintain long term survival, for up to several weeks (Meyer-Franke et al., 1995).

Retinal explants

Retinal explant cultures can be made by dissection of whole retinas into small pieces (Smalheiser et al., 1981). The explant allows the maintenance of cellular diversity and anatomic arrangements. The injury effects of dissociation are avoided, but axotomy is still present, although not as close to the cell body as with cell dissociation.

A commonly used technique for explants was first used to demonstrate that RGC neurites from both embryonic and adult mice can grow on laminin, but adding antibodies to the β1/β3 integrin blocked the laminin-dependent growth on embryonic optic fibers only (Bates and Meyer, 1997). Subsequent studies from several laboratories have established the use of this technique for assessing neurite extension from RGCs (Bahr et al., 1988; Manabe et al., 2002), measuring RGC survival (Turner, 1985; Fischer et al., 2000; Xin et al., 2007), and studying RGC differentiation (Wang et al., 2002).

Glial cultures

Glial and other supporting cell types have also been used in culture. Retinal astrocytes and Müller cells can be cultured to varying degree of purity as can optic nerve head lamina cribrosa cells (Hernandez et al., 1988).

These different cell populations can be studied with respect to neurotrophin secretion (Lambert et al., 2001), uptake of glutamate (Kawasaki et al., 2000), induction of injury signals (Neufeld et al., 1997), and other interactions with retinal neurons.

RGC-5 cells

The RGC-5 cell line is a transformed retinal ganglion cell. It was derived by transforming postnatal day 1 rat retinal cells by Krishnamoorthy et al. (2001). This cell line expresses neuronal markers' characteristic of RGCs such as Thy-1, Brn-3, neuritin, synaptophysin, NMDA-R1, and $GABA_B$ receptors. These cells are serum- and neurotrophin-dependent. They do not express the astrocyte marker GFAP.

A great advantage of using these cells is that they are uniform in phenotype, allowing good repeatability of experiments. Also, as a dividing cell line, they are in principle of limitless availability. However, a disadvantage is that these cells are mitotically active and are therefore phenotypically different from a normal postmitotic RGC. In addition, RGC-5 cells are morphologically more similar to glial cells in culture than to primary RGC, and do not express the repertoire of ion channels characteristic of RGCs.

Differentiation of RGC-5 cells

RGC-5 cells can be treated with agents that differentiate them. Differentiation with succinyl concanavalin A (sConA) makes RGC-5 cells sensitive to glutamate toxicity. This glutamate excitotoxicity is blocked by NMDA antagonists (Krishnamoorthy et al., 2001). However, RGC-5 cells differentiated with sConA do not assume a neuronal morphology, nor is proliferation halted.

Treatment of RGC-5 cells with the broad-spectrum protein kinase inhibitor staurosporine also differentiates them, but in a different way (Frassetto et al., 2006). The mechanism by which staurosporine induces RGC-5 cell differentiation is different from staurosporine differentiation of other cell types. It is unlikely to be a result of apoptosis because staurosporine, a known apoptosis inducer, does not activate apoptotic cascade in RGC-5 cells. This is an important distinction because differentiation resulting in apoptosis would not be useful for studying RGC pathophysiology. Staurosporine induces RGC-5 cells to differentiate, express neurites, and become postmitotic. Staurosporine also induces electrophysiological changes that are in the same direction as mature RGC because both cells have large voltage-gated conductance.

Staurosporine-differentiated RGC-5 cells differ in significant ways from primary cultured RGCs. Staurosporine differentiation is transcription independent and results in cells that are viable in the absence of any neurotrophic factor support, unlike normally differentiated RGCs. Neurotrophic factor dependence would be a necessary component for reproducing functional connectivity of neurons to the central nervous system, which is the goal of in vivo application of neuronal stem cells.

A third method of differentiation is with histone deacetylase (HDAC) inhibition. We have studied the relation between histone acetylation and the differentiation and survival of RGC-5 cells and compared it with the transcription-independent differentiation induced by staurosporine (Schwechter et al., 2007). Trichostatin (TSA) is a potent, specific, and well-characterized class 1 and class 2 HDAC inhibitor. TSA causes significant differentiation and neuritogenesis in RGC-5 cells. Differences between HDAC inhibition and staurosporine differentiation include the proportion of differentiated cells, cell viability, cell morphology, and transcriptional dependence. Also, treatment of RGC-5 cells with TSA resulted in RGC-5 cells that are neurotrophic factor dependent, unlike cells treated with staurosporine. Interestingly, HDAC inhibition also increases the sensitivity of RGC-5 cells to differentiation by very low concentrations of staurosporine (Schwechter et al., 2007).

Although not strictly a fourth method of differentiation, Harvey and Chintala studied the effect of plasminogen activators and their inhibition on staurosporine-treated RGC-5 cells (Harvey and Chintala, 2007). Plasmin aids in the elongation process of newly formed neurites by degrading the extracellular matrix. Plasminogen activators convert plasminogen to plasmin. However, under certain circumstances, plasminogen activators promote cell death. Undifferentiated RGC-5 cells do not express the plasminogen activators, tPA and uPA. When treated with staurosporine, RGC-5 expression of those two plasminogen activators is observed. At a high staurosporine concentration, there is an increase in tPA and uPA, but there is also an increase in cell death and shorter neurites. When RGC-5 cells are treated with staurosporine and plasminogen activator inhibitors, there is a decrease in plasminogen activator proteolytic activity and cell death, and an increase in neurite outgrowth. Thus, differentiation of RGC-5 cells with staurosporine induces the expression of tPA and uPA, and these plasminogen activators cause RGC-5 death. By inhibiting this process, the neuritic tree can be stabilized and survival enhanced (Harvey and Chintala, 2007).

RGC-5 cell neurites

Much work has been done to characterize the factors guiding axonal pathfinding in RGCs to appropriate sites in the brain (Oster et al., 2004). Studies have also characterized factors controlling the formation of the dendritic arbor and its stratification in the retina during development, including cell density (Troilo et al., 1996) and neurotrophin levels (Lom et al., 2002). However, none of these studies monitored neurite development over time in a single RGC. Purified RGCs have been cultured and their neurite outgrowth has been studied intensively, but these cells have undergone injury to their existing neuritis in the process of purification.

RGC-5 cells express what appears to be axon and dendrites (Lieven et al., 2007). Microtubule-associated proteins are a family of proteins responsible for microtubule stabilization and organization. Microtubule-associated protein 2 (MAP2) is particularly involved in cytoskeletal changes associated with neuronal differentiation (Caceres et al., 1986). MAP2 exists in several isoforms, the most prevalent being MAP2a, MAP2b, and MAP2c. MAP2a and MAP2b are present in dendrites of mature neurons (Bernhardt and Matus, 1984). MAP2c is present in developing neurons (Meichsner et al., 1993) but not in mature neurons. Tau is another microtubule-associated protein expressed in differentiated neurons, expressed exclusively in the axon in vivo (Binder et al., 1985) and in the soma of cultured cells. The expression of tau characteristically presents as a gradient, with greater amounts at the distal axon (Kempf et al., 1996). Growth-associated protein 43 (GAP43) is neuronal protein expressed in neurite growth cones, specifically those of axons (Goslin et al., 1988).

Our studies confirmed MAP2c expression in some neurites of staurosporine-treated RGC-5 cells. The expression of GAP43 in growth cones and the presence of a tau gradient confirm the development of axons and the establishment of neuronal polarity in these cells. However, the use of staurosporine-differentiated RGC-5 cells as a model for neurite formation in RGCs has potential shortcomings. It is unclear whether this staurosporine differentiation program is similar to what occurs during differentiation of primary RGCs. Also, the number of axons is low (Lieven et al., 2007).

Advantages and disadvantages of culture models

The processes studied with cell culture models appear to be so distinct from those associated with optic neuropathies as to make them less helpful for studying disease. Yet there are many features of cell culture that cannot be replicated in animal models. In cell culture models, there is an ability to control a cell's exposure to specific chemical factors, drugs, interactions with different cell types, and changes in the extracellular milieu. Multiple conditions can be studied in parallel within the same experiment when it would be near impossible to do so in situ within an animal's eye. Results of those studies may sometimes be extrapolated to in vivo conditions; identification of brain-derived neurotrophic factor (BDNF) as a survival factor for cultured RGCs is a good example.

Cell culture also allows the study of one single cell within a culture well and the possibility to observe it over time. Intracellular processes can be examined with fluorescent imaging techniques. Multiple factors can be studied this way, including intracellular calcium, magnesium, and zinc, levels of reactive oxygen species, cell pH, movements of organelles, and mitochondrial membrane potential. Electrophysiological studies can be performed with RGCs by using whole cell or patch clamping. Another advantage of cell culture is the ability of transfecting RGCs using adeno-associated virus, herpes simplex virus (HSV), and lentivirus.

The primary disadvantages of culture models are inherent in the details of culture preparation. Axotomy, which is a necessary part of culturing an RGC, is the most significant one. It is also difficult to model the effects of pressure and ischemia on the RGC axon in culture, as it is not easy to affect the axon without also affecting the cell body. Also, cell culture media and gassed atmosphere need to be used; this environment does not replicate the milieu of the RGC within the retina. Cell culture models require a substrate over which the cells extend their processes. The substrate used in models is not the same as within a retina. Models are two-dimensional while retina is a three-dimensional structure. Finally, culture models are less helpful than animal models in predicting the success of a therapy in a human disease.

References

Bahr, M., Vanselow, J. and Thanos, S. (1988) In vitro regeneration of adult rat ganglion cell axons from retinal explants. Exp. Brain Res., 73: 393–401.

Barres, B.A., Silverstein, B.E., Corey, D.P. and Chun, L.L. (1988) Immunological, morphological, and electrophysiological variation among retinal ganglion cells purified by panning. Neuron, 1: 791–803.

Bates, C.A. and Meyer, R.L. (1997) The neurite-promoting effect of laminin is mediated by different mechanisms in embryonic and adult regenerating mouse optic axons in vitro. Dev. Biol., 181: 91–101.

Bernhardt, R. and Matus, A. (1984) Light and electron microscopic studies of the distribution of microtubule-associated protein 2 in rat brain: a difference between dendritic and axonal cytoskeletons. J. Comp. Neurol., 226: 203–221.

Binder, L.I., Frankfurter, A. and Rebhun, L.I. (1985) The distribution of tau in the mammalian central nervous system. J. Cell Biol., 101: 1371–1378.

Caceres, A., Banker, G.A. and Binder, L. (1986) Immunocytochemical localization of tubulin and microtubule-associated protein 2 during the development of hippocampal neurons in culture. J. Neurosci., 6: 714–722.

Feng, G., Mellor, R.H., Bernstein, M., Keller-Peck, C., Nguyen, Q.T., Wallace, M., Nerbonne, J.M., Lichtman, J.W. and Sanes, J.R. (2000) Imaging neuronal subsets in transgenic mice expressing multiple spectral variants of GFP. Neuron, 28: 41–51.

Fischer, D., Pavlidis, M. and Thanos, S. (2000) Cataractogenic lens injury prevents traumatic ganglion cell death and promotes axonal regeneration both in vivo and in culture. Invest. Ophthalmol. Vis. Sci., 41: 3943–3954.

Frassetto, L.J., Schlieve, C.R., Lieven, C.J., Utter, A.A., Jones, M.V., Agarwal, N. and Levin, L.A. (2006) Kinase-dependent differentiation of a retinal ganglion cell precursor. Invest. Ophthalmol. Vis. Sci., 47: 427–438.

Garcia, M., Forster, V., Hicks, D. and Vecino, E. (2002) Effects of Müller glia on cell survival and neuritogenesis in adult porcine retina in vitro. Invest. Ophthalmol. Vis. Sci., 43: 3735–3743.

Goslin, K., Schreyer, D.J., Skene, J.H. and Banker, G. (1988) Development of neuronal polarity: GAP-43 distinguishes axonal from dendritic growth cones. Nature, 336: 672–674.

Harvey, R. and Chintala, S.K. (2007) Inhibition of plasminogen activators attenuates the death of differentiated retinal ganglion cells and stabilizes their neurite network in vitro. Invest. Ophthalmol. Vis. Sci., 48: 1884–1891.

Hernandez, M.R., Igoe, F. and Neufeld, A.H. (1988) Cell culture of the human lamina cribrosa. Invest. Ophthalmol. Vis. Sci., 29: 78–89.

Kawasaki, A., Otori, Y. and Barnstable, C.J. (2000) Müller cell protection of rat retinal ganglion cells from glutamate and nitric oxide neurotoxicity. Invest. Ophthalmol. Vis. Sci., 41: 3444–3450.

Kempf, M., Clement, A., Faissner, A., Lee, G. and Brandt, R. (1996) Tau binds to the distal axon early in development of polarity in a microtubule- and microfilament-dependent manner. J. Neurosci., 16: 5583–5592.

Krishnamoorthy, R.R., Agarwal, P., Prasanna, G., Vopat, K., Lambert, W., Sheedlo, H.J., Pang, I.H., Shade, D., Wordinger, R.J., Yorio, T., Clark, A.F. and Agarwal, N. (2001) Characterization of a transformed rat retinal ganglion cell line. Brain Res. Mol. Brain Res., 86: 1–12.

Lambert, W., Agarwal, R., Howe, W., Clark, A.F. and Wordinger, R.J. (2001) Neurotrophin and neurotrophin receptor expression by cells of the human lamina cribrosa. Invest. Ophthalmol. Vis. Sci., 42: 2315–2323.

Leahy, K.M., Ornberg, R.L., Wang, Y., Zhu, Y., Gidday, J.M., Connor, J.R. and Wax, M.B. (2004) Quantitative ex vivo detection of rodent retinal ganglion cells by immunolabeling Brn-3b. Exp. Eye Res., 79: 131–140.

Leifer, D., Lipton, S.A., Barnstable, C.J. and Masland, R.H. (1984) Monoclonal antibody to Thy-1 enhances regeneration of processes by rat retinal ganglion cells in culture. Science, 224: 303–306.

Levin, L.A. (2005) Retinal ganglion cells and supporting elements in culture. J. Glaucoma, 14(4): 305–307.

Lieven, C.J., Millet, L.E., Hoegger, M.J. and Levin, L.A. (2007) Induction of axon and dendrite formation during early RGC-5 cell differentiation. Exp. Eye Res., 85: 678–683.

Lom, B., Cogen, J., Sanchez, A.L., Vu, T. and Cohen-Cory, S. (2002) Local and target-derived brain-derived neurotrophic factor exert opposing effects on the dendritic arborization of retinal ganglion cells in vivo. J. Neurosci., 22: 7639–7649.

Manabe, S., Kashii, S., Honda, Y., Yamamoto, R., Katsuki, H. and Akaike, A. (2002) Quantification of axotomized ganglion cell death by explant culture of the rat retina. Neurosci. Lett., 334: 33–36.

Meichsner, M., Doll, T., Reddy, D., Weisshaar, B. and Matus, A. (1993) The low molecular weight form of microtubule-associated protein 2 is transported into both axons and dendrites. Neuroscience, 54: 873–880.

Meyer-Franke, A., Kaplan, M.R., Pfrieger, F.W. and Barres, B.A. (1995) Characterization of the signaling interactions that promote the survival and growth of developing retinal ganglion cells in culture. Neuron, 15: 805–819.

Neufeld, A.H., Hernandez, M.R. and Gonzalez, M. (1997) Nitric oxide synthase in the human glaucomatous optic nerve head. Arch. Ophthalmol., 115: 497–503.

Oster, S.F., Deiner, M., Birgbauer, E. and Sretavan, D.W. (2004) Ganglion cell axon pathfinding in the retina and optic nerve. Semin. Cell. Dev. Biol., 15: 125–136.

Schwechter, B.R., Millet, L.E. and Levin, L.A. (2007) Histone deacetylase inhibition-mediated differentiation of RGC-5 cells and interaction with survival. Invest. Ophthalmol. Vis. Sci., 48: 2845–2857.

Smalheiser, N.R., Crain, S.M. and Bornstein, M.B. (1981) Development of ganglion cells and their axons in organized cultures of fetal mouse retinal explants. Brain Res., 204: 159–178.

Tezel, G. and Wax, M.B. (2000) Increased production of tumor necrosis factor-alpha by glial cells exposed to simulated ischemia or elevated hydrostatic pressure induces apoptosis in cocultured retinal ganglion cells. J. Neurosci., 20: 8693–8700.

Troilo, D., Xiong, M., Crowley, J.C. and Finlay, B.L. (1996) Factors controlling the dendritic arborization of retinal ganglion cells. Vis. Neurosci., 13: 721–733.

Turner, J.E. (1985) Neurotrophic stimulation of fetal rat retinal explant neurite outgrowth and cell survival: age-dependent relationships. Brain Res., 350: 251–263.

Wang, S.W., Mu, X., Bowers, W.J. and Klein, W.H. (2002) Retinal ganglion cell differentiation in cultured mouse retinal explants. Methods, 28: 448–456.

Xin, H., Yannazzo, J.A., Duncan, R.S., Gregg, E.V., Singh, M. and Koulen, P. (2007) A novel organotypic culture model of the postnatal mouse retina allows the study of glutamate-mediated excitotoxicity. J. Neurosci. Methods, 159: 35–42.

CHAPTER 21

Rat models for glaucoma research

John C. Morrison[*], Elaine Johnson and William O. Cepurna

Casey Eye Institute, Oregon Health & Science University, Portland, OR, USA

Abstract: Rats are becoming an increasingly popular model system for understanding mechanisms of optic nerve injury in primary open-angle glaucoma (POAG). Although the anatomy of the rat optic nerve head (ONH) is different from the human, the ultrastructural relationships between astrocytes and axons are quite similar, making it likely that cellular processes of axonal damage in these models will be relevant to human glaucoma. All of these models rely on elevating intraocular pressure (IOP), a major risk factor for glaucoma. Methods that produce increased resistance to aqueous humor outflow at the anterior chamber angle, specifically hypertonic saline injection of aqueous outflow pathways and laser treatment of the limbal tissues, appear to produce a specific regional pattern of injury that may have a particular relevance to understanding regional injury in human glaucoma. Because increased pressure fluctuations are a characteristic of such models and the rodent ONH appears to have high susceptibility to elevated IOP, special instrumentation and measurement techniques are required to document pressure exposure in these eyes and understand the pressure levels that the eyes and the optic nerve are exposed to. With these techniques, it is possible to obtain an excellent correlation between pressure and the extent of nerve damage. Continued use of these models will lead to a better understanding of cellular mechanisms of pressure-induced optic nerve damage and POAG.

Keywords: glaucoma; intraocular pressure; optic nerve damage; astrocyte; axon; optic nerve head; animal models

Rat models for glaucoma research

Primary open-angle glaucoma (POAG) is the most common form of glaucoma in the United States and Europe. This is a slowly progressive form of optic nerve damage and blindness that begins with loss of peripheral vision and is followed by gradual shrinkage of remaining central vision and, ultimately, disappearance of even central, sharp vision. Unfortunately, blindness is irreversible, as optic nerve fibers do not regenerate.

Clinically, glaucoma is recognized by characteristic "cupping" of the optic nerve head (ONH). Cupping results from loss of retinal ganglion cell (RGC) axons, combined with collapse and posterior bowing of their supporting connective tissue sheets, or lamina cribrosa. In many patients, these physical changes are most pronounced in the superior and inferior poles of the ONH, leading to vertical enlargement of the cup and eventual undermining of the neural rim beneath the edge of the sclera in these areas.

[*]Corresponding author. Tel.: +1 503 494 3038; Fax: +1 503 494 3075; E-mail: morrisoj@ohsu.edu

This regional glaucoma injury produces a specific pattern of visual field loss (Quigley and Green, 1979; Quigley et al., 1981, 1988, 1989; Kerrigan-Baumrind et al., 2000). The most characteristic field defect in glaucoma (the arcuate defect) arches either above or below fixation, following the path of the nerve fiber bundles that pass through the superior and inferior poles of the optic nerve (Sommer et al., 1991; Tuulonen and Airaksinen, 1991). In glaucomatous optic nerve cross sections, this results in an "hour-glass" configuration of damage, with the greatest atrophy in the superior and inferior regions (Quigley and Green, 1979).

The best explanation for this injury pattern appears to lie in the structure of the ONH. In humans and other species with relatively large eyes, structural support to the optic nerve bundles is provided by the lamina cribrosa, which consists multiple "plates" of connective tissue that span the scleral opening, with pores that allow the optic nerve fiber bundles to exit the eye (Hernandez, 1992; Morrison et al., 1994). In humans, the pores of the superior and inferior laminas appear larger than elsewhere, and their connective tissue beams are thinner and more sparse (Quigley et al., 1981; Radius, 1981). This suggests that the thinner lamina cribrosa in these regions provides less adequate support for nerve fibers, increasing their risk for glaucomatous damage. This also strongly suggests that the ONH is the likely site of early injury in glaucoma, along with observations of axonal transport obstruction within the lamina in human glaucoma and animal models (Quigley et al., 1979, 1981; Pease et al., 2000; Martin et al., 2003).

Vision loss in POAG is generally slow and progressive and may take decades to develop. Clinically, it has been observed that patients with greater nerve damage and field loss can suffer progressive visual loss at levels of intraocular pressure (IOP) that would be tolerated by the eyes with less damage (Grant and Burke, 1982). This suggests that there is something unique about the glaucomatous eye that renders the remaining optic nerve fibers more susceptible to IOP.

This progressive susceptibility most likely results from gradual changes occurring in the ONH, within RGCs, or both. Understanding these changes will help explain why many patients continue to progress despite apparently successful pressure lowering and why pressure in others is never remarkably higher than the normal range. It may also lead to the development of specific treatments designed to reverse or stabilize these conditions and preserve the remaining nerve fibers.

Because we currently lack noninvasive methods for assaying cellular function in humans, it is not yet possible to study these possibilities directly in human glaucoma. Therefore, relevant animal models will remain essential in helping us understand the mechanisms of glaucomatous optic nerve damage.

Use of animal models for POAG

Most animal models of glaucoma employ experimental elevation of IOP. Although IOP is only one of several known glaucoma risk factors, large clinical trials have confirmed that aggressive IOP lowering is beneficial in a spectrum of open-angle glaucomas, including normal-tension glaucoma, ocular hypertension, and early and late glaucoma (Drance, 1999; Kass et al., 2002; Leske et al., 2003; Nouri-Mahdavi et al., 2004b). From this, it is reasonable to expect that models based on elevated IOP will be highly relevant to optic nerve damage in open-angle glaucoma.

Experimental models of pressure-induced optic nerve damage possess certain advantages over spontaneous models. First, unilateral pressure elevation leaves the fellow eye available as a control against the effects of inter-animal variability. Second, the more predictable onset of the pressure elevation makes it possible to determine sequential events of optic nerve and retina damage.

Anatomically, nonhuman primates should be the most relevant experimental glaucoma model for studying human diseases. However, these animals are expensive, making them impractical for cell biology studies and drug trials requiring a large number of animals. Without special training and experience, it can be difficult to monitor IOP frequently enough to develop a solid understanding of the pressure insult to which the eye is exposed.

For these reasons, a less-expensive, more manageable model of pressure-induced optic nerve damage is needed. We feel this need can be supplied by laboratory rats. A large body of knowledge on the cell biology of neuropathology based on rats already exists, thus providing an array of tools for studying pressure-induced optic nerve damage when produced in these animals.

This chapter will discuss the current status of rat models that can be used to study optic nerve damage in POAG. This will include methods for measuring IOP in rats, assessing damage, and a comparison of the major experimental methods used to produce elevated IOP. We will conclude with a summary of the additional advances needed to optimize the ability of these models to help us improve care of glaucoma patients.

Suitability of the rat for models of optic nerve damage in POAG

It should be noted that the rat ONH lacks a well-developed, collagenous lamina cribrosa (Morrison et al., 1995a). While this differs from the primate, it does not diminish the utility of the rat for understanding cellular mechanisms of axonal injury from elevated IOP and in glaucoma (Fig. 1).

Sparse connective tissue associated with blood vessels has been described in the rat ONH, lined with astrocytes and composed of extracellular matrix materials very similar to that of the primate lamina cribrosa (Morrison et al., 1995a). We have also found that the cellular response to elevated IOP in rats is very similar to that in human and nonhuman primate glaucomas (Hernandez et al., 1990, 2000; Morrison et al., 1990; Johnson et al., 1996). This includes disorganization of the normal columnar structure of astrocytes and aberrant deposition of collagen and laminin within spaces normally occupied by axon bundles.

The most useful feature that the rat eye offers for POAG research lies in the close association between astrocytes and axons within the ONH. Ultrastructurally, rat ONH astrocyte processes lie within axonal bundles, providing intimate contact with all axons, a situation that also exists in the primate (Morrison et al., 2005) (Fig. 2). This close

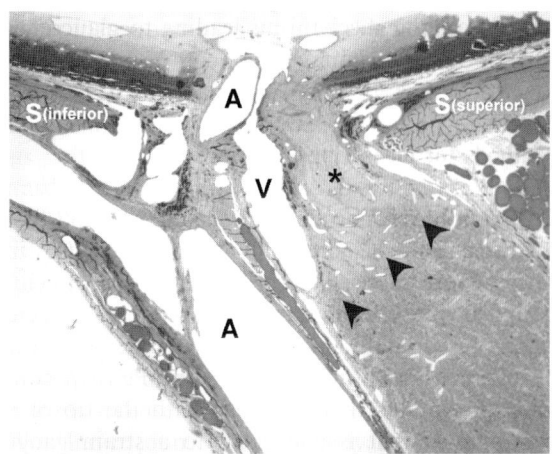

Fig. 1. Longitudinal section of the rat ONH, demonstrating close apposition of the ONH (*) with the superior sclera (S), while the inferior sclera is separated from the ONH by central artery (A) and vein (V). Arrowheads indicate the beginning of the myelinated optic nerve, posterior to the sclera (× 400).

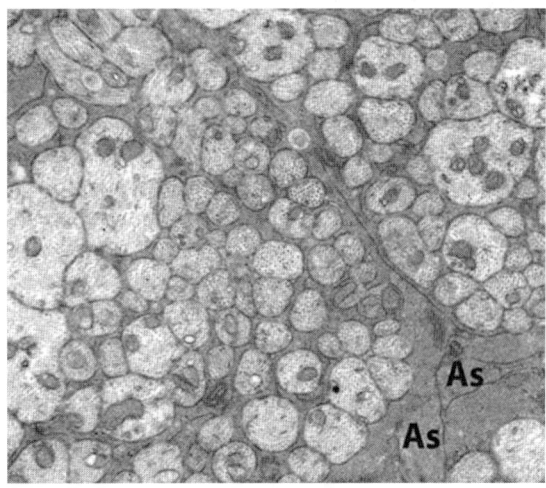

Fig. 2. TEM of the rat ONH cross section showing extension of astrocyte (As) processes into axon bundles. Note close contact between processes and nearly all axons, which are cut in cross section (× 30,000).

association between axons and astrocytes, which rest on the connective tissue lamina and peripapillary sclera of the ONH, provides a potential link by which IOP-generated forces in the load-bearing tissues of the ONH can get translated into axonal damage (Burgoyne et al., 2005). In this way, the rat eye presents an opportunity to understand the pressure measurements must be noninvasive and repeatable to provide sufficient documentation of the IOP that the eye (and the ONH) experiences during the experimental period.

While it is possible to monitor IOP repeatedly over time using a general anesthetic, measurements performed on a daily basis can result in progressive weight loss and even a reduction in IOP, presumably due to cumulative side effects of the general anesthetics (Moore et al., 1995). More importantly, general anesthetics will cause variable lowering of IOP while active, resulting in IOP measurements that are artifactually low and not reflective of the IOP actually experienced by the eye (Jia et al., 2000a). Because of these problems, and because the animal will be awake for nearly all chronic experiments, we advocate measuring IOP in awake animals, using only topical anesthesia. Fortunately, the Brown Norway rat has proven to be docile and easy to handle, and it is possible to obtain meaningful IOP measurements in these awake animals.

Obtaining an accurate assessment of IOP also requires understanding the normal, daily circadian fluctuations of IOP. When housed in a standard 12-h light:dark cycle, IOP is significantly lower in the light phase and higher in the dark (Moore et al., 1996). This appears to be a true circadian phenomenon since the IOP cycle can be reversed by inverting the light and dark phases and probably results from fluctuations in the rate of aqueous humor production (Gregory et al., 1985; Mclaren et al., 1996).

When superimposed on experimental obstruction of aqueous outflow, this circadian fluctuation can become greatly exaggerated. Over 30% of eyes with experimental outflow obstruction develop a significant IOP elevation over the fellow control eyes only in the dark phase, all of which have significant optic nerve injury (Jia et al., 2000b). In this situation, if pressure were measured only in the light phase, this nerve injury would not be explained by the recorded IOP history.

Simply measuring IOP at the same time of day does not adequately account for these problems. Using pressures measured in the nonaffected fellow eyes as a way of accounting for fluctuations is also not sufficient since IOP fluctuations following outflow obstruction can be random and highly variable.

These results suggest that, when animals with experimental outflow obstruction are housed in standard lighting conditions, IOP must be monitored in both the dark and light phases. This approach is cumbersome, and the resulting large IOP fluctuations may make it difficult to correlate pressure history with optic nerve damage. For these reasons, we feel that it is important to minimize these underlying circadian IOP fluctuations.

We have accomplished this by placing animals in a low-level constant light environment, which Rowland, studying rabbits, previously found reduces circadian IOP fluctuation (Rowland et al., 1981). We, and other laboratories, have found that, when animals are housed in constant fluorescent light conditions (40–90 lux), IOP in normal eyes will consistently measure between 27 and 28 mmHg by the Tono-Pen, regardless of the time of day. This is in contrast to light and dark pressures of 19–21 and 28–30 mmHg, when in standard lighting conditions (Moore et al., 1996; Morrison et al., 2005; Pang et al., 2005b).

In eyes with experimental aqueous outflow obstruction, IOP in constant light becomes significantly elevated over that in the fellow eyes. However, unlike a standard light:dark cycle, where the fluctuation between light and dark phases in some animals can be markedly accentuated (Jia et al., 2000b), circadian variation is not observed when animals are placed in constant light (Morrison et al., 2005). It should be emphasized that the constant light paradigm described here does not eliminate all IOP fluctuations since significant pressure variation is common to human glaucoma and glaucoma models that rely on aqueous humor outflow obstruction (Piltz et al., 1985; Asrani et al., 2000; Pena et al., 2001; Nouri-Mahdavi et al., 2004a). It reduces the overall extent of pressure fluctuation by limiting circadian variations in aqueous humor production, a significant source of marked, dark-cycle pressure elevations (Smith and Gregory, 1989). It also frees the experimenter of the need to measure IOP by a rigid schedule and monitor both light- and dark-phase pressures. The reliability of this approach has been repeatedly demonstrated by the close correlations it provides between mean IOP measurements and many different measures of injury, such as extent of optic nerve damage, RGC loss, and measures of altered cellular function in both tissues (Jia et al., 2000b; Johnson et al., 2000, 2006, 2007; Schlamp et al., 2001; Ahmed et al., 2004; Fortune et al., 2004; Morrison et al., 2005; Pang et al., 2005a).

We have observed normal Brown Norway rats in constant light for as long as 12 months and found that the pressure effects are maintained throughout the entire period. Importantly, the retinas were histologically normal, and retinal nuclear layer thickness measurements performed in collaboration with Dr. Matt LaVail were indistinguishable from retinas of age-matched animals simultaneously housed under standard light:dark conditions for the same time period (Morrison et al., 2005). It thus appears that the low level of constant light in our model does not induce photoreceptor toxicity in these pigmented rats (LaVail, 1980).

Assessing optic nerve and retina damage

Assessing damage in any glaucoma model can be done by examining either the optic nerve or the retina. It is critical that this provides an objective picture of either injury or cellular dropout in order to understand the effects of elevated IOP. This is also critical for performing reliable testing of potential neuroprotective compounds, which depend on careful documentation of IOP and damage in both experimental and control groups.

Nerve damage can be measured by evaluating a single cross section of the optic nerve, which contains the entire output of the RGCs. At the light microscopic (LM) level, standard staining of plastic-embedded sections readily allows identification of normal axons by staining the myelin sheaths. Semiautomated image analysis methods can count and determine axon density within specified areas (Quigley et al., 1987). These are then used to calculate entire nerve counts by measuring the total area of the nerve cross section. In experimental models, damage can be determined by comparing these nerve counts to counts in the normal, fellow optic nerves. Although some researchers will determine injury by comparing axon density alone, this is subject to error since nerve damage in these models is not uniform and the individual nerve areas used to count axons may not faithfully capture the true extent of axon loss.

An advantage of this semiautomated optic nerve analysis is that it requires relatively few manipulations and is objective. A disadvantage of this method is that it depends heavily on the quality of tissue fixation and, as discussed below, can produce an underestimate of actual axon counts.

Optic nerve axon counts can also be determined in cross sections by transmission electron microscopy (TEM) (Chauhan et al., 2002; Cepurna et al., 2005). Here, tissue sections from experimental and control nerves are photographed in a random, standardized fashion, and axon counts are performed manually. At this level, the axons can be readily identified by their myelin sheaths and contained microtubules.

Relatively few comparisons of this TEM method with the semiautomated LM approach detailed above have been performed, but available evidence suggests that they do not supply identical results. Two reports using the LM method in Wistar rats found total counts between 79.8 and 87.3 thousand axons (Ricci et al., 1988; Levkovitch-Verbin et al., 2002b). In contrast, studies in the same species of similar age using TEM counting methods reported 102.2 and 105.8 thousand axons (Fukuda et al., 1982; Sugimoto et al., 1984).

A direct comparison of these counting methods in the same optic nerves has provided a similar result (Cepurna et al., 2005; Morrison et al., 2005). Manual axon counts of TEM photographs encompassing approximately 65% of the optic nerve cross section in normal Brown Norway rat eyes resulted in a mean total of 126.4 ± 7.8 ($\pm$SD) thousand axons. By contrast, an LM method using an IBAS image analysis system resulted in a mean total of 86.8 ± 11.5 thousand axons. This difference was statistically significant, with the LM method undercounting axons by a mean of 39.6 ± 19.4 thousand axons, or $31 \pm 15\%$ of the TEM count. These underestimates ranged from 19 to 43%.

The precise reasons for this discrepancy are unknown. However, because TEM allows positive identification of all myelinated axons and nearly 65% of all axons were manually counted, there is a relatively less chance that the TEM method either over- or undercounted axons. By contrast, the large underestimate of axons by the LM method could have resulted from axons that were just at or smaller than the limit of resolution for the light microscope and got ignored by the software of the image analysis system.

Unfortunately, the extent of this underestimation by the LM method was variable as well as large, and one cannot simply extrapolate between these methods. This means that the difference was due to nonuniform errors, such as tissue fixation and staining, despite the use of perfusion fixation and obtaining sections for each technique from the same block of tissue. These uncertainties and potential problems must always be kept in mind whenever using the LM method for axonal counting to determine optic nerve damage.

Another way to assess optic nerve damage is to develop a qualitative LM grading scale for optic nerve degeneration (Jia et al., 2000b). This avoids the potential pitfalls of LM axon counting noted above and is more rapid than the TEM method. This approach depends on recognizing axonal degeneration at the LM level by the appearance of swollen axons that lack apparent axoplasm and dark axons due to collapsed myelin sheaths. The extent of injury is then graded, based on a stereotypical pattern of injury that we have observed in rats with elevated IOP due to aqueous humor outflow obstruction (Table 1). In this system, each optic nerve cross section is assigned a grade, based on a 5-point scale, ranging from grade 1 (normal) to 5 (degeneration involving the entire nerve cross section).

Other investigators, working with albino animals and another method of aqueous outflow obstruction, have developed and used similar qualitative damage scales (Levkovitch-Verbin et al., 2002a, b). Another group, working with chronic ischemia from endothelin administration, has described another grading scale (Chauhan et al., 2006), although it differs considerably from ours due to the different pattern of injury produced by endothelin.

To support the reliability of our grading system, a comparison to actual nerve counts by TEM in the same nerves reveals a linear relationship between the two methods, representing approximately 12,000 axons per grade (Morrison et al., 2005). This comparison also showed that some eyes with mild nerve injury (grade 2–2.5) had axon counts that were within the normal range of variation in total axons, which can be as high as 20%. This suggests that this grading system is more sensitive to mild damage than either TEM or LM counting methods.

Other investigators choose to evaluate damage by documenting loss of RGCs. This usually involves placing a tracer, such as fluorogold, into the superior colliculus or on the stump of a severed optic nerve. The tracer, which is then transported by retrograde axoplasmic flow, will accumulate in RGCs, which can then be counted, usually in flat mount preparations. Generally, these counts are performed systematically in representative regions of the retina and RGC losses determined as a reduction of RGC density or by calculating total RGCs based on the total retinal area.

Because this technique relies on axonal transport, it allows the examiner to assess viable cells. On the other hand, it depends on uniform uptake of the tracer by axons, which may not always happen and can result in an underestimate of actual RGC. It also relies on sampling several regions of the retina, which must be standardized and consistent, since RGC density varies with different regions of the retina in normal eyes. This increases the chance that regional injury may not be adequately or reproducibly detected. Automated methods of counting all labeled RGCs over

Table 1. LM grading system for optic nerve damage due to elevated IOP due to aqueous humor outflow obstruction

Grade 1	Normal
Grade 2	Focal region of degeneration (generally superior)
Grade 3	Degeneration spreading beyond the focal region
Grade 4	Degeneration involving the entire cross section of the nerve, with approximately equal numbers of apparently normal and degenerating axons
Grade 5	Apparent degeneration of nearly all axons

the entire retina have been developed to address this problem (Danias et al., 2002) but are not widely employed.

Experimental methods of producing elevated IOP

The different rat models for simulating optic nerve damage in POAG are defined based on the method used to increase eye pressure. To date, three main approaches have been developed. These include scarring the anterior chamber angle by either injecting hypertonic saline into the aqueous humor outflow pathways or using a laser. The third method, which uses cautery to close off venous outflow from the eye, may not work by the same mechanism, and the reasons for this will be presented.

Hypertonic saline injection of aqueous humor outflow pathways

Aqueous humor outflow in the rat eye has many similarities to that of the primate (Morrison et al., 1995b). These include a trabecular meshwork and a prominent Schlemm's canal, connected to the episcleral vasculature by trans-scleral collector channels. In the rat, episcleral vessels consist a plexus of aqueous veins that encircles the limbus and drains posteriorly via radial veins into the orbit.

In the hypertonic saline method, the investigator takes advantage of this anatomy by injecting hyperosmolar saline into one of the episcleral radial veins (Morrison et al., 1997). By placing a plastic ring around the equator of the eye, the injection can be directed into Schlemm's canal and driven across the trabecular meshwork. The hypertonicity of the saline damages cell membranes and produces scarring of the meshwork and the anterior chamber angle, the extent of which parallels the rise in pressure.

Elevation of IOP generally occurs 7–10 days following the injection, depending on the relative time course of the scarring process and resumption of normal aqueous humor production (Morrison et al., 1997). A second injection can be performed if pressure does not rise after 2 weeks, although some investigators have routinely performed a second injection in the same eye (Chauhan et al., 2002; Hanninen et al., 2002). Interestingly, we have found that elderly animals are unusually sensitive to this injection, possibly due to a narrower angle owing to enlargement of the lens, and the resulting IOP is often too high (Morrison et al., 2007). Modifying the ring to limit saline to specific portions of the meshwork reduces the extent of IOP elevation in these animals.

The duration of pressure elevation can last several months, although most experiments are limited to a few weeks, so that tissues can be harvested for specific histologic or cell biology studies (Morrison et al., 1997; Johnson et al., 2000; Chauhan et al., 2002; Hanninen et al., 2002; Mckinnon et al., 2002). Injections of greater than 2.0 M saline will produce higher pressures, and concentrations below this produce less severe elevations.

This method generally produces a range of pressures, as high as twofold or more above normal. With this, it is possible to detect responses that might be associated with early nerve injury (Johnson et al., 2007). Other changes will be more linearly related to the extent of injury. In general, optic nerve damage and cellular responses can be related to mean IOP, peak IOP elevation, and cumulative pressure exposure (days of pressure elevation times the mmHg above normal) (Chauhan et al., 2002; Mckinnon et al., 2002; Johnson et al., 2007).

Laser treatment of limbal tissues

This method uses external laser to the limbus, with either an argon or diode laser. While some investigators initially inject India ink into the anterior chamber to improve laser uptake and angle damage (Ueda et al., 1998), others have accomplished a similar result without injecting foreign materials into the anterior chamber (Woldemussie et al., 2001; Levkovitch-Verbin et al., 2002b).

It is possible that this treatment produces elevated IOP by coagulating the limbal vasculature and indirectly obstructing aqueous outflow. However, one study has clearly shown that angle treatment is necessary to achieve chronic IOP elevation (Levkovitch-Verbin et al., 2002b). In this manner, the mechanism of pressure rise with this treatment is very similar to hypertonic saline injection. Most likely, pressure increases following laser treatment directed at the limbal vasculature are actually due to collateral damage to the anterior chamber angle (Woldemussie et al., 2001).

This technique is generally used in nonpigmented animals since energy uptake in pigmented eyes can produce a dramatic inflammatory response that can confound interpretation of cellular responses in the retina and the ONH. Because it is difficult to measure IOP in awake albino rats, this restriction will also affect accuracy of pressure monitoring and correlations between pressure and damage.

In most reports, this technique results in a rapid IOP increase, followed by a gradual reduction to normal in a few, often 3, weeks (Levkovitch-Verbin et al., 2002b). In general, additional treatments are used as needed to achieve sustained pressure elevations (Martin et al., 2003). Some groups routinely perform a second treatment during the first 3 weeks (Schori et al., 2001; Bakalash et al., 2002; Ishii et al., 2003; Pease et al., 2006).

Episcleral vein cautery

The third method of creating elevated IOP in rats consists cauterizing large episcleral veins located posterior to the rectus muscle insertions (Shareef et al., 1995; Sawada and Neufeld, 1999). Unlike the limbal laser method, this procedure appears to work in pigmented as well as nonpigmented animals (Neufeld et al., 2002; Grozdanic et al., 2003b; Danias et al., 2006). However, there is little consensus on the effectiveness of this technique. Some groups report that IOP returns to normal after 2–4 weeks (Shareef et al., 1995; Ahmed et al., 2001). Mittag noted that IOP elevations longer than 3 weeks could only be achieved with subconjunctival injections of 5-fluorouracil (Mittag et al., 2000). This suggests that normalization of IOP results from the formation of collateral vessels, and that this is prevented by the antimetabolite. Other investigators have found that pressures will remain elevated longer than this, even up to 7 months, without retreatment (Neufeld et al., 1999, 2002).

The mechanism of pressure rise in this method is also controversial. Some authors feel that the cauterized vessels are simply episcleral veins that drain aqueous humor from the limbal plexus and elevate IOP results from this increased resistance to aqueous outflow (Shareef et al., 1995). Other investigators have determined that these vessels are most likely vortex veins (Grozdanic et al., 2003b; Pang and Clark, 2007). In rats, these veins receive blood from the choroid as well as the anterior uvea and veins at the base of the optic nerve (Morrison et al., 1987, 1999). Given these relationships, cauterization of these veins will instantaneously reduce venous outflow from nearly all parts of the eye and result in ocular congestion and a rise in IOP equivalent to arterial blood pressure (Goldblum and Mittag, 2002).

Several lines of evidence suggest that this model is not equivalent to the other two models. Rats with elevated IOP due to obstruction of aqueous humor outflow following angle closure exhibit a pattern of injury in which the superior region of the optic nerve is damaged first (Morrison et al., 1997). This is also reflected in our injury grading scale in which the focal, grade 2 lesion generally occurs within the superior optic nerve. Within the ONH, early injury is most apparent in the superior portion of the nerve at the level of the sclera (Cepurna et al., 2002) (Fig. 4). Work with the laser model has shown a similar pattern, where the greatest reduction of RGCs occurs in the superior retina (Woldemussie et al., 2001). By contrast, injury following cautery has been described as primarily occurring in peripheral regions of the retina (Sawada and Neufeld, 1999; Ko et al., 2001). Currently, little work on injury within the ONH itself has been reported with this particular model.

A recent high-frequency ultrasound study of the anterior chamber in rats following hypertonic saline treatment has demonstrated a significant deepening of the anterior chamber, suggesting a true aqueous outflow obstruction at the chamber angle (Nissirios et al., 2008) (Fig. 5). On the other hand, vein cautery did not result in any significant alteration in the angle as compared to control eyes. This suggests that elevated IOP in this model does not result from aqueous humor outflow obstruction.

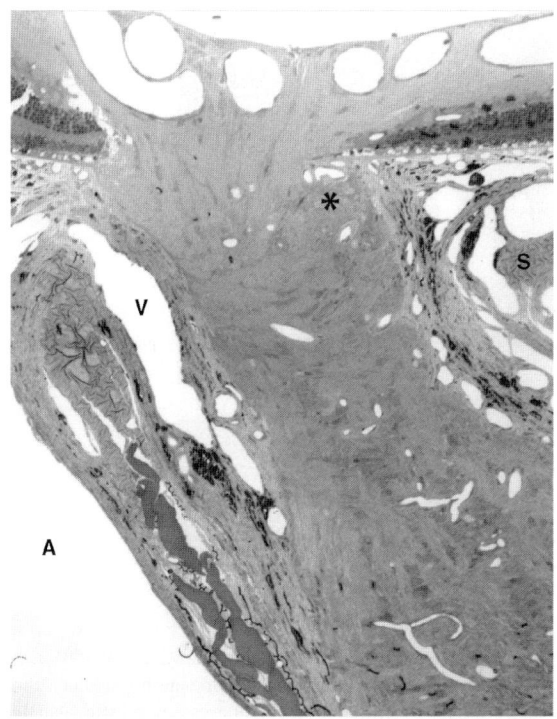

Fig. 4. Longitudinal section of the rat eye ONH with elevated IOP and early injury. Note that majority of degeneration (*) is located superiorly at the level of the sclera (S). Central retinal artery (A) and vein (V) mark the inferior aspect of the nerve and separate it from the sclera (× 400).

Additional evidence for a difference in mechanism among these models can be found in the extent of nerve and RGC damages. An assessment of the percentage of nerve damage or RGC loss as a function of the duration of IOP elevation has shown that the hypertonic saline and limbal laser models were each capable of producing injury as high as 80% over time periods of up to 11 weeks (Morrison et al., 2005). By contrast, maximum injury with the vein cautery model was no greater than 40% and frequently less, even in eyes with pressure elevations as long as 6 months (Sawada and Neufeld, 1999; Danias et al., 2006). In two studies, elevated IOP from vein cautery failed to produce any damage at all (Mittag et al., 2000; Grozdanic et al., 2003b).

Choroidal congestion may be directly responsible for this apparent reduction in optic nerve susceptibility following vein cautery (Goldblum and Mittag, 2002; Grozdanic et al., 2003b). Careful anatomic study of the rat optic nerve vasculature has shown that the choroidal veins are intimately connected to the veins surrounding the ONH and the optic nerve (Morrison et al., 1999). Thus, congestion of the choroidal vasculature is likely to increase pressure within these veins, which in turn may protect nerve fibers from the effects of elevated IOP, possibly by a splinting effect. Alternatively, since direct obstruction to aqueous humor outflow does not occur in this model, stress forces on the wall of the eye, including the peripapillary sclera, are likely to be different than with the other models. If, as suggested by Burgoyne, optic nerve damage depends on these stresses (Burgoyne et al., 2005), one would expect that damage would be different, and possibly less, following vein cautery.

Finally, while the cautery model has been used to demonstrate successful neuroprotection with the nitric oxide synthase inhibitor aminoguanidine (Neufeld et al., 1999, 2002), similar studies with the hypertonic saline model and in the DBA-2J mouse model have failed to corroborate this finding (Pang et al., 2005a; Libby et al., 2007). This further suggests that the mechanisms of optic nerve injury following cautery are different from those that occur in models that specifically rely on aqueous humor outflow obstruction.

Conclusions

The past 15 years have witnessed a remarkable expansion in our knowledge of the cellular events of optic nerve and retina injury due to elevated IOP. Much of this is due to the wide use of models based on experimentally elevated IOP in rats. As expected, these models have made it possible to study both optic nerve and retinal responses using state-of-the-art molecular biology techniques (Johnson et al., 2007; Yang et al., 2007), and they have provided opportunities for testing potentially useful compounds designed to protect optic nerves and RGCs from elevated IOP (Martin et al., 2003; Pang et al., 2005a; Zhou et al., 2005).

The advent of models in rats has quickly paved the way for the study of pressure-induced nerve damage in mice. These models, the most

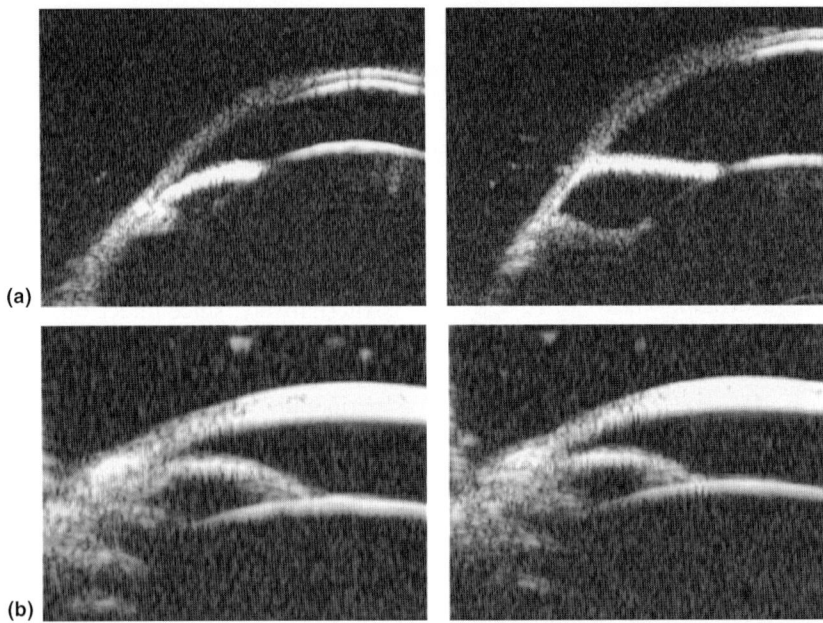

Fig. 5. Ultrasound biomicroscopy studies of the anterior chamber angle in rat eyes following experimental elevation of IOP using hypertonic saline (a) and episcleral vein cautery (b). (a) Note marked deepening of the anterior chamber following hypertonic saline (right) compared to a fellow control eye (left). (b) The anterior chambers of the experimental and control eyes are indistinguishable following vein cautery. Courtesy of John Danias, M.D., Ph.D.

commonly used of which are spontaneous and bilateral, allow genetic manipulation of potentially key pathways to help define their roles in glaucomatous damage (John et al., 1999; Libby et al., 2005; Howell et al., 2007) and are now beginning to see action for testing neuroprotective therapies (Libby et al., 2007). Regardless of these advantages, rat models will continue to be useful, as they are well suited for experimental elevation of IOP and allow specific study of this factor alone.

In spite of our enthusiasm for these models, we would like to emphasize several aspects of their application to POAG.

As mentioned above, the rat ONH lacks a collagenous lamina cribrosa that would appear substantial enough to provide either significant protection for optic nerve fibers or actively contribute to axonal damage. Yet, we, and others, have found that experimental IOP elevation can cause optic nerve damage and RGC loss in rats and, for that matter, in mice, which may have an even less robust lamina. This tells us that a collagenous lamina cribrosa is not required to produce optic nerve damage in rodents with elevated IOP. This finding also suggests that glial cells, and astrocytes in particular, may contribute to this process—thanks to their intimate relationship to axons within the ONH. ONH gene array studies now indicate that elevated IOP in rats can produce up- and down-regulation of a large number of genes (Johnson et al., 2007). Although it is unknown which exact cells are responsible for these changes, the predominance of astrocytes within the rat ONH suggests that they are worthy of specific studies, particularly for responses that might contribute directly to axonal injury.

An additional feature of rat glaucoma models produced by aqueous humor outflow obstruction is the fact that they exhibit a pattern of injury that begins regionally and then gradually spreads to the entire nerve. This is interesting since regional injury and a specific pattern of visual field loss are hallmark of human POAG. Although the pattern of injury differs between the two species, in both cases, it appears to result in unique anatomic features at the level of the ONH. We believe that a careful study of the rat ONH anatomy and how this relates to the pattern of nerve injury will provide powerful insights into the mechanisms by which elevation of IOP results in axonal injury, RGC death, and blindness.

Finally, it must be remembered that pressure elevation in POAG is only moderate in most cases, and the resulting injury occurs over many years. In this regard, animal models relevant to POAG must avoid pressures that are excessively high. With rodent models, we believe that this is particularly important. Over the years, we have found evidence that even IOPs that are only mildly above normal are capable of producing focal injury within the ONH (Morrison et al., 2002; Johnson et al., 2007) (Fig. 6). Restricting pressures to this range and developing reliable, sensitive, and noninvasive methods for measuring these pressures will thus be critical. This will not only avoid the possibility of studying eyes with pressure elevations high enough to produce optic nerve injury through mechanisms not normally present in POAG, but it will avoid catastrophic global damage and restrict injury to the slow, progressive nature more typical of POAG.

Mild IOP elevations are also needed for animal models used to test neuroprotective compounds. Again, reliable methods for measuring IOP in these lower ranges help ensure identical IOP exposure for experimental (those receiving neuroprotective agent) and control (those receiving placebo or vehicle drug) groups. This will not only allow researchers to test compounds and mechanisms relevant to POAG but also reduce the chance of overlooking potentially therapeutic effective drugs by avoiding pressures that are so high enough that no neuroprotective compound could be useful.

Abbreviations

IOP	intraocular pressure
LM	light microscopic
ONH	optic nerve head
POAG	primary open-angle glaucoma
RGC	retinal ganglion cell
TEM	transmission electron microscopy

Acknowledgements

This study was supported by NIH grant EY016866, EY010145 and unrestricted funds from Research to Prevent Blindness.

References

Ahmed, F., Brown, K.M., Stephan, D.A., Morrison, J.C., Johnson, E.C. and Tomarev, S.I. (2004) Microarray analysis of changes in mRNA levels in the rat retina after experimental elevation of intraocular pressure. Invest. Ophthalmol. Vis. Sci., 45(4): 1247–1258.

Ahmed, F.A., Hegazy, K., Chaudhary, P. and Sharma, S.C. (2001) Neuroprotective effect of alpha(2) agonist (brimonidine) on adult rat retinal ganglion cells after increased intraocular pressure. Brain Res., 913(2): 133–139.

Asrani, S., Zeimer, R., Wilensky, J., Gieser, D., Vitale, S. and Lindenmuth, K. (2000) Large diurnal fluctuations in intraocular pressure are an independent risk factor in patients with glaucoma. J. Glaucoma, 9(2): 134–142.

Bakalash, S., Kipnis, J., Yoles, E. and Schwartz, M. (2002) Resistance of retinal ganglion cells to an increase in intraocular pressure is immune-dependent. Invest. Ophthalmol. Vis. Sci., 43(8): 2648–2653.

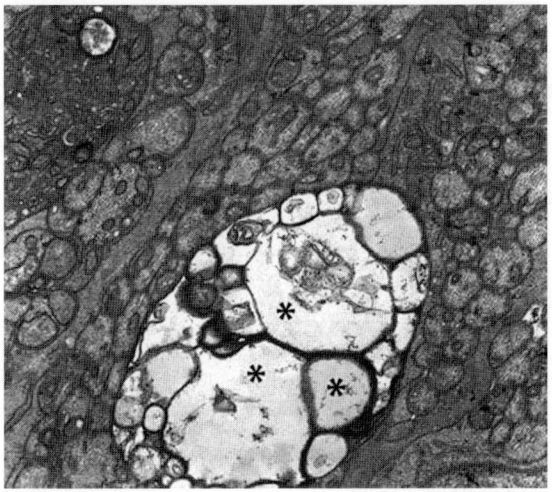

Fig. 6. TEM of the superior ONH at the level of the sclera following hypertonic saline injection showing degenerating axons (*) with swelling and loss of normal axoplasm. Mean IOP elevation in this eye was only 18% above fellow eye for 5 weeks (× 600).

Burgoyne, C.F., Downs, J.C., Bellezza, A.J., Suh, J.K. and Hart, R.T. (2005) The optic nerve head as a biomechanical structure: a new paradigm for understanding the role of IOP-related stress and strain in the pathophysiology of glaucomatous optic nerve head damage. Prog. Retin. Eye Res., 24(1): 39–73.

Cepurna, W.O., Jia, L., Johnson, E.C. and Morrison, J.C. (2002) ARVO abstract: superior scleral localization of rat optic nerve head response to mildly elevated intraocular pressure. Invest. Ophthalmol. Vis. Sci., 43: p. U1161.

Cepurna, W.O., Kayton, R.J., Johnson, E.C. and Morrison, J.C. (2005) Age related optic nerve axonal loss in adult Brown Norway rats. Exp. Eye Res., 80(6): 877–884.

Chauhan, B.C., Levatte, T.L., Garnier, K.L., Tremblay, F., Pang, I.H., Clark, A.F. and Archibald, M.L. (2006) Semiquantitative optic nerve grading scheme for determining axonal loss in experimental optic neuropathy. Invest. Ophthalmol. Vis. Sci., 47(2): 634–640.

Chauhan, B.C., Pan, J., Archibald, M.L., LeVatte, T.L., Kelly, M.E. and Tremblay, F. (2002) Effect of intraocular pressure on optic disc topography, electroretinography, and axonal loss in a chronic pressure-induced rat model of optic nerve damage. Invest. Ophthalmol. Vis. Sci., 43(9): 2969–2976.

Chui, W.S., Lam, A., Chen, D. and Chiu, R. (2008) The influence of corneal properties on rebound tonometry. Ophthalmology, 115(1): 80–84.

Cohan, B.E. and Bohr, D.F. (2001a) Goldmann applanation tonometry in the conscious rat. Invest. Ophthalmol. Vis. Sci., 42(2): 340–342.

Cohan, B.E. and Bohr, D.F. (2001b) Measurement of intraocular pressure in awake mice. Invest. Ophthalmol. Vis. Sci., 42(11): 2560–2562.

Danias, J., Kontiola, A.I., Filippopoulos, T. and Mittag, T. (2003) Method for the noninvasive measurement of intraocular pressure in mice. Invest. Ophthalmol. Vis. Sci., 44(3): 1138–1141.

Danias, J., Shen, F., Goldblum, D., Chen, B., Ramos-Esteban, J., Podos, S.M. and Mittag, T. (2002) Cytoarchitecture of the retinal ganglion cells in the rat. Invest. Ophthalmol. Vis. Sci., 43(3): 587–594.

Danias, J., Shen, F., Kavalarakis, M., Chen, B., Goldblum, D., Lee, K., Zamora, M.F., Su, Y., Podos, S.M. and Mittag, T. (2006) Characterization of retinal damage in the episcleral vein cauterization rat glaucoma model. Exp. Eye Res., 82(2): 219–228.

Drance, S.M. (1999) The collaborative normal-tension glaucoma study and some of its lessons. Can. J. Ophthalmol., 34(1): 1–6.

Filippopoulos, T., Matsubara, A., Danias, J., Huang, W., Dobberfuhl, A., Ren, L., Mittag, T., Miller, J.W. and Grosskreutz, C.L. (2006) Predictability and limitations of non-invasive murine tonometry: comparison of two devices. Exp. Eye Res., 83(1): 194–201.

Fortune, B., Bui, B.V., Morrison, J.C., Johnson, E.C., Dong, J., Cepurna, W.O., Jia, L., Barber, S. and Cioffi, G.A. (2004) Selective ganglion cell functional loss in rats with experimental glaucoma. Invest. Ophthalmol. Vis. Sci., 45(6): 1854–1862.

Fukuda, Y., Sugimoto, T. and Shirokawa, T. (1982) Strain differences in quantitative analysis of the rat optic nerve. Exp. Neurol., 75(2): 525–532.

Goldblum, D., Kontiola, A.I., Mittag, T., Chen, B. and Danias, J. (2002) Non-invasive determination of intraocular pressure in the rat eye. Comparison of an electronic tonometer (Tono-Pen), and a rebound (impact probe) tonometer. Graefes Arch. Clin. Exp. Ophthalmol., 240(11): 942–946.

Goldblum, D. and Mittag, T. (2002) Prospects for relevant glaucoma models with retinal ganglion cell damage in the rodent eye. Vision Res., 42(4): 471–478.

Grant, W.M. and Burke, J.F., Jr. (1982) Why do some people go blind from glaucoma?. Ophthalmology, 89(9): 991–998.

Gregory, D.S., Aviado, D.G. and Sears, M.L. (1985) Cervical ganglionectomy alters the circadian rhythm of intraocular pressure in New Zealand White rabbits. Curr. Eye Res., 4(12): 1273–1279.

Grozdanic, S.D., Betts, D.M., Sakaguchi, D.S., Allbaugh, R.A., Kwon, Y.H. and Kardon, R.H. (2003a) Laser-induced mouse model of chronic ocular hypertension. Invest. Ophthalmol. Vis. Sci., 44(10): 4337–4346.

Grozdanic, S.D., Betts, D.M., Sakaguchi, D.S., Kwon, Y.H., Kardon, R.H. and Sonea, I.M. (2003b) Temporary elevation of the intraocular pressure by cauterization of vortex and episcleral veins in rats causes functional deficits in the retina and optic nerve. Exp. Eye Res., 77(1): 27–33.

Hanninen, V.A., Pantcheva, M.B., Freeman, E.E., Poulin, N.R. and Grosskreutz, C.L. (2002) Activation of caspase 9 in a rat model of experimental glaucoma. Curr. Eye Res., 25(6): 389–395.

Hernandez, M.R. (1992) Ultrastructural immunocytochemical analysis of elastin in the human lamina cribrosa. Changes in elastic fibers in primary open-angle glaucoma. Invest. Ophthalmol. Vis. Sci., 33(10): 2891–2903.

Hernandez, M.R., Andrzejewska, W.M. and Neufeld, A.H. (1990) Changes in the extracellular matrix of the human optic nerve head in primary open-angle glaucoma. Am. J. Ophthalmol., 109(2): 180–188.

Hernandez, M.R., Pena, J.D., Selvidge, J.A., Salvador-Silva, M. and Yang, P. (2000) Hydrostatic pressure stimulates synthesis of elastin in cultured optic nerve head astrocytes. Glia, 32(2): 122–136.

Howell, G.R., Libby, R.T., Jakobs, T.C., Smith, R.S., Phalan, F.C., Barter, J.W., Barbay, J.M., Marchant, J.K., Mahesh, N., Porciatti, V., Whitmore, A.V., Masland, R.H. and John, S.W. (2007) Axons of retinal ganglion cells are insulted in the optic nerve early in DBA/2J glaucoma. J. Cell Biol., 179(7): 1523–1537.

Ishii, Y., Kwong, J.M. and Caprioli, J. (2003) Retinal ganglion cell protection with geranylgeranylacetone, a heat shock protein inducer, in a rat glaucoma model. Invest. Ophthalmol. Vis. Sci., 44(5): 1982–1992.

Jia, L., Cepurna, W.O., Johnson, E.C. and Morrison, J.C. (2000a) Effect of general anesthetics on IOP in rats with experimental aqueous outflow obstruction. Invest. Ophthalmol. Vis. Sci., 41(11): 3415–3419.

Jia, L., Cepurna, W.O., Johnson, E.C. and Morrison, J.C. (2000b) Patterns of intraocular pressure elevation after aqueous humor outflow obstruction in rats. Invest. Ophthalmol. Vis. Sci., 41(6): 1380–1385.

Jia, L., Cepurna, W.O., Johnson, E.C., Morrison, J.C. (2006) The TonoLab Rebound Tonometer Reliably Measures Intraocular Pressure (IOP) in Awake Rats. Invest. Ophthalmol. Vis. Sci., 47: E-Abstract 1252.

John, S.W., Anderson, M.G. and Smith, R.S. (1999) Mouse genetics: a tool to help unlock the mechanisms of glaucoma. J. Glaucoma, 8(6): 400–412.

Johnson, E.C., Cepurna, W.O., Jia, L. and Morrison, J.C. (2006) The use of cyclodialysis to limit exposure to elevated intraocular pressure in rat glaucoma models. Exp. Eye Res., 83(1): 51–60.

Johnson, E.C., Deppmeier, L.M., Wentzien, S.K., Hsu, I. and Morrison, J.C. (2000) Chronology of optic nerve head and retinal responses to elevated intraocular pressure. Invest. Ophthalmol. Vis. Sci., 41(2): 431–442.

Johnson, E.C., Jia, L., Cepurna, W.O., Doser, T.A. and Morrison, J.C. (2007) Global changes in optic nerve head gene expression after exposure to elevated intraocular pressure in a rat glaucoma model. Invest. Ophthalmol. Vis. Sci., 48(7): 3161–3177.

Johnson, E.C., Morrison, J.C., Farrell, S., Deppmeier, L., Moore, C.G. and McGinty, M.R. (1996) The effect of chronically elevated intraocular pressure on the rat optic nerve head extracellular matrix. Exp. Eye Res., 62(6): 663–674.

Kass, M.A., Heuer, D.K., Higginbotham, E.J., Johnson, C.A., Keltner, J.L., Miller, J.P., Parrish, R.K., II, Wilson, M.R. and Gordon, M.O. (2002) The ocular hypertension treatment study: a randomized trial determines that topical ocular hypotensive medication delays or prevents the onset of primary open-angle glaucoma. Arch. Ophthalmol., 120(6): 701–713, discussion 829–730.

Kerrigan-Baumrind, L.A., Quigley, H.A., Pease, M.E., Kerrigan, D.F. and Mitchell, R.S. (2000) Number of ganglion cells in glaucoma eyes compared with threshold visual field tests in the same persons. Invest. Ophthalmol. Vis. Sci., 41(3): 741–748.

Kim, C.Y., Kuehn, M.H., Anderson, M.G. and Kwon, Y.H. (2007) Intraocular pressure measurement in mice: a comparison between Goldmann and rebound tonometry. Eye, 21(9): 1202–1209.

Ko, M.L., Hu, D.N., Ritch, R., Sharma, S.C. and Chen, C.F. (2001) Patterns of retinal ganglion cell survival after brain-derived neurotrophic factor administration in hypertensive eyes of rats. Neurosci. Lett., 305(2): 139–142.

Kontiola, A. (1996) A new electromechanical method for measuring intraocular pressure. Doc. Ophthalmol., 93(3): 265–276.

Kontiola, A. and Puska, P. (2004) Measuring intraocular pressure with the Pulsair 3000 and Rebound tonometers in elderly patients without an anesthetic. Graefes Arch. Clin. Exp. Ophthalmol., 242(1): 3–7.

Kontiola, A.I. (2000) A new induction-based impact method for measuring intraocular pressure. Acta Ophthalmol. Scand., 78(2): 142–145.

Kontiola, A.I., Goldblum, D., Mittag, T. and Danias, J. (2001) The induction/impact tonometer: a new instrument to measure intraocular pressure in the rat. Exp. Eye Res., 73(6): 781–785.

LaVail, M.M. (1980) Eye pigmentation and constant light damage in the rat retina. In: Williams T.P. and Baker B.N. (Eds.), The Effects of Constant Light on Visual Processes. Plenum Press, New York, pp. 357–387.

Leske, M.C., Heijl, A., Hussein, M., Bengtsson, B., Hyman, L. and Komaroff, E. (2003) Factors for glaucoma progression and the effect of treatment: the early manifest glaucoma trial. Arch. Ophthalmol., 121(1): 48–56.

Levkovitch-Verbin, H., Martin, K.R., Quigley, H.A., Baumrind, L.A., Pease, M.E. and Valenta, D. (2002a) Measurement of amino acid levels in the vitreous humor of rats after chronic intraocular pressure elevation or optic nerve transection. J. Glaucoma, 11(5): 396–405.

Levkovitch-Verbin, H., Quigley, H.A., Martin, K.R., Valenta, D., Baumrind, L.A. and Pease, M.E. (2002b) Translimbal laser photocoagulation to the trabecular meshwork as a model of glaucoma in rats. Invest. Ophthalmol. Vis. Sci., 43(2): 402–410.

Libby, R.T., Anderson, M.G., Pang, I.H., Robinson, Z.H., Savinova, O.V., Cosma, I.M., Snow, A., Wilson, L.A., Smith, R.S., Clark, A.F. and John, S.W. (2005) Inherited glaucoma in DBA/2J mice: pertinent disease features for studying the neurodegeneration. Vis. Neurosci., 22(5): 637–648.

Libby, R.T., Howell, G.R., Pang, I.H., Savinova, O.V., Mehalow, A.K., Barter, J.W., Smith, R.S., Clark, A.F. and John, S.W. (2007) Inducible nitric oxide synthase, Nos2, does not mediate optic neuropathy and retinopathy in the DBA/2J glaucoma model. BMC Neurosci., 8(1): p. 108.

Martin, K.R., Quigley, H.A., Zack, D.J., Levkovitch-Verbin, H., Kielczewski, J., Valenta, D., Baumrind, L., Pease, M.E., Klein, R.L. and Hauswirth, W.W. (2003) Gene therapy with brain-derived neurotrophic factor as a protection: retinal ganglion cells in a rat glaucoma model. Invest. Ophthalmol. Vis. Sci., 44(10): 4357–4365.

McKinnon, S.J., Lehman, D.M., Kerrigan-Baumrind, L.A., Merges, C.A., Pease, M.E., Kerrigan, D.F., Ransom, N.L., Tahzib, N.G., Reitsamer, H.A., Levkovitch-Verbin, H., Quigley, H.A. and Zack, D.J. (2002) Caspase activation and amyloid precursor protein cleavage in rat ocular hypertension. Invest. Ophthalmol. Vis. Sci., 43(4): 1077–1087.

McLaren, J.W., Brubaker, R.F. and FitzSimon, J.S. (1996) Continuous measurement of intraocular pressure in rabbits by telemetry. Invest. Ophthalmol. Vis. Sci., 37(6): 966–975.

Mittag, T.W., Danias, J., Pohorenec, G., Yuan, H.M., Burakgazi, E., Chalmers-Redman, R., Podos, S.M. and Tatton, W.G. (2000) Retinal damage after 3 to 4 months of elevated intraocular pressure in a rat glaucoma model. Invest. Ophthalmol. Vis. Sci., 41(11): 3451–3459.

Moore, C.G., Epley, D., Milne, S.T. and Morrison, J.C. (1995) Long-term non-invasive measurement of intraocular pressure in the rat eye. Curr. Eye Res., 14(8): 711–717.

Moore, C.G., Johnson, E.C. and Morrison, J.C. (1996) Circadian rhythm of intraocular pressure in the rat. Curr. Eye Res., 15(2): 185–191.

Moore, C.G., Milne, S.T. and Morrison, J.C. (1993) Non-invasive measurement of rat intraocular pressure with the Tono-Pen. Invest. Ophthalmol. Vis. Sci., 34(2): 363–369.

Morris, C.A., Crowston, J.G., Lindsey, J.D., Danias, J. and Weinreb, R.N. (2006) Comparison of invasive and non-invasive tonometry in the mouse. Exp. Eye Res., 82(6): 1094–1099.

Morrison, J., Farrell, S., Johnson, E., Deppmeier, L., Moore, C.G. and Grossmann, E. (1995a) Structure and composition of the rodent lamina cribrosa. Exp. Eye Res., 60(2): 127–135.

Morrison, J.C., Cepurna, W.O., Jia, L., Aubert, J. and Johnson, E.C. (2002) Abstract: mechanism of focal optic nerve injury from elevated intraocular pressure. Invest. Ophthalmol. Vis. Sci., 43: p. U813.

Morrison, J.C., Cepurna, W.O., Jia, L., Aubert, J. and Johnson, E.C. (2007) Characteristics of Modeling Pressure-Induced Optic Nerve Damage in Elderly Rat Eyes ARVO. Invest. Ophthalmol. Vis. Sci., 48: E-Abstract 3662.

Morrison, J.C., DeFrank, M.P. and Van Buskirk, E.M. (1987) Comparative microvascular anatomy of mammalian ciliary processes. Invest. Ophthalmol. Vis. Sci., 28(8): 1325–1340.

Morrison, J.C., Dorman-Pease, M.E., Dunkelberger, G.R. and Quigley, H.A. (1990) Optic nerve head extracellular matrix in primary optic atrophy and experimental glaucoma. Arch. Ophthalmol., 108(7): 1020–1024.

Morrison, J.C., Fraunfelder, F.W., Milne, S.T. and Moore, C.G. (1995b) Limbal microvasculature of the rat eye. Invest. Ophthalmol. Vis. Sci., 36(3): 751–756.

Morrison, J.C., Johnson, E.C., Cepurna, W. and Jia, L. (2005) Understanding mechanisms of pressure-induced optic nerve damage. Prog. Retin. Eye Res., 24(2): 217–240.

Morrison, J.C., Johnson, E.C., Cepurna, W.O. and Funk, R.H. (1999) Microvasculature of the rat optic nerve head. Invest. Ophthalmol. Vis. Sci., 40(8): 1702–1709.

Morrison, J.C., Moore, C.G., Deppmeier, L.M., Gold, B.G., Meshul, C.K. and Johnson, E.C. (1997) A rat model of chronic pressure-induced optic nerve damage. Exp. Eye Res., 64(1): 85–96.

Morrison, J.C., Rask, P., Johnson, E.C. and Deppmeier, L. (1994) Chondroitin sulfate proteoglycan distribution in the primate optic nerve head. Invest. Ophthalmol. Vis. Sci., 35(3): 838–845.

Neufeld, A.H., Das, S., Vora, S., Gachie, E., Kawai, S., Manning, P.T. and Connor, J.R. (2002) A prodrug of a selective inhibitor of inducible nitric oxide synthase is neuroprotective in the rat model of glaucoma. J. Glaucoma, 11(3): 221–225.

Neufeld, A.H., Sawada, A. and Becker, B. (1999) Inhibition of nitric-oxide synthase 2 by aminoguanidine provides neuroprotection of retinal ganglion cells in a rat model of chronic glaucoma. Proc. Natl. Acad. Sci. U.S.A., 96(17): 9944–9948.

Nissirios, N., Goldblum, D., Rohrer, K., Mittag, T. and Danias, J. (2007) Noninvasive determination of intraocular pressure (IOP) in nonsedated mice of 5 different inbred strains. J. Glaucoma, 16(1): 57–61.

Nissirios, N., Chanis, R., Johnson, E., Morrison, J., Cepurna, W. O., Jia, L., Mittag T., Danias, J. (2008) Comparison of anterior segment structures in two rat glaucoma models. An ultrasound biomicroscopic (UBM) study. Invest. Ophthalmol. Vis. Sci., 49(6): 2478–2482.

Nouri-Mahdavi, K., Hoffman, D., Coleman, A.L., Liu, G., Li, G., Gaasterland, D. and Caprioli, J. (2004a) Predictive factors for glaucomatous visual field progression in the Advanced Glaucoma Intervention Study. Ophthalmology, 111(9): 1627–1635.

Nouri-Mahdavi, K., Hoffman, D., Gaasterland, D. and Caprioli, J. (2004b) Prediction of visual field progression in glaucoma. Invest. Ophthalmol. Vis. Sci., 45(12): 4346–4351.

Pang, I.H. and Clark, A.F. (2007) Rodent models for glaucoma retinopathy and optic neuropathy. J. Glaucoma, 16(5): 483–505.

Pang, I.H., Johnson, E.C., Jia, L., Cepurna, W.O., Shepard, A.R., Hellberg, M.R., Clark, A.F. and Morrison, J.C. (2005a) Evaluation of inducible nitric oxide synthase in glaucomatous optic neuropathy and pressure-induced optic nerve damage. Invest. Ophthalmol. Vis. Sci., 46(4): 1313–1321.

Pang, I.H., Wang, W.H. and Clark, A.F. (2005b) Acute effects of glaucoma medications on rat intraocular pressure. Exp. Eye Res., 80(2): 207–214.

Pease, M.E., Hammond, J.C. and Quigley, H.A. (2006) Manometric calibration and comparison of TonoLab and Tono-Pen tonometers in rats with experimental glaucoma and in normal mice. J. Glaucoma, 15(6): 512–519.

Pease, M.E., McKinnon, S.J., Quigley, H.A., Kerrigan-Baumrind, L.A. and Zack, D.J. (2000) Obstructed axonal transport of BDNF and its receptor TrkB in experimental glaucoma. Invest. Ophthalmol. Vis. Sci., 41(3): 764–774.

Pena, J.D., Agapova, O., Gabelt, B.T., Levin, L.A., Lucarelli, M.J., Kaufman, P.L. and Hernandez, M.R. (2001) Increased elastin expression in astrocytes of the lamina cribrosa in response to elevated intraocular pressure. Invest. Ophthalmol. Vis. Sci., 42(10): 2303–2314.

Piltz, J.R., Starita, R., Miron, M. and Henkind, P. (1985) Momentary fluctuations of intraocular pressure in normal and glaucomatous eyes. Am. J. Ophthalmol., 99(3): 333–339.

Quigley, H.A., Addicks, E.M., E. M. Green, W.R. and Maumenee, A.E. (1981) Optic nerve damage in human glaucoma. II. The site of injury and susceptibility to damage. Arch. Ophthalmol., 99(4): 635–649.

Quigley, H.A., Dunkelberger, G.R. and Green, W.R. (1988) Chronic human glaucoma causing selectively greater loss of large optic nerve fibers. Ophthalmology, 95(3): 357–363.

Quigley, H.A., Dunkelberger, G.R. and Green, W.R. (1989) Retinal ganglion cell atrophy correlated with automated perimetry in human eyes with glaucoma. Am. J. Ophthalmol., 107(5): 453–464.

Quigley, H.A. and Green, W.R. (1979) The histology of human glaucoma cupping and optic nerve damage: clinicopathologic correlation in 21 eyes. Ophthalmology, 86(10): 1803–1830.

Quigley, H.A., Guy, J. and Anderson, D.R. (1979) Blockade of rapid axonal transport. Effect of intraocular pressure elevation in primate optic nerve. Arch. Ophthalmol., 97(3): 525–531.

Quigley, H.A., Sanchez, R.M., Dunkelberger, G.R., L'Hernault, N.L. and Baginski, T.A. (1987) Chronic glaucoma selectively damages large optic nerve fibers. Invest. Ophthalmol. Vis. Sci., 28(6): 913–920.

Radius, R.L. (1981) Regional specificity in anatomy at the lamina cribrosa. Arch. Ophthalmol., 99(3): 478–480.

Ricci, A., Bronzetti, E. and Amenta, F. (1988) Effect of ageing on the nerve fibre population of rat optic nerve. Gerontology, 34(5–6): 231–235.

Rowland, J.M., Potter, D.E. and Reiter, R.J. (1981) Circadian rhythm in intraocular pressure: a rabbit model. Curr. Eye Res., 1(3): 169–173.

Sahin, A., Basmak, H., Niyaz, L. and Yildirim, N. (2007a) Reproducibility and tolerability of the ICare rebound tonometer in school children. J. Glaucoma, 16(2): 185–188.

Sahin, A., Niyaz, L. and Yildirim, N. (2007b) Comparison of the rebound tonometer with the Goldmann applanation tonometer in glaucoma patients. Clin. Exp. Ophthalmol., 35(4): 335–339.

Sawada, A. and Neufeld, A.H. (1999) Confirmation of the rat model of chronic, moderately elevated intraocular pressure. Exp. Eye Res., 69(5): 525–531.

Schlamp, C.L., Johnson, E.C., Li, Y., Morrison, J.C. and Nickells, R.W. (2001) Changes in Thy1 gene expression associated with damaged retinal ganglion cells. Mol. Vis., 7: 192–201.

Schori, H., Kipnis, J., Yoles, E., WoldeMussie, E., Ruiz, G., Wheeler, L.A. and Schwartz, M. (2001) Vaccination for protection of retinal ganglion cells against death from glutamate cytotoxicity and ocular hypertension: implications for glaucoma. Proc. Natl. Acad. Sci. U.S.A., 98(6): 3398–3403.

Shareef, S.R., Garcia-Valenzuela, E., Salierno, A., Walsh, J. and Sharma, S.C. (1995) Chronic ocular hypertension following episcleral venous occlusion in rats [letter]. Exp. Eye Res., 61(3): 379–382.

Smith, S.D. and Gregory, D.S. (1989) A circadian rhythm of aqueous flow underlies the circadian rhythm of IOP in NZW rabbits. Invest. Ophthalmol. Vis. Sci., 30(4): 775–778.

Sommer, A., Katz, J., Quigley, H.A., Miller, N.R., Robin, A.L., Richter, R.C. and Witt, K.A. (1991) Clinically detectable nerve fiber atrophy precedes the onset of glaucomatous field loss. Arch. Ophthalmol., 109(1): 77–83.

Sugimoto, T., Fukuda, Y. and Wakakuwa, K. (1984) Quantitative analysis of a cross-sectional area of the optic nerve: a comparison between albino and pigmented rats. Exp. Brain Res., 54(2): 266–274.

Tuulonen, A. and Airaksinen, P.J. (1991) Initial glaucomatous optic disk and retinal nerve fiber layer abnormalities and their progression. Am. J. Ophthalmol., 111(4): 485–490.

Ueda, J., Sawaguchi, S., Hanyu, T., Yaoeda, K., Fukuchi, T., Abe, H. and Ozawa, H. (1998) Experimental glaucoma model in the rat induced by laser trabecular photocoagulation after an intracameral injection of India ink. Jpn. J. Ophthalmol., 42(5): 337–344.

Wang, W.H., Millar, J.C., Pang, I.H., Wax, M.B. and Clark, A.F. (2005) Noninvasive measurement of rodent intraocular pressure with a rebound tonometer. Invest. Ophthalmol. Vis. Sci., 46(12): 4617–4621.

WoldeMussie, E., Ruiz, G., Wijono, M. and Wheeler, L.A. (2001) Neuroprotection of retinal ganglion cells by brimonidine in rats with laser-induced chronic ocular hypertension. Invest. Ophthalmol. Vis. Sci., 42(12): 2849–2855.

Yang, Z., Quigley, H.A., Pease, M.E., Yang, Y., Qian, J., Valenta, D. and Zack, D.J. (2007) Changes in gene expression in experimental glaucoma and optic nerve transection: the equilibrium between protective and detrimental mechanisms. Invest. Ophthalmol. Vis. Sci., 48(12): 5539–5548.

Zhou, Y., Pernet, V., Hauswirth, W.W. and Di Polo, A. (2005) Activation of the extracellular signal-regulated kinase 1/2 pathway by AAV gene transfer protects retinal ganglion cells in glaucoma. Mol. Ther., 12(3): 402–412.

CHAPTER 22

Mouse genetic models: an ideal system for understanding glaucomatous neurodegeneration and neuroprotection

Gareth R. Howell[1], Richard T. Libby[2] and Simon W.M. John[1,3,4],*

[1]The Jackson Laboratory, 600 Main Street, Bar Harbor, ME 04609, USA
[2]University of Rochester Eye Institute, University of Rochester Medical Center, Rochester, NY, USA
[3]The Howard Hughes Medical Institute, Bar Harbor, ME, USA
[4]Department of Ophthalmology, Tufts University of Medicine, Boston, MA, USA

Abstract: Here we review how mouse studies are contributing to understanding glaucoma. We include discussion of aqueous humor drainage and intraocular pressure elevation, because new treatments to avoid exposure to high pressure will indirectly protect neurons from glaucoma, and complement direct neuroprotective strategies. We describe how mouse models are adding to both the understanding of glaucomatous neurodegeneration and the development of neuroprotective strategies.

Keywords: glaucoma; neurodegeneration; neuroprotection

Introduction

Glaucoma is a heterogeneous group of diseases characterized by the dysfunction and death of retinal ganglion cells (RGCs). It is a leading neurodegenerative cause of blindness and is reported to be the second leading cause of blindness worldwide (Quigley and Broman, 2006). Approximately 70 million people are affected by glaucoma (Quigley, 1996). Elevated intraocular pressure (IOP) is one of the major risk factors for glaucoma, although high IOP is not necessary to cause glaucoma (Ritch et al., 1996). Additionally, high IOP by itself is not sufficient to cause glaucoma, as many individuals with high IOP do not develop glaucoma. Treatments for glaucoma currently center on lowering IOP

*Corresponding author. Tel.: +1 207 288 6476;
Fax: +1 207 288 6079; E-mail: simon.john@jax.org

levels, but do not prevent the development or progression of visual abnormalities in all patients (Gordon et al., 2002; Kass et al., 2002). To rationally develop better human treatments, it is vital that we better understand both pressure-dependent and pressure-independent mechanisms that contribute to glaucomatous RGC death.

The mouse as a model system

A variety of model systems are used to gain understanding into glaucoma pathology. Different species, which have their own unique mix of advantages and disadvantages, are used, including monkeys, dogs, rabbits, rats, and mice. The mouse is a powerful experimental tool for gaining understanding of the pathophysiological mechanisms underlying glaucoma (John et al., 1999; John, 2005). Along with humans, mice have one of the best characterized mammalian genomes. Mice are inexpensive to breed and are amenable to an array of genetic manipulations (see Peters et al., 2007). These properties allow the role of individual genes or molecular pathways to be assessed within a controlled genetic environment (John et al., 1999; Smithies and Maeda, 1995).

Mice are suitable models for studying IOP elevation in glaucoma

Detailed molecular understanding of aqueous humor (AqH) drainage and its abnormalities in glaucoma is an important goal for glaucoma research. This understanding will guide the development of new strategies to protect neurons from pressure-induced insults. Pressure-induced insults are key factors in the degeneration of RGCs in glaucoma. The AqH drainage structures play an important role in determining IOP and its elevation in glaucoma. The mouse is well suited for studying AqH drainage because it has similar AqH drainage structures to humans (Smith et al., 2001). The two types of outflow pathways present in humans, conventional and uveoscleral, exist in mice (see John et al., 1999; Lindsey and Weinreb, 2002). Both species have an endothelial-lined Schlemm's canal (SC) and a trabecular meshwork (TM) consisting of layers of well-organized trabecular beams covered with endothelial-like trabecular cells. These drainage structures reside in the angle of the eye (Smith et al., 2001; Gould et al., 2004b). The organizational similarity extends to drainage structure development and the genes that influence it (Gould et al., 2004b). The documented similarities between mice and humans in drainage structure anatomy, in functional responses to drugs that inhibit aqueous production and facilitate outflow and in values for various outflow parameters (Aihara et al., 2003; Avila et al., 2003), indicate that mice represent very suitable models for studying IOP and its glaucoma-associated elevation (John, 2005).

Tools for glaucoma research

The small size of the mouse eye initially hampered the use of the mouse as a glaucoma model. Various techniques of ocular analysis have been adapted for mice, including slit-lamp gonioscopy, indirect ophthalmoscopy, and IOP measurement (Smith et al., 2002). Because of its importance for all studies of glaucoma, we next discuss IOP assessment.

Accurate IOP measurements are fundamental to the study of glaucoma

In mice, the most accurate and precise measurements are achieved using invasive cannulation methods. In the microneedle system, a very fine fluid-filled glass microneedle is inserted into the anterior chamber. The needle is connected to a pressure transducer and the pressure reading is monitored on a computer (Savinova et al., 2001; John and Savinova, 2002). A second system available is an adaptation of the servo-null micropipette system (SNMS) that was developed for measuring pressure in structures smaller than 25 μm (Avila et al., 2001). Although cannulation methods have the disadvantage of penetrating the eye, limiting the number of reasonable longitudinal readings, large numbers of genetically identical mice of different ages can be evaluated. This provides an accurate record of IOP for any given population/strain.

Various tonometric devices have been adapted for use in mice based on devices for assessing IOP in humans (for recent review, see Pang and Clark, 2007). Only the optical interferometry tonometer (OIT) and impact/rebound (I/R) tonometer have been directly compared to the more accurate invasive method in the same eye in vivo (Filippopoulos et al., 2006; Morris et al., 2006) and so these are the best validated instruments. We, and others, cannot obtain accurate or reliable readings using the Tono-Pen tonometer (John et al., 1997; Reitsamer et al., 2004; Dalke et al., 2005; Pease et al., 2006).

It is clear that with appropriate care accurate noninvasive measurements are achievable in some mouse strains. However, inappropriate use in strains with corneal abnormalities or in strains for which accurate in vivo validation has not been performed can be very misleading (differing corneal properties between strains with outwardly normal corneas affect accuracy). For some of these strains that have relatively normal corneas, correction factors can be determined (Nissirios et al., 2007). The commonly used mouse model of glaucoma

(DBA/2J) has keratopathy at young ages and many mice develop corneal calcification as they age. These changes often render noninvasive measurements inaccurate. Therefore, we strongly recommend the use of invasive methods in the DBA/2J model. Currently, invasive methods are the only way to obtain accurate and reproducible IOP measurements in the majority of DBA/2J mice.

The future of IOP assessment

To truly understand the relationship between high IOP (duration, fluctuation, and magnitude) and glaucoma, new technologies are needed. An ideal system for measuring IOP would be capable of automated IOP assessment around the clock. A small implantable device that measures IOP over an extended period of time (ideally months or years) would revolutionize glaucoma research in mice and other animal models, and possibly even human patients.

Assessment of RGC function

Electrophysiological assessment of retinal function has been used to gain a better understanding of the function of many molecules in the retina, to understand disease pathogenesis, and to measure the efficacy of treatments for retinal disease in mice (Peachey and Ball, 2003; Pinto et al., 2007). Pang and Clark have reviewed the different electrophysiological tests and measures that have been used to assess retinal function in rodent models of glaucoma (Pang and Clark, 2007). Electroretinography (ERG) has been used extensively to assess retinal function after artificial elevation of IOP in rodents (Pang and Clark, 2007). The pattern electroretinogram (PERG) is a specialized kind of ERG that isolates inner retinal activity, including a large component that is generated by RGCs (Holder, 2004). In humans, the PERG appears to be an effective tool for assessing glaucomatous damage (Bach, 2001; Ventura et al., 2006). Recently, Porciatti and colleagues have refined the PERG for use in mice (Nagaraju et al., 2007; Porciatti, 2007; Porciatti et al., 2007; Saleh et al., 2007). The mouse retinal activity detected by PERG is greatly reduced early in glaucoma and thus PERG is a useful tool for assessing the efficacy of neuroprotection (Howell et al., 2007). PERG and other ERG approaches will likely prove valuable tools for assessing neuroprotective therapies as well as for helping to understand early pathophysiological events in glaucomatous neurodegeneration.

Mouse models of glaucoma

This section does not provide an exhaustive list of all mouse glaucoma models, as other reviews are available (John et al., 1999; Goldblum and Mittag, 2002; Gould et al., 2004b; John, 2005; Libby et al., 2005a; Lindsey and Weinreb, 2005; Weinreb and Lindsey, 2005; Pang and Clark, 2007). Mouse models relevant to high IOP and glaucoma can be divided into two classes, those that are inherited and those that are experimentally induced. Although experimentally induced models are valuable, inherited models are likely to more accurately model human glaucomas. We now discuss available inherited mouse models in the context of different types of human glaucoma.

Primary open-angle glaucoma

Primary open-angle glaucoma (POAG) is the most common clinically defined subtype of glaucoma. In POAG, the angle and drainage routes are clinically observed to be unimpeded. There are at least 20 loci implicated in initiating POAG and only a few genes have been identified to date (Fan et al., 2006). The pathological mechanisms of POAG remain poorly defined. Studies in mice are helping to unravel mechanisms of POAG pathogenesis, but the challenge of developing mouse models of POAG remains.

MYOC

Approximately 3–5% of late-onset POAG patients and up to 30% of juvenile open-angle glaucoma patients (an earlier and more severe form of POAG) are caused by mutations in the myocilin gene (*MYOC*) (Wiggs et al., 1998; Fingert et al., 1999). Interestingly, patients with null or early truncation alleles of *MYOC* do not develop high IOP and POAG (Lam et al., 2000; Wiggs and Vollrath, 2001), and *Myoc* null mice of different genetic backgrounds do not develop high IOP (Kim et al., 2001). This suggests that mutant MYOC causes glaucoma through a gain of function mechanism. Supporting this, overexpressing *Myoc* in mice did not alter IOP (Gould et al., 2004a; Kim et al., 2001). Surprisingly, introducing a Y423H mutation (analogous to a severe human mutation, Y437H) into the mouse *Myoc* gene did not cause elevated IOP or glaucomatous neurodegeneration (Gould et al., 2006). Similarly, although open-angle glaucoma is reported in a separate study that introduced the same mutation into mice, IOP was elevated by only 2 mmHg in the mutant mice (Senatorov et al., 2006). This magnitude of IOP elevation seems highly unlikely to cause glaucomatous degeneration as the mean value remained well below that which normally causes glaucoma in mice. The inability of the mutant mouse *Myoc* gene to substantially elevate IOP may suggest that the response of mouse drainage structure cells to mutant MYOC is different to that of humans. Alternatively, a key difference between the mouse and human genes and encoded proteins may determine whether or not mutant MYOC is pathogenic.

In support of an important difference between the mouse and human proteins, virally expressing the human *MYOC* gene with the analogous Y437H mutation in mice resulted in substantial IOP elevation (Shepard et al., 2007). In the same study, viral expression of the same mutation in a version of the human gene that did not contain the peroxisomal targeting signal type 1 receptor (PTS1R) binding sequence did not elevate IOP (Shepard et al., 2007) (Fig. 1). The authors suggest that mutations in human *MYOC* gene expose a cryptic PTS1R binding site that is ordinarily buried within the normal protein. Exposing this site causes mislocalization of mutant MYOC protein to peroxisomes. Importantly, the cryptic PTS1R binding site does not exist in the mouse MYOC protein and so mutations in mouse *Myoc* gene cannot induce targeting of mutant proteins to peroxisomes. The inability of human and mouse proteins that lack the cryptic PTS1R sequence to elevate IOP, despite intracellular accumulation of mutant proteins (Gould et al., 2006), suggests that peroxisomal targeting is a necessary component of pathogenesis.

Although not yet assessed, the model of virally expressing mutant MYOC to elevate IOP may provide a valuable and convenient model of glaucomatous neurodegeneration. If further studies show that this viral model does induce neurodegeneration, then it would be relatively easy to induce high IOP in mice with different genetic backgrounds. This approach has great potential for identifying genes that modify susceptibility to neurodegeneration. In addition, it will be important to develop a transgenic model expressing the mutant human *MYOC* gene. This will allow assessment of disease pathology without the complications and potential variability of viral infection. Transgenically humanizing the mouse would be likely to more closely model the human condition and may allow more convenient mechanistic studies.

OPTN

The optineurin (*OPTN*) gene is implicated in POAG, including patients without significant IOP elevation (normal-tension glaucoma, NTG). NTG may result from direct insults within the optic nerve and retina that are not pressure-induced. Although it is not clear if all reported *OPTN* mutations cause POAG (Alward et al., 2003; Libby et al., 2005a), the E50K mutation is clearly important (Rezaie et al., 2002; Sarfarazi and Rezaie, 2003). OPTN is expressed in RGCs (De Marco et al., 2006) and in astrocytes of the optic nerve (Obazawa et al., 2004). Overexpression of OPTN appears to have a protective effect on RGCs that are exposed to noxious stimuli (De Marco et al., 2006). In response to apoptotic stimuli, OPTN ordinarily translocates from the Golgi to the nucleus. This suggests OPTN translocation to the nucleus protects from apoptosis. The E50K mutant OPTN no longer relocates to the nucleus (De Marco et al., 2006) and this may render RGCs more susceptible to apoptotic death and glaucoma. Although some mouse strains have *Optn* mutations (Libby et al., 2005a), mice with the

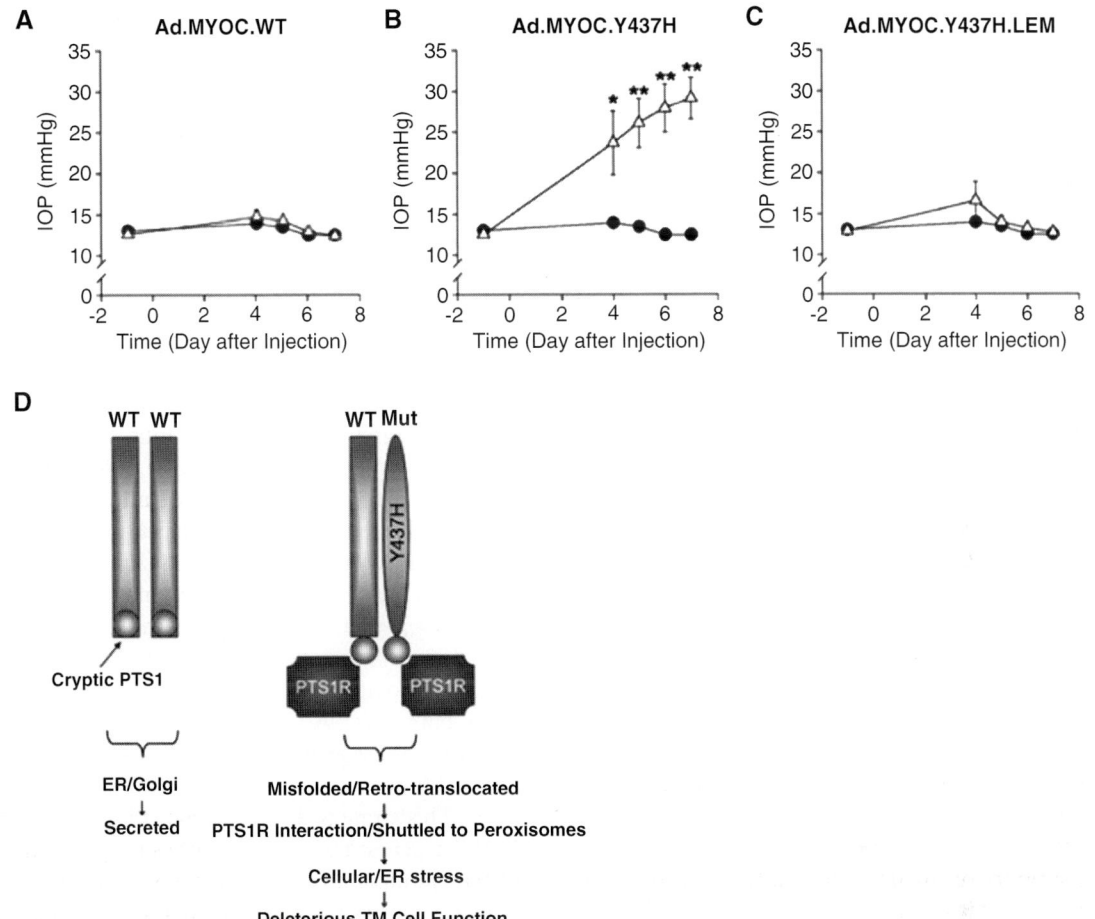

Fig. 1. Glaucoma-causing MYOC mutants require PTS1R to elevate IOP. (A) Virally expressing the wild-type form of the human *MYOC* gene (Ad.MYOC.WT) had no affect on IOP. (B) Virally expressing the human *MYOC* gene with the Y437H mutation (Ad.MYOC.Y437H) resulted in substantial IOP elevation. (C) Expressing the same mutant gene as in (B) but after disruption of the PTS1R site (Ad.MYOC.Y437H.LEM) did not elevate IOP. (D) This suggests that mutations in the human *MYOC* gene lead to a protein misfolding and exposure of an ordinarily cryptic PTS1R site. With an exposed PTS1R motif, mutant MYOC proteins can bind to PTSR1 and be shuttled to the peroxisome, a necessary component of pathogenesis. Adapted with permission from Shepard et al. (2007).

E50K mutation are not reported. Development of a mouse model either with the E50K mutation knocked into the mouse *Optn* locus or with the mutant human gene will be important for investigating the role of this gene in POAG.

WDR36

The WD40 repeat domain 36 gene (*WDR36*) was recently implicated in POAG (Monemi et al., 2005). Mutations in WDR36 do not always cause glaucoma, and current data suggest that this gene primarily modifies the severity of glaucoma induced by other glaucoma genes (Hauser et al., 2006). Currently, no mouse models with mutations in the *WDR36* gene exist.

Strategies for developing new models of POAG

As discussed above, mice are useful for modeling the effects of POAG genes, but there is still a need for POAG models with high IOP and glaucomatous neurodegeneration. A number of strategies are being used to develop these models. In addition to modeling the effects of mutations in the *MYOC*, *OPTN*, and *WDR36* genes alone, mice can be used to model and understand the combined effects of these mutant genes. Mice with combinations of these mutations may also provide valuable models of glaucomatous neurodegeneration.

A complementary strategy for producing POAG models is to alter genes whose known function suggests that they may cause glaucoma when mutant. The more we understand about the mechanisms of POAG, the smarter we can be in predicting candidate genes/pathways that may be worth perturbing to produce additional models. Candidate genes can be altered so that they either produce no proteins or produce specific mutant proteins. As discussed for myocilin, disease pathogenesis can require specific mutant proteins and even the presence of mutations in the human protein. A pitfall of the candidate gene approach is that it requires knowledge of gene function and depends on a priori assumptions about pathogenic mechanisms. Since the mechanisms are poorly understood, the candidate gene approach is often unsuccessful.

Another valuable strategy is to induce novel mutations by random mutagenesis of the mouse genome. Commonly, the genomes of founder males are mutagenized with chemical agents such as ethyl nitrosourea (ENU) and breeding and screening strategies are developed to uncover phenotypes of interest in mutant offspring (John et al., 1999; Thaung et al., 2002; O'Brien and Frankel, 2004). An ENU mutagenesis is also underway in our laboratory and is producing a series of new mouse lines with POAG relevant phenotypes. This approach requires no previous assumptions or mechanistic knowledge to identify glaucoma genes and is a powerful tool for producing new models.

Developmental glaucoma

Developmental glaucomas are caused by genes involved in ocular development. Many developmental glaucomas involve obvious dysgenesis of readily visible anterior chamber structures (anterior segment dysgenesis, ASD) such as the iris and pupil. In others (primary congenital glaucoma, PCG), the defects are subtle involving abnormal development of SC and TM drainage structures.

The mouse provides an important model system for deciphering the molecular cascades involved in ASD and developmental glaucoma (Gould and John, 2002). A number of mutant genes cause ASD (reviewed in Gould et al., 2004b), and some were first discovered to do so in mice. For example, mutation of a basement membrane collagen gene, *Col4a1*, was recently shown to cause severe ASD in mice (Van Agtmael et al., 2005; Gould et al., 2007). Soon afterwards, Sibon and colleagues found that mutations in the human orthologue of *Col4a1* are associated with Axenfeld-Rieger syndrome, a form of ASD that is often associated with glaucoma (Sibon et al., 2007).

Although some mutant genes that cause ASD-associated phenotypes also induced high IOP in mice (e.g., *Col4a1* and *Bmp4*), some have not been shown to do so. Even in cases without high IOP, mouse mutants provide valuable models for investigating ASD pathways. Additionally, breeding mutations that are not associated with high IOP into different mouse strains can uncover new high IOP phenotypes. This strategy also enables the identification of modifier genes and pathways that interact with the known mutant genes to determine whether or not IOP becomes elevated (Gould and John, 2002). The mouse is a powerful mammalian model for identifying modifier genes and unraveling complex genetic interactions. Recently, we found that genetic deficiency of tyrosinase increased the severity of the pathology of *Foxc1* and *Cyp1b1* mutations in mice (Libby et al., 2003), and that a pathway involving l-DOPA can be targeted with therapeutic benefit.

In addition to allowing studies of IOP elevation, ASD models may also provide new and early onset models of glaucomatous neurodegeneration. Although no robust models of glaucomatous neurodegeneration are reported, there are promising leads. For example, *Col4A1* mutants are reported to suffer optic nerve excavation (Van Agtmael et al., 2005) and we have confirmed this on some genetic backgrounds (unpublished data). However, further studies are needed to determine glaucoma frequency and whether or not these mutants will become a tractable model. On some but not all genetic backgrounds, *Col4A1* and *Bmp4* mutations result in developmental abnormalities of the optic nerve, which complicates their use as glaucoma models but provides a valuable opportunity to study genes that impact the health and development of RGCs. Again, mice will allow the identification of important modifier genes and the characterization of genetic interactions that modulate susceptibility to glaucoma.

Pigmentary glaucoma

Pigmentary glaucoma (PG) results secondarily to the dispersion of iris pigment into the anterior chamber of the eye. Dispersed pigment enters the drainage structures, causing damage that leads to IOP elevation and glaucomatous neurodegeneration. Forms of PG are reported in DBA/2J, DBA/2Nnia, and AKXD-28/TyJ mice (Sheldon et al., 1995; John et al., 1998; Anderson et al., 2001). DBA/2J mice have mutations in two genes that induce the pigment dispersion, the *b* mutation in the tyrosinase-related protein 1 gene ($Tyrp1^b$) and a stop-codon mutation in the glycoprotein (transmembrane) *nmb* gene ($Gpnmb^{R150X}$) (Chang et al., 1999; Anderson et al., 2002). The pigment dispersion is a consequence of both melanosomal toxicity and abnormal ocular immunity (Anderson et al., 2002; Mo et al., 2003; John, 2005). The glaucoma in DBA/2J mice has hallmarks of human glaucoma, including an age-related variable progression of optic nerve atrophy in response to elevated IOP, a regional pattern of RGC death, and optic nerve excavation. Importantly, at least in our colony, damage appears limited to RGCs and the high pressure results in direct axon damage within the lamina of the optic nerve (Danias et al., 2003; Jakobs et al., 2005; Schlamp et al., 2006; Howell et al., 2007), as discussed below. DBA/2J mice have become the most widely used mouse model to decipher mechanisms of glaucomatous neurodegeneration and for developing new neuroprotective strategies.

Experimentally induced models of glaucoma

Experimentally induced models have the advantage that IOP can be elevated conveniently and experiments conducted over a short time frame. IOP can be artificially elevated by damaging the drainage structures or the blood vessels into which they drain (reviewed in Pang and Clark, 2007). Evaluation of retinas following sustained IOP elevation (4–12 weeks) has indicated increased RGC apoptosis, decreased optic nerve cross-sectional area and axonal density (Gross et al., 2003; Grozdanic et al., 2003), preferential loss of superior axons in the optic nerve (Mabuchi et al., 2004), and sustained ERG deficits (Grozdanic et al., 2003).

Although these induced models are valuable and can provide important insights (Nakazawa et al., 2006), there is still phenotypic variability and not all groups have been able to reproduce the procedures. Additionally, due to the sudden nature of IOP elevation following induced ocular trauma, there may be differences in the neurodegenerative and remodeling mechanisms compared to the naturally occurring inherited glaucomas. Optic nerve excavation – a hallmark of glaucoma – has not been reported for any of these models. For each of these models, it is not clear if the lack of reported optic nerve excavation is due to differences in disease mechanisms, due to the strain backgrounds used, or simply because detailed optic nerve head evaluation is needed.

In addition to pressure-induced neurodegeneration, direct neuronal injury can provide important new information. Optic nerve crush is an experimentally induced model for direct optic nerve and axon injury, and apoptotic RGC death (Li et al., 1999). RGC loss after controlled optic nerve crush occurs over 3 weeks (Li et al., 1999). Although crush provides a robust and rapid system for evaluating potential role(s) of individual genes in RGC death and optic nerve degeneration, it is a more severe insult than glaucoma. There are undoubtedly important common pathogenic mechanisms between RGC death in crush and in glaucoma, but there are also differences. For example, a radiation treatment prevents glaucomatous RGC death and associated optic nerve degeneration, but does not appear to protect against crush-induced damage in DBA/2J mice (Anderson et al., 2005 and unpublished observations). Similarly, cleavage of the autoinhibitory domain of the protein phosphatase calcineurin

(CaN), which promotes apoptosis, occurs in two rodent models of glaucoma (induced rat model and DBA/2J mice), but not after optic nerve crush (Huang et al., 2005). Finally, heterozygosity for a *Bax* null allele protects RGC somas from apoptosis after a glaucomatous insult that causes axonal degeneration, but not after crush (Libby et al., 2005b).

Recently, Yang and colleagues performed a microarray analysis between RGC death induced by optic nerve transection and pressure-induced neurodegeneration (translimbal laser photocoagulation) (Yang et al., 2007). By comparing gene expression changes following crush and during experimental glaucoma and then filtering out genes that changed in both models, they report gene expression changes that are suggested to be specific to pressure-induced injury (Yang et al., 2007).

Mouse models to characterize processes involved in glaucomatous neurodegeneration

The death of RGCs and the associated degeneration of the optic nerve are unifying features of glaucoma, but the underlying mechanisms are poorly understood. In recent years, the DBA/2J strain has become the most widely used inherited mouse model to study glaucomatous neurodegeneration. In this section, we describe how the DBA/2J model (and the related strain, DBA/2NNia, collectively referred to as DBA/2) is being used to understand mechanisms involved in neurodegeneration during glaucoma.

Similar patterns of glaucomatous damage occur in humans and mice

A characteristic feature of human glaucoma is the occurrence of focal visual defects due to region-specific loss or impairment of RGCs. The most consistent regional defects are arcuate scotomas, which are detected by visual field tests (Shields, 1997). The arcuate nerve fibers originate in the temporal region of the retina and arch above or below the fovea to the optic nerve head. In the mouse, the RGC axons do not curve across the retinal surface but radiate straight toward the optic nerve. Considering this, patterns of regional damage equivalent to that in human patients occur in mice. In DBA/2 mice, RGCs and their axons are lost in "fan-shaped" or "patchy" regions (Danias et al., 2003; Jakobs et al., 2005; Schlamp et al., 2006). These fan-shaped regions of RGC loss are likely to be analogous to the arcuate scotomas seen in human glaucoma. This regional pattern of RGC loss during glaucoma is likely to result from focal damage to discrete "bundles" of axons within the lamina.

The lamina cribrosa is an important site of early glaucomatous damage

Important studies in humans and primates established that early glaucomatous damage affects the RGC axon segments within the lamina cribrosa (Anderson and Hendrickson, 1974, 1977; Quigley and Anderson, 1977; Quigley et al., 1979, 1980, 1981, 1983; Quigley and Addicks, 1980, 1981). In humans and primates, the lamina cribrosa is composed of plates of extracellular matrix (ECM) that provide support for the axons as they pass through the posterior wall of the eye (Fig. 2a). It was hypothesized that in response to elevated IOP, bowing of the ECM plates would damage axon bundles by mechanical stress. Mechanical distortion of the ECM plates has also been suggested to contribute to glaucoma by damaging blood vessels (Quigley and Addicks, 1981; Maumenee, 1983; Fechtner and Weinreb, 1994). The ECM plates are covered by astrocytes that provide neurotrophic and other forms of support to the neurons.

Historically, the mouse was not regarded as a useful model for glaucoma because it was reported to lack a lamina cribrosa (Tansley, 1956; Fujita et al., 2000; Morcos and Chan-Ling, 2000; May and Lutjen-Drecoll, 2002). However, the mouse does have an astrocyte-rich structure in the same position as the human lamina cribrosa (Xie et al., 2005; Petros et al., 2006; Schlamp et al., 2006; Howell et al., 2007). The astrocytes of this region form an enmeshing network of glial cells through which the RGC axons pass, and they are intimately associated with the axons. There is no evidence of collagenous ECM plates in this region of the mouse optic nerve. To reflect the equivalent location compared to the human lamina cribrosa,

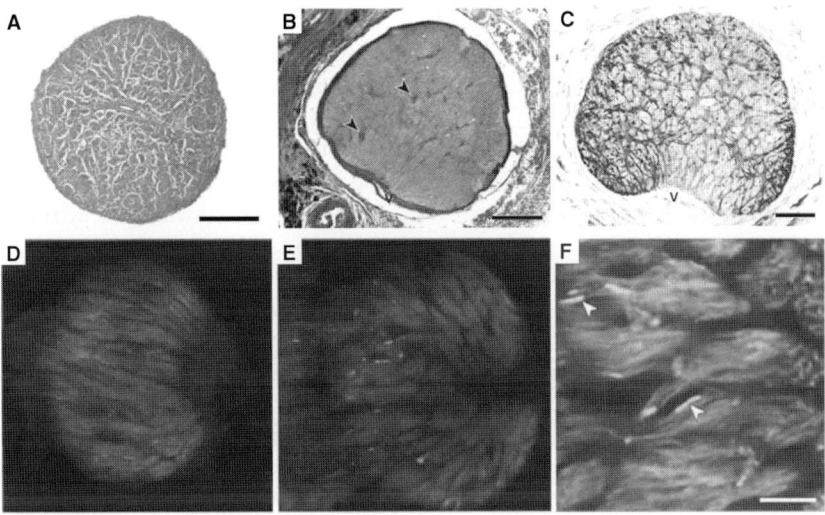

Fig. 2. (A–C) The lamina cribrosa has robust ECM plates (A, shown in blue, Masson's Trichrome) through which bundles of axons pass. In contrast, the mouse glial lamina has no collagenous plates (B), although collagens are clearly visible in blood vessel walls (arrowheads). Similar to the human lamina, the mouse glial lamina has an extensive meshwork of astrocytes (C, stained positive for GFAP). (D–F) Early focal damage at the glial lamina is visualized using DBA/2J.Thy1-CFP mice. In these mice, RGC axons appear green when viewed by confocal microscopy. Because axonal contents accumulate in regions of damage, early axon damage is evident as very brightly fluorescent axon segments (arrows). (D) No damage is seen in preglaucomatous young eyes. (E) Obvious axonal swellings were evident specifically in the lamina of eyes that were at early stages of glaucoma. (F) Focal regions of damage are clearly visible as bright, slightly swollen axonal regions (arrowheads). Scale bars: (A) 1 mm; (B, C) 50 μm; (D–F) 20 μm. (A) Adapted with permission from Karim et al. (2004). (B–F) Adapted with permission from Howell et al. (2007). (See Color Plate 22.2 in color plate section.)

the similar arrangement of glial cells, but the absence of ECM plates, this region of the mouse optic nerve has been termed the glial lamina (Howell et al., 2007).

An insult occurs to the axons of RGCs within the lamina in glaucoma

The mechanisms by which RGC somata die are different from those involved in axon degeneration. This was shown by genetically ablating the function of BAX, a proapoptotic molecule, in DBA/2J mice. BAX deficiency prevented the death of essentially all RGC soma, but the axons of these mice were still degenerated (Libby et al., 2005b). Therefore, in these mice at least, RGC soma death is a BAX-dependent apoptotic process, whereas axon degeneration is a BAX-independent process.

Similar to studies in other species (see above), we determined that the first sign of axon damage occurs within the lamina in DBA/2J mice (Fig. 2) (Howell et al., 2007). Demonstrating that the first signs of axon damage occur within the lamina is not proof that axons are insulted within the lamina. In the general case, it is well established that the first site of neuronal damage may be remote from the site of insult (reviewed in Conforti et al., 2007). For example, in transected motor axons, neuromuscular junctions that are many centimeters from the lesion degenerate first. The axons immediately adjacent to the lesion remain intact for two to three times longer than the distant terminals (Beirowski et al., 2005). We have demonstrated that RGC axons survive up to, but not into, the lamina in BAX-deficient mice that retain all of their RGC soma. This provides strong experimental evidence supporting a direct insult to RGC axons within the lamina during glaucoma (Guo et al., 2007; Howell et al., 2007). Given that DBA/2J mice develop glaucoma with regional cell death and characteristic optic nerve excavation, the lack of ECM plates in the lamina suggests mechanical distortion of ECM plates is not necessary to damage axons within the lamina during glaucoma. Although it remains possible that the ECM plates may modulate damage in human glaucoma (as the nerve is much larger), these findings strongly focus attention on other components of the lamina.

What is the nature of the insult at the lamina?

One current challenge is to identify mechanism(s) initiating axon damage at the lamina and not just the processes involved in the propagation of axon degeneration. This is of particular importance when considering improved human therapies. It is possible that the initiating events are either intrinsic to the RGC itself or extrinsic involving other cell types. Other important cell types that are present in the optic nerve are glial cells (such as astrocytes, microglia, and oligodendrocytes). These cell types ordinarily play important roles in maintaining healthy, fully functioning RGCs, but in response to stressful conditions can become harmful. Examples of pathways/molecules known to be active in these different cell types during glaucomatous injury are given below to highlight the possible importance of all of these cell types in glaucomatous neurodegeneration.

Intrinsic changes in RGCs leading to axon degeneration in glaucoma are reported. An interesting line of research relates to amyloid precursor protein (APP) and amyloid-β (Aβ). Mutations in APP cause Alzheimer's disease (AD), and recent studies imply that there is a significantly higher incidence of glaucoma among patients with AD (Bayer et al., 2002). The same abnormal Aβ peptide found in AD was found in AqH of 40% of the assessed glaucoma patients. In addition, in a rat model of glaucoma, caspase-3, a major activator of the apoptotic cascade, is activated in RGCs and cleaves APP to produce neurotoxic fragments that include Aβ (McKinnon et al., 2002; McKinnon, 2003). Supporting this, Goldblum and colleagues have shown that APP and Aβ increase in the RGC layer, lamina, and pia/dura layer in aged DBA/2 mice (15 months old) compared to young DBA/2J and C57BL/6J controls (Goldblum et al., 2007). They suggest a disruption of the homeostatic properties of secreted APP with consecutive Aβ cytotoxicity as a contributing factor to RGC loss in glaucoma. This would suggest that glaucoma and AD share common features, and suggests mechanisms identified as important in one disease should be investigated in the other. A promising study suggests that targeting different components of Aβ formation and aggregation pathway can effectively reduce glaucomatous RGC apoptosis as shown in an experimental glaucoma in rats (Guo et al., 2007).

Astrocytes form a cellular network in the glial lamina in the mouse. It has been suggested that IOP elevation can alter astrocytes so that they damage RGC axons (Hernandez, 2000). Astrocytes become reactive in glaucoma. An increase in glial fibrillary acidic protein (GFAP) is considered a hallmark of reactive astrocytes (Pekny and Nilsson, 2005). GFAP is upregulated in experimental models of glaucoma in primates and rats (Tanihara et al., 1997; Wang et al., 2000), as well as in human glaucomatous eyes (Tezel et al., 2003). Two independent microarray experiments, the first in an experimentally induced rat model, the second using DBA/2J mice, showed increases in the expression of astrocyte-related genes in response to elevated IOP (Ahmed et al., 2004; Steele et al., 2006), and we have made similar observations (unpublished data).

Microglia are another cell type present in the optic nerve and may contribute to glaucoma (Tezel and Wax, 2004; May and Mittag, 2006; Nakazawa et al., 2006). Microglia increase in numbers as glaucoma progresses, and this is true in DBA/2J mice (Inman and Horner, 2007). It is possible that microglia are necessary to initiate or propagate damage in the retina or optic nerve. Individual microglia may contribute to highly local insults within specific regions of the lamina and could conceivably underlie the fan-shaped patterns of cell death.

Oligodendrocytes were recently suggested to participate in glaucoma (Nakazawa et al., 2006). Oligodendrocytes are numerous in the optic nerve from the start of the myelinated portion (approximately 100 μm behind the glial lamina in mice). Axons within the glial lamina are unmyelinated and no myelin-producing oligodendrocytes have been shown to be present within the glial lamina. Based on findings using an induced model of high IOP, an intriguing model involving TNFα and oligodendrocyte death was suggested to damage RGCs in

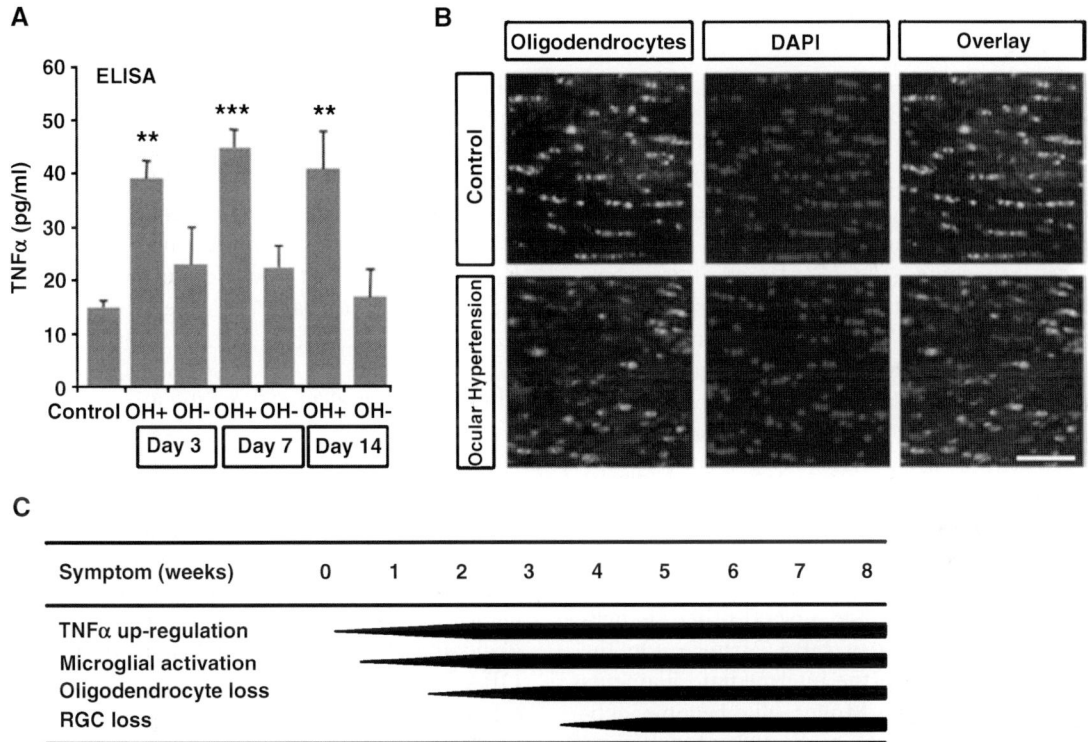

Fig. 3. Tumor necrosis factor α (TNFα) mediates oligodendrocyte loss in an experimental mouse model of glaucoma. Ocular hypertension (OH) was induced by angle closure via laser irradiation. (A) The treatment resulted in a rapid upregulation of TNFα. (B) Eight weeks after OH induction, a significant number of oligodendrocytes are lost. (C) Timeline of events in these experiments. The authors show that microglia can mediate the cytotoxic effects of TNFα. This study controlled for pressure-independent effects of the laser treatment both by using mice in which induction of OH was not successful and by using mice with laser treatment of the iris. Adapted with permission from Nakazawa et al. (2006).

glaucoma (Fig. 3) (Nakazawa et al., 2006). It is important to further evaluate and understand any pathogenic role of oligodendrocytes in glaucoma.

Other changes occur in the retina in glaucoma

PERG and complement

In addition to the optic nerve, changes within the retina must be considered in glaucoma. Recent studies show that PERG is greatly reduced early in DBA/2J glaucoma (Howell et al., 2007). Although the mechanisms of PERG attenuation are not known, they may involve direct changes within the retina. The complement pathway has been implicated in glaucoma and C1q, the initiating protein in the classical complement cascade, is upregulated in aged DBA/2J mice compared to young controls (Stasi et al., 2006; Steele et al., 2006). In an extensive microarray study, we have shown that a variety of complement genes, including C1q and C3, are upregulated early in DBA/2J glaucoma prior to significant RGC death or axon degeneration (Howell, Libby, and John, in preparation). We have also shown that C1q is localized to RGC synapses in the IPL early in glaucoma (Stevens et al., 2007). Mice deficient in C1q or C3 exhibit large sustained defects in synapse elimination during retinal development (Stevens et al., 2007). We have experiments underway to functionally test the role of the complement system in glaucoma, with DBA/2J mice deficient in C1q, C3, or C3R.

Dendrites and neurofilament accumulation

Early dendrite changes have been found in primate glaucoma (Weber et al., 1998). In glaucomatous mice, some surviving RGCs also have abnormal dendrite morphology (Jakobs et al., 2005). These RGCs show a loss of second and higher order dendrites with only a few and short third-order dendrites remaining (Fig. 4). In addition, some RGCs show the accumulation of nonphosphorylated neurofilament. These cells appear to represent a "remodeled" stable state and persist for some time prior to cell death. Similar cells would be potential targets for therapeutic rescue in human glaucoma.

Müller glia and ceruloplasmin

Although further experiments are needed to determine their importance, several other changes are reported in retinas of DBA/2J mice. On the basis of the increase in the intermediate filaments GFAP and Vimentin, Müller glia become activated in glaucoma (Xue et al., 2006; Inman and Horner, 2007). Ceruloplasmin (Cp) expression (both protein and mRNA) increases locally in Müller glia cells and in their end feet in the inner limiting membrane in glaucomatous DBA/2J eyes. Cp expression also increases in human and primate glaucomas (Stasi et al., 2007). The authors suggest Cp, a known stress response protein (Markowitz et al., 1955), is likely upregulated in response to IOP elevation, but the significance of this for glaucoma is not clear.

Müller glia and heat shock proteins

Heat shock proteins (Hsps) are upregulated in glaucoma (Park et al., 2001; Sakai et al., 2003; Tezel et al., 2004; Qing et al., 2005). For example, HSP27, particularly an activated, phosphorylated form, is upregulated in Müller glia cells in human glaucoma (Tezel et al., 2000), a rat model of glaucoma (Park et al., 2001), and more recently, DBA/2J glaucoma (Huang et al., 2007). HSP27 is a low-molecular-weight heat shock protein shown to have protective properties in response to

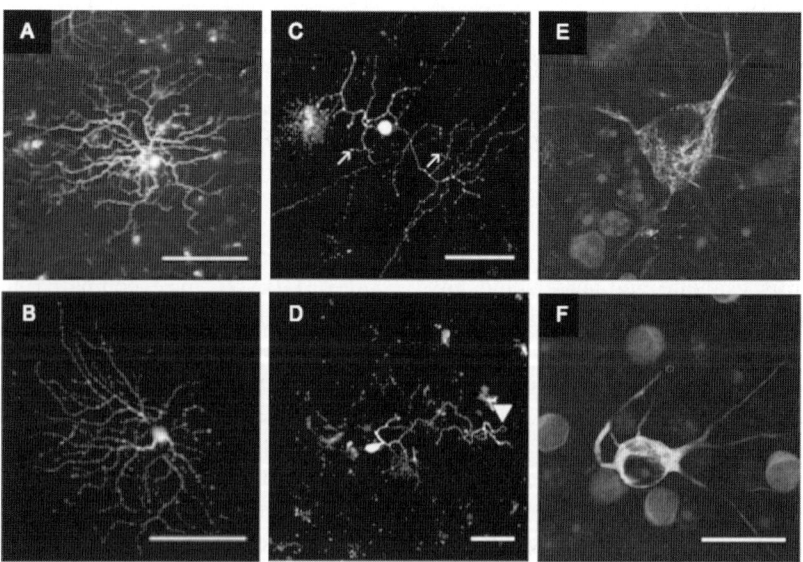

Fig. 4. Abnormal dendrite morphology in surviving RGCs from glaucomatous DBA/2J mice. (A, B) Normal dendritic arbors of healthy RGCs. (C, D) In contrast, only a few and short third-order dendrites are visible (arrows), and there are sometimes long and tortuous dendrites (arrowheads) in remodeled RGCs. (E, F) As a neurofilament marker, SMI32 (green) visualizes the cytoskeleton, which appears as a web of filaments around the cell nucleus in normal RGCs (E). In glaucomatous retinas, neurofilaments of affected RGCs stain brightly and appear condensed tightly around the nucleus (F). ChAT staining (red), and nuclei staining (blue). Scale bars: (A–D) 100 µm; (E, F) 20 µm. Adapted with permission from Jakobs et al. (2005).

apoptotic stimuli (Martin et al., 1999). The activation of Müller glia and the activation of heat shock proteins are suggested to be part of an initial protective mechanism in response to elevated IOP, one that ultimately fails in glaucoma. Understanding intrinsic protective mechanism(s) that may be occurring prior to RGC death in glaucoma models could be an important step toward developing neuroprotective strategies for human glaucomas.

Using mouse models to develop neuroprotective strategies

As described above, it is becoming evident that multiple processes are involved in the degeneration of RGCs in glaucoma. This is important when considering neuroprotective strategies. It is unlikely that targeting a single process will prevent glaucoma in either mice or humans. Combinatorial therapeutic approaches that target multiple pathogenic processes are most likely to prove the most beneficial to human patients. It is important to note that treatment regimens that are protective in one setting may be harmful in another. Various cell types have been suggested to be protective in one setting but harmful in others (Tezel and Wax, 2004).

Somal protection

The profound protective effect of BAX deficiency (saves all RGC soma) in DBA/2J mice (Libby et al., 2005b) raises the possibility that BAX inhibitors may have a similar protective effect in humans. One can imagine that such treatments might block or delay RGC soma death until regenerative treatments can be developed to regrow axons. Alternatively, combination therapies may be developed involving both BAX inhibitors and drugs that target axon degeneration. These ideas can be readily tested using mice.

Axonal protection

We recently showed that the slow Wallerian degeneration allele (Wld^S) (Lunn et al., 1989; Mack et al., 2001) protects from axon degeneration in DBA/2J mice (Howell et al., 2007). Wld^S more than doubled the number of eyes with no detectable glaucoma compared to standard DBA/2J mice and preserved PERG. Combining Wld^S with BAX deficiency tended to be more protective than Wld^S alone, but the difference was not statistically significant. Wld^S encodes a fusion protein containing 70 N-terminal amino acids of ubiquitination factor $Ube4b$ linked to full-length nicotinamide mononucleotide adenylyltransferase 1 ($Nmnat1$) (Mack et al., 2001). The mechanisms by which Wld^S protects neurons are not fully understood (Araki et al., 2004; Laser et al., 2006). It will be important to understand these process(es) and to develop neuroprotective strategies that mimic the Wld^S protection.

Erythropoietin administration

Erythropoietin (EPO) has neuroprotective properties in animal models of stroke, excitotoxic injury, and neuroinflammation (reviewed in Ehrenreich et al., 2004). EPO is thought to inhibit apoptosis because antiapoptotic genes such as $Bcl2$ have been shown to be upregulated upon EPO application (Sattler et al., 2004). EPO has a direct protective effect on RGCs in culture (Becerra and Amaral, 2002; Bocker-Meffert et al., 2002; Weishaupt et al., 2004; Yamasaki et al., 2005). Systemic EPO administration to DBA/2J mice from 4 months of age appears to prevent significant RGC death and axon degeneration by 12 months of age (Zhong et al., 2007) (Fig. 5). Based on the Zhong study, the protective effect of EPO against glaucoma is very promising and further studies evaluating its therapeutic potential are eagerly awaited. Interestingly, EPO appeared to reduce axon degeneration as well as RGC apoptosis (Zhong et al., 2007), whereas BAX deficiency prevented RGC apoptosis but axon degeneration still occurs (Libby et al., 2005b).

Radiation-based treatment

A radiation-based treatment completely prevents glaucomatous damage in DBA/2J mice. This treatment, discovered serendipitously, involves administering 1000 rads of radiation to mice in

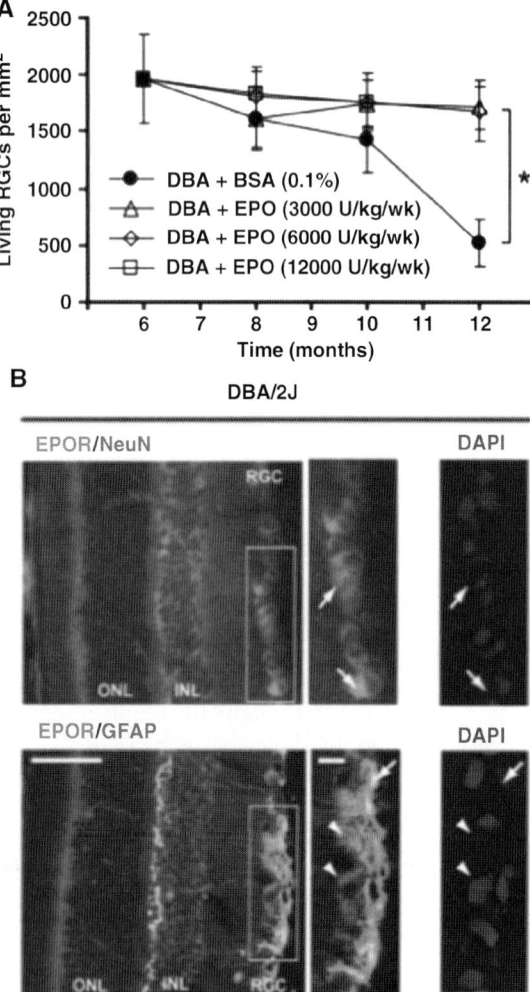

Fig. 5. (A) EPO protects from age-dependent death of RGCs in DBA/2J mice. (A) EPO profoundly protected RGC layer cells from glaucoma in DBA/2J mice. These results are convincing as the authors successfully analyzed at least 24 eyes (from 12 mice) in each treatment group (172 eyes total, personal communication, Ng). EPO was administered intraperitoneally three times per week for 2, 4, or 6 months. Compared with untreated control mice, which lost >70% of total RGCs, DBA/2J mice treated with different concentrations of EPO show no significant RGC loss up to 12 months of age. (B) Localization of EPOR expression in the retina of DBA/2J mice immunostained with antibodies to EPOR; NeuN, a marker for neurons; and GFAP, a marker for astrocytes. Top: In the DBA/2J mice, EPOR expression was colocalized with most of the NeuN-positive cells in the RGC layer (yellow, arrows). Bottom: most of the EPOR-positive (green, arrowheads) cells are located below the GFAP-positive (red) astrocytes in the RGC layer, and a few astrocytes have detectable EPOR expression (yellow, arrow). Scale bars, 100 μm; insets: 10 μm. Adapted with permission from Zhong et al. (2007).

combination with a syngeneic bone marrow transplant (Anderson et al., 2005). This treatment profoundly protects from glaucoma in the vast majority of treated animals but does not alter IOP. Importantly, this treatment is given to young mice and remains protective until old age. This protection appears to extend to humans because the incidence of glaucoma is reported to be lower in populations exposed to radiation (such as atomic bomb survivors, Yamada et al., 2004; and see Anderson et al., 2005). At this time, the mechanism(s) of neuroprotection are not known, and in its current form, the treatment is not directly transferable to humans. Further studies are needed to understand the process(es) involved in this radiation-induced neuroprotection, but there is great potential for the development of powerful new neuroprotective therapies for those at risk of glaucomatous visual loss.

References

Ahmed, F., Brown, K.M., Stephan, D.A., Morrison, J.C., Johnson, E.C. and Tomarev, S.I. (2004) Microarray analysis of changes in mRNA levels in the rat retina after experimental elevation of intraocular pressure. Invest. Ophthalmol. Vis. Sci., 45: 1247–1258.

Aihara, M., Lindsey, J.D. and Weinreb, R.N. (2003) Experimental mouse ocular hypertension: establishment of the model. Invest. Ophthalmol. Vis. Sci., 44: 4314–4320.

Alward, W.L., Kwon, Y.H., Kawase, K., Craig, J.E., Hayreh, S.S., Johnson, A.T., Khanna, C.L., Yamamoto, T., Mackey, D.A., Roos, B.R., Affatigato, L.M., Sheffield, V.C. and Stone, E.M. (2003) Evaluation of optineurin sequence variations in 1,048 patients with open-angle glaucoma. Am. J. Ophthalmol., 136: 904–910.

Anderson, D.R. and Hendrickson, A. (1974) Effect of intraocular pressure on rapid axoplasmic transport in monkey optic nerve. Invest. Ophthalmol., 13: 771–783.

Anderson, D.R. and Hendrickson, A.E. (1977) Failure of increased intracranial pressure to affect rapid axonal transport at the optic nerve head. Invest. Ophthalmol. Vis. Sci., 16: 423–426.

Anderson, M.G., Libby, R.T., Gould, D.B., Smith, R.S. and John, S.W. (2005) High-dose radiation with bone marrow transfer prevents neurodegeneration in an inherited glaucoma. Proc. Natl. Acad. Sci. U.S.A., 102: 4566–4571.

Anderson, M.G., Smith, R.S., Hawes, N.L., Zabaleta, A., Chang, B., Wiggs, J.L. and John, S.W. (2002) Mutations in genes encoding melanosomal proteins cause pigmentary glaucoma in DBA/2J mice. Nat. Genet., 30: 81–85.

Anderson, M.G., Smith, R.S., Savinova, O.V., Hawes, N.L., Chang, B., Zabaleta, A., Wilpan, R., Heckenlively, J.R., Davisson, M. and John, S.W. (2001) Genetic modification of glaucoma associated phenotypes between AKXD-28/Ty and DBA/2J mice. BMC Genet., 2: p. 1.

Araki, T., Sasaki, Y. and Milbrandt, J. (2004) Increased nuclear NAD biosynthesis and SIRT1 activation prevent axonal degeneration. Science, 305: 1010–1013.

Avila, M.Y., Carre, D.A., Stone, R.A. and Civan, M.M. (2001) Reliable measurement of mouse intraocular pressure by a servo-null micropipette system. Invest. Ophthalmol. Vis. Sci., 42: 1841–1846.

Avila, M.Y., Mitchell, C.H., Stone, R.A. and Civan, M.M. (2003) Noninvasive assessment of aqueous humor turnover in the mouse eye. Invest. Ophthalmol. Vis. Sci., 44: 722–727.

Bach, M. (2001) Electrophysiological approaches for early detection of glaucoma. Eur. J. Ophthalmol., 11(Suppl 2): S41–S49.

Bayer, A.U., Ferrari, F. and Erb, C. (2002) High occurrence rate of glaucoma among patients with Alzheimer's disease. Eur. Neurol., 47: 165–168.

Becerra, S.P. and Amaral, J. (2002) Erythropoietin — an endogenous retinal survival factor. N. Engl. J. Med., 347: 1968–1970.

Beirowski, B., Adalbert, R., Wagner, D., Grumme, D.S., Addicks, K., Ribchester, R.R. and Coleman, M.P. (2005) The progressive nature of Wallerian degeneration in wild-type and slow Wallerian degeneration (WldS) nerves. BMC Neurosci., 6: p. 6.

Bocker-Meffert, S., Rosenstiel, P., Rohl, C., Warneke, N., Held-Feindt, J., Sievers, J. and Lucius, R. (2002) Erythropoietin and VEGF promote neural outgrowth from retinal explants in postnatal rats. Invest. Ophthalmol. Vis. Sci., 43: 2021–2026.

Chang, B., Smith, R.S., Hawes, N.L., Anderson, M.G., Zabaleta, A., Savinova, O., Roderick, T.H., Heckenlively, J.R., Davisson, M.T. and John, S.W. (1999) Interacting loci cause severe iris atrophy and glaucoma in DBA/2J mice. Nat. Genet., 21: 405–409.

Conforti, L., Adalbert, R. and Coleman, M.P. (2007) Neuronal death: where does the end begin? Trends Neurosci., 30: 159–166.

Dalke, C., Pleyer, U. and Graw, J. (2005) On the use of Tono-Pen XL for the measurement of intraocular pressure in mice. Exp. Eye Res., 80: 295–296.

Danias, J., Lee, K.C., Zamora, M.F., Chen, B., Shen, F., Filippopoulos, T., Su, Y., Goldblum, D., Podos, S.M. and Mittag, T. (2003) Quantitative analysis of retinal ganglion cell (RGC) loss in aging DBA/2NNia glaucomatous mice: comparison with RGC loss in aging C57/BL6 mice. Invest. Ophthalmol. Vis. Sci., 44: 5151–5162.

De Marco, N., Buono, M., Troise, F. and Diez-Roux, G. (2006) Optineurin increases cell survival and translocates to the nucleus in a Rab8-dependent manner upon an apoptotic stimulus. J. Biol. Chem., 281: 16147–16156.

Ehrenreich, H., Aust, C., Krampe, H., Jahn, H., Jacob, S., Herrmann, M. and Siren, A.L. (2004) Erythropoietin: novel approaches to neuroprotection in human brain disease. Metab. Brain Dis., 19: 195–206.

Fan, B.J., Wang, D.Y., Lam, D.S. and Pang, C.P. (2006) Gene mapping for primary open angle glaucoma. Clin. Biochem., 39: 249–258.

Fechtner, R.D. and Weinreb, R.N. (1994) Mechanisms of optic nerve damage in primary open angle glaucoma. Surv. Ophthalmol., 39: 23–42.

Filippopoulos, T., Matsubara, A., Danias, J., Huang, W., Dobberfuhl, A., Ren, L., Mittag, T., Miller, J.W. and Grosskreutz, C.L. (2006) Predictability and limitations of non-invasive murine tonometry: comparison of two devices. Exp. Eye Res., 83: 194–201.

Fingert, J.H., Heon, E., Liebmann, J.M., Yamamoto, T., Craig, J.E., Rait, J., Kawase, K., Hoh, S.T., Buys, Y.M., Dickinson, J., Hockey, R.R., Williams-Lyn, D., Trope, G., Kitazawa, Y., Ritch, R., Mackey, D.A., Alward, W.L., Sheffield, V.C. and Stone, E.M. (1999) Analysis of myocilin mutations in 1703 glaucoma patients from five different populations. Hum. Mol. Genet., 8: 899–905.

Fujita, Y., Imagawa, T. and Uehara, M. (2000) Comparative study of the lamina cribrosa and the pial septa in the vertebrate optic nerve and their relationship to the myelinated axons. Tissue Cell, 32: 293–301.

Goldblum, D., Kipfer-Kauer, A., Sarra, G.M., Wolf, S. and Frueh, B.E. (2007) Distribution of amyloid precursor protein and amyloid-beta immunoreactivity in DBA/2J glaucomatous mouse retinas. Invest. Ophthalmol. Vis. Sci., 48: 5085–5090.

Goldblum, D. and Mittag, T. (2002) Prospects for relevant glaucoma models with retinal ganglion cell damage in the rodent eye. Vision Res., 42: 471–478.

Gordon, M.O., Beiser, J.A., Brandt, J.D., Heuer, D.K., Higginbotham, E.J., Johnson, C.A., Keltner, J.L., Miller, J.P., Parrish, R.K., 2nd, Wilson, M.R. and Kass, M.A. (2002) The ocular hypertension treatment study: baseline factors that predict the onset of primary open-angle glaucomaArch. Ophthalmol., 120: 714–720, discussion 829–830.

Gould, D.B. and John, S.W. (2002) Anterior segment dysgenesis and the developmental glaucomas are complex traits. Hum. Mol. Genet., 11: 1185–1193.

Gould, D.B., Marchant, J.K., Savinova, O.V., Smith, R.S. and John, S.W. (2007) Col4a1 mutation causes endoplasmic reticulum stress and genetically modifiable ocular dysgenesis. Hum. Mol. Genet., 16: 798–807.

Gould, D.B., Miceli-Libby, L., Savinova, O.V., Torrado, M., Tomarev, S.I., Smith, R.S. and John, S.W. (2004a) Genetically increasing Myoc expression supports a necessary pathologic role of abnormal proteins in glaucoma. Mol. Cell. Biol., 24: 9019–9025.

Gould, D.B., Reedy, M., Wilson, L.A., Smith, R.S., Johnson, R.L. and John, S.W. (2006) Mutant myocilin nonsecretion in vivo is not sufficient to cause glaucoma. Mol. Cell. Biol., 26: 8427–8436.

Gould, D.B., Smith, R.S. and John, S.W. (2004b) Anterior segment development relevant to glaucoma. Int. J. Dev. Biol., 48: 1015–1029.

Gross, R.L., Ji, J., Chang, P., Pennesi, M.E., Yang, Z., Zhang, J. and Wu, S.M. (2003) A mouse model of elevated intraocular pressure: retina and optic nerve findingsTrans. Am. Ophthalmol. Soc., 101: 163–169, discussion 169–171.

Grozdanic, S.D., Betts, D.M., Sakaguchi, D.S., Allbaugh, R.A., Kwon, Y.H. and Kardon, R.H. (2003) Laser-induced mouse model of chronic ocular hypertension. Invest. Ophthalmol. Vis. Sci., 44: 4337–4346.

Guo, L., Salt, T.E., Luong, V., Wood, N., Cheung, W., Maass, A., Ferrari, G., Russo-Marie, F., Sillito, A.M., Cheetham, M.E., Moss, S.E., Fitzke, F.W. and Cordeiro, M.F. (2007) Targeting amyloid-beta in glaucoma treatment. Proc. Natl. Acad. Sci. U.S.A., 104: 13444–13449.

Hauser, M.A., Allingham, R.R., Linkroum, K., Wang, J., LaRocque-Abramson, K., Figueiredo, D., Santiago-Turla, C., del Bono, E.A., Haines, J.L., Pericak-Vance, M.A. and Wiggs, J.L. (2006) Distribution of WDR36 DNA sequence variants in patients with primary open-angle glaucoma. Invest. Ophthalmol. Vis. Sci., 47: 2542–2546.

Hernandez, M.R. (2000) The optic nerve head in glaucoma: role of astrocytes in tissue remodeling. Prog. Retin. Eye Res., 19: 297–321.

Holder, G.E. (2004) Electrophysiological assessment of optic nerve disease. Eye, 18: 1133–1143.

Howell, G.R., Libby, R.T., Jakobs, T.C., Smith, R.S., Phalan, F.C., Barter, J.W., Barbay, J.M., Marchant, J.K., Mahesh, N., Porciatti, V., Whitmore, A.V., Masland, R.H. and John, S.W.M. (2007) Axons of retinal ganglion cells are insulted in the optic nerve early in DBA/2J glaucoma. J. Cell Biol., 179: 1523–1537.

Huang, W., Fileta, J.B., Dobberfuhl, A., Filippopolous, T., Guo, Y., Kwon, G. and Grosskreutz, C.L. (2005) Calcineurin cleavage is triggered by elevated intraocular pressure, and calcineurin inhibition blocks retinal ganglion cell death in experimental glaucoma. Proc. Natl. Acad. Sci. U.S.A., 102: 12242–12247.

Huang, W., Fileta, J.B., Filippopoulos, T., Ray, A., Dobberfuhl, A. and Grosskreutz, C.L. (2007) Hsp27 phosphorylation in experimental glaucoma. Invest. Ophthalmol. Vis. Sci., 48: 4129–4135.

Inman, D.M. and Horner, P.J. (2007) Reactive nonproliferative gliosis predominates in a chronic mouse model of glaucoma. Glia, 55: 942–953.

Jakobs, T.C., Libby, R.T., Ben, Y., John, S.W. and Masland, R.H. (2005) Retinal ganglion cell degeneration is topological but not cell type specific in DBA/2J mice. J. Cell Biol., 171: 313–325.

John, S.W. (2005) Mechanistic insights into glaucoma provided by experimental genetics the cogan lecture. Invest. Ophthalmol. Vis. Sci., 46: 2649–2661.

John, S.W., Anderson, M.G. and Smith, R.S. (1999) Mouse genetics: a tool to help unlock the mechanisms of glaucoma. J. Glaucoma, 8: 400–412.

John, S.W., Hagaman, J.R., MacTaggart, T.E., Peng, L. and Smithes, O. (1997) Intraocular pressure in inbred mouse strains. Invest. Ophthalmol. Vis. Sci., 38: 249–253.

John, S.W., Smith, R.S., Savinova, O.V., Hawes, N.L., Chang, B., Turnbull, D., Davisson, M., Roderick, T.H. and Heckenlively, J.R. (1998) Essential iris atrophy, pigment dispersion, and glaucoma in DBA/2J mice. Invest. Ophthalmol. Vis. Sci., 39: 951–962.

John, S.W.M. and Savinova, O.V. (2002) Intraocular pressure measurement in mice: technical aspects. In: Smith R.S., John S.W.M., Nishina P.M. and Sundberg J.P. (Eds.), Systematic Evaluation of the Mouse Eye: Anatomy, Pathology and Biomethods. CRC Press, Boca Raton, FL, pp. 313–320.

Karim, S., Clark, R.A., Poukens, V. and Demer, J.L. (2004) Demonstration of systematic variation in human intraorbital optic nerve size by quantitative magnetic resonance imaging and histology. Invest. Ophthalmol. Vis. Sci., 45: 1047–1051.

Kass, M.A., Heuer, D.K., Higginbotham, E.J., Johnson, C.A., Keltner, J.L., Miller, J.P., Parrish, R.K., 2nd, Wilson, M.R. and Gordon, M.O. (2002) The ocular hypertension treatment study: a randomized trial determines that topical ocular hypotensive medication delays or prevents the onset of primary open-angle glaucomaArch. Ophthalmol., 120: 701–713. discussion 829–830

Kim, B.S., Savinova, O.V., Reedy, M.V., Martin, J., Lun, Y., Gan, L., Smith, R.S., Tomarev, S.I., John, S.W. and Johnson, R.L. (2001) Targeted disruption of the myocilin gene (Myoc) suggests that human glaucoma-causing mutations are gain of function. Mol. Cell Biol., 21: 7707–7713.

Lam, D.S., Leung, Y.F., Chua, J.K., Baum, L., Fan, D.S., Choy, K.W. and Pang, C.P. (2000) Truncations in the TIGR gene in individuals with and without primary open-angle glaucoma. Invest. Ophthalmol. Vis. Sci., 41: 1386–1391.

Laser, H., Conforti, L., Morreale, G., Mack, T.G., Heyer, M., Haley, J.E., Wishart, T.M., Beirowski, B., Walker, S.A., Haase, G., Celik, A., Adalbert, R., Wagner, D., Grumme, D., Ribchester, R.R., Plomann, M. and Coleman, M.P. (2006) The slow Wallerian degeneration protein, WldS, binds directly to VCP/p97 and partially redistributes it within the nucleus. Mol. Biol. Cell, 17: 1075–1084.

Li, Y., Schlamp, C.L. and Nickells, R.W. (1999) Experimental induction of retinal ganglion cell death in adult mice. Invest. Ophthalmol. Vis. Sci., 40: 1004–1008.

Libby, R.T., Gould, D.B., Anderson, M.G. and John, S.W. (2005a) Complex genetics of glaucoma susceptibility. Annu. Rev. Genomics Hum. Genet., 6: 15–44.

Libby, R.T., Li, Y., Savinova, O.V., Barter, J., Smith, R.S., Nickells, R.W. and John, S.W. (2005b) Susceptibility to neurodegeneration in a glaucoma is modified by Bax gene dosage. PLoS Genet., 1: 17–26.

Libby, R.T., Smith, R.S., Savinova, O.V., Zabaleta, A., Martin, J.E., Gonzalez, F.J. and John, S.W. (2003) Modification of ocular defects in mouse developmental glaucoma models by tyrosinase. Science, 299: 1578–1581.

Lindsey, J.D. and Weinreb, R.N. (2002) Identification of the mouse uveoscleral outflow pathway using fluorescent dextran. Invest. Ophthalmol. Vis. Sci., 43: 2201–2205.

Lindsey, J.D. and Weinreb, R.N. (2005) Elevated intraocular pressure and transgenic applications in the mouse. J. Glaucoma, 14: 318–320.

Lunn, E.R., Perry, V.H., Brown, M.C., Rosen, H. and Gordon, S. (1989) Absence of Wallerian degeneration does not hinder regeneration in peripheral nerve. Eur. J. Neurosci., 1: 27–33.

Mabuchi, F., Aihara, M., Mackey, M.R., Lindsey, J.D. and Weinreb, R.N. (2004) Regional optic nerve damage in experimental mouse glaucoma. Invest. Ophthalmol. Vis. Sci., 45: 4352–4358.

Mack, T.G., Reiner, M., Beirowski, B., Mi, W., Emanuelli, M., Wagner, D., Thomson, D., Gillingwater, T., Court, F., Conforti, L., Fernando, F.S., Tarlton, A., Andressen, C., Addicks, K., Magni, G., Ribchester, R.R., Perry, V.H. and Coleman, M.P. (2001) Wallerian degeneration of injured axons and synapses is delayed by a Ube4b/Nmnat chimeric gene. Nat. Neurosci., 4: 1199–1206.

Markowitz, H., Gubler, C.J., Mahoney, J.P., Cartwright, G.E. and Wintrobe, M.M. (1955) Studies on copper metabolism. XIV. Copper, ceruloplasmin and oxidase activity in sera of normal human subjects, pregnant women, and patients with infection, hepatolenticular degeneration and the nephrotic syndrome. J. Clin. Invest., 34: 1498–1508.

Martin, J.L., Hickey, E., Weber, L.A., Dillmann, W.H. and Mestril, R. (1999) Influence of phosphorylation and oligomerization on the protective role of the small heat shock protein 27 in rat adult cardiomyocytes. Gene. Expr., 7: 349–355.

Maumenee, A.E. (1983) Causes of optic nerve damage in glaucoma. Robert N. Shaffer lecture. Ophthalmology., 90: 741–752.

May, C.A. and Lutjen-Drecoll, E. (2002) Morphology of the murine optic nerve. Invest. Ophthalmol. Vis. Sci., 43: 2206–2212.

May, C.A. and Mittag, T. (2006) Optic nerve degeneration in the DBA/2NNia mouse: is the lamina cribrosa important in the development of glaucomatous optic neuropathy? Acta. Neuropathol. (Berl.), 111: 158–167.

McKinnon, S.J. (2003) Glaucoma: ocular Alzheimer's disease? Front. Biosci., 8: s1140–s1156.

McKinnon, S.J., Lehman, D.M., Kerrigan-Baumrind, L.A., Merges, C.A., Pease, M.E., Kerrigan, D.F., Ransom, N.L., Tahzib, N.G., Reitsamer, H.A., Levkovitch-Verbin, H., Quigley, H.A. and Zack, D.J. (2002) Caspase activation and amyloid precursor protein cleavage in rat ocular hypertension. Invest. Ophthalmol. Vis. Sci., 43: 1077–1087.

Mo, J.S., Anderson, M.G., Gregory, M., Smith, R.S., Savinova, O.V., Serreze, D.V., Ksander, B.R., Streilein, J.W. and John, S.W. (2003) By altering ocular immune privilege, bone marrow-derived cells pathogenically contribute to DBA/2J pigmentary glaucoma. J. Exp. Med., 197: 1335–1344.

Monemi, S., Spaeth, G., Dasilva, A., Popinchalk, S., Ilitchev, E., Liebmann, J., Ritch, R., Heon, E., Crick, R.P., Child, A. and Sarfarazi, M. (2005) Identification of a novel adult-onset primary open-angle glaucoma (POAG) gene on 5q22.1. Hum. Mol. Genet., 14: 725–733.

Morcos, Y. and Chan-Ling, T. (2000) Concentration of astrocytic filaments at the retinal optic nerve junction is coincident with the absence of intra-retinal myelination: comparative and developmental evidence. J. Neurocytol., 29: 665–678.

Morris, C.A., Crowston, J.G., Lindsey, J.D., Danias, J. and Weinreb, R.N. (2006) Comparison of invasive and non-invasive tonometry in the mouse. Exp. Eye Res., 82: 1094–1099.

Nagaraju, M., Saleh, M. and Porciatti, V. (2007) IOP-dependent retinal ganglion cell dysfunction in glaucomatous DBA/2J mice. Invest. Ophthalmol. Vis. Sci., 48: 4573–4579.

Nakazawa, T., Nakazawa, C., Matsubara, A., Noda, K., Hisatomi, T., She, H., Michaud, N., Hafezi-Moghadam, A., Miller, J.W. and Benowitz, L.I. (2006) Tumor necrosis factor-alpha mediates oligodendrocyte death and delayed retinal ganglion cell loss in a mouse model of glaucoma. J. Neurosci., 26: 12633–12641.

Nissirios, N., Goldblum, D., Rohrer, K., Mittag, T. and Danias, J. (2007) Noninvasive determination of intraocular pressure (IOP) in nonsedated mice of 5 different inbred strains. J. Glaucoma, 16: 57–61.

Obazawa, M., Mashima, Y., Sanuki, N., Noda, S., Kudoh, J., Shimizu, N., Oguchi, Y., Tanaka, Y. and Iwata, T. (2004) Analysis of porcine optineurin and myocilin expression in trabecular meshwork cells and astrocytes from optic nerve head. Invest. Ophthalmol. Vis. Sci., 45: 2652–2659.

O'Brien, T.P. and Frankel, W.N. (2004) Moving forward with chemical mutagenesis in the mouse. J. Physiol., 554: 13–21.

Pang, I.H. and Clark, A.F. (2007) Rodent models for glaucoma retinopathy and optic neuropathy. J. Glaucoma, 16: 483–505.

Park, K.H., Cozier, F., Ong, O.C. and Caprioli, J. (2001) Induction of heat shock protein 72 protects retinal ganglion cells in a rat glaucoma model. Invest. Ophthalmol. Vis. Sci., 42: 1522–1530.

Peachey, N.S. and Ball, S.L. (2003) Electrophysiological analysis of visual function in mutant mice. Doc. Ophthalmol., 107: 13–36.

Pease, M.E., Hammond, J.C. and Quigley, H.A. (2006) Manometric calibration and comparison of TonoLab and TonoPen tonometers in rats with experimental glaucoma and in normal mice. J. Glaucoma, 15: 512–519.

Pekny, M. and Nilsson, M. (2005) Astrocyte activation and reactive gliosis. Glia, 50: 427–434.

Peters, L.L., Robledo, R.F., Bult, C.J., Churchill, G.A., Paigen, B.J. and Svenson, K.L. (2007) The mouse as a model for human biology: a resource guide for complex trait analysis. Nat. Rev. Genet., 8: 58–69.

Petros, T.J., Williams, S.E. and Mason, C.A. (2006) Temporal regulation of EphA4 in astroglia during murine retinal and optic nerve development. Mol. Cell. Neurosci., 32: 49–66.

Pinto, L.H., Invergo, B., Shimomura, K., Takahashi, J.S. and Troy, J.B. (2007) Interpretation of the mouse electroretinogram. Doc. Ophthalmol., 115: 127–136.

Porciatti, V. (2007) The mouse pattern electroretinogram. Doc. Ophthalmol., 115: 145–153.

Porciatti, V., Saleh, M. and Nagaraju, M. (2007) The pattern electroretinogram as a tool to monitor progressive retinal ganglion cell dysfunction in the DBA/2J mouse model of glaucoma. Invest. Ophthalmol. Vis. Sci., 48: 745–751.

Qing, G., Duan, X. and Jiang, Y. (2005) Heat shock protein 72 protects retinal ganglion cells in rat model of acute glaucoma. Yan. Ke. Xue. Bao., 21: 163–168.

Quigley, H.A. (1996) Number of people with glaucoma worldwide. Br. J. Ophthalmol., 80: 389–393.

Quigley, H.A. and Addicks, E.M. (1980) Chronic experimental glaucoma in primates. II. Effect of extended intraocular

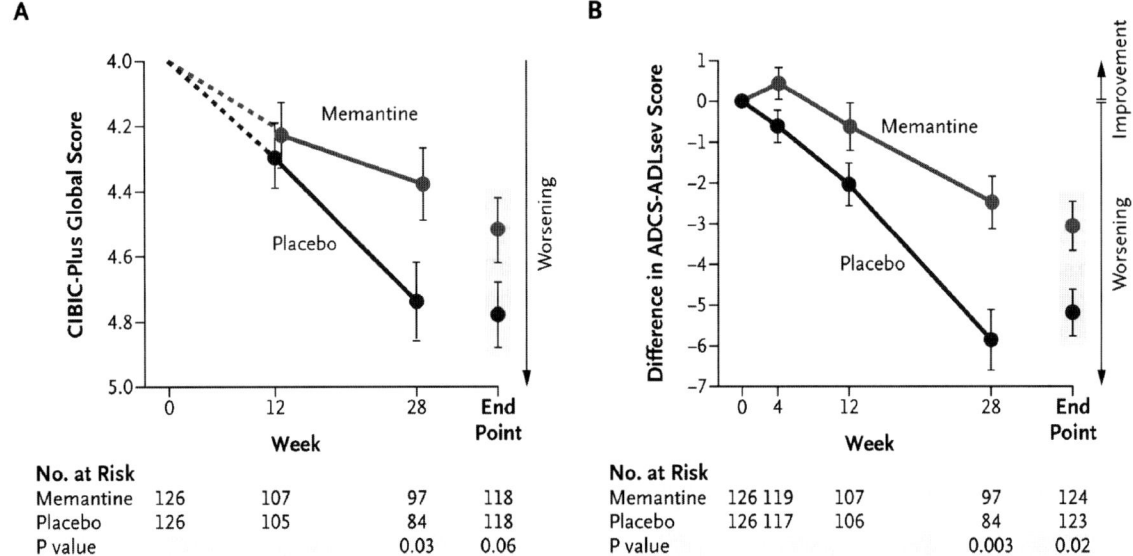

Fig. 3. Primary efficacy variables. The mean (±S.E.) scores at each specified time in the observed-cases analysis are shown. The boxes indicate the mean (±S.E.) at the endpoint in the analysis, with the last observation carried forward in the intent-to-treat population. Panel A shows the change from baseline in the Clinician's Interview-Based Impression of Change Plus Caregiver Input (CIBIC-Plus) global scores. Panel B shows the change from baseline in the Alzheimer's Disease Cooperative Study Activities of Daily Living Inventory, modified for severe dementia (ADCS-ADLsev). Reproduced with permission from Reisberg et al. (2003). Copyright © 2003 Massachusetts Medical Society. All rights reserved.

many of the strategies detailed above. The most definitive evidence will require a randomized clinical trial comparing the potential neuroprotective treatment to a control treatment or placebo. Unfortunately, there are few controlled clinical trials of neuroprotection in ophthalmology. One example is the Ischemic Optic Neuropathy Decompression Trial (IONDT) (The Ischemic Optic Neuropathy Decompression Trial Research Group, 1997). This was an National Eye Institute–sponsored multicenter clinical trial designed to assess the safety and efficacy of optic nerve decompression surgery compared with careful follow-up alone in patients with NAION. The IONDT illustrates the need for a well-controlled clinical study to adequately assess neuroprotective therapy, since the natural history of many degenerative diseases has not been well documented. Prior to this trial, improvement in visual acuity in patients with NAION was thought to be rare (less than 10%) (Boghen and Glaser, 1975; Repka et al., 1983), and several nonrandomized trials showed benefits of optic nerve decompression; however, none of these were randomized studies (Sergott et al., 1989; Kelman and Elman, 1991; Spoor et al., 1991, 1993). Interestingly, results from the IONDT showed that patients assigned to surgery did no better than patients assigned to careful follow-up (The Ischemic Optic Neuropathy Decompression Trial Research Group, 1997). Improved visual acuity of three or more lines was achieved by 23.6% of the surgery group compared with 42.7% of the careful follow-up group. In fact, patients receiving surgery had a significantly greater risk of losing three or more lines of vision at 6 months: 23.9% in the surgery group worsened compared with 12.4% in the careful follow-up group. Optic nerve decompression surgery was found to be ineffective and potentially harmful to patients with this disease.

Endpoints

A critical factor in designing clinical trials for neuroprotection in ophthalmology is endpoint selection. The primary outcome measures in the IONDT were gain or loss of three or more lines of visual acuity on the New York Lighthouse chart at 6 months after randomization (The Ischemic Optic Neuropathy Decompression Trial Research Group, 1997). This is a functional endpoint that has been used in a number of ophthalmology trials and is felt to be clinically meaningful. Visual acuity has also been used as a standard clinical endpoint for clinical trials in macular degeneration, where central acuity can be affected early in the course of the disease. Unfortunately, changes in central acuity may be an insensitive endpoint for many ophthalmic diseases. For example, central visual acuity loss occurs relatively late in the course of some diseases, including glaucoma and retinitis pigmentosa. Although visual field loss can be used as a functional endpoint, it occurs slowly and may require many years before meaningful changes occur.

Identification and validation of surrogate endpoints will improve our ability to assess neuroprotective therapies. In a sense, IOP has been used as a surrogate endpoint for glaucoma treatments. Without well-controlled data, lowering IOP was felt to reduce the risk of vision loss in patients with glaucoma. Data from randomized clinical trials that support the benefits of lowering IOP are now becoming available. In a recent report from the Advanced Glaucoma Intervention Study (AGIS), investigators showed that eyes with 100% of visits with IOP less than 18 mmHg over 6 years had mean changes from baseline in visual field defect scores close to zero during follow-up, whereas eyes with less than 50% of visits with IOP less than 18 mmHg had an estimated worsening over follow-up of 0.63 units of visual field defect score (The AGIS Investigators, 2000).

Studies have documented that rates of change differ among measures of visual function. Depending on the disease, endpoints may change at differing rates and have a large impact on length of clinical trials and the required sample size. In a trial of patients with retinitis pigmentosa, investigators assessed changes in measures of visual function in patients over time (Holopigian et al., 1996). The smallest amount of change occurred for visual acuity and hue discrimination, and the greatest amount of change occurred for visual field area.

New methods for assessing visual function may be useful for measuring endpoints in clinical trials for neuroprotection. Electrophysiological testing, newer methods of visual field testing including Frequency Doubling Technology (FDT), Short Wavelength Automated Perimetry (SWAP), and contrast sensitivity are just a few examples. The electroretinogram was used as the primary outcome measure in a randomized clinical trial of vitamin A and vitamin E supplementation for retinitis pigmentosa (Berson et al., 1993). This was a National Eye Institute–sponsored, randomized, double-masked trial to determine whether supplements of vitamin A or vitamin E alone or in combination affect the course of retinitis pigmentosa. In this study, the main outcome measure was the cone electroretinogram amplitude. Patients receiving 15,000 IU/day of vitamin A were 32% less likely to have a decline in amplitude of 50% or more from baseline than those not receiving this dosage ($P = 0.03$). Although not statistically significant, similar trends were observed for rates of decline of visual field area. These data support the potential benefit of identifying endpoints that may be more sensitive or less variable in clinical trials.

More accurate and sensitive measures of visual function will improve our ability to test neuroprotective therapies in ophthalmology. The scanning laser ophthalmoscope to assess perimetry (Sunness et al., 1995) or the multifocal electroretinogram (Hood et al., 1997) may be useful to evaluate endpoints in trials of retinal disease. Similarly, new methods of assessing glaucomatous damage such as the Heidelberg retinal tomograph (HRT), the GDx nerve fiber analyzer, and the optical coherence tomograph (OCT) may be important measures of visual function in neuroprotection trials in patients with glaucoma (Wollstein et al., 2000; Bowd et al., 2001; Miglior et al., 2001; Zangwill et al., 2001).

Measurements of visual field have become standard ways to assess the functional progression of glaucoma. There have been a number of recent developments in automated perimetry that have led to improved diagnosis and assessment of disease progression in glaucoma (Johnson, 2002).

Hood, D.C., Seiple, W., Hologigian, K. and Greenstein, V. (1997) A comparison of the components of the multifocal and full-field ERGs. Vis. Neurosci., 14(3): 533–544.

Hutchings, N., Hosking, S.L., Wild, J.M. and Flanagan, J.G. (2001) Long-term fluctuation in short-wavelength automated perimetry in glaucoma suspects and glaucoma patients. Invest. Ophthalmol. Vis. Sci., 42(10): 2332–2337.

Johnson, C.A. (2002) Recent developments in automated perimetry in glaucoma diagnosis and management. Curr. Opin. Ophthalmol., 13(2): 77–84.

Johnson, C.A. and Samuels, S.J. (1997) Screening for glaucomatous visual field loss with frequency-doubling perimetry. Invest. Ophthalmol. Vis. Sci., 38(2): 413–425.

Kass, M.A., Heuer, D.K., Higginbotham, E.J., Johnson, C.A., Keltner, J.L., Miller, J.P., Parrish, R.K., II, Wilson, M.R. and Gordon, M.O. (2002) The Ocular Hypertension Treatment Study: a randomized trial determines that topical ocular hypotensive medication delays or prevents the onset of primary open-angle glaucoma. Arch. Ophthalmol., 120(6): 701–713.

Kelman, S.E. and Elman, M.J. (1991) Optic nerve sheath decompression for nonarteritic ischemic optic neuropathy improves multiple visual function parameters. Arch. Ophthalmol., 109(5): 667–671.

Klistorner, A. and Graham, S.L. (2000) Objective perimetry in glaucoma. Ophthalmology, 107(12): 2283–2299.

Landers, J., Goldberg, I. and Graham, S. (2000) A comparison of short wavelength automated perimetry with frequency doubling perimetry for the early detection of visual loss in ocular hypertension. Clin. Experiment. Ophthalmol., 28(4): 248–252.

Lichter, P.R., Musch, D.C., Gillespie, B.W., Guire, K.E., Janz, N.K., Wren, P.A. Mills, R.P. and CIGTS Study Group. (2001) Interim clinical outcomes in the collaborative initial glaucoma treatment study comparing initial treatment randomized to medications or surgery. Ophthalmology, 108(11): 1943–1953.

Mangione, C.M., Lee, P.P., Gutierrez, P.R., Spritzer, K., Berry, S. Hays, R.D. and National Eye Institute Visual Function Questionnaire Field Test Investigators. (2001) Development of the 25-item National Eye Institute Visual Function Questionnaire. Arch. Ophthalmol., 119(7): 1050–1058.

Miglior, S., Casula, M., Guareschi, M., Marchetti, I., Iester, M. and Orzalesi, N. (2001) Clinical ability of Heidelberg retinal tomograph examination to detect glaucomatous visual field changes. Ophthalmology, 108(9): 1621–1627.

Miglior, S., Zeyen, T., Pfeiffer, N., Cunha-Vaz, J., Torri, V. Adamsons, I. and European Glaucoma Prevention Study (EGPS) Group. (2005) Results of the European Glaucoma Prevention Study. Ophthalmology, 112(3): 366–375.

Mitchell, J.D. and Borasio, G.D. (2007) Amyotrophic lateral sclerosis. Lancet, 369(9578): 2031–2041.

Mizoule, J., Meldrum, B., Mazadier, M., Croucher, M., Ollat, C., Uzan, A., Legrand, J.J., Gueremy, C. and Le Fur, G. (1985) 2-Amino-6-trifluoromethoxy benzothiazole, a possible antagonist of excitatory amino acid neurotransmission—I. Anticonvulsant properties. Neuropharmacology, 24(8): 767–773.

Pallàs, M. and Camins, A. (2006) Molecular and biochemical features in Alzheimer's disease. Curr. Pharm. Des., 12(33): 4389–4408.

Plaitakis, A. and Caroscio, J.T. (1987) Abnormal glutamate metabolism in amyotrophic lateral sclerosis. Ann. Neurol., 22(5): 575–579.

Reisberg, B., Doody, R., Stöffler, A., Schmitt, F., Ferris, S. Möbius, H.J. and Memantine Study Group. (2003) Memantine in moderate-to-severe Alzheimer's disease. N. Engl. J. Med., 348(14): 1333–1341.

Repka, M.X., Savino, P.J., Schatz, N.J. and Sergott, R.C. (1983) Clinical profile and long-term implications of anterior ischemic optic neuropathy. Am. J. Ophthalmol., 96(4): 478–483.

Rothstein, J.D., Martin, L.J. and Kuncl, R.W. (1992) Decreased glutamate transport by the brain and spinal cord in amyotrophic lateral sclerosis. N. Engl. J. Med., 326(22): 1464–1468.

Sacco, R.L., DeRosa, J.T., Haley, E.D., Jr., Levin, B., Ordronneau, P., Phillips, S.J., Rundek, T., Snipes, R.G. Thompson, J.L. and Glycine Antagonist in Neuroprotection Americas Investigators. (2001) Glycine antagonist in neuroprotection for patients with acute stroke: GAIN Americas: a randomized controlled trial. JAMA, 285(13): 1719–1728.

Sackett, D.L. (1979) Bias in analytic research. J. Chronic Dis., 32(1–2): 51–63.

Sergott, R.C., Cohen, M.S., Bosley, T.M. and Savino, P.J. (1989) Optic nerve decompression may improve the progressive form of nonarteritic ischemic optic neuropathy. Arch. Ophthalmol., 107(12): 1743–1754.

Spoor, T.C., McHenry, J.G. and Lau-Sickon, L. (1993) Progressive and static nonarteritic ischemic optic neuropathy treated by optic nerve sheath decompression. Ophthalmology, 100(3): 306–311.

Spoor, T.C., Wilkinson, M.J. and Ramocki, J.M. (1991) Optic nerve sheath decompression for the treatment of progressive nonarteritic ischemic optic neuropathy. Am. J. Ophthalmol., 111(4): 724–728.

Sunness, J.S., Schuchard, R.A., Shen, N., Rubin, G.S., Dagnelie, G. and Haselwood, D.M. (1995) Landmark-driven fundus perimetry using the scanning laser ophthalmoscope. Invest. Ophthalmol. Vis. Sci., 36(9): 1863–1874.

Tariot, P.N., Farlow, M.R., Grossberg, G.T., Graham, S.M., McDonald, S. Gergel, I. and Memanine Study Group. (2004) Memantine treatment in patients with moderate to severe Alzheimer disease already receiving donepezil: a randomized controlled trial. JAMA, 291(3): 317–324.

The AGIS Investigators. (2000) The Advanced Glaucoma Intervention Study (AGIS): 7. The relationship between control of intraocular pressure and visual field deterioration. Am. J. Ophthalmol., 130(4): 429–440.

The Ischemic Optic Neuropathy Decompression Trial Research Group. (1995) Optic nerve decompression surgery for nonarteritic anterior ischemic optic neuropathy (NAION) is not effective and may be harmful. The Ischemic Optic Neuropathy Decompression Trial Research Group. JAMA, 273(8): 625–632.

WoldeMussie, E., Ruiz, G., Wijono, M. and Wheeler, L.A. (2001) Neuroprotection of retinal ganglion cells by brimonidine in rats with laser-induced chronic ocular hypertension. Invest. Ophthalmol. Vis. Sci., 42(12): 2849–2855.

Wollstein, G., Garway-Heath, D.F., Fontana, L. and Hitchings, R.A. (2000) Identifying early glaucomatous changes. Comparison between expert clinical assessment of optic disc photographs and confocal scanning ophthalmoscopy. Ophthalmology, 107(12): 2272–2277.

Zangwill, L.M., Bowd, C., Berry, C.C., Williams, J., Blumenthal, E.Z., Sánchez-Galeana, C.A., Vasile, C. and Weinreb, R.N. (2001) Discriminating between normal and glaucomatous eyes using the Heidelberg retina tomograph, GDx Nerve Fiber Analyzer, and Optical Coherence Tomograph. Arch. Ophthalmol., 119(7): 985–993.

SECTION V

Neuroprotection: New Vistas in Pathophysiology

CHAPTER 24

Pathogenesis of ganglion "cell death" in glaucoma and neuroprotection: focus on ganglion cell axonal mitochondria

Neville N. Osborne*

Nuffield Laboratory of Ophthalmology, University of Oxford, Walton Street, Oxford OX2 6AW, UK

Abstract: Retinal ganglion cell axons within the globe are functionally specialized being richly provided with many mitochondria. The mitochondria produce the high energy requirement for nerve conduction in the unmyelinated part of the ganglion cell axons. We have proposed that in the initiation of glaucoma, an alteration in the quality of blood flow dynamics in the optic nerve head causes a compromise in the retinal ganglion cell axon energy requirement, rendering the ganglion cells susceptible to additional insults. One secondary insult might be light entering the eye to further affect ganglion cell axon mitochondrial function. Other insults to the ganglion cells might be substances (e.g., glutamate, nitric oxide, TNF-α) released from astrocytes. These effects ultimately cause ganglion cell death because of the inability of mitochondria to maintain normal function. We therefore suggest that ganglion cell apoptosis in glaucoma is both receptor and mitochondrial mediated. Agents targeted specifically at enhancing ganglion cell mitochondrial energy production should therefore be beneficial in a disease like glaucoma. Ganglion cell death in glaucoma might therefore, in principle, not be unlike the pathophysiology of numerous neurological disorders involving energy dysregulation and oxidative stress.

The trigger(s) for ganglion cell apoptosis in glaucoma is/are likely to be multifactorial, and the rationale for targeting impaired energy production as a possibility of improving a patient's quality of life is based on logic derived from laboratory studies where neuronal apoptosis is shown to occur via different mechanisms. Light-induced neuronal apoptosis is likely to be more relevant to ganglion cell death in glaucoma than, for example, neuronal apoptosis associated with Parkinson's disease. Logic suggests that enhancing mitochondrial function generally will slow down ganglion cell apoptosis and therefore benefit glaucoma patients. On the basis of our laboratory studies, we suggest that supplements such as creatine, α-lipoic acid, nicotinamide, and epigallocatechin gallate (EGCG), all of which counteract oxidative stress induced by light and other triggers, are worthy of consideration for the treatment of such patients as they can be taken orally to reach the retina without having significant side effects.

Keywords: pathogenesis of glaucoma; ganglion cells; mitochondria; apoptosis; light injury to mitochondria; neuroprotection; creatine; α-lipoic acid; nicotinamide; epigallocatechin gallate

---

*Corresponding author. Tel.: +44 1865 248996;
Fax: +44 1865 794508; E-mail: neville.osborne@eye.ox.ac.uk

DOI: 10.1016/S0079-6123(08)01124-2

Introduction

Glaucoma, or glaucomatous optic neuropathy, is a chronic neurodegenerative disease characterized by a progressive loss of retinal ganglion cells (Quigley, 1999; Whitmore et al., 2005). The disease is associated with a specific remodeling of the optic nerve head. Primary open-angle glaucoma (POAG) constitutes the majority of all forms of glaucoma where the iris position is not affected. Traditionally, glaucoma has been viewed as a disease of elevated intraocular pressure (IOP). Excessive elevation of IOP can cause compression of retinal ganglion cell axons at the optic nerve head to affect axonal transport and alter the appropriate nutritional requirements for ganglion cell survival. Blood flow in the optic nerve head is also reduced because of compression of blood vessels and/or altered perfusion occurring when IOP is moderately elevated. Compelling evidence therefore exists to show that raised IOP can be the cause of visual loss in glaucoma. This is supported by the finding that the lowering IOP is often linked with the prevention of visual loss (Weinreb and Khaw, 2004).

A substantial number of glaucoma patients do not have raised IOP, however, and often lowering of elevated IOP does not result in the prevention of visual loss (Flammer and Orgül, 1998; Weinreb and Khaw, 2004). Moreover, not all ocular hypertensive patients have glaucoma. Clearly, raised IOP is not synonymous with loss of vision in all glaucoma patients. Factors other than raised IOP therefore contribute to loss of vision in glaucoma. However, unequivocal proof of risk factors other than raised IOP causing loss of vision in glaucoma remains a problem. A number of potential risk factors (Fig. 1) have been identified that include fluctuation of IOP, aging, family history, severe myopia, central cornea thickness, hypertension, hypotension, vasospasm, hemorheology, immune system, diabetes mellitus, sleep disturbances, family history, and light (Flammer et al., 2002; Pache and Flammer, 2006, Osborne et al., 2001, 2006). In addition, a number of candidate genes have also been associated with the risk of developing glaucoma (Abu-Amero et al., 2006).

Retinal ganglion cells and mitochondria

Retinal ganglion cell axons are unmyelinated within the ocular globe and have now been shown to have varicosities that are rich in mitochondria. These axons make desmosome- and hemidesmosome-like junctions with other axons and glial cells (Wang et al., 2003; Carelli et al., 2004). The mitochondria-enriched axon varicosities have been interpreted as functional sites with local

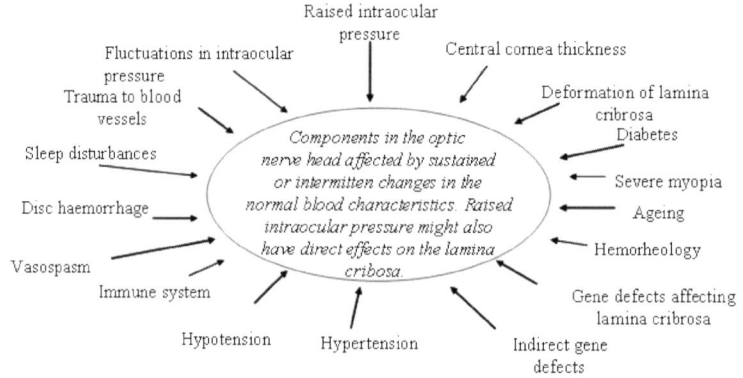

Fig. 1. Potential causes for an ischemic insult to the optic nerve head as might occur in glaucoma.

high-energy demand, possibly relevant for signal transmission. Significantly, immunocytochemical labeling of cytochrome c oxidase, a marker related to the intensity of oxidative phosphorylation, is greatest in the nerve fiber layer and prelaminar–laminar ganglion cell axons within the globe. It also decreases drastically in the myelin posterior to the lamina. Thus, there is an asymmetrical distribution of mitochondria along the ganglion cell axons, with the unmyelinated regions within the globe containing an enriched population of mitochondria (Bristow et al., 2002; Carelli et al., 2004). Clearly, an alteration in the functional status of these ganglion cell axon mitochondria will influence ganglion cell survival, and the question arises as to whether this phenomenon plays a part in the pathogenesis of optic neuropathies such as glaucoma.

Mitochondria are constantly produced in neuronal soma and transported to their terminals where they participate in synaptic transmission and undergo fusion and fission (Chan, 2006). One explanation for the different number of mitochondria in the prelaminar and postlaminar regions of a ganglion cell axon is that mitochondria are transported more slowly in the prelaminar region as they are required to provide a lot of energy to propagate an action potential along an unmyelinated axon. If this is the case, then prelaminar axonal transport would vary between different ganglion cells depending on the length of unmyelinated axon within the globe. A reduction in nutrient/oxygen supply to the optic nerve head, for example, caused by raised IOP, will ultimately affect the efficiency of energy production by mitochondria in the prelaminar region, and as a consequence axonal transport is affected. There is good evidence to suggest that ganglion cell axonal transport is attenuated in glaucoma (Levy, 1976). Moreover, reduced nutrient/oxygen supply to the optic nerve head region might not affect all ganglion cells in the same way, with some being energetically compromised to a greater extent than others. This might be partially dependent on the length and number of mitochondria in a ganglion cell axon within the globe.

Recent studies have proven that mitochondrial respiratory chain enzymes such as favin and cytochrome oxidases are able to absorb light maximally around 440–450 nm, and as a consequence, contribute to the generation of reactive oxygen species (ROS) in cells (Chen et al., 1992, 2003; King et al., 2004; Godley et al., 2005; Osborne et al., 2006). The interaction of light with ganglion cell axon mitochondria in healthy cells probably results in a generation of ROS, which is removed by the appropriate scavenging mechanisms. However, when ganglion cells are in an energetic compromised state, normal scavenging mechanisms might not be adequate. The proposition has therefore been made that light becomes a risk factor to energetic compromised cells in glaucoma. Laboratory studies on cells in culture provide support for this notion (Osborne et al., 2006, 2008; Lascaratos et al., 2007).

Possible causes for ganglion cell death in glaucoma

It is now clear that the ganglion cell death in glaucoma is not solely caused by raised IOP. It is also evident that, in order to develop additional methods of treatment other than the lowering of IOP, it is necessary to understand the pathogenesis of the disease. From our present knowledge, it is tempting to conclude that impairment in the quality of blood flow in the microcirculation associated with the optic nerve head region is the factor initiating retinal changes that leads to loss of vision in glaucoma (Fig. 1), although it remains to be unequivocally demonstrated that optic nerve head blood flow is compromised in glaucoma. Present logic suggests that sustained or intermittent changes in the microcirculation in the optic nerve head region is likely to occur because of raised or fluctuation changes in IOP, aging, vasospasm, hemorheology, hypotension, or hypertension. This could result in variable ischemia/reperfusion and/or hypoxic insults to ganglion cell axons and glial cells in the optic nerve head region. This idea is supported by the finding that HIF-1α, activated by a reduced oxygen supply, is upregulated in pathological retinas from glaucoma subjects (Tezel and Wax, 2004).

How might ischemic insults to retinal ganglion cell axons and glia in the optic nerve head region cause the death of ganglion cells at different times in the life of a glaucoma patient? We have hypothesized that in the initial stages of the disease, ganglion cells will be forced to survive at a lower energetic status because of the altered optic nerve head blood flow (Fig. 2) and that as the disease progresses, individual ganglion cells will be at risk of a combination of different types of additional secondary insults. As a result, ganglion cells will die at different times. We have also proposed that secondary insults originate from light entering the eye (Osborne et al., 2006) and from altered function to glial cells (astrocytes, Müller cells, and microglia), originating from abnormal blood flow in the optic nerve head region that is also the cause for ganglion cells being energetically compromised (Figs. 1 and 2). This is based partly on experimental studies that have shown that a variety of potentially toxic substances (glutamate, TNF-α, serine, nitric oxide, and potassium) become elevated in the extracellular retinal spaces when retinal glial cell function is affected (see Tezel, 2006). Moreover, in pathological tissues, an upregulation of substances like nitric oxide synthase (involved in the formation of nitric oxide) (Neufeld et al., 1999) and TNF-α (see Tezel, 2006) has been reported. Also, an elevation of glutamate occurs in the vitreous humor of glaucoma patients (Dryer et al., 1996). Elevation of these substances will affect the survival of all retinal neurones by having membrane effects that may be receptor mediated. However, their effects on retinal ganglion cells will be particularly detrimental because these cells are in an already compromised energetic state. For example, elevated extracellular levels of glutamate will depolarize excessively all cells containing NMDA-type receptors, but the ganglion cells, being at a compromised energetic state, will not be able to compensate as efficiently as other neurones for this excessive depolarization, causing them to become dysfunctional and eventually die by apoptosis. This hypothesis therefore predicts that the trigger(s) for different ganglion cells dying by

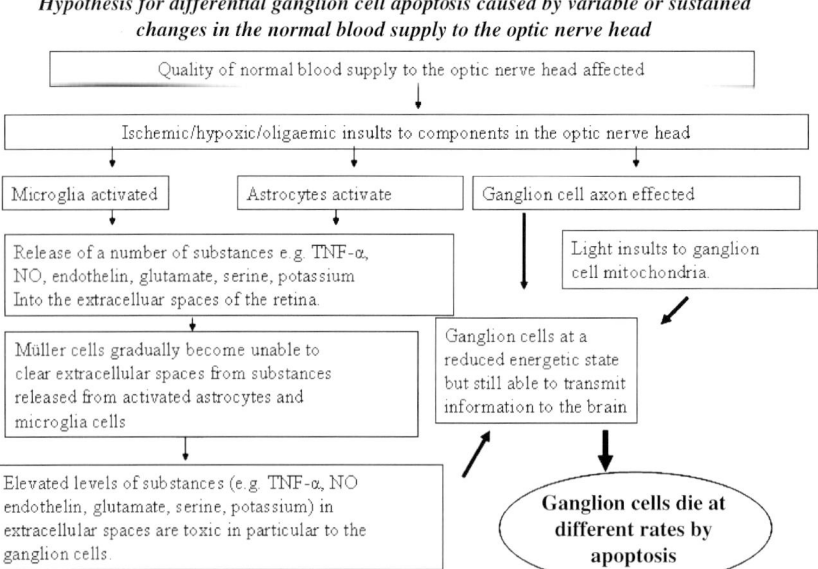

Fig. 2. Hypothetical cascade of events that lead to ganglion cells dying at different times during the lifetime of a glaucoma patient. The initial insult in the optic nerve head region affects all components (astrocytes, microglia, Müller cell end feet, ganglion cell axons, and lamina cribosa), resulting in ganglion cells existing at a reduced energetic state and an elevation of various substances (glutamate, TNF-α, NO, endothelin, d-serine, and potassium) in the extracellular space caused by activated astrocyte and microglia cells. These substances together with light impinging on mitochondria in ganglion cell axons eventually cause apoptotic insults to the energetically compromised ganglion cells at different times, depending on the axonal length and receptor profile of individual ganglion cells.

apoptosis in glaucoma will not necessarily be the same. This implies that the death pathways for different ganglion cells dying at various stages in glaucoma may not be the same.

Mitochondrial functions and apoptosis

Mitochondria are involved in numerous metabolic functions that include oxidative energy metabolism, the related production of ROS, and a role in promoting and regulating apoptotic death of the cell (Schon and Manfredi, 2003; Chan, 2006). Among the numerous stresses known to participate in mitochondrion-mediated apoptosis, at least in vitro, bioenergetic failure and elevated ROS figure prominently (Kroemer and Reed, 2000; Schon and Manfredi, 2003). Moreover, it is now established that light can be absorbed by mitochondrial photosensitizers, such as cytochrome oxidase and flavin-containing oxidases, causing a production of ROS (King et al., 2004; Godley et al., 2005). It is also known that the rate of ROS production from mitochondria is increased in a variety of pathologic conditions that include aging, hypoxia, and ischemia (Osborne et al., 1999; Chan, 2006; Morin and Simon, 2006).

Of key importance is the role of mitochondria in oxidative energy metabolism. Oxidative phosphorylation generates most of the cell's ATP, and any impairment of the organelle's ability to produce energy can have catastrophic consequences, not only due to the primary loss of ATP, but also due to indirect impairment of downstream functions, such as maintenance of organellar and cellular calcium homeostasis. Moreover, deficient mitochondrial metabolism will generate ROS that can wreak havoc in the cell. It is for such reasons that all evidence points to inadequate mitochondrial function being linked to apoptosis. This is supported by laboratory findings showing that substances that can maintain mitochondrial function (e.g., to allow ATP to be generated efficiently) or scavenge excessive ROS production blunt the process of apoptosis.

The hallmarks of apoptosis are condensation of nuclear and cytoplasmic contents, nuclear DNA fragmentation, cell blebbing, and autophagy of membrane-bound bodies. Apoptosis appears to occur via different pathways. A mitochondrion-mediated (intrinsic) pathway has been described where an external insult — elevated cytosolic calcium, to cite one example — acts to cause a release of cytochrome c, located in the inner membrane space of mitochondria. Cytosolic cytochrome c can then bind apoptotic protease-activating factor 1 (APAF-1), which then binds to the inactive form of caspase-9 (see Fig. 3). This complex "apoptosome" can then activate a cascade of events, which includes caspase-3, ultimately resulting in the hallmarks of apoptosis. Mitochondrion-mediated activation of caspase-9 can also occur via extracellular receptor-mediated signals (extrinsic pathway) to target various ligands (growth factor deprivation) — e.g., Bad, Bax, and Bik — to the mitochondrion, thereby causing cytochrome c release. Under certain circumstances, a separate mitochondrial-independent pathway also operates, which involves the activation of caspase-8 and caspase-3. In addition, a caspase-independent/mitochondrial-dependent pathway exists where apoptosis-inducing factor (AIF), confined to mitochondria, is released and translocates to the nucleus to cause chromatin condensation and DNA degradation (see Fig. 3).

Mitochondrial function enhancement and the attenuation of ganglion cell death

Mitochondria provide the bulk of a neurone's energy by oxidation of reducing equivalents (e.g., NADH and $FADH_2$), via the electron transport chain, to ultimately yield ATP. In addition, mitochondria contribute to cytosolic calcium buffering (Steeghs et al., 1997), apoptosis (Green and Reed, 1998; Kroemer and Reed, 2000; Schon and Manfredi, 2003), excitotoxicity (Peng et al., 1998), and generation of superoxide (McLennan and Degli Esposti, 2000). Also, mitochondrial flavin and cytochrome oxidases are affected by light to stimulate a production of ROS. Thus, an alteration in any aforementioned process or a combination of the processes maybe a cause of retinal ganglion cells dying in glaucoma.

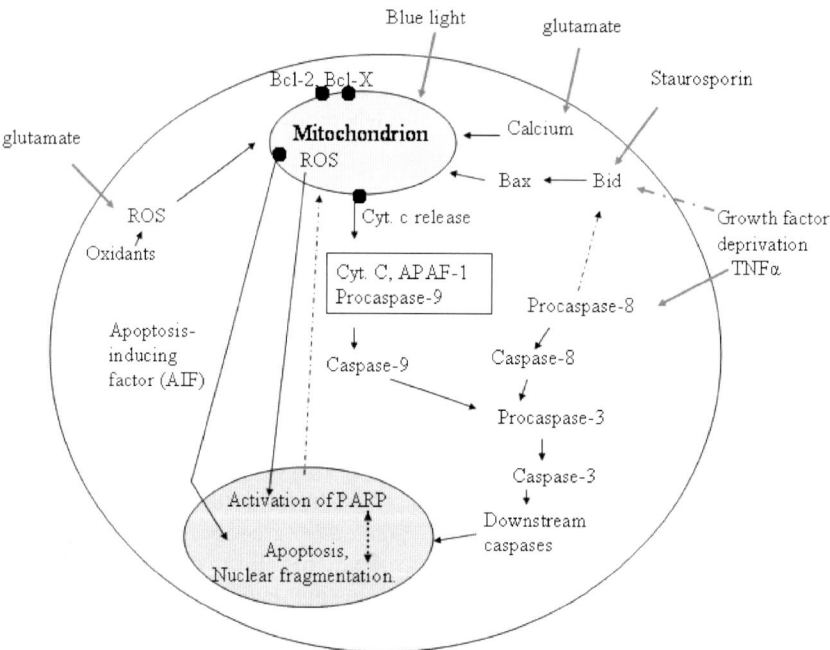

Fig. 3. Schematic view of the main mitochondrial-dependent and -independent (involves activation of caspase-8) apoptotic pathways. Mitochondrial-dependent apoptosis often involves the release of cytochrome c and the activation of caspase-9 but can also occur through mitochondrial release of cleaved AIF or ROS. AIF, apoptosis-inducing factor; PARP, poly(ADP-ribose) polymerase 1; ROS, reactive oxygen species; APAF-1, apoptotic protease-activating factor 1; Cyt. c, cytochrome c.

Substances that enhance mitochondrial function might therefore benefit glaucoma patients because ganglion cell axons have many mitochondria and these organelles are implicated in their apoptosis. Targeting impaired energy production is a realistic possibility. This will necessitate the use of an agent that can reach the retina, enhance the metabolic state of energetically compromised ganglion cells, and not affect healthy retinal cells adversely. Substances that might fulfil these criteria include creatine, α-lipoic acid, nicotinamide, and epigallocatechin gallate (EGCG), and each will be considered in turn. All four substances can be taken orally at regular intervals by humans without any obvious detrimental affects.

Creatine

Creatine is a guanidine compound that is ubiquitous among mammalian cells. Its concentration appears to reflect the energy requirements of various tissues, being highest in retinal photoreceptors with intermediate levels in the brain (Wyss and Kaddurah-Daouk, 2000). Creatine is released from the liver or consumed in the diet and transported into cells by specific transporters (Wyss and Kaddurah-Daouk, 2000). Within the cell, ATP and ADP and creatine are metabolized to form phosphocreatine by mitochondrial creatine kinase. Phosphocreatine then serves not only as an intermediate temporary energy buffer, but also as an energy shuttle from subcellular sites of energy production (mitochondria) to sites of energy consumption (Brewer and Wallimann, 2000). Thus, the phosphocreatine/creatine kinase system has been suggested to be of physiological importance in tissues with high energy and fluctuating energy requirements. This should apply to the intraretinal ganglion cell axons in particular because of their many mitochondria and presumed high level of creatine kinase. It has also been hypothesized that supplementation with creatine will result in cells making more phosphocreatine (Brewer and

Wallimann, 2000), and this has been supported by experimental studies. For example, creatine supplementation attenuates accumulation of oxidative stress markers and is shown to scavenge superoxide and peroxynitrite (Lawler et al., 2002). In cell culture experiments, creatine protects against toxicity induced by glutamate or β-amyloid (Brewer and Wallimann, 2000). Creatine supplementation has also been shown to exhibit remarkable neuroprotection in animal models of amyotropic lateral sclerosis (Klivenyi et al., 1998), Huntington's disease, Parkinson's disease, and traumatic brain injury (Zhu et al., 2004). Thus, the idea that creatine supplementation is beneficial for the treatment of glaucoma by attenuating the death process to ganglion cells seems to be a real possibility.

In preliminary studies, rats were injected subcutaneously twice a day with creatine (1 g/kg body weight/day) for six days, which was followed a day later for another five days (because of Sunday break from the laboratory). Controls received saline. On the sixth day, NMDA (concentration within the vitreous estimated to be 100 μM) was injected into the vitreous humor of one eye while the other received vehicle. On the 12th day, the animals were killed. Retinas were fixed and optic nerves analyzed by Western blotting for neurofilament light (NF-L) protein content relative to actin. Sections of fixed retinas were cut in areas of similar eccentricities. Eyes injected with NMDA showed a clear reduction of amacrine (ChAT) and ganglion (Thy-1) cell markers when compared with vehicle-injected eyes. However, in those animals that had been prophylactically treated with creatine, the effect of the NMDA injection was less evident (Fig. 4). Moreover, analysis of NF-L content in optic nerves showed that the reduction caused by NMDA treatment was significantly less in rats treated with creatine. These preliminary studies therefore provide support for the idea that creatine supplementation can attenuate an NMDA insult to inner retinal neurones, which include ganglion cells.

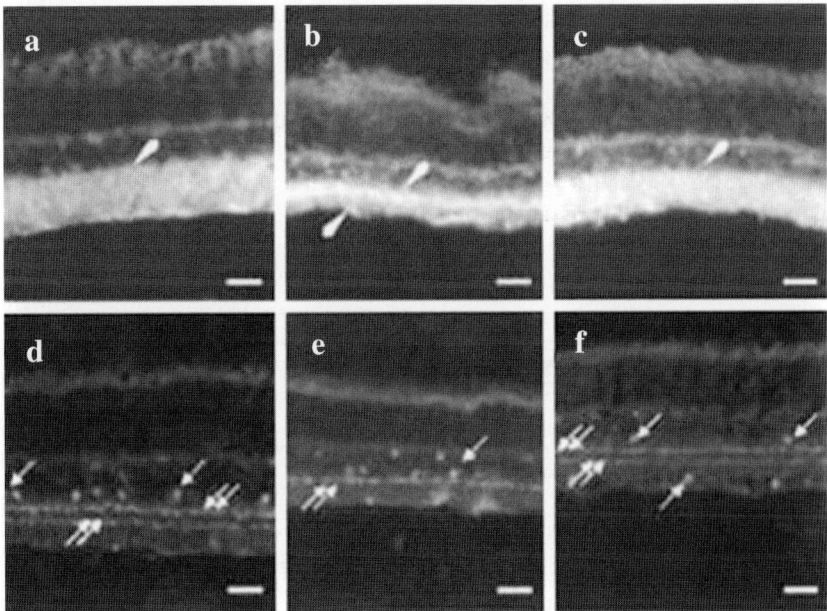

Fig. 4. Influence of creatine administration on the negative effects to the rat retina following an intraocular injection of NMDA. The normal thick band of Thy-1 immunoreactivity associated with the ganglion cells (a, arrow heads) and the typical localization of choline acetyltransferase (ChAT) associated with amacrine cells (small arrows) and their two processes in the inner plexiform layer (d) are much affected by NMDA injection (b, e). However, in animals treated with creatine (c, f), the influence of NMDA injection on Thy-1 and ChAT immunoreactivity is much reduced.

α-Lipoic acid

α-Lipoic acid (1,2-dithiolane-3-pentanoic acid) is a mitochondrial dithiol compound that functions as a coenzyme for pyruvate dehydrogenase and α-ketoglutarate dehydrogenase. It is a cofactor in the mitochondrial dehydrogenase complex that catalyzes the oxidative decarboxylation of α-keto acids such as pyruvate and α-ketoglutarate. In the process, α-lipoic acid is reduced to dihydrolipoic acid and the two substances operate as a redox couple. Besides quenching of free radicals, lipoic acid/dihydrolipoic acid chelate transition metals and also assist in the regeneration of other antioxidants, such as glutathione, α-tocopherol, and ascorbate (Biewenga et al., 1997). Administration of α-lipoic acid to rodents has been demonstrated to reduce the damage that occurs after ischemic-reperfusion injuries in the cerebral cortex (Packer et al., 1997), heart (Freisleben, 2000), and peripheral nerve (Mitsui et al., 1999). α-Lipoic acid has also been found to exert protective effects in cell culture models of hypoxia and excitotoxicity (Tirosh et al., 1999). Despite such positive results, the effectiveness of α-lipoic acid as a neuroprotectant in the retina has largely been ignored to date. It is known that in the early diabetic rat retina, decreased free cytosolic and mitochondrial $NAD^+/NADH$ ratios and increased levels of 4-hydroxyalkenals, which are indicative of retinal hypoxia and increased lipid peroxidation, respectively, are reversed by treatment with α-lipoic acid (Obrosova et al., 2001). It is also important to note that α-lipoic acid is well tolerated by humans with no known side effects reported when taken daily (500–600 mg).

In order to test whether α-lipoic acid might be a candidate drug for the prophylactic use in attenuating ganglion cell death in glaucoma, we examined this possibility in a rat model given ischemia (Chidlow et al., 2002). Rats were injected intraperitoneally with either vehicle or α-lipoic acid (100 mg/kg) once daily for 11 days. On the third day, ischemia was delivered to the rat retina by raising the intraocular pressure above systolic blood pressure for 45 min. The electroretinogram (ERG) was measured prior to ischemia and five days after reperfusion. Rats were killed five or eight days after reperfusion, and the retinas were processed for immunohistochemistry and for determination of mRNA levels by RT-PCR. Ischemia–reperfusion caused a significant reduction in the a- and b-wave amplitudes of the ERG (Fig. 5), a decrease in nitric oxide synthase and Thy-1 immunoreactivities, a decrease in retinal ganglion cell-specific mRNAs (Thy-1 and NF-L), and an increase in bFGF and CNTF mRNA levels. All of these changes were clearly counteracted by α-lipoic acid. Moreover, in mixed rat retinal cultures, α-lipoic acid partially counteracted the loss of GABA-immunoreactive neurons induced by anoxia. The results of the study demonstrate that α-lipoic acid provides

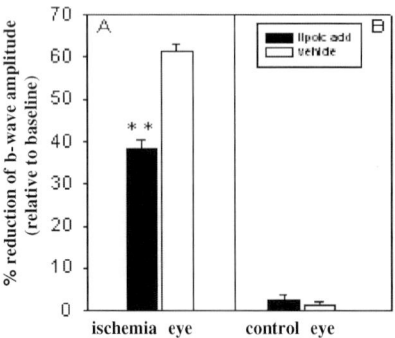
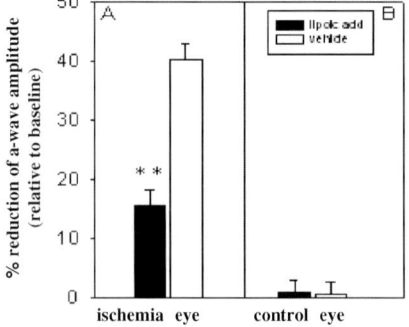

Fig. 5. Amplitude recordings of the a- and b-waves of electroretinograms from rats following ischemia/reperfusion and treated with either vehicle or α-lipoic acid. It can be seen that ischemia/reperfusion caused clear reductions in the a- and b-wave amplitudes, and these were significantly attenuated by α-lipoic acid. Results are means ± SEM, where $n = 16$.

protection from ischemia-reperfusion injuries to the retina as a whole and to ganglion cells in particular.

Nicotinamide

Nicotinamide is a precursor of NADH, which is a substrate for complex 1 (NADH-CoQ oxidoreductase). It is also an inhibitor of poly-ADP-ribose polymerase (PARP), an enzyme which is activated by DNA damage and in turn depletes both NADH and ATP (Eliasson et al., 1997). Activation of PARP plays a role in neuronal injury induced by both ischemia and MPTP toxicity (Mandir et al., 1999). Inhibition of PARP has also been shown to attenuate neuronal injury and ATP depletion produced by MTPT (Cosi and Marien, 1998). Appropriate nicotinamide pretreatment, in a middle cerebral artery occlusion model of stroke, reduced infarct volume, augumented cerebral blood flow, and increased brain NAD^+ levels in ischemic penumbra (Sadanaga-Akiyoshi et al., 2003). Moreover, in human studies, nicotinamide administration improved phosphocreatine/ATP ratios in certain patients. Recent studies on retinal cell cultures and rats support the supposition that appropriate nicotinamide supplementation might benefit glaucoma patients. In one series of experiments, nicotinamide (500 mg/kg) was administered intraperitoneally just before ischemia and into the vitreous (5 μl of a 200 μM solution) immediately

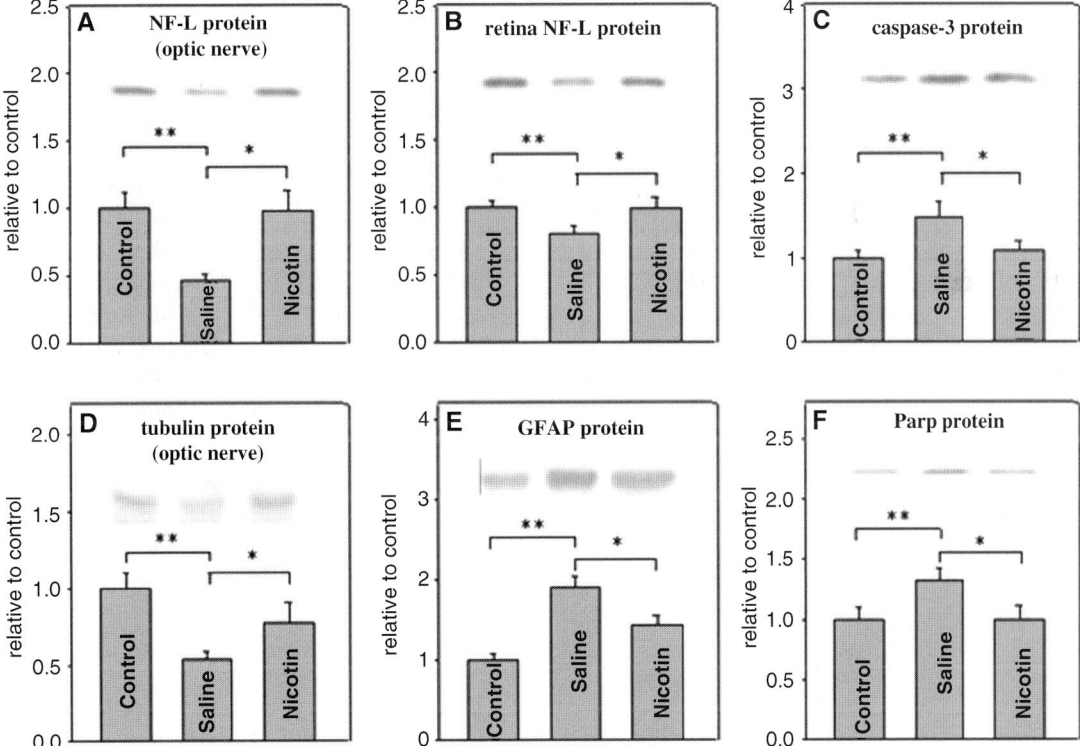

Fig. 6. Protein levels in whole retinas (NF-L, caspase-3, GFAP, and PARP) and optic nerves (NF-L and tubulin) relative to their individual amount of actin of eyes given ischemia where saline ($n = 6$) or nicotinamide ($n = 6$) was administered, and compared with the retina and optic nerve of the eye that in each case had not received any treatment (control). Results expressed as mean values ± SEMs. It can be seen that significant (**$p < 0.01$) changes in all proteins were found in retinas given ischemia (saline) relative to the untreated control retinas. These changes in the retinas given ischemia are significantly blunted (*$p < 0.05$) when the animals are treated with nicotinamide.

after the insult (Ji et al., 2008). Seven days after ischemia (IOP raised to 120 mmHg for 50 min), retinas were analyzed for the localization of various antigens as well as the content of certain proteins and mRNAs. In addition, optic nerves were analyzed for their NF-L content. Ischemia/reperfusion affected a number of parameters in the retina that included a decrease in ganglion cell proteins (Thy-1 and NF-L) and mRNAs (NF-L and melanopsin) (see Fig. 6). Also, ischemia/reperfusion caused clear changes in the localization of nitric oxide synthase and ChAT and an activation of a certain protein/mRNAs (caspase-3, PARP) involved in apoptosis. In addition, ischemia/reperfusion caused a drastic loss of NF-L protein in the optic nerve. All these effects induced by ischemia/reperfusion were significantly blunted by nicotinamide treatment (Fig. 7).

Epigallocatechin gallate

A number of in vivo and in vitro reports have shown that catechins, a class of flavonoid present in large amounts in green tea, can exert neuroprotective actions in several models of neurodegenerative disorders (Mitscher et al., 1997; Mandel et al., 2005). The major catechin component of green tea, (−)-epigallocatechin-3-gallate (EGCG), at low micromolar and submicromolar concentrations, attenuates a variety of insults by what appear to be different mechanisms. For example, EGCG acts as an antioxidant, increases endogenous antioxidant defenses, counteracts inflammation-mediated neuronal injury, and causes a downregulation of pro-apoptotic genes. In relation to the eye, the antioxidant property of EGCG has been reported to account for its protective action of the lens against photooxidative stress

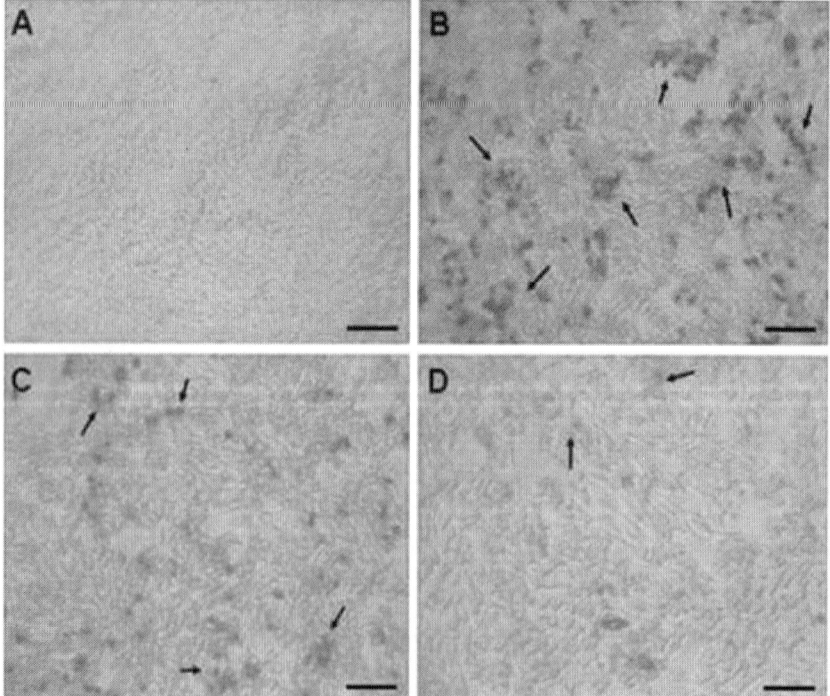

Fig. 7. Staining of RGC-5 cells exposed to light for 96 h (B) compared with cells maintained in the dark (A) by the APOPercentage™ methodology for apoptosis (arrows). Inclusion of 0.5 mm nicotinamide (C) or 100 μM NU1025, a known PARP inhibitor (D), clearly reduced the positive staining of cell membranes as an indication for apoptosis induced by light. Scale bar = 20 μm. (See Color Plate 24.7 in color plate section.)

(Thiagarajan et al., 2001; Vinson and Zhang, 2005). Moreover, our recent studies have shown that photoreceptor degeneration induced by an intraocular injection of sodium nitroprusside (nitric oxide generator) can be attenuated by co-injection with EGCG (Zhang and Osborne, 2006) and that a systemic administration of EGCG attenuates retinal damage caused by ischemia/reperfusion (Zhang et al., 2007). These studies provide clear evidence that EGCG is a powerful antioxidant and can act as a neuroprotectant, making its potential use for the treatment of retinal degenerative diseases worthy of consideration.

Analysis of the bioavailability of EGCG in rats (Suganuma et al., 1998) shows orally administered EGCG to reach various tissues including the central nervous system, within 6 h. Moreover, a second administration after a 6 h interval enhances tissue levels four to six times above that of a single administration. In addition, all the evidence suggests that when a large amount of EGCG is orally administered, it is well tolerated with no ophthalmic abnormalities detected (Isbrucker et al., 2006). It has also been estimated that the half-life of green tea catechins after high oral dosing is between 451 and 479 min in albino rats, which is approximately 1–10 times longer than when administered by an intravenous route (Zhu et al., 2000). In humans it has been shown that the pharmacokinetic parameters for EGCG are similar when present as decaffeinated green tea or as a pure form. Oral administration of EGCG at a high daily dose of 800 mg is known to be safe and without side effects in healthy individuals. Recently, we therefore examined whether orally administrated EGCG to rats can attenuate retinal ischemia (Zhang et al., 2008). Ischemia was delivered to one eye of a number of rats by raising the intraocular pressure, where EGCG (0.5%, each rat intake was approximately 200 ml/day) was present in the drinking water of half the animals three days before ischemia and also during the next five days of reperfusion. The ERGs of both eyes from all rats were recorded before ischemia and five days following ischemia. Seven days after ischemia, either retinas from both eyes of all rats were analyzed for the localization of various antigens or extracts were prepared for analysis for the level of specific proteins and mRNAs. Ischemia/reperfusion to the retina affected a number of parameters. These included the localization of Thy-1 and choline acetyltransferase, the a- and b-wave amplitudes of the ERG, the content of certain retinal and optic nerve proteins, and various mRNAs. Significantly, EGCG statistically blunted many of the effects induced by ischemia/reperfusion (see Figs. 8 and 9). This included ischemia-induced alterations to proteins associated with ganglion cells (NF-L and tubulin), Müller cells (GFAP), photoreceptors (rhodopsin kinase), and apoptosis (caspase-3, PARP, and AIF).

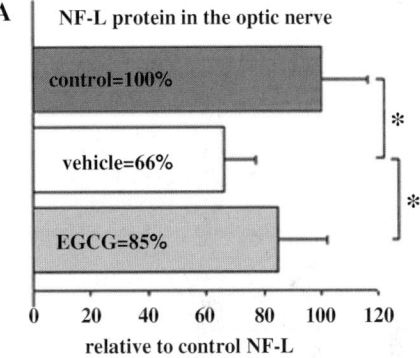

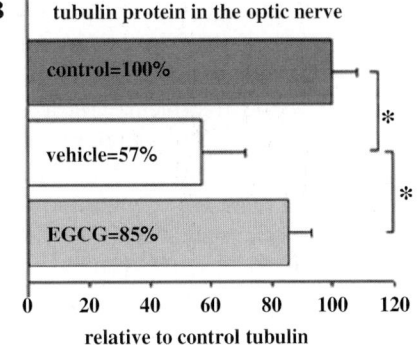

Fig. 8. NF-L (A) and tubulin (B) proteins relative to actin in the nonischemic optic nerves (control) and following ischemia/reperfusion with vehicle treatment (vehicle) or EGCG treatment (EGCG). Ischemia/reperfusion caused a significant decrease in the amounts of both NF-L and tubulin proteins. EGCG treatment significantly attenuated this reduction. Data are means with error bars indicating ±SEMs. $n = 8$ in each case, $^*p < 0.05$, and $^{**}p < 0.01$.

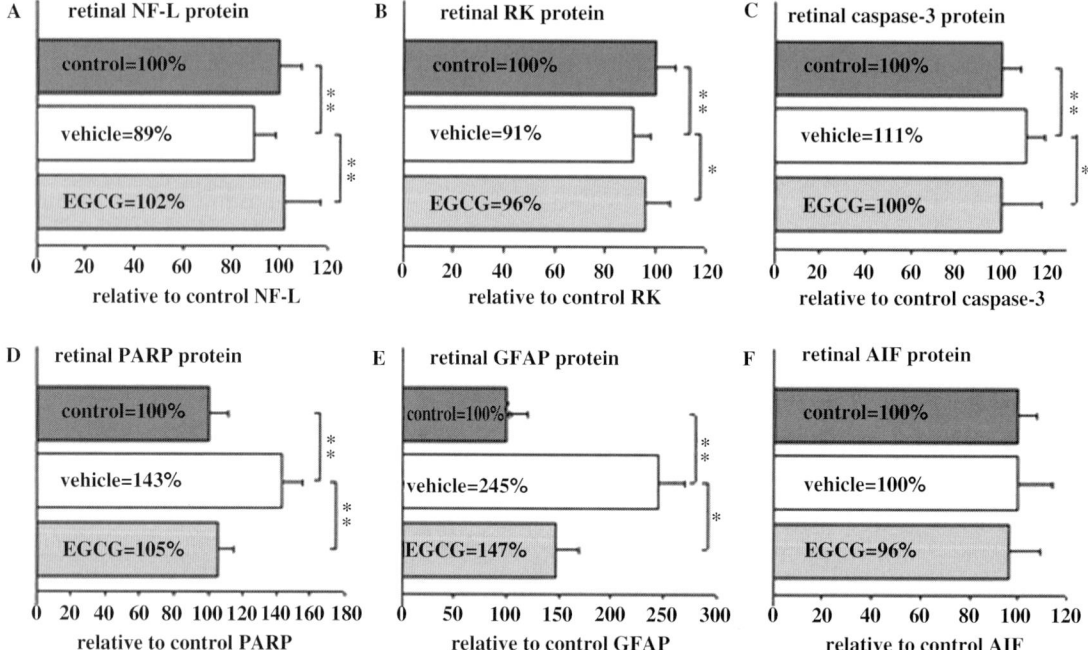

Fig. 9. NF-L (A), rhodopsin kinase (B), caspase-3 (C), PARP (D), GFAP (E), and AIF (F) proteins relative to actin in the retina in the nonischemic ischemia (control) and following ischemia/reperfusion with vehicle treatment (vehicle) or EGCG treatment (EGCG). Ischemia/reperfusion caused a significant change in the amounts of all proteins (except for AIF), and treatment with EGCG significantly counteracted this effect. Data are means with error bars indicating ±SEMs. $n = 6$ in each case, $*p < 0.05$, and $**p < 0.01$.

Conclusion

This article makes a case for the use of substances such as creatine, α-lipoic acid, nicotinamide, and EGCG as adjunctive therapies in the treatment of glaucoma. It is argued that, on diagnosis of glaucoma, immediate treatment with a substance that enhances mitochondrial function will increase the survival of still functional ganglion cells. It is postulated that the process of ganglion cell death in glaucoma occurs in a number of stages, initiated by an ischemic insult to their axons. In this state, ganglion cells can still survive for many years, but they are nevertheless more susceptible to further insults. The final demise of a particular ganglion cell occurs when the cell's energy metabolism reaches a critical point caused by additional insults. Potential additional insults include light or raised extracellular levels of glutamate, NO, or TNF-α. It is suggested that should a substance such as creatine, α-lipoic acid, nicotinamide, and EGCG be available to ganglion cells in their susceptible stage, mitochondrial function will be potentially enhanced and in the process prolong ganglion survival. The potential advantage of daily intake of such nontoxic compounds by glaucoma patients to slow down visual loss is a theoretical one based on animal studies but one that makes sense. It is not suggested as an alternative to the use of IOP-lowering agents but rather as a form of adjunct treatment.

References

Abu-Amero, K.K., Morales, J. and Bosley, T.M. (2006) Mitochondrial abnormalities in patients with primary open-angle glaucoma. Invest. Ophthalmol. Vis. Sci., 47: 2533–2541.

Biewenga, G.P., Haenen, G.R. and Bast, A. (1997) The pharmacology of the antioxidant lipoic acid. Gen. Pharmacol., 29: 315–331.

Brewer, G.J. and Wallimann, T.W. (2000) Protective effect of the energy precursor creatine against toxicity of glutamate and β-amyloid in rat hippocampal neurons. J. Neurochem., 74: 1968–1979.

Bristow, E.A., Griffiths, P.G., Andrews, R.M., Johnson, M.A. and Turnbull, D.M. (2002) The distribution of mitochondrial activity in relation to optic nerve structure. Arch. Ophthalmol., 120: 791–796.

Carelli, V., Ross-Cisneros, F.N. and Sadun, A.A. (2004) Mitochondrial dysfunction as a cause of optic neuropathies. Prog. Retin. Eye Res., 24: 53–90.

Chan, D.C. (2006) Mitochondria: dynamic organelles in disease, ageing, and development. Cell, 125: 1241–1252.

Chen, E., Soderberg, P.G. and Lindstrom, B. (1992) Cytochrome oxidase activity in rat retina after exposure to 404 nm blue light. Curr. Eye Res., 11: 825–831.

Chen, Q., Vazquez, E.J., Moghaddas, S., Hoppel, C.L. and Lesnefsky, E.J. (2003) Production of reactive oxygen species by mitochondria: central role of complex III. J. Biol. Chem., 278(38): 36027–36031.

Chidlow, G., Schmidt, K.-G., Wood, J.P.M., Melena, J. and Osborne, N.N. (2002) α-Lipoic acid protects the retina against ischemia-reperfusion. Neuropharmacology, 43: 1015–1025.

Cosi, C. and Marien, M. (1998) Decreases in mouse brain NAD+ and ATP induced by 1-methyl-4-phenyl-1,2,3,6-tetrahydropyridine (MPTP): prevention by the poly (ADP-ribose) polymerase inhibitor, benzamide. Brain Res., 809: 58–67.

Dryer, E.B., Zurakowski, D., Schumer, R.A., Podos, S.M. and Lipton, S.A. (1996) Elevated glutamate levels in the vitreous body of humans and monkeys with glaucoma. Arch. Ophthalmol., 114: 299–305.

Eliasson, M.J.L., Sampei, K., Madir, A.S., Hurn, P.D., Traystman, R.J., Bao, J., Pieper, A., Wang, Z.Q., Dawson, T.M., Snyder, S.H. and Dawson, V.L. (1997) Poly(ADP-ribose) polymerase gene disruption renders mice resistant to cerebral ischemia. Nat. Med., 3: 1089–1095.

Flammer, J. and Orgül, S. (1998) Optic nerve blood-flow abnormalities in glaucoma. Prog. Retin. Eye Res., 17: 267–289.

Flammer, J., Orgül, S., Costa, V.P., Orzalesi, N., Krieglstein, G.K., Serra, L.M., Renard, J.-P. and Stefansson, E. (2002) The impact of ocular blood flow in glaucoma. Prog. Retin. Eye Res., 21: 359–393.

Freisleben, H.J. (2000) Lipoic acid reduces ischemia-reperfusion injury in animal models. Toxicology, 148: 159–171.

Godley, B.F., Shamsi, F.A., Liang, F.Q., Jarrett, S.G., Davies, S. and Boulton, M. (2005) Blue light induces mitochondrial DNA damage and free radical production in epithelial cells. J. Biol. Chem., 280: 21061–21066.

Green, D.R. and Reed, J.C. (1998) Mitochondria and apoptosis. Science, 281: 1309–1312.

Isbrucker, R.A., Bausch, J., Edwards, J.A. and Wolz, E. (2006) Safety studies on epigallocatechin gallate (EGCG) preparations. Part 1: genotoxicity. Food Chem. Toxicol., 44: 626–635.

Ji, D., Li, G.-U. and Osborne, N.N. (2008) Nicotinamide attenuates retinal ischemia and light insults to neurones. Neurochem. Int., 52: 786–798.

King, A., Gottlieb, E., Brooks, D.G., Murphy, M.P. and Dunaief, J.L. (2004) Mitochondria-derived reactive oxygen species mediate blue light-induced death of retinal pigment epithelial cells. Photochem. Photobiol., 79: 470–475.

Klivenyi, P., Ferrante, R.J. and Matthews, R.T. (1998) Neuroprotective effects of creatine in a transgenic animal model of ALS. Nat. Med., 5: 347–350.

Kroemer, G. and Reed, J.C. (2000) Mitochondrial control of cell death. Nat. Med., 6: 513–519.

Lascaratos, G., Ji, D., Wood, J.P.M. and Osborne, N.N. (2007) Visible light affects mitochondrial function and induces neuronal death in retinal cell cultures. Vision Res., 47: 1191–1201.

Lawler, J.M., Barnes, W.S., Wu, G., Song, W. and Demaree, S. (2002) Direct antioxidant properties of creatine. Biochem. Biophys. Res. Commun., 290: 47–52.

Levy, N.S. (1976) The effect of interruption of the short ciliary arteries on slow axoplasmic transport and histology within the optic nerve of the rhesus monkey. Invest. Ophthalmol., 15: 405–410.

Mandel, S.A., Avramovich-Tirosh, Y., Reznichenko, L., Zheng, H., Weinreb, O., Amit, T. and Youndim, M.B.H. (2005) Multifunctional activities of green tea catechins in neuroprotection. Neurosignals, 14: 46–60.

Mandir, A.S., Przedborski, S., Jackson-Lewis, V., Wang, Z.Q., Simbulan-Rosenthal, C.M., Smulson, M.E., Hoffman, B.E., Guastella, D.B., Dawson, V.L. and Dawson, T.M. (1999) Poly (ADP-ribose) polymerase activation mediates 1-methyl-4-phenyl-1,2,3,6-tetrahydropyridine (MPTP)-induced parkinsonism. Proc. Natl. Acad. Sci. U.S.A., 96: 5774–5779.

McLennan, H.R. and Degli Esposti, M. (2000) The contribution of mitochondrial respiratory complexes to the production of reactive oxygen species. J. Bioenerg. Biomembr., 32: 153–162.

Mitscher, L.A., Jung, M., Shankel, D., Dou, J.-H., Steele, L. and Pillai, S.P. (1997) Chemoprotection: a review of the potential therapeutic antioxidant properties of green tea (*Camellia sinensis*) and certain of its constituents. Med. Res. Rev., 17: 327–365.

Mitsui, Y., Schmelzer, J.D., Zollman, P.J., Mitsui, M., Tritschler, H.J. and Low, P.A. (1999) Alpha-lipoic acid provides neuroprotection from ischemia-reperfusion injury of peripheral nerve. J. Neurol. Sci., 163: 11–16.

Morin, C. and Simon, N. (2006) Mitochondria: a target for neuroprotective intervention in cerebral ischemia-reperfusion. Curr. Pharm. Des., 12: 739–757.

Neufeld, A.H., Sawada, A. and Becker, B. (1999) Inhibition of nitric-oxide synthase 2 by aminoguanidine provides neuroprotection of retinal ganglion cells in a rat model of chronic glaucoma. Proc. Nat. Acad. Sci. U.S.A., 96: 9944–9948.

Obrosova, I.G., Stevens, M.J. and Lang, H.J. (2001) Diabetes-induced changes in retinal NAD-redox status: pharmacological modulation and implications for pathogenesis of diabetic retinopathy. Pharmacology, 62: 172–180.

Osborne, N.N., Lascaratos, G., Bron, A.J., Chidlow, G. and Wood, J.P.M. (2006) A hypothesis to suggest that light is a risk factor in glaucoma and the mitochondrial optic neuropathies. Br. J. Ophthalmol., 90: 237–241.

Osborne, N.N., Li, G.-Y., Ji, D., Mortiboys, H.J. and Jackson, S. (2008) Light affects mitochondria to cause apoptosis to cultured cells: possible relevance to ganglion cell death in certain optic neuropathies. J. Neurochem., 105: 2013–2028.

Osborne, N.N., Melena, J., Chidlow, G. and Wood, J.P.M. (2001) A hypothesis to explain cell death caused by vascular insults to the optic nerve head: possible implication for the treatment of glaucoma. Br. J. Ophthalmol., 85: 1252–1259.

Osborne, N.N., Ugarte, M., Chao, M., Chidlow, G., Bae, J.H., Wood, J.P.M. and Nash, M.S. (1999) Neuroprotection in relation to retinal ischemia and relevance to glaucoma. Surv. Ophthalmol., 43(Suppl 1): S102–S128.

Pache, M. and Flammer, J. (2006) A sick eye in a sick body? Systemic findings in patients with primary open-angle glaucoma. Surv. Ophthalmol., 51: 179–212.

Packer, L., Tritschler, H.J. and Wessel, K. (1997) Neuroprotection by the metabolic antioxidant alpha-lipoic acid. Free Radic. Biol. Med., 22: 359–378.

Peng, T.I., Jou, M.G., Sheu, S.S. and Greenamyre, J.T. (1998) Visualization of NMDA receptor-induced mitochondrial calcium accumulation in striatal neurons. Exp. Neurol., 149: 1–12.

Quigley, H.A. (1999) Neuronal death in glaucoma. Prog. Retin. Eye Res., 18: 39–57.

Sadanaga-Akiyoshi, F., Yao, H., Tanuma, S., Nakahara, T., Hong, J.S., Ibayashi, S., Uchimura, H. and Fujishima, M. (2003) Nicotinamide attenuates focal ischemic brain injury in rats: with special reference to changes in nicotinamide and NAD+ levels in ischemic core and penumbra. Neurochem. Res., 28: 1227–1234.

Schon, E.A. and Manfredi, G. (2003) Neuronal degeneration and mitochondrial function. J. Clin. Invest., 111: 303–312.

Steeghs, K., Benders, A., Oerlemans, F., de Haan, A., Heerschap, A., Ruitenbeek, W., Jost, C., van Deursen, J., Perryman, B., Pette, D., Brückwilder, M., Koudijs, J., Jap, P., Veerkamp, J. and Wieringa, B. (1997) Altered Ca^{2+} responses in muscles with combined mitochondrial and cystolic creatine kinase deficiencies. Cell, 89: 93–103.

Suganuma, M., Okabe, S., Oniyama, M., Tada, Y., Ito, H. and Fujiki, H. (1998) Wide distribution of [^{3}H](-)- epigallocatechin gallate, a cancer preventive tea polyphend, in mouse tissue. Carcinogenesis, 10: 1771–1776.

Tezel, G. (2006) Oxidative stress in glaucomatous neurodegeneration: mechanisms and consequences. Prog. Retin. Eye Res., 25: 490–513.

Tezel, G. and Wax, M.B. (2004) Hypoxia-inducible factor 1-alpha in the glaucomatous retina and optic nerve head. Arch. Ophthalmol., 122: 1348–1356.

Thiagarajan, G., Chandani, S., Sundari, C.S., Rao, S.H., Kulkarni, A.V. and Balasubramanian, D. (2001) Antioxidant properties of green and black tea, and their potential ability to retard the progression of eye lens cataract. Exp. Eye Res., 73: 393–401.

Tirosh, O., Sen, C.K., Roy, S., Kobayashi, M.S. and Packer, L. (1999) Neuroprotective effects of alpha-lipoic acid and its positively charged amide analogue. Free Radic. Biol. Med., 26: 1418–1426.

Vinson, J.A. and Zhang, J. (2005) Black and green teas equally inhibit diabetic cataracts in a streptozotocin-induced rat model of diabetes. J. Agric. Food Chem., 53: 3710–3713.

Wang, L., Dong, J., Cull, G., Fortune, B. and Cioffi, G.A. (2003) Varicosities of intraretinal ganglion cell axons in human and nonhuman primates. Invest. Ophthalmol. Vis. Sci., 44: 2–9.

Weinreb, R.N. and Khaw, P.T. (2004) Primary open-angle glaucoma. Lancet, 363: 1711–1720.

Whitmore, A.V., Libby, R.T. and John, S.W.M. (2005) Glaucoma: thinking in new ways – a role for autonomous axonal self-destruction and other compartmentalised processes?. Prog. Retin. Eye Res., 24: 639–662.

Wyss, M. and Kaddurah-Daouk, R. (2000) Creatine and creatine metabolism. Physiol. Rev., 80: 1107–1213.

Zhang, B. and Osborne, N.N. (2006) Oxidative-induced retinal degeneration is attenuated by epigallocatechin gallate. Brain Res., 1124: 176–187.

Zhang, B., Rusciano, D. and Osborne, N.N. (2008) Orally administered epigallocatechin gallate attenuates retinal neuronal death in vivo and light-induced apoptosis in vitro. Brain Res., 1198: 141–152.

Zhang, B., Safa, R., Rusciano, D. and Osborne, N.N. (2007) Epigallocatechin gallate, an active ingredient from green tea, attenuates damaging influences to the retina caused by ischemia/reperfusion. Brain Res., 1159: 40–53.

Zhu, M., Chen, Y. and Li, R.C. (2000) Oral absorption and bioavailability of tea catechins. Planta Med., 66: 444–447.

Zhu, S., Mingwei, Li., Figueroa, B.E., Liu, A., Stavrovskaya, I.G., Pasinelli, P., Beal, M.F., Brown, R.H., Kristal, B.S., Ferrante, R.J. and Friedlander, R.M. (2004) Prophylactic creatine administration mediates neuroprotection in cerebral ischemia in mice. J. Neurosci., 24: 5909–5912.

CHAPTER 25

Astrocytes in glaucomatous optic neuropathy

M. Rosario Hernandez[1],*, Haixi Miao[1] and Thomas Lukas[2]

[1]*Department of Ophthalmology, Feinberg School of Medicine, Northwestern University, Chicago, Illinois, USA*
[2]*Department of Molecular Pharmacology and Biological Chemistry, Feinberg School of Medicine, Northwestern University, Chicago, Illinois, USA*

Abstract: Glaucoma, the second most prevalent cause of blindness worldwide, is a degenerative disease characterized by loss of vision due to loss of retinal ganglion cells. There is no cure for glaucoma, but early intervention with drugs and/or surgery may slow or halt loss of vision. Increased intraocular pressure (IOP), age, and genetic background are the leading risk factors for glaucoma. Our laboratory and other investigators have provided evidence that astrocytes are the cells responsible for many pathological changes in the glaucomatous optic nerve head (ONH). Over several years, in vivo and in vitro techniques characterized the changes in quiescent astrocytes that lead to the reactive phenotype in glaucoma. Reactive astrocytes alter the homeostasis and integrity of the neural and connective tissues in the ONH of human and experimental glaucoma in monkeys. During the transition of quiescent astrocytes to the reactive phenotype altered astrocyte homeostatic functions such as cell–cell communication, migration, growth factor pathway activation, and responses to oxidative stress may impact pathological changes in POAG. Our data also suggests that the creation of a non-supportive environment for the survival of RGC axons through remodeling of the ONH by reactive astrocytes leads to progression of glaucomatous optic neuropathy.

Keywords: glaucoma; optic nerve head; astrocytes; signal transduction; migration; oxidative stress; extracellular matrix; growth factors

Introduction

Primary open angle glaucoma (POAG) is a retinal disease characterized by optic neuropathy associated with optic cupping and loss of visual field (Johnson et al., 2000; Weinreb and Khaw, 2004). Typically affecting older adults, the intraocular pressure (IOP) exceeds the level that is tolerated by an individual optic nerve. Although genes for some types of glaucoma including POAG have been characterized, the underlying cause of POAG remains unknown (Hewitt et al., 2006; Wiggs, 2007). A likely target of the mechanical stress generated by elevated IOP is the lamina cribrosa in the optic nerve head (ONH) (Bellezza et al., 2003). The signals generated by IOP-related stress are sensed by hypothetical "mechanosensors" in the lamina cribrosa of which the astrocyte is the main cellular component (Hernandez, 2000).

Astrocytes are the major glial cell type in the non-myelinated ONH in most mammals and provide cellular support functions to the axons

*Corresponding author. Tel.: 312-503-1064; Fax: 312-503-1062; E-mail: m-hernandez-neufeld@northwestern.edu

while interfacing between connective tissue surfaces and surrounding blood vessels. In addition to astrocytes, a second cell type exists in the lamina cribrosa of humans and non-human primates, the lamina cribrosa cell. Lamina cribrosa cells can be distinguished from astrocytes because they do not express glial fibrillary acidic protein (GFAP) in vivo or in vitro and they do not express vascular specific markers or microglial markers (Hernandez et al., 1988). In the normal ONH, astrocytes are considered to be quiescent (Bachoo et al., 2004). Astrocytes become reactive in response to injury or disease and participate in the formation of a glial scar, which does not support axonal survival or growth (Liu et al., 2006).

Quiescent astrocytes

Quiescent astrocytes are terminally differentiated and in the CNS there are several subpopulations with regional specialization. In the lamina cribrosa, astrocytes form lamellae oriented perpendicular to the axons surrounding a core of fibroelastic extracellular matrix (Hernandez, 2000) (Fig. 1A, B, D). ONH astrocytes have many of the same functions as astrocytes in white matter (Araque et al., 2001; Hansson and Ronnback, 2003). Astrocytes supply energy substrate to axons in the optic nerve and maintain extracellular pH and ion homeostasis in the periaxonal space (Magistretti, 2006; Obara et al., 2008). Sodium channels in astrocytes participate in Na^+ homeostasis, providing a path for Na^+ entry into the cytoplasm (Bowman et al., 1984). The level of intracellular Na^+ regulates the activity of various transporters, particularly Na^+/K^+ ATPase and the Na^+/glutamate transporter (Anderson and Swanson, 2000). Astrocytes regulate water exchange between the brain and vascular space through expression of the water channel AQP4 in membrane domains in their end-feet around vessels (Tait et al., 2008). In astrocytes, voltage-gated calcium channels deliver Ca^{2+} into the cytoplasm and participate in generation of glial Ca^{2+} signals. Astrocytes maintain the scant periaxonal ECM consisting of glycoproteins, such as laminin and proteoglycans. In the normal CNS, astrocytes express a wide variety of growth factors and receptors, many of which serve as trophic and survival factors for neurons. ONH astrocytes and lamina cribrosa cells express bone morphogenetic proteins (Wordinger et al., 2002), neurotrophins, and receptors (Lambert et al., 2004a, b).

Reactive astrocytes in glaucoma

Adult, quiescent astrocytes become "reactive" after injury or disease and participate in formation of a glial scar, which does not support axonal survival or growth (Hernandez and Pena, 1997; Liu et al., 2006). The major hallmarks of a reactive astrocyte are an enlarged cell body and a thick network of processes with increased expression of GFAP and vimentin (Yang and Hernandez, 2003). Reactive astrocytes increase expression of various cell surface molecules that play important roles in cell–cell recognition and in cell adhesion to substrates, as well as various growth factors, cytokines, and receptors (Ricard et al., 2000; Hernandez et al., 2002). Reactive astrocytes express many new ECM proteins such as laminin, tenascin C, and proteoglycans (Pena et al., 1999b). The expression of TGF-α (Junier, 2000) and TGF-β (Pena et al., 1999a), ciliary neurotrophic factor (CNTF) (Liu et al., 2007), fibroblast growth factor 2 (Smith et al., 2001), platelet-derived growth factor (PDGF) (Gris et al., 2007), and their receptors has been reported to induce the transition of quiescent astrocytes into the reactive phenotype or to modulate the function of reactive astrocytes. Participating in the different pathogenic mechanisms of Alzheimer's disease (Abraham, 2001), amyotrophic lateral sclerosis (ALS) (Pehar et al., 2005), Parkinson's disease (Croisier and Graeber, 2006), and multiple sclerosis (Kielian and Esen, 2004), quiescent astrocytes become reactive in response to a wide variety of stimuli (Ridet et al., 1997; Heales et al., 2004). Reactive astrocytes in the glaucomatous ONH are large rounded cells with many thick processes which expresses increased amounts of GFAP, vimentin, and HSP27, that is motile, and that is located either at the edge of the laminar plates or inside the nerve bundles (Hernandez, 2000) (Fig. 1C).

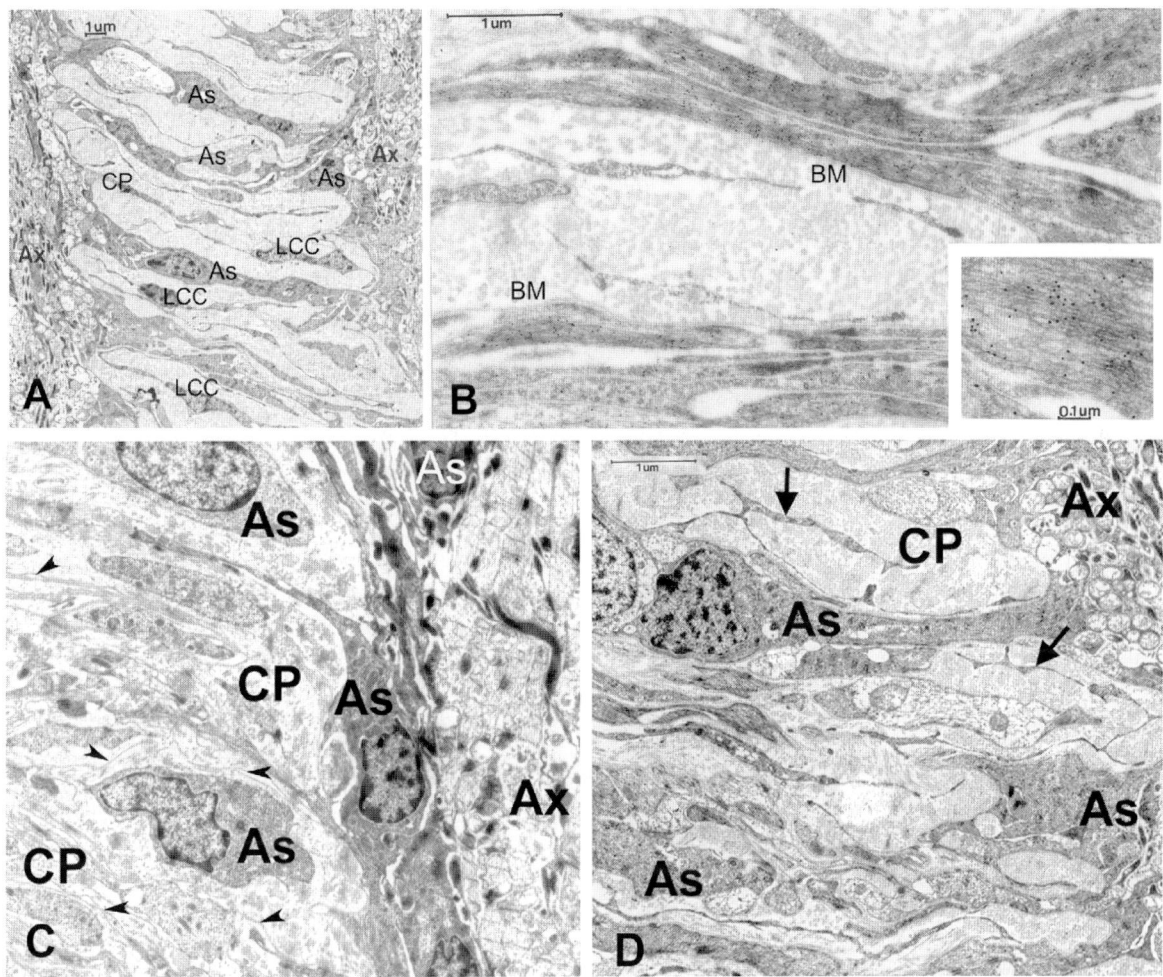

Fig. 1. Electron micrographs of sagittal sections of the monkey optic nerve head. (A) Normal lamina cribrosa. Low power view of the cribriform plates (CP) oriented perpendicular to the axons (Ax) of the retinal ganglion cells. Astrocytes (As) appear elongated and extend thin processes into the ECM of the plates. Lamina cribrosa cells (LCC) are star-shaped cells inside the ECM of the plates. Bar 1 μm. (B) Higher magnification of the cribriform plates showing astrocytes (As) forming lamellae. Note the basement membrane (BM) that surrounds the astrocyte cell body and processes. Bar 1 μm. The inset shows the cytoplasm of an astrocyte stained with anti-human GFAP and immunogold labeling. Bar 0.1 μm. (C) Lamina cribrosa in experimental glaucoma. Eye with elevated IOP (mean IOP 45.5 ± 9.2 mmHg for 7 weeks). In a sagittal view of the lamina cribrosa, note that astrocytes (As) appear rounded and that the ECM of the cribriform plates (CP) is disorganized. Astrocyte cell bodies appear migrating or are located in the axon bundles. (D) Normal contralateral eye. Note that astrocytes are elongated and have thin processes (arrows) into the ECM of the plates. Ax: axons. Bar 1 μm.

Astrocyte cell–cell communication in the optic nerve head

The cell processes of astrocytes are connected to each other via gap junctions (Rose and Ransom, 1997) forming a functional syncytium that allows astrocytes to communicate and maintain control of the ionic and metabolic homeostasis in the ONH. Astrocytic gap junctions are built mainly of connexin-43 (Cx43; GJA1). Connexins form a special class of ion channels that mediate cell-to-cell passage of ions and small molecules, which is thought to help coordinate the activity of astrocytes (Cusato et al., 2003; Nagy et al., 2004). The gap junctional intercellular communication (GJIC) exhibited by cells is highly regulated (Giaume and McCarthy, 1996). One mechanism for regulation is connexin phosphorylation, which has been extensively studied in cells coupled by Cx43 (Bruzzone et al., 1996; Goodenough et al., 1996). Cx43 phosphorylation is correlated with alterations in gap junction uncoupling (Warn-Cramer et al., 1998) and its dephosphorylation is correlated with either decreased GJIC (Oelze et al., 1995) or increased channel conductance (Moreno et al., 1994).

Immunohistochemistry of Cx43 in the monkey normal and glaucomatous optic nerve shows that Cx43 is associated with astrocyte membranes in the lamina cribrosa in normal ONH (Fig. 2A). In contrast Cx43 is intracellular in astrocytes in the glaucomatous ONH (Fig. 2B). Real time RT-PCR comparing primary cultures of astrocytes from normal and glaucomatous human eyes indicate that gene expression of GJA1 is higher in glaucomatous astrocytes compared to normal age matched controls (Fig. 2C). Next we examined the effects of elevated hydrostatic pressure on gap junction intracellular communication GJIC in human ONH astrocytes in vitro providing evidence that astrocytes decrease intercellular communication upon exposure to elevated HP, a mechanical stress (Fig. 3) (Malone et al., 2007). The data further demonstrates that activation of epidermal growth factor receptor (EGFR) mediates long-term phosphorylation of Cx43 on tyrosine in response to pressure (Fig. 4). Activation of EGFR by pressure induces phosphorylation of Cx43 on tyrosine residues and subsequent closure of gap junction communication between neighboring astrocytes. In the normal optic nerve, astrocyte GJIC is maintained so that astrocyte–astrocyte cellular coupling in the synctitium is continuous, allowing propagation of extracellular cues to obtain coordinated responses. Under exposure to elevated pressure in vitro and perhaps in vivo, gap junctions close via activation of the EGFR pathway. Closure of gap junctions will interrupt the continuity of astrocyte intercellular communication causing loss of cell–cell contact and loss of homeostatic regulation. Under these conditions, astrocytes may adopt the reactive astrocyte phenotype characteristic of response to injury in the CNS. In summary, if astrocyte–astrocyte gap junction communication is interrupted in vivo by chronic elevated IOP, as in glaucoma, the homeostasis of the RGC non-myelinated axons may be altered and the blood nerve barrier may be impaired leading to axonal loss and optic disc remodeling characteristic of glaucomatous neuropathy.

Signal transduction in glaucomatous astrocytes

Protein tyrosine kinases (PTKs)

PTKs are the primary mediators of the signaling network that transmit extracellular signals into the cell. PTK signaling activates several small G proteins, including Ras, Rap-1, and the cdc42-Rac-Rho family, as well as pathways regulated by mitogen-activated protein kinases (MAPKs), phosphatidylinositol 3-kinase (PI3K), and phospholipase C. Receptor-type PTKs are activated by ligand binding and directly transduce the extracellular information into intracellular tyrosine phosphorylation events, whereas nonreceptor tyrosine kinases (nrPTK) function as signal transducers in concert with receptor-like molecules that lack tyrosine kinase activity. Members of the receptor PTK family include the EGFR (Wieduwilt and Moasser, 2008), the FGF receptor (FGFR) (Rusnati and Presta, 2007), the PDGF receptor (PDGFR) (Fruttiger et al., 1996), neurotrophin receptors (Lykissas et al., 2007), and ephrin receptors (Himanen et al., 2007). NrPTKs include Src, Fyn, TEC, TXK, JAK1-3, and FAK. In mouse brain astrocytes, members of Src family tyrosine regulate cell adhesion via a FAK-dependent mechanism (Beggs et al., 2003). Among many stimuli, PTKs are activated by mechanical stress (Sawada et al., 2006). Inhibitors of PTKs, such as genistein, block the proliferation through an autocrine response of release of soluble factors in response to transmural compression in astrocytoma cells. Recently, Liu and Neufeld (2007) demonstrated that phosphorylation of EGFR is a necessary step for induction of nitric oxide synthase (iNOS) in astrocytes exposed to HP.

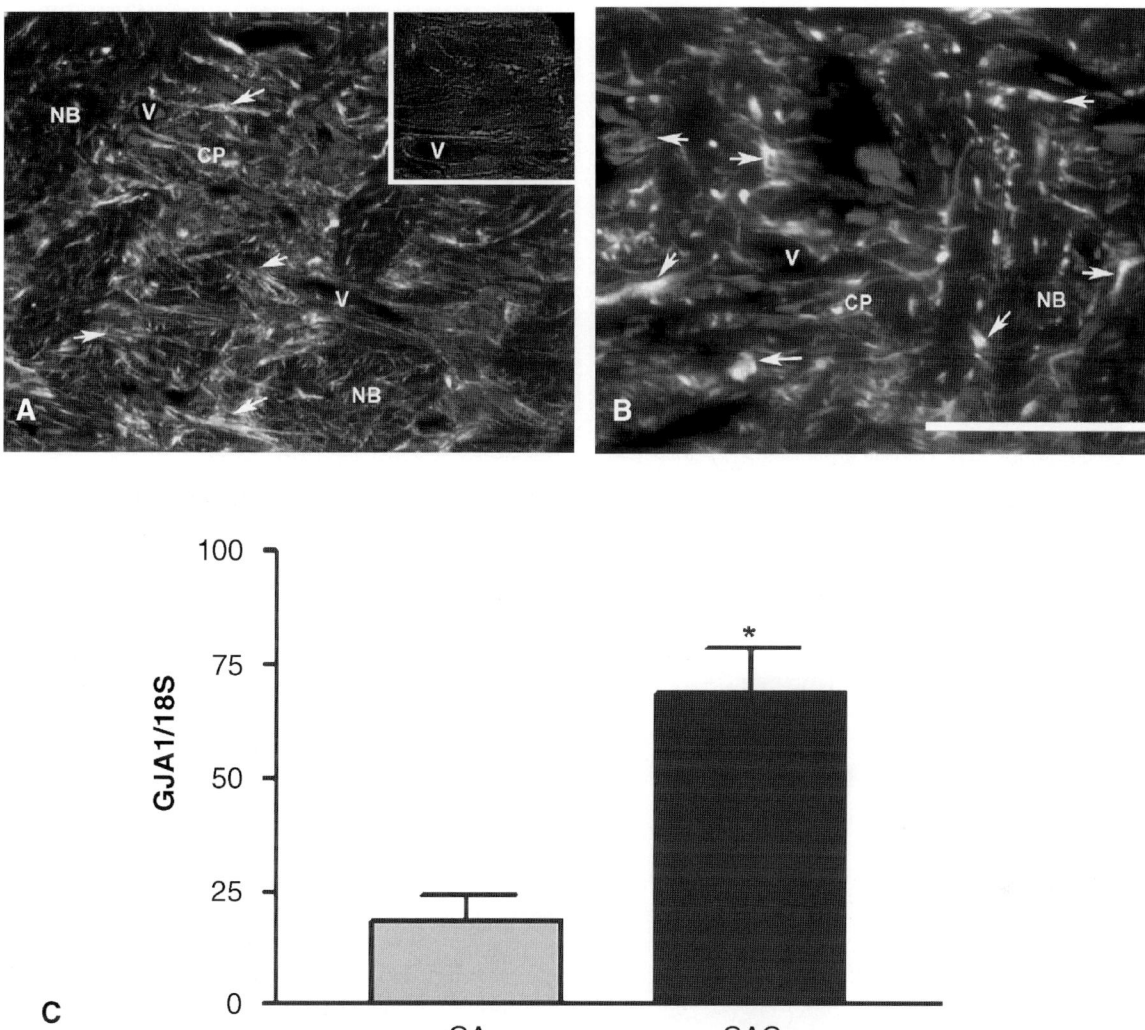

Fig. 2. Expression of connexin-43 in the monkey optic nerve head. (A) Immunofluorescent staining for GFAP shows astrocytes (green) forming lamellae in the cribriform plates of the lamina cribrosa (CP) stained with Cx43 (red). Fine lines of colocalization of Cx43 on astrocytes membranes (yellow, arrows), V: blood vessel, NB: nerve bundles. Inset: Cx43 immunoreactivity following the fine lamellar processes of astrocytes in the normal lamina cribrosa. (B) In the contralateral eye with experimental glaucoma, astrocytes (green) are disorganized and appear rounded with intracellular Cx43 immunoreactivity (yellow, arrows) in the lamina cribrosa. A and B: Magnification bar = 35 μm. For methods see (Agapova et al., 2006a). (C) Relative amount of Cx43 (GJA1) mRNA in human normal and glaucomatous ONH astrocytes measured by quantitative RT-PCR. Bar graphs represent relative expression of Cx43 mRNA normalized to 18S in normal ($n = 8$) and glaucomatous ($n = 8$) ONH astrocyte cultures. Two-tailed t test was used. *Indicates $p < 0.0002$. (See Colour Plate 25.2 in the colour plate section.)

Serine/threonine protein mitogen-activated kinases (MAPKs)

MAPKs phosphorylate specific serines and threonines of target proteins and regulate gene expression, mitosis, movement, metabolism, and programmed cell death. MAPK-catalyzed phosphorylation of proteins functions as a switch to turn on or off activity. Substrates include other protein kinases, phospholipases, transcription

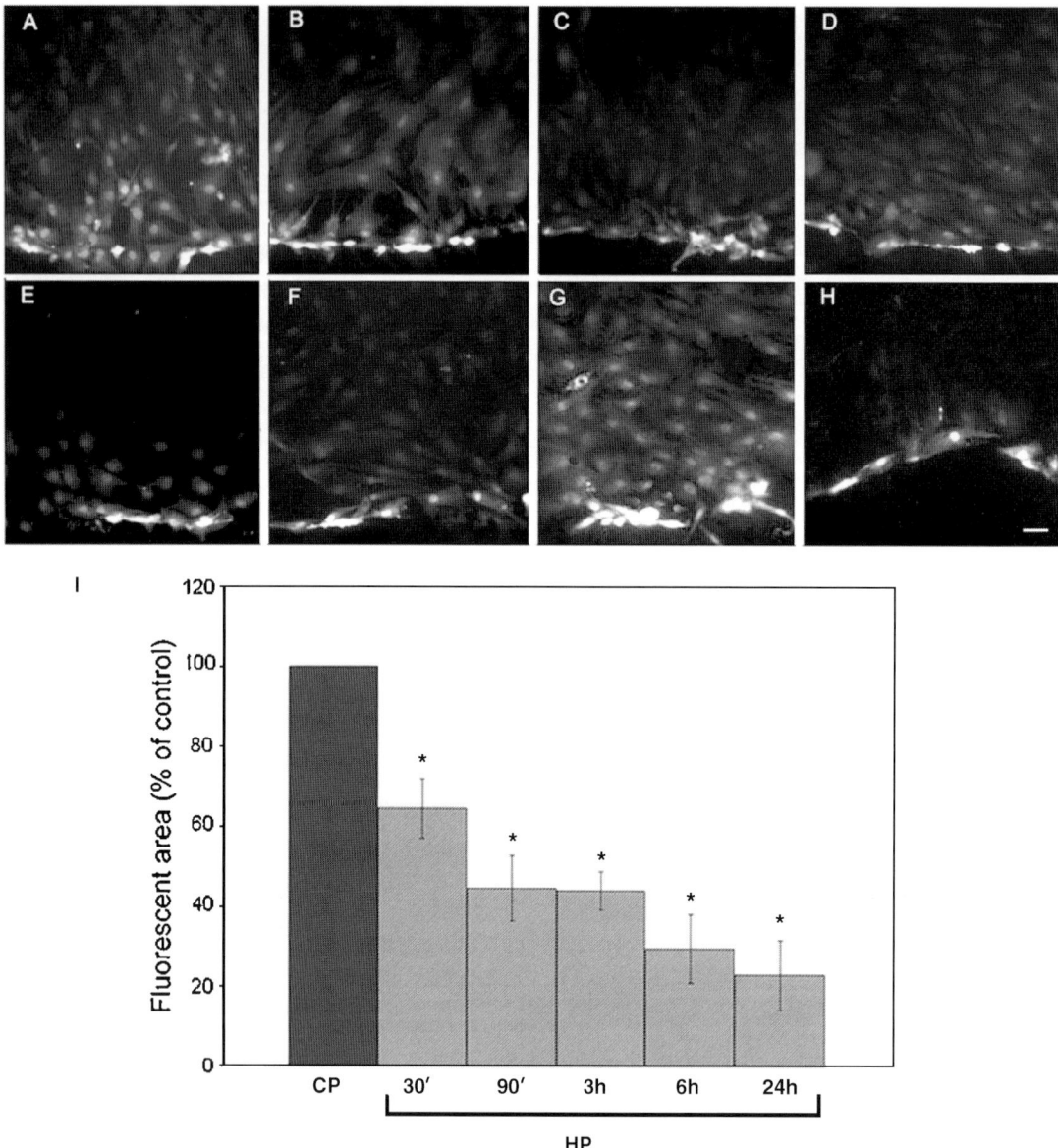

Fig. 3. Effects of hydrostatic pressure (HP) on GJIC measured by the SLDT technique. Fluorescent micrographs of cultured ONH astrocytes exposed to (A) CP or (B) 30 min, (C) 90 min, (D) 3 h, (E) 6 h, and (F) 24 h HP conditions. (G) GJIC in control astrocytes exposed to CP. (H) Carbenoxolone (CBX) inhibited GJIC in ONH astrocytes exposed to the same conditions (Magnification bar = 30 μm). (I) Exposure to HP induces a decrease in astrocyte GJIC determined by the SLDT technique. Quantitation of the LY fluorescent area was determined using OPTIMAS software and expressed as a percentage of the control (100%) exposed to CP (solid bar) and to HP (gray bar) for 30 min, 90 min, 3 h, 6 h, and 24 h (mean ± standard deviation). HP induced a decrease in GJIC by 36 ± 7, 56 ± 8, 57 ± 4, 71 ± 7, and 78% ± 9, respectively, compared to controls. *Indicates statistical significance at $p < 0.05$ from three independent experiments.

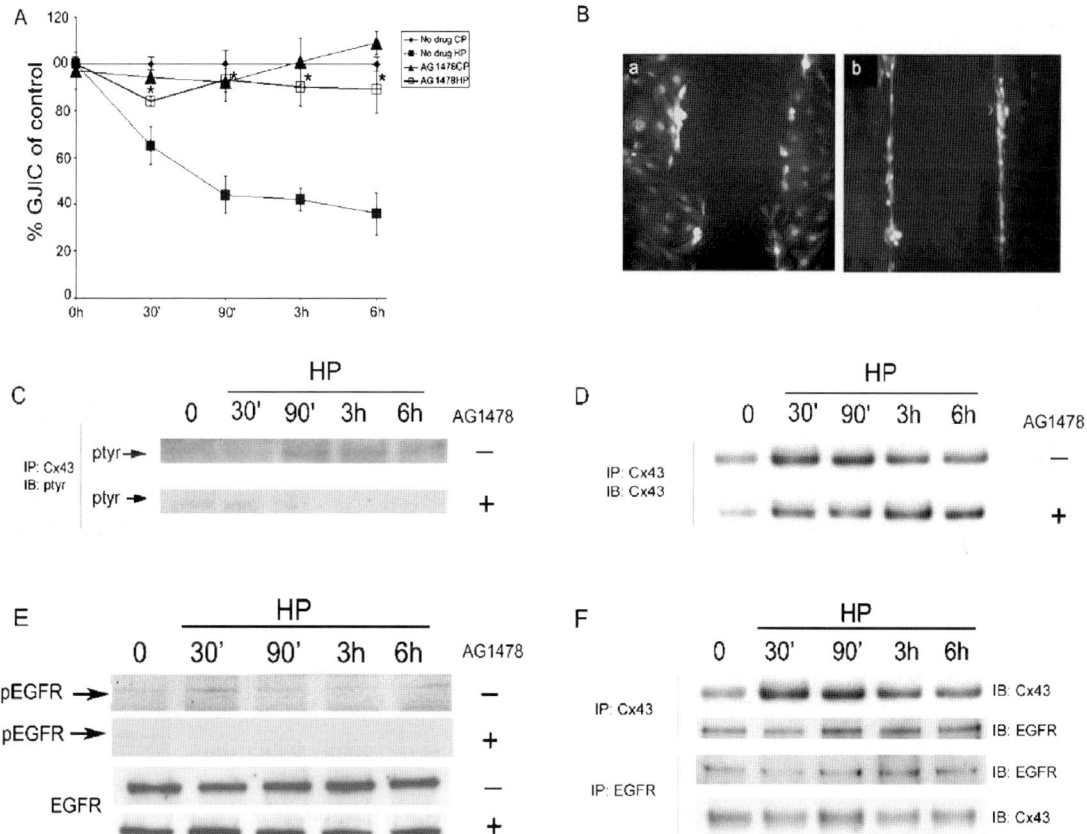

Fig. 4. EGFR mediates GJIC and Cx43 tyrosine phosphorylation in response to HP. (A) GJIC analysis of ONH astrocytes under HP conditions with AG1478 (3 nM), an EGFR inhibitor, results in complete opening of gap junctions from 0–6 h compared to astrocytes exposed to HP without the inhibitor ($^*p<0.05$ from three different experiments). (B) Micrographs of a scratch assay in astrocytes exposed to HP for 90 min with (a) and without (b) AG1478 demonstrate reversal of inhibition of GJIC due to HP in the presence of EGFR inhibitor. (C) Immunoblot analysis of Cx43 immunoprecipitates from ONH astrocytes exposed to HP with and without AG1478. Astrocytes were preincubated with AG1478 for 30 min and protein lysates were collected at 30 min, 90 min, 3 h, and 6 h. Lysates from untreated controls were collected at time 0. Immunoblots on Cx43 immunoprecipitates using phosphotyrosine antibody show a unique band of Cx43, which disappears in the presence of AG1478, indicating tyrosine phosphorylation of Cx43 under HP. (D) Immunoblot with monoclonal anti-Cx43 (BD) on immunoprecipitates of polyclonal anti-Cx43 (Sigma) showed an increase in Cx43 in astrocytes exposed to HP. Ag1478 treatment did not inhibit the increase in Cx43 under pressure. (E) Western blot of phosphorylated EGFR (pEGFR) from astrocytes lysates exposed to HP with and without AG1478. Note that a single band at 175 kDa corresponding to pEGFR appears in lysates from ONH astrocytes exposed to HP, which disappears in the presence of AG1478. Total EGFR levels were not affected. (F) Upper blots: detection of EGFR in Cx43 immunoprecipitates using a polyclonal antibody against human EGFR (Cell Signaling #2232). Note a band at 175 kDa corresponding to EGFR associated with Cx43 that appears in samples exposed to HP. Lower blots: detection of Cx43 in EGFR immunoprecipitates using a polyclonal antibody against human Cx43 (Sigma). Note a darker band at 43 kDa corresponding to Cx43 associated with EGFR in samples exposed to HP for 90 min (from Malone et al., 2007).

factors, and cytoskeletal proteins. There are three well-characterized subfamilies of MAPKs: extracellular signal-regulated kinases, ERK1 and ERK2; c-Jun NH$_2$-terminal kinases, JNK-1, JNK-2, and JNK-3; and the four p38 enzymes — p38α, p38β, p38γ, and p38δ. Studies of tissue samples from patients with various CNS diseases have demonstrated activation of the ERK/MAPK pathway, suggesting a role for the triggering and/or persistence of reactive astrogliosis (Tichauer et al., 2007). Mechanical trauma induces rapid activation of ERK/MAPK pathway in astrocytes in vitro

(Mandell et al., 2001). Activation of ERK/MAPK pathway has been shown in ONH astrocytes in vitro and in vivo in glaucoma and in astrocytes exposed to hydrostatic pressure (Tezel et al., 2003; Hashimoto et al., 2005).

G protein-coupled receptors

Signaling through G protein-coupled receptors (GPCRs) mediates numerous cellular functions including mechanotransduction (Chachisvilis et al., 2006). The effects of activation of G protein-coupled metabotropic glutamate receptors, GABA receptors, nucleotides, and catecholamines in brain astrocytes are under intense study. Endothelin-1 (ET-1) has been proposed to play role in astrogliosis that occurs in the glaucomatous ONH. G protein-coupled endothelin-1 receptors ET_A and ET_B are abundantly expressed and widely distributed in ocular tissues, including the retina, optic nerve, and ONH astrocytes as described recently (Prasanna et al., 2002; Chauhan et al., 2004; Wang et al., 2006). In the CNS, ET-1 regulates astrocyte cell shape through tyrosine phosphorylation of FAK and paxillin (Koyama et al., 2000). Recently, it was shown that the endothelin receptor B is expressed in ONH astrocytes in human glaucoma (Wang et al., 2006). ONH astrocytes and lamina cribrosa cells express functional endothelin receptor A and endothelin receptor B (ETB) in culture (Rao et al., 2007). Moreover, ET-1 regulates the expression of matrix metalloproteinases (MMPs) and tissue inhibitors in ONH astrocytes further suggesting that the ET pathway may be involved in tissue remodeling in glaucoma (He et al., 2007).

The regulators of G protein signaling (RGS) are an important way to modulate signal transduction (RGS5 ocular). RGS5 protein and mRNA are abundantly expressed in human ONH astrocytes in vivo and in vitro (Fig. 5). Current studies suggest that the function of RGS5 is to modulate astrocyte responses in response to pressure. Among the known functions of RGS5 the modulation of G protein-coupled receptors such as prostaglandin EP3 and EP4 which affect the transcription of genes such as NOS2 (Hu et al., 2005). Figure 5 illustrates preferential localization of RGS5 to laminar astrocytes in the glaucomatous human ONH of Caucasian donors compared to normal donors (Fig. 5B) compared with low levels of RGS5 in astrocytes in the normal ONH (Fig. 5A). Consistent with the immunohistochemistry, primary cultures of ONH astrocytes from glaucomatous donors exhibit higher levels of RGS5 protein compared with normal donors by Western blot analysis (Fig. 5C). Real time quantitative RT-PCR detected significant higher mRNA levels for RGS5 in primary cultures of ONH astrocytes from Caucasian donors with POAG compared to normal donors (Fig. 5D).

Ras superfamily of small G proteins

The Ras superfamily members participate in many cellular processes including cell movement and act as signal transducers and/or regulators of membrane trafficking (Seachrist and Ferguson, 2003). Ras proteins cycle between a GTP-bound state and a GDP-bound state because of their GTP hydrolysis activity and a higher affinity for GTP than GDP. In vivo, this cycle is regulated by GTPase activator proteins, which increase the rate of GTP hydrolysis, and guanine nucleotide exchange factors, which stimulate the exchange of GDP for GTP. Members of the Rho, Rab, and Ran families are also regulated by GDP dissociation inhibitors, which, by binding to the GDP-bound form, inhibit nucleotide exchange. Rac1 and Cdc42 regulate migration in astrocytes because of their ability to regulate actin cytoskeletal dynamics by signaling through effectors of the Ras-activated kinases (Wheeler et al., 2006; Bourguignon et al., 2007). CDC42, a small Rho GTPase that participate in polarized motility in astrocytes (Osmani et al., 2006) was upregulated in ONH astrocytes exposed to hydrostatic pressure at the mRNA and protein levels. We also demonstrated early increased Rho activity in ONH astrocytes exposed to hydrostatic pressure (Yang et al., 2004).

Astrocyte migration in the glaucomatous optic nerve head

Migration of astrocytes occurs during normal development, in neurodegenerative diseases, after

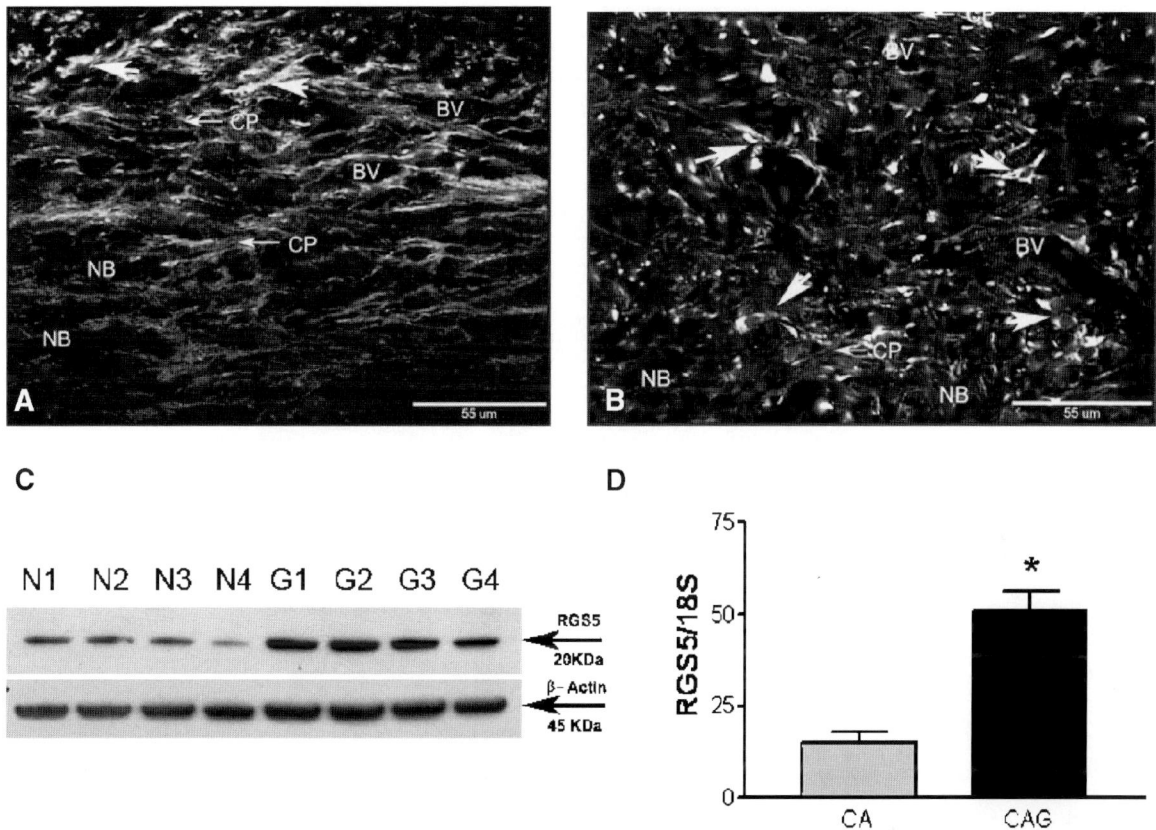

Fig. 5. Expression of RGS5 in the human optic nerve head. (A) Immunofluorescent staining for GFAP shows astrocytes (green) forming lamellae in the cribriform plates of the lamina cribrosa (CP) (yellow, arrows), few astrocytes stained with RGS5 antibody (yellow) V: blood vessel, NB: nerve bundles. (B) In an eye with POAG, there is an increase in RGS5 immunoreactivity in laminar astrocytes (yellow, arrows). A and B: Magnification bar = 35 μm. For methods see (Agapova et al., 2006a). (C) Western blot analysis. (D) Relative amount of RGS5 mRNA in human normal and glaucomatous ONH astrocytes measured by quantitative RT-PCR. Bar graphs represent relative expression of SOD2 mRNA normalized to 18S in normal ($n = 8$) and glaucomatous ($n = 8$) ONH astrocyte cultures. (See Colour Plate 25.5 in the colour plate section.)

injury, and during tumor invasion in the CNS. Cell migration is a biochemical and mechanical response that can originate via in response to ECM molecules such as laminin, fibronectin, or environmental cues such as growth factors or cytokines including the TGFβ family (Yu et al., 2007), and EGF (Liu and Neufeld, 2007). Most importantly, in glaucoma, reactive astrocytes have been shown to migrate from the cribriform plates into the nerve bundles (Liu and Neufeld, 2004) and synthesize neurotoxic mediators such as nitric oxide (NO) and TNF-α, which may be released near axons causing neuronal damage (Liu and Neufeld, 2000; Neufeld and Liu, 2003; Tezel et al., 2004). These findings suggest that migration of astrocytes likely precedes the remodeling in the optic neuropathy.

Exposure of ONH astrocytes to elevated hydrostatic pressure in vitro increased cell migration (Fig. 6E, F) through activation of at least two signaling pathways: PI-3K activation and pressure induced erbB2, EGFR1, and PDGFR kinases activation. Interestingly, inhibition of tyrosine kinase receptors did not affect astrocyte basal migration induced by the cell-free area. On the other hand, mechanical denudation induced migration by a rapid and sustained increase in COX2 activity (Salvador-Silva et al., 2004). The

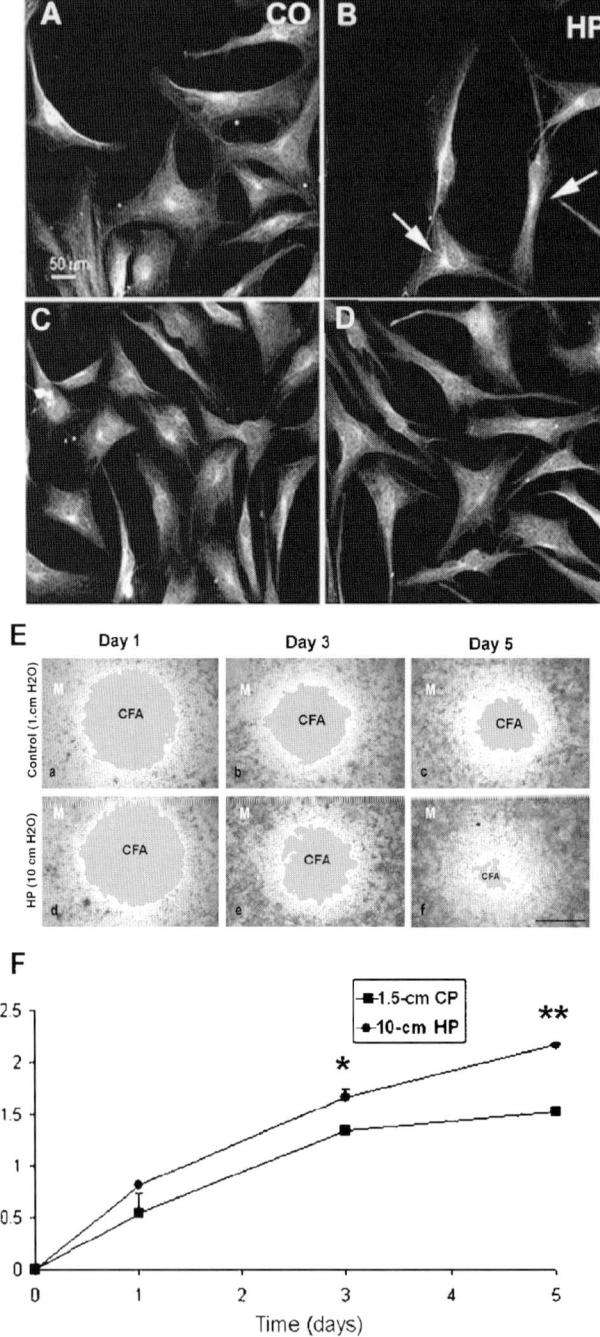

importance of these findings is that in a model of reactive astrocytes in vitro, separate mechanisms govern migration in response to mechanical injury and in response to a physical stress such as elevated pressure.

Previous work in our laboratory demonstrated changes in astrocytes cytoskeleton after exposure to hydrostatic pressure in the intact monolayer (Wax et al., 2000; Salvador-Silva et al., 2001). These changes included redistribution of actin to the cell margins and localization of GFAP and vimentin intermediate filaments to the perinuclear region. In these experiments, astrocytes in the intact monolayer appeared flat and polygonal in shape when exposed to control pressure. GFAP filaments were organized as a diffuse network running radially toward the edges of the cell border (Fig. 6C). In astrocytes migrating into the cell-free area under ambient pressure, GFAP (Fig. 6A) staining was diffuse but evenly distributed throughout the cell. Astrocytes in the intact monolayer exposed to HP appeared elongated in response to HP (Fig. 6D). GFAP staining was stronger around the nuclear region. In migrating astrocytes exposed to HP, GFAP (Fig. 6B) staining was strong around the nucleus towards the rear of the cell.

Cell adhesion of ONH astrocytes

Reactive astrocytes express several cell adhesion molecules including cadherins, integrins, and neural cell adhesion molecules (NCAM) in CNS degenerations (Charles et al., 2000). Three NCAM isoforms result from alternative splicing of a single gene: two major transmembrane isoforms, denoted $NCAM_{140}$ and $NCAM_{180}$, and $NCAM_{120}$ a glycosyl-phosphatidylinositol-linked isoform (Maness and Schachner, 2007). The adhesive properties of NCAM depend on alternative splicing of the primary transcript or post-translational modifications, such as polysialylation (Maness and Schachner, 2007). Activation by extracellular cues of cellular signaling pathways regulates alternative splicing. Recent studies suggest that c-Jun, a transcription factor, regulates alternative splicing of NCAM pre-mRNA and the synthesis of $NCAM_{140}$ (Feng et al., 2002). In the adult human ONH, quiescent astrocytes and lamina cribrosa cells express $NCAM_{140}$ as the predominant isoform (Ricard et al., 2000) $NCAM_{120}$ and $NCAM_{180}$ are not expressed by quiescent ONH astrocytes in vivo or in vitro (Ricard et al., 2000; Hernandez et al., 2002). In the human glaucomatous ONH, reactive astrocytes differentially express $NCAM_{180}$ mRNA and protein (Hernandez et al., 2002). Injury to the mouse optic nerve increases expression of $NCAM_{180}$ in the axons and in reactive astrocytes (Becker et al., 2001).

Integrin receptors cluster at focal adhesion complexes, where cells anchor into the ECM. Binding at these sites can activate intracellular signaling pathways to modify cell behavior, including cell migration, differentiation, adhesion to substrates, and target recognition. Integrins $\alpha 2\alpha 1$, $\alpha 3\alpha \beta 1$, $\alpha 6\beta 1$, and $\alpha 6\beta 4$ provide attachment for ONH astrocytes to basement membranes via

Fig. 6. Migration of ONH astrocytes: (A—D) Changes in intermediate filaments in ONH astrocytes exposed to hydrostatic pressure. Immunofluorescence staining of ONH astrocytes with glial fibrillary acidic protein (GFAP) in a migration assay after exposure to 10 cm H_2O hydrostatic pressure (HP) or ambient pressure (CO) for 3 days: Migrating astrocytes in the cell-free area (CFA) under control (CO) (A) and hydrostatic pressure (HP) (B) conditions: GFAP localizes to the perinuclear region and the rear of the cell in migrating cells under HP (B). Astrocytes under ambient pressure (C) and hydrostatic pressure (D) astrocytes in the intact monolayer appear elongated in response to HP (D), and show stronger GFAP staining around the perinuclear region. Under control pressure, GFAP appears as a diffuse network running radially toward the edges of the cell border and the cells remain flat and polygonal in shape (C). (E) Representative micrographs of the closure of the cell-free area (CFA) under control pressure (CP) (a–c) or under hydrostatic pressure (HP) (d–f). Note that at day 1 there is no difference between HP and CP. At days 3 (b, e) and 5 (c, f), the CFA is smaller in HP than in CP. Scale bar = 600 μm in f (refers to a–f). (F) Exposure to HP increases closure of the CFA. Data represent closure of the CFA as distance migrated (in mm) from the border of the original CFA to the leading edge of the repopulated area after 1, 3, and 5 days in astrocytes exposed to CP (1.5 cm) and HP (10 cm). Exposure to HP for 3 and 5 days resulted in a significant increase in migration compared to CP (at 3 days, $^*p<0.005$; at 5 days, $^*p<0.0005$). Increase in cell migration, observed at day 1 was not significant. All data points represent mean $\pm$SD of three independent experiments (Fig. 6E and F from Salvador-Silva et al., 2004, with permission).

laminin to sense changes in stress and strain within the prelaminar region and the lamina cribrosa. Vascular endothelial cell stress may be mediated by integrins α3αβ1, α6β1, and α6β4, along with α5β1 and αVβ1 (Morrison, 2006). In glaucomatous ONH astrocytes there are low expression levels of α-integrins, suggesting that other adhesion mechanisms may facilitate migration in glaucoma (Hernandez et al., 2002).

Connective tissue changes in the glaucomatous optic nerve head

Extracellular matrix synthesis by ONH astrocytes

CNS axons fail to regenerate beyond a lesion site, even in the absence of a recognizable glial scar, suggesting that reactive astrocytes establish a local biochemical barrier (Zhang et al., 2006; Galtrey and Fawcett, 2007). Tenascin expressed by reactive astrocytes in the glaucomatous ONH may represent such a barrier (Pena et al., 1999b). Also, proteoglycans, such as neurocan and phosphacan secreted by reactive astrocytes, inhibit neurite outgrowth from different populations of neurons and sequester growth factors such as TGFβ, preventing neurite growth-promoting effects (McKeon et al., 1999).

There is substantial evidence that ONH astrocytes are responsible for the normal maintenance of the ECM in normal and that reactive astrocytes remodel the ECM in response to elevated IOP in human and experimental glaucoma (Hernandez, 2000; Tezel et al., 2001). Reactive astrocytes in the ONH express large amounts of elastin, leading to elastotic degeneration of the ECM in glaucoma and loss of resiliency and deformability in response to elevated IOP (Hernandez et al., 2000; Pena et al., 2001). Synthesis of elastin by astrocytes has not been reported in other regions of the normal CNS. However, recent reports indicate that astrocytomas, the most common form of primary brain tumors, express elastin and elastin binding protein (EBP) or elastin laminin receptors (ELRs) in vivo and in vitro, indicating that synthesis of new elastin participates in tumor proliferation and invasion (Lapis and Timar, 2002) and may facilitate ONH astrocyte migration in glaucoma (Varela and Hernandez, 1997).

We recently reported that astrocytes from African American (AA) normal donors expressed high levels of elastin mRNA and protein, and decreased levels of LOXL2 (Urban et al., 2007). Patients of Scandinavian ancestry with pseudoexfoliation glaucoma exhibit variations in the LOXL1 gene sequence that may predispose this group to elevated IOP and glaucomatous optic neuropathy due to accumulation of pseudoexfoliation material (Thorleifsson et al., 2007). We previously published marked elastosis in the ONHs of patients with pseudoexfoliation glaucoma (Netland et al., 1995). The downregulation of LOXL2 in AA astrocytes may confer a similar susceptibility to elevated IOP in this population. Synthesis of elastin and related ECM proteins by ONH astrocytes makes ONH astrocytes unique among glia and demonstrate the plasticity of this cell type to perform diverse functions in the CNS.

Extracellular matrix degradation by reactive astrocytes

MMPs, or matrixins, degrade ECM components such as collagens, proteoglycans, elastin, laminin, fibronectin, and glycoproteins in normal and pathologic conditions (Clark et al., 2007). Some MMPs are expressed constitutively in most cell types, and others are inducible and tissue-specific. Specific proteins known as tissue inhibitors of MMP (TIMP 1-4) are the physiologic regulators of these enzymes. Expression of MMPs and TIMPs by reactive astrocytes in the CNS depends on the type of injury or disease (Clark et al., 2007). Reactive astrocytes express MMP3 and MMP9 in neural inflammation in response to cytokine stimulation. In mice, deficiency in MMP9 protects against RGC death after optic nerve ligation (Chintala et al., 2002), whereas, in mice deficient in MMP2 there was no neuroprotection of RGC (Asahi et al., 2001) Reactive astrocytes are the main source of MMP9 activity, and ERK and p38 MAP kinases mediate secretion of MMP9 after mechanical injury (Wang et al., 2002). In the glaucomatous optic nerve, reactive astrocytes express increased MMP1 and MT1-MMP but do not express MMP9, MMP3, or MMP7, suggesting highly regulated proteolytic activity (Hernandez et al., 2002; Agapova et al., 2003a).

TGFβ signaling in ONH astrocytes in glaucoma

TGFβ signaling has long been associated with fibrous scar formation in the injured CNS by enhancing the deposition of laminin, fibronectin, and CSPGs (Logan et al., 1999a, b; McKeon et al., 1999; Asher et al., 2000). Furthermore, although TGFβ1 and -β2 are believed to signal through the same molecular pathways (Hartsough and Mulder, 1997), a role for scar production after spinal cord injury (SCI) has been attributed specifically to TGFβ2 that is expressed in multiple cell types including astrocytes in the spinal lesion during scar formation (Lagord et al., 2002). TGFβ2 which acts via TGFBR1 and TGFBR2 receptors and downstream signaling proteins (SMADs), is the primary form of TGFβ produced by ONH astrocytes (Pena et al., 1999a). The receptors TGFBR1 and TGFBR2, are upregulated in glaucomatous astrocytes from Caucasian donors (Fig. 7). Similarly, TGFBR3, another TGFβ receptor, is also upregulated (2.1-fold) in glaucomatous astrocytes. TGFBR3 serves as a co-receptor that alters the ligand binding properties of TGFβ receptors so as to favor interaction with TGFβ2 (Esparza-Lopez et al., 2001). Thus, in glaucomatous astrocytes, there may be increased sensitivity to TGFβ2. Downstream to the TGFBR2 are SMAD2 and SMAD3 which function as transcriptional regulators in ONH astrocytes (Fuchshofer et al., 2005) and other cells in the CNS (Flanders et al., 1998). These proteins are also coupled to Rho, Cdc42, and Rac1/Rac2 signaling pathways that control astrocyte polarity and migration (Kalman et al., 1999; Etienne-Manneville and Hall, 2001). Analysis of microarray data from glaucomatous ONH astrocytes compared to controls indicated that LM04, a LIM domain protein that modulates SMAD3 transcriptional activity (Lu et al., 2006), is upregulated 1.8-fold in glaucomatous astrocytes. SMAD3 was upregulated in ONH astrocytes exposed to hydrostatic pressure in vitro suggesting that pressure modulates the TGFβ pathway (Yang et al., 2004). Countering SMAD signaling is ubiquitin-linked degradation by SMURF2. Although SMURF2 expression is not altered in glaucomatous astrocytes, it is down-regulated by an increase in hydrostatic pressure (Yang et al., 2004). Thus, there may be a potentiation of TGFβ signaling in glaucomatous astrocytes with changes in IOP. Although there are reports of increased levels of TGFβ2 in the aqueous humor collected from POAG eyes (reviewed in Lutjen-Drecoll, 2005), our data suggests that changes in TGFβ signaling occurs at the level of the receptors and associated signaling molecules in astrocytes from glaucomatous donors.

Expression of several TGFβ2-regulated genes associated with extracellular matrix in glaucomatous astrocytes has been explored by several groups including ours. In lamina cribrosa cells TGFβ1 induced expression and release of ECM components (Kirwan et al., 2004, 2005), elastin deposition and gene expression is upregulated in astrocytes in experimental glaucoma in monkeys (Pena et al., 2001) and in human ONH astrocytes

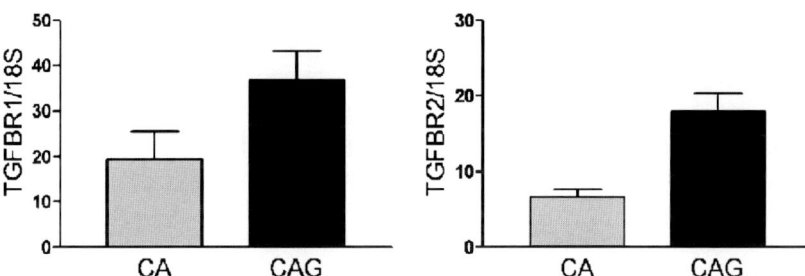

Fig. 7. Real time quantitative RT-PCR of TGFBR1 and TGFBR2, members of the TGFβ signaling pathway in human normal ONH astrocytes and glaucomatous astrocytes from Caucasian donors. Genes were normalized to 18S. Graphical representation of the relative mRNA levels in normal and glaucomatous astrocytes ($n = 8$, respectively, two-tailed t test was used. *Indicates $p < 0.05$).

under hydrostatic pressure (Hernandez et al., 2000). In cultured human optic nerve astrocytes TGFβ2 increased expression of collagen type 4 and transglutaminase 2 (Fuchshofer et al., 2005) so the alterations in TGFβ2 levels and TGFβ receptors noted above are consistent with enhanced expression of matrix protein. Interestingly, TGFβ2 induces expression of TGM2 and fibronectin in cultured human trabecular meshwork cells (Zhao et al., 2004) and ECM proteins in organ culture. Therefore, alterations in TGFβ signaling in POAG may affect not only the ONH but the aqueous outflow facility as well.

There are also alterations in the bone morphogenic protein (BMP) signaling pathways in glaucomatous astrocytes. BMPs are similar to TGFB in that the BMP receptors are coupled to SMAD proteins (SMAD1, SMAD4, SMAD5) and induce transcriptional activity (Zode et al., 2007). BMP4 is secreted by ONH astrocytes (Wordinger et al., 2002) and is upregulated 2.3-fold in cells from glaucoma donors (Hernandez et al., 2002). SMAD6 is one of the SMADs that inhibits the transcriptional activity of SMADs 1 and 4 (Chen et al., 2004). Expression of SMAD6 is upregulated in glaucomatous astrocytes (Hernandez et al., 2002) so that the elevated expression of BMP4 by glaucomatous ONH astrocytes may not affect these cells, but could affect the neighboring lamina cribrosa cells that also have BMP4 receptors (Zode et al., 2007).

Oxidative stress in ONH astrocytes

Hydrogen peroxide (H_2O_2), which is abundantly produced during CNS injury and ischemia, promotes pathological excitatory amino acid release and swelling of reactive astrocytes (Hirrlinger et al., 2000). The antioxidant glutathione (GSH) plays an important role in protecting the mitochondrial electron transport chain (ETC) from damage by oxidative stress in astrocytes and neurons. GSH represents the major CNS thiol and is essential for prevention of oxidative stress. In the CNS, astrocytes use either glutamate or glutamine as precursors for GSH synthesis. Damage to the axons of RGC in glaucoma may occur by production of reactive oxygen species (ROS) by reactive astrocytes (Tezel, 2006). Our microarray data indicates that glaucomatous astrocytes may counteract the effects of ROS on RGC axons by increasing production of GSH through increased expression of glutamate-cysteine ligase (GCL), the rate-limiting enzyme of the synthesis of GSH (Hernandez et al., 2002). GSH plays an important role in protecting the mitochondrial ETC from damage by oxidative stress in astrocytes and neurons. The antioxidant enzyme mitochondrial superoxide dismutase (SOD2) catalyzes the dismutation reaction involving the conversion of superoxide anion (O_2^-) to oxygen (O_2) and H_2O_2. Data from our laboratory shows that in glaucomatous astrocytes increased SOD2 expression (Fig. 8) is under transcriptional regulation of NF-κB via the androgen receptor (AR) (Agapova et al., 2006b).

Human OHN astrocytes treated with HNE showed immediate decreased cellular levels of GSH, which in turn induced expression of GCLC, the rate-limiting enzyme in the synthesis of GSH (Malone and Hernandez, 2007). The strong induction of GCLC and new synthesis of GSH after a 24 h recovery from the HNE treatment indicates that ONH astrocytes can manage oxidative stress. These results are consistent with similar experiments performed with Madin-Darby canine kidney II cells (MDCK II) (Ji et al., 2004), transformed human bronchial epithelial (HBE1) cells (Dickinson et al., 2002), and rat astrocytes (Ahmed et al., 2002).

Glaucomatous ONH astrocytes compared to normal ONH astrocytes have lower basal levels of GSH, which may imply a compromised oxidation–reduction system or a depleted antioxidant response in the glaucomatous optic nerve (Malone and Hernandez, 2007). This observation is consistent with a study that assessed the levels of plasma GSH in patients with untreated POAG and found lower levels of circulating GSH in glaucoma patients when compared to control subjects (Gherghel et al., 2005). However, unlike the normal ONH astrocytes, glaucomatous ONH astrocytes exhibited increased levels of GSH immediately following treatment with HNE. This result may be consistent with findings by Ferreira

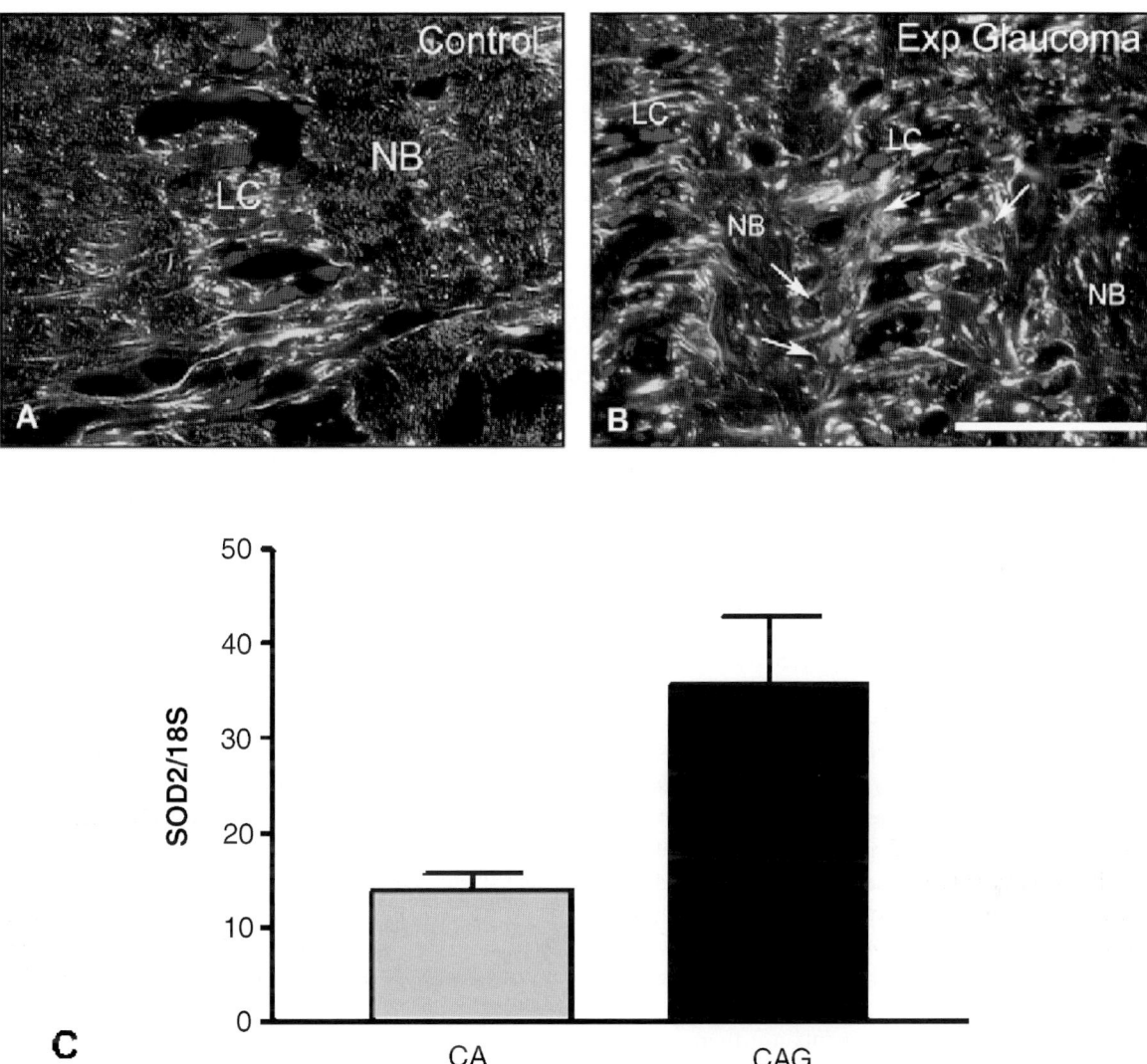

Fig. 8. Expression of SOD-2 in the monkey optic nerve head. (A) Immunofluorescent staining shows SOD-2 (red) localized to abundant mitochondria in the axons of the RGC in the lamina cribrosa (LC) (arrows) and very few in GFAP+ astrocytes (green) in a normal monkey eye. (B) In experimental glaucoma (ExpG), there is an increase in SOD-2 immunoreactivity in astrocytes (arrows) and a marked decrease in axons in the nerve bundles. A and B: Magnification bar = 35 μm. For methods see (Agapova et al., 2003a). (C) Relative amount of SOD2 mRNA in normal and glaucomatous ONH astrocytes measured by quantitative RT-PCR. Bar graphs represent relative expression of SOD2 mRNA normalized to 18S in normal ($n = 8$) and glaucomatous ($n = 8$) ONH astrocyte cultures ($n = 8$, respectively, two-tailed t test was used. *Indicates $p < 0.05$). (See Colour Plate 25.8 in the colour plate section.)

(Ferreira et al., 2004) who found 3-fold higher glutathione peroxidase activity in the aqueous humor of patients with glaucoma compared to patients with cataracts. This is further evidence of alterations in antioxidant enzymes in glaucoma.

AKR1C1, an aldo–keto reductase involved in androgen and neurosteroid metabolism, has been shown to be over-expressed in glaucomatous optic nerves, suggesting a constitutive response to chronic oxidative stress (Agapova et al., 2003b). ONH astrocytes upregulate expression AKR1C1 following treatment with HNE, which can inactivate HNE. AKR1C1 acts as a reductase by converting HNE to 1,4-dihydroxy-2-nonene (an inactive metabolite), thereby eliminating the ability of HNE to bind to protein or DNA (Burczynski et al., 2001). Enzymes such as AKR1C1 may play a more important role in the inactivation of HNE when the primary stores of GSH are depleted in the cell, as exhibited immediately following treatment with HNE in normal ONH astrocytes, because AKR1C1 can reduce HNE independent of GSH.

Glial cells protect neurons and are involved in the antioxidant defense in the brain by synthesizing GSH, mostly located in glial cells in the brain (Pearce et al., 1997) and by providing precursors that neurons require to make their own GSH (Sagara et al., 1993; Dringen et al., 1999). Since astrocytes have high levels of antioxidants (Makar et al., 1994; Dringen and Hamprecht, 1997), they can sustain the presence of ROS and therefore prevent their neurotoxicity (Wilson, 1997) which we demonstrated after HNE treatment.

Astrocytes can transport GSH conjugated to HNE out of the cell via the multidrug-resistance protein (MRP1) (Hirrlinger et al., 2002) or through the RLIP76 (Ral binding protein) (Yang et al., 2003). Once outside of the cell, the released GSH cannot be taken up by neurons, so gamma-glutamyl transpeptidase (GGT) present on the plasma membrane of astrocytes cleaves GSH into components that can be further cleaved and taken up by neurons. These basic components are then utilized by the neuron to synthesize GSH (Heales et al., 2004). Therefore, ONH astrocytes may offer neuroprotection in the optic nerve by releasing GSH and antioxidant enzymes to eliminate the products of chronic oxidative stress that may be contributing to the progression of neurodegeneration in POAG.

Conclusions

This review intends to raise many questions that will require future investigations. Astrocytes serve protective and trophic functions to the axons of the retinal ganglion cells in the optic nerve. Glaucomatous ONH astrocytes share many characteristics of reactive astrocytes in the CNS; certain properties may be specific to the pathophysiology of glaucoma. Our own data indicates that the microenvironment supported by astrocytes in the normal and glaucomatous ONH has characteristics that may impact susceptibility to glaucomatous optic neuropathy in certain individuals. Future directions include identification and characterization of signaling pathways involved in astrocyte function and further exploration on the role of selected identified genes in experimental animal and in vitro models of glaucoma.

Acknowledgments

This work was supported by National Institutes of Health Grant EY-06416 and by National Research to Prevent Blindness (RPB).

References

Abraham, C.R. (2001) Reactive astrocytes and alpha1-antichymotrypsin in Alzheimer's disease. Neurobiol. Aging, 22: 931–936.

Agapova, O.A., Kaufman, P.L. and Hernandez, M.R. (2006a) Androgen receptor and NFkB expression in human normal and glaucomatous optic nerve head astrocytes in vitro and in experimental glaucoma. Exp. Eye Res., 82: 1053–1059.

Agapova, O.A., Kaufman, P.L., Lucarelli, M.J., Gabelt, B.T. and Hernandez, M.R. (2003a) Differential expression of matrix metalloproteinases in monkey eyes with experimental glaucoma or optic nerve transection. Brain Res., 967: 132–143.

Agapova, O.A., Malone, P.E. and Hernandez, M.R. (2006b) A neuroactive steroid 5alpha-androstane-3alpha, 17beta-diol regulates androgen receptor level in astrocytes. J. Neurochem., 98: 355–363.

Agapova, O.A., Yang, P., Wang, W.H., Lane, D.A., Clark, A.F., Weinstein, B.I. and Hernandez, M.R. (2003b) Altered expression of 3alpha-hydroxysteroid dehydrogenases in human glaucomatous optic nerve head astrocytes. Neurobiol. Dis., 14: 63–73.

Ahmed, I., John, A., Vijayasarathy, C., Robin, M.A. and Raza, H. (2002) Differential modulation of growth and glutathione metabolism in cultured rat astrocytes by 4-hydroxynonenal and green tea polyphenol, epigallocatechin-3-gallate. Neurotoxicology, 23: 289–300.

Anderson, C.M. and Swanson, R.A. (2000) Astrocyte glutamate transport: review of properties, regulation, and physiological functions. Glia, 32: 1–14.

Araque, A., Carmignoto, G. and Haydon, P.G. (2001) Dynamic signaling between astrocytes and neurons. Annu. Rev. Physiol., 63: 795–813.

Asahi, M., Sumii, T., Fini, M.E., Itohara, S. and Lo, E.H. (2001) Matrix metalloproteinase 2 gene knockout has no effect on acute brain injury after focal ischemia. Neuroreport, 12: 3003–3007.

Asher, R.A., Morgenstern, D.A., Fidler, P.S., Adcock, K.H., Oohira, A., Braistead, J.E., Levine, J.M., Margolis, R.U., Rogers, J.H. and Fawcett, J.W. (2000) Neurocan is upregulated in injured brain and in cytokine-treated astrocytes. J. Neurosci., 20: 2427–2438.

Bachoo, R.M., Kim, R.S., Ligon, K.L., Maher, E.A., Brennan, C., Billings, N., Chan, S., Li, C., Rowitch, D.H., Wong, W.H. and DePinho, R.A. (2004) Molecular diversity of astrocytes with implications for neurological disorders. Proc. Natl. Acad. Sci. U.S.A., 101: 8384–8389.

Becker, C.G., Becker, T. and Meyer, R.L. (2001) Increased NCAM-180 immunoreactivity and maintenance of L1 immunoreactivity in injured optic fibers of adult mice. Exp. Neurol., 169: 438–448.

Beggs, H.E., Schahin-Reed, D., Zang, K., Goebbels, S., Nave, K.A., Gorski, J., Jones, K.R., Sretavan, D. and Reichardt, L.F. (2003) FAK deficiency in cells contributing to the basal lamina results in cortical abnormalities resembling congenital muscular dystrophies. Neuron, 40: 501–514.

Bellezza, A.J., Rintalan, C.J., Thompson, H.W., Downs, J.C., Hart, R.T. and Burgoyne, C.F. (2003) Deformation of the lamina cribrosa and anterior scleral canal wall in early experimental glaucoma. Invest. Ophthalmol. Vis. Sci., 44: 623–637.

Bourguignon, L.Y., Gilad, E., Peyrollier, K., Brightman, A. and Swanson, R.A. (2007) Hyaluronan-CD44 interaction stimulates Rac1 signaling and PKN gamma kinase activation leading to cytoskeleton function and cell migration in astrocytes. J. Neurochem., 101: 1002–1017.

Bowman, C.L., Kimelberg, H.K., Frangakis, M.V., Berwald-Netter, Y. and Edwards, C. (1984) Astrocytes in primary culture have chemically activated sodium channels. J. Neurosci., 4: 1527–1534.

Bruzzone, R., White, T.W. and Goodenough, D.A. (1996) The cellular Internet: on-line with connexins. Bioessays, 18: 709–718.

Burczynski, M.E., Sridhar, G.R., Palackal, N.T. and Penning, T.M. (2001) The reactive oxygen species and Michael acceptor-inducible human aldo-keto reductase AKR1C1 reduces the alpha, beta-unsaturated aldehyde 4-hydroxy-2-nonenal to 1,4-dihydroxy-2-nonene. J. Biol. Chem., 276: 2890–2897.

Chachisvilis, M., Zhang, Y.L. and Frangos, J.A. (2006) G protein-coupled receptors sense fluid shear stress in endothelial cells. Proc. Natl. Acad. Sci. U.S.A., 103: 15463–15468.

Charles, P., Hernandez, M.P., Stankoff, B., Aigrot, M.S., Colin, C., Rougon, G., Zalc, B. and Lubetzki, C. (2000) Negative regulation of central nervous system myelination by polysialylated-neural cell adhesion molecule. Proc. Natl. Acad. Sci. U.S.A., 97: 7585–7590.

Chauhan, B.C., LeVatte, T.L., Jollimore, C.A., Yu, P.K., Reitsamer, H.A., Kelly, M.E., Yu, D.Y., Tremblay, F. and Archibald, M.L. (2004) Model of endothelin-1-induced chronic optic neuropathy in rat. Invest. Ophthalmol. Vis. Sci., 45: 144–152.

Chen, D., Zhao, M. and Mundy, G.R. (2004) Bone morphogenetic proteins. Growth Factors, 22: 233–241.

Chintala, S.K., Zhang, X., Austin, J.S. and Fini, M.E. (2002) Deficiency in matrix metalloproteinase gelatinase B (MMP-9) protects against retinal ganglion cell death after optic nerve ligation. J. Biol. Chem., 277: 47461–47468.

Clark, I.M., Swingler, T.E., Sampieri, C.L. and Edwards, D.R. (2007) The regulation of matrix metalloproteinases and their inhibitors. Int. J. Biochem. Cell Biol., 40: 1362–1378.

Croisier, E. and Graeber, M.B. (2006) Glial degeneration and reactive gliosis in alpha-synucleinopathies: the emerging concept of primary gliodegeneration. Acta Neuropathol., 112: 517–530.

Cusato, K., Bosco, A., Rozental, R., Guimaraes, C.A., Reese, B.E., Linden, R. and Spray, D.C. (2003) Gap junctions mediate bystander cell death in developing retina. J. Neurosci., 23: 6413–6422.

Dickinson, D.A., Iles, K.E., Watanabe, N., Iwamoto, T., Zhang, H., Krzywanski, D.M. and Forman, H.J. (2002) 4-hydroxynonenal induces glutamate cysteine ligase through JNK in HBE1 cells. Free Radic. Biol. Med., 33: p. 974.

Dringen, R. and Hamprecht, B. (1997) Involvement of glutathione peroxidase and catalase in the disposal of exogenous hydrogen peroxide by cultured astroglial cells. Brain Res., 759: 67–75.

Dringen, R., Pfeiffer, B. and Hamprecht, B. (1999) Synthesis of the antioxidant glutathione in neurons: supply by astrocytes of CysGly as precursor for neuronal glutathione. J. Neurosci., 19: 562–569.

Esparza-Lopez, J., Montiel, J.L., Vilchis-Landeros, M.M., Okadome, T., Miyazono, K. and Lopez-Casillas, F. (2001) Ligand binding and functional properties of betaglycan, a co-receptor of the transforming growth factor-beta superfamily. Specialized binding regions for transforming growth factor-beta and inhibin A. J. Biol. Chem., 276: 14588–14596.

Etienne-Manneville, S. and Hall, A. (2001) Integrin-mediated activation of Cdc42 controls cell polarity in migrating astrocytes through PKCzeta. Cell, 106: 489–498.

Feng, Z., Li, L., Ng, P.Y. and Porter, A.G. (2002) Neuronal differentiation and protection from nitric oxide-induced apoptosis require c-Jun-dependent expression of NCAM140. Mol. Cell Biol., 22: 5357–5366.

Ferreira, S.M., Lerner, S.F., Brunzini, R., Evelson, P.A. and Llesuy, S.F. (2004) Oxidative stress markers in aqueous humor of glaucoma patients. Am. J. Ophthalmol., 137: 62–69.

Flanders, K.C., Ren, R.F. and Lippa, C.F. (1998) Transforming growth factor-betas in neurodegenerative disease. Prog. Neurobiol., 54: 71–85.

Fruttiger, M., Calver, A.R., Kruger, W.H., Mudhar, H.S., Michalovich, W.D., Takakura, N., Nishikawa, S. and

Richardson, W.D. (1996) PDGF mediates a neuron-astrocyte interaction in the developing retina. Neuron, 17: 1117–1131.

Fuchshofer, R., Birke, M., Welge-Lussen, U., Kook, D. and Lutjen-Drecoll, E. (2005) Transforming growth factor-beta 2 modulated extracellular matrix component expression in cultured human optic nerve head astrocytes. Invest. Ophthalmol. Vis. Sci., 46: 568–578.

Galtrey, C.M. and Fawcett, J.W. (2007) The role of chondroitin sulfate proteoglycans in regeneration and plasticity in the central nervous system. Brain Res. Rev., 54: 1–18.

Gherghel, D., Griffiths, H.R., Hilton, E.J., Cunliffe, I.A. and Hosking, S.L. (2005) Systemic reduction in glutathione levels occurs in patients with primary open-angle glaucoma. Invest. Ophthalmol. Vis. Sci., 46: 877–883.

Giaume, C. and McCarthy, K.D. (1996) Control of gap-junctional communication in astrocytic networks. Trends Neurosci., 19: 319–325.

Goodenough, D.A., Goliger, J.A. and Paul, D.L. (1996) Connexins, connexons, and intercellular communication. Annu. Rev. Biochem., 65: 475–502.

Gris, P., Tighe, A., Levin, D., Sharma, R. and Brown, A. (2007) Transcriptional regulation of scar gene expression in primary astrocytes. Glia, 55: 1145–1155.

Hansson, E. and Ronnback, L. (2003) Glial neuronal signaling in the central nervous system. FASEB J., 17: 341–348.

Hartsough, M.T. and Mulder, K.M. (1997) Transforming growth factor-beta signaling in epithelial cells. Pharmacol. Ther., 75: 21–41.

Hashimoto, K., Parker, A., Malone, P., Gabelt, B.T., Rasmussen, C., Kaufman, P.S. and Hernandez, M.R. (2005) Long-term activation of c-Fos and c-Jun in optic nerve head astrocytes in experimental ocular hypertension in monkeys and after exposure to elevated pressure in vitro. Brain Res., 1054: 103–115.

He, S., Prasanna, G. and Yorio, T. (2007) Endothelin-1-mediated signaling in the expression of matrix metalloproteinases and tissue inhibitors of metalloproteinases in astrocytes. Invest. Ophthalmol. Vis. Sci., 48: 3737–3745.

Heales, S.J., Lam, A.A., Duncan, A.J. and Land, J.M. (2004) Neurodegeneration or neuroprotection: the pivotal role of astrocytes. Neurochem. Res., 29: 513–519.

Hernandez, M.R. (2000) The optic nerve head in glaucoma: role of astrocytes in tissue remodeling. Prog. Retin. Eye Res., 19: 297–321.

Hernandez, M.R., Agapova, O.A., Yang, P., Salvador-Silva, M., Ricard, C.S. and Aoi, S. (2002) Differential gene expression in astrocytes from human normal and glaucomatous optic nerve head analyzed by cDNA microarray. Glia, 38: 45–64.

Hernandez, M.R., Igoe, F. and Neufeld, A.H. (1988) Cell culture of the human lamina cribrosa. Invest. Ophthalmol. Vis. Sci., 29: 78–89.

Hernandez, M.R. and Pena, J.D. (1997) The optic nerve head in glaucomatous optic neuropathy. Arch. Ophthalmol., 115: 389–395.

Hernandez, M.R., Pena, J.D., Selvidge, J.A., Salvador-Silva, M. and Yang, P. (2000) Hydrostatic pressure stimulates synthesis of elastin in cultured optic nerve head astrocytes. Glia, 32: 122–136.

Hewitt, A.W., Craig, J.E. and Mackey, D.A. (2006) Complex genetics of complex traits: the case of primary open-angle glaucoma. Clin. Experiment. Ophthalmol., 34: 472–484.

Himanen, J.P., Saha, N. and Nikolov, D.B. (2007) Cell-cell signaling via Eph receptors and ephrins. Curr. Opin. Cell Biol., 19: 534–542.

Hirrlinger, J., Gutterer, J.M., Kussmaul, L., Hamprecht, B. and Dringen, R. (2000) Microglial cells in culture express a prominent glutathione system for the defense against reactive oxygen species. Dev. Neurosci., 22: 384–392.

Hirrlinger, J., Schulz, J.B. and Dringen, R. (2002) Glutathione release from cultured brain cells: multidrug resistance protein 1 mediates the release of GSH from rat astroglial cells. J. Neurosci. Res., 69: 318–326.

Hu, R.G., Sheng, J., Qi, X., Xu, Z., Takahashi, T.T. and Varshavsky, A. (2005) The N-end rule pathway as a nitric oxide sensor controlling the levels of multiple regulators. Nature, 437: 981–986.

Ji, B., Ito, K. and Horie, T. (2004) Multidrug resistance-associated protein 2 (MRP2) enhances 4-hydroxynonenal-induced toxicity in Madin-Darby canine kidney II cells. Chem. Res. Toxicol., 17: 158–164.

Johnson, C.A., Cioffi, G.A., Liebmann, J.R., Sample, P.A., Zangwill, L.M. and Weinreb, R.N. (2000) The relationship between structural and functional alterations in glaucoma: a review. Semin. Ophthalmol., 15: 221–233.

Junier, M.P. (2000) What role(s) for TGFalpha in the central nervous system?. Prog. Neurobiol., 62: 443–473.

Kalman, D., Gomperts, S.N., Hardy, S., Kitamura, M. and Bishop, J.M. (1999) Ras family GTPases control growth of astrocyte processes. Mol. Biol. Cell, 10: 1665–1683.

Kielian, T. and Esen, N. (2004) Effects of neuroinflammation on glia-glia gap junctional intercellular communication: a perspective. Neurochem. Int., 45: 429–436.

Kirwan, R.P., Crean, J.K., Fenerty, C.H., Clark, A.F. and O'Brien, C.J. (2004) Effect of cyclical mechanical stretch and exogenous transforming growth factor-beta1 on matrix metalloproteinase-2 activity in lamina cribrosa cells from the human optic nerve head. J. Glaucoma., 13: 327–334.

Kirwan, R.P., Leonard, M.O., Murphy, M., Clark, A.F. and O'Brien, C.J. (2005) Transforming growth factor-beta-regulated gene transcription and protein expression in human GFAP-negative lamina cribrosa cells. Glia, 52: 309–324.

Koyama, Y., Yoshioka, Y., Hashimoto, H., Matsuda, T. and Baba, A. (2000) Endothelins increase tyrosine phosphorylation of astrocytic focal adhesion kinase and paxillin accompanied by their association with cytoskeletal components. Neuroscience, 101: 219–227.

Lagord, C., Berry, M. and Logan, A. (2002) Expression of TGFbeta2 but not TGFbeta1 correlates with the deposition of scar tissue in the lesioned spinal cord. Mol. Cell Neurosci., 20: 69–92.

Lambert, W.S., Clark, A.F. and Wordinger, R.J. (2004a) Effect of exogenous neurotrophins on Trk receptor phosphorylation, cell proliferation, and neurotrophin secretion by cells isolated from the human lamina cribrosa. Mol. Vis., 10: 289–296.

Lambert, W.S., Clark, A.F. and Wordinger, R.J. (2004b) Neurotrophin and Trk expression by cells of the human lamina cribrosa following oxygen-glucose deprivation. BMC Neurosci., 5: p. 51.

Lapis, K. and Timar, J. (2002) Role of elastin-matrix interactions in tumor progression. Semin. Cancer Biol., 12: 209–217.

Liu, B., Chen, H., Johns, T.G. and Neufeld, A.H. (2006) Epidermal growth factor receptor activation: an upstream signal for transition of quiescent astrocytes into reactive astrocytes after neural injury. J. Neurosci., 26: 7532–7540.

Liu, B. and Neufeld, A.H. (2000) Expression of nitric oxide synthase-2 (NOS-2) in reactive astrocytes of the human glaucomatous optic nerve head. Glia, 30: 178–186.

Liu, B. and Neufeld, A.H. (2004) Activation of epidermal growth factor receptor causes astrocytes to form cribriform structures. Glia, 46: 153–168.

Liu, B. and Neufeld, A.H. (2007) Activation of epidermal growth factor receptors in astrocytes: from development to neural injury. J. Neurosci. Res., 85: 3523–3529.

Liu, X., Clark, A.F. and Wordinger, R.J. (2007) Expression of ciliary neurotrophic factor (CNTF) and its tripartite receptor complex by cells of the human optic nerve head. Mol. Vis., 13: 758–763.

Logan, A., Baird, A. and Berry, M. (1999a) Decorin attenuates gliotic scar formation in the rat cerebral hemisphere. Exp. Neurol., 159: 504–510.

Logan, A., Green, J., Hunter, A., Jackson, R. and Berry, M. (1999b) Inhibition of glial scarring in the injured rat brain by a recombinant human monoclonal antibody to transforming growth factor-beta2. Eur. J. Neurosci., 11: 2367–2374.

Lu, Z., Lam, K.S., Wang, N., Xu, X., Cortes, M. and Andersen, B. (2006) LMO4 can interact with Smad proteins and modulate transforming growth factor-beta signaling in epithelial cells. Oncogene, 25: 2920–2930.

Lutjen-Drecoll, E. (2005) Morphological changes in glaucomatous eyes and the role of TGFbeta2 for the pathogenesis of the disease. Exp. Eye Res., 81: 1–4.

Lykissas, M.G., Batistatou, A.K., Charalabopoulos, K.A. and Beris, A.E. (2007) The role of neurotrophins in axonal growth, guidance, and regeneration. Curr. Neurovasc. Res., 4: 143–151.

Magistretti, P.J. (2006) Neuron-glia metabolic coupling and plasticity. J. Exp. Biol., 209: 2304–2311.

Makar, T.K., Nedergaard, M., Preuss, A., Gelbard, A.S., Perumal, A.S. and Cooper, A.J. (1994) Vitamin E, ascorbate, glutathione, glutathione disulfide, and enzymes of glutathione metabolism in cultures of chick astrocytes and neurons: evidence that astrocytes play an important role in antioxidative processes in the brain. J. Neurochem., 62: 45–53.

Malone, P., Miao, H., Parker, A., Juarez, S. and Hernandez, M.R. (2007) Pressure induces loss of gap junction communication and redistribution of connexin 43 in astrocytes. Glia, 55: 1085–1098.

Malone, P.E. and Hernandez, M.R. (2007) 4-Hydroxynonenal, a product of oxidative stress, leads to an antioxidant response in optic nerve head astrocytes. Exp. Eye Res., 84: 444–454.

Mandell, J.W., Gocan, N.C. and Vandenberg, S.R. (2001) Mechanical trauma induces rapid astroglial activation of ERK/MAP kinase: evidence for a paracrine signal. Glia, 34: 283–295.

Maness, P.F. and Schachner, M. (2007) Neural recognition molecules of the immunoglobulin superfamily: signaling transducers of axon guidance and neuronal migration. Nat. Neurosci., 10: 19–26.

McKeon, R.J., Jurynec, M.J. and Buck, C.R. (1999) The chondroitin sulfate proteoglycans neurocan and phosphacan are expressed by reactive astrocytes in the chronic CNS glial scar. J. Neurosci., 19: 10778–10788.

Moreno, A.P., Rook, M.B., Fishman, G.I. and Spray, D.C. (1994) Gap junction channels: distinct voltage-sensitive and insensitive conductance states. Biophys. J., 67: 113–119.

Morrison, J.C. (2006) Integrins in the optic nerve head: potential roles in glaucomatous optic neuropathy (an American Ophthalmological Society thesis). Trans. Am. Ophthalmol. Soc., 104: 453–477.

Nagy, J.I., Dudek, F.E. and Rash, J.E. (2004) Update on connexins and gap junctions in neurons and glia in the mammalian nervous system. Brain Res. Brain Res. Rev., 47: 191–215.

Netland, P.A., Ye, H., Streeten, B.W. and Hernandez, M.R. (1995) Elastosis of the lamina cribrosa in pseudoexfoliation syndrome with glaucoma. Ophthalmology, 102: 878–886.

Neufeld, A.H. and Liu, B. (2003) Comparison of the signal transduction pathways for the induction of gene expression of nitric oxide synthase-2 in response to two different stimuli. Nitric Oxide, 8: 95–102.

Obara, M., Szeliga, M. and Albrecht, J. (2008) Regulation of pH in the mammalian central nervous system under normal and pathological conditions: facts and hypotheses. Neurochem. Int., 52: 905–919.

Oelze, I., Kartenbeck, J., Crusius, K. and Alonso, A. (1995) Human papillomavirus type 16 E5 protein affects cell-cell communication in an epithelial cell line. J. Virol., 69: 4489–4494.

Osmani, N., Vitale, N., Borg, J.P. and Etienne-Manneville, S. (2006) Scrib controls Cdc42 localization and activity to promote cell polarization during astrocyte migration. Curr. Biol., 16: 2395–2405.

Pearce, R.K., Owen, A., Daniel, S., Jenner, P. and Marsden, C.D. (1997) Alterations in the distribution of glutathione in the substantia nigra in Parkinson's disease. J. Neural Transm., 104: 661–677.

Pehar, M., Vargas, M.R., Cassina, P., Barbeito, A.G., Beckman, J.S. and Barbeito, L. (2005) Complexity of astrocyte-motor neuron interactions in amyotrophic lateral sclerosis. Neurodegener. Dis., 2: 139–146.

Pena, J.D., Agapova, O., Gabelt, B.T., Levin, L.A., Lucarelli, M.J., Kaufman, P.L. and Hernandez, M.R. (2001) Increased elastin expression in astrocytes of the lamina cribrosa in response to elevated intraocular pressure. Invest. Ophthalmol. Vis. Sci., 42: 2303–2314.

Pena, J.D., Taylor, A.W., Ricard, C.S., Vidal, I. and Hernandez, M.R. (1999a) Transforming growth factor beta isoforms in human optic nerve heads. Br. J. Ophthalmol., 83: 209–218.

Pena, J.D., Varela, H.J., Ricard, C.S. and Hernandez, M.R. (1999b) Enhanced tenascin expression associated with reactive astrocytes in human optic nerve heads with primary open angle glaucoma. Exp. Eye Res., 68: 29–40.

Prasanna, G., Krishnamoorthy, R., Clark, A.F., Wordinger, R.J. and Yorio, T. (2002) Human optic nerve head astrocytes as a target for endothelin-1. Invest. Ophthalmol. Vis. Sci., 43: 2704–2713.

Rao, V.R., Krishnamoorthy, R.R. and Yorio, T. (2007) Endothelin-1, endothelin A and B receptor expression and their pharmacological properties in GFAP negative human lamina cribrosa cells. Exp. Eye Res., 84: 1115–1124.

Ricard, C.S., Kobayashi, S., Pena, J.D., Salvador-Silva, M., Agapova, O. and Hernandez, M.R. (2000) Selective expression of neural cell adhesion molecule (NCAM)-180 in optic nerve head astrocytes exposed to elevated hydrostatic pressure in vitro. Brain Res. Mol. Brain Res., 81: 62–79.

Ridet, J.L., Malhotra, S.K., Privat, A. and Gage, F.H. (1997) Reactive astrocytes: cellular and molecular cues to biological function. Trends Neurosci., 20: 570–577.

Rose, C.R. and Ransom, B.R. (1997) Gap junctions equalize intracellular Na^+ concentration in astrocytes. Glia, 20: 299–307.

Rusnati, M. and Presta, M. (2007) Fibroblast growth factors/fibroblast growth factor receptors as targets for the development of anti-angiogenesis strategies. Curr. Pharm. Des., 13: 2025–2044.

Sagara, J.I., Miura, K. and Bannai, S. (1993) Maintenance of neuronal glutathione by glial cells. J. Neurochem., 61: 1672–1676.

Salvador-Silva, M., Aoi, S., Parker, A., Yang, P., Pecen, P. and Hernandez, M.R. (2004) Responses and signaling pathways in human optic nerve head astrocytes exposed to hydrostatic pressure in vitro. Glia, 45: 364–377.

Salvador-Silva, M., Ricard, C.S., Agapova, O.A., Yang, P. and Hernandez, M.R. (2001) Expression of small heat shock proteins and intermediate filaments in the human optic nerve head astrocytes exposed to elevated hydrostatic pressure in vitro. J. Neurosci. Res., 66: 59–73.

Sawada, Y., Tamada, M., Dubin-Thaler, B.J., Cherniavskaya, O., Sakai, R., Tanaka, S. and Sheetz, M.P. (2006) Force sensing by mechanical extension of the Src family kinase substrate p130Cas. Cell, 127: 1015–1026.

Seachrist, J.L. and Ferguson, S.S. (2003) Regulation of G protein-coupled receptor endocytosis and trafficking by Rab GTPases. Life Sci., 74: 225–235.

Smith, C., Berry, M., Clarke, W.E. and Logan, A. (2001) Differential expression of fibroblast growth factor-2 and fibroblast growth factor receptor 1 in a scarring and nonscarring model of CNS injury in the rat. Eur. J. Neurosci., 13: 443–456.

Tait, M.J., Saadoun, S., Bell, B.A. and Papadopoulos, M.C. (2008) Water movements in the brain: role of aquaporins. Trends Neurosci., 31: 37–43.

Tezel, G. (2006) Oxidative stress in glaucomatous neurodegeneration: mechanisms and consequences. Prog. Retin. Eye Res., 25: 490–513.

Tezel, G., Chauhan, B.C., LeBlanc, R.P. and Wax, M.B. (2003) Immunohistochemical assessment of the glial mitogen-activated protein kinase activation in glaucoma. Invest. Ophthalmol. Vis. Sci., 44: 3025–3033.

Tezel, G., Hernandez, M.R. and Wax, M.B. (2001) In vitro evaluation of reactive astrocyte migration, a component of tissue remodeling in glaucomatous optic nerve head. Glia, 34: 178–189.

Tezel, G., Yang, X., Yang, J. and Wax, M.B. (2004) Role of tumor necrosis factor receptor-1 in the death of retinal ganglion cells following optic nerve crush injury in mice. Brain Res., 996: 202–212.

Thorleifsson, G., Magnusson, K.P., Sulem, P., Walters, G.B., Gudbjartsson, D.F., Stefansson, H., Jonsson, T., Jonasdottir, A., Jonasdottir, A., Stefansdottir, G., Masson, G., Hardarson, G.A., Petursson, H., Arnarsson, A., Motallebipour, M., Wallerman, O., Wadelius, C., Gulcher, J.R., Thorsteinsdottir, U., Kong, A., Jonasson, F. and Stefansson, K. (2007) Common sequence variants in the LOXL1 gene confer susceptibility to exfoliation glaucoma. Science, 317: 1397–1400.

Tichauer, J., Saud, K. and von Bernhardi, R. (2007) Modulation by astrocytes of microglial cell-mediated neuroinflammation: effect on the activation of microglial signaling pathways. Neuroimmunomodulation, 14: 168–174.

Urban, Z., Agapova, O., Hucthagowder, V., Yang, P., Starcher, B.C. and Hernandez, M.R. (2007) Population differences in elastin maturation in optic nerve head tissue and astrocytes. Invest. Ophthalmol. Vis. Sci., 48: 3209–3215.

Varela, H.J. and Hernandez, M.R. (1997) Astrocyte responses in human optic nerve head with primary open-angle glaucoma. J. Glaucoma., 6: 303–313.

Wang, L., Fortune, B., Cull, G., Dong, J. and Cioffi, G.A. (2006) Endothelin B receptor in human glaucoma and experimentally induced optic nerve damage. Arch. Ophthalmol., 124: 717–724.

Wang, X., Mori, T., Jung, J.C., Fini, M.E. and Lo, E.H. (2002) Secretion of matrix metalloproteinase-2 and -9 after mechanical trauma injury in rat cortical cultures and involvement of MAP kinase. J. Neurotrauma., 19: 615–625.

Warn-Cramer, B.J., Cottrell, G.T., Burt, J.M. and Lau, A.F. (1998) Regulation of connexin-43 gap junctional intercellular communication by mitogen-activated protein kinase. J. Biol. Chem., 273: 9188–9196.

Wax, M.B., Tezel, G., Kobayashi, S. and Hernandez, M.R. (2000) Responses of different cell lines from ocular tissues to elevated hydrostatic pressure. Br. J. Ophthalmol., 84: 423–428.

Weinreb, R.N. and Khaw, P.T. (2004) Primary open-angle glaucoma. Lancet, 363: 1711–1720.

Wheeler, A.P., Wells, C.M., Smith, S.D., Vega, F.M., Henderson, R.B., Tybulewicz, V.L. and Ridley, A.J. (2006) Rac1 and Rac2 regulate macrophage morphology but are not essential for migration. J. Cell Sci., 119: 2749–2757.

Wieduwilt, M.J. and Moasser, M.M. (2008) The epidermal growth factor receptor family: biology driving targeted therapeutics. Cell Mol. Life Sci., 65: 1566–1584.

Wiggs, J.L. (2007) Genetic etiologies of glaucoma. Arch. Ophthalmol., 125: 30–37.

Wilson, J.X. (1997) Antioxidant defense of the brain: a role for astrocytes. Can. J. Physiol. Pharmacol., 75: 1149–1163.

Wordinger, R.J., Agarwal, R., Talati, M., Fuller, J., Lambert, W. and Clark, A.F. (2002) Expression of bone morphogenetic proteins (BMP), BMP receptors, and BMP associated proteins in human trabecular meshwork and optic nerve head cells and tissues. Mol. Vis., 8: 241–250.

Yang, P., Agapova, O., Parker, A., Shannon, W., Pecen, P., Duncan, J., Salvador-Silva, M. and Hernandez, M.R. (2004) DNA microarray analysis of gene expression in human optic nerve head astrocytes in response to hydrostatic pressure. Physiol. Genomics, 17: 157–169.

Yang, P. and Hernandez, M.R. (2003) Purification of astrocytes from adult human optic nerve heads by immunopanning. Brain Res. Brain Res. Protoc., 12: 67–76.

Yang, Y., Sharma, R., Sharma, A., Awasthi, S. and Awasthi, Y.C. (2003) Lipid peroxidation and cell cycle signaling: 4-hydroxynonenal, a key molecule in stress mediated signaling. Acta. Biochim. Pol., 50: 319–336.

Yu, A.L., Fuchshofer, R., Birke, M., Priglinger, S.G., Eibl, K.H., Kampik, A., Bloemendal, H. and Welge-Lussen, U. (2007) Hypoxia/reoxygenation and TGF-beta increase alphaB-crystallin expression in human optic nerve head astrocytes. Exp. Eye Res., 84: 694–706.

Zhang, H., Uchimura, K. and Kadomatsu, K. (2006) Brain keratan sulfate and glial scar formation. Ann. N.Y. Acad. Sci., 1086: 81–90.

Zhao, X., Ramsey, K.E., Stephan, D.A. and Russell, P. (2004) Gene and protein expression changes in human trabecular meshwork cells treated with transforming growth factor-beta. Invest. Ophthalmol. Vis. Sci., 45: 4023–4034.

Zode, G.S., Clark, A.F. and Wordinger, R.J. (2007) Activation of the BMP canonical signaling pathway in human optic nerve head tissue and isolated optic nerve head astrocytes and lamina cribrosa cells. Invest. Ophthalmol. Vis. Sci., 48: 5058–5067.

CHAPTER 26

Glaucoma as a neuropathy amenable to neuroprotection and immune manipulation

Michal Schwartz* and Anat London

Department of Neurobiology, The Weizmann Institute of Science, Rehovot, Israel

Abstract: Glaucoma, once thought as a single disease, is actually a group of diseases of the optic nerve involving loss of retinal ganglion cells. The process of cell death occurs in a characteristic pattern of optic neuropathy, a broad term for a certain pattern of damage to the optic nerve (the bundle of nerve fibers that carries information from the eye to the brain). Untreated glaucoma leads to permanent damage of the optic nerve and resultant visual field loss, which can progress to blindness. Worldwide, it is estimated that about 66.8 million people have visual impairment as a result of glaucoma, with 6.7 million suffering from blindness.

Keywords: glaucoma; neuroprotection; protective autoimmunity; monocytes; optic neuropathy; therapeutic vaccination; neurodegenerative diseases; secondary degeneration

Glaucoma as a neurodegenerative disease

Traditionally, elevation in intraocular pressure (IOP) has been considered to be the main cause of glaucoma (Weinreb and Khaw, 2004); IOP is determined by the balance between secretion and drainage of aqueous humor. In glaucoma, this balance is interrupted, as insufficient fluid drains out of the eye, leading to increased IOP. As a result, the retina and the optic nerve heads are subjected to mechanical (Burgoyne et al., 2005; Sigal et al., 2005), hypoxic (Tezel and Wax, 2004), and oxidative tissue stress (Tezel et al., 2000).

Over the past decades, scientists have focused on the elevated IOP as a primary therapeutic target, trying to diminish this major risk factor (Quigley and Maumenee, 1979; Kass et al., 2002; Leske et al., 2003; Johnson et al., 2006; Nickells et al., 2007), while totally disregarding the process of damage that derives from it. Consequently, the current approved glaucoma medications and surgical therapies are directed at lowering IOP, and indeed there are evidences from several clinical trials for a significant attenuation of progressive visual field loss among the treated patients (Quigley and Maumenee, 1979; Heijl et al., 2002; Kass et al., 2002; Leske et al., 2003).

However, some patients continue to suffer from an ongoing visual field loss even after their IOP was effectively controlled (Jay and Allan, 1989; Nouri-Mahdavi et al., 1995; Brubaker, 1996). Even more confusing is the case of normal tension glaucoma (NTG) in which progressive retinal ganglion cell (RGC) death and subsequent glaucomatous damage occurs in the absence of any elevated IOP. Moreover, some studies have reported a negligible relationship between mean IOP and vision loss in glaucoma (Richler et al., 1982; Schulzer et al., 1990;

*Corresponding author. Tel.: +972 8 934 2467;
Fax: +972 8 934 6018; E-mail: Michal.schwartz@weizmann.ac.il

Chauhan and Drance, 1992). These observations indicate the possible contribution of IOP-independent mechanisms to disease progression.

It seems, therefore, that glaucoma is a complex multivariate disease, initiated by several risk factors (with elevated IOP as only one of them), whose individual contributions to glaucomatous destruction have not yet been fully elucidated. As a result, the efforts of researchers have shifted toward understanding and subsequently preventing the disease progression, regardless of the primary cause. Thus, the major goal of glaucoma treatment is moving to neuroprotection, preventing the spread of damage, and protection from the progressive loss of the nerve fiber layers (Schwartz et al., 1996; Schwartz, 2003).

There are many molecular and cellular elements that contribute to the pathological progression and neuronal loss in glaucoma, even after the primary risk factor no longer exists. Following the initial insult, there is a progressive self-perpetuating secondary degeneration of neurons that were spared from the primary injury. This secondary damage is an outcome of the hostile environment produced by the degenerating neurons. The noxious extracellular environment includes mediators of oxidative stress and free radicals, excessive amounts of glutamate and excitotoxicity, increased calcium concentration, deprivation of neurotrophins and growth factors, abnormal accumulation of proteins, and apoptotic signals (Scheme 1), all of which are universal features of many neurodegenerative diseases (Schwartz,

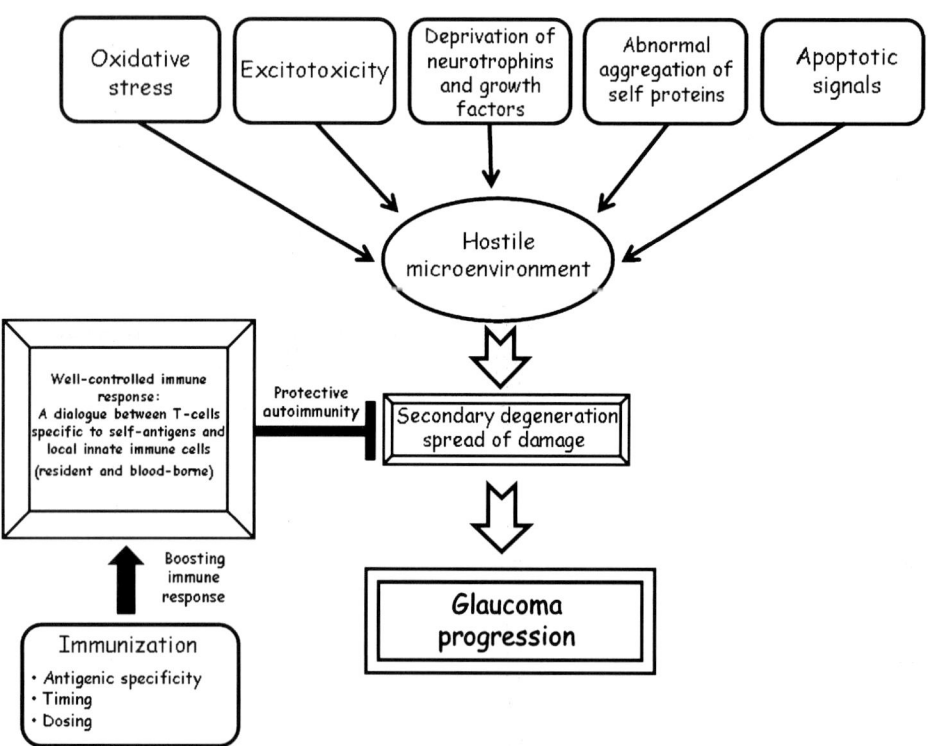

Scheme 1. Immune protection in glaucoma. Degenerating neurons create a noxious milieu, which consist of oxidative stress and free radicals, excessive amounts of glutamate and excitotoxicity, increased calcium concentration, deprivation of neurotrophins and growth factors, abnormal accumulation of proteins, and apoptotic signals. These features are characteristics of a hostile microenvironment to the remaining neurons that leads to secondary degeneration and further loss of neurons. The immune system plays a key role in the ability of the optic nerve and the retina to withstand these threatening conditions, by recruiting both innate (resident and blood-borne macrophages) and adaptive (self-antigens specific T cells) cells that together create a protective niche and thereby halt disease progression. The spontaneous immune response might not be sufficient, and therefore boosting it by immunization (with the appropriate antigen, in specific timing and dosing) may be a suitable therapeutic vaccination to treat glaucoma.

2005). These characteristics place glaucoma among the common neurodegenerative disorders.

Oxidative stress and free radicals

Oxidative stress is involved in the pathogenesis of many neurodegenerative disorders (Beckman et al., 1993; Abe et al., 1995; Giasson et al., 2000; Castegna et al., 2003; Andersen, 2004; Potashkin and Meredith, 2006; Sultana et al., 2006). The central nervous system (CNS) has a unique sensitivity to oxidative stress. Its function requires electrical excitability, transsynaptic chemical connections, and a high metabolic rate, which entail the augmented use of O_2 and ATP synthesis. In addition, the CNS lacks an appropriate defense system against the elevated levels of reactive oxygen species (ROS), produced in these tissues. These ROS, accumulating in cells that undergo oxidative stress, react with nitric oxide to produce free radicals, leading to a chain of reactions that result in mitochondrial dysfunction, DNA degradation, and eventually cell death. As in Alzheimer's disease (Castegna et al., 2003; Sultana et al., 2006), Parkinson's disease (Giasson et al., 2000), and amyotrophic lateral sclerosis (ALS) (Beckman et al., 1993; Abe et al., 1995), the association of oxidative stress with neurodegeneration has been increasingly reported in glaucoma. Free radicals cause extensive damage to the RGCs and their axons (Oku et al., 1997; Levkovitch-Verbin et al., 2000; Tezel, 2006); they contribute to the secondary degeneration either by a direct neurotoxic effect or indirectly through the induction of glial dysfunction (Tezel and Wax, 2003), oxidative modification of proteins (Tezel et al., 2005), and activation of apoptotic pathways (Martindale and Holbrook, 2002). Oxygen-derived free radicals are therefore an important therapeutic target for treating glaucoma. A variety of antioxidants (Ritch, 2000; Siu et al., 2006) and nitric oxide synthase (NOS) inhibitors (Neufeld et al., 1999) are currently being investigated as potential therapeutic agents.

Excessive glutamate, increased calcium levels, and excitotoxicity

Glutamate is an essential neurotransmitter, participating in a variety of neurological processes in the CNS (Sahai, 1990; Lipton and Rosenberg, 1994). It is also the main excitatory neurotransmitter in the retina and is involved in phototransduction (Tsacopoulos et al., 1998).

Excessive levels of glutamate are toxic and detrimental to neurons; an excess of glutamate can hyperactivate the *N*-methyl-D-aspartate (NMDA) receptor, resulting in a poisonous influx of calcium (Sucher et al., 1997) — a phenomenon termed excitotoxicity (Siliprandi et al., 1992; Dreyer, 1998). In glaucoma, the initially degenerating neurons expel their glutamate stores into the extracellular environment, thereby damaging their still healthy, neighboring neurons. Moreover, Müller glial cells, which normally take up glutamate, fail to do so in glaucoma (Napper et al., 1999), and thus glutamate levels continue to escalate, leading to RGC death (Olney, 1969; Olney et al., 1986). Excitotoxicity is also common in other neurodegenerative diseases and neurological disorders, including ALS (Van Den Bosch et al., 2006), Alzheimer's disease (Riederer and Hoyer, 2006; Lipton, 2007), Parkinson's disease (Beal, 1998; Lancelot and Beal, 1998), stroke, and Huntington's disease (Choi, 1988b; Lipton and Rosenberg, 1994). Blocking NMDA receptors by a glutamate antagonist can prevent the glaucomatous excitatory damage (Stuiver et al., 1996). However, since glutamate is a fundamental neurotransmitter, vital for the normal maintenance of the retina and essential to many CNS functions (Sahai, 1990), the blockage of its receptor is accompanied by many side effects.

Another approach is to focus on the increased influx of calcium caused by the excess of glutamate and the hyperstimulation of voltage-gated calcium channels (Choi, 1988a, b). Indeed, some calcium channel blockers have been shown to reduce retinal damage (Takahashi et al., 1992; Bath et al., 1996).

Deprivation of neurotrophins and growth factors

Neurotrophins are crucial for the normal maintenance of the CNS. These factors are required by all types of neurons including the RGCs. They are produced in the superior colliculus and lateral geniculate nucleus in the brain and are delivered along the optic nerve to the RGCs. Any interference with this neurotrophin supply could lead to neuronal damage. Ganglion cells are supported by brain-derived neurotrophic factor (BDNF), which delays apoptosis and prevents secondary degeneration (Mansour-Robaey et al., 1994; Peinado-Ramon et al., 1996; Gao et al., 1997; Klocker et al., 1998; Rocha et al., 1999). Moreover, retinal cells respond to ciliary neurotrophic factor (CNTF) and fibroblast growth factor (FGF) (Unoki and LaVail, 1994). In glaucoma, retrograde and anterograde axonal transport are defective, often resulting in an insufficient supply of neurotrophins and growth factors to the RGCs, leading to their apoptotic breakdown. Administration of neurotrophins can protect the RGCs and prevent their programmed cell death (Mansour-Robaey et al., 1994; Peinado-Ramon et al., 1996; Gao et al., 1997; Klocker et al., 1998; Rocha et al., 1999).

Abnormal accumulation of proteins

A shared feature among many of the neurodegenerative diseases is the abnormal increased accumulation of certain self-proteins during disease progression. The accumulation of β-amyloid protein in the senile plaques in Alzheimer's disease and of α-synuclein protein in Parkinson's disease is among the main characteristics of each of these diseases (Lansbury and Lashuel, 2006). The abnormal processing of amyloid precursor protein has also been reported in glaucoma (McKinnon et al., 2002). This is in addition to the oxidatively modified proteins that are produced during glaucomatous neurodegeneration (Tezel et al., 2005). Drug candidates that inhibit aggregation are now being tested in the clinic for the treatment of Alzheimer's and Parkinson's diseases, and might also serve in the future as a therapy for glaucoma (Lansbury and Lashuel, 2006; Weydt and La Spada, 2006).

Apoptotic signals

The final outcome of disease progression is the enhanced activation of programmed cell death pathways among the RGCs. In a normal cell, a balance is maintained between pro-apoptotic and anti-apoptotic proteins. During disease progression, this balance is interrupted and there is an increase in the proportion of pro-apoptotic signals. This is manifested by the increased expression of pro-apoptotic genes, such as *bax* (Oltvai and Korsmeyer, 1994), in parallel to the downregulation of the anti-apoptotic genes, such as *bcl-xL* (Levin et al., 1997). Following the initial death signal, there is a rapid cascade of caspase activation that eventually results in a noninflammatory cell death (Thornberry, 1998; Thornberry and Lazebnik, 1998; Quigley, 1999). The programmed cell death cycle of retinal cells is an interesting potential therapeutic target, though caution should be taken when considering this approach because any anti-apoptotic agent can also serve as a potential carcinogen.

Pharmacological neuroprotection for glaucoma

The features described above, presenting the toxic environment that is one of the well-known characteristics of neurodegenerative diseases, are also associated with glaucoma. This hostile environment might explain the fact that in glaucoma, as in other neurodegenerative disorders, primary cell death is followed by the secondary degeneration of surrounding neurons that were affected by the toxic microenvironment caused by the initial dying cells. This spread of damage, as a part of disease progression, is one of the hallmarks of neurodegenerative diseases and can be clearly seen in glaucoma. We, as well as others, have proposed that the factors contributing to this ongoing degeneration are physiological compounds emerging in toxic quantities from the injured fibers or their cell bodies. Studies along these lines have revealed that some of the compounds identified in the pathogenesis of glaucoma are already known to be active in other neurodegenerative diseases. The many common features between the neurodegenerative diseases and glaucoma have led to the recognition of glaucoma as a neurodegenerative disease, rather than simply a disease of elevated ocular pressure (Gupta and Yucel, 2007; Mozaffarieh and Flammer, 2007; Nickells, 2007). This raises the possibility of utilizing similar therapeutic approaches for glaucoma that are currently being used in other neurodegenerative disorders (Schwartz, 2005). Indeed, treatments used for Alzheimer's and Parkinson's diseases are becoming more relevant to glaucoma; for example, memantine, an NMDA channel blocker, used as a therapy for Alzheimer's and Parkinson's diseases, is currently being evaluated for glaucoma treatment. It is likely that as our recognition of glaucoma as a neurodegenerative disease increases, more of the newly developed therapies will be based on neuroprotection (Hartwick, 2001; Chidlow et al., 2007), fighting against the spread of damage and degeneration of the RGCs.

Protection of the retinal ganglion cells involves the immune system

The CNS was always viewed as an "immune privileged" site, in which any immune response was considered harmful, and was usually associated with a disease or other malfunction. Our own studies of acute and chronic injuries to the rodent optic nerve led us to the unexpected discovery that the immune system plays a key role in the ability of the optic nerve and the retina to withstand injurious conditions.

Our first observations that the immune system (in the form of T cells directed to certain self-antigens) can protect injured neurons from death came from studies in rodents showing that passive transfer of T cells specific to myelin basic protein (MBP) reduces the loss of RGCs after traumatic optic nerve injury (Moalem et al., 1999). We found that these T cells are also effective when directed to either cryptic or pathogenic epitopes of MBP, as well as toward other myelin-derived antigens or their epitopes (Fisher et al., 2001; Mizrahi et al., 2002). These findings raised a number of critical questions. For example, are myelin antigens capable of protecting the visual system from all types of acute or chronic insult? Is the observed neuroprotective activity of immune cells merely an artifact of our experimental system, or does it indicate the critical participation of the immune system in fighting off injurious conditions in the visual system and in the CNS in general? If the latter is true, does that mean that glaucoma is a systemic disease? And if so, can this finding be translated into a systemic therapy that would protect the eye?

In a series of experiments carried out over almost a decade, we have learned much about the role of the immune response in neuroprotection. We first learned that the injury-induced response of T cells, reactive to specific self-antigens residing in the site of stress (eye or brain), is a spontaneous physiological response that protects the nerve against the degenerative effects of the hostile environment — a concept that was established in our lab and named "protective autoimmunity" (Yoles et al., 2001). Unfortunately, this protective response might not always be sufficient or properly controlled, which might explain the minimal spontaneous recovery following many severe CNS insults, including the damage that occurs in glaucoma. Secondly, we discovered that the specificity of such protective T cells depends on the site of the insult and type of cells damaged. Thus, for example, the protective effect of vaccination with myelin-associated antigens is restricted to injuries of the white matter, that is, to myelinated axons (Schori et al., 2001; Mizrahi et al., 2002; Avidan et al., 2004). If the insult is to the retina, which contains no myelin, myelin-related antigens have no effect.

We also sought to identify the phenotype of the beneficial autoimmune T cells, and to understand what determines the balance between a beneficial (neuroprotective) outcome of the T cell–mediated response to a CNS injury and a destructive effect causing autoimmune disease. Finally, we examined ways of translating the beneficial response into a therapy for glaucoma.

Some critical aspects of this approach had to be addressed along the way: (1) we verified that the loss of RGCs in a rat model of high IOP, simulating some types of glaucoma, is T cell dependent (Bakalash et al., 2002); (2) we attempted to determine whether the specific self-antigens that are harnessed by the protective autoimmune T cells in our rat model of chronic glaucoma, and which can be boosted for therapeutic purposes, reside in the retina or the optic nerve (Bakalash et al., 2003); and finally, (3) we searched for an antigen that would be able to safely boost the physiological response without causing autoimmune disease.

Using a rat model of elevated IOP, we showed that a protective response could be obtained only with an antigen residing in the retina, suggesting that, at least in this model, the site of self-perpetuating degeneration, and therefore the site in need of protection, is not the optic nerve but the retina (Schori et al., 2001; Bakalash et al., 2003). We further determined that in immune-deficient animals, the number of surviving RGCs following an insult of elevated IOP is significantly lower than that in matched controls with an intact immune system. This suggests that the ability to withstand insult to the optic nerve or to the retina depends on the integrity of the immune system, and specifically on the specific cell population within the immune system that recognizes the site-specific self-antigens. Interestingly, treatment with steroids, which have an immunosuppressant effect, caused a significant loss of RGCs (Bakalash et al., 2003). Moreover, in a model of RGC loss induced by non-IOP conditions, for example, uveitis, steroids, which alleviate the inflammatory manifestations of the uveitis, not only failed to protect RGCs but even caused further death. These results prompted us to suggest that the well-controlled boosting of the T cell response might protect RGCs, even under conditions of NTG as well as in uveitis.

T cells specific to antigens residing in the site of damage help clean and heal

To manifest their protective effects, anti-self T cells should home to the site of damage and be locally activated. This is why only those antigens that are present at the site of lesion can be used for the vaccination. Once activated, the T cells provide a source of cytokines and growth factors, which create the proper niche for microglia and infiltrating blood-borne monocytes, so as to make them active protective cells that the eye can tolerate (Scheme 1). Namely, such activated microglia/macrophages can take up glutamate, remove debris, and produce growth factors; additionally, they do not produce agents that are part of their cytotoxic mechanism, such as TNF-α, which the eye, like the brain, tolerates poorly (Moalem et al., 2000; Butovsky et al., 2001; Barouch and Schwartz, 2002; Butovsky et al., 2005; Shaked et al., 2005). Such T cells are constitutively controlled by regulatory T cells that are part of the physiological immune network and are themselves amenable to control upon need (Kipnis et al., 2002, 2004).

Searching for an antigen for potential glaucoma therapy

Among the many immunomodulatory compounds that we tested for therapy of glaucoma was glatiramer acetate (GA), also known as Copaxone®, a synthetic 4-amino-acid copolymer, currently used as a treatment (administered according to a daily regimen) for multiple sclerosis. We chose to test this compound because of its low-affinity cross-reaction with a wide range of self-antigens residing in the CNS. In the rat model of chronically high IOP, vaccination with GA significantly reduces RGC loss even if the IOP remains high. Vaccination does not prevent disease onset, but rather slows down the progression of the RGCs loss by controlling the milieu of the nerve and retina, making it less hostile to neuronal survival and allowing the RGCs to better withstand the stress (Kipnis et al., 2000, 2002; Schori et al., 2001; Benner et al., 2004; Schwartz, 2007). In our initial studies, we used GA emulsified in complete Freund's adjuvant (CFA) (Schori et al., 2001). We subsequently found that in models of optic nerve insults and in models of spinal cord injury, regardless of the choice of the antigens, the type of adjuvant (amount of mycobacteria), the timing of the vaccination, and the dosing critically affect the outcome (Hauben et al., 2001; Bakalash et al., 2005; Blair et al., 2005). In subsequent studies in a model of IOP, we tested adjuvant-free GA, and found that GA is effective without adjuvant, but the onset of the treatment, the frequency, and the dosing are critical (Bakalash et al., 2005). Given that a self-perpetuating degeneration is a multiparameter disease in which numerous factors are participating, it seems that a therapy involving the well-controlled activation of the immune system can provide a multidimensional effectual therapy. For chronic treatment, daily, weekly, and monthly regimens were tested;

no benefit was found in daily injection, while weekly and monthly treatment were effective (Bakalash et al., 2005).

Concluding remarks

During the past decade, scientists and clinicians began to accept that glaucoma should not be viewed as a distinct syndrome with its own peculiar features, but as one of the large group of neurodegenerative disorders of the CNS. Regardless of the primary source of tissue injury, degeneration may continue due to the loss of equilibrium that exists between the initial insult and the ability of the eye to withstand it. As described above, this stage of equilibrium is generally managed by the immune system. According to our view, rather than or in addition to fighting off the risk factor(s), there is a need to protect the tissue from the ongoing spread of damage — an approach collectively termed "neuroprotection" (Schwartz et al., 1996). This view of glaucoma has led to major changes in the nature of glaucoma research, the way in which clinicians perceive the disease, and the approach to therapy. One comprehensive approach is to harness the immune system, which if properly controlled can be used to modulate the local milieu, so as to become protective rather than hostile to the RGCs (Scheme 1). Such selective immune activation can address the multiple risk factors contributing to the glaucomatous degeneration process. Currently, more specific antigens are being tested as possible candidates for specific immune therapy of glaucoma.

References

Abe, K., Pan, L.H., Watanabe, M., Kato, T. and Itoyama, Y. (1995) Induction of nitrotyrosine-like immunoreactivity in the lower motor neuron of amyotrophic lateral sclerosis. Neurosci. Lett., 199: 152–154.

Andersen, J.K. (2004) Oxidative stress in neurodegeneration: cause or consequence?. Nat. Med., 10(Suppl): S18–S25.

Avidan, H., Kipnis, J., Butovsky, O., Caspi, R.R. and Schwartz, M. (2004) Vaccination with autoantigen protects against aggregated beta-amyloid and glutamate toxicity by controlling microglia: effect of CD4+CD25+ T cells. Eur. J. Immunol., 34: 3434–3445.

Bakalash, S., Kessler, A., Mizrahi, T., Nussenblatt, R. and Schwartz, M. (2003) Antigenic specificity of immunoprotective therapeutic vaccination for glaucoma. Invest. Ophthalmol. Vis. Sci., 44: 3374–3381.

Bakalash, S., Kipnis, J., Yoles, E. and Schwartz, M. (2002) Resistance of retinal ganglion cells to an increase in intraocular pressure is immune-dependent. Invest. Ophthalmol. Vis. Sci., 43: 2648–2653.

Bakalash, S., Shlomo, G.B., Aloni, E., Shaked, I., Wheeler, L., Ofri, R. and Schwartz, M. (2005) T-cell-based vaccination for morphological and functional neuroprotection in a rat model of chronically elevated intraocular pressure. J. Mol. Med., 83: 904–916.

Barouch, R. and Schwartz, M. (2002) Autoreactive T cells induce neurotrophin production by immune and neural cells in injured rat optic nerve: Implications for protective autoimmunity. FASEB J., 16: 1304–1306.

Bath, C.P., Farrell, L.N., Gilmore, J., Ward, M.A., Hicks, C.A., O'Neill, M.J. and Bleakman, D. (1996) The effects of ifenprodil and eliprodil on voltage-dependent Ca^{2+} channels and in gerbil global cerebral ischaemia. Eur. J. Pharmacol., 299: 103–112.

Beal, M.F. (1998) Excitotoxicity and nitric oxide in Parkinson's disease pathogenesis. Ann. Neurol., 44: S110–S114.

Beckman, J.S., Carson, M., Smith, C.D. and Koppenol, W.H. (1993) ALS, SOD and peroxynitrite. Nature, 364: p. 584.

Benner, E.J., Mosley, R.L., Destache, C.J., Lewis, T.B., Jackson-Lewis, V., Gorantla, S., Nemachek, C., Green, S.R., Przedborski, S. and Gendelman, H.E. (2004) Therapeutic immunization protects dopaminergic neurons in a mouse model of Parkinson's disease. Proc. Natl. Acad. Sci. U.S.A., 101: 9435–9440.

Blair, M., Pease, M.E., Hammond, J., Valenta, D., Kielczewski, J., Levkovitch-Verbin, H. and Quigley, H. (2005) Effect of glatiramer acetate on primary and secondary degeneration of retinal ganglion cells in the rat. Invest. Ophthalmol. Vis. Sci., 46: 884–890.

Brubaker, R.F. (1996) Delayed functional loss in glaucoma. LII Edward Jackson Memorial Lecture. Am. J. Ophthalmol., 121: 473–483.

Burgoyne, C.F., Downs, J.C., Bellezza, A.J., Suh, J.K. and Hart, R.T. (2005) The optic nerve head as a biomechanical structure: a new paradigm for understanding the role of IOP-related stress and strain in the pathophysiology of glaucomatous optic nerve head damage. Prog. Retin. Eye Res., 24: 39–73.

Butovsky, O., Hauben, E. and Schwartz, M. (2001) Morphological aspects of spinal cord autoimmune neuroprotection: colocalization of T cells with B7-2 (CD86) and prevention of cyst formation. FASEB J., 15: 1065–1067.

Butovsky, O., Talpalar, A.E., Ben-Yaakov, K. and Schwartz, M. (2005) Activation of microglia by aggregated beta-amyloid or lipopolysaccharide impairs MHC-II expression and renders them cytotoxic whereas IFN-gamma and IL-4 render them protective. Mol. Cell. Neurosci., 29: 381–393.

Castegna, A., Thongboonkerd, V., Klein, J.B., Lynn, B., Markesbery, W.R. and Butterfield, D.A. (2003) Proteomic identification of nitrated proteins in Alzheimer's disease brain. J. Neurochem., 85: 1394–1401.

Chauhan, B.C. and Drance, S.M. (1992) The relationship between intraocular pressure and visual field progression in glaucoma. Graefes Arch. Clin. Exp. Ophthalmol., 230: 521–526.

Chidlow, G., Wood, J.P. and Casson, R.J. (2007) Pharmacological neuroprotection for glaucoma. Drugs, 67: 725–759.

Choi, D.W. (1988a) Calcium-mediated neurotoxicity: relationship to specific channel types and role in ischemic damage. Trends Neurosci., 11: 465–469.

Choi, D.W. (1988b) Glutamate neurotoxicity and diseases of the nervous system. Neuron, 1: 623–634.

Dreyer, E.B. (1998) A proposed role for excitotoxicity in glaucoma. J. Glaucoma, 7: 62–67.

Fisher, J., Levkovitch-Verbin, H., Schori, H., Yoles, E., Butovsky, O., Kaye, J.F., Ben-Nun, A. and Schwartz, M. (2001) Vaccination for neuroprotection in the mouse optic nerve: implications for optic neuropathies. J. Neurosci., 21: 136–142.

Gao, H., Qiao, X., Hefti, F., Hollyfield, J.G. and Knusel, B. (1997) Elevated mRNA expression of brain-derived neurotrophic factor in retinal ganglion cell layer after optic nerve injury. Invest. Ophthalmol. Vis. Sci., 38: 1840–1847.

Giasson, B.I., Duda, J.E., Murray, I.V., Chen, Q., Souza, J.M., Hurtig, H.I., Ischiropoulos, H., Trojanowski, J.Q. and Lee, V.M. (2000) Oxidative damage linked to neurodegeneration by selective alpha-synuclein nitration in synucleinopathy lesions. Science, 290: 985–989.

Gupta, N. and Yucel, Y.H. (2007) Glaucoma as a neurodegenerative disease. Curr. Opin. Ophthalmol., 18: 110–114.

Hartwick, A.T. (2001) Beyond intraocular pressure: neuroprotective strategies for future glaucoma therapy. Optom. Vis. Sci., 78: 85–94.

Hauben, E., Agranov, E., Gothilf, A., Nevo, U., Cohen, A., Smirnov, I., Steinman, L. and Schwartz, M. (2001) Posttraumatic therapeutic vaccination with modified myelin self-antigen prevents complete paralysis while avoiding autoimmune disease. J. Clin. Invest., 108: 591–599.

Heijl, A., Leske, M.C., Bengtsson, B., Hyman, L., Bengtsson, B. and Hussein, M. (2002) Reduction of intraocular pressure and glaucoma progression: results from the Early Manifest Glaucoma Trial. Arch. Ophthalmol., 120: 1268–1279.

Jay, J.L. and Allan, D. (1989) The benefit of early trabeculectomy versus conventional management in primary open angle glaucoma relative to severity of disease. Eye, 3(Pt 5): 528–535.

Johnson, E.C., Cepurna, W.O., Jia, L. and Morrison, J.C. (2006) The use of cyclodialysis to limit exposure to elevated intraocular pressure in rat glaucoma models. Exp. Eye Res., 83: 51–60.

Kass, M.A., Heuer, D.K., Higginbotham, E.J., Johnson, C.A., Keltner, J.L., Miller, J.P., Parrish II, R.K., Wilson, M.R. and Gordon, M.O. (2002) The Ocular Hypertension Treatment Study: a randomized trial determines that topical ocular hypotensive medication delays or prevents the onset of primary open-angle glaucoma. Arch. Ophthalmol., 120: 701–713, discussion 829–730

Kipnis, J., Cardon, M., Avidan, H., Lewitus, G.M., Mordechay, S., Rolls, A., Shani, Y. and Schwartz, M. (2004) Dopamine, through the extracellular signal-regulated kinase pathway, downregulates CD4+CD25+ regulatory T-cell activity: implications for neurodegeneration. J. Neurosci., 24: 6133–6143.

Kipnis, J., Mizrahi, T., Hauben, E., Shaked, I., Shevach, E. and Schwartz, M. (2002) Neuroprotective autoimmunity: Naturally occurring CD4+CD25+ regulatory T cells suppress the ability to withstand injury to the central nervous system. Proc. Natl. Acad. Sci. U.S.A., 99: 15620–15625.

Kipnis, J., Yoles, E., Porat, Z., Cohen, A., Mor, F., Sela, M., Cohen, I.R. and Schwartz, M. (2000) T cell immunity to copolymer 1 confers neuroprotection on the damaged optic nerve: possible therapy for optic neuropathies. Proc. Natl. Acad. Sci. U.S.A., 97: 7446–7451.

Klocker, N., Cellerino, A. and Bahr, M. (1998) Free radical scavenging and inhibition of nitric oxide synthase potentiates the neurotrophic effects of brain-derived neurotrophic factor on axotomized retinal ganglion cells in vivo. J. Neurosci., 18: 1038–1046.

Lancelot, E. and Beal, M.F. (1998) Glutamate toxicity in chronic neurodegenerative disease. Prog. Brain Res., 116: 331–347.

Lansbury, P.T. and Lashuel, H.A. (2006) A century-old debate on protein aggregation and neurodegeneration enters the clinic. Nature, 443: 774–779.

Leske, M.C., Heijl, A., Hussein, M., Bengtsson, B., Hyman, L. and Komaroff, E. (2003) Factors for glaucoma progression and the effect of treatment: the early manifest glaucoma trial. Arch. Ophthalmol., 121: 48–56.

Levin, L.A., Schlamp, C.L., Spieldoch, R.L., Geszvain, K.M. and Nickells, R.W. (1997) Identification of the bcl-2 family of genes in the rat retina. Invest. Ophthalmol. Vis. Sci., 38: 2545–2553.

Levkovitch-Verbin, H., Harris-Cerruti, C., Groner, Y., Wheeler, L.A., Schwartz, M. and Yoles, E. (2000) RGC death in mice after optic nerve crush injury: oxidative stress and neuroprotection. Invest. Ophthalmol. Vis. Sci., 41: 4169–4174.

Lipton, S.A. (2007) Pathologically-activated therapeutics for neuroprotection: mechanism of NMDA receptor block by memantine and S-nitrosylation. Curr. Drug Targets, 8: 621–632.

Lipton, S.A. and Rosenberg, P.A. (1994) Excitatory amino acids as a final common pathway for neurologic disorders. N. Engl. J. Med., 330: 613–622.

Mansour-Robaey, S., Clarke, D.B., Wang, Y.C., Bray, G.M. and Aguayo, A.J. (1994) Effects of ocular injury and administration of brain-derived neurotrophic factor on survival and regrowth of axotomized retinal ganglion cells. Proc. Natl. Acad. Sci. U.S.A., 91: 1632–1636.

Martindale, J.L. and Holbrook, N.J. (2002) Cellular response to oxidative stress: signaling for suicide and survival. J. Cell Physiol., 192: 1–15.

McKinnon, S.J., Lehman, D.M., Kerrigan-Baumrind, L.A., Merges, C.A., Pease, M.E., Kerrigan, D.F., Ransom, N.L., Tahzib, N.G., Reitsamer, H.A., Levkovitch-Verbin, H.,

Quigley, D.J. and Zack, D.J. (2002) Caspase activation and amyloid precursor protein cleavage in rat ocular hypertension. Invest. Ophthalmol. Vis. Sci., 43: 1077–1087.

Mizrahi, T., Hauben, E. and Schwartz, M. (2002) The tissue-specific self-pathogen is the protective self-antigen: the case of uveitis. J. Immunol., 169: 5971–5977.

Moalem, G., Gdalyahu, A., Shani, Y., Otten, U., Lazarovici, P., Cohen, I.R. and Schwartz, M. (2000) Production of neurotrophins by activated T cells: implications for neuroprotective autoimmunity. J. Autoimmun., 15: 331–345.

Moalem, G., Leibowitz-Amit, R., Yoles, E., Mor, F., Cohen, I.R. and Schwartz, M. (1999) Autoimmune T cells protect neurons from secondary degeneration after central nervous system axotomy. Nat. Med., 5: 49–55.

Mozaffarieh, M. and Flammer, J. (2007) Is there more to glaucoma treatment than lowering IOP? Surv. Ophthalmol., 52(Suppl 2): S174–S179.

Napper, G.A., Pianta, M.J. and Kalloniatis, M. (1999) Reduced glutamate uptake by retinal glial cells under ischemic/hypoxic conditions. Vis. Neurosci., 16: 149–158.

Neufeld, A.H., Sawada, A. and Becker, B. (1999) Inhibition of nitric-oxide synthase 2 by aminoguanidine provides neuroprotection of retinal ganglion cells in a rat model of chronic glaucoma. Proc. Natl. Acad. Sci. U.S.A., 96: 9944–9948.

Nickells, R.W. (2007) From ocular hypertension to ganglion cell death: a theoretical sequence of events leading to glaucoma. Can. J. Ophthalmol., 42: 278–287.

Nickells, R.W., Schlamp, C.L., Li, Y., Kaufman, P.L., Heatley, G., Peterson, J.C., Faha, B. and VerHoeve, J.N. (2007) Surgical lowering of elevated intraocular pressure in monkeys prevents progression of glaucomatous disease. Exp. Eye Res., 84: 729–736.

Nouri-Mahdavi, K., Brigatti, L., Weitzman, M. and Caprioli, J. (1995) Outcomes of trabeculectomy for primary open-angle glaucoma. Ophthalmology, 102: 1760–1769.

Oku, H., Yamaguchi, H., Sugiyama, T., Kojima, S., Ota, M. and Azuma, I. (1997) Retinal toxicity of nitric oxide released by administration of a nitric oxide donor in the albino rabbit. Invest. Ophthalmol. Vis. Sci., 38: 2540–2544.

Olney, J.W. (1969) Glutaate-induced retinal degeneration in neonatal mice. Electron microscopy of the acutely evolving lesion. J. Neuropathol. Exp. Neurol., 28: 455–474.

Olney, J.W., Price, M.T., Samson, L. and Labruyere, J. (1986) The role of specific ions in glutamate neurotoxicity. Neurosci. Lett., 65: 65–71.

Oltvai, Z.N. and Korsmeyer, S.J. (1994) Checkpoints of dueling dimers foil death wishes. Cell, 79: 189–192.

Peinado-Ramon, P., Salvador, M., Villegas-Perez, M.P. and Vidal-Sanz, M. (1996) Effects of axotomy and intraocular administration of NT-4, NT-3, and brain-derived neurotrophic factor on the survival of adult rat retinal ganglion cells. A quantitative in vivo study. Invest. Ophthalmol. Vis. Sci., 37: 489–500.

Potashkin, J.A. and Meredith, G.E. (2006) The role of oxidative stress in the dysregulation of gene expression and protein metabolism in neurodegenerative disease. Antioxid. Redox Signal., 8: 144–151.

Quigley, H.A. (1999) Neuronal death in glaucoma. Prog. Retin. Eye Res., 18: 39–57.

Quigley, H.A. and Maumenee, A.E. (1979) Long-term follow-up of treated open-angle glaucoma. Am. J. Ophthalmol., 87: 519–525.

Richler, M., Werner, E.B. and Thomas, D. (1982) Risk factors for progression of visual field defects in medically treated patients with glaucoma. Can. J. Ophthalmol., 17: 245–248.

Riederer, P. and Hoyer, S. (2006) From benefit to damage. Glutamate and advanced glycation end products in Alzheimer brain. J. Neural Transm., 113: 1671–1677.

Ritch, R. (2000) Potential role for Ginkgo biloba extract in the treatment of glaucoma. Med. Hypotheses, 54: 221–235.

Rocha, M., Martins, R.A. and Linden, R. (1999) Activation of NMDA receptors protects against glutamate neurotoxicity in the retina: evidence for the involvement of neurotrophins. Brain Res., 827: 79–92.

Sahai, S. (1990) Glutamate in the mammalian CNS. Eur. Arch. Psychiatry Clin. Neurosci., 240: 121–133.

Schori, H., Kipnis, J., Yoles, E., WoldeMussie, E., Ruiz, G., Wheeler, L.A. and Schwartz, M. (2001) Vaccination for protection of retinal ganglion cells against death from glutamate cytotoxicity and ocular hypertension: implications for glaucoma. Proc. Natl. Acad. Sci. U.S.A., 98: 3398–3403.

Schulzer, M., Drance, S.M., Carter, C.J., Brooks, D.E., Douglas, G.R. and Lau, W. (1990) Biostatistical evidence for two distinct chronic open angle glaucoma populations. Br. J. Ophthalmol., 74: 196–200.

Schwartz, M. (2003) Neurodegeneration and neuroprotection in glaucoma: development of a therapeutic neuroprotective vaccine: the Friedenwald lecture. Invest. Ophthalmol. Vis. Sci., 44: 1407–1411.

Schwartz, M. (2005) Lessons for glaucoma from other neurodegenerative diseases: can one treatment suit them all? J. Glaucoma, 14: 321–323.

Schwartz, M. (2007) Modulating the immune system: a vaccine for glaucoma?. Can. J. Ophthalmol., 42: 439–441.

Schwartz, M., Belkin, M., Yoles, E. and Solomon, A. (1996) Potential treatment modalities for glaucomatous neuropathy: neuroprotection and neuroregeneration. J. Glaucoma, 5: 427–432.

Shaked, I., Tchroesh, D., Gersner, R., Meiri, G., Mordechai, S., Xiao, X., Hart, R.P. and Schwartz, M. (2005) Protective autoimmunity: interferon-gamma enables microglia to remove glutamate without evoking inflammatory mediators. J. Neurochem., 92: 997–1009.

Sigal, I.A., Flanagan, J.G. and Ethier, C.R. (2005) Factors influencing optic nerve head biomechanics. Invest. Ophthalmol. Vis. Sci., 46: 4189–4199.

Siliprandi, R., Canella, R., Carmignoto, G., Schiavo, N., Zanellato, A., Zanoni, R. and Vantini, G. (1992) N-Methyl-D-aspartate-induced neurotoxicity in the adult rat retina. Vis. Neurosci., 8: 567–573.

Siu, A.W., Maldonado, M., Sanchez-Hidalgo, M., Tan, D.X. and Reiter, R.J. (2006) Protective effects of melatonin in experimental free radical-related ocular diseases. J. Pineal Res., 40: 101–109.

the eye, such as age-related macular degeneration, cataract, and glaucoma.

Many findings lead us to believe that these diseases are the cytopathological consequence of an unfavorable state between the intracellular concentration of free radicals and the cells' capacity to neutralize them through an increase in endogenous production of free radicals, a reduction of antioxidant molecules and/or a decrease of the capacity to repair the oxidative damage on cellular macromolecules.

The main eye diseases have a common pathogenetic mechanism: a long period of latency between induction and clinical manifestation with a multifactorial etiology resulting from the interaction between environmental risk factors and heightened individual genetic susceptibility.

In particular, in glaucoma, free radicals have different target tissues and, especially, in the anterior segment of the eye, they affect the trabecular meshwork (TM) and, more specifically, its endothelium. Contemporarily, in the posterior segment, free radicals affect/involve retinal ganglion cells (RGCs) and optic nerve head (ONH) in its extracellular component, leading to a series of events defined as "the glaucomatous cascade."

Oxidative stress

The loss of electrons from the outer orbit of atoms is defined as "oxidation" and leads to the formation of highly reactive molecules. A molecular species whose atoms contain one or more unpaired electrons in their outer orbits is defined as a free radical. This structure gives rise to the fundamental property of such molecules: instability. Indeed, in order to restore the equilibrium of their magnetic fields, atoms containing unpaired electrons tend to capture electrons from other atoms of nearby molecules, which in turn become free radicals, thus a chain reaction is triggered.

Organisms use oxygen to oxidize food; nutrients composed of carbohydrates, proteins, and fats are oxidized to carbon dioxide and water. The energy released during the oxidation process is stored in the form of adenosine triphosphate (ATP) and is subsequently used in numerous metabolic reactions.

While oxygen is essential to animal metabolism, it can also be harmful. Indeed, numerous uncontrolled oxidation reactions (auto-oxidation) may take place in the presence of oxygen and can cause cell damage. Free radicals are oxygen compounds resulting from the numerous metabolic reactions involving oxygen. They tend to readdress their imbalance by "attacking" nearby molecules in order to recover their missing electron, thereby making other molecules unstable. The ensuing chain reaction gives rise to the formation of new compounds, some of which may be toxic.

Although many atoms and molecules can form free radicals in vivo, the most important for biological systems seem to be the radical ions associated with oxygen reduction. The diatomic molecules of oxygen (O_2) can be reduced to form the superoxide (O_2^-) and hydroxyl (OH^-) radicals. These oxygen free radicals have an enormous potential to harm living organisms. The superoxide is not particularly reactive in aqueous environments, while it is highly destructive in the lipophilic linings of biological membranes. The hydroxyl form is the most reactive and the most dangerous of all free radicals: it survives only momentarily before combining with one of the molecules nearby, such as DNA, proteins, and other macromolecules.

In physiological conditions, there is a balance between the endogenous production of free radicals and their neutralization by antioxidant defense mechanisms. When the production of radicals exceeds the organism's capacity to neutralize them ("scavenger" activity) or when antioxidant substances activity decreases, damage ensues.

Free radicals are neutralized both by a range of enzymes, such as superoxide dismutase (SOD), glutathione peroxidase, or catalase, and by numerous molecules that are either endogenously produced, such as glutathione (GSH), or dietary introduced, such as flavonoids, vitamins C, E, and others. These molecules are able to capture free radicals and accept the unpaired electron and "pass it on." Molecules with the most effective action are those which have aromatic rings, particularly those with several hydroxyl groups, such as the polyphenols.

The most well-known free radical chain reaction is lipid peroxidation. In this process, a free radical removes a hydrogen atom from a lateral chain of a polyunsaturated fatty acid. Consequently, the carbon atom from which the hydrogen atom has been removed is left with an unpaired electron. In other words, the lateral chain of the fatty acid is transformed into a reactive free radical. These carbon atom radicals (lipids) normally end up combining with molecular oxygen (O_2) in the membrane to produce a peroxyl radical lipid (O_2). O_2 is extremely reactive and triggers a chain reaction within the lateral chains of the polyunsaturated fatty acids. Over time, peroxidation produces enough peroxidated lipids to damage the structure and fluidity of the membrane. Through a similar mechanism, peroxidation can cause severe damage to the essential proteins of the membrane.

The lateral chains of the polyunsaturated fatty acids, which guarantee the necessary fluidity of the membrane lipids, are particularly sensitive to attack by free radicals. If the free radicals of the chains are not deactivated, their chemical reactivity can damage all types of cellular macromolecules. The main targets of peroxidation reactions are proteins, cell membranes, and nucleic acids (DNA and RNA), including mitochondrial DNA (mtDNA). Indeed, mtDNA is less protected than nuclear DNA, and is therefore more sensitive to free radical attack (De Grey, 1997). mtDNA damage is another possible mechanism involved in the etiopathogenesis of degenerative diseases.

Peroxidation phenomena in the organism are countered by antioxidant compounds, which inhibit the formation of free radicals. These compounds include water-soluble antioxidants (e.g., ascorbic acid, cysteine, GSH), lipid-insoluble antioxidants (e.g., tocopherols and retinols), and enzymes such as SOD, which catalyzes the transformation of free radicals into hydrogen peroxide. Although hydrogen peroxide is also an active oxygen compound, it can be further transformed into oxygen and water by catalase. Also, several metal-binding proteins (e.g., transferrin) (Babizhayev and Costa, 1994; Rose et al., 1998) and flavonoids (e.g., genistein, diazine, glycyrrhizin, etc.) (Kapiotis et al., 1997; Wang et al., 1998; Tang et al., 2007) have an antioxidant activity.

Table 1. Many studies underscore the role of endogenous oxidative damage in the pathogenesis of glaucoma

ROS effects in the eye

Induction	Inhibition
Protein synthesis	HTM cellularity
H_2O_2	HTM mobility
Catalase	
Vascular permeability factor	Glutamate synthase
Vascular endothelial growth factor	Na^+-dependent glutamate transporter
Ischemia	Glutamate reuptake by astroglial cells
Heat shock proteins	Ionic imbalance
Matrix components	Glutathione
Peroxynitrite	Expression of neural cell adhesion molecule
Endothelins	Na^+, K^+-ATPase
ECM remodeling	Ascorbate
TGF	Antioxidant activity

Note: This interpretation is in agreement with both vascular and mechanical pathogenic theories. The set of processes that make up the glaucomatous cascade triggered by free radicals results in the progressive apoptotic degeneration of trabecular meshwork, retina, and optic nerve.

The formation of metal-mediated free radicals causes various modifications of DNA nucleotides, enhanced lipid peroxidation, and altered calcium and sulfhydryl homeostasis. Lipid peroxides, formed by the attack of free radicals on polyunsaturated fatty acid residues of phospholipids, can further react with redox metals, finally producing mutagenic and other exocyclic DNA adducts (Valko et al., 2005), such as 8-hydroxy-2′-deoxyguanosine (8-OH-dG), which is an indicator of oxidative DNA damage.

Concerning eye diseases, all this phenomena are particularly important, above all in the pathogenesis of glaucoma (Table 1), where age and oxidative stress appear to play a fundamental pathogenic role (Saccà et al., 2007).

Trabecular meshwork

Glaucoma is a group of optic neuropathies characterized by a progressive degeneration of RGCs and visual-field damage, which represents the final common pathway resulting from a number of different conditions that can affect the eye. Even if this disease has been known from the time of Hippocrates (Nathan, 2000), its pathogenesis is still misunderstood.

The chambers of the eye are filled with aqueous humor (AH), a fluid with an ionic composition very similar to blood plasma, with two main functions: to provide nutrients to eye tissues (e.g., cornea, iris, and lens) and to maintain intraocular pressure (IOP). Therefore, the anterior chamber of the eye can be regarded as a highly specialized vascular compartment whose inner walls are composed of the endothelia of iris, cornea, and TM (Brandt and O'Donnell, 1999).

In most cases, glaucoma is accompanied by an increase of the IOP. Ocular hypertension is one of the major risk factors for the development and progression of primary open-angle glaucoma (POAG), a leading cause of blindness. TM is a tissue located in the anterior chamber angle of the eye, and it is a crucial determinant of IOP even if, at the moment, many aspects of the regulation of AH outflow remain unclear. Nevertheless, it is known that the region of maximal resistance to AH outflow resides at the peripheral juxtacanalicular TM, which connects the TM to the Schlemm's canal (Johnson and Johnson, 2001). TM is directly involved in the regulation of AH outflow (Wiederholt et al., 1995). The subendothelial region of Schlemm's canal does not form a continuous fluid system, and the pathways through the connective tissue of the cribriform region are responsible for outflow facility and determine the filtration area of the inner wall of Schlemm's canal (Lutjen-Drecoll, 1973). This tissue has unique morphologic and functional properties involved in the regulation of AH outflow. The conventional outflow pathway is organized with a plumbing arrangement for maintaining a fluid barrier to prevent the passage of AH, consisting of trabecular lamellae covered with HTM cells, in front of a resistor, consisting of juxtacanalicular HTM cells and the inner wall of Schlemm's canal. The outermost juxtacanalicular or cribriform region has no collagenous beams, but rather several cell layers which some authors claim to be immersed in loose extracellular material/matrix (Tian et al., 2000). TM cells also regulate the formation and turnover of extracellular matrix (ECM) (Yue, 1996). A disproportionate accretion of ECM occurs in the TM region of POAG eyes, and this buildup is responsible for the development of greater resistance to AH outflow, resulting in increased IOP (Rohen and Witmer, 1972; Lee and Grierson, 1974). The main resistance to the AH outflow is located in the TM directly underneath the inner wall of Schlemm's canal (Maepea and Bill, 1992). The inner wall of Schlemm's canal is unique, sharing extraordinary characteristics with both types of specialized endothelia: lymphatic and blood capillary endothelia (Ramos et al., 2007). In this layer, there are the "giant vacuoles." They are really outpouchings of the endothelium into Schlemm's canal, caused by the pressure drop across inner wall endothelial cells (Brilakis and Johnson, 2001). The distal openings, or pores, in these vacuoles are a second feature of the inner wall endothelium (Johnson, 2006). The majority of these pores are transcellular. The transcellular pores do not connect the extracellular fluid with the cellular cytoplasm. These pores usually form on giant vacuoles because it is in this region in which the cell is greatly attenuated and the cytoplasm becomes thin (Johnson, 2006). When cell thickness is reduced below a critical value, vascular endothelium can form transcellular pores involved in transport processes (Neal and Michel, 1996; Savla et al., 2002). The region immediately underlying the inner wall and basement membrane is the juxtacanalicular connective tissue having many large spaces filled with an ECM gel (Ethier et al., 1986; Ten Hulzen and Johnson, 1996). Furthermore, the ECM may generate significant aqueous outflow resistance (Bradley et al., 1998). Studies found that some inner wall pores may be artifacts of the fixation process (Sit et al., 1997; Ethier et al., 1998). Actually, TM pores contribute only 10% of the aqueous outflow resistance (Sit et al., 1997). As stated by Johnson, "a fundamental reassessment of the mechanism by which AH crosses the inner wall endothelium is necessary" (Johnson et al., 2002). The concept of the outflow system as a passive filter has been surpassed; the structures through which AH leaves the anterior chamber may be objects of physical or pharmacologic manipulation for therapeutic purposes. In any case, endothelial cells of TM seem to have a leading role in outflow: probably, their tridimensional architecture and allocation on the trabecular beams considerably increases the filtration surface whose degeneration, resulting in the decay of HTM cellularity, causes IOP increase and triggers glaucoma. Important alterations in the cellular component and in the TM in toto, during POAG or in aging, occur in the inner layers near the anterior chamber (Alvarado et al., 1984a, b). Furthermore, in the juxtacanalicular–Schlemm's canal region, a marker of cell senescence stains the endothelia cells more than controls (Liton et al., 2005).

The analysis of POAG patients' eyes revealed thickened trabeculae, an increased amount of plaque-material deposits, and a collagen abundance. The uveal meshwork is partly deprived of cells. The cribriform layer often contained numerous enlarged, light cells with many small mitochondria and lysosomes (Gabelt and Kaufman, 2005).

Human outflow facility decreases with age. The age-related decrease is about 30% from the youngest (<40 years) to the oldest (>60 years) (Gaasterland et al., 1978). This phenomenon could be explained by the decrease in pseudofacility alone which could, in part, be due to the increase in ocular rigidity with age (Toris et al., 1999). The aqueous production shows an age-related decline that approximately amounts to 15–35% over the age range of 20–80 years (Gaasterland et al., 1978; Brubaker et al., 1981; Toris et al., 1999).

IOP increase and free radicals

Anterior chamber endothelial cells are always in contact with free radicals and are involved in antioxidant activities that counteract the toxic effects of oxidative stress. Aging is an inevitable biological process characterized by a general decline in physiological function that manifests itself in the anterior chamber with an endothelial cellular loss in TM (Alvarado et al., 1981, 1984a, b) involving, above all, the filtering area (Alvarado et al., 1981). This phenomenon can explain why IOP slightly increases with age in many Western populations (Brubaker et al., 1981).

In POAG, TM cell loss corresponds to a decrease of the AH flow: 65% of inner-layer trabecular cells die. In addition, in the cribriform layer, there is also an increase of the amount of extracellular material and of thickness in the elastic fiber sheaths (Gabelt et al., 2003).

Increased incidence of cells detaching from trabeculae and a progressively greater incidence of fusion between adjacent trabeculae occur with age (Grierson et al., 1982).

HTM is a heterogeneous tissue where the endothelial cells drive a mechanism that controls the permeability of the TM. These endothelial cells release vasoactive cytokines and other factors able to increase the permeability of the endothelial barrier of the Schlemm's canal. Therefore, cytokines alterations may cause outflow shift and, subsequently, IOP may rise to abnormal levels (Alvarado et al., 2005).

It is possible that toxic substances present in AH could in some way contribute to the appearance of these burden pathogenetic alterations of the TM. For example, the AH is capable of producing levels of H_2O_2 (Spector et al., 1998), yet HTM cells are in contact with relatively high concentrations of H_2O_2 and the exposure to H_2O_2 causes a decrease in facility in eyes with the GSH-depleted TM (Kahn et al., 1983).

Free radicals are the triggering agents in cellular decay, and AH antioxidant activity in patients with POAG is reduced (Ferreira et al., 2004). Indeed, the H_2O_2 effect on the adhesion of HTM cells to ECM proteins results in rearrangement of cytoskeletal structures that may induce a decrease of TM cell adhesion, cell loss, and compromise HTM integrity (Zhou et al., 1999).

In AH, there are many factors that have a protective role for endothelial cells, as GSH, which protects anterior segment tissues from high levels of H_2O_2 (Costarides et al., 1991). Likewise, a 28-amino acid neurotrophic factor present in human AH protects corneal endothelial cells from H_2O_2 and from other oxidative insults (Koh and Waschek, 2000).

Both GSH and ascorbate have been detected in AH (Rose et al., 1998). In humans, ascorbic acid is the major water-soluble antioxidant (Reiss, 1986), and is 15 times higher in concentration in the AH than in plasma (Becker, 1957), even if its concentration is closely correlated to the level of ascorbic acid in serum (Haung et al., 1997). Both vitamin C and GSH operate in fluid outside the cell, and GSH participates directly in the neutralization of free radicals and reactive oxygen compounds, being essential to protect protein thiol groups in order to maintain enzyme activity; moreover, GSH maintains exogenous antioxidants such as vitamins C and E in their reduced (active) forms (Cardoso et al., 1998). The GSH redox system is believed to protect ocular tissues from H_2O_2 low concentrations, whereas catalase is thought to protect ocular tissues from the damage induced by higher H_2O_2 concentrations (Riley, 1990; Costarides et al., 1991).

These antioxidant agents seem to play a particularly important role in glaucomatous disease. Indeed, patients with glaucoma exhibit low levels of circulating GSH, suggesting a general impairment of antioxidative defenses (Gherghel et al., 2005). In particular, in AH, oxidative stress may lead to an induction of antioxidant enzymes and contribute to decreasing the activities of the antioxidant enzymes SOD, catalase, and glutathione peroxidase (Ferreira et al., 2004). Indeed, POAG patients' AH contains lipid peroxides in higher concentrations than in normal eyes (Babizhayev and Bunin, 1989).

When mitochondrial GSH and vitamin E levels are reduced to 20% of the normal level, lipid peroxidation occurs (Augustin et al., 1997). In any case, GSH decrease and H_2O_2 increase in AH may contribute to collagen deposition in POAG TM (Veach, 2004). Yet, DNA damage of TM cells in POAG patients shows statistically significant correlations with IOP peaks (Saccà et al., 2005). Besides, a recent experimental study shows that oxidative damage and IOP increase contribute to glaucomatous optic neuropathy (Liu et al., 2007). Therefore, insufficient GSH combined with exogenous H_2O_2 may induce collagen matrix remodeling and trabecular cell apoptosis independently of mitochondria (Saccà et al., 2007).

H_2O_2 activates both trabecular cells and astrocytes to produce heat shock proteins (HSP) (Tamm et al., 1996; Takuma et al., 2002). HSP are produced in response to stress from reactive oxygen species (ROS), and in glaucomatous eyes, there is an increased concentration of HSP, both in the lamina cribrosa and TM (Tamm et al., 1996; Lutjen-Drecoll et al., 1998; Tezel et al., 2000). This phenomenon appeared to be the consequence of oxidative stress in ECM tissues (Tamm et al., 1996; Takuma et al., 2002). From a biomolecular point of view, mitochondria are selective targets for the protective effects of HSP against oxidative injury, still, HSP prevented lipid peroxidation of the mitochondrial membrane and H_2O_2-induced cell death (Polla et al., 1996) (Fig. 1).

Trabecular meshwork-inducible glucocorticoid response (TIGR), also known as myocilin (MYOC), may have a role in the regulation of aqueous outflow and, therefore, may have an extracellular function. The enhanced matrix deposition mediated by ascorbate in the AH could lead to the increased deposition of TIGR/MYOC from the AH into the ECM of TM in vivo (Filla et al., 2002). How extracellular TIGR/MYOC may function in the TM is unknown; moreover, the specificity of a TIGR/MYOC interaction with matrix proteins is confirmed and specifically interacts with the Hep II domain of fibronectin (Filla et al., 2002). This fibronectin domain is well known for its biological role in cellular adhesion, cytoskeleton organization, signal transduction, and phagocytosis (Gong et al., 1996; Tian et al., 2000).

Furthermore, in glaucoma patients' TM, the expression of the endothelial-leukocyte adhesion molecule (ELAM-1) is abundant and it is mediated by an autocrine feedback mechanism of activation by interleukin-1 through the nuclear transcription factor kappa B (NF-κB) (Wang et al., 2001).

The influence of genetic polymorphism

Moreover, genetic polymorphisms have been detected for GSH transferase (GST) isoenzyme, and the *GSTM1*-null genotype has been found to be significantly more common in patients with POAG than in controls (Izzotti et al., 2003). This genetic risk factor for glaucoma development is actually controversial even if other authors have confirmed this data (Izzotti and Saccà, 2004; Yildirim et al., 2005). For example, Unal et al. (2007) have reported

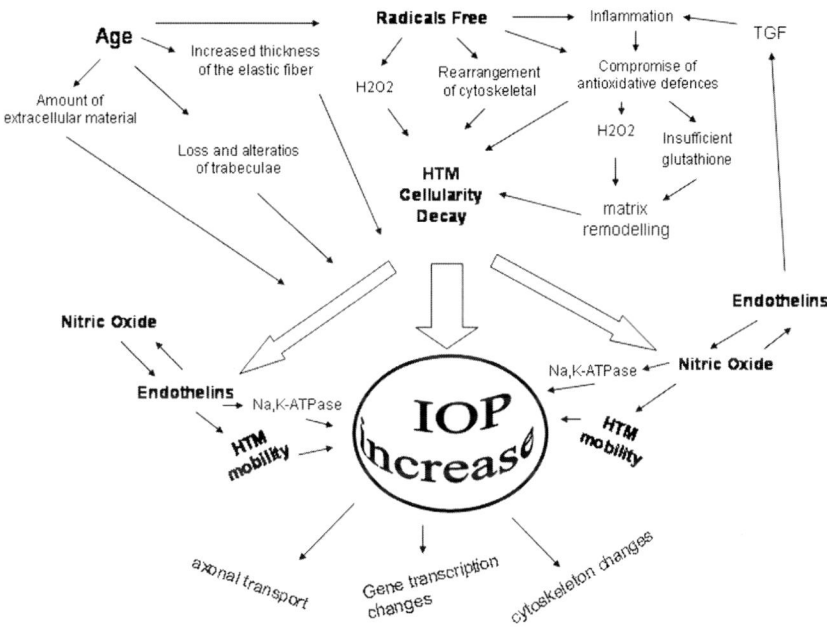

Fig. 1. IOP increase plays an important role in glaucoma. This factor is strongly connected with oxidative stress. Elevated IOP can, during the disease process, trigger all glaucomatous targets points. Particular importance is covered by the loss of endothelial cells that occurs in HTM. It is presumable that this cell injury may start the glaucomatous cascade acting at the same time in anterior and posterior segment.

that the *GSTM1*-positive genotype and *GSTT1*-null genotype or the combination of both may be associated with an increased risk of development of POAG in the Turkish population.

Free radicals probably play a fundamental role in the decrease of HTM cells: indeed, oxidative DNA damage is significantly increased in the TM of glaucoma patients than in controls and a statistically significant correlation has been found among HTM DNA oxidative damage, visual field damage, and IOP (Izzotti et al., 2003; Saccà et al., 2005). These data confirm weakened antioxidant defenses and the elevation of oxidative stress in glaucoma patients.

Glaucomatous cascade

Nitric oxide and endothelins

HTM is a complex organ that is stably engaged in the AH and it is able to respond to vasoactive substances, including vasoconstrictors, such as endothelins (ET), and vasodilators, such as nitric oxide (NO). A balance between vasoconstrictors and vasodilators is necessary for the maintenance of the physiological structure and function of endothelia (Wiederholt et al., 2000). Whenever this balance is disrupted, as in glaucoma, the outcome is endothelial dysfunction and injury, triggering the glaucomatous pathogenic cascade (Gibbons, 1997).

NO is produced by NO synthase (NOS) enzyme. This enzyme can have three different forms: neuronal NOS (NOS1), macrophage — or inducible — NOS (NOS2), and endothelial NOS (NOS3). NOS2 seems to be the most important isoenzyme involved in glaucoma pathogenesis (Liu and Neufeld, 2000). Indeed, NOS2 is expressed only in the ONH of patients with glaucoma (Neufeld et al., 1997). The induction of NOS2 expression generates high levels of NO, which have been associated with neural tissue toxicity (Dawson et al., 1993, 1994).

Vitamin C protects NO by scavenging superoxide and peroxynitrite within endothelial cells, which enhances NO exit from the cells, and spares NOS or tetrahydrobiopterin from oxidative modification (May, 2000).

NO production increases after elevation of the pressure gradient over the TM (Schneemann et al., 2003). In an experimental animal model of glaucoma, increased IOP appears to be a major causative factor for the overproduction of NO through inducible NOS (NOS2) activation, resulting in RGC death and optic nerve damage (Shareef et al., 1999). The presence of free radicals may induce NO to generate toxic products interacting with oxygen, iron, and/or copper: this can aggravate the metabolic conditions of the TM and alter its mobility (Tamm and Lutjen Drecoll, 1998; Wiederholt, 1998; Haefliger et al., 1999). NO has multiple effects: it is a neural messenger having a fundamental role in the nervous system, and it may modulate the sodium pump that regulates the neuronal energy metabolism of the brain (Royes et al., 2005). Through this mechanism, NO, glutamate, and certain other intercellular messengers bring about a marked and prolonged alteration in Na^+, K^+-ATPase activity and, so, causing an alteration of cellular energy usage, form a focal point for the action of several cellular messengers that have been implicated in neuronal viability in certain degenerative diseases and under conditions of stress (Ellis and Nathanson, 1998).

Moreover, the increase of the NO concentration, reacting with anion superoxide to form $ONOO^-$ (peroxynitrite) or other substances derived from oxidative stress, can lead to neurotoxicity (Lipton, 1999).

Also, ET can participate in the regulation of IOP, and despite TM mobility, ET levels in the glaucomatous AH are more elevated than in controls (Wiederholt, 1998; Haefliger et al., 1999; Orgul et al., 1999). From this point of view, ET-1 may induce TM contraction, and this contraction increases outflow resistance, whereas TM relaxation increases outflow facility (Wiederholt et al., 2000). In any case, ET-1 has a variety of pathophysiological ocular functions, depending on receptor subtype and tissue involved: ET induces an increase in intracellular calcium (Marsden and Brenner, 1991) and vasoconstriction (Marsault et al., 1993). Therefore, ET-1 would be the main effector of the glaucoma ischemia (Orgul et al., 1999). Surely, optic nerve ischemia could contribute to visual field loss and neuropathy; for this reason, ET-1 has a primary role in the vascular theory of glaucoma (Flammer, 1994). ET may be involved in vasoconstriction and/or vasospasms from abnormal autoregulation of the retinal microcirculation (Cioffi et al., 1995; Haefliger et al., 1999). Besides, chronic administration of low doses of ET into the perineural region in primates produces damage to the ONH that is characteristic of that seen in glaucoma and independent of IOP changes (Cioffi and Sullivan, 1999). Indeed, ET-1 induces ECM remodeling in the ONH, influencing the regulation of matrix metalloproteinases (MMPs) (He et al., 2007).

Furthermore, the effect of ET-1 is related to a reduction of Na^+, K^+-ATPase activity, underlining its vasoconstrictive properties, and it may contribute to the decrease of AH formation (Prasanna et al., 2001; Petzold et al., 2003) (Fig. 2).

In any case, free radicals play an important role in the development of ischemia/reperfusion (I/R) injury. There is not only a dysregulation of the vascular response to increased levels of ET, but there are also direct effects of ET on target tissues, depending on the expression and distribution of their ET receptor (Yorio et al., 2002). NO can modulate the expression, sensitivity, and signal termination of ET receptors (Redmond et al., 1996).

ET-1 and NO have been considered important mediators of apoptotic RGCs death in glaucoma, where lamina cribrosa appears to be the primary site of injury of the ONH (Haefliger et al., 1999; Quigley, 2005).

In animal models of glaucoma, RGCs die via apoptosis (Kerrigan et al., 1997); apoptosis is a genetically predetermined program of cell death, which can be activated by many different factors (Levin, 1999; Kikuchi et al., 2000; Tatton et al., 2001). There are different stimuli for apoptotic RGCs death: hypoxia, neurotrophin withdrawal, and glutamate-mediated toxicity, and in this context, oxidative stress seems to play an important role (Meldrum and Garthwaite, 1990;

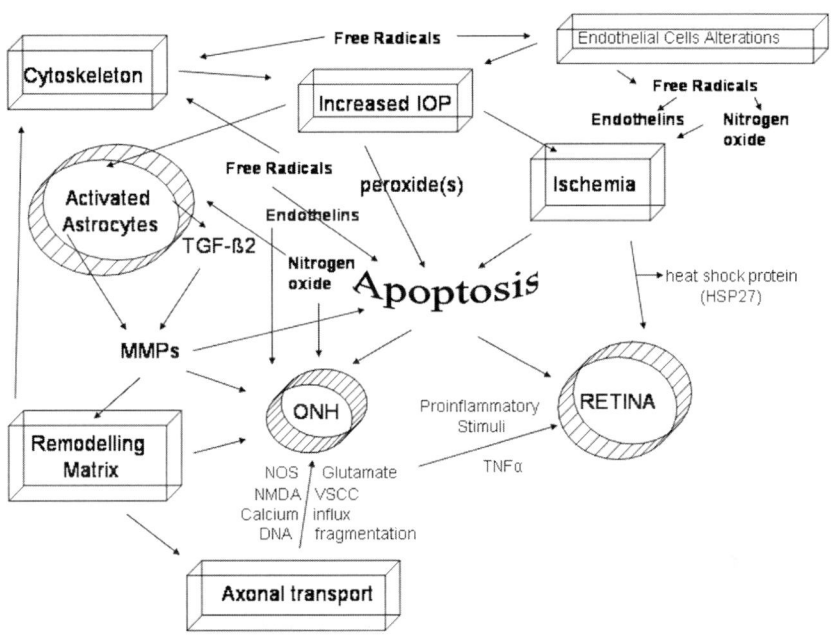

Fig. 2. Retinal ganglion cells death and axon degeneration represent the principal event in glaucoma course. The ways that manage the loss of this specific neuron population remain unclear. From a pathophysiological point of view, main steps of glaucoma course are intraocular pressure increase, ischemia, glial cells activation, apoptosis, and optic nerve head changes. The common denominator is the oxidative damage that is influencing these glaucomatous targets.

Lambert et al., 2004; Moreno et al., 2004; Tezel and Yang, 2004). Therefore, mechanical and vascular factors might work synergistically, leading to the same end result (Prasanna et al., 2005). IOP increase may induce factors such as mechanical stretching on the TM; mechanical strain on glial cells supporting ganglion cell axons; the scleral mechanical properties that should have such a large influence on ONH biomechanics, where the lamina cribrosa appears to be the primary site of injury (Burgoyne et al., 2005; Quigley, 2005; Sigal et al., 2005).

Lamina cribrosa is a specific connective region through which RGC axons exit the eye (Birch et al., 1997). In normal conditions, lamina cribrosa provides metabolic and mechanical support to nerve fibers against a pressure gradient, during course of glaucoma, and a remarkable overthrow of its architecture is observable: deformation, collapse, and ECM reorganization are manifest (Miller and Quigley, 1988). Elevated IOP induces both axonal transport and cytoskeleton changes in the ONH. Changes to the cytoskeleton may contribute to the axonal transport abnormalities that occur in elevated IOP (Balaratnasingam et al., 2007). Furthermore, this axonal transport blockage results in RGCs death and optic nerve degeneration (Quigley et al., 1983).

Extracellular matrix

The expression of a variety of TM genes is significantly affected by mechanical stretching and age-related variations. These are involved in apoptosis, cellular proliferation, and in other major aspects of cellular metabolism (Vittal et al., 2005). A particularly interesting data is provided by the involvement of the ECM: indeed, several ECM proteins may contribute to homeostatic modifications of AH outflow resistance, being up- or downregulated (Vittal et al., 2005). ECM remodeling occurring in TM during POAG is similar to the remodeling of other tissues, in particular the same pathological process occurs in atherosclerosis: trabecular cells are endothelial-like cells similar to the fibrogenic cells in these disorders (Veach, 2004). In any case, lower concentrations of oxidized low-density lipids have stimulated ECM remodeling (Bachem et al., 1999).

A recent study indicates that increased fibronectin synthesis could result in concomitant increase in IOP (Fleenor et al., 2006). Increased laminin and collagen type IV synthesis by TM cells exposed to ascorbic acid was also demonstrated (Zhou et al., 1998). A tight correlation between reduced permeability and increased expression of the ECM components has been observed, suggesting a possible link between excess matrix deposition and potential blockage in aqueous outflow (Tane et al., 2007). In vitro researches on human eyes have found a decrease in glycosaminoglycans (GAGs) synthesis, particularly hyaluronic acid, in glaucomatous eyes compared with normal eyes, and an increase in chondroitin sulfate content (Knepper et al., 1996b; Navajas et al., 2005).

Transforming growth factors (TGFs) are a family of cytokines that control a large variety of cellular processes like inflammation, wound healing, and ECM accumulation (Sporn and Roberts, 1992; Saika, 2004; Kottler et al., 2005). TGF regulates production of a wide variety of ECM genes, including elastin, collagens, fibrillin, laminin, and fibulin. Three structurally similar isoforms have been identified: TGF-β_1, TGF-β_2, and TGF-β_3 (Massague, 1990). TGF-β_2 levels are elevated in glaucomatous human AH (Tripathi et al., 1994) and alter ECM metabolism (Wordinger et al., 2007).

Bone morphogenetic proteins (BMP) are groups of growth factors known for their ability to induce the formation of bone and cartilage, which are expressed in the human TM and ONH (Wordinger et al., 2002). Recently, it has been discovered that BMP-7, which is expressed in the adult TM, modulates and antagonizes the effects of TGF-β_2 signaling on tissues of the outflow pathways in vivo and leads to increased ECM deposition and elevated IOP (Kane et al., 2005; Fuchshofer et al., 2007). In any case, elevated levels of TGF-β_2 in the AH may have the dual effect of both a direct increase of ECM components production in TM (e.g., fibronectin) and an enhanced production of gene products inhibiting ECM degradation (Fleenor et al., 2006). Furthermore, TGF-β_2 increases the expression of plasminogen activator inhibitor-1 (PAI-1). However, elevated PAI-1 levels have been shown to be linked to both decreased adhesion and increased detachment of a variety of cell types (Czakey and Loskutoff, 2004). This phenomenon may act with oxidative stress in cellular decay in TM.

TGF in the AH is also responsible for anterior chamber-associated immune deviation, a mechanism that protects the eye from inflammation and immune-related tissue damage (Wilbanks et al., 1992). Indeed, TGF-β_2 is one of the most important immunosuppressive cytokines in the anterior chamber of the eye and has a fibrogenic effect in trabecular cells (Alexander et al., 1998). A direct correlation between oxidative stress and TGF-β_2 expression has been demonstrated (Poli and Parola, 1997).

Moreover, TGF rules the expression levels of hyaluronan synthases (Usui et al., 2000). Among the GAGs of TM, hyaluronan is the most abundant, and it has been suggested to be an important modulator of aqueous outflow resistance and TM cell survival (Acott et al., 1985; Knepper et al., 1996a; Lerner et al., 1997). It represents a significant factor in outflow resistance in POAG, particularly during elevated IOP (Knepper et al., 2005).

ECM production in the TM may be mediated by vitamin C (Epstein et al., 1990; Sawaguchi et al., 1992). Ascorbic acid is reported to stimulate increased hyaluronic acid synthesis in glaucomatous trabecular cells compared with normal human trabecular cells (Schachtschabel and Binninger, 1993). Also, ascorbate reduces the viscosity of hyaluronic acid, thus increasing outflow through the trabeculum (McCarty, 1998). Researchers seem to be in disagreement about the trend of glaucoma patients having an ascorbate deficiency (Asregadoo, 1979; Lane, 1980; Beit-Yannai et al., 2007). However, there is compelling research on its effectiveness in treating glaucoma: high doses of vitamin C, first in animals and then in humans, showed that ascorbate decreases IOP (Virno et al., 1966). Other authors have confirmed this capacity of vitamin C, used both orally and topically (Linner, 1996). It is possible that ascorbic acid reduces IOP by the depolymerization of the TM's hyaluronic acid component (Linner, 1996). In any case, this vitamin is fundamental in ECM homeostasis: vitamin C neutralizes oxygen free radicals and is a reductant of oxidized vitamin E (Varma, 1991). It is also important in protecting cell membranes from lipid peroxidation (Kang et al., 2003).

The amount of vitamin E is low in the anterior chamber of the eye and it is not significantly associated with the risk of POAG. Vitamin E also prevents endogenous mitochondrial production of ROS (Southam et al., 1991).

The ECM degradation may lead to axonal loss in the ONH, and ROS stimulate astrocyte secretions and degrade tropoelastin (Hernandez and Pena, 1997; Hayashi et al., 1998; Tanaka et al., 1999; Hernandez, 2000).

Metalloproteinases

Other molecules that seem to play a very important role on collagen remodeling are the MMPs. MMPs are a family of calcium- and zinc-dependent extracellular endoproteinases that degrade ECM proteins (Nagase and Woessner, 1999). MMPs are secreted by cells as proenzymes and are in balance with their tissue inhibitors, which directly downregulate MMPs (Visse and Nagase, 2003). It is interesting to observe that MMPs may also be useful biomarkers of atherosclerotic risk and they serve as predictors of coronary and cerebrovascular disease recurrence (Rodriguez et al., 2007). In any case, in POAG eyes, MMPs levels are high, especially MMP-1 and MMP-3 levels that have a broad range of substrates including collagens, elastin, fibronectin, gelatin, tenascin, and laminin (Ronkko et al., 2007).

Increased MMP activity decreases collagen deposition, and AH outflow facility is increased by stimulating MMP activity. It is possible that elevated MMP levels in the AH of glaucomatous eyes may be produced by inflammatory cells and/or by the trabecular cells (Ronkko et al., 2007). The biologic activity of MMPs is regulated by their activation state, and their conversion to functionally active forms requires a specific multi-step activation process involving the proteolytic removal of part of the molecule (Nagase, 1997).

PAI is a potent, fast-acting, and irreversible inhibitor of active forms of tissue plasminogen representing the real inhibitor of plasmin that is involved in the activation of MMPs (Rondeau et al., 1995).

The elevated levels of TGF-β_2 in AH increase the production of ECM components in TM (e.g., fibronectin) and PAI-1 expression in HTM (Fuchshofer et al., 2003). PAI-1 is able to modulate the association between factors such as vitronectin and urokinase-like plasminogen activator with adhesion receptors (Fleenor et al., 2006). Hence, PAI-1 may be involved in TM cell loss, modulating TM cell migration and/or adhesion, including phagocytosis, oxidative stress, MYOC, and the presence of chemoattractants within the AH (Zhou et al., 1999; Hogg et al., 2000; Wentz-Hunter et al., 2004; Fleenor et al., 2006). Thus, it seems possible that an agent that increases one or more MMPs within the TM might have beneficial effects on IOP in patients with glaucoma. Blockage of the endogenous activity of the MMPs reduces outflow facility, probably because ECM turnover, initiated by one or more MMPs, appears to be essential to maintain IOP homeostasis (Bradley et al., 1998). An imbalance in the protease/antiprotease system is also important and may play a role in the pathogenesis of glaucoma (Ronkko et al., 2007).

Other factors of interest

Through the induction of oxidative damage, mechanical and vascular factors lead to the same final pathologic consequence (Prasanna et al., 2005). Among these mechanical factors, neurotrophic factor deprivation deserves to be mentioned. Elevated IOP blocks axonal transport at the level of the lamina cribrosa. One of the molecules delivered to the retina by a retrograde way of axonal transport is brain-derived neurotrophic factor (BDNF). Its importance has been shown by an experimental study on glaucomatous animals in which its intravitreous administration increases the number of surviving RGCs in comparison with untreated eyes (Ko et al., 2001). Another factor that may influence RGCs death in glaucoma is the glial cell activation, which is dopaminergic cell cultures, significantly enhancing the survival of dopaminergic neurons in primary cultures from embryonic mouse mesencephala exposed to neurotoxic concentrations of glutamate (Radad et al., 2004).

Memantine and its derivates

Memantine treatment has been found to significantly reduce lipid peroxidation (Ozsuer et al., 2005). In rats, systemic treatment with memantine has proved efficacious in experimental glaucoma, significantly reducing the glaucoma-induced loss of retinal ganglion cells (Hare et al., 2001). However, this potential neuroprotective agent, which blocks virtually all NMDA-receptor activity, is reported to have unacceptable clinical side effects (Lipton, 2003). Nevertheless, systemic treatment with memantine in monkeys has proved both safe and effective in reducing the functional loss and histologic changes associated with experimental glaucoma (Hare et al., 2001, 2004; Yucel et al., 2006). Orgogozo et al. (2002) evaluated the efficacy of memantine in patients with symptomatic mild-to-moderate vascular dementia of 6 months duration; the drug was well tolerated, with the most common adverse effects being agitation, confusion, and dizziness. Memantine has recently been approved by the European Union and the United States Food and Drug Administration (FDA) for clinical use in moderate-to-severe Alzheimer's disease to improve symptoms and reduce the rate of clinical deterioration. A multicenter phase-3 trial is currently underway to evaluate the effect of memantine in retarding the progression of glaucoma.

A series of second-generation memantine derivatives are currently being developed and may prove to have even greater neuroprotective properties than memantine. Adamantane, a derivative of memantine, blocks only the excessive NMDA receptor excitotoxic activity of glutamate in glaucoma without disrupting normal activity (Lipton, 2003). Another memantine derivative, amantadine, is a noncompetitive NMDA receptor antagonist that is thought to work both presynaptically and postsynaptically by increasing the amount of dopamine in the striatum (Peeters et al., 2003).

While all these drug types may provide relief from glaucoma damage, their long-term efficacy remains to be investigated through studies with longer follow-up periods. Finally, these second-generation drugs take advantage of the fact that the NMDA receptor has other modulatory sites, in addition to its ion channel, which could also be used for safe, effective clinical intervention (Lipton, 2005).

Conclusions

Despite the fact that glaucoma is one of the leading causes of blindness in the world, much remains to be discovered about the mechanisms triggering this disease and about its molecular pathogenesis (Quigley and Broman, 2006). Oxidative stress probably plays a primary role in the pathogenesis of glaucoma and seems to manifest its effects both downstream, on the cells of the TM, and upstream, on the ganglion cells of the retina. Several factors are at play in these processes, and many more remain to be elucidated.

Increasing evidence seems to suggest the role of genetic factors in POAG development, but only few selected genes, such as GLC1A, GSTM1, and ELAM-1, have been examined for their possible role in POAG pathogenesis (Alward et al., 1998; Wang et al., 2001; Izzotti and Saccà, 2004). More specifically, the following questions are still open: (a) Is deletion of the GSTM1 gene, which results in decreased GSTM1 expression, a risk factor for POAG? (b) Do stress-response genes involved in the NF-κB cascade have any role in POAG? Are these genes the targets of various drugs that exert anti-inflammatory and antioxidant activities (Barnes and Karin, 1997; Wiederholt, 1998; Totan et al., 2001)? (c) Is the expression of the genes involved in blood vessel regulation altered? (These results could support the "vascular theory" of POAG pathogenesis.) (d) Are genes involved in the defense against oxidation dysregulated in POAG? (These results could support the use of antioxidant administration in POAG therapy and prevention.)

In the light of all these considerations, it should be remembered that laser treatment for glaucoma is also closely connected with oxidative stress and the metabolism of TM (Fig. 3).

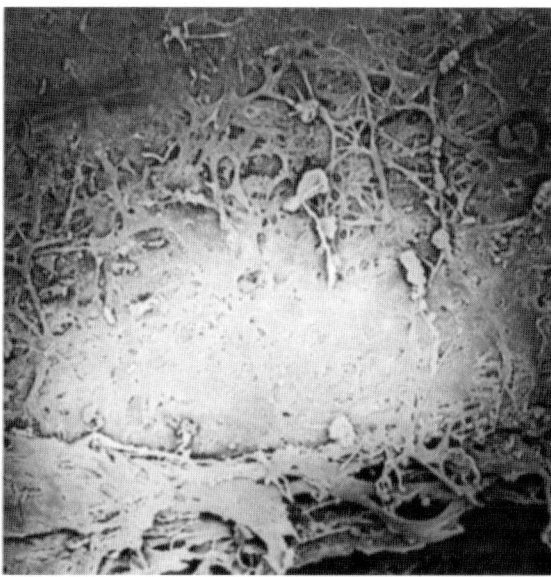

Fig. 3. Scanning electron microscope photograph of the human sclerocorneal trabecular meshwork (magnification 1500 ×). The argon laser spot is well evident. It has produced a large area of disruption and coagulation of the trabecular lamellae, and an impermeable appearing scar. The mechanisms of how laser works are unknown. Wise theorized that laser spots caused a thermal burn to the tissue resulting in shrinkage of the meshwork at the site of the laser application, opening the intertrabecular spaces between the laser sites (Wise and Witter, 1979). A second theory or the "repopulation theory" says that laser would stimulate the repopulation of the meshwork with fresh trabecular cells, which may result in the formation of healthy TM (Bylsma et al., 1988; Dueker et al., 1990). The biologic theory (third theory) suggests that the mechanism of action of the ALT or SLT is the result of the thermal energy from the laser-stimulating cellular activity that involves the release of cytokines from trabecular cells and matrix metalloprotease enzyme synthesis, resulting in increased turnover of the ECM (Parshley et al., 1995, 1996). It is our convincement that biologic effects may be more important than physical process. In AH of rabbit eyes treated with selective laser trabeculoplasty, increase in lipid peroxidase level and free oxygen radicals formed in the pigmented TM due to photodisruption (Guzey et al., 2001) may initiate anti-inflammatory cells to clean up the whole TM and facilitate aqueous outflow (Latina and de Leon, 2005).

Shedding light on the central role of free radicals in this context would help us to understand the pathophysiology of glaucoma and to develop new approaches for its prevention and treatment.

Abbreviations

AH	aqueous humor
BDNF	brain-derived neurotrophic factor
BMP	bone morphogenetic proteins
ECM	extracellular matrix
ET	endothelins
GBE	Ginkgo biloba extract
GSH	glutathione
GST	glutathione transferase
HSP	heat shock proteins
HTM	human trabecular meshwork
IOP	intraocular pressure
MMP	matrix metalloproteinase
MYOC	myocilin
NMDA	N-methyl-D-aspartate receptors
NO	nitric oxide
NOS	nitric oxide synthase
ONH	optic nerve head
PAI	plasminogen activator inhibitor
ROS	reative oxygen species
TGF	transforming growth factor
TIGR	trabecular meshwork-inducible glucocorticoid response
TM	trabecular meshwork
TNF-α	tumor necrosis factor-alpha
VDCC	voltage-dependent calcium channels

References

Acott, T.S., Westcott, M., Passo, M.S. and Van Buskirk, E.M. (1985) Trabecular meshwork glycosaminoglycans in human and cynomolgus monkey eye. Invest. Ophthalmol. Vis. Sci., 26: 1320–1329.

Agostinho, P., Duarte, C.B., Carvalho, A.P. and Oliveira, C.R. (1997a) Oxidative stress affects the selective ion permeability of voltage-sensitive Ca^{2+} channels in cultured retinal cells. Neurosci. Res., 27: 323–334.

Agostinho, P., Duarte, C.B. and Oliveira, C.R. (1996) Intracellular free Na^+ concentration increases in cultured retinal cells under oxidative stress conditions. Neurosci. Res., 25: 343–351.

Agostinho, P., Duarte, C.B. and Oliveira, C.R. (1997b) Impairment of excitatory amino acid transporter activity by oxidative stress conditions in retinal cells: effect of antioxidants. FASEB J., 11: 154–163.

Alexander, J.P., Samples, J.R. and Acott, T.S. (1998) Growth factor and cytokine modulation of trabecular meshwork matrix metalloproteinase and TIMP expression. Curr. Eye Res., 17: 276–285.

Alvarado, J.A., Alvarado, R.G., Yeh, R.F., Franse-Carman, L., Marcellino, G.R. and Brownstein, M.J. (2005) A new insight into the cellular regulation of aqueous outflow: how trabecular meshwork endothelial cells drive a mechanism that regulates the permeability of Schlemm's canal endothelial cells. Br. J. Ophthalmol., 89: 1500–1505.

Alvarado, J.A., Murphy, C.G. and Juster, R. (1984a) Trabecular meshwork cellularity in primary open-angle glaucoma and nonglaucomatous normals. Ophthalmology, 91: 564–579.

Alvarado, J.A., Murphy, C.G., Polansky, J.R. and Juster, R. (1981) Age-related changes in trabecular meshwork cellularity. Invest. Ophthalmol. Vis. Sci., 21: 714–727.

Alvarado, J.A., Murphy, C.G., Polansky, J.R. and Juster, R. (1984b) Studies on pathogenesis of primary open angle glaucoma: regional analyses of trabecular meshwork cellularity and dense collagen. Int. Congr. Ser., 636: 3–8.

Alward, W.L., Fingert, J.H., Coote, M.A., Johnson, A.T., Lerner, S.F., Junqua, D., Durcan, F.J., McCartney, P.J., Mackey, D.A., Sheffield, V.C. and Stone, E.M. (1998) Clinical features associated with mutations in the chromosome 1 open-angle glaucoma gene (GLC1A). N. Engl. J. Med., 338(15): 1022–1027.

Asregadoo, E.R. (1979) Blood levels of thiamine and ascorbic acid in chronic open-angle glaucoma. Ann. Ophthalmol., 11: 1095–1100.

Augustin, W., Wiswedel, I., Noack, H., Reinheckel, T. and Reichelt, O. (1997) Role of endogenous and exogenous antioxidants in the defence against functional damage and lipid peroxidation in rat liver mitochondria. Mol. Cell. Biochem., 174: 199–205.

Babizhayev, M.A. and Bunin, A. (1989) Lipid peroxidation in open angle glaucoma. Acta Ophthalmol. (Copenh.), 67: 371–377.

Babizhayev, M.A. and Costa, E.B. (1994) Lipid peroxide and reactive oxygen species generating systems of the crystalline lens. Biochim. Biophys. Acta, 1225: 326–337.

Bachem, M.G., Wendelin, D., Schneiderhan, W., Haug, C., Zorn, U., Gross, H.J., Schmid-Kotsas, A. and Grunert, A. (1999) Depending on their concentration oxidized low density lipoproteins stimulate extracellular matrix synthesis or induce apoptosis in human coronary artery smooth muscle cells. Clin. Chem. Lab. Med., 37: 319–326.

Balaratnasingam, C., Morgan, W.H., Bass, L., Matich, G., Cringle, S.J. and Yu, D.Y. (2007) Axonal transport and cytoskeletal changes in the laminar regions after elevated intraocular pressure. Invest. Ophthalmol. Vis. Sci., 48: 3632–3644.

Barnes, P.J. and Karin, M. (1997) Nuclear factor-kappaB: a pivotal transcription factor in chronic inflammatory disease. N. Engl. J. Med., 336(15): 1066–1071.

Becker, B. (1957) Chemical composition of human aqueous humor: effects of acetazoleamide. Arch. Ophthalmol., 57: 793–800.

Beit-Yannai, E., Trembovler, V. and Solomon, A.S. (2007) Decrease in reducing power of aqueous humor originating from glaucomatous rabbits. Eye, 21: 658–664.

Birch, M., Brotchie, D., Roberts, N. and Grierson, I. (1997) The three-dimensional structure of the connective tissue in the lamina cribrosa of the human optic nerve head. Ophthalmologica, 211: 183–191.

Bradley, J.M., Vranka, J., Colvis, C.M., Conger, D.M., Alexander, J.P., Fish, A.S., Samples, J.R. and Acott, T.S. (1998) Effect of matrix metalloproteinases activity on outflow in perfused human organ culture. Invest. Ophthalmol. Vis. Sci., 39: 2649–2658.

Brandt, J.D. and O'Donnell, M.E. (1999) How does the trabecular meshwork regulate outflow? Clues from the vascular endothelium. J. Glaucoma, 8: 328–339.

Brilakis, H.S. and Johnson, D.H. (2001) Giant vacuole survival time and implications for aqueous humor outflow. J. Glaucoma, 10: 277–283.

Brubaker, R.F., Nagataki, S., Townsend, D.J., Burns, R.R., Higgins, R.G. and Wentworth, W. (1981) The effect of age on aqueous humor formation in man. Ophthalmology, 88: 283–288.

Burgoyne, C.F., Downs, J.C., Bellezza, A.J., Suh, J.K. and Hart, R.T. (2005) The optic nerve head as a biomechanical structure: a new paradigm for understanding the role of IOP-related stress and strain in the pathophysiology of glaucomatous optic nerve head damage. Prog. Retin. Eye Res., 24: 39–73.

Bylsma, S.S., Samples, J.R., Acott, T.S. and van Buskirk, E.M. (1988) Trabecular cell division after argon laser trabeculoplasty. Arch. Ophthalmol., 106: 544–547.

Cardoso, S.M., Pereira, C. and Oliveira, C.R. (1998) The protective effect of vitamin E, idebenone and reduced glutathione on free radical mediated injury in rat brain synaptosomes. Biochem. Biophys. Res. Commun., 246: 703–710.

Choi, D.W. (1988) Glutamate neurotoxicity and diseases of the nervous system. Neuron, 1: 623–634.

Choo, M.K., Park, E.K., Han, M.J. and Kim, D.H. (2003) Antiallergic activity of ginseng and its ginsenosides. Planta Med., 69: 518–522.

Cioffi, G.A., Orgul, S., Onda, E., Bacon, D.R. and Van Buskirk, E.M. (1995) An in vivo model of chronic optic nerve ischemia: the dose-dependent effects of endothelin-1 on the optic nerve microvasculature. Curr. Eye Res., 14: 1147–1153.

Cioffi, G.A. and Sullivan, P. (1999) The effect of chronic ischemia on the primate optic nerve. Eur. J. Ophthalmol., 9: S34–S36.

Costarides, A.P., Riley, M.V. and Green, K. (1991) Roles of catalase and the glutathione redox cycle in the regulation of the anterior-chamber hydrogen peroxide. Ophthalmic Res., 23: 284–294.

Czakey, R.P. and Loskutoff, D.J. (2004) Unexpected role of plasminogen activator inhibitor 1 in cell adhesion and detachment. Exp. Biol. Med., 229: 1090–1096.

Dawson, V.L., Brahmbhatt, H.P., Mong, J.A. and Dawson, T.M. (1994) Expression of inducible nitric oxide synthase causes delayed neurotoxicity in primary mixed neuronalglial cortical cultures. Neuropharmacology, 33: 1425–1430.

Dawson, V.L., Dawson, T.M., Bartley, D.A., Uhl, G.R. and Snyder, S.H. (1993) Mechanisms of nitric oxide-mediated neurotoxicity in primary brain cultures. J. Neurosci., 13: 2651–2661.

Defeudis, F.V. (2002) Bilobalide and neuroprotection. Pharmacol. Res., 46: 565–568.

De Grey, A.D. (1997) A proposed refinement of the mitochondrial free radical theory of aging. Bioessays, 19: 161–166.

Dreyer, E.B., Zhang, D. and Lipton, S.A. (1995) Transcriptional or translational inhibition blocks low dose NMDA-mediated cell death. Neuroreport, 6: 942–944.

Dueker, D.K., Norberg, M., Johnson, D.H., Tschumper, R.C. and Feeney-Burns, L. (1990) Stimulation of cell division by argon and Nd:YAG laser trabeculoplasty in cynomolgus monkeys. Invest. Ophthalmol. Vis. Sci., 31: 115–124.

Dugan, L.L., Sensi, S.L., Canzoniero, L.M., Handran, S.D., Rothman, S.M., Lin, T.S., Goldberg, M.P. and Choi, D.W. (1995) Mitochondrial production of reactive oxygen species in cortical neurons following exposure to N-methyl-D-aspartate. J. Neurosci., 15: 6377–6388.

Ellis, D. and Nathanson, J. (1998) Nitric oxide in the human eye: sites of synthesis and physiologic actions on intraocular pressure, blood flow, sodium transport and neuronal viability. In: Haefliger I.O. and Flammer J. (Eds.), Nitric Oxide and Endothelin in the Pathogenesis of Glaucoma. Lippincott-Raven, Philadelphia, pp. 178–204.

Epstein, D.L., De Kater, A.W., Lou, M. and Patel, J. (1990) Influences of glutathione and sulfhydryl containing compounds on aqueous humor outflow function. Exp. Eye Res., 50: 785–793.

Ethier, C.R., Coloma, F.M., Sit, A.J. and Johnson, A.M. (1998) Two pore types in the inner wall endothelium of Schlemm's canal. Invest. Ophthalmol. Vis. Sci., 39: 2041–2048.

Ethier, C.R., Kamm, R.D., Palaszewski, B.A., Johnson, M.C. and Richardson, T.M. (1986) Calculations of flow resistance in the juxtacanalicular meshwork. Invest. Ophthalmol. Vis. Sci., 27: 1741–1750.

Ferreira, S.M., Lerner, S.F., Brunzini, R., Evelson, P.A. and Llesuy, S.F. (2004) Oxidative stress markers in aqueous humor of glaucoma patients. Am. J. Ophthalmol., 137: 62–69.

Filla, M.S., Liu, X., Nguyen, T.D., Polansky, J.R., Brandt, C.R., Kaufman, P.L. and Peters, D.M. (2002) In vitro localization of TIGR/MYOC in trabecular meshwork extracellular matrix and binding to fibronectin. Invest. Ophthalmol. Vis. Sci., 43: 151–161.

Flammer, J. (1994) The vascular concept of glaucoma. Surv. Ophthalmol., 38: 3–6.

Fleenor, D.L., Shepard, A.R., Hellberg, P.E., Jacobson, N., Pang, I.H. and Clark, A.F. (2006) TGF-β_2 induced changes in human trabecular meshwork: implications for intraocular pressure. Invest. Ophthalmol. Vis. Sci., 47: 226–234.

Fuchshofer, R., Welge-Lussen, U. and Lutjen-Drecoll, E. (2003) The effect of TGF-beta2 on human trabecular meshwork extracellular proteolytic system. Exp. Eye Res., 77: 757–765.

Fuchshofer, R., Yu, A.H., Welge-Lussen, U. and Tamm, E.R. (2007) Bone morphogenetic protein-7 is an antagonist of transforming growth factor-beta2 in human trabecular meshwork cells. Invest. Ophthalmol. Vis. Sci., 48: 715–726.

Gaasterland, D., Kupfer, C., Milton, R., Ross, K., McCain, L. and MacLellan, H. (1978) Studies of aqueous humour dynamics in man. VI. Effect of age upon parameters of intraocular pressure in normal human eyes. Exp. Eye Res., 26: 651–656.

Gabelt, B.T., Gottanka, J., Lutjen-Drecoll, E. and Kaufman, P.L. (2003) Aqueous humor dynamics and trabecular meshwork and anterior ciliary muscle morphologic changes with age in rhesus monkeys. Invest. Ophthalmol. Vis. Sci., 44: 2118–2125.

Gabelt, B.T. and Kaufman, P.L. (2005) Changes in aqueous humor dynamics with age and glaucoma. Prog. Retin. Eye Res., 24: 612–637.

Gherghel, D., Griffiths, H.R., Hilton, E.J., Cunliffe, I.A. and Hosking, S.L. (2005) Systemic reduction in glutathione levels occurs in patients with primary open-angle glaucoma. Invest. Ophthalmol. Vis. Sci., 46: 877–883.

Gibbons, G.H. (1997) Endothelial function as a determinant of vascular structure and function: a new therapeutic target. Am. J. Cardiol., 79: 3–8.

Gong, H., Tripathi, R.C. and Tripathi, B. (1996) Morphology of the aqueous outflow pathway. Microsc. Res. Tech., 33: 336–367.

Grierson, I., Wang, Q., McMenamin, P.G. and Lee, W.R. (1982) The effects of age and antiglaucoma drugs on the meshwork cell population. Res. Clin. Forums, 4: p. 69.

Guzey, M., Vural, H., Satici, A., Karadede, S. and Dogan, Z. (2001) Increase of free oxygen radicals in aqueous humour induced by selective Nd:YAG laser trabeculoplasty in the rabbit. Eur. J. Ophthalmol., 11(1): 47–52.

Haefliger, I.O., Dettmann, E., Liu, R., Meyer, P., Prünte, C., Messerli, J. and Flammer, J. (1999) Potential role of nitric oxide and endothelin in the pathogenesis of glaucoma. Surv. Ophthalmol., 43(Suppl 1): S51–S58.

Hare, W., WoldeMussie, E., Lai, R., Ton, H., Ruiz, G., Feldmann, B., Wijono, M., Chun, T. and Wheeler, L. (2001) Efficacy and safety of memantine, an NMDA-type open-channel blocker, for reduction of retinal injury associated with experimental glaucoma in rat and monkey. Surv. Ophthalmol., 45: S284–S289.

Hare, W.A., WoldeMussie, E., Weinreb, R.N., Ton, H., Ruiz, G., Chun, T. and Wheeler, L. (2004) Efficacy and safety of memantine treatment for reduction of changes associated with experimental glaucoma in monkey. II. Structural measures. Invest. Ophthalmol. Vis. Sci., 45: 2640–2651.

Haung, W., Koralewska-Makar, A., Bauer, B. and Akesson, B. (1997) Extracellular glutathione peroxidase and ascorbic acid in aqueous humor and serum of patients operated on for cataract. Clin. Chim. Acta, 261: 117–130.

Hayashi, A., Ryu, A., Suzuki, T., Kawada, A. and Tajima, S. (1998) In vitro degradation of tropoelastin by reactive oxygen species. Arch. Dermatol. Res., 290: 497–500.

He, S., Prasanna, G. and Yorio, T. (2007) Endothelin-1-mediated signaling in the expression of matrix metalloproteinases and tissue inhibitors of metalloproteinases in astrocytes. Invest. Ophthalmol. Vis. Sci., 48: 3737–3745.

Henning, S.M., Fajardo-Lira, C., Lee, H.W., Youssefian, A.A., Go, V.L.W. and Heber, D. (2003) Catechin content of 18 teas and a green tea extract supplement correlated with the antioxidant capacity. Nutr. Cancer, 45: 226–235.

Hernandez, M.R. (2000) The optic nerve head in glaucoma: role of astrocytes in tissue remodeling. Prog. Retin. Eye Res., 19: 297–321.

Hernandez, M.R. and Pena, J.D. (1997) The optic nerve head in glaucomatous optic neuropathy. Arch. Ophthalmol., 115: 389–395.

Hirooka, K., Tokuda, M., Miyamoto, O., Itano, T., Baba, T. and Shiraga, F. (2004) The Ginkgo biloba extract (EGB 761) provides a neuroprotective effect on retinal ganglion cells in a rat model of chronic glaucoma. Curr. Eye Res., 28: 153–157.

Hofmann, F., Biel, M. and Flockerzi, V. (1994) Molecular basis for Ca^{2+} channel diversity. Annu. Rev. Neurosci., 17: 399–418.

Hogg, P., Calthorpe, M., Batterbury, M. and Grierson, I. (2000) Aqueous humor stimulates the migration of human trabecular meshwork cells in vitro. Invest. Ophthalmol. Vis. Sci., 41: 1091–1098.

Izzotti, A. and Saccà, S.C. (2004) Glutathione S-transferase M1 and its implications in glaucoma pathogenesis: a controversial matter. Exp. Eye Res., 79: 141–142.

Izzotti, A., Saccà, S.C., Cartiglia, C. and De Flora, S. (2003) Oxidative deoxyribonucleic acid damage in the eyes of glaucoma patients. Am. J. Med., 114: 638–646.

Johnson, M. (2006) What controls aqueous humour outflow resistance? Exp. Eye Res., 82: 545–557.

Johnson, M., Chan, D., Read, A.T., Christensen, C., Sit, A. and Ethier, C.R. (2002) The pore density in the inner wall endothelium of Schlemm's canal of glaucomatous eyes. Invest. Ophthalmol. Vis. Sci., 43: 2950–2955.

Johnson, D.H. and Johnson, M. (2001) How does nonpenetrating glaucoma surgery work? Aqueous outflow resistance and glaucoma surgery. J. Glaucoma, 10: 55–67.

Joo, S.S., Won, T.J. and Lee do, I. (2005) Reciprocal activity of ginsenosides in the production of proinflammatory repertoire, and their potential roles in neuroprotection in vivo. Planta Med., 71: 476–481.

Kahn, M.G., Giblin, F.J. and Epstein, D.L. (1983) Glutathione in calf trabecular meshwork and its relation to aqueous humor outflow facility. Invest. Ophthalmol. Vis. Sci., 24: 1283–1287.

Kane, R., Stevenson, L., Godson, C., Stitt, A.W. and O'Brien, C. (2005) Gremlin gene expression in bovine retinal pericytes exposed to elevated glucose. Br. J. Ophthalmol., 89: 1638–1642.

Kang, J.H., Pasquale, L.R., Willett, W., Rosner, B., Egan, K.M., Faberowski, N. and Hankinson, S.E. (2003) Antioxidant intake and primary open-angle glaucoma: a prospective study. Am. J. Epidemiol., 158: 337–346.

Kapiotis, S., Hermann, M., Held, I., Seelos, C., Ehringer, H. and Gmeiner, B.M. (1997) Genistein, the dietary-derived angiogenesis inhibitor, prevents LDL oxidation and protects endothelial cells from damage by atherogenic LDL. Arterioscler. Thromb. Vasc. Biol., 17: 2868–2874.

Kerrigan, L.A., Zack, D.J., Quigley, H.A., Smith, S.D. and Pease, M.E. (1997) TUNEL-positive ganglion cells in human primary open-angle glaucoma. Arch. Ophthalmol., 115: 1031–1035.

Kikuchi, M., Tenneti, L. and Lipton, S.A. (2000) Role of p38 mitogen-activated protein kinase in axotomy-induced apoptosis of rat retinal ganglion cells. J. Neurosci., 1: 5037–5044.

Kim, J.H., Hong, Y.H., Lee, J.H., Kim, D.H., Jeong, S.M., Lee, B.H., Lee, S.M. and Nah, S.Y. (2005) A role for the carbohydrate portion of ginsenoside Rg3 in Na^+ channel inhibition. Mol. Cells, 19: 137–142.

Kim, S., Ahn, K., Oh, T.H., Nah, S.Y. and Rhim, H. (2002) Inhibitory effect of ginsenosides on NMDA receptor-mediated signals in rat hippocampal neurons. Biochem. Biophys. Res. Commun., 296: 247–254.

Knepper, P.A., Fadel, J.R., Miller, A.M., Goossens, W., Choi, J., Nolan, M.J. and Whitmer, S. (2005) Reconstitution of trabecular meshwork GAGs: influence of hyaluronic acid and chondroitin sulfate on flow rates. J. Glaucoma, 14: 230–238.

Knepper, P.A., Goossens, W., Hvizd, M. and Palmberg, P.F. (1996a) Glycosaminoglycans of the human trabecular meshwork in primary open-angle glaucoma. Invest. Ophthalmol. Vis. Sci., 37: 1360–1367.

Knepper, P.A., Goossens, W. and Palmberg, P.F. (1996b) Glycosaminoglycan stratification of the juxtacanalicular tissue in normal and primary open-angle glaucoma. Invest. Ophthalmol. Vis. Sci., 37: 2414–2425.

Ko, M.L., Hu, D.N., Ritch, R., Sharma, S.C. and Chen, C.F. (2001) Patterns of retinal ganglion cell survival after brain-derived neurotrophic factor administration in hypertensive eyes of rats. Neurosci. Lett., 305: 139–142.

Kobuchi, H., Droy-Lefaix, M.T., Christen, Y. and Packer, L. (1997) Ginkgo biloba extract (EGb 761): inhibitory effect on nitric oxide production in the macrophage cell line RAW 264.7. Biochem. Pharmacol., 53: 897–904.

Koh, S.W. and Waschek, A. (2000) Corneal endothelial cell survival in organ cultures under acute oxidative stress: effect of VIP. Invest. Ophthalmol. Vis. Sci., 41: 4085–4092.

Köse, K. and Dogan, P. (1995) Lipoperoxidation induced by hydrogen peroxide in human erythrocyte membranes. Comparison of the antioxidant effect of Ginkgo biloba extract (EGb 761) with those of water-soluble and lipid-soluble antioxidants. J. Int. Med. Res., 23: 9–18.

Kottler, U.B., Junemann, A.G.M., Aigner, T., Zenkel, M., Rummelt, C. and Schlotzer-Schrehardt, U. (2005) Comparative effects of TGF-β_1 and TGF-β_2 on extracellular matrix production, proliferation, migration, and collagen contraction of human Tenon's capsule fibroblasts in pseudoexfoliation and primary open-angle glaucoma. Exp. Eye Res., 80: 121–134.

Kuehn, M.H., Fingert, J.H. and Kwon, Y.H. (2005) Retinal ganglion cell death in glaucoma: mechanisms and neuroprotective strategies. Ophthalmol. Clin. North Am., 18: 383–395.

Lambert, W.S., Clark, A.F. and Wordinger, R.J. (2004) Neurotrophin and Trk expression by cells of the human lamina cribrosa following oxygen-glucose deprivation. BMC Neurosci., 5(3): p. 51.

Lane, B.C. (1980) Evaluation of intraocular pressure with daily, sustained closework stimulus to accommodation, lowered tissue chromium and dietary deficiency of ascorbic acid. Doc. Ophthalmol., 28: 149–155.

Latina, M.A. and de Leon, J.M. (2005) Selective laser trabeculoplasty. Ophthalmol. Clin. North Am., 18: 409–419.

Lee, W.R. and Grierson, I. (1974) Relationships between intraocular pressure and the morphology of the outflow apparatus. Trans. Ophthalmol. Soc. U.K., 94: 430–449.

Lee, H.C. and Wei, Y.H. (2001) Mitochondrial alterations, cellular response to oxidative stress and defective degradation of proteins in aging. Biogerontology, 2: 231–244.

Lerner, L.E., Polansky, J.R., Howes, E.L. and Stern, R. (1997) Hyaluronan in the human trabecular meshwork. Invest. Ophthalmol. Vis. Sci., 38: 1222–1228.

Levin, L.A. (1999) Direct and indirect approaches to neuroprotective therapy of glaucomatous optic neuropathy. Surv. Ophthalmol., 43(Suppl 1): S98–S101.

Linner, E. (1996) The effect of ascorbic acid on intraocular pressure. In: Paterson G., Miller S.J.H. and Paterson G.D. (Eds.), Drug Mechanisms in Glaucoma. J & A Churchill, London, pp. 153–164.

Lipton, S.A. (1999) Neuronal protection destruction by NO. Cell Death Differ., 6: 943–951.

Lipton, S.A. (2003) Possible role for memantine in protecting retinal ganglion cells from glaucomatous damage. Surv. Ophthalmol., 48: S38–S46.

Lipton, S.A. (2005) The molecular basis of memantine action in Alzheimer's disease and other neurologic disorders: low-affinity, uncompetitive antagonism. Curr. Alzheimer Res., 2: 155–165.

Liton, P.B., Challa, P., Stinnett, S., Luna, C., Epstein, D.L. and Gonzalez, P. (2005) Cellular senescence in the glaucomatous outflow pathway. Exp. Gerontol., 40: 745–748.

Liu, B. and Neufeld, A.H. (2000) Expression of nitric oxide synthase-2 (NOS-2) in reactive astrocytes of the human glaucomatous optic nerve head. Glia, 30: 178–186.

Liu, Q., Ju, W.K., Crowston, J.G., Xie, F., Perry, G., Smith, M.A., Lindsey, J.D. and Weinreb, R.N. (2007) Oxidative stress is an early event in hydrostatic pressure induced retinal ganglion cell damage. Invest. Ophthalmol. Vis. Sci., 48: 4580–4589.

Lucas, D.R. and Newhouse, J.P. (1957) The toxic effect of sodium L-glutamate on the inner layers of the retina. AMA Arch. Ophthalmol., 58193–58201.

Lutjen-Drecoll, E. (1973) Structural factors influencing outflow facility and its changeability under drugs. A study in macaca arctoides. Invest. Ophthalmol. Vis. Sci., 12: 280–294.

Lutjen-Drecoll, E., May, C.A., Polansky, J.R., Johnson, D.H., Bloemendal, H. and Nguyen, T.D. (1998) Localization of the stress proteins alpha B-crystallin and trabecular meshwork inducible glucocorticoid response protein in normal and glaucomatous trabecular meshwork. Invest. Ophthalmol. Vis. Sci., 39: 517–525.

Maepea, O. and Bill, A. (1992) Pressures in the juxtacanalicular tissue and Schlemm's canal in monkeys. Exp. Eye Res., 65: 879–883.

Mandel, S., Amit, T., Reznichenko, L., Weinreb, O. and Youdim, M.B. (2006) Green tea catechins as brain-permeable, natural iron chelators-antioxidants for the treatment of neurodegenerative disorders. Mol. Nutr. Food Res., 50: 229–234.

Marcocci, L., Maguire, J.J., Droy-Lefaix, M.T. and Packer, L. (1994) The nitric oxide-scavenging properties of Ginkgo biloba extract (EGb 761). Biochem. Biophys. Res. Commun., 201: 748–755.

Marsault, R., Feolde, E. and Frelin, C. (1993) Receptor externalization determines sustained contractile responses to endothelin-1 in the rat aorta. Am. J. Physiol., 264: C687–C693.

Marsden, P.A. and Brenner, B.M. (1991) Nitric oxide and endothelins: novel autocrine/paracrine regulators of the circulation. Semin. Nephrol., 11: 169–185.

Martin, K.R., Levkovitch-Verbin, H., Valenta, D., Baumrind, L., Pease, M.E. and Quigley, H.A. (2002) Retinal glutamate transporter changes in experimental glaucoma and after optic nerve transection in the rat. Invest. Ophthalmol. Vis. Sci., 43: 2236–2243.

Massague, J. (1990) The transforming growth factor-beta family. Annu. Rev. Cell Biol., 6: 597–641.

May, J.M. (2000) How does ascorbic acid prevent endothelial dysfunction?. Free Radic. Biol. Med., 28: 1421–1429.

McCarty, M.F. (1998) Primary open-angle glaucoma may be a hyaluronic acid deficiency disease: potential for glucosamine in prevention and therapy. Med. Hypotheses, 51: 483–484.

Meldrum, B. and Garthwaite, J. (1990) Excitatory amino acid neurotoxicity and neurodegenerative disease. Trends Pharmacol. Sci., 11: 379–387.

Miller, K.M. and Quigley, H.A. (1988) The clinical appearance of the lamina cribrosa as a function of the extent of glaucomatous optic nerve damage. Ophthalmology, 95: 135–138.

Moreno, M.C., Campanelli, J., Sande, P., Sanez, D.A., Keller Sarmiento, M.I. and Rosenstein, R.E. (2004) Retinal oxidative stress induced by high intraocular pressure. Free Radic. Biol. Med., 37: 803–812.

Nagase, H. (1997) Activation mechanisms of matrix metalloproteinases. Biol. Chem., 378: 151–160.

Nagase, H. and Woessner, J.F., Jr. (1999) Matrix metalloproteinases. J. Biol. Chem., 274: 21491–21494.

Nathan, J. (2000) Hippocrates to Duke-Elder: an overview of the history of glaucoma. Clin. Exp. Optom., 83: 116–118.

Navajas, E.V., Martins, J.R., Melo, L.A., Jr., Saraiva, V.S., Dietrich, C.P., Nader, H.B. and Belfort, R., Jr. (2005) Concentration of hyaluronic acid in primary open-angle glaucoma aqueous humor. Exp. Eye Res., 80: 853–857.

Neal, C.R. and Michel, C.C. (1996) Openings in frog microvascular endothelium induced by high intravascular pressures. J. Physiol., 492: 39–52.

Neufeld, A.H. (1999) Nitric oxide: a potential mediator of retinal ganglion cell damage in glaucoma. Surv. Ophthalmol., 43: S129–S135.

Neufeld, A.H., Hernandez, M.R. and Gonzalez, M. (1997) Nitric oxide synthase in the human glaucomatous optic nerve head. Arch. Ophthalmol., 115: 497–503.

Orgogozo, J.M., Rigaud, A.S., Stoffler, A., Möbius, H.J. and Forette, F. (2002) Efficacy and safety of memantine in patients with mild to moderate vascular dementia. A randomized, placebo-controlled trial (MMM300). Stroke, 33: 1834–1838.

Orgul, S., Gugleta, K. and Flammer, J. (1999) Physiology of perfusion as it relates to the optic nerve head. Surv. Ophthalmol., 43(Suppl 1): S17–S26.

Oyama, Y., Fuchs, P.A., Katayama, N. and Noda, K. (1994) Myricetin and quercetin, the flavonoid constituents of Ginkgo biloba extract, greatly reduce oxidative metabolism in both resting and Ca (2+)-loaded brain neurons. Brain Res., 635: 125–129.

Ozsuer, H., Gorgulu, A., Kiris, T. and Cobanoglu, S. (2005) The effects of memantine on lipid peroxidation following closed-head trauma in rats. Neurosurg. Rev., 28: 143–147.

Parshley, D.E., Bradley, J.M., Fish, A., Hadaegh, A., Samples, J.R., Van Buskirk, E.M. and Acott, T.S. (1996) Laser trabeculoplasty induces stromelysin expression by trabecular juxtacanalicular cells. Invest. Ophthalmol. Vis. Sci., 37: 795–804.

Parshley, D.E., Bradley, J.M., Samples, J.R., Van Buskirk, E.M. and Acott, T.S. (1995) Early changes in MMPs and inhibitors after in vitro laser treatment to the trabecular meshwork. Curr. Eye Res., 14: 537–544.

Peeters, M., Maloteaux, J.M. and Hermans, E. (2003) Distinct effects of amantadine and memantine on dopaminergic transmission in the rat striatum. Neurosci. Lett., 343: 205–209.

Petzold, G.C., Einhaupl, K.M., Dirnagl, U. and Dreier, J.P. (2003) Ischemia triggered by spreading neuronal activation is induced by endothelin-1 and hemoglobin in the subarachnoid space. Ann. Neurol., 54: 591–598.

Poli, G. and Parola, M. (1997) Oxidative damage and fibrogenesis. Free Radic. Biol. Med., 22(1–2): 287–305.

Polla, B.S., Kantengwa, S., Francois, D., Salvioli, S., Franceschi, C., Marsac, C. and Cossarizza, A. (1996) Mitochondria are selective targets for the protective effects of heat shock against oxidative injury. Proc. Natl. Acad. Sci. U.S.A., 93: 6458–6463.

Prasanna, G., Hulet, C., Desai, D., Krishnamoorthy, R.R., Narayan, S., Brun, A.M., Suburo, A.M. and Yorio, T. (2005) Effect of elevated intraocular pressure on endothelin-1 in a rat model of glaucoma. Pharmacol. Res., 51: 41–50.

Quaranta, L., Bettelli, S., Uva, M.G., Semeraro, F., Turano, R. and Gandolfo, E. (2003) Effect of Ginkgo biloba extract on preexisting visual field damage in normal tension glaucoma. Ophthalmology, 110: 359–362.

Quigley, H.A. (2005) Glaucoma: macrocosm to microcosm: the Friedenwald lecture. Invest. Ophthalmol. Vis. Sci., 46: 2662–2670.

Quigley, H.A. and Broman, A.T. (2006) The number of people with glaucoma world-wide in 2010 and 2020. Br. J. Ophthalmol., 90(3): 253–254.

Quigley, H.A., Hohman, R.M., Addicks, E.M., Massof, R.W. and Green, W.R. (1983) Morphologic changes in the lamina cribrosa correlated with neural loss in open-angle glaucoma. Am. J. Ophthalmol., 95: 673–691.

Radad, K., Gille, G., Moldzio, R., Saito, H. and Rausch, W.D. (2004) Ginsenosides Rb1 and Rg1 effects on mesencephalic dopaminergic cells stressed with glutamate. Brain Res., 1021: 41–53.

Ramos, R.F., Hoying, J.B., Witte, M.H. and Daniel Stamer, W. (2007) Schlemm's canal endothelia, lymphatic, or blood vasculature?. J. Glaucoma, 16: 391–405.

Redmond, E.M., Cahill, P.A., Hodges, R., Zhang, S. and Sitzmann, J.V. (1996) Regulation of endothelin receptors by nitric oxide in cultured rat vascular smooth muscle cells. J. Cell. Physiol., 166: 469–479.

Rego, A.C., Santos, M.S. and Oliveira, C.R. (1996) Oxidative stress, hypoxia, and ischemia-like conditions increase the release of endogenous amino acids by distinct mechanisms in cultured retinal cells. J. Neurochem., 66: 2506–2516.

Reiss, G.R. (1986) Ascorbic acid levels in the aqueous humor of nocturnal and diurnal mammals. Arch. Ophthalmol., 104: 753–755.

Riley, M.V. (1990) Physiologic neutralization mechanisms and the response of the corneal endothelium to hydrogen peroxide. CLAO J., 16: S16–S21.

Ritch, R. (2000) Potential role for Ginkgo biloba extract in the treatment of glaucoma. Med. Hypotheses, 54: 221–235.

Ritch, R. (2005) Complementary therapy for the treatment of glaucoma: a perspective. Ophthalmol. Clin. North Am., 18: 597–609.

Rodriguez, J.A., Orbe, J. and Paramo, J.A. (2007) Metalloproteases, vascular remodeling and atherothrombotic syndromes. Rev. Esp. Cardiol., 60: 959–967.

Rohen, J.W. and Witmer, R. (1972) Electron microscopic studies on the trabecular meshwork in glaucoma simplex. Graefes. Arch. Klin. Exp. Ophthalmol., 183: 251–266.

Rondeau, E., Sraer, J. and Schleuning, W. (1995) The renal plasminogen activating system. In: Schlondorff D. and Bonventre J. (Eds.), Molecular Nephrology: Kidney Function in Health and Diseases. Marcel Dekker, New York, pp. 699–715.

Ronkko, S., Rekonen, P., Kaarniranta, K., Puustjarvi, T., Teräsvirta, M. and Uusitalo, H. (2007) Matrix metalloproteinases and their inhibitors in the chamber angle of normal eyes and patients with primary open-angle glaucoma and exfoliation glaucoma. Graefes. Arch. Clin. Exp. Ophthalmol., 245: 697–704.

Rose, R.C., Richer, S.P. and Bode, A.M. (1998) Ocular oxidants and antioxidant protection. Proc. Soc. Exp. Biol. Med., 217: 397–407.

Royes, L.F., Fighera, M.R., Furian, A.F., Oliveira, M.S., Fiorenza, N.G., de Carvalho Myskiw, J., Frussa-Filho, R. and Mello, C.F. (2005) Involvement of NO in the convulsive behavior and oxidative damage induced by the intrastriatal injection of methylmalonate. Neurosci. Lett., 376(2): 116–120.

Saccà, S.C., Izzotti, A., Rossi, P. and Traverso, C. (2007) Glaucomatous outflow pathway and oxidative stress. Exp. Eye Res., 84: 389–399.

Saccà, S.C., Pascotto, A., Camicione, P., Capris, P. and Izzotti, A. (2005) Oxidative DNA damage in the human trabecular meshwork: clinical correlation in patients with primary open-angle glaucoma. Arch. Ophthalmol., 123: 458–463.

Saika, S. (2004) TGF-beta signal transduction in corneal wound healing as a therapeutic target. Cornea, 23: S25–S30.

Savla, U., Neal, C.R. and Michel, C. (2002) Openings in frog microvascular endothelium at different rates of increase in pressure and at different temperatures. J. Physiol., 539: 285–293.

Sawaguchi, S., Yue, B.Y., Chang, I.L., Wong, F. and Higginbotham, E.J. (1992) Ascorbic acid modulates collagen type I gene expression by cells from an eye tissue-trabecular meshwork. Cell. Mol. Biol., 38: 587–604.

Schachtschabel, D.O. and Binninger, E. (1993) Stimulatory effects of ascorbic acid on hyaluronic acid synthesis of in vitro cultured normal and glaucomatous trabecular meshwork cells of the human eye. Gerontology, 26: 243–246.

Schneemann, A., Leusink-Muis, A., van den Berg, T., Hoyng, P.F. and Kamphuis, W. (2003) Elevation of nitric oxide production in human trabecular meshwork by increased pressure. Graefes. Arch. Clin. Exp. Ophthalmol., 241: 321–326.

Scott, G.I., Colligan, P.B., Ren, B.H. and Ren, J. (2001) Ginsenosides Rb1 and Re decrease cardiac contraction in adult rat ventricular myocytes: role of nitric oxide. Br. J. Pharmacol., 134: 1159–1165.

Shareef, S., Sawada, A. and Neufeld, A.H. (1999) Isoforms of nitric oxide synthase in the optic nerves of rat eyes with chronic moderately elevated intraocular pressure. Invest. Ophthalmol. Vis. Sci., 40(12): 2884–2891.

Sigal, I.A., Flanagan, J.G. and Ethier, C.R. (2005) Factors influencing optic nerve head biomechanics. Invest. Ophthalmol. Vis. Sci., 46: 4189–4199.

Sit, A.J., Coloma, F.M., Ethier, C.R. and Johnson, M. (1997) Factors affecting the pores of the inner wall endothelium of Schlemm's canal. Invest. Ophthalmol. Vis. Sci., 38: 1517–1525.

Southam, E., Thomas, P.K., King, R.H., Goss-Sampson, M.A. and Muller, D.P. (1991) Experimental vitamin E deficiency in rats. Morphological and functional evidence of abnormal axonal transport secondary to free radical damage. Brain, 114: 915–936.

Spector, A., Ma, W. and Wang, R.R. (1998) The aqueous humor is capable of generating and degrading H_2O_2. Invest. Ophthalmol. Vis. Sci., 39: 1188–1197.

Sporn, M.B. and Roberts, A.B. (1992) Transforming growth factor-β_2: recent progress and new challenges. J. Cell Biol., 119: 1017–1021.

Takuma, K., Mori, K., Lee, E., Enomoto, R., Baba, A. and Matsuda, T. (2002) Heat shock inhibits hydrogen peroxide-induced apoptosis in cultured astrocytes. Brain Res., 946: 232–238.

Tamm, E.R. and Lutjen Drecoll, E. (1998) Nitric oxide in the outflow pathways of the aqueous humor. In: Haefliger I.O. and Flammer J. (Eds.), Nitric Oxide and Endothelin in the pathogenesis of Glaucoma. Lippincott-Raven, New York, pp. 158–167.

Tamm, E.R., Russell, P., Johnson, D.H. and Piatigorsky, J. (1996) Human and monkey trabecular meshwork accumulate alpha B-crystallin in response to heat shock and oxidative stress. Invest. Ophthalmol. Vis. Sci., 37: 2402–2413.

Tanaka, J., Toku, K., Zhang, B., Ishihara, K., Sakanaka, M. and Maeda, N. (1999) Astrocytes prevent neuronal death induced by reactive oxygen and nitrogen species. Glia, 28: 85–96.

Tane, N., Dhar, S., Roy, S., Pinheiro, A., Ohira, A. and Roy, S. (2007) Effect of excess synthesis of extracellular matrix components by trabecular meshwork cells: possible consequence on aqueous outflow. Exp. Eye Res., 84: 832–842.

Tang, B., Qiao, H., Meng, F. and Sun, X. (2007) Glycyrrhizin attenuates endotoxin-induced acute liver injury after partial hepatectomy in rats. Braz. J. Med. Biol. Res., 40(12): 1637–1646.

Tatton, N.A., Tezel, G., Insolia, S.A., Nandor, S.A., Edward, P.D. and Wax, M.B. (2001) In situ detection of apoptosis in normal pressure glaucoma: a preliminary examination. Surv. Ophthalmol., 45(Suppl 3): S268–S272.

Ten Hulzen, R.D. and Johnson, D.H. (1996) Effect of fixation pressure on juxtacanalicular tissue and Schlemm's canal. Invest. Ophthalmol. Vis. Sci., 37: 114–124.

Tezel, G., Hernandez, R. and Wax, M.B. (2000) Immunostaining of heat shock proteins in the retina and optic nerve head of normal and glaucomatous eyes. Arch. Ophthalmol., 118: 511–518.

Tezel, G., Li, L.Y., Patil, R.V. and Wax, M.B. (2001) TNF-alpha and TNF-alpha receptor-1 in the retina of normal and glaucomatous eyes. Invest. Ophthalmol. Vis. Sci., 42: 1787–1794.

Tezel, G. and Yang, X. (2004) Caspase-independent component of retinal ganglion cell death, in vitro. Invest. Ophthalmol. Vis. Sci., 45: 4049–4059.

The AGIS Investigators. The Advanced Glaucoma Intervention Study (AGIS): 7. (2000) The relationship between control of intraocular pressure and visual field deterioration. Am. J. Ophthalmol., 130: 429–440.

Tian, B., Geiger, B., Epstein, D.L. and Kaufman, P.L. (2000) Cytoskeletal involvement in the regulation of aqueous humor outflow. Invest. Ophthalmol. Vis. Sci., 41: 619–623.

Toris, C.B., Yablonski, M.E., Wang, Y.L. and Camras, C.B. (1999) Aqueous humor dynamics in the aging human eye. Am. J. Ophthalmol., 127: 407–412.

Totan, Y., Cecik, O., Borazan, M., Uz, E., Sogut, S. and Akyol, O. (2001) Plasma malondialdehyde and nitric oxide levels in age related macular degeneration. Br. J. Ophthalmol., 85(12): 1426–1428.

Tripathi, R.C., Li, J., Chan, W.F. and Tripathi, B.J. (1994) Aqueous humor in glaucomatous eyes contains an increased level of TGF-beta 2. Exp. Eye Res., 59: 723–727.

Ullian, E.M., Barkis, W.B., Chen, S., Diamond, J.S. and Barres, B.A. (2004) Invulnerability of retinal ganglion cells to NMDA excitotoxicity. Mol. Cell. Neurosci., 26: 544–557.

Unal, M., Guven, M., Devranoglu, K., Ozaydin, A., Batar, B., Tamçelik, N., Görgün, E.E., Uçar, D. and Sarici, A. (2007) Glutathione S-transferase M1 and T1 genetic polymorphisms are related to the risk of primary open-angle glaucoma: a study in a Turkish population Br J. Ophthalmol., 91: 527–530.

Usui, T., Amano, S., Oshika, T., Suzuki, K., Miyata, K., Araie, M., Heldin, P. and Yamashita, H. (2000) Expression regulation of hyaluronan synthase in corneal endothelial cells. Invest. Ophthalmol. Vis. Sci., 41: 3261–3267.

Valko, M., Morris, H. and Cronin, M.T. (2005) Metals, toxicity and oxidative stress. Curr. Med. Chem., 12: 1161–1208.

Van Jaarsveld, H., Kuyl, J.M., Schulenburg, D.H. and Wiid, N.M. (1996) Effect of flavonoids on the outcome of myocardial mitochondrial ischemia/reperfusion injury. Res. Commun. Mol. Pathol. Pharmacol., 91: 65–75.

Varma, S.D. (1991) Scientific basis for medical therapy of cataracts by antioxidants. Am. J. Clin. Nutr., 53(Suppl): 335S–345S.

Veach, J. (2004) Functional dichotomy: glutathione and vitamin E in homeostasis relevant to primary open-angle glaucoma. Br. J. Nutr., 91: 809–829.

Virno, M., Bucci, M.G., Pecori-Giraldi, J. and Cantore, G. (1966) Intravenous glycerol-vitamin C (sodium salt) as osmotic agents to reduce intraocular pressure. Am. J. Ophthalmol., 62: 824–833.

Visse, R. and Nagase, H. (2003) Matrix metalloproteinases and tissue inhibitors of metalloproteinases: structure, function, and biochemistry. Circ. Res., 92(8): 827–839.

Vittal, V., Rose, A., Gregory, K.E., Kelley, M.J. and Acott, T.S. (2005) Changes in gene expression by trabecular meshwork cells in response to mechanical stretching. Invest. Ophthalmol. Vis. Sci., 46: 2857–2868.

Wang, N., Chintala, S.K., Fini, M.E. and Schuman, J.S. (2001) Activation of a tissue-specific stress response in the aqueous outflow pathway of the eye defines the glaucoma disease phenotype. Nat. Med., 7: 304–309.

Wang, Y., Zhang, X., Lebwohl, M., DeLeo, V. and Wei, H. (1998) Inhibition of ultraviolet B (UVB)-induced c-fos and c-jun expression in vivo by a tyrosine kinase inhibitor genistein. Carcinogenesis, 19: 649–654.

Wentz-Hunter, K., Kubota, R., Shen, X. and Yue, B.Y. (2004) Extracellular myocilin affects activity of human trabecular meshwork cells. J. Cell Physiol., 200: 45–52.

Wiederholt, M. (1998) Nitric oxide and endothelin in aqueous humor outflow regulation. In: Haefliger I.O. and Flammer J. (Eds.), Nitric Oxide and Endothelin in the Pathogenesis of Glaucoma. Lippincott-Raven, New York, pp. 168–177.

Wiederholt, M., Bielka, S., Schweig, F., Lutjen-Drecoll, E. and Lepple-Wienhues, A. (1995) Regulation of outflow rate and resistance in the perfused anterior segment of the bovine eye. Exp. Eye Res., 61: 223–234.

Wiederholt, M., Thieme, H. and Stumpff, F. (2000) The regulation of trabecular meshwork and ciliary muscle contractility. Prog. Retin. Eye Res., 19: 271–295.

Wilbanks, G.A., Mammolenti, M. and Streilein, J.W. (1992) Studies on the induction of anterior chamber-associated immune deviation (ACAID). III. Induction of ACAID upon intraocular transforming growth factor-beta. Eur. J. Immunol., 22: 165–173.

Wise, J.B. and Witter, S.L. (1979) Argon laser therapy for open angle glaucoma: a pilot study. Arch. Ophthalmol., 97: 319–322.

Wordinger, R.J., Agarwal, R., Talati, M., Fuller, J., Lambert, W. and Clark, A.F. (2002) Expression of bone morphogenetic proteins (BMP), BMP receptors, and BMP associated proteins in human trabecular meshwork and optic nerve head cells and tissues. Mol. Vis., 8: 241–250.

Wordinger, R.J., Fleenor, D.L., Hellberg, P.E., Pang, I.H., Tovar, T.O., Zode, G.S., Fuller, J.A. and Clark, A.F. (2007) Effects of TGF-beta2, BMP-4, and gremlin in the trabecular meshwork: implications for glaucoma. Invest. Ophthalmol. Vis. Sci., 48: 1191–1200.

Yang, C.S., Chung, J.Y., Yang, G., Chhabra, S.K. and Lee, M.J. (2000) Tea and tea polyphenols in cancer prevention. J. Nutr., 130: 472S–478S.

Yildirim, O., Ates, N.A., Tamer, L., Oz, O., Yilmaz, A., Atik, U. and Camdeviren, H. (2005) May glutathione S-transferase M1 positive genotype afford protection against primary open-angle glaucoma? Graefes. Arch. Clin. Exp. Ophthalmol., 243: 327–333.

Yorio, T., Krishnamoorthy, R. and Prasanna, G. (2002) Endothelin: is it a contributor to glaucoma pathophysiology? J. Glaucoma, 11: 259–270.

Yucel, Y.H., Gupta, N., Zhang, Q., Mizisin, A.P., Kalichman, W. and Weinreb, R.N. (2006) Memantine protects neurons from shrinkage in the lateral geniculate nucleus in experimental glaucoma. Arch. Ophthalmol., 124: 217–225.

Yue, B.Y. (1996) The extracellular matrix and its modulation in the trabecular meshwork. Surv. Ophthalmol., 40: 379–390.

Zeevalk, G.D., Bernard, L.P., Sinha, C., Ehrhart, J. and Nicklas, W.J. (1998) Excitotoxicity and oxidative stress during inhibition of energy metabolism. Dev. Neurosci., 20: 444–453.

Zhou, L., Higginbotham, E.J. and Yue, B.Y. (1998) Effects of ascorbic acid on levels of fibronectin, laminin and collagen type 1 in bovine trabecular meshwork in organ culture. Curr. Eye Res., 17: 211–217.

Zhou, L., Li, Y. and Yue, B.Y. (1999) Oxidative stress affects cytoskeletal structure and cell-matrix interactions in cells from an ocular tissue: the trabecular meshwork. J. Cell. Physiol., 180: 182–189.

CHAPTER 28

TNF-α signaling in glaucomatous neurodegeneration

Gülgün Tezel*

Departments of Ophthalmology & Visual Sciences and Anatomical Sciences & Neurobiology, University of Louisville School of Medicine, Louisville, KY, USA

Abstract: Growing evidence supports the role of tumor necrosis factor-alpha (TNF-α) as a mediator of neurodegeneration in glaucoma. Glial production of TNF-α is increased, and its death receptor is upregulated on retinal ganglion cells (RGCs) and optic nerve axons in glaucomatous eyes. This multifunctional cytokine can induce RGC death through receptor-mediated caspase activation, mitochondrial dysfunction, and oxidative stress. In addition to direct neurotoxicity, potential interplay of TNF-α signaling with other cellular events associated with glaucomatous neurodegeneration may also contribute to spreading neuronal damage by secondary degeneration. Opposing these cell death–promoting signals, binding of TNF receptors can also trigger the activation of survival signals. A critical balance between a variety of intracellular signaling pathways determines the predominant in vivo bioactivity of TNF-α as best exemplified by differential responses of RGCs and glia. This review focuses on the present evidence supporting the involvement of TNF-α signaling in glaucomatous neurodegeneration and possible treatment targets to provide neuroprotection.

Keywords: glaucoma; neurodegeneration; retinal ganglion cell; glia; tumor necrosis factor-alpha

Tumor necrosis factor-alpha (TNF-α) is a pro-inflammatory cytokine with multiple functions in the immune response. Since its initial discovery as a serum factor causing tumor necrosis, it has been clear that besides its overwhelming functions in the regulation of inflammatory processes, this potent immunomediator is also a major mediator of apoptosis. Thousands of studies over the last 30 years have implicated TNF-α in the pathogenesis of a wide spectrum of human diseases, including sepsis, diabetes, cancer, collagen tissue diseases, and neurodegenerative diseases (Locksley et al., 2001; Chen and Goeddel, 2002). In addition to macrophages, lymphoid cells, mast cells, endothelial cells, and fibroblasts, TNF-α is also produced by glial cells and may lead to neuronal cell death (Downen et al., 1999). Cytotoxic effects of TNF-α are largely associated with its ability to induce apoptosis signaling as well as its function as a potent activator of neurotoxic substances such as nitric oxide and excitotoxins. In addition, picogram concentrations of TNF-α known to be noncytotoxic induce neuronal cell death through the silencing of survival signals (Venters et al., 2000). Several studies also implicate sphingomyelin hydrolysis and ceramide generation in TNF-α-induced cell death (Dressler et al., 1992). TNF-α is rapidly upregulated in the brain after injury, suggesting an important role of this cytokine in modifying the neurodegenerative process. Its excessive synthesis after brain

---

*Corresponding author. Tel.: +1 502 852 7395;
Fax: +1 502 852 3811; E-mail: gulgun.tezel@louisville.edu

injury has been correlated with poor outcome, and its inhibition has been associated with reduced neuronal damage in various diseases of the central nervous system (Shohami et al., 1999; McCoy et al., 2006). Regarding the optic nerve, intravitreal injections of TNF-α have resulted in degeneration of optic nerve axons and delayed loss of retinal ganglion cell (RGC) bodies (Kitaoka et al., 2006).

Evidence supporting the role of TNF-α as a mediator of retinal ganglion cell death in glaucoma

Growing evidence supports that TNF-α, through the binding of TNF receptor-1 (TNF-R1), a death receptor, is involved in mediating RGC death during glaucomatous neurodegeneration. Glial production of TNF-α is increased in the retina and optic nerve head, and TNF-R1 is upregulated in RGCs and their axons in glaucomatous human donor eyes (Yan et al., 2000; Yuan and Neufeld, 2000; Tezel et al., 2001). Findings of ongoing in vivo studies support that TNF-α and TNF-R1 are also upregulated following experimental elevation of intraocular pressure (IOP) in experimental animal models, and TNF-α signaling is involved in the RGC death process during neurodegeneration in ocular hypertensive eyes. TNF-α secreted by stressed glial cells in glaucomatous tissues can induce RGC death through receptor-mediated caspase cascade, mitochondrial dysfunction, and oxidative damage (Tezel and Wax, 2000a; Tezel and Yang, 2004).

An exciting finding of our earlier in vitro studies was the activation of retinal caspase-8 in response to different stress stimuli evident in glaucomatous eyes (Tezel and Wax, 1999, 2000b). Since caspase-8 is associated with the apoptosis pathway triggered by TNF death receptor binding (Hsu et al., 1996), this finding stimulated interest in the role of TNF-α signaling in glaucoma. Following this initial observation, a series of experiments provided evidence that the TNF-α signaling is indeed involved in RGC death in different experimental paradigms. An important evidence comes from primary co-culture experiments. These experiments proved that following exposure to different glaucomatous stimuli, glial cells adversely affect the survival of co-cultured RGCs through increased production of TNF-α. A similar neurotoxic effect was replicated by transferring the conditioned medium obtained from stressed glia cultures, and inhibition of TNF-α bioactivity by treatment of co-cultures with a neutralizing antibody resulted in a decreased rate of apoptosis in RGCs (Tezel and Wax, 2000a).

Another series of in vitro experiments also revealed that in addition to caspase activity, mitochondrial dysfunction accompanies RGC death induced by TNF-α (Tezel and Yang, 2004). These studies demonstrated that the inhibition of caspase activity is not adequate to block RGC death in primary cultures exposed to TNF-α if the mitochondrial membrane potential is lost and mitochondrial cell death mediators, cytochrome c and apoptosis-inducing factor, are released. RGCs exposed to TNF-α also accumulated reactive oxygen species (ROS) over time, and when combined with caspase inhibition, a free radical scavenger treatment reduced ROS and provided an additional increase in RGC survival (Tezel and Yang, 2004). These findings support that in addition to receptor-mediated caspase cascade, RGC death induced by TNF-α also involves caspase-dependent and caspase-independent components of the mitochondrial cell death pathway and oxidative injury (Fig. 1).

Parallel studies through immunohistochemical analysis of human donor eyes revealed an increased immunolabeling for TNF-α and TNF-R1 in the optic nerve head (Yan et al., 2000) and retina (Tezel et al., 2001) of glaucomatous eyes compared to age-matched controls. Using in situ hybridization, mRNA signals for TNF-α or TNF-R1 were also found to be similarly more intense in glaucomatous eyes relative to controls (Tezel et al., 2001). Upregulation of TNF-α in the glaucomatous optic nerve head and retina was mostly detectable in glial cells. However, TNF-R1 upregulation was also prominent on nerve bundles and RGC bodies. The upregulation of TNF-α and TNF-R1 in glaucomatous tissues supports the association of TNF-α signaling with glaucomatous neurodegeneration. The predominant localization of TNF death receptor to RGCs and their axons

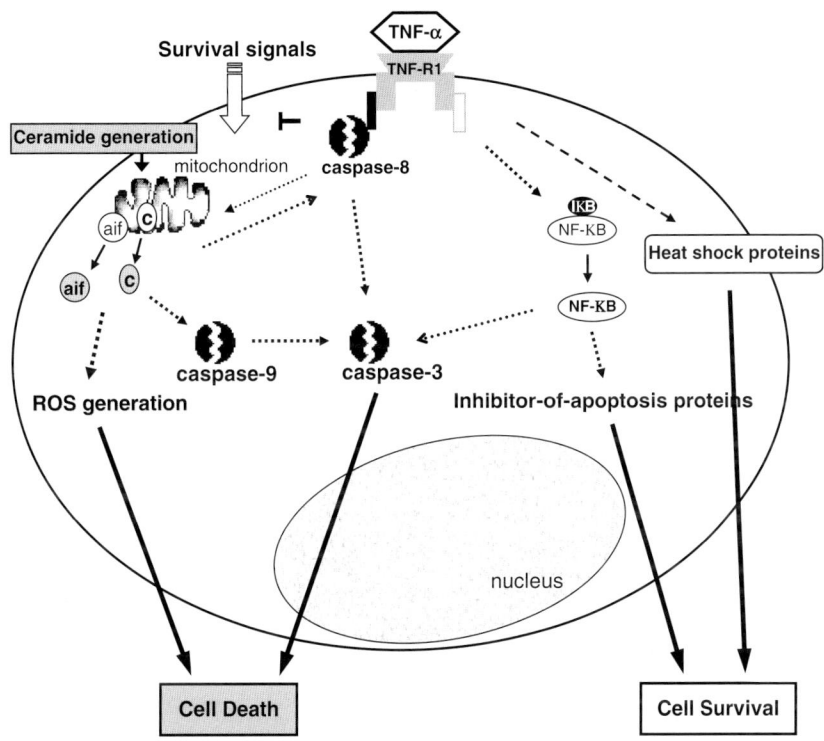

Fig. 1. Binding of TNF-α to TNF-R1, a death receptor, can induce RGC death through receptor-mediated caspase activation, and caspase-dependent and caspase-independent components of the mitochondrial cell death pathway, which includes increased generation of ROS leading to oxidative damage. In addition, ceramide generation and silencing of survival signals have been associated with TNF-α-mediated cell death. Opposing these cell death–promoting signals, binding of TNF-R1 can also trigger the activation of survival signals, which include the activation of a transcription factor, NF-κB, whose target genes include inhibitor-of-apoptosis proteins. Another intrinsic protection mechanism activated after TNF receptor binding involves heat shock proteins. A critical balance between a variety of intracellular signaling pathways determines whether an RGC will die or survive the exposure to TNF-α (TNF-α, tumor necrosis factor-alpha; TNF-R1, TNF death receptor; aif, apoptosis-inducing factor; c, cytochrome c; ROS, reactive oxygen species; NF-κB, nuclear factor-kappaB; IKB, NF-κB-inhibiting protein).

indicates that they are sensitive targets for the cytotoxic effects of TNF-α.

A series of in vivo observations also support the involvement of TNF-α signaling in the neurodegenerative process of glaucoma. Our in vivo studies utilizing a chronic pressure-induced rat model of glaucoma demonstrated retinal caspase activation in ocular hypertensive eyes, which includes the activation of caspase-8 (available at www.iovs.org, 2005; 46:E-Abstract 3772). Caspase-8 activation after IOP elevation was also detected by another study using two different rat glaucoma models (McKinnon et al., 2002). Activation of caspase-8 is an important early step following death receptor binding (Hsu et al., 1996), although there is in vitro evidence suggesting that caspase-8 activation may also occur downstream of mitochondria (Slee et al., 1999). The activation of not only the receptor-mediated caspase cascade but also the TNF-α-mediated mitochondrial cell pathway has been associated with caspase-8 activation, since caspase-8 cleaves a pro-apoptotic member of the Bcl-2 family of proteins, Bid, which then participates in the activation of the mitochondrial cell death pathway (Luo et al., 1998). Further studies also indicate Bid-independent mechanisms for the involvement of mitochondria during TNF-α-mediated cell death (Chen et al., 2007).

In vivo experiments also determined the expression and cellular localization of TNF-α and

TNF-R1, and IOP-dependent regulation of these molecules in ocular hypertensive rat eyes. Findings of these experiments demonstrated an upregulation of TNF-α signaling in ocular hypertensive eyes relative to controls. This upregulation in gene and protein expression exhibited a close association with the cumulative IOP exposure and neuronal damage (available at www.iovs.org, 2005; 46:E-Abstract 3772). Similar to glaucomatous human eyes (Yan et al., 2000; Tezel et al., 2001), increased TNF-α immunolabeling in ocular hypertensive rat eyes was mainly localized to glial cells. However, TNF-R1 immunolabeling in the retina and optic nerve head of ocular hypertensive rats was most prominent on RGCs and their axons. These findings further support the association of TNF-α signaling with the experimental paradigm of glaucomatous neurodegeneration.

Increased gene expression for TNF-α and TNF-R1 in ocular hypertensive eyes is consistent with other studies using gene microarray analysis in experimental glaucoma, which have also identified differential regulation of genes associated with TNF-α signaling. For example, a study detected upregulation of a transcription factor regulating TNF-α gene expression, Litaf, in the ocular hypertensive rat retina (Ahmed et al., 2004). Another study using Affymetrix analysis of rat retinal RNA identified multiple genes differentially regulated in eyes with ocular hypertension or optic nerve transaction. Findings of this study also support that the participation of TNF-α in glaucomatous injury is associated with JNK signaling (available at www.iovs.org, 2007; 48: E-Abstract 3279).

To assess the specific role of TNF death receptor signaling in the induction of RGC death, in vivo studies also utilized another experimental model, the optic nerve crush injury model, in TNF-R1 knockout mice (Tezel et al., 2004). Although not a perfect simulation of glaucomatous conditions, optic nerve degeneration in the crush injury model mimics many of the features of glaucomatous optic nerve degeneration. Most importantly, spreading of damage by secondary degeneration of RGCs is likely similar in crush injury and glaucomatous injury of the optic nerve. Counts of RGCs and their axons 6 weeks after the injury demonstrated that their loss was significantly less in TNF-R1 knockout mice compared with the controls. The most prominent decrease in neuronal loss detected in these animals was beyond the initial 2-week period after injury. This time period was correlated with the period of glial activation and increased glial immunolabeling for TNF-α in these eyes. No further protection against neuronal loss was detectable in TNF-R1 knockout mice treated with D-JNKI1, which is a specific peptide inhibitor of JNK. However, anti-JNK treatment of control animals provided significant protection against neuronal loss during the same secondary degeneration period. Phospho-JNK immunolabeling of RGCs in control mice subjected to optic nerve crush significantly decreased following their treatment with D-JNKI1, and anti-JNK treatment protected RGCs from degeneration, similar to the lack of TNF-R1 (Tezel et al., 2004). These findings provide evidence that TNF death receptor signaling is involved in the secondary degeneration of RGCs following optic nerve injury and is associated with JNK signaling.

TNF-α has also been suggested to mediate oligodendrocyte death and delayed RGC loss in a mouse model of glaucoma (Nakazawa et al., 2006). Intravitreal TNF-α injections in normal mice mimicked these effects. Conversely, treatment with an antibody neutralizing TNF-α activity or deleting the genes encoding TNF-α or TNF-R2 blocked the deleterious effects of ocular hypertension.

Optineurin gene mutations detected in glaucoma patients (Sarfarazi and Rezaie, 2003) provide another line of evidence supporting the role of TNF-α signaling in glaucoma. In addition to TNF-α gene polymorphism detected in different ethnic populations (Lin et al., 2003), a possible interaction between polymorphisms in optineurin and TNF-α genes has been suggested to increase the risk of glaucoma in the Japanese population (Funayama et al., 2004). Optineurin, expressed by RGCs (Wang et al., 2007), has been proposed to be associated with TNF-α signaling pathway leading to RGC death based on its direct interaction with adenovirus E3-14.7 kDa protein, which utilizes TNF receptor pathways to mediate apoptosis (Sarfarazi and Rezaie, 2003). However, a mutated form of optineurin identified in normal-tension glaucoma patients loses its ability for nuclear translocation and when overexpressed, compromises the mitochondrial membrane integrity, thereby resulting in decreased cell survival (De Marco et al., 2006). In a recent study, a glaucoma-associated mutation of optineurin has been shown to selectively induce oxidative stress-mediated RGC death, and optineurin has been suggested to be a likely component of the TNF-α signaling pathway leading to RGC death (Chalasani et al., 2007). More recently, microRNA silencing of optineurin has resulted in a prominent increase in TNF-α-induced nuclear factor kappaB (NF-κB) activity. These findings support a physiologic role of optineurin in dampening TNF-α signaling in association with glaucoma (Zhu et al., 2007).

Interplay of TNF-α signaling with other cellular events associated with glaucomatous neurodegeneration

Multiple pathogenic mechanisms, possibly sequentially functioning and interplaying with each other, have been associated with RGC death and optic nerve degeneration in glaucoma. Various intrinsic adaptive/protective mechanisms are also known to be activated in response to glaucomatous stress, and life and death decisions are made by integrating a variety of cell death and survival signals in RGCs. Potential interplay of TNF-α signaling with other cellular events associated with glaucomatous neurodegeneration further contributes to the complexity of the picture (Fig. 2).

Besides caspase activation through receptor-mediated apoptosis signaling and increased generation of ROS through mitochondrial dysfunction, a neurodestructive role of TNF-α in glaucoma may also be associated with its ability to induce glial nitric oxide production (Goureau et al., 1997). This highly potent secondary oxidant has been implicated in glaucomatous injury (Neufeld et al., 1999).

In addition, known interaction of cytokines with the glutamate system is of great importance among different mechanisms possibly involved in TNF-α-induced neurodegeneration. The role of TNF-α in various pathophysiological conditions of the central nervous system has been associated with its interaction with ion channels, as this cytokine can modulate ion channel activity, thereby regulating neuronal excitability, synaptic plasticity, and excitotoxic injury (Pickering et al., 2005). Similar to several pathological conditions that are largely dependent on excessive glutamate release and subsequent over-stimulation of the NMDA receptor, glutamate excitoxicity has been associated with glaucomatous neurodegeneration (Vorwerk et al., 1999).

TNF-α also activates matrix metalloproteinases (Gottschall and Yu, 1995), which are not only involved in tissue remodeling in the glaucomatous optic nerve head (Yan et al., 2000) but have also been associated with neurotoxicity. Retinal matrix metalloproteinase activity has been related to RGC death induced by IOP elevation (Guo et al., 2005). Although direct pathogenic role of matrix metalloproteinases in glaucoma is unclear, these proteases have been suggested to play a role in glutamate receptor processing thereby predisposing RGCs to damage. More relevant to TNF-α-mediated RGC death, the release of TNF-α from its membrane-bound precursor is also a matrix metalloproteinase-dependent process. TNF-α, which is produced as a biologically inactive transmembrane precursor, pro-TNF-α, must be enzymatically cleaved to release the 17 kDa active form of TNF-α. Matrix metalloproteinases that catalyze the normal turnover of extracellular matrix molecules can also function as TNF-α-converting enzyme to release the soluble TNF-α from its membrane-bound precursor (Black et al., 1997).

Endothelin-1 (ET-1), a potent vasoactive peptide, which can produce optic nerve damage analogous to that in glaucoma, has been associated with glaucomatous neurodegeneration, since ET-1 levels are increased in the glaucomatous aqueous and vitreous humor (Prasanna et al., 2003). TNF-α is a potent stimulator of ET-1 synthesis and secretion in several ocular cell types, including optic nerve head astrocytes (Desai et al., 2004).

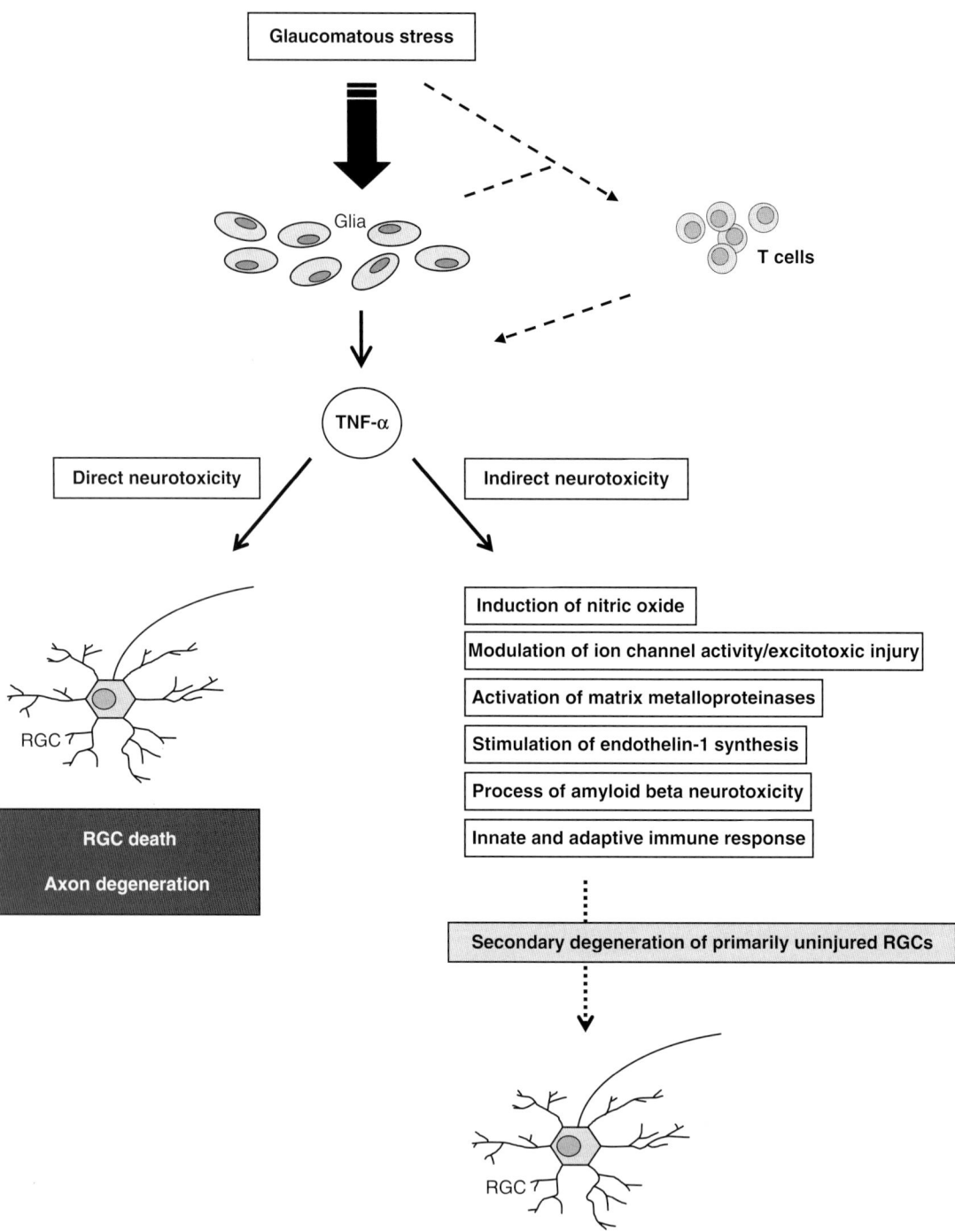

Fig. 2. Glial production of TNF-α is increased in the glaucomatous retina and optic nerve head. In addition to direct neurotoxicity to RGCs and their axons, potential interplay of TNF-α signaling with other cellular events associated with glaucomatous neurodegeneration likely contributes to secondary degeneration of primarily uninjured RGCs. In addition, glaucomatous tissue stress and glial reactivity may also initiate an immune response, and dysregulation of innate and adaptive immunity may facilitate neuronal injury. TNF-α is one of the most abundant cytokines produced by reactive T cells and appears to play a major role in T cell–mediated neurotoxicity.

Proposed involvement of amyloid-β in glaucomatous neurodegeneration, mimicking the Alzheimer's disease at the molecular level, has also been linked with the TNF death receptor signaling. Activation of amyloid-β requires caspase-8 activity by cross-linking with TNF family death receptors (Ivins et al., 1999). Based on the observation of retinal caspase-8 activation and abnormal processing of retinal amyloid precursor protein in ocular hypertensive rats (McKinnon et al., 2002), chronic amyloid-β neurotoxicity has been associated with RGC death in glaucoma.

Another association of TNF-α with glaucomatous neurodegeneration appears to be related to its key role in immune response with many effects ranging from inflammation to apoptosis. Immune priviledge of the central nervous system makes it crucial that glial cells are capable of responding rapidly to any injurious condition or damage. It is now generally accepted that these resident cells with immunoregulatory functions, particularly the microglia, play a key role in local immune response, which includes increased production of cytokines like TNF-α. Pertinent to its originally described function in the antitumoral response of the immune system, TNF-α is an important mediator of the innate immunity. Cytokine response may initially provide an intrinsic mechanism facilitating tissue repair. However, local production of cytokines, especially during injury and disease, makes the neuronal tissue more vulnerable to harmful effects of the innate immune response. Although innate immunity may initially participate in healing the injured tissue by eliminating cell debris and potentially toxic protein aggregates, uncontrolled and continued activity may lead to neurodestructive events. It is evident that the immunoregulatory function of optic nerve head and retinal glia is altered in glaucomatous eyes (Yang et al., 2001; Tezel et al., 2003), which can lead to an aberrant activity of the immune system, thereby facilitating primary and/or secondary degeneration of RGCs in glaucoma (Tezel and Wax, 2004). In addition to direct neurototoxicity of cytokines secreted by reactive glia in glaucomatous eyes, cytokine-stimulated microglia also generate copious amounts of reactive oxygen and reactive nitrogen species, creating additional stress on neurons. Conversely, oxidants can stimulate pro-inflammatory gene transcription in glia as well as serving as co-stimulatory molecules required for antigen presentation to the immune system (Tezel et al., 2007). TNF-α appears to be one of the key cytokines involved in the vicious cycle of the local immune response with neurodestructive consequences. In respect to adaptive immune response, also evident in glaucoma patients, this most abundant cytokine produced by reactive T cells may also participate in T cell–mediated neuronal injury. There is evidence supporting that any neurotoxicity through an adaptive immune response mediated by T cells is likely be associated with TNF receptor family signaling, although related evidence needs further verification (available at www.iovs.org, 2006; 47: E-Abstract 1828).

Involvement of TNF-α in innate immune response may also have implications in axonal degeneration in glaucomatous eyes. Although the initial site of injury is unclear, glaucomatous neurodegeneration exhibits widespread damage through a set of compartmentalized subcellular processes, which eventually involve RGC soma and dendrites in the retina and axons in the optic nerve. Axonal degeneration in glaucoma may evidently extend from the eye to synapses in the brain, possibly through an active and regulated self-destruction program, such as Wallerian degeneration. There is evidence supporting that TNF-α signaling may also be associated with axonal dysfunction and loss by disintegrating axons to initiate the innate immune response that characterizes Wallerian degeneration. TNF-α and its receptors are differentially regulated during Wallerian degeneration (Shamash et al., 2002). One of the functions of TNF-α during Wallerian degeneration has been suggested to be the induction of macrophage recruitment for debris removal (Liefner et al., 2000). TNF-α-induced matrix metalloproteinase activity has also been shown to promote macrophage recruitment into injured nerves during delayed axonal degeneration (Shubayev et al., 2006). Indeed, TNF-α knockout mice exhibit slower macrophage migration into the injured nerve and delayed myelin clearance during Wallerian degeneration (Liefner et al., 2000). However, another study utilizing TNF-α knockout mice has shown a significantly higher number of preserved axons within the degenerating distal nerve stump (Siebert and Bruck, 2003), thereby suggesting that despite debris removal function of invading macrophages, TNF-α secreted by these cells may also mediate axonal damage. It is highly warranted to determine the pathogenic importance of TNF-α signaling for axon degeneration in glaucoma, since in addition to increased TNF-α production of neighboring cells, optic nerve axons prominently express TNF-R1, which is upregulated in glaucoma (Yan et al., 2000; Yuan and Neufeld, 2000; Tezel et al., 2001).

Signaling pathways promoting neurodestructive or neuroprotective effects of TNF-α

TNF-α is generally assumed to be a cytokine exacerbating and sustaining neurodegenerative processes, while many other cytokines mostly promote regeneration, protection, and cell survival. Although TNF-R1 is able to signal cell death via its cytoplasmic death domain in a variety of cell types, including neurons, and despite massive evidence supporting numerous neurotoxic consequences of TNF-α signaling, this fascinating cytokine does not always cause neuronal cell death but can also exert protective functions (Rath and Aggarwal, 1999; Wajant et al., 2003).

The prominent paradoxical role of TNF-α in the central nervous system has generally been attributed to neuromodulatory function of this cytokine in the normal brain, but neurotoxic features under injurious conditions. Bioactivity of this cytokine can be influenced considerably by the functional state and type of cells, and the concentration and duration of exposure. Diverse cellular effects of TNF-α may also be dependent on the combination of other cytokines present at the same time, and even the temporal sequence of several cytokines acting on the same cell. Besides these parameters determining the net in vivo effect of TNF-α, diverse actions of TNF-α have also been associated with the composition of TNF receptors. Constitutive expression of TNF receptors, which are present on most cell types throughout the brain, exhibits rapid upregulation in response to injury, and the dual role of TNF-α has been associated with the opposing actions of TNF receptors. Two main subgroups of the TNF receptor superfamily, TNF-R1 (CD120a; p55/60) and TNF-R2 (CD120b; p75/80), bind both the membrane-bound and soluble forms of TNF-α, as well as another secreted molecule, lymphotoxin-α. However, soluble TNF-α predominantly stimulates TNF-R1 and has limited signaling capacities on TNF-R2. TNF-R1 is constitutively expressed in most tissues and appears to be the key mediator of TNF-α signaling. However, TNF-R2 is typically found and highly regulated in immune system cells, and seems to play a major role in lymphoid tissue. Extracellular domains of these receptors can be proteolytically cleaved, yielding soluble receptor fragments, which are considered to be physiological mechanisms that scavenge excess TNF-α. Soluble TNF receptors have been detected in the vitreous of normal subjects as well as of patients with retinal diseases (Sippy et al., 1996). Regarding intracellular domains, TNF-R1 contains a protein–protein interaction domain, called death domain, which recruits different signaling proteins involved in the apoptotic cascade. After TNF-α-induced formation of a multiprotein signaling complex at the cell membrane, cytoplasmic tail of this complex binds to a death domain adaptor protein [TNF receptor–associated death domain protein (TRADD)], which then binds to a receptor-interacting protein (FLICE) containing a protease domain (caspase-8) and initiates the apoptotic caspase cascade directly. TNF-R1 is also a potent activator of gene expression via indirect recruitment of the members of the TNF receptor–associated factor (TRAF) family. Interestingly, TNF-R2 directly recruits TRAF2, thereby inducing gene expression through a cross-talk relation with TNF-R1.

On the basis of current understanding of TNF-α signaling, it is generally accepted that TNF-R1 signaling promotes neuronal cell death while TNF-R2 mediates proliferative and regulatory signals promoting cell survival (Shohami et al., 1999). Counteracting functions of these TNF receptors in neuronal tissues are also evident in the retina. For example, a recent study utilizing a murine model of retinal ischemia has demonstrated that TNF-R1 aggravates neuronal cell death, whereas TNF-R2 promotes neuroprotection via activation of PKB/Akt (Fontaine et al., 2002). Growing evidence from many studies elucidating the mechanisms of TNF-α signaling pathways over the past three decades now supports that the diverse bioactivities of TNF-α can actually be mediated through each of these receptors. Evidently, TNF-α through TNF-R1 and also through the nondeath domain-containing receptor, TNF-R2, can promote cell death, while binding any of these receptors can promote cell survival (Rath and Aggarwal, 1999; Wajant et al., 2003).

It is fascinating how the signaling through TNF-R2 not containing death domain can induce cell death. Many evidences now support that the stimulation of TNF-R2 does not directly engage the apoptotic program, but the induction of endogenously produced membrane-bound TNF-α can subsequently activate TNF-R1. In addition to such a TRADD-independent signaling mechanism, a TNF-R2-dependent process can lead to depletion of TRAF2 and TRAF2-associated protective factors. TNF-R2 may also compete with TNF-R1 for the recruitment of newly synthesized TRAF2-bound anti-apoptotic factors, thereby promoting caspase activation rather than producing anti-apoptotic factors (Wajant et al., 2003). Furthermore, TNF-R2 signaling can potentiate programmed necrosis via TNF-R1 and caspase inhibition (Chan et al., 2003); however, the relevance of such a cell death mechanism to glaucomatous neurodegeneration is unclear.

Opposing these cell death–promoting signals, binding of two TNF receptors can also trigger the activation of survival signals (Fig. 1). Although many details are yet unclear, an important transcription factor, NF-κB, appears to be a regulator of the neuronal survival programs induced by TNF-α signaling (Baeuerle and Baltimore, 1996). Target genes of this important transcription factor include those that encode inhibitor-of-apoptosis proteins (IAPs), which provide an intrinsic protection mechanism against caspase activity (Deveraux et al., 1997). NF-κB may also induce cell survival through anti-apoptotic members of the Bcl-2 family and through JNK inhibition (Tang et al., 2001).

Another cell survival–promoting consequence of TNF-α signaling is the induction of heat shock proteins (HSPs) in cross-talk with NF-κB signaling (Nakano et al., 1996). A signaling mechanism similar to TNF-R1 and TNF-R2 cross-talk in inducing cell death has also been proposed for the induction of HSPs as a survival-promoting consequence of TNF-α signaling. TNF-R1 has been shown to be required for stress-induced HSP70 expression, and an intracellular role of TNF-R1 through a TNF-α-independent pathway has been described for the production of HSP70 (Heimbach et al., 2001).

Thus, bioactivities of TNF-α are exerted by activating a cross-talk relationship between multiple signaling pathways (Shohami et al., 1999; Wajant et al., 2003). While the activation of caspases promotes cell death, activation of NF-κB-dependent genes inhibits apoptosis and promotes the survival and proliferation effects of TNF-α (Beg and Baltimore, 1996). These opposing pathways are inhibitory to each other, and a critical balance determines the predominant in vivo bioactivity of TNF-α.

Diverse consequences of TNF-α signaling in glaucoma

Differential cellular response to TNF-α is best exemplified by differential susceptibilities of RGC and glia to glaucomatous injury. Primary co-culture studies demonstrate that RGCs undergo TNF-α-mediated apoptosis following exposure to glaucomatous stimuli, but glial cells survive the same condition (Tezel and Wax, 2000a). This is also in agreement with in vivo and clinical observations that glaucoma is a selective disease of neurons, specifically the RGCs and their axons. Despite a selective vulnerability of RGCs to primary and/or secondary degeneration in glaucoma, glial cells not only survive-glaucomatous stress but also exhibit an activated phenotype (Tezel et al., 2003). However, current understanding of factors determining diverse cellular responses to chronic and widespread tissue stress in glaucoma, such as the death of neurons but the sparing and activation of glia, are limited.

One explanation for the relative protection of glia against TNF-α cytotoxicity could be differential expression of different receptor subtypes. However, although the expression level of TNF receptors may vary between RGCs and glial cells, they are expressed by both cell types (Tezel et al., 2001). Considerable evidence rather suggests that diverse responses of RGCs and glial cells to TNF-α are associated with differential regulation of many signaling molecules involved in TNF-α signaling between these cell types. Since the functional activation of several adaptive/protective or pathogenic proteins involved in TNF-α signaling is known to require phosphorylation, kinase activity appears to be crucial in determining the ultimate cell fate in response to TNF-α. For example, similar to other cell types, functional activation of TNF-α-induced survival signaling through NF-κB in RGCs and glia is phosphorylation dependent (Tezel and Yang, 2005). On the other hand, JNK activity following optic nerve injury has been associated with the amplification of TNF-α-mediated cell death signaling, thereby switching the life balance toward cell death (Tezel et al., 2004). Findings of a recent histopathological study in human donor eyes support a prominent and persistent activation of glial ERK signaling in glaucomatous retinas, although the most prominent immunolabeling for phospho-JNK or phospho-p38 in these eyes is localized to RGCs (Tezel et al., 2003). Such a differential kinase activity between RGCs and glia appears to be associated with the different susceptibility of these cell types to TNF-α, since a critical balance between the survival-promoting ERK pathway and the death-promoting JNK and p38 pathways regulates cell fate (Xia et al., 1995).

Consistent with immunohistochemical observations, a recent comparative gene array analysis revealed differential regulation of MAPK or NF-κB signaling pathways between RGCs and glial cells exposed to TNF-α (Tezel and Yang, 2005). Findings of this in vitro study support that phosphorylation cascades are important components of TNF-α signaling and that a cross-talk between death- or survival-promoting signals determines whether an RGC should live or die in response to TNF-α. Besides the differential activity of MAPK and NF-κB signaling between RGCs and glial cells exposed to TNF-α, differentially regulated genes in RGCs and glial cells exposed to TNF-α also included HSP27. Upregulation and phosphorylation of this chaperonin protein selectively in glial cells suggests that various intrinsic adaptive/protective mechanisms, including HSPs (as well as MAPK and NF-κB signaling pathways), are important for relative protection of these cells against TNF-α-mediated cell death. Better understanding of the factors determining diverse responses of RGCs and glia to TNF-α can provide clues for therapeutic manipulation of TNF-α signaling for the gain of RGC survival, similar to that naturally seen in glia.

Conclusions on neuroprotective treatment targets in glaucoma

Growing evidence supports that TNF-α-mediated neurotoxicity is a component of the neurodegenerative injury in glaucoma. It is clear that a critical balance between diverse signaling pathways determines RGC fate after increased exposure to TNF-α in glaucomatous eyes. Based on our current understanding of diverse bioactivities of TNF-α signaling, which can promote both cell death and survival, specific inhibition of cell death signaling and/or amplification of survival signaling, rather than the inhibition of death receptor binding, is expected to accomplish RGC protection. Since such strategies are expected to not impede the survival-promoting signaling triggered by TNF-α, their neuroprotective outcome should be superior. Current understanding of TNF-α-mediated cell death signaling in RGCs suggests that neuroprotective strategies targeting TNF-α signaling for RGC rescue should include tools to block the caspase cascade and also improve the ability of these neurons to survive cytotoxic consequences of mitochondrial dysfunction. In addition, several molecules involved in TNF-α signaling, such as MAPKs, NF-κB, and HSPs, appear to be promising treatment targets for manipulation of the life balance between cell death and survival signals. Neuroprotective ability of anti-TNF-α strategies against axonal injury also needs to be clarified. Because the main source of TNF-α is glial cells, targeting key glial activation pathways should also be determined for attenuation of the glia-associated component of neurodegeneration, while maintaining glial neurosupportive functions. Evidently, multiple neurodegenerative processes are involved in RGC death, and current strategies are not expected to individually provide complete neuroprotection in glaucoma patients. Elucidation of specific signaling pathways is therefore crucial to define new treatment targets for effective neuroprotective interventions. Ongoing efforts utilizing targeted proteomic approaches are expected to identify signaling molecules associated with glaucomatous neurodegeneration. RNA interference technology also offers a powerful tool through which specific siRNAs can be used to determine functional significance of newly identified molecules as treatment targets. This technology can also serve as an intervention strategy along with other genomic or pharmacologic treatments to provide neuroprotection in glaucoma.

Acknowledgments

Dr. Tezel's work is supported by National Eye Institute (2R01 EY013813, 1R01 EY017131, R24 EY015636), Bethesda, MD, and an unrestricted grant to University of Louisville Department of Ophthalmology & Visual Sciences from Research to Prevent Blindness Inc., New York, NY.

References

Ahmed, F., Brown, K.M., Stephan, D.A., Morrison, J.C., Johnson, E.C. and Tomarev, S.I. (2004) Microarray analysis of changes in mRNA levels in the rat retina after experimental elevation of intraocular pressure. Invest. Ophthalmol. Vis. Sci., 45: 1247–1258.

Baeuerle, P.A. and Baltimore, D. (1996) NF-kappa B: ten years after. Cell, 87: 13–20.

Beg, A.A. and Baltimore, D. (1996) An essential role for NF-kappaB in preventing TNF-alpha-induced cell death. Science, 274: 782–784.

Black, R.A., Rauch, C.T., Kozlosky, C.J., Peschon, J.J., Slack, J.L., Wolfson, M.F., Castner, B.J., Stocking, K.L., Reddy, P., Srinivasan, S., Nelson, N., Boiani, N., Schooley, K.A., Gerhart, M., Davis, R., Fitzner, J.N., Johnson, R.S., Paxton, R.J., March, C.J. and Cerretti, D.P. (1997) A metalloproteinase disintegrin that releases tumour-necrosis factor-alpha from cells. Nature, 385: 729–733.

Chalasani, M.L., Radha, V., Gupta, V., Agarwal, N., Balasubramanian, D. and Swarup, G. (2007) A glaucoma-associated mutant of optineurin selectively induces death of retinal ganglion cells which is inhibited by antioxidants. Invest. Ophthalmol. Vis. Sci., 48: 1607–1614.

Chan, F.K., Shisler, J., Bixby, J.G., Felices, M., Zheng, L., Appel, M., Orenstein, J., Moss, B. and Lenardo, M.J. (2003) A role for tumor necrosis factor receptor-2 and receptor-interacting protein in programmed necrosis and antiviral responses. J. Biol. Chem., 278: 51613–51621.

Chen, G. and Goeddel, D.V. (2002) TNF-R1 signaling: a beautiful pathway. Science, 296: 1634–1635.

Chen, X., Ding, W.X., Ni, H.M., Gao, W., Shi, Y.H., Gambotto, A.A., Fan, J., Beg, A.A. and Yin, X.M. (2007) Bid-independent mitochondrial activation in tumor necrosis factor alpha-induced apoptosis and liver injury. Mol. Cell. Biol., 27: 541–553.

De Marco, N., Buono, M., Troise, F. and Diez-Roux, G. (2006) Optineurin increases cell survival and translocates to the nucleus in a Rab8-dependent manner upon an apoptotic stimulus. J. Biol. Chem., 281: 16147–16156.

Desai, D., He, S., Yorio, T., Krishnamoorthy, R.R. and Prasanna, G. (2004) Hypoxia augments TNF-alpha-mediated endothelin-1 release and cell proliferation in human optic nerve head astrocytes. Biochem. Biophys. Res. Commun., 318: 642–648.

Deveraux, Q.L., Takahashi, R., Salvesen, G.S. and Reed, J.C. (1997) X-linked IAP is a direct inhibitor of cell-death proteases. Nature, 388: 300–304.

Downen, M., Amaral, T.D., Hua, L.L., Zhao, M.L. and Lee, S.C. (1999) Neuronal death in cytokine-activated primary human brain cell culture: role of tumor necrosis factor-alpha. Glia, 28: 114–127.

Dressler, K.A., Mathias, S. and Kolesnick, R.N. (1992) Tumor necrosis factor-alpha activates the sphingomyelin signal transduction pathway in a cell-free system. Science, 255: 1715–1718.

Fontaine, V., Mohand-Said, S., Hanoteau, N., Fuchs, C., Pfizenmaier, K. and Eisel, U. (2002) Neurodegenerative and neuroprotective effects of tumor necrosis factor (TNF) in retinal ischemia: opposite roles of TNF receptor 1 and TNF receptor 2. J. Neurosci., 22: p. RC216.

Funayama, T., Ishikawa, K., Ohtake, Y., Tanino, T., Kurosaka, D., Kimura, I., Suzuki, K., Ideta, H., Nakamoto, K., Yasuda, N., Fujimaki, T., Murakami, A., Asaoka, R., Hotta, Y., Tanihara, H., Kanamoto, T., Mishima, H., Fukuchi, T., Abe, H., Iwata, T., Shimada, N., Kudoh, J., Shimizu, N. and Mashima, Y. (2004) Variants in optineurin gene and their association with tumor necrosis factor-{alpha} polymorphisms in Japanese patients with glaucoma. Invest. Ophthalmol. Vis. Sci., 45: 4359–4367.

Gottschall, P.E. and Yu, X. (1995) Cytokines regulate gelatinase A and B (matrix metalloproteinase 2 and 9) activity in cultured rat astrocytes. J. Neurochem., 64: 1513–1520.

Goureau, O., Amiot, F., Dautry, F. and Courtois, Y. (1997) Control of nitric oxide production by endogenous TNF-alpha in mouse retinal pigmented epithelial and Muller glial cells. Biochem. Biophys. Res. Commun., 240: 132–135.

Guo, L., Moss, S.E., Alexander, R.A., Ali, R.R., Fitzke, F.W. and Cordeiro, M.F. (2005) Retinal ganglion cell apoptosis in glaucoma is related to intraocular pressure and IOP-induced effects on extracellular matrix. Invest. Ophthalmol. Vis. Sci., 46: 175–182.

Heimbach, J.K., Reznikov, L.L., Calkins, C.M., Robinson, T.N., Dinarello, C.A., Harken, A.H. and Meng, X. (2001) TNF receptor I is required for induction of macrophage heat shock protein 70. Am. J. Physiol. Cell Physiol., 281: C241–C247.

Hsu, H., Shu, H.B., Pan, M.G. and Goeddel, D.V. (1996) TRADD-TRAF2 and TRADD-FADD interactions define two distinct TNF receptor 1 signal transduction pathways. Cell, 84: 299–308.

Ivins, K.J., Thornton, P.L., Rohn, T.T. and Cotman, C.W. (1999) Neuronal apoptosis induced by beta-amyloid is mediated by caspase-8. Neurobiol. Dis., 6: 440–449.

Kitaoka, Y., Kwong, J.M., Ross-Cisneros, F.N., Wang, J., Tsai, R.K., Sadun, A.A. and Lam, T.T. (2006) TNF-alpha-induced optic nerve degeneration and nuclear factor-kappaB p65. Invest. Ophthalmol. Vis. Sci., 47: 1448–1457.

Liefner, M., Siebert, H., Sachse, T., Michel, U., Kollias, G. and Bruck, W. (2000) The role of TNF-alpha during Wallerian degeneration. J. Neuroimmunol., 108: 147–152.

Lin, H.J., Tsai, F.J., Chen, W.C., Shi, Y.R., Hsu, Y. and Tsai, S.W. (2003) Association of tumour necrosis factor alpha -308 gene polymorphism with primary open-angle glaucoma in Chinese. Eye, 17: 31–34.

Locksley, R.M., Killeen, N. and Lenardo, M.J. (2001) The TNF and TNF receptor superfamilies: integrating mammalian biology. Cell, 104: 487–501.

Luo, X., Budihardjo, I., Zou, H., Slaughter, C. and Wang, X. (1998) Bid, a Bcl2 interacting protein, mediates cytochrome c release from mitochondria in response to activation of cell surface death receptors. Cell, 94: 481–490.

McCoy, M.K., Martinez, T.N., Ruhn, K.A., Szymkowski, D.E., Smith, C.G., Botterman, B.R., Tansey, K.E. and Tansey, M.G. (2006) Blocking soluble tumor necrosis factor signaling with dominant-negative tumor necrosis factor inhibitor attenuates loss of dopaminergic neurons in models of Parkinson's disease. J. Neurosci., 26: 9365–9375.

McKinnon, S.J., Lehman, D.M., Kerrigan-Baumrind, L.A., Merges, C.A., Pease, M.E., Kerrigan, D.F., Ransom, N.L., Tahzib, N.G., Reitsamer, H.A., Levkovitch-Verbin, H., Quigley, H.A. and Zack, D.J. (2002) Caspase activation and amyloid precursor protein cleavage in rat ocular hypertension. Invest. Ophthalmol. Vis. Sci., 43: 1077–1087.

Nakano, M., Knowlton, A.A., Yokoyama, T., Lesslauer, W. and Mann, D.L. (1996) Tumor necrosis factor-alpha-induced expression of heat shock protein 72 in adult feline cardiac myocytes. Am. J. Physiol., 270: H1231–H1239.

Nakazawa, T., Nakazawa, C., Matsubara, A., Noda, K., Hisatomi, T., She, H., Michaud, N., Hafezi-Moghadam, A., Miller, J.W. and Benowitz, L.I. (2006) Tumor necrosis factor-alpha mediates oligodendrocyte death and delayed retinal ganglion cell loss in a mouse model of glaucoma. J. Neurosci., 26: 12633–12641.

Neufeld, A.H., Sawada, A. and Becker, B. (1999) Inhibition of nitric-oxide synthase 2 by aminoguanidine provides neuroprotection of retinal ganglion cells in a rat model of chronic glaucoma. Proc. Natl. Acad. Sci. U.S.A., 96: 9944–9948.

Pickering, M., Cumiskey, D. and O'Connor, J.J. (2005) Actions of TNF-alpha on glutamatergic synaptic transmission in the central nervous system. Exp. Physiol., 90: 663–670.

Prasanna, G., Narayan, S., Krishnamoorthy, R.R. and Yorio, T. (2003) Eyeing endothelins: a cellular perspective. Mol. Cell. Biochem., 253: 71–88.

Rath, P.C. and Aggarwal, B.B. (1999) TNF-induced signaling in apoptosis. J. Clin. Immunol., 19: 350–364.

Sarfarazi, M. and Rezaie, T. (2003) Optineurin in primary open angle glaucoma. Ophthalmol. Clin. North Am., 16: 529–541.

Shamash, S., Reichert, F. and Rotshenker, S. (2002) The cytokine network of Wallerian degeneration: tumor necrosis factor-alpha, interleukin-1alpha, and interleukin-1beta. J. Neurosci., 22: 3052–3060.

Shohami, E., Ginis, I. and Hallenbeck, J.M. (1999) Dual role of tumor necrosis factor alpha in brain injury. Cytokine Growth Factor Rev., 10: 119–130.

Shubayev, V.I., Angert, M., Dolkas, J., Campana, W.M., Palenscar, K. and Myers, R.R. (2006) TNFalpha-induced MMP-9 promotes macrophage recruitment into injured peripheral nerve. Mol. Cell. Neurosci., 31: 407–415.

Siebert, H. and Bruck, W. (2003) The role of cytokines and adhesion molecules in axon degeneration after peripheral nerve axotomy: a study in different knockout mice. Brain Res., 960: 152–156.

Sippy, B.D., Hofman, F.M., Wright, A.D., He, S., Ryan, S.J. and Hinton, D.R. (1996) Soluble tumor necrosis factor receptors are present in human vitreous and shed by retinal pigment epithelial cells. Exp. Eye Res., 63: 311–317.

Slee, E.A., Harte, M.T., Kluck, R.M., Wolf, B.B., Casiano, C.A., Newmeyer, D.D., Wang, H.G., Reed, J.C., Nicholson, D.W., Alnemri, E.S., Green, D.R. and Martin, S.J. (1999) Ordering the cytochrome c-initiated caspase cascade: hierarchical activation of caspases-2, -3, -6, -7, -8, and -10 in a caspase-9-dependent manner. J. Cell Biol., 144: 281–292.

Tang, G., Minemoto, Y., Dibling, B., Purcell, N.H., Li, Z., Karin, M. and Lin, A. (2001) Inhibition of JNK activation through NF-kappaB target genes. Nature, 414: 313–317.

Tezel, G., Chauhan, B.C., LeBlanc, R.P. and Wax, M.B. (2003) Immunohistochemical assessment of the glial mitogen-activated protein kinase activation in glaucoma. Invest. Ophthalmol. Vis. Sci., 44: 3025–3033.

Tezel, G., Li, L.Y., Patil, R.V. and Wax, M.B. (2001) Tumor necrosis factor-alpha and its receptor-1 in the retina of normal and glaucomatous eyes. Invest. Ophthalmol. Vis. Sci., 42: 1787–1794.

Tezel, G. and Wax, M.B. (1999) Inhibition of caspase activity in retinal cell apoptosis induced by various stimuli in vitro. Invest. Ophthalmol. Vis. Sci., 40: 2660–2667.

Tezel, G. and Wax, M.B. (2000a) Increased production of tumor necrosis factor-alpha by glial cells exposed to simulated ischemia or elevated hydrostatic pressure induces apoptosis in cocultured retinal ganglion cells. J. Neurosci., 20: 8693–8700.

Tezel, G. and Wax, M.B. (2000b) The mechanisms of hsp27 antibody-mediated apoptosis in retinal neuronal cells. J. Neurosci., 20: 3552–3562.

Tezel, G. and Wax, M.B. (2004) The immune system and glaucoma. Curr. Opin. Ophthalmol., 15: 80–84.

Tezel, G. and Yang, X. (2004) Caspase-independent component of retinal ganglion cell death, in vitro. Invest. Ophthalmol. Vis. Sci., 45: 4049–4059.

Tezel, G. and Yang, X. (2005) Comparative gene array analysis of TNF-alpha-induced MAPK and NF-kappaB signaling pathways between retinal ganglion cells and glial cells. Exp. Eye Res., 81: 207–217.

Tezel, G., Yang, X., Luo, C., Peng, Y., Sun, S.L. and Sun, D. (2007) Mechanisms of immune system activation in glaucoma: oxidative stress-stimulated antigen presentation by the retina and optic nerve head glia. Invest. Ophthalmol. Vis. Sci., 48: 705–714.

Tezel, G., Yang, X., Yang, J. and Wax, M.B. (2004) Role of tumor necrosis factor receptor-1 in the death of retinal ganglion cells following optic nerve crush injury in mice. Brain Res., 996: 202–212.

Venters, H.D., Dantzer, R. and Kelley, K.W. (2000) A new concept in neurodegeneration: TNFalpha is a silencer of survival signals. Trends Neurosci., 23: 175–180.

Vorwerk, C.K., Gorla, M.S. and Dreyer, E.B. (1999) An experimental basis for implicating excitotoxicity in glaucomatous optic neuropathy. Surv. Ophthalmol., 43(Suppl 1): S142–S150.

Wajant, H., Pfizenmaier, K. and Scheurich, P. (2003) Tumor necrosis factor signaling. Cell Death Differ., 10: 45–65.

Wang, J.T., Kunzevitzky, N.J., Dugas, J.C., Cameron, M., Barres, B.A. and Goldberg, J.L. (2007) Disease gene candidates revealed by expression profiling of retinal ganglion cell development. J. Neurosci., 27: 8593–8603.

Xia, Z., Dickens, M., Raingeaud, J., Davis, R.J. and Greenberg, M.E. (1995) Opposing effects of ERK and JNK-p38 MAP kinases on apoptosis. Science, 270: 1326–1331.

Yan, X., Tezel, G., Wax, M.B. and Edward, D.P. (2000) Matrix metalloproteinases and tumor necrosis factor alpha in glaucomatous optic nerve head. Arch. Ophthalmol., 118: 666–673.

Yang, J., Yang, P., Tezel, G., Patil, R.V., Hernandez, M.R. and Wax, M.B. (2001) Induction of HLA-DR expression in human lamina cribrosa astrocytes by cytokines and simulated ischemia. Invest. Ophthalmol. Vis. Sci., 42: 365–371.

Yuan, L. and Neufeld, A.H. (2000) Tumor necrosis factor-alpha: a potentially neurodestructive cytokine produced by glia in the human glaucomatous optic nerve head. Glia, 32: 42–50.

Zhu, G., Wu, C.J., Zhao, Y. and Ashwell, J.D. (2007) Optineurin negatively regulates TNFalpha-induced NF-kappaB activation by competing with NEMO for ubiquitinated RIP. Curr. Biol., 17: 1438–1443.

CHAPTER 29

Involvement of the *Bcl2* gene family in the signaling and control of retinal ganglion cell death

Robert W. Nickells*, Sheila J. Semaan and Cassandra L. Schlamp

Department of Ophthalmology and Visual Sciences University of Wisconsin, Madison, WI 53706, USA

Abstract: Retinal ganglion cell death by apoptosis is a well-established outcome in the glaucomatous pathology of the retina. Extensive research into the molecular events underlying this process show us that members of the *Bcl2* gene family play a critical role in the activation and control of ganglion cell death. Perhaps the most critical molecule at play is the pro-apoptotic protein BAX. Without BAX, ganglion cell somas appear to survive an optic nerve insult indefinitely. Once BAX is activated, however, the cell death program reaches an irreversible point, where the process cannot be blocked. Interacting with BAX are other members of this larger gene family, including the anti-apoptotic protein BCL-X, and several members of the BH3-only proteins that serve as sensors and activators of the cell death program. A hypothetical model of how all these molecules interact in glaucoma is presented.

Keywords: apoptosis; retinal ganglion cells; *Bcl2* gene family; BAX; glaucoma

Introduction

Retinal ganglion cell (RGC) death is the end-stage pathology of all glaucomas. In this review, we summarize the current level of knowledge of the molecular events associated with this process, particularly the involvement of a class of genes related to the *Bcl2* gene family. Much of the research that has lead to our understanding of the molecular events associated with ganglion cell death has originated from the study of rodent models of both experimental glaucoma and chronic ocular hypertension (Weinreb and Lindsey, 2005). For this reason, the genetic nomenclature that we use here is restricted to convention for rats and mice.

More than a decade ago, several laboratories reported that RGCs died by a process similar to developmental programmed cell death, known as apoptosis. This modality of cell death was evident in models of acute optic nerve lesion, such as axotomy and crush (Berkelaar et al., 1994; Garcia-Valenzuela et al., 1994; Quigley et al., 1995; Li et al., 1999), and in models of experimental glaucoma (Garcia-Valenzuela et al., 1995; Quigley et al., 1995; Libby et al., 2005b), which is now believed to be a milder version of the crush lesion (Nickells, 2007). Similar observations have also been reported for ganglion cell death in human glaucoma and ischemic optic neuropathy (Levin and Louhab, 1996; Kerrigan et al., 1997).

Morphologically, dying ganglion cells exhibit many of the characteristic features of apoptosis, including dendritic tree and soma shrinkage (Misantone et al., 1984; Weber et al., 1998; Morgan

*Corresponding author. Tel.: 608-265-6037; Fax: 608-262-1479;
E-mail: nickells@wisc.edu

et al., 2000; Morgan, 2002), chromatin condensation, nuclear envelope dissolution, and the formation of apoptotic bodies (Quigley et al., 1995), and DNA fragmentation. More recent observations now suggest that ganglion cells death is subdivided into regional compartments (Whitmore et al., 2005), which can occur as independent autonomous self-destruct pathways. In this newer model, ganglion cell death begins with degeneration of the axon in the optic nerve and then spreads to the dendrites and soma in the retina. This model is consistent with the evidence that the initial site of damage in glaucoma is the lamina cribrosa (Quigley et al., 1981; Howell et al., 2007) and suggests a process that links elevated intraocular pressure (IOP) with the activation and eventual execution of the entire ganglion cell (Nickells, 2007). Technically, apoptosis is the term used to describe the autonomous self-destruct pathway executed by cell somas.

Intrinsic apoptosis vs. extrinsic apoptosis

There are two basic pathways of apoptosis (Adams and Cory, 2007), both of which culminate in the activation of a cascade of cysteine proteases called caspases (Salvesen and Dixit, 1997; Slee et al., 1999; Adams and Cory, 2002). The caspases function to literally digest the cellular contents from within, thus allowing for the elimination of the cell without adversely imparting significant stress on the surrounding tissue of the whole organism. Caspases are present as pro-proteins in healthy cells and they become activated when the prodomain is cleaved off. There is a great deal of evidence that caspase activation is present in dying ganglion cells, supporting the early observations that these cells die by apoptosis with more molecular evidence (Kermer et al., 1998; McKinnon et al., 2002; Huang et al., 2005a). Caspases can be activated by two distinct apoptotic pathways. Extrinsic apoptosis, also referred to as the death receptor pathway, relies on the interaction of an extracellular ligand of the Tumor Necrosis Factor (TNF) family with a death receptor on the surface of the target cell. Once bound, the receptor engages adaptor proteins resulting in the direct activation of the caspase cascade through the proteolytic cleavage of procaspase 8. Alternatively, caspases can also be activated by an internal signaling pathway called intrinsic apoptosis. The intrinsic pathway is significantly more complicated and relies heavily on mitochondrial involvement. In this pathway, signaling molecules from the *Bcl2* gene family are recruited to the mitochondrial outer membrane causing disruption of mitochondrial function and the release of the electron transporter cytochrome c. Once released, cytochrome c complexes with a molecule called apoptosis inducing factor-1 (Apaf-1) and procaspase 9, creating a structure known as the apoptosome (Adams and Cory, 2002). This complex facilitates the cleavage of procaspase 9, which then is able to activate the caspase cascade. An important consideration when comparing the two pathways is that, by itself, mitochondrial dysfunction is lethal to the cell. This is principally due to disruption of the electron transport chain, resulting in the loss of ATP production. In addition, because molecular oxygen is no longer converted to water by the transfer of electron free radicals, mitochondrial dysfunction also leads to the generation of excessive reactive oxygen species. Thus, involvement of mitochondria is often considered an irreversible step in the apoptotic pathway (Chang et al., 2002; Nickells, 2004). This is experimentally evidenced by studies using caspase inhibitors, which can provide only a transient protective effect to cells undergoing intrinsic apoptosis (Chang et al., 2002). A caveat to the concept of distinct apoptotic pathways is that the two can share common elements in some cells. One process of the extrinsic pathway is to recruit the intrinsic pathway via the activation of signaling molecules that affect mitochondria. A more detailed discussion of these molecules and their function will be given in a subsequent section of this review.

The majority of experimental evidence points to RGCs executing the intrinsic apoptotic pathway. This evidence is partially based on studies of mitochondrial changes in dying ganglion cells (Mittag et al., 2000; Tatton et al., 2001), and the activation of caspase 9 in these cells (Kermer et al., 2000). In addition, as will be discussed below, genetic studies using mice mutated for different members of the *Bcl2* gene family, show dramatic effects in preventing ganglion cell soma death. Lastly, even though intrinsic apoptosis is likely the dominant mechanism of ganglion cell death in glaucoma, there is also compelling evidence that downstream activation of extrinsic apoptosis may play a role in the overall pathology of this disease.

The *Bcl2* family of proteins

Both the extrinsic and intrinsic apoptotic pathways involve proteins with structurally similar amino acid domains. Originally, these domains were identified in a protein called BCL-2, the product of a gene found in a t(14;18) chromosomal translocation common to human B-cell lymphomas (Bakhshi et al., 1985). Members of the *Bcl2* gene family share homology with at least one of four conserved amino acid domains, hence the name Bcl Homology (BH) domains (Fig. 1). The BH1, BH2, and BH3 domains of a single protein can organize into a hydrophobic groove, which may allow family members containing all three to form pore structures in lipid bilayers. More importantly, however, proteins containing the BH3-only domain are able to interact with the groove by virtue of an amphipathic α-helical domain (Adams and Cory, 1998). Additionally, most members of the *Bcl2* gene family have the ability to anchor, or insert, themselves into membranes, which has led to the speculation that they function to create membranous pore structures or destabilize lipid bilayers.

Within the *Bcl2* gene family, there are three subfamilies of proteins. The first group contains related proteins that are anti-apoptotic and promote cell survival. The major representatives of this subfamily are *Bcl2* and *BclX*, although at least 10 other similar proteins have been identified. The second group contains related proteins that are pro-apoptotic and promote cell death. This subfamily is smaller, containing three known members, with the most well characterized being *Bax* and *Bak*. Finally, a third group has also been classified in the *Bcl2* family by virtue of containing a single BH3 domain. Whereas genes within the first two subfamilies tend to be homologs of each other, the BH3-only proteins are generally unrelated except for the BH3 domain. There are at least eight known members of this subfamily and they also play a role in activating apoptosis.

Gene expression studies of RGCs indicate that members of all three subfamilies are transcribed in these cells. Quantitative mRNA analysis and localization experiments indicate that *BclX* is the most prevalent member of the anti-apoptotic genes in ganglion cells, exhibiting at least a 16-fold greater abundance over *Bcl2* transcripts in the whole retina (Levin et al., 1997). This does not preclude that *Bcl2* is also expressed in the ganglion cells, since sensitive PCR-based methods of mRNA analysis are also able to detect transcripts from this gene in whole retinal extracts (Chaudhary et al., 1999). Overexpression of anti-apoptotic proteins has a dramatic impact on ganglion cell survival after optic nerve trauma. Transgenic mice that selectively upregulate *Bcl2* in neurons exhibit prolonged RGC survival after optic nerve lesions (Bonfanti et al., 1995; Bonfanti et al., 1996; Cenni et al., 1996), while introduction of exogenous BCL-X by gene therapy or permeable fusion proteins prevents ganglion cell loss in rat axotomy models (Liu et al., 2001; Kretz et al., 2004; Malik et al., 2005).

Of the pro-apoptotic subfamily, *Bax* appears to be the most relevant molecule for ganglion cell death. This observation comes from genetic experiments using mice with engineered mutations in the *Bax* gene (Mosinger Ogilvie et al., 1998). Mice homozygous for the *Bax* mutation exhibit complete abrogation of ganglion cell soma death in models of acute and chronic optic nerve damage (Li et al., 2000; Libby et al., 2005b). This is generally true for many neuronal cell-types, while non-neuronal cells appear to express comparable amounts of *Bax* and *Bak* and both genes must be disabled to block cell death (Wei et al., 2001). There is also some controversy regarding the transcriptional regulation of *Bax* in damaged ganglion cells. Some studies have suggested that *Bax* levels increase early in the apoptotic pathway (Isenmann et al., 1997; Näpänkangas et al., 2003), while others have not been able to confirm this finding (McKinnon et al., 2002) (S. Semaan and R. Nickells, unpublished data). Historically, *Bax*

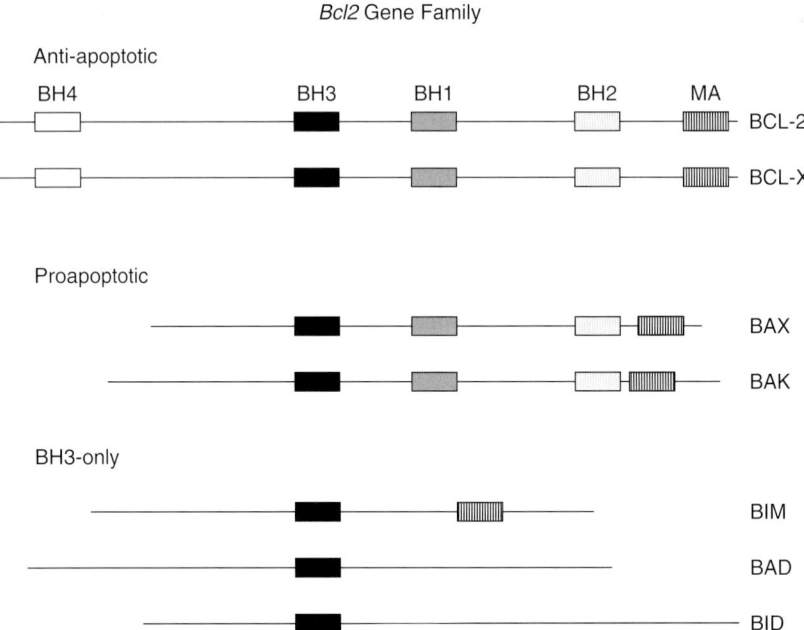

Fig. 1. Structural similarities of the *Bcl2* gene family. A diagram of several proteins of the *Bcl2* gene family discussed in this review. Proteins of this family share amino acid motifs called Bcl Homology (BH) Domains. BCL-2 and BCL-X are homologs and contain four structurally related domains. BH1, BH2, and BH3 domains interact and form a globular structure with a hydrophobic groove. In addition to this region, these proteins contain a C-terminal membrane anchor (MA). BCL-2 and BCL-X function as anti-apoptotic proteins. A second group of proteins in this family function as pro-apoptotic proteins. The most well characterized members of this family are BAX and BAK, both of which contain BH1, BH2, and BH3 domains and a membrane anchor. The third subfamily in this group is the BH3-only proteins. These proteins are unrelated except for the presence of a BH3 domain. Like BAX and BAK, BH3-only proteins also play a role in mediating apoptosis in cells. Retinal ganglion cells have been reported to express BCL-X, BAX, BIM, BAD, and BID.

mRNA and proteins are thought to exist as latent molecules in neurons (Putcha et al., 2003; Adams and Cory, 2007). A comprehensive discussion of *Bax* function in RGCs is presented in a later section of this review.

Ganglion cells also express several BH3-only proteins, including genes called *Bim*, *Bad*, and *Bid*. The interplay of the proteins from these genes is a critical control over the apoptotic program in most cells, including ganglion cells. While BH3-only proteins are unable to activate cell death by themselves, they play a role in modulating the interaction between pro-apoptotic and anti-apoptotic members. In our current understanding of these proteins, they appear to act as both activators and sensors of the apoptotic process within cells, such that as cell death is initiated, different BH3-only proteins are recruited to augment the process and ensure successful execution of the pathway. In a separate section of this review, we present a model of how BH3-only proteins may interact in dying cells of a glaucomatous retina.

The requirement of BAX for RGC soma death

In healthy cells, BAX resides in an inactive state, principally localized in the cytosol, or loosely associated with intracellular membranes. BAX molecules that are resident in or near membranes may interact specifically with anti-apoptotic proteins, thereby preventing "accidental" permealization of these organelles. Upon activation of apoptosis, evidence shows that BAX is translocated primarily to the mitochondrial outer membrane (Wolter et al., 1997; Putcha et al., 1999) (Fig. 2). Once there, BAX is thought to undergo a conformational change, allowing it to insert into the membrane bilayer, resulting in the release of cytochrome c. Although the exact mechanism of this release is currently not known (Danial and Korsmeyer, 2004), a popular model is that BAX can oligomerize and form pores in the mitochondrial outer membrane, enabling the release of cytochrome c. A caveat to this model is that proteinaceous pores have never been identified either in vivo or in vitro. An alternate model suggests that BAX proteins interact specifically with mitochondrial lipids, causing a decrease in the stability of the lipid bilayer and increasing permeability (Polster and Fiskum, 2004). Of note in this whole process is the suspected antagonistic function of the anti-apoptotic proteins. BCL-X is typically already localized to mitochondrial membranes by virtue of a specific mitochondrial targeting sequence at its carboxy terminus (Kaufmann et al., 2003) (Fig. 2). Cells expressing BCL-X are able to block apoptosis even with active BAX molecules present, suggesting that there is a direct antagonism between the two. One model suggests that these two proteins bind to each other and that this prevents BAX from affecting membrane permeability, but not necessarily its ability to insert into membranes (Putcha et al., 1999). This model is challenged, however, by some evidence that anti-apoptotic BCL proteins can exert a protective effect without associating with BAX (Liu et al., 2006). The true mechanism (or mechanisms) of action of BCL-X is still an unanswered question.

Much of the knowledge of *Bax* involvement in neuronal death comes from studies using mice with an engineered mutation in the *Bax* gene. Mice carrying mutant *Bax* genes exhibit a marked decrease in developmental neuronal death in a variety of sites such as the brainstem, cerebellum, and hippocampus (White et al., 1998). This is also true of the retina, which exhibits supernumerary neurons in both the inner nuclear layer and ganglion cell layer (Mosinger Ogilvie et al., 1998). Using this mouse model, researchers found that ganglion cell death after acute optic nerve trauma (crush lesion), was absolutely dependent on *Bax* function, but other stimuli of ganglion cell death, such as high levels of excitotoxins, elicited a *Bax*-independent program of apoptosis (Li et al., 2000) (Fig. 3). The reason underlying the alternative cell death pathways is still not clear, but probably lies in which BH3-only sentinel molecules are activated under varying circumstances.

The mutant *Bax* allele was also used to establish *Bax* function in ganglion cell death in glaucoma. In these experiments, the mutant allele was congenically cross-bred onto the genetic background of the DBA/2J line of inbred mouse. During the last decade, studies on this line have shown that they exhibit abnormalities in melanocyte formation and melanin synthesis. These defects cause a breakdown in the iris stroma and dispersion of pigment debris that becomes trapped in the mouse trabecular meshwork leading to pathology of the angle tissues and outflow pathway (John et al., 1998; Chang et al., 1999; Anderson et al., 2002; Mo et al., 2003). These anterior chamber defects are clinically evident in mice at 6 months of age and very prominent in mice older than 8 months. Concomitant with these defects, the DBA/2J mouse develops elevated IOP followed shortly after by degeneration of the optic nerve and the subsequent loss of RGCs (Libby et al., 2005a; Schlamp et al., 2006). DBA/2J mice congenic for the *Bax* mutant allele revealed three important features of RGC loss in response to elevated IOP (Libby et al., 2005b). First,

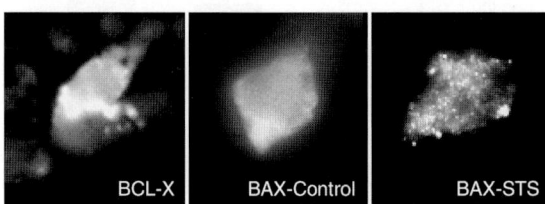

Fig. 2. Localization of BCL-X ad BAX in HEK 293 tissue culture cells. Proteins are visualized by immunofluoresence. In unstressed cells, BCL-X staining is punctate, typical of its association with cellular organelles, principally mitochondria (left panel). BAX labeling in unstressed cells (center panel) is more diffuse, but becomes punctate within 24 h after treatment with staurosporine (STS), which activates apoptosis in these cells (right panel). The punctate labeling pattern is associated with the translocation of BAX from the cytoplasm to the mitochondrial outer membrane.

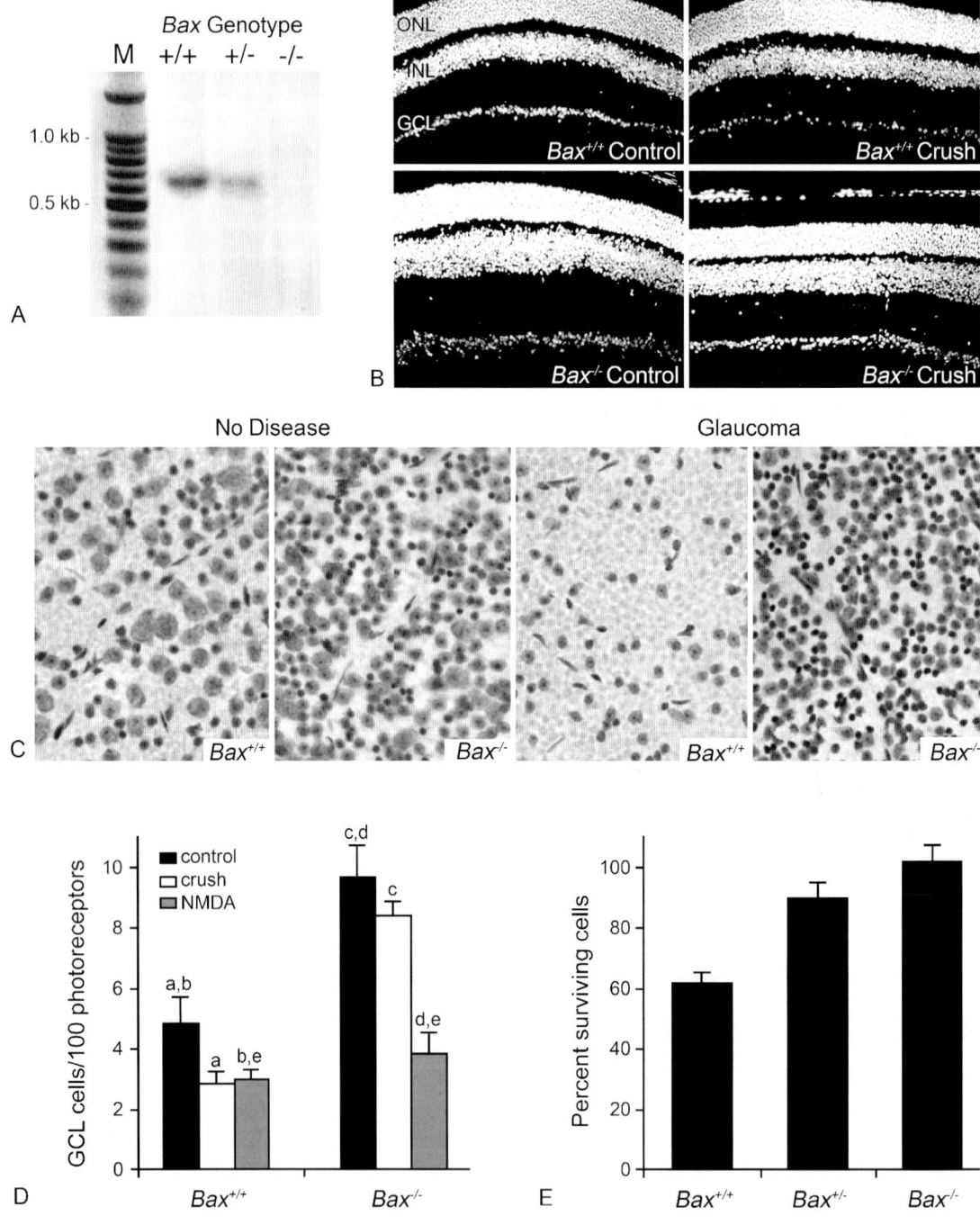

glaucomatous $Bax-/-$ mice exhibited complete abrogation of ganglion cell soma loss (Fig. 3). Thus, regardless of what combinations of apoptotic pathways are activated in ganglion cells, the initial pathology in the retina must occur through a Bax-dependent, intrinsic pathway. Second, even though ganglion cell somas were spared, glaucomatous $Bax-/-$ mice still exhibited maximal damage to the optic nerve, including the complete loss of axons. This observation has helped to formulate the concept that RGCs die using compartmentalized self-destruct pathways (Whitmore et al., 2005). In this model, axons, dendrites, somas, and synapses may all die independent of each other. Such a model may help explain the link between elevated IOP and the initial activation of ganglion cell loss in glaucoma (Whitmore et al., 2005; Nickells, 2007). Third, this study also revealed that Bax gene dosage could markedly affect cell survival. Previous studies using Bax mutant mice had suggested that only a single functioning Bax allele was sufficient for normal activation of neuronal death (Deckwerth et al., 1996). In the DBA/2J genetic background, however, $Bax+/-$ mice exhibited the same attenuated ganglion cell death observed in completely Bax-deficient mice. Recent studies using the Bax mutant allele in different genetic strains have also shown that reducing Bax gene dosage reduces transcript levels from this gene (Fig. 3). Strain differences in the cell death response to Bax gene dosage may be related to polymorphisms in the Bax gene promoter that affect transcriptional activity of this gene (S. Semaan and R. Nickells, unpublished). These studies suggest that therapeutically reducing BAX levels by just 50% could have a pronounced effect on the survival of ganglion cells.

BH3-only proteins and the early signaling of ganglion cell apoptosis

As noted in an earlier section of this review, the $Bcl2$ gene family is divided into three main subgroups based on the presence of common amino acid domains. We have already discussed the potential functions of the $BclX$ and Bax genes and products, but the critical element for the activation of intrinsic apoptosis is the interactions of these proteins with the members of the BH3-only family

Fig. 3. The role of BAX in retinal ganglion cell death in mice. (A) Semi-quantitative reverse transcriptase PCR analysis of Bax mRNA in mice carrying mutant alleles of the Bax gene. Mice heterozygous for the Bax mutant ($Bax+/-$) exhibit about 50% of the normal Bax mRNA, while homozygous mutant mice ($-/-$) have no detectable Bax transcripts. (B) Loss of Bax expression blocks retinal ganglion cell death after acute optic nerve crush. The images shown are sections of mouse retinas stained with 4,6-diamidino-2-phenylindole. Panels on the left are from unoperated eyes of wild type and Bax-deficient mice. $Bax-/-$ animals exhibit a thickened inner nuclear layer (INL) and approximately twice the number of retinal ganglion cells in the ganglion cell layer (GCL) compared to wild type littermates. No change is evident in the outer nuclear layer (ONL). The right panels show retinas from eyes 2 weeks after optic nerve crush. Wild type mice exhibit a dramatic loss of cells in the GCL, while Bax-deficient animals show no significant loss of cells. (C) Loss of Bax expression blocks retinal ganglion cell death in glaucomatous DBA/2J mice. Panels show Nissl-stained retinal wholemounts of wild type and Bax-deficient DBA/2J mice. In this experimental paradigm, glaucomatous damage was defined as an increase in intraocular pressure and degeneration of the axons in the optic nerve. Young DBA/2J mice, with no evidence of disease are shown in the 2 panels on the left. Bax-deficient mice tend to have smaller ganglion cells, possibly restricted in size because of the higher density of these cells. As DBA/2J mice age, they develop ocular hypertension and an optic neuropathy. Wild type mice with severe optic nerve disease (>90% axon loss) also show a loss of ganglion cell somas in the retina (panel 3). Conversely, $Bax-/-$ mice with ocular hypertension and severe optic nerve disease exhibit complete abrogation of ganglion cell soma death (panel 4). (D) Quantitative analysis ganglion cell density in mice after optic nerve crush or intravitreal injection of the glutamate analog N-methyl-d-aspartate (NMDA). Wild type mice exhibit significant cell loss after both crush and NMDA-injections ($^{a,\ b}P<0.005$). Bax-deficient mice, however, exhibit minimal cell loss after crush ($^{c}P>0.10$), but nearly maximal cell loss after NMDA injections ($^{d}P<0.05$), although statistically more than $Bax+/+$ mice ($^{e}P=0.01$). Cell density in the ganglion cell layer was measured from sections of mouse retinas. (E) Quantitative analysis of ganglion cell loss in DBA/2J glaucomatous mice. Wild type mice exhibit the loss of ~40% of the total number of neurons in the ganglion cell layer ($P<0.001$ compared to mice with no disease). $Bax+/-$, and $Bax-/-$ mice, however, show no appreciable loss of cells ($P=0.207$ and 0.426, respectively). In this experiment, cell loss was measured from Nissl-stained wholemounts and compared to retinas from young mice with no evidence of disease. Images in (B) and (D) are adapted with permission from Li et al. (2000). Images in (C) and (E) are adapted with permission from Libby et al. (2005b), which is an open access publication.

of molecules. Evidence exists for the function of BIM, BAD, and proteolytically cleaved BID (tBID) peptides in the process of RGC death. In one respect, evidence for an active role of each may seem contradictory to a common final pathway because the known activation events of these BH3-only proteins are dramatically different. In this section, we will suggest a model that incorporates all three of these peptides in the overall process of ganglion cell death. This model is based on a recent hypothesis that BH3-only proteins are really sentinels of the apoptotic program, which are recruited to help amplify and accelerate the process (Adams and Cory, 2007).

The actual function of the BH3-only proteins is controversial. By virtue of the BH3 domain, they are able to interact with the globular BH domains of other *Bcl2* family proteins. In many cells, the presence of BH3-only proteins dramatically facilitates the cell death process. This lead to early hypotheses that these small proteins interacted directly with pro-apoptotic *Bcl2* family proteins, like BAX, possibly targeting them to organellar membranes or allowing these proteins to change conformation and insert into lipid bilayers. Current experimental evidence, however, suggests that this model is not correct. In vivo, BH3-only proteins show a distinct affinity for anti-apoptotic members, such as BCL2 and BCL-X (Adams and Cory, 2007). A new model of indirect activation has emerged that suggests that the activation of BH3-only proteins allows them to interact and neutralize anti-apoptotic proteins, thus limiting their ability to antagonize their pro-apoptotic counterparts.

Studies using *Bax−/−* mice clearly show that a *Bax*-mediated program of cell death must be initiated in order to achieve complete retinal pathology (Li et al., 2000; Libby et al., 2005b). In the compartmentalized cell death model of glaucoma, the initiating event of ganglion cell soma death is the deficit of neurotrophins due to the blockade of axoplasmic transport in the lamina cribrosa. This process is biochemically similar to serum deprivation in cultured neurons. Studies using primary cultures of neurons showed that this traumatic episode activated both the transcriptional upregulation, and then the post-translational phosphorylation, of the BIM BH3-only protein (Putcha et al., 2003). This study also showed that the activating phosphorylation of BIM was mediated by c-Jun N-Terminal Kinases (JNKs) localized to the mitochondria. Several studies have identified BIM regulation and activation as one of the critical molecules participating in ganglion cell death after experimental trauma to the optic nerve (Näpänkangas et al., 2003; McKernan and Cotter, 2007). The role of BIM in glaucoma is less clear, but several investigators have observed JNK upregulation and activation in rat models of experimental glaucoma (Kwong and Caprioli, 2006; Levkovitch-Verbin et al., 2007), suggesting a link with BIM activation.

Other studies have also found associations between the activation of the BH3-only protein BAD and RGC death. Like BAX, the pro-apoptotic action of BAD requires interaction with charged lipids in the mitochondrial outer membrane. BAD is often present as a latent phosphorylated molecule in cells. Phosphorylation appears to block its interaction with lipids and facilitates an interaction with 14-3-3 inhibitory peptides (Hekman et al., 2006). The activation of BAD generally requires the activity of a protein phosphatase such as Calcineurin, which removes these inhibitory phosphates. Recently, Huang and colleagues showed that activated Calcineurin was present in ganglion cells of rat models of optic nerve trauma, including experimental glaucoma, and in aged DBA/2J mice (Huang et al., 2005b). This led to a decrease in the level of phosphorylated BAD and was temporally associated with increased ganglion cell death. In addition, pharmacologic inhibition of Calcineurin using FK506 was able to attenuate the process of cell death, at least in a short-term analysis of 10 days post ocular hypertension. Overall, the activation of Calcineurin, by proteolytic cleavage, is linked to increased levels of intracellular calcium ions and the activation of calmodulin. In dying cells, the most likely source of free calcium ions is from stores in the endoplasmic reticulum (ER), which are extruded during phases of ER stress. Intrinsically, the mediator of this release is the secondary integration of activated BAX protein into the ER lipid bilayer (Nutt et al., 2002; Breckenridge et al.,

2003). In $Bax-/-$ neurons, for example, Ca^{2+} ions are not released from ER stores. Thus, even though the activity of BAD is to neutralize anti-apoptotic proteins, presumably allowing BAX to function, the activation of BAD may be reliant on the initial activation of BAX by some other mechanism.

The third BH3-only protein that has been found to play a role in ganglion cell death is BID. Like BIM and BAD, the putative function of BID is to neutralize anti-apoptotic proteins. Unlike its two counterparts, however, BID predominantly plays a role in amplifying the extrinsic apoptotic pathway (Adams and Cory, 2007). As discussed above, extrinsic apoptosis initiates from an extracellular interaction between a death-inducing ligand, such as TNFα, with a death receptor located on the membrane of the target cell. This interaction directly activates the caspase cascade, initially through the proteolytic cleavage of procaspase 8. One of the proteolytic targets of caspase 8 is inactive BID, which when cleaved to become tBID, is able to interact with multiple anti-apoptotic *Bcl2* family members (Adams and Cory, 2007). This sequence of events is not essential for the full activation of the caspase cascade and cell death, but instead helps amplify and accelerate the process by recruiting the intrinsic pathway as well. There is a great deal of interest in this pathway in the process of RGC death and glaucoma. Several studies have helped establish that TNFα levels are elevated in eyes after optic nerve trauma, including glaucoma (Tezel et al., 2001; Tezel et al., 2004), and there is an increase in the level of tBID in ganglion cells (Huang et al., 2005a). The relative importance of the TNFα pathway has been established using mice carrying mutations in the TNFα receptor-1 and 2 genes. TNFα-R1 mutants, after optic nerve crush, show no immediate change in the rate of ganglion cell loss, but a significant attenuation of loss 2 weeks after the injury (Tezel et al., 2004), supporting a role for TNF-mediated cell death in secondary degeneration. Surprisingly, TNFα-R1 mutants show no effect in a mouse model of ocular hypertension, but TNFα-R2 mutants do (Nakazawa et al., 2006). The delayed reliance on TNF-mediated ganglion cell death suggests that it plays a more significant role in late stages of retinal pathology. This is even more likely since the principal source of TNFα is activated glia, particularly microglia, in the retina (Tezel and Wax, 2000; Tezel et al., 2001; Nakazawa et al., 2006).

Given the available evidence, the following hypothetical model of BH3-only protein mediated activation of ganglion cell death arises (Fig. 4). Initially, neurotrophin deprivation stimulates activating kinases that lead to the upregulation and modification of BIM. BIM neutralizes some of the BCL-X in ganglion cells, allowing for BAX, which has been translocated and inserted into the mitochondrial membrane, to facilitate an increase in membrane permeability. If this step is blocked, such as in $Bax-/-$ or $Bim-/-$ mice, there is complete and long-term rescue of cell death (Li et al., 2000; Libby et al., 2005b; McKernan and Cotter, 2007). If BAX does become active in this initial stage, then not only can it insert into mitochondrial membranes, but it can also insert into ER membranes, precipitating ER stress and the release of calcium ions. The release of Ca^{2+} is potentially cytotoxic by itself. If calcium levels become too high, cell death can ensue by a variety of mechanisms, many of which are not as controlled as apoptosis and thus deleterious to the overall health of the organism or surrounding tissue. Since apoptosis is in the best interest of the organism, elevated calcium levels are sensed by another BH3-only protein, BAD, which becomes activated via a calcium-calmodulin dependent proteolytic mechanism. Activated BAD then collaborates with BIM by also competing for and neutralizing BCL-X. This could be considered as a second gear in the apoptotic machinery. As ganglion cell somas begin to die, the overall pathology of the retina is sensed by resident glial cells. In some cases, glia, such as astrocytes may try to prevent further ganglion cell death. Other cells, however, may exacerbate the pathology. It is suspected that TNFα release principally originates from activated microglia. The presence of TNFα activates concomitant receptors on ganglion cells, thus activating extrinsic apoptosis and the tBID BH3-only sensor. As with BAD, tBID likely functions to help neutralize BCL-X along with BIM. An interesting caveat to the TNFα-mediated

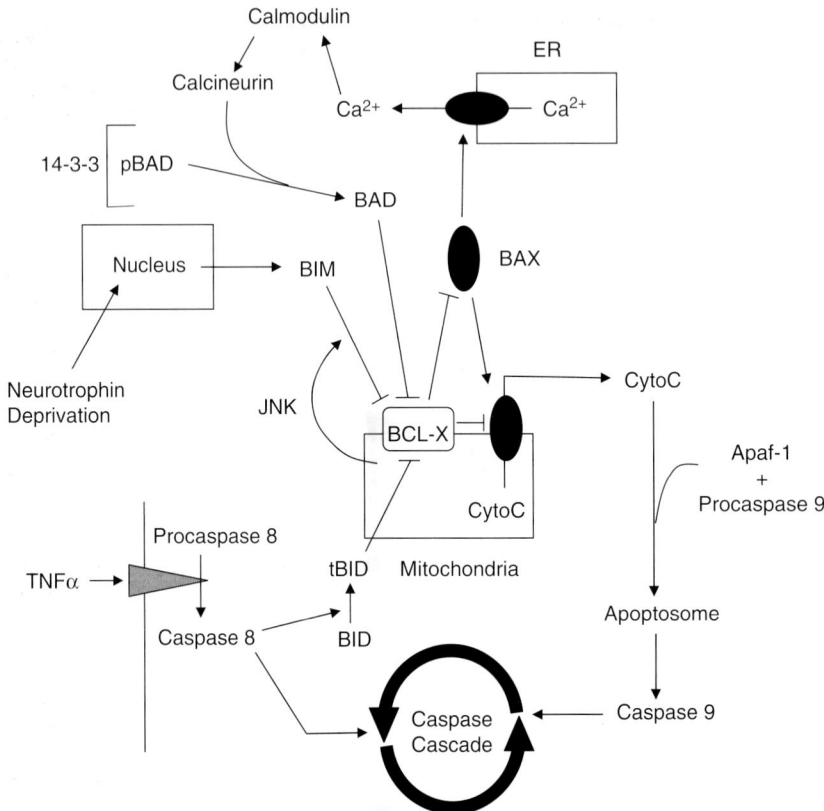

Fig. 4. Flow diagram of the activation of apoptosis in retinal ganglion cells. The hypothetical initiating event for apoptosis is damage to ganglion cell axons in the optic nerve head. This precipitates the loss of neurotrophic support to ganglion cell somas, which is normally derived from retrograde axoplasmic transport of factors via the optic nerve. Loss of support stimulates a series of signaling events resulting in the transcriptional activation of the BH3-only protein BIM. After modification of BIM by a mitochondrial c-Jun N-Terminal Kinase (JNK), BIM antagonizes the anti-apoptotic protein BCL-X, which is localized to the surface of the mitochondria. With the function of BCL-X impaired, cytoplasmic BAX is able to translocate to the mitochondrial inner membrane and facilitate the release of cytochrome C (CytoC). CytoC binds with procaspase 9 and Apaf-1 to form the apoptosome, which actives caspase 9 and initiates the caspase cascade. Secondarily, BAX is also able to insert into the membrane of the endoplasmic reticulum, mediating the release of calcium ion stores. This release affects the activation of the protein phosphatase Calcineurin, which dephosphorylates the BH3-only protein BAD. Once active, BAD can also antagonize BCL-X possibly allowing BAX to function more efficiently. At some point downstream of the initial damage to optic nerve axons, cells in the retina (presumably microglia) begin to secrete Tumor Necrosis Factor α (TNFα). TNFα interacts with either R1 or R2 receptors on the surface of ganglion cells, leading to the direct activation of caspase 8, which activates the caspase cascade. Caspase 8 can also cleave the BH3-only protein BID, forming tBID, which then antagonizes BCL-X and allows the activation of BAX. Thus, TNFα can activate both the extrinsic and intrinsic apoptotic programs. Studies on genetically engineered mice reveal that activation of BAX is likely the most critical step in allowing this whole process to occur. Mice lacking a functioning *Bax* gene exhibit no ganglion cell loss in either acute or chronic models of optic nerve damage, similar to mice that overexpress *Bax* antagonists like *Bcl2*. This finding suggests that downstream activation of TNFα-mediated pathways first requires some initial period of ganglion cell damage that is controlled by the BIM/BAX pathway of intrinsic cell death.

pathway, however, is that unlike BAD, tBID activation may occur in cells that do not have already activated BAX protein in them. Thus, TNFα could be a critical component of the suspected secondary degeneration of healthy ganglion cells not directly affected by pathology at the lamina cribrosa (Schwartz, 2005; Tezel, 2006; Nickells, 2007). The evidence that there is no ganglion cell loss in *Bax−/−* mice may seem to contradict this hypothesis, since once extrinsic apoptosis is activated, BAX is not critical for cell death and is recruited only to help the process. If TNFα-mediated cell death plays a significant role in glaucoma related cell death, why, then, is there no cell loss in Bax−/− mice? One plausible explanation is that secondary degeneration mediated by TNFα release first requires the initial Bax-dependent loss of ganglion cells. This phenomenon will likely be the subject of investigation in the future.

Conclusion

There have been many exciting and important advances in the field of glaucoma research in the last 15 years. As we learn more about the molecular pathways activated in dying ganglion cells, we are beginning to understand many of the complexities of this disease as a whole. More than ever, a comprehensive understanding of initial activating events and subsequent downstream effects will allow future developments of a variety of therapeutic interventions targeted at blocking the progressive loss of neurons associated with this disease.

Abbreviations

Apaf-1	apoptosis activating factor-1
BH	Bcl homology
CytoC	cytochrome C
ER	endoplasmic reticulum
GCL	ganglion cell layer
INL	inner nuclear layer
IOP	intraocular pressure
JNK	c-Jun N-terminal kinase
MA	membrane anchor
NMDA	N-methyl-d-aspartate
ONL	outer nuclear layer
TNF	tumor necrosis factor

Acknowledgments

This work was supported by grants from the National Eye Institute (R01 EY12223 to RWN) and an unrestricted research grant from Research to Prevent Blindness, Inc., to the Department of Ophthalmology and Visual Sciences of the University of Wisconsin.

References

Adams, J.M. and Cory, S. (1998) The bcl-2 protein family: arbiters of cell survival. Science, 281: 1322–1326.

Adams, J.A. and Cory, S. (2002) Apoptosomes: engines for caspase activation. Curr. Opin. Cell Biol., 14: 715–720.

Adams, J.M. and Cory, S. (2007) The Bcl-2 apoptotic switch in cancer development and therapy. Oncogene, 26: 1324–1337.

Anderson, M.G., Smith, R.S., Hawes, N.L., Zabaleta, A., Chang, B., Wiggs, J.L. and John, S.W.M. (2002) Mutations in genes encoding melanosomal proteins cause pigmentary glaucoma in DBA/2J mice. Nat. Genet., 30: 81–85.

Bakhshi, A., Jensen, J.P., Goldman, P., Wright, J.J., McBride, O.W., Epstein, A.L. and Korsmeyer, S.J. (1985) Cloning the chromosomal breakpoint of t (14;18) human lymphomas: clustering around JH on chromosome 14 and near a transcriptional unit on 18. Cell, 41: 899–906.

Berkelaar, M., Clarke, D.B., Wang, Y.-C. and Aguayo, A.J. (1994) Axotomy results in delayed death and apoptosis of retinal ganglion cells in adult rats. J. Neurosci., 14: 4368–4374.

Bonfanti, L., Chierzi, S., Cenni, M.C., Strettoi, E., Liu, X.-H., Martinou, J.C., Maffei, L. and Rabacchi, S.A. (1995) Survival of retinal ganglion cells over-expressing Bcl-2 protein in neonatal transgenic mice following optic nerve section. Soc. Neurosci. Abstr., 21: p. 1556.

Bonfanti, L., Strettoi, E., Chierzi, S., Cenni, M.C., Liu, X.-H., Martinou, J.C., Maffei, L. and Rabacchi, S.A. (1996) Protection of retinal ganglion cells from natural and axotomy-induced cell death in neonatal transgenic mice overexpressing bcl-2. J. Neurosci., 16: 4186–4194.

Breckenridge, D.G., Germain, M., Mathai, J.P., Nguyen, M. and Shore, G.C. (2003) Regulation of apoptosis by endoplasmic reticulum pathways. Oncogene, 22: 8608–8618.

Cenni, M.C., Bonfanti, L., Martinou, J.-C., Ratto, G.M., Strettoi, E. and Maffei, L. (1996) Long-term survival of retinal ganglion cells following optic nerve section in adult bcl-2 transgenic mice. Eur. J. Neurosci., 8: 1735–1745.

Chang, B., Smith, R.S., Hawes, N.L., Anderson, M.G., Zabaleta, A., Savinova, O., Roderick, T.H., Heckenlively, J.R., Davisson, M.T. and John, S.W.M. (1999) Interacting loci cause severe iris atrophy and glaucoma in DBA/2J mice. Nat. Genet., 21: 405–409.

Chang, L.K., Putcha, G.V., Deshmukh, M. and Johnson, E.M., Jr. (2002) Mitochondrial involvement in the point of no return in neuronal apoptosis. Biochimie, 84: 223–231.

Chaudhary, P., Ahmed, F., Quebada, P. and Sharma, S.C. (1999) Caspase inhibitors block the retinal ganglion cell death following optic nerve transection. Mol. Brain Res., 67: 36–45.

Danial, N.N. and Korsmeyer, S.J. (2004) Cell death: critical control points. Cell, 116: 205–219.

Deckwerth, T.L., Elliot, J.L., Knudson, C.M., Johnson, E.M., Jr., Snider, W.D. and Korsmeyer, S.J. (1996) BAX is required for neuronal death after trophic factor deprivation and during development. Neuron, 17: 401–411.

Garcia-Valenzuela, E., Gorczyca, W., Darzynkiewicz, Z. and Sharma, S.C. (1994) Apoptosis in adult retinal ganglion cells after axotomy. J. Neurobiol., 25: 431–438.

Garcia-Valenzuela, E., Shareef, S., Walsh, J. and Sharma, S.C. (1995) Programmed cell death of retinal ganglion cells during experimental glaucoma. Exp. Eye Res., 61: 33–44.

Hekman, M., Albert, S., Galmiche, A., Rennefahrt, U.E.E., Fueller, J., Fischer, A.J., Puehringer, D., Wiese, S. and Rapp, U.R. (2006) Reversible membrane interaction of BAD requires two C-terminal lipid binding domains in conjunction with 14-3-3 protein binding. J. Biol. Chem., 281: 17321–17336.

Howell, G.R., Libby, R.T., Jakobs, T.C., Smith, R.S., Phalan, F.C., Barter, J.W., Barbay, J.M., Marchant, J.K., Mahesh, N., Porciatti, V., Whitmore, A.V., Masland, R.H. and John, S.W.M. (2007) Axons of retinal ganglion cells are insulted in the optic nerve early in DBA/2J glaucoma. J. Cell Biol., 179: 1523–1537.

Huang, W., Dobberfuhl, A., Filippopoulos, T., Ingelsson, M., Fileta, J.B., Poulin, N.R. and Grosskreutz, C.L. (2005a) Transcriptional up-regulation and activation of initiating caspases in experimental glaucoma. Am. J. Pathol., 167: 673–681.

Huang, W., Fileta, J.B., Dobberfuhl, A., Filippopoulos, T., Guo, Y., Kwon, G. and Grosskreutz, C.L. (2005b) Calcineurin cleavage is triggered by elevated intraocular pressure, and calcineurin inhibition blocks retinal ganglion cell death in experimental glaucoma. Proc. Natl. Acad. Sci. U.S.A., 102: 12242–12247.

Isenmann, S., Wahl, C., Krajewski, S., Reed, J.C. and Bähr, M. (1997) Up-regulation of Bax protein in degenerating retinal ganglion cells precedes apoptotic cell death after optic nerve lesion in the rat. Eur. J. Neurosci., 9: 1763–1772.

John, S.W.M., Smith, R.S., Savinova, O.V., Hawes, N.L., Chang, B., Turnbull, D., Davisson, M., Roderick, T.H. and Heckenlively, J.R. (1998) Essential iris atrophy, pigment dispersion, and glaucoma in DBA/2J mice. Invest. Ophthalmol. Vis. Sci., 39: 951–962.

Kaufmann, T., Schlipf, S., Sanz, J., Neubert, K., Stein, R. and Borner, C. (2003) Characterization of the signal that directs Bcl-XL, but not Bcl-2, to the mitochondrial outer membrane. J. Cell Biol., 160: 53–64.

Kermer, P., Ankerhold, R., Klocker, N., Krajewski, S., Reed, J.C. and Bähr, M. (2000) Caspase-9 involvement in secondary death of axotomized rat retinal ganglion cells in vivo. Brain Res. Mol. Brain Res., 28: 144–150.

Kermer, P., Klöcker, N., Labes, M. and Bähr, M. (1998) Inhibition of CPP32-like proteases rescues axotomized retinal ganglion cells from secondary death in vivo. J. Neurosci., 18: 4656–4662.

Kerrigan, L.A., Zack, D.J., Quigley, H.A., Smith, S.D. and Pease, M.E. (1997) TUNEL-positive ganglion cells in human primary open-angle glaucoma. Arch. Ophthalmol., 115: 1031–1035.

Kretz, A., Kügler, S., Happold, C., Bähr, M. and Isenmann, S. (2004) Excess BclXL increases the intrinsic growth potential of adult CNS neurons in vitro. Mol. Cell Neurosci., 26: 63–74.

Kwong, J.M. and Caprioli, J. (2006) Expression of phosphorylated c-Jun N-terminal kinase (JNK) in experimental glaucoma in rats. Exp. Eye Res., 82: 576–582.

Levin, L.A. and Louhab, A. (1996) Apoptosis of retinal ganglion cells in anterior ischemic optic neuropathy. Arch. Ophthalmol., 114: 488–491.

Levin, L.A., Schlamp, C.L., Spieldoch, R.L., Geszvain, K.M. and Nickells, R.W. (1997) Identification of bcl-2 family genes in the rat retina. Invest. Ophthalmol. Vis. Sci., 38: 2545–2553.

Levkovitch-Verbin, H., Harizman, N., Dardik, R., Nisgav, Y., Vander, S. and Melamed, S. (2007) Regulation of cell death and survival pathways in experimental glaucoma. Exp. Eye Res., 85: 250–258.

Li, Y., Schlamp, C.L. and Nickells, R.W. (1999) Experimental induction of retinal ganglion cell death in adult mice. Invest. Ophthalmol. Vis. Sci., 40: 1004–1008.

Li, Y., Schlamp, C.L., Poulsen, K.P. and Nickells, R.W. (2000) Bax-dependent and independent pathways of retinal ganglion cell death induced by different damaging stimuli. Exp. Eye Res., 71: 209–213.

Libby, R.T., Anderson, M.G., Pang, I.-H., Robinson, Z.H., Savinova, O.V., Cosma, I.M., Snow, A., Wilson, L.A., Smith, R.S., Clark, A.F. and John, S.W.M. (2005a) Inherited glaucoma in DBA/2J mice: pertinent disease features for studying the neurodegeneration. Vis. Neurosci., 22: 637–648.

Libby, R.T., Li, Y., Savinova, O.V., Barter, J., Smith, R.S., Nickells, R.W. and John, S.W.M. (2005b) Susceptibility to neurodegeneration in glaucoma is modified by Bax gene dosage. PLoS Genet., 1: 17–26.

Liu, X., Zhu, Y., Dai, S., White, J., Peyerl, F., Kappler, J.W. and Marrack, P. (2006) Bcl-XL does not have to bind to Bax to protect T cells from death. J. Exp. Med., 203: 2953–2961.

Liu, X.H., Collier, R.J. and Youle, R.J. (2001) Inhibition of axotomy-induced neuronal apoptosis by extracellular delivery of a Bcl-XL fusion protein. J. Biol. Chem., 276: 46326–46332.

Malik, J.M.I., Shevtsova, Z., Bahr, M. and Kugler, S. (2005) Long-term in vivo inhibition of CNS neurodegeneration by BclXL gene transfer. Mol. Ther., 11: 373–381.

McKernan, D.P. and Cotter, T.G. (2007) A critical role for Bim in retinal ganglion cell death. J. Neurochem., 102: 922–930.

McKinnon, S.J., Lenhman, D.M., Kerrigan-Baumrind, L.A., Merges, C.A., Pease, M.E., Kerrigan, D.F., Ransom, N.L., Tahzib, N.G., Reitsamer, H.A., Levkovitch-Verbin, H., Quigley, H.A. and Zack, D.J. (2002) Caspase activation and amyloid precursor protein cleavage in rat ocular hypertension. Invest. Ophthalmol. Vis. Sci., 43: 1077–1087.

Misantone, L.J., Gershenbaum, M. and Murray, M. (1984) Viability of retinal ganglion cells after optic nerve crush in rats. J. Neurocytol., 13: 449–465.

Mittag, T.W., Danias, J., Pohorenec, G., Yuan, H.M., Burakgazi, E., Chalmers-Redman, R., Podos, S.M. and Tatton, W.G. (2000) Retinal damage after 3 to 4 months of elevated intraocular pressure in a rat glaucoma model. Invest. Ophthalmol. Vis. Sci., 41: 3451–3459.

Mo, J.S., Anderson, M.G., Gregory, M., Smith, R.S., Savinova, O.V., Sereze, D.V., Ksander, B.R., Streilein, J.W. and John, S.W.M. (2003) By altering ocular immune privilege, bone marrow-derived cells pathogenetically contribute to DBA/2J pigmentary glaucoma. J. Exp. Med., 197: 1335–1344.

Morgan, J.E. (2002) Retinal ganglion cell shrinkage in glaucoma. J. Glaucoma, 11: 365–370.

Morgan, J.E., Uchida, H. and Caprioli, J. (2000) Retinal ganglion cell death in experimental glaucoma. Br. J. Ophthalmol., 84: 303–310.

Mosinger Ogilvie, J., Deckwerth, T.L., Knudson, C.M. and Korsmeyer, S.J. (1998) Suppression of developmental retinal cell death but not photoreceptor degeneration in Bax-deficient mice. Invest. Ophthalmol. Vis. Sci., 39: 1713–1720.

Nakazawa, T., Nakazawa, C., Matsubara, A., Noda, K., Hisatomi, T., She, H., Michaud, N., Hafezi-Moghadam, A., Miller, J.W. and Benowitz, L.I. (2006) Tumor necrosis factor-a mediates oligodendrocyte death and delayed retinal ganglion cell loss in a mouse model of glaucoma. J. Neurosci., 26: 12633–12641.

Nickells, R.W. (2004) The molecular biology of retinal ganglion cell death: caveats and controversies. Brain Res. Bull., 62: 439–446.

Nickells, R.W. (2007) From ocular hypertension to ganglion cell death: a theoretical sequence of events leading to glaucoma. Can. J. Ophthalmol., 42: 278–287.

Nutt, L.K., Chandra, J., Pataer, A., Fang, B., Roth, J.A., Swisher, S.G., O'Neil, R.G. and McConkey, D.J. (2002) Bax-mediated Ca^{2+} mobilizatoin promotes cytochrome C release during apoptosis. J. Biol. Chem., 277: 20301–20308.

Näpänkangas, U., Lindqvist, N., Lindholm, D. and Hallböök, F. (2003) Rat retinal ganglion cells upregulate the pro-apoptotic BH3-only protein Bim after optic nerve transection. Mol. Brain Res., 120: 30–37.

Polster, B.M. and Fiskum, G. (2004) Mitochondrial mechanisms of neuronal cell apoptosis. J. Neurochem., 90: 1281–1289.

Putcha, G.V., Deshmukh, M. and Johnson, E.M., Jr. (1999) BAX translocation is a critical event in neuronal apoptosis: regulation by neuroprotectants, BCL-2, and caspases. J. Neurosci., 19: 7476–7485.

Putcha, G.V., Le, S., Frank, S., Besirli, C.G., Clark, K., Chu, B., Alix, S., Youle, R.J., LaMarche, A., Maroney, A.C. and Johnson, E.M., Jr. (2003) JNK-mediated BIM phosphorylation potentiates BAX-dependent apoptosis. Neuron, 38: 899–914.

Quigley, H.A., Addicks, E.M., Green, W.R. and Maumenee, A.E. (1981) Optic nerve damage in human glaucoma: II. The site of injury and susceptibility to damage. Arch. Ophthalmol., 99: 635–649.

Quigley, H.A., Nickells, R.W., Kerrigan, L.A., Pease, M.E., Thibault, D.J. and Zack, D.J. (1995) Retinal ganglion cell death in experimental glaucoma and after axotomy occurs by apoptosis. Invest. Ophthalmol. Vis. Sci., 36: 774–786.

Salvesen, G.S. and Dixit, V.M. (1997) Caspases: intracellular signaling by proteolysis. Cell, 91: 443–446.

Schlamp, C.L., Li, Y., Dietz, J.A., Janssen, K.T. and Nickells, R.W. (2006) Progressive ganglion cell loss and optic nerve degeneration in DBA/2J mice is variable and asymmetric. BMC Neurosci., 7: p. 66.

Schwartz, M. (2005) Lessons for glaucoma from other neurodegenerative diseases: can one treatment suit them all?. J. Glaucoma, 14: 321–323.

Slee, E.A., Harte, M.T., Kluck, R.M., Wolf, B.B., Casiano, C.A., Newmeyer, D.D., Wang, H.-G., Reed, J.C., Nicholson, D.W., Alnemri, E.S., Green, D.R. and Martin, S.J. (1999) Ordering the cytochrome c-initiated caspase cascade: hierarchical activation of caspases-2, -3, -6, -7, -8, and -10 in a caspase-9-dependent manner. J. Cell Biol., 144: 281–292.

Tatton, W.G., Chalmers-Redman, R.M., Sud, A., Podos, S.M. and Mittag, T.W. (2001) Maintaining mitochondrial membrane impermeability: an opportunity for new therapy in glaucoma?. Surv. Ophthalmol., 45(suppl.): S295–S296.

Tezel, G. (2006) Oxidative stress in glaucomatous neurodegeneration: mechanisms and consequences. Prog. Retin. Eye Res., 25: 490–513.

Tezel, G. and Wax, M.B. (2000) Increased production of tumor necrosis factor-a by glial cells exposed to simulated ischemia or elevated hydrostatic pressure induces apoptosis in cocultured retinal ganglion cells. J. Neurosci., 20: 8693–8700.

Tezel, G., Li, L.Y., Patil, R.V. and Wax, M.B. (2001) TNF-alpha and TNF-alpha receptor-1 in the retina of normal and glaucomatous eyes. Invest. Ophthalmol. Vis. Sci., 42: 1787–1794.

Tezel, G., Yang, X., Yang, J. and Wax, M.B. (2004) Role of tumor necrosis factor receptor-1 in the death of retinal ganglion cells following optic nerve crush injury in mice. Brain Res., 996: 202–212.

Weber, A.J., Kaufman, P.L. and Hubbard, W.C. (1998) Morphology of single ganglion cells in the glaucomatous primate retina. Invest. Ophthalmol. Vis. Sci., 39: 2304–2320.

Wei, M.C., Zong, W.X., Cheng, E.H.Y., Lindsten, T., Panoutsakopoulou, V., Ross, A.J., Roth, K.A., MacGregor, G.R., Thompson, C.B. and Korsmeyer, S.J. (2001) Proapoptotic BAX and BAK: a requisite gateway to mitochondrial dysfunction and death. Science, 292: 727–730.

Weinreb, R.N. and Lindsey, J.D. (2005) The importance of models in glaucoma research. J. Glaucoma, 14: 302–304.

White, F.A., Keller-Peck, C.R., Knudson, C.M., Korsmeyer, S.J. and Snider, W.D. (1998) Widespread elimination of naturally occurring neuronal death in bax-deficient mice. J. Neurosci., 18: 1428–1439.

Whitmore, A.V., Libby, R.T. and John, S.W.M. (2005) Glaucoma: thinking in new ways — a role for autonomous axonal self-destruction and compartmentalised processes? Prog. Retin. Eye Res., 24: 639–662.

Wolter, K.G., Hsu, Y.T., Smith, C.L., Nechushtan, A., Xi, X.G. and Youle, R.J. (1997) Movement of bax from the cytosol to mitochondria during apoptosis. J. Cell Biol., 139: 1281–1292.

CHAPTER 30

Assessment of neuroprotection in the retina with DARC

Li Guo[1] and M. Francesca Cordeiro[1,2,*]

[1] Glaucoma & Retinal Degeneration Research Group, UCL Institute of Ophthalmology, London, UK
[2] The Glaucoma Research Group, Western Eye Hospital, London, UK

Abstract: Currently, assessment of new drug efficacy in glaucoma relies on conventional perimetry to monitor visual field changes. However, visual field defects cannot be detected until 20–40% of retinal ganglion cells (RGCs), the key cells implicated in the development of irreversible blindness in glaucoma, have been lost. We have recently developed a new, noninvasive real-time imaging technology, which is named DARC (detection of apoptosing retinal cells), to visualize single RGC undergoing apoptosis, the earliest sign of glaucoma. Utilizing fluorescently labeled annexin 5 and confocal laser scanning ophthalmoscopy, DARC enables evaluation of treatment effectiveness by monitoring RGC apoptosis in the same living eye over time. Using DARC, we have assessed different neuroprotective therapies in glaucoma-related animal models and demonstrated DARC to be a useful tool in screening neuroprotective strategies. DARC will potentially provide a meaningful clinical end point that is based on the direct assessment of the RGC death process, not only being useful in assessing treatment efficacy, but also leading to the early identification of patients with glaucoma. Clinical trials of DARC in glaucoma patients are due to start in 2008.

Keywords: DARC; RGC apoptosis; glaucoma; neuroprotection; glutamate modulation; targeting Aβ pathway; coenzyme Q10

Introduction

Glaucoma is the major cause of irreversible blindness worldwide and visual loss is attributed to retinal ganglion cell (RGC) death — a hallmark of glaucoma. Glaucoma is commonly linked with raised intraocular pressure (IOP), which has previously been implicated as a major cause of RGC death. Lowering IOP is currently the only clinical therapy available for glaucoma treatment, with an estimated cost of $5 billion annually in America alone by 2011 (Lee et al., 2006). However, pressure-lowering strategies have been shown to be inadequate in the prevention of progressive glaucomatous damage (Collaborative Normal-Tension Glaucoma Study Group, 1998; Oliver et al., 2002). This has provoked much research in non-IOP-lowering strategies, i.e., neuroprotective approaches in glaucoma management.

Currently, the most widely advocated neuroprotective agents in the prevention of RGC death are modifiers of the glutamate pathway, as this, somewhat controversially, is implicated in the development of glaucomatous RGC death (Dreyer et al., 1996; Lipton, 2004b; Guo et al., 2006). Memantine, an NMDA antagonist that has FDA

*Corresponding author. Tel./Fax: +44 207 608 6938;
E-mail: M.Cordeiro@ucl.ac.uk

approval for the treatment of Alzheimer's disease (AD), is currently the only neuroprotective drug in clinical trial (phase IV) in glaucoma patients (Hare et al., 2004a, b; Greenfield et al., 2005; Yucel et al., 2006). With the increasing association of glaucoma as a neurodegenerative disorder, there is a trend to use existing neuroprotective strategies that have been shown effective in central nerve system (CNS) diseases (Lipton, 2001). Indeed our recent study has revealed that the amyloid-β drugs used in the treatment of AD patients are also effective in prevention of RGC death in experimental glaucoma (Guo et al., 2007).

At present, monitoring new drug efficacy in glaucoma patients relies on the detection of visual field changes, which has accounted for a long period of follow-up (5 years) necessary for a clinical trial, such as the ongoing memantine clinical trial. Additionally, visual field testing is not a sensitive and accurate method to detect glaucomatous damage as it has been estimated that up to 20–40% of RGCs are lost before visual field defects are detected (Kerrigan-Baumrind et al., 2000). Moreover, given that RGC loss plays a key role in glaucoma with RGC apoptosis being recognized as early event (Quigley et al., 1995; Cordeiro et al., 2004), it would be a fundamental advance if RGC apoptosis could be monitored in evaluating therapeutic efficiency. There is currently no established clinical end point for the assessment of neuroprotective strategies in glaucoma.

We have recently devised an imaging technology called DARC (detection of apoptosing retinal cells) to detect RGC apoptosis in vivo. In this review, we will be focusing on assessment of neuroprotective strategies in the prevention of glaucomatous RGC apoptosis with DARC. DARC will be clinically trialed in 2008, with a view to assessing its role in glaucoma management and neuroprotective drug monitoring.

DARC

Introducing the DARC technique

We have recently established a novel, noninvasive real-time imaging technology to visualize individual RGC apoptosis in vivo in glaucoma-related experimental models. This method, which we have given the acronym DARC (detection of apoptosing retinal cells), employs flurorescently labeled annexin 5 and confocal laser scanning ophthalmoscopy. For the first time, we have been able to image changes occurring in RGC apoptosis over hours, days, and months in vivo (Cordeiro et al., 2004). This technique has recently been demonstrated to be a useful tool in screening neuroprotective strategies in glaucoma-related rat models (Guo et al., 2006, 2007).

Annexin 5-labeled apoptosis and ophthalmoloscopy

Apoptosis plays an important role in both physiology and pathology, particularly glaucoma, where vision loss is attributed to RGC apoptosis and death (Quigley et al., 1995; Kerrigan et al., 1997). Apoptosis is a regulated process of cell suicide, so called programmed cell death (PCD), characterized by cell shrinkage, DNA fragmentation, and chromatin condensation with a lack of inflammatory responses.

The idea of using annexin 5 to detect apoptosis was derived from a cellular phenomenon, first described by Fadok et al. in 1992. They observed that phosphatidylserine (PS), presenting in the inner leaflet of the plasma membrane, becomes exposed on the surface of apoptotic cells, and acts as an "eat me" signal for phagocytes (Fadok et al., 1992). It was further found that annexin 5, a human protein, is able to bind to PS selectively with a high affinity in the presence of Ca^{2+} (Koopman et al., 1994). As the externalization of PS occurs early in the apoptotic process, well before DNA fragmentation and nuclear condensation can be detected microscopically, annexin 5 binds to the exposed PS thereby identifying early stages of apoptosis.

Because of its properties, annexin 5 has been used in the detection of cells undergoing apoptosis using fluorescein isothiocyanate-labeled annexin 5 (FITC-annexin 5) (Koopman et al., 1994; Dumont et al., 2001). This annexin 5 affinity assay was further developed by labeling annexin 5 with biotin or with several radionuclides to facilitate various protocols for measuring apoptosis both in vitro and in vivo animal models (Blankenberg et al.,

1998; Vriens et al., 1998; Dumont et al., 2000; Post et al., 2002; Kenis et al., 2004; Murakami et al., 2004). The availability of 99Tc-labeled annexin 5 produced under good manufacturing practice (GMP) regulations led to the first studies of noninvasive detection of apoptosis in patients (Hofstra et al., 2001; Narula et al., 2001; Boersma et al., 2003; van de Wiele et al., 2003; Kietselaer et al., 2004). Annexin 5 imaging has been demonstrated to be clinically feasible and safe in patients. However, these techniques have never been used in the eye.

Using a related but nonradioactive approach, we have developed an imaging technology that uses fluorescently labeled annexin 5 with direct visualization of the retina to detect the dynamic process of individual cells undergoing apoptosis (Cordeiro et al., 2004). This is possible because the presence of clear optical media in the eye, compared to other organs in the body, allows direct visualization of labeled disease processes as they occur in the eye. The technique uses an imaging device such as a confocal scanning laser ophthalmoscope (cSLO) with an argon laser of 488 nm necessary to excite the administered annexin 5-bound fluorophore, and a photodetector system with a cutoff 521 nm filter to detect the fluorescent emitted light. We have to date used cSLOs with specialized imaging software to compensate for eye movements and improve the signal to noise ratio (von Ruckmann et al., 1995; Wade and Fitzke, 1998).

For imaging, animals under anesthesia are held in a stereotaxic frame, and their pupils dilated. Videos of scanned retinal areas are recorded and assessed for fluorescence using a method we have previously described (Fitzke, 2000). For each eye, a retinal montage is constructed from images captured at the same time point. It is then possible to count a total number of apoptosing RGCs for each time point in vivo (Maass et al., 2007) (Fig. 1).

Detection of RGC apoptosis in glaucoma-related animal models with DARC

To evaluate DARC's sensitivity and accuracy in monitoring RGC apoptosis, we have assessed its performance in vivo using several different glaucoma-related animal models and shown that

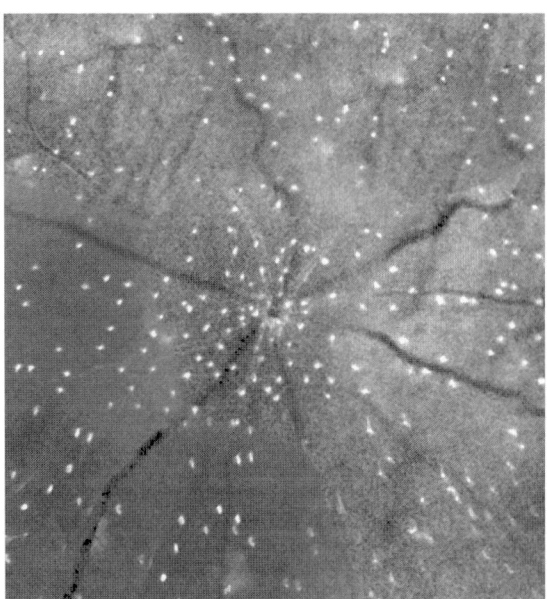

Fig. 1. DARC in vivo imaging shows retinal ganglion cell (RGC) apoptosis on a retina montage of an OHT rat at 3 weeks after IOP elevation. The white spots represent apoptotic RGCs.

DARC enables the real-time imaging of apoptotic changes occurring in RGCs longitudinally with histological validations.

In a well-established rat model of chronic ocular hypertension (OHT) (Cordeiro et al., 2004; Guo et al., 2005a, b, 2006, 2007), using DARC we revealed that RGC apoptosis occurs in vivo, accounting for 1, 15, 13, 7, and 2% of total RGCs, with RGC losses of 17, 22, 36, 45, and 60% of the original population at 2, 3, 4, 8, and 16 weeks, respectively, after surgical elevation of IOP. In an optic nerve transection rat model, we recorded RGC apoptosis in 0.3, 1, 8, and 4% of total RGCs, with RGC losses of 0, 3, 49, and 76% at 0, 3, 7, and 12 days, respectively. DARC imaging is also applicable to the mouse eye (Maass et al., 2007). Our results were obtained from a large cohort of animals — with minimal intra- and intervariability reinforcing the reproducibility and repeatability of the technique (Cordeiro et al., 2004; Guo et al., 2005a, b, 2006, 2007).

Additionally, we have developed a model of drug-induced RGC apoptosis using intravitreal staurosporine (SSP), a nonselective protein kinase inhibitor and a well-known potent inducer of neuronal apoptosis (Koh et al., 1995; Belmokhtar et al., 2001; Thuret et al., 2003). Assessment of the SSP model with DARC has shown that it generates rapid and extensive RGC apoptosis with peak levels visualized at 2 h in the rat, and is a useful tool in the assessment of neuroprotective strategies, with strong data attainable within a relatively short time (Cordeiro et al., 2004; Guo et al., 2006).

Assessment of glutamate modulation with DARC

Glutamate at synaptic endings

Glutamate is a predominant excitatory amino acid and important neurotransmitter in the CNS and the retina, involved in a variety of physiological processes and pathophysiological states. During synaptic activity, glutamate, triggered by nerve impulses is released from the presynaptic cell into the synaptic cleft and subsequently binds to glutamate receptors on the postsynaptic membrane to start intracellular signaling cascades. To allow for efficient signaling to occur, glutamate levels in the synaptic cleft have to be maintained at very low levels. This process is regulated by glutamate transporters, which rapidly remove excess glutamate from the extracellular space via a sodium–potassium coupled uptake mechanism.

High levels of extracellular glutamate stimulates glutamate receptors, allowing excessive Ca^{2+} influx into the cells triggering a multitude of intracellular events in the postsynaptic neuron, which ultimately results in neuronal cell death. This phenomenon is known as excitotoxicity, which has been implicated as an important mechanism of a number of neurodegenerative diseases, such as Alzheimer's (AD), Parkinson's (PD), and Huntington's (HD) diseases.

Glutamate excitotoxicity in glaucoma

There has been increasing evidence that glutamate excitotoxicity has been involved in RGC death in glaucoma although its exact role is controversial. Glutamate-related RGC death was first reported in the late 1950s when Lucas and Newhouse (1957) found that subcutaneous injection of glutamate selectively damaged the inner layer of the retina. A decade ago, Dreyer et al. (1996) reported an increase of glutamate concentration in the vitreous in glaucoma patients and experimental glaucoma monkeys and this finding was further supported by studies using dog (Brooks et al., 1997) and quail (Dkhissi et al., 1999). Although these results have been recently challenged by other researchers (Carter-Dawson et al., 2002; Levkovitch-Verbin et al., 2002; Honkanen et al., 2003; Wamsley et al., 2005), this does not exclude an excitotoxicity mechanism in glaucoma as there are the inherent difficulties in measuring in vivo glutamate levels (Salt and Cordeiro, 2006).

In addition to the possibility of increased release of glutamate from damaged RGCs in glaucoma, a reduced clearance of glutamate may also occur as a result of inefficient uptake by glutamate transporters present in the membranes of neurons and glial cells. In support of this, a significant reduction of the glutamate transporters GLAST (EAAT1) and GLT-1 (EAAT2) has been observed in experimental rat glaucoma retina (Martin et al., 2002). However, instead of a reduction, an increased expression of GLAST in the retinal Muller cells was reported using a similar model (Woldemussie et al., 2004), where the authors suggested a compensatory mechanism. Although other studies have failed to demonstrate a glaucoma-induced defect in retinal glutamate clearance mechanisms (Hartwick et al., 2005), a specific glutamate transporter GLT-1c, which is normally only expressed by photoreceptors, has been identified in RGCs in experimental glaucoma, and it may be indicative of an anomaly in glutamate homeostasis (Sullivan et al., 2006).

Perhaps the most effective exploration of the role of glutamate excitotoxicity in glaucoma is to examine the effect of blocking RGC glutamate receptors. This may be done via ion channel-associated (ionotropic) or G protein-coupled (metabotropic) receptors in the CNS and the retina. RGCs express multiple subtypes of these receptors (Hamassaki-Britto et al., 1993; Brandstatter et al., 1994; Hartveit et al., 1995; Fletcher et al., 2000). The ionotropic (iGlu) receptors include N-methyl-D-aspartate (NMDA), α-amino-3-hydroxy-5-methyl-4-isoxazolepropionate (AMPA), and KA (kainite) subtypes; and among them, NMDA receptors, which are the most permeable to Ca^{2+}, have been extensively studied.

Postsynaptic NMDA receptors are heteromeric ion channel complexes that consist of two NR1 and two NR2 subunits that can be either of the NR2A, -2B, -2C, or -2D type. The NR1 contains a specific glutamate-binding site, whereas the NR2 binds to glycine, which acts as a co-agonist for receptor activation. When both glutamate and glycine bind onto the receptors, the cells are depolarized, resulting in the release of Mg^{2+}, which blocks the channel. The opening of the channel is transient under normal physiological condition, allowing appropriate amounts of extracellular Ca^{2+} and Na^+ to flood into the cell to initiate intracellular signaling cascades. However, overexpression of extracellular glutamate causes the channel to stay open longer than needed; as a result, excessive Ca^{2+} flows into the cell triggering cell death.

Blocking NMDA receptors by specific antagonists has been shown to be effective in reducing RGC death in experimental glaucoma. Memantine, an uncompetitive NMDA antagonist, is perhaps the best-known glutamate modifier and has recently been shown to be clinically useful in the treatment of moderate to severe AD (Lleo et al., 2006). Memantine has been demonstrated to be a highly effective neuroprotective agent in various animal models of RGC death (Lagreze et al., 1998; Osborne, 1999; Kim et al., 2002; WoldeMussie et al., 2002). Application of memantine significantly increased RGC survival in experimental rat glaucoma (WoldeMussie et al., 2002), and a similar result was reported using the DBA/2J transgenic mouse model of glaucoma (Schuettauf et al., 2002). Additionally, memantine has shown to be preventive of not only structural damage but also functional loss in experimental monkey glaucoma (Hare et al., 2004a, b). Memantine is currently in a phase IV clinical trial of glaucoma, the results of which are eagerly awaited.

Assessment of glutamate modulation in glaucoma-related models with DARC

Using DARC, we have assessed the effects of different glutamate modulation strategies, including a nonselective (MK801) and a selective (ifenprodil) NMDA receptor antagonist and a metabotropic glutamate receptor agonist (mGluR Group II, LY354740), in glaucoma-related rat models in vivo. We have shown that DARC is a useful tool in screening neuroprotective therapies and monitoring RGC apoptosis in experimental glaucoma (Guo et al., 2006).

Like memantine, MK801 is a broadspectrum NMDA antagonist and has been reported to protect retina neurons from NMDA- and pressure-induced toxicity (el-Asrar et al., 1992; Chaudhary et al., 1998). Ifenprodil, an NR2B subunit-selective NMDA antagonist, is specific against SSP-induced cell death (Williams et al., 2002). LY354740 is a highly potent and selective group II metabotropic receptor (mGluR) agonist, which has been documented to be neuroprotective against NMDA- or SSP-induced neuronal death in rat CNS (Monn et al., 1997; Bond et al., 1998; Kingston et al., 1999). Metabotropic receptors have been classified into three groups. There is evidence that activation of group I mGluRs increases neuronal excitation, whereas that of group II and III mGluRs reduces synaptic transmission; therefore, group II and III mGluR agonists and group I antagonists can be thought to be neuroprotective (Nicoletti et al., 1996). All mGluRs have been found to be expressed in the retina (Hartveit et al., 1995; Shen and Slaughter, 1998; Tehrani et al., 2000; Robbins et al., 2003).

In this study (Guo et al., 2006), we first assessed the effects of each single agent on RGC apoptosis in our SSP model. Using DARC, we showed that all treatments reduced RGC apoptosis in a dose-dependent manner, and that MK801 was more effective than ifenprodil. However, MK801 has been reported to be neurotoxic due to its high affinity for the NMDA receptor channel, causing its accumulation in the channels and blocking critical normal functions (Lipton, 1993, 2004a). To minimize its toxicity but still take advantage of the neuroprotective properties, we also looked at the effect of combining low-dose MK801 with LY354740 and found this combination to be most effective in preventing RGC apoptosis compared to either agent alone, which may be attributed to their different pharmacological properties in modulating glutamate excitatory transmission.

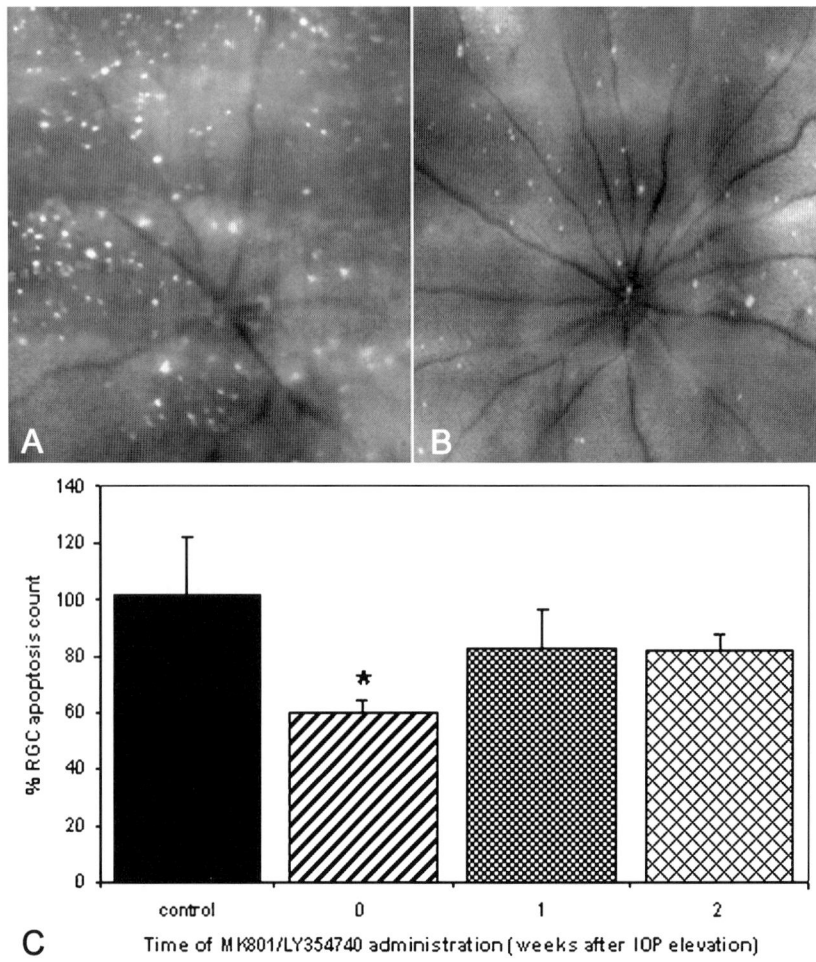

Fig. 2. DARC reveals combination effects of glutamate modulators on reduction of RGC apoptosis in an OHT model. Pressure-induced RGC apoptosis at 3 weeks (A) was significantly reduced by combination treatment of MK801 and LY354740 (B). Timing administration showed that the most effective treatment application was at 0 weeks (the time of IOP elevation, C).

We then applied the most optimal combination regimens of MK801 and LY354740 to an OHT model at different time points (0, 1, and 2 weeks after IOP elevation), and our DARC results revealed that the most effective timing of the treatment application was at 0 weeks (the time of IOP elevation) ($p<0.05$, Fig. 2). We believe this is due to the application of glutamate modulators at the time of the primary insult (IOP elevation) leading to maximal inhibition of glutamate release from primary injured RGCs, resulting in the prevention of secondary degeneration — a process in which RGCs that survive the initial injury are subsequently damaged by the toxic effects of the primary degenerating neurons (Levkovitch-Verbin et al., 2001, 2003; Kaushik et al., 2003). Our DARC data strongly supports the involvement of glutamate in glaucoma (Guo et al., 2006; Salt and Cordeiro, 2006).

Assessment of targeting the amyloid-β pathway with DARC

Amyloid-β neurotoxicity in AD

The Alzheimer protein amyloid-β (Aβ) is the major constituent of amyloid plaques in AD. Aβ is an amino acid peptide derived from the proteolytic cleavage of amyloid precursor protein (APP), a transmembrane protein. There are two catabolic pathways being identified for APP processing: the nonamyloidogenic pathway relying on α-secretase activity and resulting in secretion of soluble forms of APP, functioning as a cell surface signaling molecule to modulate neurite outgrowth, synaptogenesis and cell survival (Li et al., 1997; Pastorino and Lu, 2006), and the amyloidogenic pathway where β- and γ-secretase activities lead to Aβ generation. Aggregation of Aβ in the brain is believed to be the primary driver of neurodegeneration and cognitive decline leading to dementia (Auld et al., 2002), and Aβ neurotoxicity has been involved in AD's neuropathology, although the proximate cause of the neurodegeneraton responsible for cognitive impairment is not clear (Hardy and Selkoe, 2002; Pepys, 2006). Blocking the APP amyloidogenic pathway has been shown to be a promising approach to prevent neuronal damage in AD.

Amyloid-β implication in glaucoma

Amyloid-β has recently been reported to be implicated in the development of RGC apoptosis in glaucoma, showing caspase-3-mediated abnormal processing of APP with increased expression of Aβ in RGCs in the OHT rat (McKinnon et al., 2002) and DBA/2J transgenic mice (Goldblum et al., 2007). In addition, decreased vitreous levels of Aβ have been found in patients with glaucoma compared to controls, suggesting the deposition of Aβ in the retinal (Yoneda et al., 2005). Further evidence of a link between glaucoma and AD has emerged from studies showing that patients with AD have RGC loss and reduced thickness of the retinal nerve fiber layer (RNFL) associated with typical glaucomatous changes, such as optic neuropathy and visual functional impairment (Blanks et al., 1996a, b; Parisi et al., 2001; Danesh-Meyer et al., 2006; Iseri et al., 2006; Tamura et al., 2006; Berisha et al., 2007), as is also the case in PD (Bayer et al., 2002). This indicates similar pathological mechanisms involving Aβ leading to RGC loss as implicated in the brain (Loffler et al., 1995; Vickers et al., 1995; Archer et al., 1998; Johnson et al., 2002).

Using DARC, we have recently provided further evidence from experimental glaucoma supporting the involvement of Aβ in development of glaucomatous RGC apoptosis (Guo et al., 2007). We observed a significant increase of Aβ deposition in the RGC layer in OHT model compared to control, with Aβ colocalized to apoptotic RGCs. DARC imaging data further showed that application of exogenous Aβ peptide induced significant RGC apoptosis in vivo in a dose- and time-dependent manner.

Assessment of targeting Aβ pathway in experimental glaucoma with DARC

The finding of increased Aβ deposition in experimental glaucoma and its induction of RGC apoptosis in vivo suggests that Aβ could be a factor mediating the apoptotic changes in RGC cells in our glaucoma model. To examine our hypothesis that targeting Aβ formation and aggregation might reduce RGC apoptosis, we examined three different agents, including a β-secretase inhibitor (βSI), an anti-Aβ antibody (Aβab), and Congo red (CR) (Guo et al., 2007) (Fig. 3).

β-Secretase, a membrane-anchored aspartic protease, is responsible for the initial step of APP cleavage in the amyloidogenic pathway leading to the generation of Aβ; βSIs therefore inhibit the generation of Aβ, with evidence of in vitro and in vivo efficacy in AD-related models (Citron, 2002; Yamamoto et al., 2004). CR, a dye commonly used to stain amyloid-β histologically, has been shown to completely block Aβ aggregation and toxicity in rat hippocampal neuron cultures (Lorenzo and Yankner, 1994), and it is believed that its inhibitory effects are the result of an interference with Aβ protein misfolding, fibril formation (Lorenzo and Yankner, 1994; Hirakura et al., 1999). Aβabs are thought to work by not only blocking Aβ aggregation (Bard et al., 2000) but also increasing Aβ clearance in AD-related animal models (Vasilevko and Cribbs, 2006) (Fig. 3).

Our DARC data showed that all three treatments altered the profile of RGC apoptosis in a

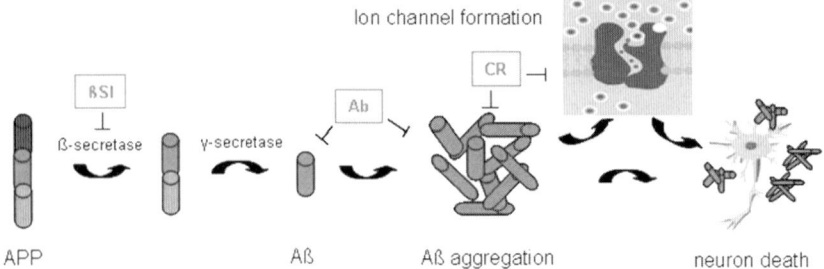

Fig. 3. The amyloid-beta (Aβ) pathway and its targeting. Abnormal processing of APP causes the generation of Aβ, leading to neuronal death. Using DARC, we assessed the effects of targeting Aβ formation and aggregation on prevention of glaucomatous RGC apoptosis by a β-secretase inhibitor (βSI), an anti-Aβ antibody (Aβab), and Congo red (CR).

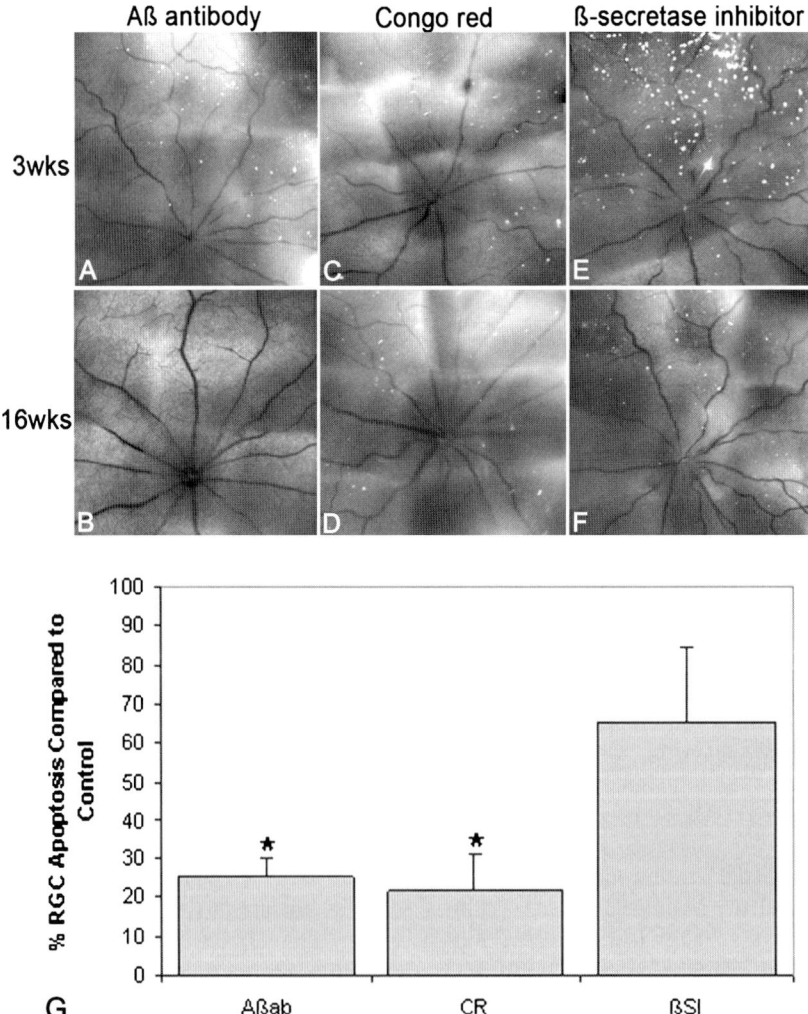

Fig. 4. DARC in vivo images show the effects of single agents including an Aβ antibody (Aβab, A and B), Congo red (CR, C and D), and a β-secretase inhibitor (βSI, E and F) on RGC apoptosis at 3 (A, C, and E) and 16 (B, D, and F) weeks after IOP elevation. Aβab and CR caused significant percentage reduction of RGC apoptosis at 3 weeks compared to control (G).

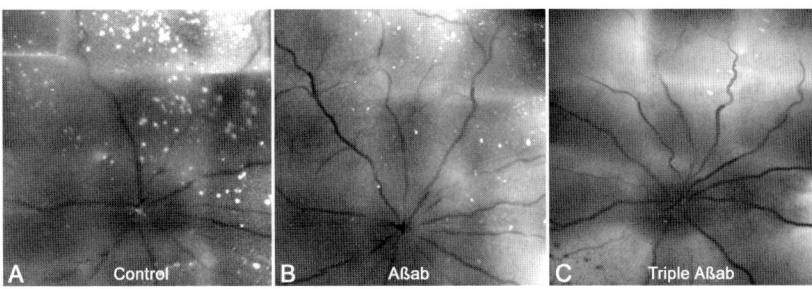

Fig. 5. Effects of combination Aβ-targeting therapy on RGC apoptosis. Compared to nontreatment control (A), DARC imaging shows triple therapy (C, Aβab + CR + βSI) was more effective than Aβab alone (B) in reduction of RGC apoptosis in an OHT model.

temporal manner by delaying the development of peak RGC apoptosis as well as reducing the peak level of RGC apoptosis. Among them, the anti-Aβ ab appeared the most effective in the prevention of RGC apoptosis compared with CR and the βSI. The single application of the Aβab showed a prolonged effect up to 16 weeks following IOP elevation. In comparison, CR appeared to have a shorter window of RGC protection and βSI showed no significant effect on glaucomatous RGC apoptosis (Fig. 4).

Perhaps the most significant finding of the work with DARC has been combination therapy, targeting three different aspects of the Aβ pathway. In combination, the neuroprotective effect of all three agents (triple therapy) was significantly improved at 3 weeks after IOP elevation compared with Aβab alone (Fig. 5). The combination therapy produced the maximal reduction of RGC apoptosis (>80%).

Using DARC, we have highlighted that Aβ is a likely mediator of pressure-induced RGC death and that targeting multiple stages in the Aβ pathway may provide a potential neuroprotective strategy in glaucoma management.

Assessment of coenzyme Q10 in glaucoma-related models with DARC

Coenzyme Q10 (CoQ10) is an important insoluble component of the mitochondrial respiratory chain where it transports electrons from complexes I and II to complex III, by which ATP is produced. CoQ10 has been reported to afford neuroprotection, preventing neuronal death (Papucci et al., 2003). CoQ10 acts as a potent antioxidant by maintaining the mitochondrial membrane potential and inhibiting ROS generation when neuronal cells are subject to oxidative stress (McCarthy et al., 2004; Somayajulu et al., 2005). CoQ10 has been shown to prevent RGC apoptosis in a pressure-induced transient ischemia model by minimizing synaptic glutamate increase and inhibiting the MPTP formation (Nucci et al., 2007). CoQ10 has been safely used in patients with neurodegenerative disorders (Levy et al., 2006; Liu, 2007).

We have investigated the effects of topical CoQ10 on reducing RGC apoptosis in vivo using our SSP rat model (Cordeiro et al., 2007). Our DARC imaging results showed that CoQ10 0.1% significantly reduced SSP-induced RGC apoptosis compared to CoQ10 0.05% and carrier. The most effective timing administration was at 1 h after SSP application. The effects of CoQ10 on preventing RGC apoptosis may be attributed to its property of inhibiting mitochondrial depolarization, cytochrome *c* release, and caspase-9 activation (Papucci et al., 2003).

Summary

Currently, there is no good and quick method for assessing neuroprotection in glaucoma. The only neuroprotective drug that has undergone large scale clinical trial in glaucoma, memantine, has relied on visual field status as a defined end point, which has led to an expensive 6-year period of follow-up — with still no published outcome. We

(2004b) Efficacy and safety of memantine treatment for reduction of changes associated with experimental glaucoma in monkey, II: Structural measures. Invest. Ophthalmol. Vis. Sci., 45(8): 2640–2651.

Hartveit, E., Brandstatter, J.H., Enz, R. and Wassle, H. (1995) Expression of the mRNA of seven metabotropic glutamate receptors (mGluR1 to 7) in the rat retina. An in situ hybridization study on tissue sections and isolated cells. Eur. J. Neurosci., 7(7): 1472–1483.

Hartwick, A.T., Zhang, X., Chauhan, B.C. and Baldridge, W.H. (2005) Functional assessment of glutamate clearance mechanisms in a chronic rat glaucoma model using retinal ganglion cell calcium imaging. J. Neurochem., 94(3): 794–807.

Hirakura, Y., Lin, M.C. and Kagan, B.L. (1999) Alzheimer amyloid abeta1-42 channels: effects of solvent, pH, and Congo Red. J. Neurosci. Res., 57(4): 458–466.

Hofstra, L., Dumont, E.A., Thimister, P.W., Heidendal, G.A., DeBruine, A.P., Elenbaas, T.W., Boersma, H.H., van Heerde, W.L. and Reutelingsperger, C.P. (2001) In vivo detection of apoptosis in an intracardiac tumor. JAMA, 285(14): 1841–1842.

Honkanen, R.A., Baruah, S., Zimmerman, M.B., Khanna, C.L., Weaver, Y.K., Narkiewicz, J., Waziri, R., Gehrs, K.M., Weingeist, T.A., Boldt, H.C., Folk, J.C., Russell, S.R. and Kwon, Y.H. (2003) Vitreous amino acid concentrations in patients with glaucoma undergoing vitrectomy. Arch. Ophthalmol., 121(2): 183–188.

Iseri, P.K., Altinas, O., Tokay, T. and Yuksel, N. (2006) Relationship between cognitive impairment and retinal morphological and visual functional abnormalities in Alzheimer disease. J. Neuroophthalmol., 26(1): 18–24.

Johnson, L.V., Leitner, W.P., Rivest, A.J., Staples, M.K., Radeke, M.J. and Anderson, D.H. (2002) The Alzheimer's A beta-peptide is deposited at sites of complement activation in pathologic deposits associated with aging and age-related macular degeneration. Proc. Natl. Acad. Sci. U.S.A., 99(18): 11830–11835.

Kaushik, S., Pandav, S.S. and Ram, J. (2003) Neuroprotection in glaucoma. J. Postgrad. Med., 49(1): 90–95.

Kenis, H., van Genderen, H., Bennaghmouch, A., Rinia, H.A., Frederik, P., Narula, J., Hofstra, L. and Reutelingsperger, C.P. (2004) Cell surface-expressed phosphatidylserine and annexin A5 open a novel portal of cell entry. J. Biol. Chem., 279(50): 52623–52629.

Kerrigan, L., Zack, D., Quigley, H., Smith, S.D. and Pease, M. (1997) TUNEL-positive ganglion cells in human primary open-angle glaucoma. Arch. Ophthalmol., 115: 1031–1035.

Kerrigan-Baumrind, L., Quigley, H., Pease, M., Kerrigan, D. and Mitchell, R. (2000) Number of ganglion cells in glaucoma eyes compared with threshold visual field tests in the same persons. Invest. Ophthalmol. Vis. Sci., 41: 741–748.

Kietselaer, B.L., Reutelingsperger, C.P., Heidendal, G.A., Daemen, M.J., Mess, W.H., Hofstra, L. and Narula, J. (2004) Noninvasive detection of plaque instability with use of radiolabeled annexin A5 in patients with carotid-artery atherosclerosis. N. Engl. J. Med., 350(14): 1472–1473.

Kim, T.W., Kim, D.M., Park, K.H. and Kim, H. (2002) Neuroprotective effect of memantine in a rabbit model of optic nerve ischemia. Korean J. Ophthalmol., 16(1): 1–7.

Kingston, A.E., O'Neill, M.J., Lam, A., Bales, K.R., Monn, J.A. and Schoepp, D.D. (1999) Neuroprotection by metabotropic glutamate receptor glutamate receptor agonists: LY354740, LY379268 and LY389795. Eur. J. Pharmacol., 377(2–3): 155–165.

Koh, J.Y., Wie, M.B., Gwag, B.J., Sensi, S.L., Canzoniero, L.M., Demaro, J., Csernansky, C. and Choi, D.W. (1995) Staurosporine-induced neuronal apoptosis. Exp. Neurol., 135(2): 153–159.

Koopman, G., Reutelingsperger, C.P., Kuijten, G.A., Keehnen, R.M., Pals, S.T. and van Oers, M.H. (1994) Annexin V for flow cytometric detection of phosphatidylserine expression on B cells undergoing apoptosis. Blood, 84(5): 1415–1420.

Lagreze, W.A., Knorle, R., Bach, M. and Feuerstein, T.J. (1998) Memantine is neuroprotective in a rat model of pressure-induced retinal ischemia. Invest. Ophthalmol. Vis. Sci., 39(6): 1063–1066.

Lee, P.P., Walt, J.G., Doyle, J.J., Kotak, S.V., Evans, S.J., Budenz, D.L., Chen, P.P., Coleman, A.L., Feldman, R.M., Jampel, H.D., Katz, L.J., Mills, R.P., Myers, J.S., Noecker, R.J., Piltz-Seymour, J.R., Ritch, R.R., Schacknow, P.N., Serle, J.B. and Trick, G.L. (2006) A multicenter, retrospective pilot study of resource use and costs associated with severity of disease in glaucoma. Arch. Ophthalmol., 124(1): 12–19.

Levkovitch-Verbin, H., Quigley, H.A., Kerrigan-Baumrind, L.A., D'Anna, S.A., Kerrigan, D. and Pease, M.E. (2001) Optic nerve transection in monkeys may result in secondary degeneration of retinal ganglion cells. Invest. Ophthalmol. Vis. Sci., 42(5): 975–982.

Levkovitch-Verbin, H., Martin, K.R., Quigley, H.A., Baumrind, L.A., Pease, M.E. and Valenta, D. (2002) Measurement of amino acid levels in the vitreous humor of rats after chronic intraocular pressure elevation or optic nerve transection. J. Glaucoma, 11(5): 396–405.

Levkovitch-Verbin, H., Quigley, H.A., Martin, K.R., Zack, D.J., Pease, M.E. and Valenta, D.F. (2003) A model to study differences between primary and secondary degeneration of retinal ganglion cells in rats by partial optic nerve transection. Invest. Ophthalmol. Vis. Sci., 44(8): 3388–3393.

Levy, G., Kaufmann, P., Buchsbaum, R., Montes, J., Barsdorf, A., Arbing, R., Battista, V., Zhou, X., Mitsumoto, H., Levin, B. and Thompson, J.L. (2006) A two-stage design for a phase II clinical trial of coenzyme Q10 in ALS. Neurology, 66(5): 660–663.

Li, H.L., Roch, J.M., Sundsmo, M., Otero, D., Sisodia, S., Thomas, R. and Saitoh, T. (1997) Defective neurite extension is caused by a mutation in amyloid beta/A4 (A beta) protein precursor found in familial Alzheimer's disease. J. Neurobiol., 32(5): 469–480.

Lipton, S.A. (1993) Prospects for clinically tolerated NMDA antagonists: open-channel blockers and alternative redox states of nitric oxide. Trends Neurosci., 16(12): 527–532.

Lipton, S.A. (2001) Retinal ganglion cells, glaucoma and neuroprotection. Prog. Brain Res., 131: 712–718.

Lipton, S.A. (2004a) Failures and successes of NMDA receptor antagonists: molecular basis for the use of open-channel blockers like memantine in the treatment of acute and chronic neurologic insults. Neurorx, 1(1): 101–110.

Lipton, S.A. (2004b) Paradigm shift in NMDA receptor antagonist drug development: molecular mechanism of uncompetitive inhibition by memantine in the treatment of Alzheimer's disease and other neurologic disorders. J. Alzheimers Dis., 6(6 Suppl): S61–S74.

Liu, J. (2007) The effects and mechanisms of mitochondrial nutrient alpha-lipoic acid on improving age-associated mitochondrial and cognitive dysfunction: an overview. Neurochem. Res.

Lleo, A., Greenberg, S.M. and Growdon, J.H. (2006) Current pharmacotherapy for Alzheimer's disease. Annu. Rev. Med., 57: 513–533.

Loffler, K.U., Edward, D.P. and Tso, M.O. (1995) Immunoreactivity against tau, amyloid precursor protein, and beta-amyloid in the human retina. Invest. Ophthalmol. Vis. Sci., 36(1): 24–31.

Lorenzo, A. and Yankner, B.A. (1994) Beta-amyloid neurotoxicity requires fibril formation and is inhibited by congo red. Proc. Natl. Acad. Sci. U.S.A., 91(25): 12243–12247.

Lucas, D.R. and Newhouse, J.P. (1957) The toxic effect of sodium L-glutamate on the inner layers of the retina. AMA Arch. Ophthalmol., 58(2): 193–201.

Maass, A., von Leithner, P.L., Luong, V., Guo, L., Salt, T.E., Fitzke, F.W. and Cordeiro, M.F. (2007) Assessment of rat and mouse RGC apoptosis imaging in vivo with different scanning laser ophthalmoscopes. Curr. Eye Res., 32(10): 851–861.

Martin, K.R., Levkovitch-Verbin, H., Valenta, D., Baumrind, L., Pease, M.E. and Quigley, H.A. (2002) Retinal glutamate transporter changes in experimental glaucoma and after optic nerve transection in the rat. Invest. Ophthalmol. Vis. Sci., 43(7): 2236–2243.

McCarthy, S., Somayajulu, M., Sikorska, M., Borowy-Borowski, H. and Pandey, S. (2004) Paraquat induces oxidative stress and neuronal cell death; neuroprotection by water-soluble coenzyme Q10. Toxicol. Appl. Pharmacol., 201(1): 21–31.

McKinnon, S.J., Lehman, D.M., Kerrigan-Baumrind, L.A., Merges, C.A., Pease, M.E., Kerrigan, D.F., Ransom, N.L., Tahzib, N.G., Reitsamer, H.A., Levkovitch-Verbin, H., Quigley, H.A. and Zack, D.J. (2002) Caspase activation and amyloid precursor protein cleavage in rat ocular hypertension. Invest. Ophthalmol. Vis. Sci., 43(4): 1077–1087.

Monn, J.A., Valli, M.J., Massey, S.M., Wright, R.A., Salhoff, C.R., Johnson, B.G., Howe, T., Alt, C.A., Rhodes, G.A., Robey, R.L., Griffey, K.R., Tizzano, J.P., Kallman, M.J., Helton, D.R. and Schoepp, D.D. (1997) Design, synthesis, and pharmacological characterization of (+)-2-aminobicyclo[3.1.0]hexane-2,6-dicarboxylic acid (LY354740): a potent, selective, and orally active group 2 metabotropic glutamate receptor agonist possessing anticonvulsant and anxiolytic properties. J. Med. Chem., 40(4): 528–537.

Murakami, Y., Takamatsu, H., Taki, J., Tatsumi, M., Noda, A., Ichise, R., Tait, J.F. and Nishimura, S. (2004) 18F-Labelled annexin V: a PET tracer for apoptosis imaging. Eur. J. Nucl. Med. Mol. Imaging, 31(4): 469–474.

Narula, J., Acio, E.R., Narula, N., Samuels, L.E., Fyfe, B., Wood, D., Fitzpatrick, J.M., Raghunath, P.N., Tomaszewski, J.E., Kelly, C., Steinmetz, N., Green, A., Tait, J.F., Leppo, J., Blankenberg, F.G., Jain, D. and Strauss, H.W. (2001) Annexin-V imaging for noninvasive detection of cardiac allograft rejection. Nat. Med., 7(12): 1347–1352.

Nicoletti, F., Bruno, V., Copani, A., Casabona, G. and Knopfel, T. (1996) Metabotropic glutamate receptors: a new target for the therapy of neurodegenerative disorders? Trends Neurosci., 19(7): 267–271.

Nucci, C., Tartaglione, R., Cerulli, A., Mancino, R., Spano, A., Cavaliere, F., Rombola, L., Bagetta, G., Corasaniti, M.T. and Morrone, L.A. (2007) Retinal damage caused by high intraocular pressure-induced transient ischemia is prevented by coenzyme Q10 in rat. Int. Rev. Neurobiol., 82: 397–406.

Oliver, J.E., Hattenhauer, M.G., Herman, D., Hodge, D.O., Kennedy, R., Fang-Yen, M. and Johnson, D.H. (2002) Blindness and glaucoma: a comparison of patients progressing to blindness from glaucoma with patients maintaining vision. Am. J. Ophthalmol., 133(6): 764–772.

Osborne, N.N. (1999) Memantine reduces alterations to the mammalian retina, in situ, induced by ischemia. Vis. Neurosci., 16(1): 45–52.

Papucci, L., Schiavone, N., Witort, E., Donnini, M., Lapucci, A., Tempestini, A., Formigli, L., Zecchi-Orlandini, S., Orlandini, G., Carella, G., Brancato, R. and Capaccioli, S. (2003) Coenzyme q10 prevents apoptosis by inhibiting mitochondrial depolarization independently of its free radical scavenging property. J. Biol. Chem., 278(30): 28220–28228.

Parisi, V., Restuccia, R., Fattapposta, F., Mina, C., Bucci, M.G. and Pierelli, F. (2001) Morphological and functional retinal impairment in Alzheimer's disease patients. Clin. Neurophysiol., 112(10): 1860–1867.

Pastorino, L. and Lu, K.P. (2006) Pathogenic mechanisms in Alzheimer's disease. Eur. J. Pharmacol., 545(1): 29–38.

Pepys, M.B. (2006) Amyloidosis. Annu. Rev. Med., 57: 223–241.

Post, A.M., Katsikis, P.D., Tait, J.F., Geaghan, S.M., Strauss, H.W. and Blankenberg, F.G. (2002) Imaging cell death with radiolabeled annexin V in an experimental model of rheumatoid arthritis. J. Nucl. Med., 43(10): 1359–1365.

Quigley, H.A., Nickells, R.W., Kerrigan, L.A., Pease, M.E., Thibault, D.J. and Zack, D.J. (1995) Retinal ganglion cell death in experimental glaucoma and after axotomy occurs by apoptosis. Invest. Ophthalmol. Vis. Sci., 36(5): 774–786.

Robbins, J., Reynolds, A.M., Treseder, S. and Davies, R. (2003) Enhancement of low-voltage-activated calcium currents by group II metabotropic glutamate receptors in rat retinal ganglion cells. Mol. Cell. Neurosci., 23(3): 341–350.

Salt, T.E. and Cordeiro, M.F. (2006) Glutamate excitotoxicity in glaucoma: throwing the baby out with the bathwater? Eye, 20(6): 730–731, author reply 731–732.

Schuettauf, F., Quinto, K., Naskar, R. and Zurakowski, D. (2002) Effects of anti-glaucoma medications on ganglion cell survival: the DBA/2J mouse model. Vision Res., 42(20): 2333–2337.

Shen, W. and Slaughter, M.M. (1998) Metabotropic and ionotropic glutamate receptors regulate calcium channel currents in salamander retinal ganglion cells. J. Physiol., 510(Pt 3): 815–828.

Somayajulu, M., McCarthy, S., Hung, M., Sikorska, M., Borowy-Borowski, H. and Pandey, S. (2005) Role of mitochondria in neuronal cell death induced by oxidative stress; neuroprotection by coenzyme Q10. Neurobiol. Dis., 18(3): 618–627.

Sullivan, R.K., Woldemussie, E., Macnab, L., Ruiz, G. and Pow, D.V. (2006) Evoked expression of the glutamate transporter GLT-1c in retinal ganglion cells in human glaucoma and in a rat model. Invest. Ophthalmol. Vis. Sci., 47(9): 3853–3859.

Tamura, H., Kawakami, H., Kanamoto, T., Kato, T., Yokoyama, T., Sasaki, K., Izumi, Y., Matsumoto, M. and Mishima, H.K. (2006) High frequency of open-angle glaucoma in Japanese patients with Alzheimer's disease. J. Neurol. Sci., 246(1–2): 79–83.

Tehrani, A., Wheeler-Schilling, T.H. and Guenther, E. (2000) Coexpression patterns of mGLuR mRNAs in rat retinal ganglion cells: a single-cell RT-PCR study. Invest. Ophthalmol. Vis. Sci., 41(1): 314–319.

Thuret, G., Chiquet, C., Herrag, S., Dumollard, J.M., Boudard, D., Bednarz, J., Campos, L. and Gain, P. (2003) Mechanisms of staurosporine induced apoptosis in a human corneal endothelial cell line. Br. J. Ophthalmol., 87(3): 346–352.

van de Wiele, C., Lahorte, C., Vermeersch, H., Loose, D., Mervillie, K., Steinmetz, N.D., Vanderheyden, J.L., Cuvelier, C.A., Slegers, G. and Dierck, R.A. (2003) Quantitative tumor apoptosis imaging using technetium-99m-HYNIC annexin V single photon emission computed tomography. J. Clin. Oncol., 21(18): 3483–3487.

Vasilevko, V. and Cribbs, D.H. (2006) Novel approaches for immunotherapeutic intervention in Alzheimer's disease. Neurochem. Int., 49(2): 113–126.

Vickers, J.C., Lazzarini, R.A., Riederer, B.M. and Morrison, J.H. (1995) Intraperikaryal neurofilamentous accumulations in a subset of retinal ganglion cells in aged mice that express a human neurofilament gene. Exp. Neurol., 136(2): 266–269.

von Ruckmann, A., Fitzke, F.W. and Bird, A.C. (1995) Distribution of fundus autofluorescence with a scanning laser ophthalmoscope. Br. J. Ophthalmol., 79(5): 407–412.

Vriens, P.W., Blankenberg, F.G., Stoot, J.H., Ohtsuki, K., Berry, G.J., Tait, J.F., Strauss, H.W. and Robbins, R.C. (1998) The use of technetium Tc 99m annexin V for in vivo imaging of apoptosis during cardiac allograft rejection. J. Thorac. Cardiovasc. Surg., 116(5): 844–853.

Wade, A.R. and Fitzke, F.W. (1998) A fast, robust pattern recognition system for low light level image registration and its application to retinal imaging. Opt. Express, 3(5): 190–197.

Wamsley, S., Gabelt, B.T., Dahl, D.B., Case, G.L., Sherwood, R.W., May, C.A., Hernandez, M.R. and Kaufman, P.L. (2005) Vitreous glutamate concentration and axon loss in monkeys with experimental glaucoma. Arch. Ophthalmol., 123(1): 64–70.

Williams, A.J., Dave, J.R., Lu, X.M., Ling, G. and Tortella, F.C. (2002) Selective NR2B NMDA receptor antagonists are protective against staurosporine-induced apoptosis. Eur. J. Pharmacol., 452(1): 135–136.

WoldeMussie, E., Yoles, E., Schwartz, M., Ruiz, G. and Wheeler, L.A. (2002) Neuroprotective effect of memantine in different retinal injury models in rats. J. Glaucoma, 11(6): 474–480.

Woldemussie, E., Wijono, M. and Ruiz, G. (2004) Muller cell response to laser-induced increase in intraocular pressure in rats. Glia, 47(2): 109–119.

Yamamoto, R., Yoneda, S. and Hara, H. (2004) Neuroprotective effects of beta-secretase inhibitors against rat retinal ganglion cell death. Neurosci. Lett., 370(1): 61–64.

Yoneda, S., Hara, H., Hirata, A., Fukushima, M., Inomata, Y. and Tanihara, H. (2005) Vitreous fluid levels of beta-amyloid((1-42)) and tau in patients with retinal diseases. Jpn. J. Ophthalmol., 49(2): 106–108.

Yucel, Y.H., Gupta, N., Zhang, Q., Mizisin, A.P., Kalichman, M.W. and Weinreb, R.N. (2006) Memantine protects neurons from shrinkage in the lateral geniculate nucleus in experimental glaucoma. Arch. Ophthalmol., 124(2): 217–225.

CHAPTER 31

Potential roles of (endo)cannabinoids in the treatment of glaucoma: from intraocular pressure control to neuroprotection

Carlo Nucci[1,2,*], Monica Bari[3], Arnoldo Spanò[1], MariaTiziana Corasaniti[4], Giacinto Bagetta[5], Mauro Maccarrone[6,a] and Luigi Antonio Morrone[5,a]

[1] Ophthalmological Unit, Department of Biopathology and Diagnostic Imaging, University of Rome "Tor Vergata," Rome, Italy
[2] Experimental Neuropharmacology Center, "Mondino-Tor Vergata," Fondazione C. Mondino-IRCCS, Rome, Italy
[3] Department of Experimental Medicine and Biochemical Sciences, University of Rome "Tor Vergata," Rome, Italy
[4] Department of Pharmacobiological Sciences, University Magna Graecia, Catanzaro, Italy
[5] Department of Pharmacobiology, University of Calabria and UCHAD, Arcavacata di Rende (Cosenza) Italy
[6] Department of Biomedical Sciences, University of Teramo, Teramo, Italy

Abstract: Recent evidence shows that the endocannabinoid system is involved in the pathogenesis of numerous neurodegenerative diseases of the central nervous system. Pharmacologic modulation of cannabinoid receptors or the enzymes involved in the synthesis, transport, or breakdown of endogenous cannabinoids has proved to be a valid alternative to conventional treatment of these diseases. In this review, we will examine recent findings that demonstrate the involvement of the endocannabinoid system in glaucoma, a major neurodegenerative disease of the eye that is a frequent cause of blindness. Experimental findings indicate that the endocannabinoid system contributes to the control of intraocular pressure (IOP), by modulating both production and drainage of aqueous humor. There is also a growing body of evidence of the involvement of this system in mechanisms leading to the death of retinal ganglion cells, which is the end result of glaucoma. Molecules capable of interfering with the ocular endocannabinoid system could offer valid alternatives to the treatment of this disease, based not only on the reduction of IOP but also on neuroprotection.

Keywords: cannabinoids; intraocular pressure; glaucoma; neuroprotection; retinal ganglion cells

Introduction

Glaucoma is an optic neuropathy characterized by apoptotic death of the retinal ganglion cells (RGCs) and loss of the axons that make up the optic nerve (Weinreb and Khaw, 2004). The changes provoked by glaucoma are progressive, and if left untreated can produce severe visual-field deficits. The pathogenesis of glaucoma is complex and multifactorial, but elevated intraocular pressure (IOP) is one of the risk factors most closely associated with both onset and progression of the disease. Increases in the IOP are believed to be responsible for blood-flow alterations that ultimately lead to hypoxia and ischemia of the retina

*Corresponding author. Tel.: 39 06 7259 6147;
Fax: 39 06 2026232; E-mail: nucci@med.uniroma2.it
[a] Equally senior authors

fact, porcine ocular tissues synthesize and degrade AEA (Matsuda et al., 1997), as do bovine (Bisogno et al., 1999) and rat retina (Glaser et al., 2005) and various human eye tissues (Chen et al., 2005). In addition, CB1 receptors, FAAH, and TRPV1 have been shown to be widespread in the retina of rats and other mammals by immunocytochemical methods (Yazulla and Studholme, 2004). Recently, it has been demonstrated that elevated hydrostatic pressure induces a TRPV1-dependent influx of extracellular Ca^{2+} in retinal microglia that produce and release IL-6 (Sappington and Calkins, 2008). The presence of a functional endocannabinoid system in the eye supports a role for eCBs in ocular physiology. Experimental evidence has accumulated in the last few years to support the role of AEA in the CNS, while it acts as neuroprotective or neurotoxic agent. In fact, eCBs have been shown to possess protective activity in an experimental model of allergic uveitis (Pryce et al., 2003), and to regulate photoreception and neurotransmission in the retina (Fan and Yazulla, 2005; Struik et al., 2006). Several studies have shown that topical administration of CB1 agonists lowers IOP in rabbits, nonhuman primates, and glaucomatous humans (Devane et al., 1992; Laine et al., 2001, 2002a, b; Chien et al., 2003), possibly due to an increase in aqueous humor outflow (Chien et al., 2003; Njie et al., 2006). More recently, plant-derived and synthetic CBs have been shown to exert neuroprotective actions in the eye, with potential implications for the treatment of glaucoma (Tomida et al., 2006).

The IOP-lowering effects of endocannabinoids

The first studies on the IOP-reducing effects of CBs date back to the 1970s. Hepler and Frank (1971) found that eating or smoking marijuana reduced IOP by 5–45% (mean, approximately 25%). However, since the effect lasted for only 3–4 h, patients would have to use cannabis 8–9 times a day to keep their IOP under control. In a review of the literature published some 20 years later, Green (1998) noted that marijuana produces detectable IOP-lowering effects in only 60–65% of healthy subjects or volunteers with glaucoma, and the efficiency of treatment seemed to be dose-dependent. Similar results were observed when CBs were administered intravenously (Perez-Reyez et al., 1976; Cooler and Gregg, 1977), although preliminary data on topical administration yielded negative results (Jay and Green, 1983). The beneficial effects of marijuana on ocular hypertension were also accompanied by several toxic effects, including orthostatic hypotension, increased heart rate, emphysema-like lung changes, diverse alterations of the mental state (euphoria, reduced attention, short-term memory deficits, and altered motor coordination), and at the ocular level they included conjunctival hyperemia associated with a 50% reduction in tear secretion (Brown et al., 1977). These findings prompted the American Academy of Ophthalmology's Complementary Therapies Task Force to carry out a systematic review of the peer-reviewed literature on the IOP-lowering effects of CBs (see website www.aao.orgeyecaretreatmentalternative-therapies marijuana-glaucoma.cfm). The conclusions were that there was no scientific evidence to support the hypothesis that marijuana was more effective or safer for the treatment of glaucoma than the drugs and surgical procedures already widely used to reduce IOP.

At the same time, however, various studies revealed that there was a complex network of receptors for eCBs at the ocular level, and these findings led to a new wave of studies to determine how this endogenous system might be related to the control of IOP. These efforts identified a series of (endo)cannabinoids (cannabis derivatives and endogenous or synthetic molecules that interact specifically with CBRs), that were able to produce selective IOP-lowering effects without provoking systemic toxicity. Hosseini et al. (2006), for example, showed that topical administration (three times a day) of WIN-55-212-2, a synthetic CB that binds both CB1 and CB2 receptors, reduces the IOP by as much as 47% in rats with experimentally induced glaucoma, and this effect was maintained over a period of treatment of 4 weeks. The treatment was not associated with any psychotropic manifestations or symptoms of systemic or local toxicity. These findings were consistent with previous observations in primates

(Chien et al., 2003), and they have recently been confirmed by additional studies in a rat model of glaucoma (Oltmanns et al., 2008). Porcella et al. (2001) reported that WIN-55-212-2 decreases IOP in patients with a type of glaucoma that was refractory to conventional treatment. eCBs like AEA (Pate et al., 1998) or noladin ether (Laine et al., 2002a) have also been found to reduce IOP without producing systemic toxic effects. In addition, the IOP-lowering effects of AEA have been shown to be potentiated by simultaneous administration of phenylmethylsulfonyl fluoride, a compound that inhibits AEA breakdown (Laine et al., 2002b). The latter finding highlights the potential therapeutic perspectives offered by drug-induced modulation of the endocannabinoid tone in the control of IOP. In a recent pilot study, sublingual administration of THC reduced IOP in patients with glaucoma, without producing any significant systemic side effects (Tomida et al., 2004). It seems clear that, while the use of marijuana for the treatment of glaucoma is not supported by scientific evidence, other molecules — natural and synthetic — that interact with the ocular endocannabinoid system are offering new perspectives for the control of IOP.

The mechanisms underlying the effects of (e)CBs on IOP have yet to be completely defined. Straiker's demonstration in 1999 of the presence of CB1 receptors in the pigmented epithelium of the ciliary body, the trabecular meshwork, the Schlemm canal, and the ciliary muscle suggested that CB1 receptor agonists might influence both the production and drainage of aqueous humor. This hypothesis was subsequently confirmed by studies in which pretreatment with CB1 receptor antagonists prevented the IOP-lowering effects normally observed with the metabolically stable analog of anandamide, methanandamide (Pate et al., 1997), or with the synthetic CBs CP55,940 (Pate et al., 1998) and WIN-55,212-2 (Song and Slowey, 2000; Hosseini et al., 2006; Oltmanns et al., 2008). In addition, Lograno and Romano (2004) showed that activation of the CB1 receptor by AEA or CP55,940 caused contraction of the ciliary muscle, an event known to promote outflow of the aqueous humor through the trabecular meshwork. In this context, activation of CB1 receptors present in blood vessels within the ciliary body is thought to reduce the production of aqueous humor by inducing vasodilatation. Chien et al. (2003) confirmed this hypothesis by demonstrating that the pressure-lowering effects of WIN-55212-2 in monkeys with experimentally induced glaucoma are caused by a reduction of approximately 18% in aqueous humor production, an effect that is mediated by the CB1 receptor.

It has recently been suggested that eCBs might also have IOP-lowering effects that are not receptor-mediated. Rosch et al. (2006) found that treatment with AEA, its stable analog methanandamide, or with THC increases cyclooxygenase-2 (COX-2) expression in cultured cells from the nonpigmented epithelium of the ciliary body. As a result, the supernatants from these cultures contained higher levels of the COX-2 product prostaglandin E_2 (PGE_2) and of matrix metalloproteinases-1, -3, and -9. These mediators are known to be involved in remodeling of the aqueous humor outflow pathways, and thus contribute to the regulation of IOP (Weinreb and Lindsay, 2002). This finding was consistent with a report by Maihofner et al. (2001), who showed that COX-2 expression in patients with advanced glaucoma is considerably lower than that observed in healthy individuals (Maihofner et al., 2001). Patients with chronic glaucoma or steroid-induced glaucoma also have lower aqueous humor levels of PGE_2 than those of patients undergoing surgery for cataract (Maihofner et al., 2001). Therefore, the eCBs might lower IOP via activation of cyclooxygenases as well as through receptor-dependent mechanisms. In line with this, studies in animals models of glaucoma have shown that the IOP-lowering effects of these agents are attenuated by drugs that block cyclooxygenases, such as indomethacin and steroids (Pate et al., 1996; Green et al., 2001).

Endocannabinoids and neuroprotection

Experimental findings support the view that drugs capable of interacting with the endocannabinoid system exert specific neuroprotective effects (Van der Stelt and Di Marzo, 2005). These have been reported in experimental models of excitotoxic

CNS damage, including stroke (Nagayama et al., 1999; Amantea et al., 2007), head trauma (Panikashvili et al., 2001), epilepsy (Marsicano et al., 2003), multiple sclerosis (Pryce et al., 2003; Centonze et al., 2007), and other neurodegenerativeneuroinflammatory diseases (Maccarrone et al., 2007). There is also a growing body of experimental data supporting a role for excitotoxicity in the pathogenesis of glaucomatous neuron injury (Nucci et al., 2005a). Subcutaneous or intravitreal administration of glutamate was first shown to produce toxic effects on retinal cells in the 1950s (Lucas and Newhouse, 1957; Sisk and Kuwabara, 1985). Later studies showed that even slight increases in glutamate levels cause retinal damage (Samy et al., 1994), primarily in the large RGCs (Glovinsky et al., 1993), which are also the first cells to display signs of glaucoma-related injury.

Recently, using a microdialysis technique, we have reported acutely increased concentrations of glutamate in the retina of rat in which high IOP-induced retinal ischemia is accompanied by delayed RGC death (Nucci et al., 2005b). These neurochemical data lend support to similar data obtained from human glaucomatous eyes (Dreyer et al., 1996; but see also Carter-Dawson et al., 2002; Honkanen et al., 2003) and from experimental animals (Adachi et al., 1998; Louzada-Júnior et al., 1992; but see also Muller et al., 1997; Kwon et al., 2005). Involvement of the excitotoxic cascade in glaucoma has been confirmed by studies in which glutamate receptor antagonists conferred neuroprotection in in vitro and in vivo models of RGC death (Sucher et al., 1997; Adachi et al., 1998). Under the experimental conditions of excitotoxicity, RGC might occur via apoptosis and this can be prevented by antagonists of both NMDA and non-NMDA glutamate receptors (i.e., MK801 and GIKI52466, respectively) (Nucci et al., 2005b). Similar neuroprotection is afforded by L-NAME, an inhibitor of nitric oxide synthase, or by free-radical scavengers such as coenzyme Q10 and vitamin E (Nucci et al., 2007a). These findings strengthen the hypothesis that excessive accumulation of extracellular glutamate plays a role in glaucoma. Via activation of NMDA and non-NMDA glutamate receptors, this excitatory neurotransmitter increases intracellular levels of calcium, which activate nitric oxide synthase and lead to the release of free nitrogen radicals with subsequent death of RGCs.

Interestingly, El-Remessy et al. (2003) found that systemic administration of THC or cannabidiol (CBD), a major nonpsychotropic constituent of cannabis, prevents RGC death triggered by intravitreal administration of NMDA in rat, and this effect was associated with reduced formation of peroxynitrites. The neuroprotective effects of these CBs was partially inhibited by SR141716A, a selective CB1 receptor antagonist. These findings are consistent with the recent report by Crandall et al. (2007), who found that 20 weeks of treatment with THC lowers IOP and reduces RGC death by approximately 75% in animals with chronic experimentally induced glaucoma.

As summarized above, the endocannabinoid system present in the human retina includes proteins that synthesize, transport, and hydrolyze anandamide, along with CB1 and TRPV1 receptors that are activated by AEA. Our research group has investigated the role of this endogenous system in the neuronal damage that follows acute ocular hypertension (Nucci et al., 2007b). In our experimented model, retinal ischemia induced by ocular hypertension was associated with a 25% reduction in intraretinal levels of AEA (Fig. 1). This effect seems to be the result of an altered endocannabinoid metabolism in the retina. Indeed, as early as 3 h after the acute hypertonic insult, we demonstrated progressive increases in the expression and activity of FAAH, the enzyme that hydrolyzes anandamide. Six and twelve hours after the insult, FAAH activity displayed increases of 150 and 230%, respectively, over baseline values. In contrast, the increased IOP did not have any significant effect on the activities of other enzymes involved in the metabolism of AEA, such NAPE-PLD, which is mainly responsible for the biosynthesis of AEA, or AMT, which transports this substance across cell membranes (Fig. 1).

Collectively, our study seems to indicate that acute elevation of IOP is associated with a reduction in retinal endocannabinoid tone, secondary to increased degradation of anandamide. To

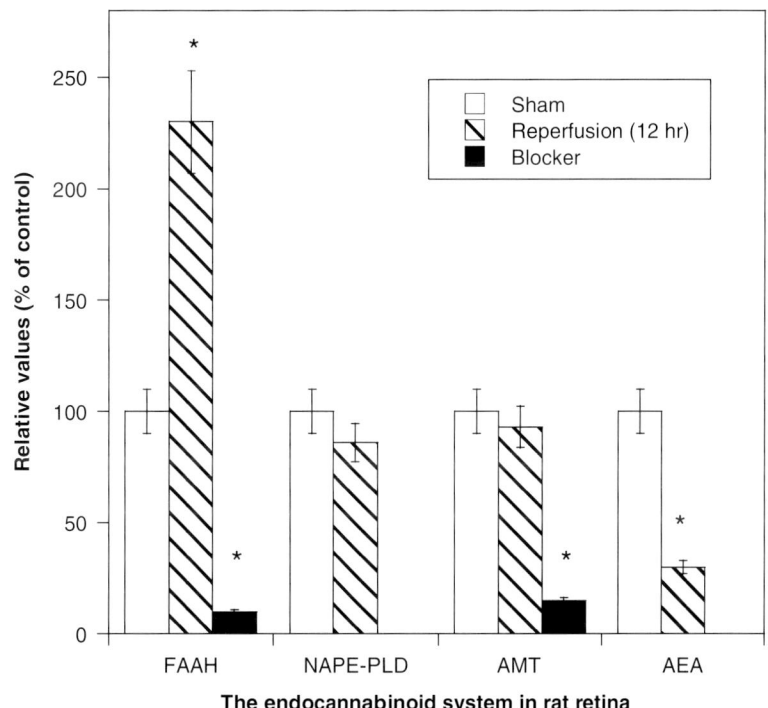

Fig. 1. Activity of FAAH, NAPE-PLD, and AMT, and endogenous levels of AEA, in the retina of rats subjected to high IOP-induced ischemia for 45 min followed by 12 h reperfusion. Sham operated animals were exposed to the same surgical procedure without ischemiareperfusion (100% = 161±20 pmol/min per mg protein, for FAAH; 39±5 pmol/min per mg protein, for NAPE-PLD; 34±5 pmol/min per mg protein, for AMT; 20±4 pmol per mg protein, for AEA). The activity of FAAH and that of AMT was assayed also in the presence of specific blockers, i.e. 10 nM URB597 and 5 μM OMDM1, respectively. Data were expressed as mean±S.D. ($n = 3$) and were analyzed by the Mann–Whitney U test. *Denotes $p<0.01$ versus sham (adapted with permission from Nucci et al., 2007b).

determine whether these events contributed to RGC death in our model, we pretreated animals with URB597, a selective FAAH inhibitor, and evaluated the retinal damage provoked by the ocular hypertensive insult in terms of total number of cells in the ganglion layer and levels of mRNA for THY-1, a specific marker of RGCs. URB597 pretreatment prevented the increase in FAAH activity triggered by acute ocular hypertension and diminished RGC loss, compared with that observed in untreated controls (Fig. 2, Table 1), suggesting that reduced levels of AEA caused by enhanced FAAH activity do indeed play a role in the retinal cell loss provoked by acute ocular hypertension. This hypothesis is further supported by our observation that intravitreal administration of methanandamide (Fig. 2, Table 1), prevents ganglion-cell death in rats exposed to retinal ischemia caused by acute ocular hypertension. The neuroprotective effects exerted by methanandamide seem to be related to activation of CB1 and TRPV1 receptors, since they are abolished by treatment with antagonists of these receptors, like SR141716A and capsazepine, respectively (Fig. 2, Table 1). Collectively, our findings seem to indicate that the retinal endo cannabinoid system provides a form of neuroprotection that can be weakened under certain conditions, leading to the activation of cell death cascades. Restoration of physiological levels of anandamide with agents that inhibit its enzymatic degradation or act as CB1 or vanilloid receptor agonists (e.g., methanandamide) appears to be a promising strategy for strengthening the protective effect of endocannabinoids in the retina, and thus for preventing cell loss.

Hosseini, A., Lattanzio, F.A., Williams, P.B., Tibbs, D., Samudre, S.S. and Allen, R.C. (2006) Chronic topical administration of WIN-55-212-2 maintains a reduction in IOP in a rat glaucoma model without adverse effects. Exp. Eye Res., 82: 753–759.

Howlett, A.C., Barth, F., Bonner, T.I., Cabral, G., Casellas, P., Devane, W.A., Felder, C.C., Herkenham, M., Mackie, K., Martin, B.R., Mechoulam, R. and Pertwee, R.G. (2002) Classification of cannabinoid receptors. Pharmacol. Rev., 54: 161–202.

Howlett, A.C., Breivogel, C.S., Childers, S.R., Deadwyler, S.A., Hampson, R.E. and Porrino, L.J. (2004) Cannabinoid physiology and pharmacology: 30 years of progress. Neuropharmacology, 47: 345–358.

Jackson, S.J., Diemel, L.T., Pryce, D. and Baker, D. (2005) Cannabinoids and neuroprotection in CNS inflammatory disease. J. Neurol. Sci., 233: 21–25.

Jay, W.M. and Green, K. (1983) Multiple drop study of topically applied of 1% Δ9 tetrahydrocannabinol in human eye. Arch. Ophthalmol., 101: 591–593.

Jordt, S.E. and Julius, D. (2002) Molecular basis for species-specific sensitivity to "hot" chili peppers. Cell, 108(3): 421–430.

Kass, M.A., Heuer, D.K., Higginbotham, E.J., Johnson, C.A., Keltner, J.L., Miller, J.P., Parrish, R.K., 2nd., Wilson, M.R. and Goedon, M.O. (2002) The ocular hypertension treatment study: a randomized trial determines that topical ocular hypotensive medication delays or prevent the onset of primary open-angle glaucoma. Arch. Ophthalmol., 120(6): 701–713.

Kim, S.H., Kim, J.Y., Kim, D.M., Ko, H.S., Kim, S.Y., Yoo, T., Hwang, S.S. and Park, S.S. (2006) Investigations on the association between normal tension glaucoma and single nucleotide polymorphisms of the endothelin-1 and endothelin receptor genes. Mol. Vis., 12: 1016–1021.

Kim, D.J. and Thayer, S.A. (2000) Activation of CB1 cannabinoid receptors inhibits neurotransmitter release from identified synaptic sites in rat hippocampal cultures. Brain Res., 852: 398–405.

Kwon, Y.H., Rickman, D.W., Baruah, S., Zimmerman, M.B., Kim, C.S., Boldt, H.C., Russell, S.R. and Hayreh, S.S. (2005) Vitreous and retinal amino acid concentrations in experimental central retinal artery occlusion in the primate. Eye, 19(4): 455–463.

Laine, K., Jarvinen, K., Mechoulam, R., Breuer, A. and Jarvinen, T. (2002a) Comparison of the enzymatic stability and intraocular pressure effects of 2-arachidonylglycerol and noladin ether, a novel putative endocannabinoid. Invest. Ophthalmol. Vis. Sci., 43: 3216–32122.

Laine, K., Jarvinen, K., Pate, D.W., Urtti, A. and Jarvinen, T. (2002b) Effect of the enzyme inhibitor, phenylmethylsulfonyl fluoride, on the IOP profiles of topical anandamides. Invest. Ophthalmol. Vis. Sci., 43: 393–397.

Laine, K., Jarvinen, T., Savinainen, J., Laitinen, J.T., Pate, D.W. and Jarvinen, K. (2001) Effects of topical anandamide-transport inhibitors, AM404 and olvanil, on intraocular pressure in normotensive rabbits. Pharm. Res., 18: 494–499.

Lastres-Becker, I., Molina-Holgado, F., Ramos, J.A., Mechoulam, R. and Fernandez-Ruiz, J. (2005) Cannabinoids provide neuroprotection against 6-hydroxydopamine toxicity in vivo and in vitro: relevance to Parkinson's disease. Neurobiol. Dis., 19: 96–107.

Lauckner, J.E., Jensen, J.B., Chen, H.Y., Lu, H.C., Hille, B. and Mackie, K. (2008) GPR55 is a cannabinoid receptor that increases intracellular calcium and inhibits M current. Proc. Natl. Acad. Sci., 105: 2699–2704.

Ligresti, A., Cascio, M.G. and Di Marzo, V. (2005) Endocannabinoid metabolic pathways and enzymes. Curr. Drug Targets CNS Neurol. Disord., 4: 615–623.

Lipton, S.A. (2003) Possible role for memantine in protecting retinal ganglion cells from glaucomatous damage. Surv. Ophthalmol., 48(Suppl 1): S38–S46.

Lograno, M.D. and Romano, M.R. (2004) Cannabinoid agonists induce contractile responses through Gi/o-dependent activation of phospholipase C in the bovine ciliary muscle. Eur. J. Pharmacol., 494: 55–62.

Louzada-Júnior, P., Dias, J.J., Santos, W.F., Lachat, J.J., Bradford, H.F. and Coutinho-Netto, J. (1992) Glutamate release in experimental ischaemia of the retina: an approach using microdialysis. J. Neurochem., 59(1): 358–363.

Lucas, D.R. and Newhouse, J.P. (1957) The toxic effect of sodium L-glutamate on the inner layers of the retina. Am. Med. Assoc. Arch. Ophthalmol., 58.: 193–201.

Maccarrone, M., Battista, N. and Centonze, D. (2007) The endocannabinoid pathway in Huntington's disease: a comparison with other neurodegenerative diseases. Prog. Neurobiol., 81: 349–379.

Maccarrone, M., Rossi, S., Bari, M., De Chiara, V., Fezza, F., Musella, A., Gasperi, V., Prosperetti, C., Bernardi, G., Finazzi-Agrò, A., Cravatt, B.F. and Centonze, D. (2008) Anandamide inhibits metabolism and physiological actions of 2-arachidonoylglycerol in the striatum. Nat. Neurosci., 11: 152–159.

Maihofner, C., Schlotzer-Schrehardt, U., Guhring, H., Zeilhofer, H.U., Naumann, G.O.H., Pahl, A., Mardin, C., Tamm, E.R. and Brune, K. (2001) Expression of ciclooxigenase-1 and -2 in normal and glaucomatous human eyes. Invest. Ophthalmol. Vis. Sci., 42: 2616–2624.

Marinelli, S., Di Marzo, V., Berretta, N., Matias, I., Maccarrone, M., Bernardi, G. and Mercuri, N.B. (2003) Presynaptic facilitation of glutamatergic synapses to dopaminergic neurons of the rat substantia nigra by endogenous stimulation of vanilloid receptors. J. Neurosci., 23(8): 3136–3144.

Marsicano, G., Goodenough, S., Monory, K., Hermann, H., Eder, M., Cannich, A., Azad, S.C., Cascio, M.G., Gutierrez, S.O. and V. Stelt, O. (2003) CB1 cannabinoid receptors and on-demand defense against excitotoxicity. Science, 302: 84–88.

Matsuda, S., Kanemitsu, N., Nakamura, A., Mimura, Y., Ueda, N., Kurahashi, Y. and Yamamoto, S. (1997) Metabolism of anandamide, an endogenous cannabinoid receptor ligand, in porcine ocular tissues. Exp. Eye Res., 64(5): 707–711.

McIntosh, B.T., Hudson, B., Yegorova, S., Jollimore, C.A.B. and Kelly, M.E.M. (2007) Agonist-dependent cannabinoid receptor signalling in human trabecular meshwork k cells. Br. J. Pharmacol., 152: 1111–1120.

McKinney, M.K. and Cravatt, B.F. (2005) Structure and function of fatty acid amide hydrolase. Annu. Rev. Biochem., 74: 411–432.

Mechoulam, R., Panikashvili, D. and Shohami, E. (2002) Cannabinoids and brain injury: therapeutic implications. Trends Mol. Med., 8: 58–61.

Muller, A., Villain, M. and Bonne, C. (1997) The release of amino acids from ischemic retina. Exp Eye Res., 64(2): 291–293.

Nagayama, T., Sinor, A.D., Simon, R.P., Chen, J., Graham, S.H., Jin, K. and Greenberg, D.A. (1999) Cannabinoids and neuroprotection in global and focal cerebral ischemia and in neuronal cultures. J. Neurosci., 19: 2987–2995.

Njie, Y.F., Kumar, A., Qiao, Z., Zhong, L. and Song, Z.H. (2006) Noladin ether acts on trabecular meshwork cannabinoid (CB1) receptors to enhance aqueous humor outflow facility. Invest. Ophthalmol. Vis. Sci., 47: 1999–2005.

Nucci, C., Cerulli, A., Turturo, A., Tartaglione, R., Morrone, LA., Bagetta, G. and Corasaniti, M.T. (2005a) Apoptosis and glaucoma: new insight from basic research to the development of novel strategies for neuroprotection. In: Reece S.M. (Ed.), Focus on Glaucoma Research. Nova Science Publisher, New York, USA, pp. 1–26.

Nucci, C., Gasperi, V., Tartaglione, R., Cerulli, A., Terrinoni, A., Bari, M., De Simone, C., Finazzi Agrò, A., Morrone, L.A., Corasaniti, M.T., Bagetta, G. and Maccarrone, M. (2007b) Involvement of the endocannabinoid system in retinal damage following high intraocular pressure-induced ischemia in rat. Invest. Ophthalmol. Vis. Sci., 48: 2997–3002.

Nucci, C., Tartaglione, R., Cerulli, A., Mancino, R., Spanò, A., Cavaliere, F., Rombolà, L., Bagetta, G., Corasaniti, M.T. and Morrone, L.A. (2007a) Retinal damage caused by high intraocular pressure (IOP)-induced transient ischemia is prevented by coenzyme Q10 in rat. Int. Rev. Neurobiol., 82: 397–406.

Nucci, C., Tartaglione, R., Rombola, L., Morrone, L.A., Fazzi, E. and Bagetta, G. (2005b) Neurochemical evidence to implicate elevated glutamate in the mechanisms of high intraocular pressure (IOP)-induced retinal ganglion cell death in rat. Neurotoxicology, 26: 935–941.

Okamoto, Y., Morishita, J., Tsuboi, K., Tonai, T. and Ueda, N. (2004) Molecular characterization of a phospholipase D generating anandamide and its congeners. J. Biol. Chem., 279: 5298–5305.

Oltmanns, M.H., Samudre, S.S., Castillo, I.G., Hosseni, A., Lichtman, A.H., Allen, R.C., Lattanzio, F.A. and Williams, P.B. (2008) Topical WIN 55212-2 alleviates intraocular hypertension in rats through a CB1 receptor-mediated mechanism of actionJ. Ocul. Pharmacol., Jan 17 Epub ahead of print.

Osborne, N.N., Casson, R.J., Wood, J.P.M., Childlow, G., Graham, M. and Melena, J. (2004) Retinal ischemia: mechanisms of damage and potential therapeutic strategies. Prog. Retin. Eye Res., 23: 91–147.

Osborne, N.N., Ugarte, M., Chao, M., Chidlow, G., Bae, J.H., Wood, J.P. and Nash, M.S. (1999) Neuroprotection in relation to retinal ischemia and relevance to glaucoma. Surv. Ophthalmol., 43: S102–S128.

Panikashvili, D., Simeonidou, C., Ben-Shabat, S., Hanus, L., Breuer, A., Mechoulam, R. and Shohami, E. (2001) An endogenous cannabinoid (2AG) is neuroprotective after brain injury. Nature, 413: 527–531.

Pate, D.W., Järvinen, K., Urtii, A., Jarho, P., Fich, M., Mahadevan, V. and Jarvinen, T. (1996) Effects of topical anandamides on intraocular pressure in normotensive rabbits. Life Sci., 58: 1849–1860.

Pate, D.W., Jarvinen, K., Urtii, A., Jarho, P., Mahadevan, V. and Jarvinen, T. (1997) Effects of topical alpha-substituted anandamides on intraocular pressure in normotensive rabbits. Pharm. Res., 14: 1738–1743.

Pate, D.W., Järvinen, K., Urtii, A., Mahadevan, V. and Jarvinen, T. (1998) Effect of the CB_1 receptor antagonist SR 141716A on cannabinoids induced ocular hypotension in normotensive rabbits. Life Sci., 63: 2181–2188.

Perez-Reyez, M., Wagner, D., Wall, M.E. and Davis, K.H. (1976) Intravenous administration of cannabinoids and intraocular pressure. In: Braude M.C. and Szara S. (Eds.), The Pharmacology of Marihuana. Raven Press, New York, NY, pp. 832–839.

Pertwee, R.G. and Ross, R.A. (2002) Cannabinoid receptors and their ligands. Prostaglandins Leukot Essent Fatty Acids, 66(2-3): 101–121.

Piomelli, D. (2003) The molecular logic of endocannabinoid signalling. Nat. Rev. Neurosci., 4(11): 873–884.

Plange, N., Arend, K.O., Kaup, M., Doehmen, B. and Adams, H. (2007) Dronabinol and retinal hemodynamics in human. Am. J. Ophthalmol., 143.: 173–174.

Porcella, A., Maxia, C.H., Gessa, G.L. and Pani, L. (2001) The synthetic cannabinoids WIN-55,212-2 decreases the intraocular pressure in human glaucoma resistant to conventional therapies. Eur. J. Neurosci., 13: 409–412.

Pryce, G., Ahmed, Z., Hankey, D.J., Jackson, S.J., Croxford, J.L., Pocock, J.M., Ledent, C., Petzold, A., Thompson, A.J., Giovannoni, G., Cuzner, M.L. and Maker, D. (2003) Cannabinoids inhibit neurodegeneration in models of multiple sclerosis. Brain, 126: 2191–2202.

Ramirez, B.G., Blazquez, C., Gomez, D.P., Guzman, M. and De Ceballos, M.L. (2005) Prevention of Alzheimer's disease pathology by cannabinoids: neuroprotection mediated by blockade of microglial activation. J. Neurosci., 25: 1904–1913.

Romano, M.R. and Lograno, M.D. (2006) Cannabinoid agonists induce relaxation in the bovine ophthalmic artery: evidences for CB1 receptors, nitric oxide and potassium channels. Br. J. Pharmacol., 147(8): 917–925.

Ronco, A.M., Llanos, M., Tamayo, D. and Hirsch, S. (2007) Anandamide inhibits endothelin-1 production by human cultured endothelial cells: a new vascular action of this endocannabinoid. Pharmacology, 79(1): 12–16.

Rosch, S., Ramer, R., Brune, K. and Hinz, B. (2006) R(+)-methanandamide and other cannabinoids induce the expression of cyclooxygenase-2 and matrix metalloproteinases in human nonpigmented ciliary epithelial cells. J. Pharmacol. Exp. Ther., 316(3): 1219–1228.

Samy, C.N., Lui, C.J., Kaiser, P.K., Lipton, S.A. and Dreyer, E.B. (1994) Toxicity of chronic glutamate administration to the retina. Invest. Ophthalmol. Vis. Sci., 35: p. 497.

Sappington, R.M. and Calkins, D. (2008) TRPV1 contributes to microglia-derived IL-6 and NF{kappa}B translocation with elevated hydrostatic pressure. Invest. Ophthalmol. Vis. Sci.

Sawzdargo, M., Nguyen, T., Lee, D.K., Lynch, K.R., Cheng, R., Heng, H.H., George, S.R. and O'Dowd, B.F. (1999) Identification and cloning of three novel human G protein-coupled receptor genes GPR52, PsiGPR53 and GPR55: GPR55 is extensively expressed in human brain. Brian Res. Mol. Brain Res., 64: p. 193.

Shen, M. and Thayer, S.A. (1998) The cannabinoid agonist win 55,212-2 inhibits calcium channels by receptor-mediated and direct pathways in cultured rat hippocampal neurons. Brain Res., 783: 77–84.

Shen, M. and Thayer, S.A. (1999) D9-tetrahydrocannabinol act as a partial agonist to modulate glutamatergic synaptic transmission between rat hippocampal neurons in culture. Mol. Pharmacol., 55: 459–462.

Sisk, D.R. and Kuwabara, T. (1985) Histologic changes in the inner retina of albino rats following intravitreal injection of monosodium L-glutamate. Graefes Arch. Clin. Exp. Ophthalmol., 223: 250–258.

Song, Z.H. and Slowey, C.A. (2000) Involvement of cannabinoids receptors in the intraocular pressure lowering effects of WIN-55,212-2. Pharmacol. Exp. Ther., 292: 136–139.

Starowicz, K., Nigam, S. and Di Marzo, V. (2007) Biochemistry and pharmacology of endovanilloids. Pharmacol. Ther., 114: 13–33.

Straiker, A., Maguire, G., Makie, K. and Lindsey, J (1999) Localization of cannabinoid CB1 receptors in the human anterior eye and retina. Invest. Ophthalmol. Vis. Sci., 40: 2442–2448.

Struik, M.L., Yazulla, S. and Kamermans, M. (2006) Cannabinoid agonist WIN 55212-2 speeds up the cone response to light offset in goldfish retina. Vis. Neurosci., 23: 285–293.

Sucher, N.J., Lipton, S.A. and Dreyer, E.B. (1997) Molecular basis of glutamate toxicity in retinal ganglion cells. Vision Res., 37: 3483–3493.

Tomida, I., Azuara-Blanco, A., House, H., Flint, M., Pertwee, R.G. and Robson, P.J. (2006) Effect of sublingual application of cannabinoids on intraocular pressure: a pilot study. J. Glaucoma, 15: 349–353.

Tomida, I., Pertwee, R.G. and Azuara-Blanco, A. (2004) Cannabinoids and glaucoma. Br. J. Ophthalmol., 88: 708–713.

Tsuboi, K., Sun, Y.X., Okamoto, Y., Araki, N., Tonai, T. and Ueda, N.J. (2005) Molecular characterization of N-acylethanolamine-hydrolyzing acid amidase, a novel member of the choloylglycine hydrolase family with structural and functional similarity to acid ceramidase. Biol. Chem., 280(12): 11082–11092.

Twitchell, W., Brown, S. and Mackie, K. (1997) Cannabinoid inhibit N- and P/Q-type calcium channels in cultured rat hippocampal neurons. J. Neurophysiol., 78: 43–50.

van der Stelt, M. and Di Marzo, V. (2005) Cannabinoid receptors and their role in neuroprotection. Neuromolecular Med., 7(1–2): 37–50.

Walter, L. and Stella, N. (2004) Cannabinoids and neuroinflammation. Br. J. Pharmacol., 141: 775–785.

Wang, L., Fortune, B., Cull, G., Dong, J. and Cioffi, G.A. (2006) Endothelin B receptor in human glaucoma and experimentally induced optic nerve damage. Arch. Ophthalmol., 124(5): 717–724.

Wei, B.Q., Mikkelsen, T.S., McKinney, M.K., Lander, E.S. and Cravatt, B.F. (2006) A second fatty acid amide hydrolase with variable distribution among placental mammals. J. Biol. Chem., 281(48): 36569–36578.

Weinreb, R.N. and Khaw, P.T. (2004) Primary open-angle glaucoma. Lancet, 363: 1711–1720.

Weinreb, R.N. and Lindsay, J.D. (2002) Metalloproteinase gene transcription in human ciliary muscle cells with latanoprost. Invest. Ophthalmol. Vis. Sci., 43: 716–722.

Yazulla, S. and Studholme, K.M. (2004) Vanilloid receptor like 1 (VRL1) immunoreactivity in mammalian retina: colocalization with somatostatin and purinergic $P2 \times 1$ receptors. J. Comp. Neurol., 474(3): 407–418.

CHAPTER 32

Glaucoma of the brain: a disease model for the study of transsynaptic neural degeneration

Yeni Yücel[1,2,*] and Neeru Gupta[1,2]

[1]*Department of Ophthalmology and Vision Sciences, Keenan Research Centre at the Li Ka Shing, Knowledge Institute of St. Michael's Hospital, University of Toronto, Toronto, Ontario, Canada*
[2]*Department of Laboratory Medicine and Pathobiology, Keenan Research Centre at the Li Ka Shing, Knowledge Institute of St. Michael's Hospital, University of Toronto, Toronto, Ontario, Canada*

Abstract: The identification of mechanisms precipitating neuronal death and injury is an intense area of investigation requiring reliable models to assess the effects of neuroprotective agents. Most are suboptimal since the effects of initial damage are diffuse and may not be reproducible or easily quantifiable. The ideal laboratory model should have the ability to (a) clearly detect evidence of neuronal injury and recovery, (b) accurately measure morphologically the extent of these changes, and (c) provide functional evidence for damage and recovery. Glaucoma is a disease of visual neurons in the eye and brain. In the visual system, neuroanatomical pathways and retinotopic organization are exquisitely defined, functional modalities are highly characterized and can be dissected physiologically, visual input parameters can be modified, visual functional output can be readily tested and measured, changes in the eye and the visual brain can be directly visualized and imaged, and pathological and compensatory changes in brain centers of vision can be examined and measured specifically. For these reasons, the glaucoma disease model is ideal for the study of response and recovery to injury in the central nervous system due to anterograde and retrograde degeneration from the eye to the brain and the brain to the eye, respectively. The study of this glaucoma model of transsynaptic brain injury may be relevant to understanding more complex pathways and point to new strategies to prevent disease progression in other neurodegenerative diseases.

Keywords: Alzheimer's disease; neurodegenerative diseases; glaucoma; models; cell atrophy; cell death; blood flow; retinal ganglion cell

Glaucoma is a leading cause of irreversible world blindness estimated to affect more than 79.6 million persons by year 2020 (Quigley and Broman, 2006). Vision loss is caused by cell death and injury to retinal ganglion cells (RGCs) (Quigley, 1999; Weinreb and Khaw, 2004), and the diagnosis of glaucoma is based on characteristic patterns of optic nerve fiber loss revealed by ophthalmological examination. It is often associated with elevated intraocular pressure (IOP), a major risk factor for the development of the disease and one that is modifiable by current IOP-lowering treatments (Weinreb and Khaw, 2004). Reducing IOP medically and surgically has been shown to slow the progression of vision loss in glaucoma, including in those patients without elevated IOP, so-called

*Corresponding author. Tel.: +1 416 864 6060, Ext. 3168; Fax: +1 416 864 5208; E-mail: yeni.yucel@utoronto.ca

low-tension glaucoma cases (Heijl et al., 2002; Leske et al., 2007). However, a significant proportion of glaucoma patients continue to lose sight in spite of IOP-lowering treatment (Heijl et al., 2002; Leske et al., 2007), and in this context, novel adjuvant strategies to prevent RGC loss/injury continue to be an area of intense investigation. This article will critically review our current understanding of neural degeneration in glaucoma, its spread within central visual pathways, and its relevance to understanding transsynaptic degeneration in neurodegenerative disease.

Retinal ganglion cells, retino-geniculate neurons

The RGC is the principal cell type injured in glaucoma. Its cell body, dendrites, and unmyelinated axon component are located within the eye. There are more than 10 RGC types in nonhuman primates (Dacey et al., 2003). The myelinated axon component of the RGC lies outside of the globe, forming the intraorbital, intracanalicular, and intracranial components of the optic nerve, optic chiasm, and optic tract (Fig. 1). The majority of RGC terminals convey visual information to the lateral geniculate nucleus (LGN), a nucleus in the thalamus (Perry et al., 1984). The remaining 10% of RGCs target structures that include superior colliculus responsible for eye movements, pretectum for pupillary reflex/eye movements, accessory optic system nuclei for optokinetic nystagmus, and suprachiasmatic nucleus for circadian rhythm retinal input (Goebel et al., 2004). RGC axons exiting the eye convey visual information to these centers while maintaining bidirectional axoplasmic transport of molecules and organelles critical to the function of RGCs and targets neurons.

Lateral geniculate nucleus

The LGN is the major relay station between the retina and the primary visual cortex in humans and nonhuman primates. This structure is composed of neuronal cell bodies arranged into six anatomically segregated layers. The two most ventral layers called magnocellular layers receive input from M RGCs conveying motion information. Four dorsal

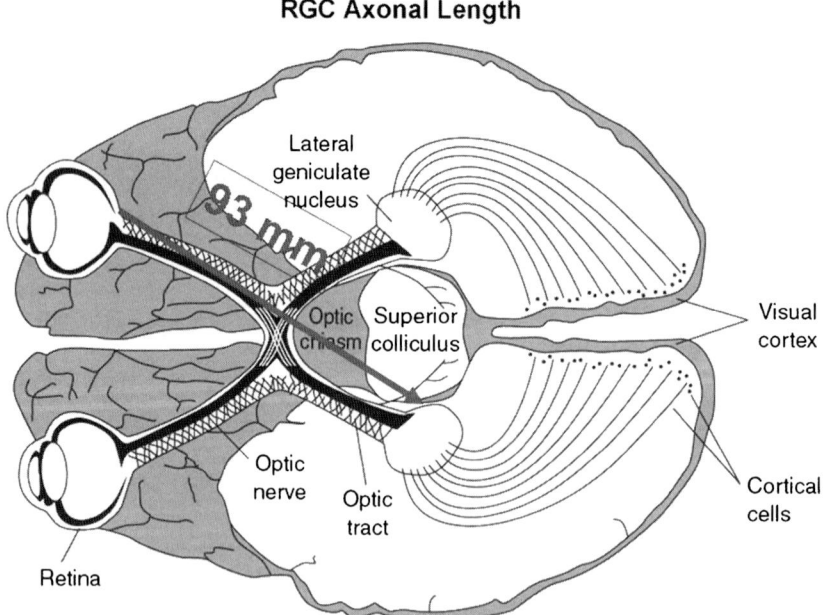

Fig. 1. Principal visual stations of the brain are apparent. Retinal ganglion cell axons measure approximately 93 mm spanning intraorbital, intracanalicular, and intracranial portions of the visual pathways including optic nerve, optic chiasm, and optic tract.

layers, called parvocellular layers, receive input from P RGCs conveying red–green color information. The koniocellular neurons located between principal layers receive input from K RGCs conveying blue–yellow color information (Kaplan, 2004). Each LGN layer receives input from one eye only: Layers 2, 3, and 5 receive input from the ipsilateral eye, and layers 1, 4, and 6 from the contralateral eye. Within the LGN, approximately 20% of the LGN neurons are GABAergic interneurons remaining in the LGN (Montero and Zempel, 1986), while 80% are relay neurons with their axons forming optic radiations that convey visual information to the primary visual cortex. Only a minority of synaptic inputs onto geniculate relay cells derive from retina (less than 10%), with the majority of synaptic input coming from GABAergic interneurons, cortical, and subcortical inputs (Van Horn et al., 2000).

Mechanisms of RGC injury in glaucoma

The death or loss of RGCs is accepted as the pathological correlate of irreversible vision loss in glaucoma. The loss of RGCs is slow and partial, and accompanied by a specific visual field loss pattern that extends in a progressive manner. Some evidence suggests that RGC death is apoptotic in nature (Quigley, 1999), and primary mechanisms leading to programmed cell death have been reviewed elsewhere (Weinreb and Khaw, 2004). Our understanding of this process is based on experimental work performed in glaucoma models with elevated IOP and optimized over the years to study the sequence of pathological events triggered by IOP elevation (Morrison et al., 1997; Gabelt et al., 2003; Lindsey and Weinreb, 2005; Weinreb and Lindsey, 2005). These pathological events associated with RGC death include glial cell alteration at the optic nerve head, disruption of intracellular transport mechanisms leading to growth factor deprivation, oxidative stress, glutamate excitotoxicity, immune alterations, and vascular pathology (Weinreb and Khaw, 2004; Nickells, 2007; Feilchenfeld et al., 2008) (Fig. 2). However, careful examination of the retina and optic nerve in these models and human glaucoma indicates that a significant proportion of RGCs are neither dead nor completely healthy and that surviving RGCs show subtle morphological changes including shrinkage of cell body, dendrites, and axons (Weber et al., 1998; Morgan et al., 2000). Morphological changes correlated with altered function (Weber and Harman, 2005), and it is possible that these early changes contribute to visual dysfunction in glaucoma. The relationship between IOP elevation and these changes is not completely elucidated, and there is no clear evidence to suggest that dysfunctional cells are destined to die. Factors implicated in RGC death may also be relevant to early injury and cell changes.

Transsynaptic degeneration of the lateral geniculate nucleus in glaucoma

In several neurodegenerative diseases, central nervous system (CNS) injury spreads from a population of neurons to other neuronal populations along anatomical and functional connections (Saper et al., 1987; Kume et al., 1993; Suzuki et al., 1995; Su et al., 1997). This process, called transsynaptic or transneuronal degeneration, is critical to disease progression in many neurological disorders, such as neurons involved in cognition in Alzheimer's disease (Su et al., 1997). A striking example of this process is in the visual system where transsynaptic degeneration has been observed in the LGN in humans following removal of the eye, a total deafferentation severing all RGC axons (Goldby, 1957). Systematic studies of this process following unilateral enucleation in nonhuman primates demonstrated atrophy of LGN neurons in all animals (Matthews et al., 1960) and loss of LGN neurons in those with longer survival time (Matthews, 1964).

Studies of transsynaptic degeneration in glaucoma come mainly from the experimental monkey model of glaucoma. In this model, the fluid drainage pathway of the eye, the trabecular meshwork, is lasered excessively by argon laser trabeculoplasty, inducing scarring and restricting fluid exit from the eye (Gaasterland and Kupfer, 1974). This caused IOP elevation, typically induced in one eye, after which eye pressure is monitored in both

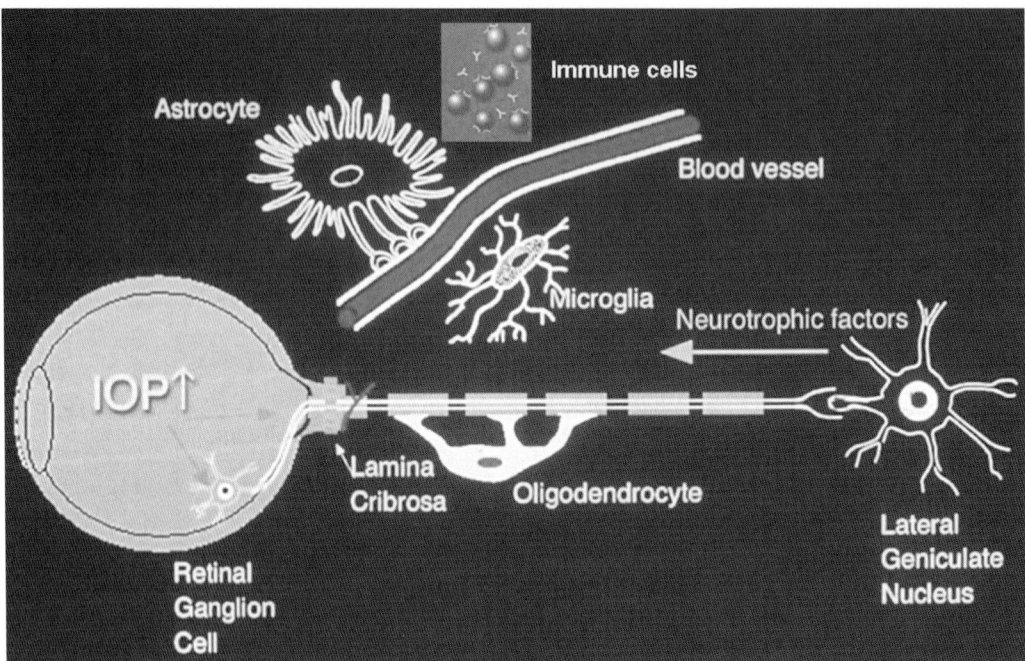

Fig. 2. Diagram illustrating multiple cell types and mechanisms implicated in neural degeneration in glaucoma.

experimental and fellow non-glaucoma eyes (Fig. 3). The experimental primate model of glaucoma is highly relevant to human disease due to similar anatomy and physiology of central visual pathways (Kaplan, 2004); mimicking of characteristic glaucomatous optic nerve changes observed in human glaucoma (Quigley et al., 1981; Yücel et al., 1999; Gabelt et al., 2003) and monkeys trained to perform reliable visual field testing show visual deficits similar to those observed in patients with glaucoma (Harwerth et al., 2002). In addition to IOP, optic disk changes (Yücel et al., 1998; Burgoyne et al., 2002; Hare et al., 2004) and visual electrophysiological changes (Frishman et al., 1996; Hare et al., 1999) can be monitored in vivo. After exposure to elevated IOP, postmortem retina and optic nerve are studied to assess pathological events in these tissues. Chronic elevated IOP in the primate model induces blocked anterograde transport to the LGN and retrograde transport at the level of the lamina cribrosa (Quigley and Anderson, 1976; Gaasterland, 1978; Quigley and Addicks, 1980; Radius and Anderson, 1981a, b), leading to RGC death (Quigley, 1999), and atrophy of surviving cell bodies and dendrites (Weber et al., 1998; Morgan et al., 2000). Although axon size measurement suggested selective loss of larger neurons (Glovinsky et al., 1991), presumably M RGCs, atrophy of RGC axons (Morgan et al., 2000) was not considered.

In experimental primate glaucoma, varying degrees of RGC loss measured by established histomorphometric techniques can be observed (Fig. 4). This includes no evidence of RGC loss with elevated IOP, or partial loss of RGCs, and even total RGC loss with replacement by glial scar (Yücel et al., 1999).

Glaucomatous RGC loss and atrophy spread by anterograde degeneration to major visual pathways in the brain (Gupta and Yücel, 2003).

Neural degeneration in magno-, parvo-, and koniocellular LGN layers

Following elevated IOP in the monkey eye, metabolic changes have been noted in the LGN layers connected to the glaucomatous eye (Vickers

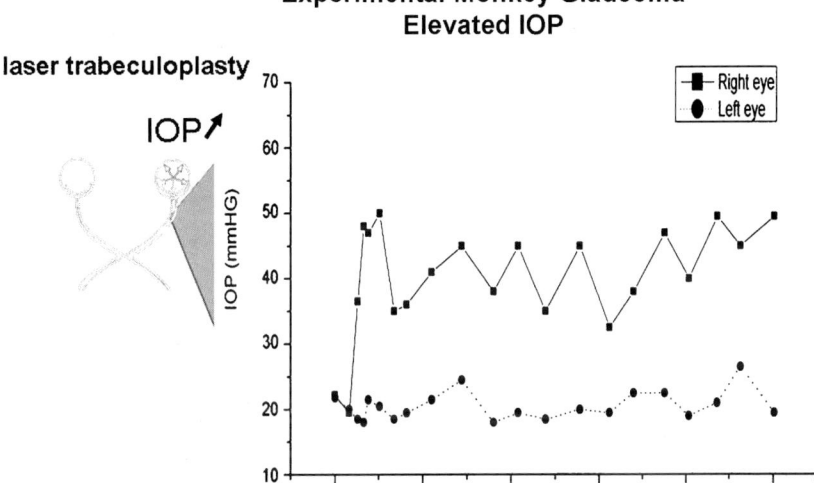

Fig. 3. Laser-induced injury to one eye induces elevated intraocular pressure that can be monitored in the experimental right and fellow non-glaucoma eyes over time.

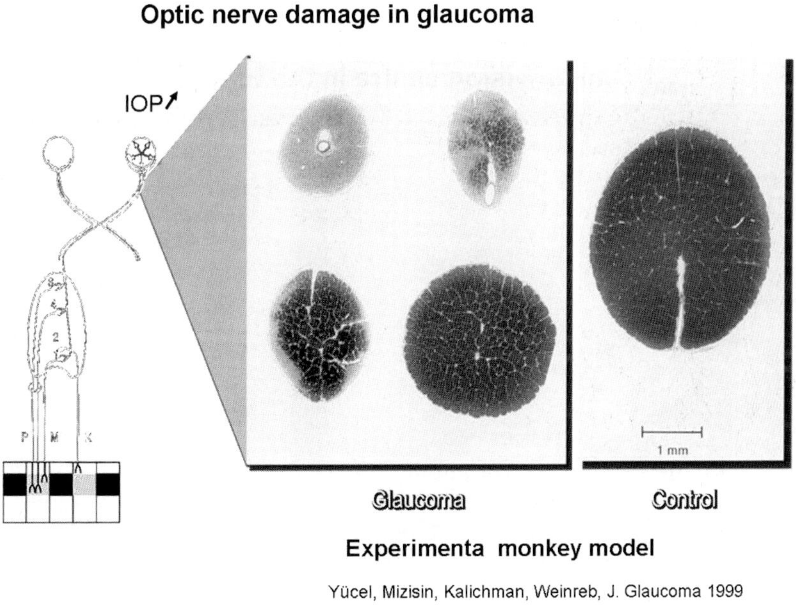

Fig. 4. Following experimentally induced elevated intraocular pressure in monkey, glaucomatous optic nerves show overall atrophy and varying degree of optic nerve fiber loss compared to the control optic nerve on the right (myelin stain in black). The bar indicates 1 mm. Adapted with permission from Yücel et al. (1999).

et al., 1997; Crawford et al., 2000). Loss of magno- and parvocellular relay neurons in this model occurred (Weber et al., 2000; Yücel et al., 2000) and increased with optic nerve fiber loss and mean IOP (Yücel et al., 2003) (Fig. 5). Surviving neurons showed atrophy that also increased with these parameters (Yücel et al., 2001), more pronounced in parvocellular layers (Yücel et al., 2001). The latter was even observed in monkeys with ocular hypertension without detectable optic nerve fiber loss (Yücel et al., 2001). In fact, this group also showed reduced LGN dendrite complexity and field area (Gupta et al., 2007), indicating that not all changes are attributable to deafferentation of the visual system.

In the koniocellular pathway, the alpha subunit of type II calmodulin-dependent protein kinase (CaMK-II alpha), a selective marker for these neurons and a major postsynaptic density protein (Hendry and Reid, 2000), was reduced in primate glaucoma (Yücel et al., 2003). Decreased immunoreactivity was observed also in ocular hypertensive monkeys without evidence of optic nerve fiber loss (Yücel et al., 2003). This finding suggests that neurochemical alterations at the synaptic level occur at early stages of LGN injury to this blue–yellow pathway. Thus, transsynaptic degeneration of the LGN in primate glaucoma appears to affect the three major vision channels, namely the magno-, parvo-, and koniocellular pathways (Yücel et al., 2003).

While changes to relay neurons in experimental glaucoma are described above, it is not known whether GABAergic interneurons in the LGN are also altered in glaucoma as has been observed following enucleation and monocular deprivation (Hendry, 1991).

Glial cells including astrocytes and NG2 cells also appeared altered in experimental glaucoma involving the optic tract and LGN (Yücel et al., unpublished data). Further studies are needed to characterize the involvement of other cell types such as microglia (Wang et al., 2000), vascular (Luthra et al., 2005), and perivascular cells in the LGN in glaucoma.

Mechanisms related to transsynaptic degeneration in glaucoma may be relevant to strategies

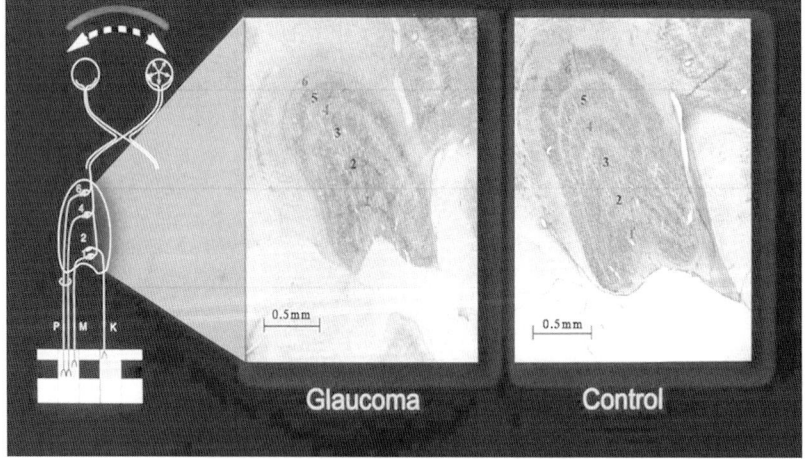

Yücel, Zhang, Gupta, Kaufman & Weinreb, Arch Ophthalmol, 2000

Fig. 5. Cross-sections of the right lateral geniculate nucleus in control (right) and glaucomatous monkeys (left) show six distinct neuronal layers. Compared to the control, in glaucoma, overall atrophy is observed. Parvalbumin stains relay neurons, and immunoreactivity in layers 1, 4, and 6 connected to the glaucomatous right eye is decreased. Adapted with permission from Yücel et al. (2000). Copyright © (2000), American Medical Association.

aimed at preventing the spread of disease to visual centers in the brain and presumably disease progression.

Visual cortex in glaucoma

In the primary visual cortex, neurons are arranged into six layers subdivided into sublayers. The M, P, and K geniculate axons terminate in sublayers of layer 4 and superficial layers in eye-specific columns (Callaway, 2004). In monkeys with unilateral glaucoma, a relative decrease in metabolic activity was detected with cytochrome oxidase activity in ocular dominance columns driven by the glaucomatous eye, compared to those driven by the fellow non-glaucoma eye (Vickers et al., 1997; Crawford et al., 2000, 2001; Yücel et al., 2003) (Fig. 6). Additional evidence of neuroplasticity of the visual system in primate glaucoma is suggested by neurochemical changes in the visual cortex involving the presynaptic molecule, growth-cone-associated protein-43 (GAP43), and inhibitory neurotransmitter receptor GABA A receptor subtype (Lam et al., 2003). Inspite of observed metabolic and neurochemical changes, evidence of neuronal loss in the primary visual cortex in this model is lacking. However, relative changes in metabolic activity of ocular dominance columns appeared more pronounced with increasing optic nerve fiber loss (Yücel et al., 2003).

Neuropathology of glaucoma in the visual pathways in the human brain

In a case of advanced human glaucoma compared to age-matched normal controls, neuropathological analysis of visual pathways revealed degeneration in the intracranial optic nerve, LGN, and visual cortex, corresponding to the visual field defects (Gupta et al., 2006a). In the presence of vision loss in glaucoma patients, pathology in vision centers of the brain may exist. In a previous report of human glaucoma, LGN neuron density appeared decreased (Chaturvedi et al., 1993). Findings of degenerative changes in the central visual pathways in human glaucoma are consistent

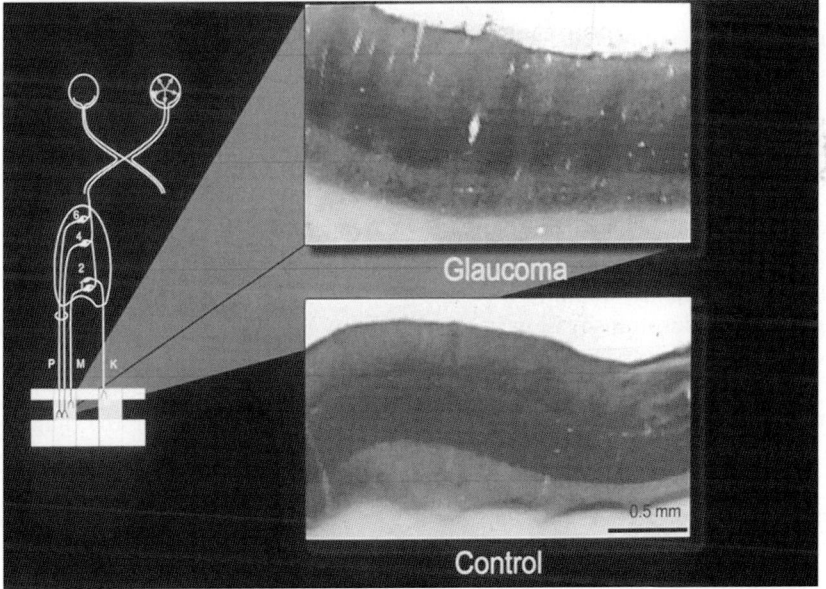

Fig. 6. Normal primate visual cortex section stained with cytochrome oxidase, an activity marker, shows continuous and homogeneous dark staining. In contrast, glaucomatous visual cortex shows alternating light and dark bands corresponding to ocular dominance columns driven by the glaucoma and non-glaucomatous fellow eyes, respectively.

with changes observed in experimental primate glaucoma (Gupta et al., 2006a).

Mechanisms of glaucoma damage in the central visual pathways

Among a number of proposed initiating mechanisms of RGC and optic nerve degeneration (Weinreb and Khaw, 2004), there is evidence that oxidative injury (Luthra et al., 2005) and glutamate excitotoxicity (Yücel et al., 2006) are implicated in transsynaptic degeneration in experimental primate glaucoma.

In oxidative injury, the accumulation of reactive oxygen species alters cellular and molecular pathways to induce cell death (Chong et al., 2005). Oxygen species can react with nitric oxide to form peroxynitrite, which mediates protein nitration to produce nitrotyrosine (Bian et al., 2006). In neurodegenerative diseases, the finding of nitrotyrosine is considered a footprint of peroxynitrate-mediated oxidative injury (Nunomura et al., 2001). Nitrotyrosine has been found in the neural parenchyma and blood vessels of the LGN in experimental primate glaucoma, suggesting a role for oxidative injury in the LGN in the blood vessels and neuronal layers connected to glaucomatous eye (Luthra et al., 2005). Oxidative injury in LGN layers connected to the fellow non-glaucoma eye might suggest differential metabolic needs between LGN layers driven by the glaucoma eye and the intercalated layers driven by the non-glaucoma fellow eye, challenging the autoregulatory capacity of common supplying blood vessels.

The glutamatergic system is ubiquitous and responsible for excitatory neurotransmission in the brain. Under conditions of excessive stimulation, neuron toxicity can occur, leading to intracellular calcium overload and neuron death (Hynd et al., 2004). This pathological process called glutamate excitotoxicity is implicated in a number of neurodegenerative diseases (Hynd et al., 2004) as in glaucomatous neural degeneration in the retina and optic nerve (Hare et al., 2001). Using pharmacological agents such as memantine, an open-channel blocker of the N-methyl-D-aspartic acid (NMDA) subtype, overstimulation of the glutamatergic system can be blocked. Memantine crosses the blood–brain barrier in monkey glaucoma and may block NMDA receptors at the levels of LGN (Tighilet et al., 1998), retina (Grunert et al., 2002), and visual cortex (Rosier et al., 1993). Monkeys with glaucoma, given daily doses of memantine and compared to vehicle-treated glaucoma monkeys, showed attenuated neuronal shrinkage of LGN relay neurons (Yücel et al., 2006). However, the same glaucoma animals treated with memantine did not show statistically significant differences in neuronal numbers compared to vehicle-treated glaucoma animals, suggesting that blockage of excitotoxicity by memantine (4 mg/kg) had no significant effect on neuronal death in LGN in experimental glaucoma.

LGN and visual cortex damage by glaucoma and/or neurodegenerative diseases may increase the susceptibility of surviving RGCs to ongoing injury by reducing their LGN trophic support and contribute to glaucomatous progression. Although the exact role of trophic factors in transsynaptic degeneration is not known, ocular delivery of BDNF has been shown to have a protective effect on RGCs (Martin et al., 2003). BDNF might be an anterograde trophic factor for survival of target neurons as seen during development (Caleo et al., 2000). Neurotrophic factors may play their effects by acting on different neural targets, such as LGN (Riddle et al., 1995), intracortical circuitry, and subcortical afferents (Berardi and Maffei, 1999).

Implications of central visual system injury in glaucoma

Glaucoma is a disease affecting the entire RGC with intraocular and intracranial components. Understanding within geniculo-cortical pathways has major implications for the diagnosis and management of the disease to slow progressive loss of sight.

In Alzheimer's disease, disease progression has been studied by measuring hippocampal atrophy (Mungas et al., 2005). Similarly, assessment of visual pathways using modern neuroimaging such as magnetic resonance imaging (MRI) to visualize the LGN (Horton et al., 1990; Fujita et al., 2001) may be useful to assess structural changes in vision centers along the components within retino-geniculo-cortical pathway in the brain. Infact, a recent 1.5 Tesla MRI study showed LGN atrophy in glaucoma patients compared to controls (Gupta et al., 2008a). Functional neuroimaging (fMRI), using the blood oxygen level dependent (BOLD) fMRI response, was decreased in human primary visual cortex in patients with primary open-angle glaucoma (Duncan et al., 2007). In future, modern functional and structural neuroimaging technologies may prove useful in assessing the spread of glaucomatous damage within the CNS. Multifocal and evoked electrophysiological measurements may be relevant to the detection of dysfunction along visual pathways (Korth et al., 1994; Klistorner and Graham, 1999; Graham et al., 2000; Hood and Greenstein, 2003). Optimization of the techniques to identify and measure structural and functional parameters in vivo in glaucoma patients is needed to better characterize brain changes in glaucoma. The choice of a reliable biomarker would help detect progression and/or a population at a high risk of progression. Non-geniculo-cortical pathways involved in eye movements and reflexes may be worth exploring in glaucoma. For example, the superior colliculus is an important visual structure for eye movements. In glaucomatous monkeys, reduced RGC terminals expressing vesicular glutamate transporter-2, a presynaptic marker of RGC terminals, were observed in superficial layers of the superior colliculus compared to controls (Alarcon-Martinez et al., unpublished data).

Cortical binocular functions such as stereovision (Bassi and Galanis, 1991; Essock et al., 1996; Gupta et al., 2006b) may also be useful to assess visual dysfunction in glaucoma.

The loss of visual field in moderate-to-advanced disease is a representation of damage to central visual pathways in glaucoma. Teasing apart visual functions selective for specific central vision pathways provides an opportunity to understand their relative contributions to visual disturbances in patients with glaucoma, and to functionally assess potential effects of candidate neuroprotective drugs in pathways with preferential functions such as magno-, parvo-, and koniocellular pathways.

Increased susceptibility of RGCs to ongoing glaucomatous injury has been described as a determinant in progression of the disease (Abedin et al., 1982). We suggest that degeneration with neuronal loss and atrophy of target neurons in the LGN may alter the normal function of surviving RGCs in glaucoma. In fact, degenerative changes in RGCs are observed following the loss of target cells in the LGN (Payne et al., 1984; Pearson and Stoffler, 1992) and lesions of the striate cortex (Cowey et al., 1989; Johnson and Cowey, 2000). Changes in the visual stations may deplete growth factor sources to be transported back to the retina, contributing to the susceptibility of surviving RGCs to ongoing glaucomatous injury and progression, and studies are needed to test this hypothesis.

Some cellular mechanisms in glaucoma appear similar to neurodegenerative diseases. Abnormal tau protein, a pathological hallmark of neurodegenerative diseases such as Alzheimer's, has also been detected in human glaucoma retina, in horizontal cells (Gupta et al., 2008b). A susceptibility gene for Alzheimer's disease complicated in neural repair, apolipoprotein epsilon 4 allele, is reported to be associated with low-tension glaucoma in a population (Vickers et al., 2002), but could not be confirmed in other populations (Lam et al., 2006; Zetterberg et al., 2007). The coexistence of glaucoma and Alzheimer's disease might not be purely coincidental (Bayer et al., 2002; McKinnon, 2003; Gupta and Yücel, 2007).

Comparing brain changes in experimental glaucoma and human glaucoma to those seen in other types of injury such as ischemic optic neuropathy (Orgül et al., 1996; Brooks et al., 2004), optic nerve transection, enucleation, and monocular visual deprivation (Hendry, 1991; Riddle et al., 1995) is needed. These experiments may help to elucidate how altered retinal input to relay cells with less than 10% of total synaptic input can cause such atrophy. Can this model of glaucoma help us to better understand the role of the thalamus beyond a relay station between retina and cortex (Sherman and Guillery, 2004)? Some of these questions might be best answered by well-characterized rodent models of glaucoma (Morrison et al., 1997; Lindsey and Weinreb, 2005; Morrison, 2005; Weinreb and Lindsey, 2005).

Recent epidemiologic evidence showed that ocular perfusion pressure, a parameter that is based on the difference between blood pressure and IOP, is a risk factor for incident glaucoma (Leske et al., 2008) and for the progression of the disease (Leske et al., 2007). The relationship between glaucoma and cerebrovascular diseases requires further elucidation. For example, small blood vessels with small silent infarcts and white matter lesions may occur in watershed areas of terminal cerebral arteries supplying posterior visual pathways. It is unclear how this might alter visual function in glaucoma patients and contribute to progression of vision loss (Yücel and Gupta, 2008).

Treatment to lower IOP prior to significant RGC loss is likely an important strategy to prevent the transsynaptic spread of damage to target visual neurons. In addition to lowering IOP, future strategies to treat glaucoma may consider protecting visual neurons in the retina and brain. Since retinal cell survival is dependent on its ability to effectively connect to target neurons, assessment of the health of target neurons (Yücel and Gupta, 2007) is critical for new treatment strategies such as ocular neuroprotection (Weinreb, 2007), gene therapy (Sieving et al., 2006; Liu et al., 2007; MacDonald et al., 2007), and stem cell transplant strategies (Levin et al., 2004; Wallace, 2007). In this context, aberrant axonal outgrowth may alter visual function while adequate axonal outgrowth may restore it. New strategies to improve axonal outgrowth such as graft after peripheral nerve injury may be relevant (Lundborg, 2003).

Further studies of plasticity in the brain following LGN degeneration in glaucoma are needed.

Conclusion

Elevated IOP and injury to RGCs can trigger degeneration in distant connected neurons in major vision centers of the brain. Lowering IOP is an important strategy to prevent RGC death in the eye and may reduce the risk of CNS degeneration in glaucoma. In patients with progressive vision loss despite adequate IOP control, secondary pathological changes occur in visual centers of the brain. Thus, IOP-lowering strategies combined with therapies to protect retina and central visual system neurons offer new opportunities to prevent blindness from glaucoma.

Glaucomatous injury to the visual system is an ideal model to study transsynaptic degeneration, as it relates to neurodegenerative diseases such as Alzheimer's. The effects of initial damage are readily apparent, reproducible, and measurable, as are the effects of subsequent injury along well-defined anatomical and functional connections within vision pathways of the brain. Mechanisms elucidated from this glaucoma model of transsynaptic brain injury may be highly relevant to other neurodegenerative diseases, helping to understand complex pathways and to identify future strategies to prevent progressive functional decline in disease states.

Acknowledgments

This work was supported in part by Canadian Institutes of Health Research, Glaucoma Research Society of Canada and The Fred Jarvis Fund.

References

Abedin, S., Simmons, R.J. and Grant, W.M. (1982) Progressive low-tension glaucoma: treatment to stop glaucomatous cupping and field loss when these progress despite normal intraocular pressure. Ophthalmology, 89(1): 1–6.

Bassi, C.J. and Galanis, J.C. (1991) Binocular visual impairment in glaucoma. Ophthalmology, 98(9): 1406–1411.

Bayer, A.U., Ferrari, F. and Erb, C. (2002) High occurrence rate of glaucoma among patients with Alzheimer's disease. Eur. Neurol., 47(3): 165–168.

Berardi, N. and Maffei, L. (1999) From visual experience to visual function: roles of neurotrophins. J. Neurobiol., 41(1): 119–126.

Bian, K., Ke, Y., Kamisaki, Y. and Murad, F. (2006) Proteomic modification by nitric oxide. J. Pharmacol. Sci., 101(4): 271–279.

Brooks, D.E., Kallberg, M.E., Cannon, R.L., Komaromy, A.M., Ollivier, F.J., Malakhova, O.E., Dawson, W.W., Sherwood, M.B., Kuekuerichkina, E.E. and Lambrou, G.N. (2004) Functional and structural analysis of the visual system in the rhesus monkey model of optic nerve head ischemia. Invest. Ophthalmol. Vis. Sci., 45(6): 1830–1840.

Burgoyne, C.F., Mercante, D.E. and Thompson, H.W. (2002) Change detection in regional and volumetric disc parameters using longitudinal confocal scanning laser tomography. Ophthalmology, 109(3): 455–466.

Caleo, M., Menna, E., Chierzi, S., Cenni, M.C. and Maffai, L. (2000) Brain-derived neurotrophic factor is an anterograde survival factor in the rat visual system. Curr. Biol., 10(19): 1155–1161.

Callaway, E.M. (2004) Cell types and local circuits in primary visual cortex of the macaque monkey. In: Chalupa L.M. and Werner J.S. (Eds.), The Visual Neurosciences. Massachusetts Institute of Technology Press, Cambridge, MA, pp. 680–694.

Chaturvedi, N., Hedley-Whyte, E.T. and Dreyer, E.B. (1993) Lateral geniculate nucleus in glaucoma. Am. J. Ophthalmol., 116(2): 182–188.

Chong, Z.Z., Li, F. and Maiese, K. (2005) Oxidative stress in the brain: novel cellular targets that govern survival during neurodegenerative disease. Prog. Neurobiol., 75(3): 207–246.

Cowey, A., Stoerig, P. and Perry, V.H. (1989) Transneuronal retrograde degeneration of retinal ganglion cells after damage to striate cortex in macaque monkeys: selective loss of P beta cells. Neuroscience, 29(1): 65–80.

Crawford, M.L., Harwerth, R.S., Smith, E.L., 3rd, Mills, S. and Ewing, B. (2001) Experimental glaucoma in primates: changes in cytochrome oxidase blobs in V1 cortex. Invest. Ophthalmol. Vis. Sci., 42(2): 358–364.

Crawford, M.L., Harwerth, R.S., Smith, E.L., 3rd, Shen, F. and Carter-Dawson, L. (2000) Glaucoma in primates: cytochrome oxidase reactivity in parvo- and magnocellular pathways. Invest. Ophthalmol. Vis. Sci., 41(7): 1791–1802.

Dacey, D.M., Peterson, B.B., Robinson, F.R. and Gamlin, P.D. (2003) Fireworks in the primate retina: in vitro photodynamics reveals diverse LGN-projecting ganglion cell types. Neuron, 37(1): 15–27.

Duncan, R.O., Sample, P.A., Weinreb, R.N., Bowd, C. and Zangwill, L.M. (2007) Retinotopic organization of primary visual cortex in glaucoma: comparing fMRI measurements of cortical function with visual field loss. Prog. Retin. Eye Res., 26(1): 38–56.

Essock, E.A., Fechtner, R.D., Zimmerman, T.J., Krebs, W.K. and Nussdorf, J.D. (1996) Binocular function in early glaucoma. J. Glaucoma, 5(6): 395–405.

Feilchenfeld, Z., Yücel, Y.H. and Gupta, N. (2008) Oxidative injury to blood vessels and glia of the pre-laminar optic nerve head in human glaucoma. Exp. Eye Res. Published online August 5, 2008, doi: 10.1016/j.exer.2008.07.011.

Frishman, L.J., Shen, F.F., Du, L., Robson, J.G., Harwerth, R.S., Smith, E.L., 3rd, Carter-Dawson, L. and Crawford, M.L. (1996) The scotopic electroretinogram of macaque after retinal ganglion cell loss from experimental glaucoma. Invest. Ophthalmol. Vis. Sci., 37(1): 125–141.

Fujita, N., Tanaka, H., Takanashi, M., Hirabuki, N., Abe, K., Yoshimura, H. and Nakamura, H. (2001) Lateral geniculate nucleus: anatomic and functional identification by use of MR imaging. AJNR Am. J. Neuroradiol., 22(9): 1719–1726.

Gaasterland, D. and Kupfer, C. (1974) Experimental glaucoma in the rhesus monkey. Invest. Ophthalmol., 13(6): 455–457.

Gaasterland, D., Tanishima, T. and Kuwabara, T. (1978) Axoplasmic flow during chronic experimental glaucoma. 1. Light and electron microscopic studies of the monkey optic nervehead during development of glaucomatous cupping. Invest. Ophthalmol. Vis. Sci., 17(9): 838–846.

Gabelt, B.T., Ver Hoeve, J.N. and Kaufman, P.L. (2003) Intraocular pressure elevation: Laser photocoagulation of the trabecular meshwork. In: Levin L. and DiPolo A. (Eds.), Ocular Neuroprotection. Marcel Dekker, New York, pp. 47–84.

Glovinsky, Y., Quigley, H.A. and Dunkelberger, G.R. (1991) Retinal ganglion cell loss is size dependent in experimental glaucoma. Invest. Ophthalmol. Vis. Sci., 32(3): 484–491.

Goebel, R., Muckli, L. and Kim, D-S. (2004) Visual system. In: Paxinos G. and Mai J.K. (Eds.), The Human Nervous System (2nd edn.). Elsevier, London, UK, pp. 1280–1305.

Goldby, F. (1957) A note on transneuronal atrophy in the human lateral geniculate body. J. Neurol. Neurosurg. Psychiatr., 20(3): 202–207.

Graham, S.L., Klistorner, A.I., Grigg, J.R. and Billson, F.A. (2000) Objective VEP perimetry in glaucoma: asymmetry analysis to identify early deficits. J. Glaucoma, 9(1): 10–19.

Grunert, U., Haverkamp, S., Fetcher, E.L. and Wassle, H. (2002) Synaptic distribution of ionotropic glutamate receptors in the inner plexiform layer of the primate retina. J. Comp. Neurol., 447(2): 138–151.

Gupta, N., Ang, L.C., Noel de Tilly, L., Bidaisee, L. and Yücel, Y.H. (2006a) Human glaucoma and neural degeneration in intracranial optic nerve, lateral geniculate nucleus, and visual cortex. Br. J. Ophthalmol., 90(6): 674–678.

Gupta, N., Fong, J., Ang, L.C. and Yücel, Y.H. (2008b) Retinal tau pathology in human glaucomas. Can. J. Ophthalmol., 43(1): 53–60.

Gupta, N., Greenberg, G., Noel de Tilly, L., Gray, B., Polemidiotis, M. and Yücel, Y.H. (2008a) Atrophy of the lateral geniculate nucleus in human glaucoma by magnetic resonance imaging. Br. J. Ophthalmol., BJO Published Online First: 12 August, 2008. doi:10.1136/bjo.2008.138172.

Gupta, N., Krishnadev, N., Hamstra, S.J. and Yücel, Y.H. (2006b) Depth perception deficits in glaucoma suspects. Br. J. Ophthalmol., 90(8): 979–981.

Gupta, N., Ly, T., Zhang, Q., Kaufman, P.L., Weinreb, R.N. and Yücel, Y.H. (2007) Chronic ocular hypertension induces dendrite pathology in the lateral geniculate nucleus of the brain. Exp. Eye Res., 84(1): 176–184.

Gupta, N. and Yücel, Y.H. (2003) Brain changes in glaucoma. Eur. J. Ophthalmol., 13(Suppl 3): S32–S35.

Gupta, N. and Yücel, Y.H. (2007) Glaucoma as a neurodegenerative disease. Curr. Opin. Ophthalmol., 18(2): 110–114.

Hare, W., Ton, H., WoldeMussie, E., Ruiz, G., Feldmann, B. and Wijono, M. (1999) Electrophysiological and histological measures of retinal injury in chronic ocular hypertensive monkeys. Eur. J. Ophthalmol., 9(Suppl 1): S30–S33.

Hare, W., WoldeMussie, E., Lai, R., Ton, H., Ruiz, G., Feldmann, B., Wijono, M., Chun, T. and Wheeler, L. (2001)

Efficacy and safety of memantine, an NMDA-type open-channel blocker, for reduction of retinal injury associated with experimental glaucoma in rat and monkey. Surv. Ophthalmol., 45(Suppl 3): S284–S289. discussion S295–S296.

Hare, W.A., WoldeMussie, E., Weinreb, R.N., Ton, H., Ruiz, G., Wijono, M., Feldmann, B., Zangwill, L. and Wheeler, L. (2004) Efficacy and safety of memantine treatment for reduction of changes associated with experimental glaucoma in monkey, II: structural measures. Invest. Ophthalmol. Vis. Sci., 45(8): 2640–2651.

Harwerth, R.S., Crawford, M.L., Frishman, L.J., Viswanathan, S., Smith, E.L., 3rd and Carter-Dawson, L. (2002) Visual field defects and neural losses from experimental glaucoma. Prog. Retin. Eye Res., 21(1): 91–125.

Heijl, A., Leske, M.C., Bengtsson, B., Hyman, L. and Hussein, M. (2002) Reduction of intraocular pressure and glaucoma progression: results from the Early Manifest Glaucoma Trial. Arch. Ophthalmol., 120(10): 1268–1279.

Hendry, S.H. (1991) Delayed reduction in GABA and GAD immunoreactivity of neurons in the adult monkey dorsal lateral geniculate nucleus following monocular deprivation or enucleation. Exp. Brain Res., 86(1): 47–59.

Hendry, S.H. and Reid, R.C. (2000) The koniocellular pathway in primate vision. Annu. Rev. Neurosci., 23: 127–153.

Hood, D.C. and Greenstein, V.C. (2003) Multifocal VEP and ganglion cell damage: applications and limitations for the study of glaucoma. Prog. Retin. Eye Res., 22(2): 201–251.

Horton, J.C., Landau, K., Maeder, P. and Hyot, W.F. (1990) Magnetic resonance imaging of the human lateral geniculate body. Arch. Neurol., 47(11): 1201–1206.

Hynd, M.R., Scott, H.L. and Dodd, P.R. (2004) Glutamate-mediated excitotoxicity and neurodegeneration in Alzheimer's disease. Neurochem. Int., 45(5): 583–595.

Johnson, H. and Cowey, A. (2000) Transneuronal retrograde degeneration of retinal ganglion cells following restricted lesions of striate cortex in the monkey. Exp. Brain Res., 132(2): 269–275.

Kaplan, E. (2004) The M, P and K pathways in the primate visual system. In: Chalupa L.M. and Werner J.S. (Eds.), The Visual Neurosciences. Massachusetts Institute of Technology Press, Cambridge, MA, pp. 481–493.

Klistorner, A.I. and Graham, S.L. (1999) Multifocal pattern VEP perimetry: analysis of sectoral waveforms. Doc. Ophthalmol., 98(2): 183–196.

Korth, M., Nguyen, N.X., Junemann, A., Martus, P. and Jonas, J.B. (1994) VEP test of the blue-sensitive pathway in glaucoma. Invest. Ophthalmol. Vis. Sci., 35(5): 2599–2610.

Kume, A., Takahashi, A. and Hashizume, Y. (1993) Neuronal cell loss of the striatonigral system in multiple system atrophy. J. Neurol. Sci., 117(1–2): 33–40.

Lam, D.Y., Fan, B.J., Wang, D.Y., Tam, P.O., Yung Tham, C.C., Leung, D.Y., Ping Fan, D.S., Chiu Lam, D.S. and Pang, C.P. (2006) Association of apolipoprotein E polymorphisms with normal tension glaucoma in a Chinese population. J. Glaucoma, 15(3): 218–222.

Lam, D.Y., Kaufman, P.L., Gabelt, B.T., To, E.C. and Matsubara, J.A. (2003) Neurochemical correlates of cortical plasticity after unilateral elevated intraocular pressure in a primate model of glaucoma. Invest. Ophthalmol. Vis. Sci., 44(6): 2573–2581.

Leske, M.C., Heijl, A., Hyman, L., Bengtsson, B., Dong, L. and Yang, Z. (2007) Predictors of long-term progression in the early manifest glaucoma trial. Ophthalmology, 114(11): 1965–1972.

Leske, M.C., Wu, S.Y., Hennis, A., Honkanen, R. and Nemesure, B. (2008) Risk factors for incident open-angle glaucoma: the Barbados eye studies. Ophthalmology, 115(1): 85–93.

Levin, L.A., Ritch, R., Richards, J.E. and Borras, T. (2004) Stem cell therapy for ocular disorders. Arch. Ophthalmol., 122(4): 621–627.

Lindsey, J.D. and Weinreb, R.N. (2005) Elevated intraocular pressure and transgenic applications in the mouse. J. Glaucoma, 14(4): 318–320.

Liu, X., Brandt, C.R., Rasmussen, C.A. and Kaufman, P.L. (2007) Ocular drug delivery: molecules, cells, and genes. Can. J. Ophthalmol., 42(3): 447–454.

Lundborg, G. (2003) Richard P. Bunge Memorial Lecture. Nerve injury and repair — a challenge to the plastic brain. J. Peripher. Nerv. Syst., 8(4): 209–226.

Luthra, A., Gupta, N., Kaufman, P.L., Weinreb, R.N. and Yücel, Y.H. (2005) Oxidative injury by peroxynitrite in neural and vascular tissue of the lateral geniculate nucleus in experimental glaucoma. Exp. Eye Res., 80(1): 43–49.

MacDonald, I.M., Sauvé, Y. and Sieving, P.A. (2007) Preventing blindness in retinal disease: ciliary neurotrophic factor intraocular implants. Can. J. Ophthalmol., 42(3): 399–402.

Martin, K.R., Quigley, H.A., Zack, D.J., Levkovitch-Verbin, H., Kielczewski, J., Valenta, D., Baumrind, L., Pease, M.E., Klein, R.L. and Huswirth, W.W. (2003) Gene therapy with brain-derived neurotrophic factor as a protection: retinal ganglion cells in a rat glaucoma model. Invest. Ophthalmol. Vis. Sci., 44(10): 4357–4365.

Matthews, M.R. (1964) Further observations on transneuronal degeneration in the lateral geniculate nucleus of the macaque monkey. J. Anat., 98: 255–263.

Matthews, M.R., Cowan, W.M. and Powell, T.P. (1960) Transneuronal cell degeneration in the lateral geniculate nucleus of the macaque monkey. J. Anat., 94(Pt 2): 145–169.

McKinnon, S.J. (2003) Glaucoma: ocular Alzheimer's disease? Front. Biosci., 8: s1140–s1156.

Montero, V.M. and Zempel, J. (1986) The proportion and size of GABA-immunoreactive neurons in the magnocellular and parvocellular layers of the lateral geniculate nucleus of the rhesus monkey. Exp. Brain Res., 62(1): 215–223.

Morgan, J.E., Uchida, H. and Caprioli, J. (2000) Retinal ganglion cell death in experimental glaucoma. Br. J. Ophthalmol., 84(3): 303–310.

Morrison, J.C. (2005) Elevated intraocular pressure and optic nerve injury models in the rat. J. Glaucoma, 14(4): 315–317.

Morrison, J.C., Moore, C.G., Deppmeier, L.M., Gold, B.G., Meshul, C.K. and Johnson, E.C. (1997) A rat model of chronic pressure-induced optic nerve damage. Exp. Eye Res., 64(1): 85–96.

Mungas, D., Harvey, D., Reed, B.R., Jagust, W.J., DeCarli, C., Beckett, L., Mack, W.J., Kramer, J.H., Weiner, M.W., Schuff, N. and Chiu, H.C. (2005) Longitudinal volumetric MRI change and rate of cognitive decline. Neurology, 65(4): 565–571.

Nickells, R.W. (2007) From ocular hypertension to ganglion cell death: a theoretical sequence of events leading to glaucoma. Can. J. Ophthalmol., 42(2): 278–287.

Nunomura, A., Perry, G., Aliev, G., Hirai, K., Takeda, A., Balraj, E.K., Jones, P.K., Ghanbari, H., Wataya, T., Shimohama, S., Chiba, S., Atwood, C.S., Petersen, R.B. and Smith, M.A. (2001) Oxidative damage is the earliest event in Alzheimer disease. J. Neuropathol. Exp. Neurol., 60(8): 759–767.

Örgül, S., Cioffi, G.A., Bacon, D.R. and Van Buskirk, E.M. (1996) An endothelin-1-induced model of chronic optic nerve ischemia in rhesus monkeys. J. Glaucoma, 5(2): 135–138.

Payne, B.R., Pearson, H.E. and Cornwell, P. (1984) Transneuronal degeneration of beta retinal ganglion cells in the cat. Proc. R. Soc. Lond. B. Biol. Sci., 222(1226): 15–32.

Pearson, H.E. and Stoffler, D.J. (1992) Retinal ganglion cell degeneration following loss of postsynaptic target neurons in the dorsal lateral geniculate nucleus of the adult cat. Exp. Neurol., 116(2): 163–171.

Perry, V.H., Oehler, R. and Cowey, A. (1984) Retinal ganglion cells that project to the dorsal lateral geniculate nucleus in the macaque monkey. Neuroscience, 12(4): 1101–1123.

Quigley, H.A. (1999) Neuronal death in glaucoma. Prog. Retin. Eye Res., 18(1): 39–57.

Quigley, H.A. and Addicks, E.M. (1980) Chronic experimental glaucoma in primates. II. Effect of extended intraocular pressure elevation on optic nerve head and axonal transport. Invest. Ophthalmol. Vis. Sci., 19(2): 137–152.

Quigley, H.A., Addicks, E.M., Green, W.R. and Maumenee, A.E. (1981) Optic nerve damage in human glaucoma. II. The site of injury and susceptibility to damage. Arch. Ophthalmol., 99(4): 635–649.

Quigley, H.A. and Anderson, D.R. (1976) The dynamics and location of axonal transport blockade by acute intraocular pressure elevation in primate optic nerve. Invest. Ophthalmol., 15(8): 606–616.

Quigley, H.A. and Broman, AT. (2006) The number of people with glaucoma worldwide in 2010 and 2020. Br. J. Ophthalmol., 90(3): 262–267.

Radius, R.L. and Anderson, D.R. (1981a) Rapid axonal transport in primate optic nerve. Distribution of pressure-induced interruption. Arch. Ophthalmol., 99(4): 650–654.

Radius, R.L. and Anderson, D.R. (1981b) Morphology of axonal transport abnormalities in primate eyes. Br. J. Ophthalmol., 65(11): 767–777.

Riddle, D.R., Lo, D.C. and Katz, L.C. (1995) NT-4-mediated rescue of lateral geniculate neurons from effects of monocular deprivation. Nature, 378(6553): 189–191.

Rosier, A.M., Arckens, L., Orban, G.A. and Vandesande, F. (1993) Laminar distribution of NMDA receptors in cat and monkey visual cortex visualized by [3H]-MK-801 binding. J. Comp. Neurol., 335(3): 369–380.

Saper, C.B., Wainer, B.H. and German, D.C. (1987) Axonal and transneuronal transport in the transmission of neurological disease: potential role in system degenerations, including Alzheimer's disease. Neuroscience, 23(2): 389–398.

Sherman, S.M. and Guillery, R.W. (2004) The visual relays in the thalamus. In: Chalupa L.M. and Werner J.S. (Eds.), The Visual Neurosciences. Massachusetts Institute of Technology Press, Cambridge, MA, pp. 565–591.

Sieving, P.A., Caruso, R.C., Tao, W., Coleman, H.R., Thompson, D.J., Fullmer, K.R. and Bush, R.A. (2006) Ciliary neurotrophic factor (CNTF) for human retinal degeneration: phase I trial of CNTF delivered by encapsulated cell intraocular implants. Proc. Natl. Acad. Sci. USA, 103(10): 3896–3901.

Su, J.H., Deng, G. and Cotman, C.W. (1997) Transneuronal degeneration in the spread of Alzheimer's disease pathology: immunohistochemical evidence for the transmission of tau hyperphosphorylation. Neurobiol. Dis., 4(5): 365–375.

Suzuki, H., Oyanagi, K., Takahashi, H. and Ikuta, F. (1995) Evidence for transneuronal degeneration in the spinal cord in man: a quantitative investigation of neurons in the intermediate zone after long-term amputation of the unilateral upper arm. Acta Neuropathol., 89(5): 464–470.

Tighilet, B., Hashikawa, T. and Jones, E.G. (1998) Cell-specific expression of type II calcium/calmodulin-dependent protein kinase isoforms and glutamate receptors in normal and visually deprived lateral geniculate nucleus of monkeys. J. Comp. Neurol., 390(2): 278–296.

Van Horn, S.C., Erişir, A. and Sherman, S.M. (2000) Relative distribution of synapses in the A-laminae of the lateral geniculate nucleus of the cat. J. Comp. Neurol., 416(4): 509–520.

Vickers, J.C., Craig, J.E., Stankovich, J., McCormack, G.H., West, A.K., Dickinson, J.L., McCartney, P.J., Coote, M.A., Healey, D.L. and Mackey, D.A. (2002) The apolipoprotein epsilon4 gene is associated with elevated risk of normal tension glaucoma. Mol. Vis., 8: 389–393.

Vickers, J.C., Hof, P.R., Schumer, R.A., Wang, R.F., Podos, S.M. and Morrison, J.H. (1997) Magnocellular and parvocellular visual pathways are both affected in a macaque monkey model of glaucoma. Aust. N.Z. J. Ophthalmol., 25(3): 239–243.

Wallace, V.A. (2007) Stem cells: a source for neuron repair in retinal disease. Can. J. Ophthalmol., 42(3): 442–446.

Wang, X., Sam-Wah Tay, S. and Ng, Y.K. (2000) Nitric oxide, microglial activities and neuronal cell death in the lateral geniculate nucleus of glaucomatous rats. Brain Res., 878 (1–2): 136–147.

Weber, A.J., Chen, H., Hubbard, W.C. and Kaufman, P.L. (2000) Experimental glaucoma and cell size, density, and number in the primate lateral geniculate nucleus. Invest. Ophthalmol. Vis. Sci., 41(6): 1370–1379.

Weber, A.J. and Harman, C.D. (2005) Structure-function relations of parasol cells in the normal and glaucomatous primate retina. Invest. Ophthalmol. Vis. Sci., 46(9): 3197–3207.

Weber, A.J., Kaufman, P.L. and Hubbard, W.C. (1998) Morphology of single ganglion cells in the glaucomatous primate retina. Invest. Ophthalmol. Vis. Sci., 39(12): 2304–2320.

Weinreb, R.N. (2007) Glaucoma neuroprotection: what is it? Why is it needed? Can. J. Ophthalmol., 42(3): 396–398.

Weinreb, R.N. and Khaw, P.T. (2004) Primary open-angle glaucoma. Lancet, 363(9422): 1711–1720.

Weinreb, R.N. and Lindsey, J.D. (2005) The importance of models in glaucoma research. J. Glaucoma, 14(4): 302–304.

Yücel, Y.H. and Gupta, N. (2007) Should we treat the brain in glaucoma? Can. J. Ophthalmol., 42: 409–413.

Yücel, Y.H. and Gupta, N. (2008) Paying attention to the cerebrovascular system in glaucoma. Can. J. Ophthalmol., 43(3): 342–346.

Yücel, Y.H., Gupta, N., Kalichman, M.W., Mizisin, A.P., Hare, W., de Souza Lima, M., Zangwill, L. and Weinreb, R.N. (1998) Relationship of optic disc topography to optic nerve fiber number in glaucoma. Arch. Ophthalmol., 116(4): 493–497.

Yücel, Y.H., Gupta, N., Zhang, Q., Mizisin, A.P., Kalichman, M.W. and Weinreb, R.N. (2006) Memantine protects neurons from shrinkage in the lateral geniculate nucleus in experimental glaucoma. Arch. Ophthalmol., 124(2): 217–225.

Yücel, Y.H., Kalichman, M.W., Mizisin, A.P., Powell, H.C. and Weinreb, R.N. (1999) Histomorphometric analysis of optic nerve changes in experimental glaucoma. J. Glaucoma, 8(1): 38–45.

Yücel, Y.H., Zhang, Q., Gupta, N., Kaufman, P.L. and Weinreb, R.N. (2000) Loss of neurons in magnocellular and parvocellular layers of the lateral geniculate nucleus in glaucoma. Arch. Ophthalmol., 118(3): 378–384.

Yücel, Y.H., Zhang, Q., Weinreb, R.N., Kaufman, P.L. and Gupta, N. (2001) Atrophy of relay neurons in magno- and parvocellular layers in the lateral geniculate nucleus in experimental glaucoma. Invest. Ophthalmol. Vis. Sci., 42(13): 3216–3222.

Yücel, Y.H., Zhang, Q., Weinreb, R.N., Kaufman, P.L. and Gupta, N. (2003) Effects of retinal ganglion cell loss on magno-, parvo-, koniocellular pathways in the lateral geniculate nucleus and visual cortex in glaucoma. Prog. Retin. Eye Res., 22(4): 465–481.

Zetterberg, M., Tasa, G., Palmer, M.S., Juronen, E., Teesalu, P., Blennow, K. and Zetterberg, H. (2007) Apolipoprotein E polymorphisms in patients with primary open-angle glaucoma. Am. J. Ophthalmol., 143(6): 1059–1060.

CHAPTER 33

Changes of central visual receptive fields in experimental glaucoma

S.C. Sharma*

Departments of Ophthalmology, Cell Biology and Anatomy, New York Medical College, Valhalla, NY 10595, USA

Abstract: Retinal ganglion cell apoptotic death in experimental glaucoma is protracted over several months and it leads to the visual dysfunction. In the rat with increased intraocular pressure (IOP), the lack of visual scotoma was observed where visual field was determined electrophysiologically on the contralateral optic tectum in the early stages of the disease. Increases in the sizes of receptive fields on the periphery represented early stage of glaucomatous dysfunction. The relationship of duration and magnitude of IOP elevation had a significant correlation between percentages of receptive field sizes in the tectum. Large increases in receptive field sizes noted in the glaucomatous retinal terminal areas suggest the ability of the remaining retinal axons to compete and compensate for the loss of retinal axons. This compensatory adaptation leads to the degradation of the visual acuity and visual thresholds when measured psychophysically.

Keywords: glaucoma; rat; superior colliculus; optic tectum; receptive fields; electrophysiology; intraocular pressure; scotoma; visual acuity; spatial frequency; contrast sensitivity; neurodegenerative disease

Glaucoma is a heterogeneous group of disorders with a resultant common denominator: optic neuropathy leading eventually to blindness. Glaucoma is the second leading cause of blindness worldwide and is characterized by loss of retinal ganglion cells (RGCs) through a process whose pathophysiology is not fully understood. The most common subtype of glaucoma in the United States is open-angle glaucoma, a chronic disease exhibiting progressive loss of RGCs and their axons. Its clinical manifestations include excavation of the optic disc and progressive loss of visual field. Several models of glaucoma in monkeys (Quigley and Hohman, 1983) and rats (Shareef et al., 1995;

Morrison et al., 1997) have been used experimentally to increase intraocular pressure (IOP), which leads to RGC death and generates features similar to those of human glaucoma.

RGCs in chronic glaucoma do not die at the same time; rather, RGC death is protracted over the course of several months or years. Therefore, the progression of optic neuropathy over the time provides avenues in which to study the state of still-living cells and to save them from death. RGC death also occurs when the optic nerve is transected. RGC death in the glaucomatous retina or following optic nerve transection is apoptotic and not necrotic (Garcia-Valenzuela et al., 1994, 1995; Quigley et al., 1995). It is reasonable to assume that the signal for apoptotic cell death is received approximately at the same time in all RGCs in the animals with optic nerve transection,

*Corresponding author. Tel.: +1 914 594 4382;
Fax: +1 914 594 4653; E-mail: Sansar_Sharma@nymc.edu

whereas in glaucomatous eyes either the signal is not received by all RGCs at the same time or they respond differentially to the signals (Laquis et al., 1998).

RGC death may have two modes. One is the direct insult of a primary pathological process in glaucoma and the second is the effect of dying RGCs on the surrounding RGCs that lead to death of normal uninjured cells, which is termed secondary degeneration (Levkovitch-Verbin et al., 2001). Convincing evidence of secondary degeneration in glaucoma is lacking. Specific initiators of apoptosis in glaucoma may include blockage of axonal transport leading to neurotrophin depletion (Quigley et al., 1979), glutamate excitotoxicity (Dreyer at al., 1996), antibodies to heat-shock proteins (Tezel and Yang, 2004), ischemia (Osborne et al., 1999), and nitric oxide synthase (NOS) upregulation with reactive oxygen species formation (Neufeld et al., 1999).

Clinical relevance of optic neuropathy

The relationship between elevated IOP and glaucomatous optic neuropathy has been extensively studied. Elevated IOP remains the etiological factor toward which all current therapeutic efforts are directed, although other approaches are now being entertained (Levin, 2003; Osborne et al., 2004).

The prevailing hypothesis regarding the mechanism of elevated IOP mediated optic nerve damage includes compression of optic axons at the lamina cribrosa, ischemia in the lamina cribrosa (Kolker and Hetherington, 1983), and expression of NOS in the optic nerve head (Shareef et al., 1999; Liu and Neufeld, 2001). Animal models became a necessity to study the pathology of the disease process and the effect of pharmacological intervention on death cascade. The resemblance between primate and human glaucoma stimulated the use of primates for research during the last couple of decades (Quigley and Hohman, 1983). However, economical and ethical issues led to a decrease in use of primates and partial replacements by rodents in glaucoma research. Rats have anatomical similarities with primates, regarding the anterior segment blood supply and aqueous humor drainage (Morrison et al., 1995). No specific model yet exists that has direct homology to human glaucoma. However, the rat model of elevated IOP developed in our laboratory (Shareef et al., 1995) compares closely with the secondary form of human glaucoma and this animal model has been replicated in several laboratories. In a recent study, Urcola et al. (2006) compared three experimental approaches to induce elevation of IOP in rats: injection of latex microspheres in the anterior chamber, injection of latex microspheres plus hydroxypropyl methyl cellulose in the anterior chamber, and episcleral venous occlusions. Although RGCs death in temporal and spatial order was similar using all methods, the episcleral venous occlusion method was preferred by these authors as it had fewer complications to the anterior chamber.

In order to obtain massive RGC death, optic nerve transection or IOP elevation has been used to study the apoptotic pathway (Quigley et al., 1979; Berkelaar et al., 1994; Garcia-Valenzuela and Sharma, 1996). In adult rats, 90% of RGCs die within the first 2–3 weeks after optic nerve transection. The apoptotic RGC death after axotomy and/or elevated IOP has been established in rat (Garcia-Valenzuela et al., 1993, 1994, 1995; Berkelaar et al., 1994) as well as in monkey (Quigley et al., 1995). The induction of apoptosis in RGCs can also be initiated by ischemia/reperfusion, withdrawal of trophic factors, and radiation. Apoptosis is the mechanism through which superfluous, ectopic, damaged, or mutated cells are eliminated. RGC survival is enhanced by intravitreal injection of neurotrophic factors (Mey and Thanos, 1993), such as brain-derived neurotrophic factor (BDNF) (Krueger-Naug et al., 2003), ciliary neurotrophic factor (CNTF) (van Adel et al., 2003), and NT3/NT4 (Peinado-Ramon et al., 1996). Numerous studies showed the survival effect of neurotrophic factors on axotomized RGCs (reviewed in Chaudhary and Sharma, 2001). A combination of trophic factors, such as glia-derived neurotrophic factor (GDNF) and BNDF, leads to further attenuation of RGC death in 75% of injured RGCs surviving up to 2 weeks following axotomy (Yan et al., 1999). Induction of

CNTF through viral vector has resulted in increased survival and regeneration of injured RGCs (van Adel et al., 2003).

In the past decade or so, great strides have been made in understanding the molecular cascades of RGCs apoptosis in experimental glaucoma; however, many questions remain unanswered. No single mechanism under one experimental condition has been elucidated to account for the apoptosis. Why some RGCs die first and why others survive for a prolonged exposure of elevated IOP stress remains a mystery. Rescuing RGCs by intervening in the death cascade has been partially successful in ameliorating cell death in experimental glaucoma. However, the functional consequences of the rescued RGCs remain unclear.

Is there a remodeling of retinal circuitry?

Extensive evidence has emerged in the past decade that some retinal diseases whose progression has been correlated with photoreceptor degeneration lead to the remodeling of retinal circuitry (reviewed in Jones et al., 2005; Marc et al., 2006). The loss of photoreceptors and, therefore, the glutamatergic loss to the bipolar cells lead to retinal remodeling apparent in migration of neurons, retinal rewiring, and creation of microneuromas. These authors reported Müller cell hypertrophy and axonal sprouting in ganglion, bipolar, and amacrine cells as a consequence of photoreceptor death. The evidence of neuronal plasticity in the adult retina is real.

Plasticity of RGCs has been the focus of earlier studies where neurons react to lesions by structural reorganization of somas, axons, and dendrites. During development, RGC morphology is influenced by either increasing or decreasing cell densities in the retina using experimental manipulations (Rappaport and Stone, 1983; Peichl and Bolz, 1984; Eysel et al., 1985; Kirby and Chalupa, 1986; Leventhal et al., 1988; Bahr et al., 1992). In glaucomatous eyes the densities of RGCs change. We have shown earlier that following increase in the IOP and subsequent death at the rate of 3% per week, the soma diameter of remaining ganglion cells became significantly larger in all cell types. Ganglion cell types I and III showed the increase in the area of dendritic field (Ahmed et al., 2001). Rodent data are surprising as the primate glaucomatous eye showed no increase in RGC soma or dendrites (Weber et al., 1998, 2000). On the contrary, Weber and Harman (2005) concluded that ganglion cells in the glaucomatous eye retain normal intrinsic electrical properties yet they have decreased dendritic field sizes. Some controversies still exist, for example, Smith et al. (1993) had proposed earlier that visual deficit in long-term glaucoma results from ganglion cell loss and not reduction in the functional capacity of the surviving neurons. One could conclude that plasticity in RGCs persisted in adult rats.

In the glaucomatous eye of the primates, data to date suggest that following loss of RGCs there is continuous atrophy of the surviving cells that leads to visual dysfunction. Data accumulated in the past several years point to transneuronal changes resulting either in the loss of some and/or reduction of other soma sizes of the lateral geniculate nucleus (LGN). No studies exist as yet showing similar changes in the rodent visual system.

Functional assessment of visual changes in glaucoma patients was made possible by recent studies of Duncan et al. (2007) using functional magnetic resonance imaging (fMRI) method. The spatial pattern of activity in the glaucomatous eye was correlated on a retinotopic map with that of the fellow eye. The fMRI signals in visual I area of the human cortex showed a loss of visual function that correlated nicely with the loss of visual field in the eye. The fMRI technique has opened a new vista to quantify glaucomatous changes in the neural activity. The fMRI method allows assessment of the changes in the visual brain centers following damaged retina; however, it is not clear whether this method can assess changes in the early stages of glaucoma before the death of either RGCs or the changes in the function of RGCs where dendrites have begun to shrink yet the axons are still properly connected to the visual centers. This problem is further compounded by the fact that fMRI studies in macular degeneration patients with losses of foveal vision show extensive activity in the cortical foveal area (Ngugen et al., 2004a, b; Baker et al., 2005). These studies point to the extensive reorganization of visual terminal areas that would account for the results. It is reasonable to assume that in macular degeneration some foveal retinal cells survived and these RGCs expanded their projections in the cortex. It is further conceivable that the parafoveal RGC axons invaded the territory in the foveal area of the cortex. Either possibility will lead one to believe that many limitations persist in assessing the early functional changes in glaucomatous eyes using fMRI methods.

Optic nerve and RGC damage produces changes in the visually evoked potentials in patients with glaucoma, as the scotoma enlarges with the progression of the disease. The enlargement of the visual field deficits is due to an increase in dysfunction or dying RGCs in the focal areas of the retina. In order to determine the functional consequences of RGC pathology in glaucomatous eyes, visual receptive fields were mapped onto the superior colliculus contralateral to the experimental eye and in control eye in elevated IOP rats (King et al., 2006).

Retinotopic organization in the normal rat was generated by recording from specific locations on the contralateral superior colliculus. Visual receptive field sizes at each recording site on the superior colliculus were in the order of 5–15°. Essentially, the nasal visual field projects to the contralateral rostral superior colliculus and the temporal visual fields project to the caudal tectum, whereas the dorsal visual field projects to the medial and the ventral visual field projects to the lateral edges of the tectum. This topographic pattern is highly ordered (Fig. 1).

To assess the effects of localized scotoma, we first created a small lesion in the retina via the insertion of a 22-gauge needle into the nasal retina. We then mapped electrophysiologically the retinotopic projection to the contralateral superior colliculus. A well-demarcated silent area in the contralateral temporal visual field was obtained (Fig. 2). An electrophysiological mapping of visual field projections in a rat that had elevated IOP (28–30 mmHg) was generated. Normal pressure before surgery and in the control eye was 16 mmHg. Following the IOP elevation via episcleral venous occlusion, the animal was allowed to recover and maintain at a 12-h light and 12-h dark cycle. IOP was measured once a week for up to 4 weeks and was within 25–27 mmHg immediately prior to recording. We assumed that RGC death would be in the range of 12–15% in 5 weeks following IOP elevation. In normal animals, the sizes of receptive fields recorded at their terminal on the superior colliculus are in the range of 5–15° of the visual space. However, in these experiments with elevated IOP for up to 5 weeks, some receptive fields on the peripheral visual area were in the range of 25–30° of the visual space. The visual field map was not interrupted. All electrode locations provided visually evoked responses

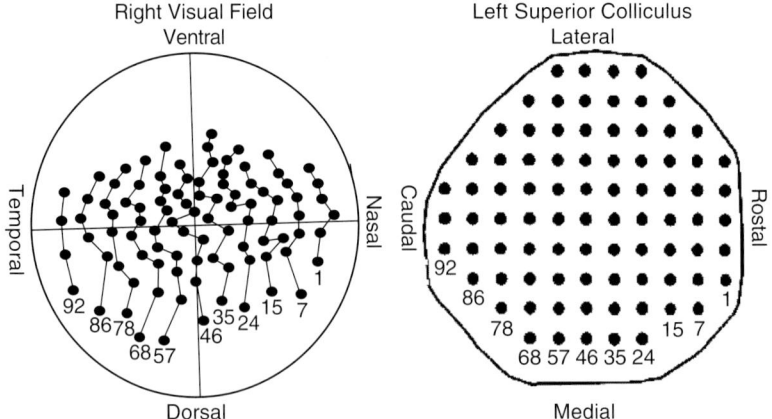

Fig. 1. The retinotopic projection in a normal rat on the contralateral superior colliculus. The numbered positions on the superior colliculus represent the electrode sites that correspond to the visual field position at which maximum responses were recorded.

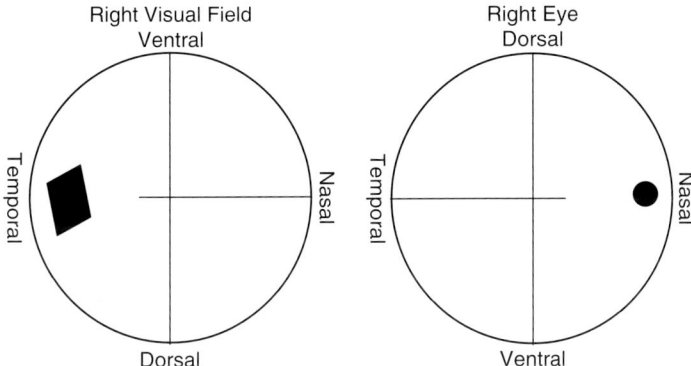

Fig. 2. The location of the needle lesion in the retina is shown in the right eye (black dot). The visual projection of this eye onto the contralateral left tectum was recorded. The right visual field map shows a scotoma (black box) in the temporal visual field. Corresponding locations on the caudal tectum were silent (data not shown). All electrophysiological parameters were the same as those used for the experiment shown in Fig. 1.

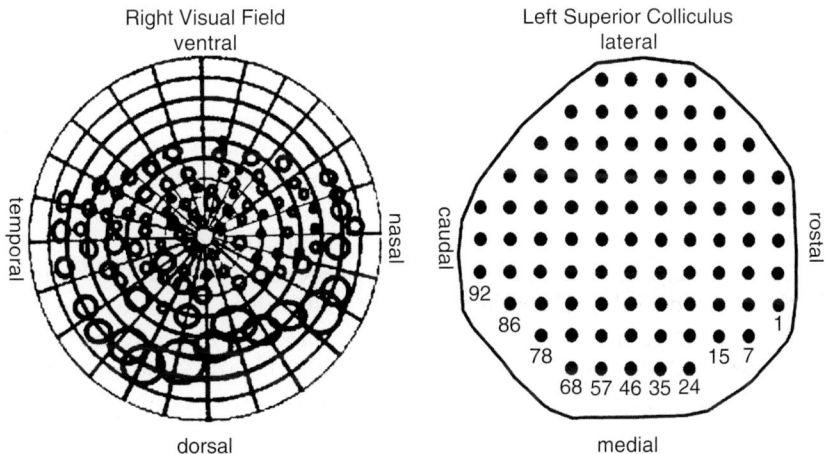

Fig. 3. Retinotopic map of the right eye onto the contralateral superior colliculus in a rat with elevated IOP in the right eye (28 mmHg). Different sized circles on the right visual field map represent receptive fields recorded from the corresponding tectal locations. Notice the enlarged receptive field sizes from the peripheral retinal locations. The conventions for recording were similar to those described for Fig. 1.

without any silent area that may correspond to the retinal scotomas. The enlarged receptive fields were definable. The recordings were done in a mesopic luminance level. The dots and numbers on the left superior colliculus indicate electrode positions, 200 μm apart. The receptive fields on the right visual field map are of various sizes; some enlarged receptive fields overlap with each other. These were noticed primarily in the dorsal areas of the visual field (Fig. 3).

These results indicated that although loss of the RGCs continued, the visual field scotoma was not apparent in the early period following elevation of IOP. Additionally, larger receptive fields on the periphery may represent the early signs of altered geometry of the retina as large RGCs die in the periphery of the retina in early stages of glaucoma (Shareef et al., 1995). It would be tempting to assume that following the death of larger cells on the periphery, the remaining ganglion cells expanded their axonal arbors in the tectum leading to the enlarged receptive fields. This may represent the earliest changes in visual receptive fields in the tectum following death of selective RGCs.

Expansion of visual receptive fields in glaucoma

As pointed out above, the sprouting of axons and dendrites in other retinal cells has been reported in adult mammals (Peichl and Bolz, 1984; Lewis et al., 1998). The glaucomatous eye's retinotopic map onto the superior colliculus was recorded in order to assess the project patterns of the remaining RGCs following significant death of RGCs due to elevated IOP (King et al., 2006). IOP was elevated in the eye via episcleral venous occlusion (Garcia-Valenzuela et al., 1995) and it remained elevated to 150% of the control eye for the duration of the experiment. The sizes of the visual receptive fields recorded from the superior colliculus, contralateral to the glaucomatous eyes, were significantly large (e.g., 236 degree^2 for normal vs. 849 degree^2 for the glaucomatous eye). This increase in area of visual receptive fields was proportional to both the degree and the duration of IOP elevation. Scotomas due to RGCs loss, although present in the eye following prolonged cell death, were undetectable due perhaps to the massive overlap of visual receptive fields in the superior colliculus generated by the remaining RGCs (Fig. 4).

The ramifications of the above-described studies are the following:

1. Change in the dendritic arbor of RGCs in glaucomatous eyes (Ahmed et al., 2001) cannot solely describe the compensatory effect of the changes in the receptive fields in the colliculus. Ahmed et al. (2001) reported changes in only two cell types in the retina whose dendritic arbors expanded following death of RGCs. Type I and III cells expand their dendritic fields for up to 6 weeks following IOP elevation, whereas the soma diameter of all three cell types was enlarged. These observations stand in contrast to that seen in the primate retina. Experimentally induced RGC death by partial optic nerve cut or other means leads to the vacated sites. This reduction in the density of RGCs in cats led to the increases in soma sizes of the remaining ganglion cells (Rappaport and Stone, 1983; Kirby and Chalupa, 1986). Observations of Rousseau et al. (1999) in adult rat showing the compensatory soma size changes following 50% RGC death supported our observations in glaucomatous rats (Ahmed et al., 2001). It is important to keep in mind that following partial optic nerve crush, cells that are destined to die (within 2 weeks) show soma swelling. The remaining cells increase their soma size similar to physiological parameters of recovery of function (Schmitt et al., 1996). It is therefore conceivable that early increase in soma sizes may represent the cells that are destined to die. It is clear that surviving RGCs have differential ability to adapt in glaucomatous retina. Many questions remain unanswered, e.g., (a) what are the differences between primate and rat retina in glaucomatous eye; (b) are the increases in dendritic arbor owing to new synaptic connections created by vacant bipolar and amacrine cells in rat and not in primates; (c) what are the differential time sequences of RGC death; (d) what are the functional consequences of remaining RGCs?

2. The relationship of duration and magnitude of IOP elevation showed a significant positive correlation between percentages of receptive field size increases in glaucomatous animal (Fig. 4D). This correlation becomes evident after 6 months of sustained elevated IOP. A greater increase in visual receptive field size correlates reasonably well with a greater increase in the IOP. When the IOP level increases over 50–60%, a definite increase in the visual receptive field is encountered. Hence, an increase in the visual receptive field may represent a very long-term effect of the remaining ganglion cells in the glaucomatous eyes in which ganglion cell axons try to compensate for the loss of the fields by occupying a larger than normal territory of

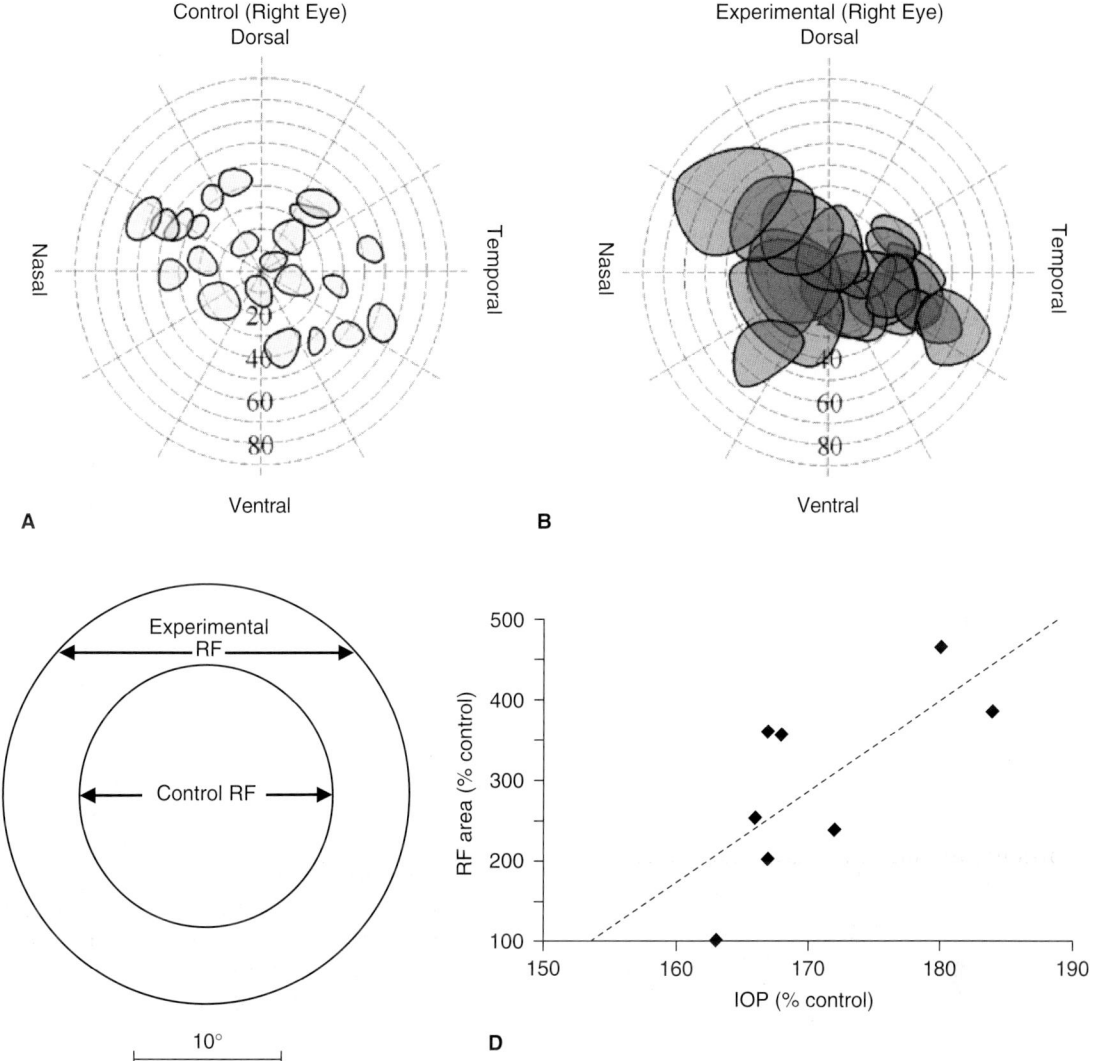

Fig. 4. Representative examples of the visually driven receptive field sizes from the contralateral superior colliculus in a normal rat (A) and in a glaucomatous experimental eye (B). The increase in receptive field size from a normal control animal is compared to that from the experimental glaucomatous animal (C). The correlation of increase in intraocular pressure with increase in visual receptive field size is shown in (D). Each point represents values from one animal. $p < 0.025$.

the colliculus. Since these changes represent the long-term effect in elevated IOP rats, one wonders how much it may take to manifest similar changes in primate glaucomatous animals.

3. The magnification factor, or the degree of visual field per millimeter square of the tectal surface, is variable in the rat superior colliculus. The comparison of the recorded visual receptive field from the experimental and mirror image in the control tecta provided the best avenue to measure the relative changes. At every point recorded on the experimental tectum, the size of the visual receptive field was significantly larger. These data suggest that every tectal point in the experimental animal independent of the magnification factor showed an increase in

the receptive field. Since electrophysiological recordings were made in the superficial layer of the tectum and the extracellular multiunit recordings were generated in the area of the receptive field, we are confident that recordings were acquired from the terminal areas of the optic axons. We must therefore assume that optic axonal terminal arbors increase to cover larger than the normal terminal area of the tectum. We have recently observed (unpublished preliminary data) a qualitative increase in the optic axonal terminal areas in the tectum of glaucomatous animals when compared to the terminal areas (of similar type) seen on the mirror image location of contralateral normal tectum.

Behavioral consequences of glaucoma

Sauve et al. (2004) reported a reasonable correlation of full-field ERG and visual field thresholds recorded from the superior colliculus in rat. These authors showed that magnitude of b-wave could be used to predict the computed area and degree of visual field preservation in the central nervous system (CNS). Grozdanic et al. (2003) reported a similarly functional characterization of retina after ischemia. By contrast, a smaller number of studies have been undertaken to measure changes in visual function after glaucoma treatment (Greve et al., 1975; Flammer and Drance, 1983; Yoshikawa et al., 1989; Rolando et al., 1991). These authors showed moderate improvements in retinal function in humans. However, in animals with experimental glaucoma, hardly any studies have dealt with RGC survival and recovery of function. In rodent's glaucoma, functional analysis has been hampered due to limited and tedious methodologies to test visual behavior. Recently, Prusky and his colleagues have developed a rapid quantification method for testing spatial vision in rats. The spatial vision (grating acuity and contrast sensitivity) can be measured by a virtual reality optomotor system in which spatial visual thresholds can be measured rapidly and without specific reinforcement training. The profiles of acuity and contrast sensitivity present a dynamic picture of the functioning of the visual system, and the ability of the rat to reflexively track a moving grating can easily be assessed. Using this procedure, one can characterize the loss of vision in glaucomatous animals and evaluate the effects of various manipulations aimed at saving RGC in glaucoma. The other complimentary aspect of vision is the measurement of visual perceptual thresholds that can be used to follow changes in visual processing in glaucoma. The perceptual threshold is measured by visual water task method. Both tests are necessary to evaluate potential assessment of visual dysfunction.

Slow horizontal head and body rotation occurs in rats when the visual field is rotated around them. These optomotor responses can be readily produced. If one eye is closed, only motion in the temporal-to-nasal direction for the contralateral eye evokes the tracking response. When the maximal spatial frequency capable of driving the response (acuity) was measured under monocular and binocular viewing conditions, the monocular acuity was identical to binocular acuity measured with the same rotation direction. Thus, the visual capabilities through each eye can be measured under binocular conditions simply by changing the direction of rotation. The spatial frequency, contrast, and velocity of stimulus pattern are generated by the computer and can be changed instantaneously.

The grating thresholds of Long-Evans rats are clustered around 1.0 cycle/degree. Using spatial frequencies already described for Long-Evans rats (Prusky et al., 2002), visual thresholds were obtained in normal rat. As described by Prusky et al. (2000), for each eye, motion in the temporal-to-nasal direction evokes tracking, whereas motion in the nasal-to-temporal direction does not. We used frequency of 0.103 cycle/degree sine wave gratings. As grating was rotated, rat ceased body movements and began to track grating with head movements in relation to the rotation. The contrast was increased or decreased to achieve the threshold. In experimental animals where only one eye had elevated IOP and the contralateral eye served as a control, the tests for contrast sensitivity and visual acuity began at 5 weeks after the surgery. The mean IOP of the control left eye was 16.5 mmHg, whereas the experimental right eye had a mean IOP of 25.15 mmHg. Visual acuity refers to the maximal spatial frequency that evoked an optomotor response. Acuity thresholds were measured for both eyes of normal and experimental animals in which one eye had elevated IOP. The eyes of normal animals had comparable mean acuity thresholds of 0.52; a paired t-test indicates no significant difference between the eyes of normal animals ($t = 1$, df $= 2$, $p = 0.42$). For elevated IOP eyes, the experimental eye had a lower mean acuity value (0.15 ± 0.013 SEM) than did the contralateral normal eye (0.52 ± 0.001 SEM). A paired t-test indicated that the acuity values for the experimental and normal eye differed significantly ($t = 26.9$, df $= 6$, $p < 0.0005$).

These animals were further tested for contrast sensitivity as a function of spatial frequency. The experiment formed a 2×6 design (eye vs. spatial frequency) with repeated measures on both variables. A 2×6 repeated-measures ANOVA indicated that the main effect of eye (normal vs. experimental elevated IOP eye) was significant ($f = 22,471.6$, df $= 1$, $p < 0.0005$).

The main effect of spatial frequency was also significant ($f = 626.7$, df $= 5$, $p < 0.0005$). Examination of the data indicates that the interaction was due to the flattening of the curve for the experimental eye relative to the curve for the normal eye. This flattening was due to a floor effect created by the poor contrast sensitivity of the experimental eye. A two-tailed, paired t-test was done at each spatial frequency to compare the mean contrast sensitivities of the experimental and normal eyes. All six t-tests were significant and the t values ranged from 33.9 to 193.4 (df $= 6$, $p < 0.0005$). Therefore, the contrast sensitivity of the experimental eye differed from that of the normal eye at all six spatial frequencies (Fig. 5).

Once again, the above-described preliminary (unpublished) data in glaucomatous rat stand in contrast to the published data from the primates. Primate glaucomatous eyes showed visual behavior changes where approximately 50% RGC have already died, but in glaucomatous rats the earliest changes in visual acuity were evident within a month following IOP elevation. It seems that about 15% RGC death was sufficient to measure the earliest changes in acuity and contrast

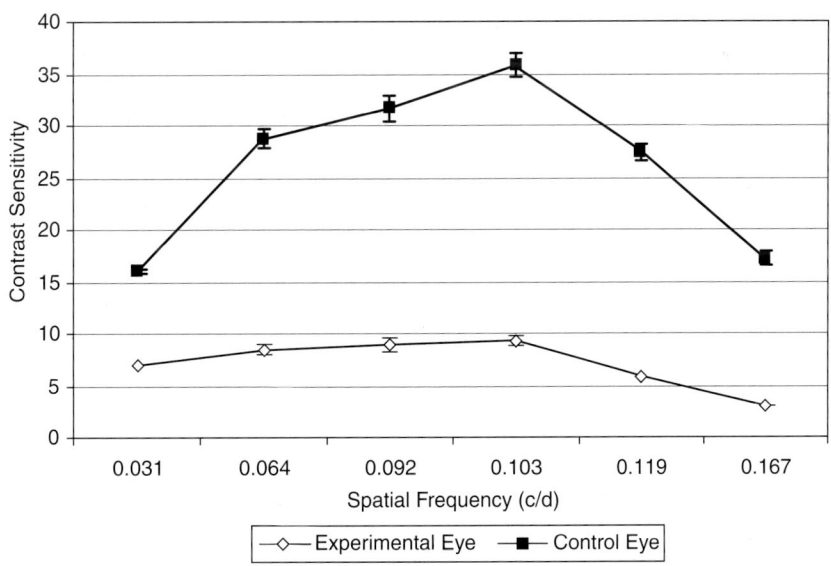

Fig. 5. Changes in contrast sensitivity at six spatial frequencies of a glaucomatous eye when measured at 5 weeks after sustained elevation of IOP (25 mmHg).

sensitivity in the rat and these changes may be due to very sensitive measuring equipment.

Glaucoma as a neurodegenerative disease versus neuroplasticity and adaptive changes

Dr. Yucel and his colleagues in Canada have shed light on the effects of elevated IOP in primate and human glaucomatous visual centers. Elevated IOP induces RGC death and as expected leads to the changes in the LGN where cell loss and the shrinkage of LGN was reported (Weber et al., 2000; Yucel et al., 2000, 2001, 2003). Finally, changes in the primary visual cortex were reported. Cytochrome oxidase activity decreased in the ocular dominant columns subserving the glaucomatous eye (Crawford et al., 2001; Lam et al., 2003). These transsynaptic changes in the visual pathways following glaucoma led the authors to describe glaucoma as a neurodegenerative disease. Evidence has further accumulated that IOP elevation induces changes in the LGN without optic nerve axonal loss. It is conceivable that initiating changes both physical (dendritic withdrawal preceding RGC death) and chemical (decrease in cytochrome oxidase activity in LGN) may be linked. Hence, one can conclude that initiation of the elevated IOP may trigger changes in the retina and CNS whose effect continues for a very long time. To date, emphasis has been placed on lowering IOP as a means to reduce RGC death in glaucomatous eyes. However, in many cases RGC death continues even after normal pressure is maintained.

Future directions

The continuous debate on the "dogma" of glaucoma, i.e., whether ganglion cells die first or optic axons are damaged first, is yet to be resolved. In both situations, changes commence in the termination sites of the visual pathways. The dramatic change seen in the visual centers in glaucomatous animal and human raise the question that the functional relationship between the loss or degradation of the visual field and the changes in visual acuity and visual thresholds must be further explored.

The factors that initiate RGCs death in glaucoma are incompletely defined. Loss of RGCs via apoptotic mechanisms induced by the elevation of IOP leads to degenerative morphological changes and cell death in the LGN and primary visual cortex. Some reorganization of retinotopic cortical maps in adult mammals has been reported following lesions of the retina. Conditions that might lead to cortical neuroplasticity following IOP elevation in monkeys have been reported (Lam et al., 2003).

Rescued RGCs in the optic nerve transection model do not provide any means to assess their functionality, as cut optic axons do not regenerate ordinarily. Furthermore, rescued or still surviving and perhaps uninjured RGCs in glaucoma maintain their nerve connections to the brain centers allowing evaluations of their functionality. At what stage(s) in the progression of glaucomatous RGCs death is there degradation of the central visual field maps? It is important to acquire such information to allow better understanding of the progression of visual loss in glaucoma, and it will provide avenues to assess the effects of neuroprotective agents on the consequence of rescuing RGCs in glaucoma.

The ability of the neuronal system to undergo remodeling and repair following injury is influenced by interactions between the remaining viable RGCs in glaucomatous retina as well as by the response of individual cells to the reduction in IOP and the administration of neuroprotective agents. Another factor that ought to be considered is that the reduction in the overall number of RGC may induce the dendrodendritic interactions of the remaining RGCs in delimiting the size of the dendritic tree. If by reducing the IOP and/or administration of neuroprotective agents one can determine the functional consequences of retina leading to either partial or complete restitution of function, it may offer a new avenue for improving visual function in glaucoma patients.

Future experiments should be directed to explore the effect of lowering the IOP, either surgically or medically, on the receptive field sizes.

Is the degradation of visual receptive field ever restored? Experiments should also be directed to evaluate the progression of changes in visual function by recording from the visual cortex.

Is there functional reorganization in the primary visual cortex in response to glaucoma in human? If the emergence of degradation of visual fields takes a few months of sustained elevated IOP in rats and it happens when ganglion cell death cascade has slowed down, it might take much longer time in humans when IOP in glaucomatous eye has been managed.

Finally, as it is becoming acceptable that glaucomatous changes are well pronounced in central visual nuclei, the question arises that treating the visual centers by neuroprotective agents so that death of LGN and cortical neurons is reduced or controlled might benefit patients with glaucoma. Of course, these treatments should be coupled with lowering IOP treatments of the eye to have any real effect in glaucomatous patients.

Acknowledgment

The work reported has been supported by NIH/NEI grants.

References

Ahmed, F.A., Chaudhary, P. and Sharma, S.C. (2001) Effects of increased intraocular pressure on rat retinal ganglion cells. Int. J. Dev. Neurosci., 19: 209–218.

Bahr, M., Wizenmann, A. and Thanos, S. (1992) Effect of bilateral tectum lesions on retinal ganglion cell morphology in rats. J. Comp. Neurol., 320: 370–380.

Baker, C.I., Peli, E., Knouf, N. and Kanwisher, N.G. (2005) Reorganization of visual processing in macular degeneration. J. Neurosci., 25: 614–618.

Berkelaar, M., Clarke, D.B., Wang, Y.C., Bray, G.M. and Aguayo, A.J. (1994) Axotomy results in delayed death and apoptosis of retinal ganglion cells in adult rats. J. Neurosci., 14: 4368–4374.

Chaudhary, P. and Sharma, S.C. (2001) Apoptotic cell death following axotomy. In: Ingoglia N.A. and Murray M. (Eds.), Axonal Regeneration in the Central Nervous System. Marcel Dekker, New York, pp. 185–218.

Crawford, M.L., Harwerth, R.S., Smith, E.L., Mills, S. and Ewing, B. (2001) Experimental glaucoma in primates: changes in cytochrome oxidase blobs in V1 cortex. Invest. Ophthalmol. Vis. Sci., 42: 358–364.

Dreyer, E.B., Zurakowski, D., Schumer, R.A., Podos, S.M. and Lipton, S.A. (1996) Elevated glutamate levels in the vitreous body of humans and monkeys with glaucoma. Arch. Ophthalmol., 114: 299–305.

Duncan, R.O., Sample, P.A., Weinbel, R.W., Bowd, C. and Zangwill, L.M. (2007) Retinotopic organization of primary visual cortex in glaucoma: comparing fMRI measurements of cortical function with visual loss. Prog. Retin. Eye Res., 26: 38–56.

Eysel, U.T., Peichl, L. and Wassle, H. (1985) Dendritic plasticity in the early postnatal feline retina: quantitative characteristics and sensitive period. J. Comp. Neurol., 242: 134–145.

Flammer, J. and Drance, S.M. (1983) Reversibility of a glaucomatous visual field defect after acetazolamide therapy. Can. J. Ophthalmol., 18: 139–141.

Garcia-Valenzuela, E., Fahmey, A. and Sharma, S.C. (1993) Axotomy induces apoptotic cell death in adult retinal ganglion cells. Eur. J. Neurosci., (Suppl 6): p. 499.

Garcia-Valenzuela, E., Gorczyca, W., Darzynkiewicz, Z. and Sharma, S.C. (1994) Apoptosis in adult retinal ganglion cells after axotomy. J. Neurobiol., 25: 431–438.

Garcia-Valenzuela, E., Shareef, S., Walsh, J. and Sharma, S.C. (1995) Programmed cell death of retinal ganglion cells during experimental glaucoma. Exp. Eye Res., 61: 33–44.

Garcia-Valenzuela, E. and Sharma, S.C. (1996) Optic nerve regeneration. In: Rich R., Shields B. and Krupin T. (Eds.), The Glaucomas. St. Louis, Mosby, pp. 199–212.

Greve, E.L., Dake, C.L. and Verduin, W.M. (1975) Proceedings: reversibility of visual field defects in glaucoma. Exp. Eye Res., 20: p. 183.

Grozdanic, S.D., Sakaguchi, D.S., Kwon, Y.H., Kardon, R.H. and Sonea, I.M. (2003) Functional characterization of retina and optic nerve after acute ocular ischemia in rats. Invest. Ophthalmol. Vis. Sci., 44: 2597–2605.

Jones, B.W., Watts, C.W. and Marc, R.E. (2005) Retinal remodelling. Clin. Exp. Optom., 88: 282–291.

King, W.M., Sarup, V., Sauve, Y., Carpenter, D.O. and Sharma, S.C. (2006) Expansion of visual receptive fields in experimental glaucoma. Vis. Neurosci., 23: 137–142.

Kirby, M.A. and Chalupa, L.M. (1986) Retinal crowding alters the morphology of alpha ganglion cells. J. Comp. Neurol., 251: 532–541.

Kolker, A.E. and Hetherington, J. (1983) Primary open-angle glaucoma. In: Mosby, C.V. (Ed.), Becker-Shafers Diagnosis and Therapy of the Glaucoma. St. Louis, Mosby, pp. 55–63.

Krueger-Naug, A.M., Emsley, J.G., Myers, T.L., Currie, R.W. and Clarke, D.B. (2003) Administration of brain-derived neurotrophic factor suppresses the expression of heat shock protein 27 in rat retinal ganglion cells following axotomy. Neuroscience, 116: 49–58.

Lam, D.Y., Kaufman, P.L. and Gabelt, B.T. (2003) Neurochemical correlates of cortical plasticity after unilateral elevated intraocular pressure in primate model of glaucoma. Invest. Ophthalmol. Vis. Sci., 44: 2573–2581.

Laquis, S., Chaudhary, P. and Sharma, S.C. (1998) The patterns of retinal ganglion cell death in hypertensive eyes. Brain Res., 784: 100–104.

Leventhal, A.G., Schall, J.D. and Ault, S.J. (1988) Extrinsic determinants of retinal ganglion cell structure in the cat. J. Neurosci., 8(6): 2028–2038.

Levin, L.A. (2003) Retinal ganglion cells and neuroprotection for glaucoma. Surv. Ophthalmol., 48(1): S21–S24.

Levkovitch-Verbin, H., Quigley, H.A., Kerrigan-Baumrind, L.A., D'Anna, S.A., Kerrigan, D. and Pease, M.E. (2001) Optic nerve transection in monkeys may result in secondary degeneration of retinal ganglion cells. Invest. Ophthalmol. Vis. Sci., 42: 975–982.

Lewis, G.P., Linberg, K.A. and Fisher, S.K. (1998) Neurite outgrowth from bipolar and horizontal cells after experimental retinal detachment. Invest. Ophthalmol. Vis. Sci., 39: 424–434.

Liu, B. and Neufeld, A.H. (2001) Nitric oxide synthase-2 in human optic nerve head astrocytes induced by elevated pressure in vitro. Arch. Ophthalmol., 119: 240–245.

Marc, R.E., Jones, B.W. and Watt, C.B. (2006) Retinal remodeling: circuitry revision triggered by photoreceptor degeneration. In: Pinaud E., Tremere R. and De-Weerd E. (Eds.), Plasticity in the Visual System: From Genes to Circuits. Spring Science, New York.

Mey, J. and Thanos, S. (1993) Intravitreal injections of neurotrophic factors support the survival of axotomized retinal ganglion cells in adult rats in vivo. Brain Res., 602: 304–317.

Morrison, J.C., Fraunfelder, F.W., Milne, S.T. and Moore, C.G. (1995) Limbal microvasculature of the rat eye. Invest. Ophthalmol. Vis. Sci., 36: 751–756.

Morrison, J.C., Moore, C.G., Deppmeier, L.M., Gold, B.G., Meshul, C.K. and Johnson, E.C. (1997) A rat model of chronic pressure-induced optic nerve damage. Exp. Eye Res., 64: 85–96.

Neufeld, A.H., Sawada, A. and Becker, B. (1999) Inhibition of nitric-oxide synthase 2 by aminoguanidine provides neuroprotection of retinal ganglion cells in a rat model of chronic glaucoma. Proc. Natl. Acad. Sci. U.S.A., 96: 9944–9948.

Ngugen, T.H., Stievenart, J.L., Saucet, J.C., Gargasson, J.F., Cohen, Y.S., Pelegrini-Issac, M., Burnod, Y., Iba-Zizen, M.T. and Cabanis, E.A. (2004a) Cortical response to age-related macular degeneration. Functional MRI study Part II. J. Fr. Ophtalmol., 27: 3572–3586.

Ngugen, T.H., Stievenart, J.L., Saucet, J.C., Gargasson, J.F., Cohen, Y.S., Pelegrini-Issac, M., Burnod, Y., Iba-Zizen, M.T. and Cabanis, E.A. (2004b) Cortical response to age-related macular degeneration. Functional MRI study Part I. J. Fr. Ophtalmol., 27: 3565–3571.

Osborne, N.N., Chidlow, G., Layton, C.J., Wood, J.P., Casson, R.J. and Melena, J. (2004) Optic nerve and neuroprotection strategies. Eye, 18: 1075–1084.

Osborne, N.N., Ugarte, M., Chao, M., Chidlow, G., Bae, J.H., Wood, J.P. and Nash, M.S. (1999) Neuroprotection in relation to retinal ischemia and relevance to glaucoma. Surv. Ophthalmol., 43(1): S102–S128.

Peichl, L. and Bolz, J. (1984) Kainic acid induces sprouting of retinal neurons. Science, 223: 503–504.

Peinado-Ramon, P., Salvador, M., Villegas-Perez, M.P. and Vidal-Sanz, M. (1996) Effects of axotomy and intraocular administration of NT-4, NT-3, and brain-derived neurotrophic factor on the survival of adult rat retinal ganglion cells. A quantitative in vivo study. Invest. Ophthalmol. Vis. Sci., 37: 489–500.

Prusky, G.T., Harker, K.T., Douglas, R.M. and Wishaw, I.Q. (2002) Variation in visual acuity within pigmented and non-pigmented and albino strains of rats. Behav. Brain Res., 136: 339–348.

Prusky, G.T., West, P.W. and Douglas, D.M. (2000) Behavioral assessment of visual acuity in mice and rats. Vision Res., 40: 2201–2209.

Quigley, H.A., Guy, J. and Anderson, D.R. (1979) Blockade of rapid axonal transport. Effect of intraocular pressure elevation in primate optic nerve. Arch. Ophthalmol., 97: 525–531.

Quigley, H.A. and Hohman, R.M. (1983) Laser energy levels for trabecular meshwork damage in the primate eye. Invest. Ophthalmol. Vis. Sci., 24: 1305–1307.

Quigley, H.A., Nickells, R.W., Kerrigan, L.A., Pease, M.E., Thibault, D.J. and Zack, D.J. (1995) Retinal ganglion cell death in experimental glaucoma and after axotomy occurs by apoptosis. Invest. Ophthalmol. Vis. Sci., 36: 774–786.

Rappaport, D.H. and Stone, J. (1983) Time course of morphological differentiation of cat retinal ganglion cells: influences on soma size. J. Comp. Neurol., 221: 42–52.

Rolando, M., Facino, M., Rathschuler, F., Bocca, L. and Zingirian, M. (1991) [Reversibility of visual field defects in primary open-angle glaucoma]. J. Fr. Ophtalmol., 14: 291–294.

Rousseau, V., Engelmann, R. and Sabel, B.A. (1999) Restoration of vision III: soma swelling dynamics predicts neuronal death or survival after optic nerve crush in vivo. Neuroreport, 10: 3387–3391.

Sauve, Y., Lu, B. and Lund, R.D. (2004) The relationship between full field electroretinogram and perimetry-like visual thresholds in RCS rats during photoreceptor degeneration and rescue by cell transplants. Vision Res., 44: 9–18.

Schmitt, U., Cross, R., Pazdernik, T.L. and Sabel, B.A. (1996) Loss and subsequent recovery of local cerebral glucose use in visual targets after controlled optic nerve crush in adult rats. Exp. Neurol., 13: 17–24.

Shareef, S., Sawada, A. and Neufeld, A.H. (1999) Isoforms of nitric oxide synthase in the optic nerves of rat eyes with chronic moderately elevated intraocular pressure. Invest. Ophthalmol. Vis. Sci., 40: 2884–2891.

Shareef, S.R., Garcia-Valenzuela, E., Salierno, A., Walsh, J. and Sharma, S.C. (1995) Chronic ocular hypertension following episcleral venous occlusion in rats. Exp. Eye Res., 61: 379–382.

Smith, E.L., Chino, Y.M., Harwerth, R.S., Ridder, R.S., Crawford, M.L.J. and DeSatnis, L. (1993) Retinal inputs to the monkey lateral geniculate nucleus in experimental glaucoma. Clin. Vis. Sci., 8: 113–139.

Tezel, G. and Yang, X. (2004) Caspase-independent component of retinal ganglion cell death, in vitro. Invest. Ophthalmol. Vis. Sci., 45: 4049–4059.

Urcola, J.H., Garcia, M., Hernandez, M. and Vecino, E. (2006) Three experimental glaucoma models in rats: comparison of the effects of intraocular pressure elevation on retinal ganglion cell size and death. Exp. Eye Res., 83: 429–437.

van Adel, B.A., Kostic, C., Deglon, N., Ball, A.K. and Arsenijevic, Y. (2003) Delivery of ciliary neurotrophic factor via lentiviral-mediated transfer protects axotomized retinal ganglion cells for an extended period of time. Hum. Gene Ther., 14: 103–115.

Weber, A.J., Chen, H., Hubbard, W.C. and Kaufman, P.L. (2000) Experimental glaucoma and cell size, density, and number in the primate lateral geniculate nucleus. Invest. Ophthalmol. Vis. Sci., 41: 1370–1379.

Weber, A.J. and Harman, C.D. (2005) Structure–function relations of parasol cells in the normal and glaucomatous primate retina. Invest. Ophthalmol. Vis. Sci., 46: 3197–3207.

Weber, A.J., Kaufman, P.L. and Hubbard, W.C. (1998) Morphology of single ganglion cells in the glaucomatous primate retina. Invest. Ophthalmol. Vis. Sci., 39: p. 2304.

Yan, Q., Wang, J., Matheson, C.R. and Urich, J.L. (1999) Glial cell line-derived neurotrophic factor (GDNF) promotes the survival of axotomized retinal ganglion cells in adult rats: comparison to and combination with brain-derived neurotrophic factor (BDNF). J. Neurobiol., 38: 382–390.

Yoshikawa, K., Baba, H., Ochi, T., Inoue, T. and Inoue, Y. (1989) Reversibility of glaucomatous defects of the central visual field. Nippon Ganka Gakkai Zasshi, 93: 405–411.

Yucel, Y.H., Zhang, Q., Gupta, N., Kaufman, P.L. and Weinreb, R.N. (2000) Loss of neurons in magnocellular and parvocellular layers of the lateral geniculate nucleus in glaucoma. Arch. Ophthalmol., 118: 378–384.

Yucel, Y.H., Zhang, Q., Weinreb, R.N., Kaufman, P.L. and Gupta, N. (2001) Atrophy of relay neurons in mango- and parvocellular layers in the lateral geniculate nucleus in experimental glaucoma. Invest. Ophthalmol. Vis. Sci., 42: 3216–3222.

Yucel, Y.H., Zhang, Q., Weinreb, R.N., Kaufman, P.L. and Gupta, N. (2003) Effects of retinal ganglion cell loss on mango-, parvo-, konicellular pathways in the lateral geniculate nucleus and visual cortex in glaucoma. Porg. Retin. Eye Res., 22: 465–481.

SECTION VI

Neuroprotection: Evidence for Future Strategies

CHAPTER 34

Targeting excitotoxic/free radical signaling pathways for therapeutic intervention in glaucoma

Masaaki Seki and Stuart A. Lipton*

Center for Neuroscience, Aging, and Stem Cell Research, Burnham Institute for Medical Research, La Jolla, CA, USA

Abstract: Glaucoma is a visual disorder characterized by progressive loss of retinal ganglion cells (RGCs), which is often associated with high intraocular pressure. However, mechanisms of RGC death in glaucoma still remain a mystery. Two theories have been proposed as pathogeneses of glaucoma: mechanical and vascular. We demonstrate that glutamate excitotoxicity triggered by overactivation of the N-methyl-D-aspartate (NMDA)-type glutamate receptors may contribute according to both theories to RGC death in glaucoma and other retinal diseases such as ischemia. From a therapeutic standpoint, NMDA receptors and downstream signaling pathways, triggered by p38 mitogen-activated protein kinase (MAPK) and caspases, are potential targets of intervention to prevent RGC death. Glutamate, however, mediates synaptic transmission essential for normal function of the nervous system. Hence, complete blockade of NMDA receptor activity causes unacceptable side effects. Studies in our laboratory have shown that an open-channel blocker of the NMDA receptors, memantine, blocks only excessive NMDA receptor activity while leaving normal function relatively intact. This characteristic endows memantine with clinical tolerability, as demonstrated by its approval for treatment of Alzheimer's disease and vascular dementia, and clinical trials for glaucoma. In this review, we discuss improved memantine derivatives, p38 MAPK, and caspase inhibitors as plausible therapeutics to prevent RGC death.

Keywords: neuroprotective agents; retinal ganglion cell; memantine; S-nitrosylation; reactive oxygen species; nitric oxide; p38 mitogen-activated protein kinase; caspase

Introduction

Glutamate is the predominant excitatory neurotransmitter in the central nervous system. However, the presence of glutamate at excessive concentration or for excessive periods of time can excite neurons to death. This phenomenon was first discovered in the retina (Lucas and Newhouse, 1957) and later named "excitotoxicity" (Olney and Ho, 1970). Excitotoxicity has been thought to participate in etiology of various neurological disorders, ranging from acute insults (e.g., stroke, hypoglycemia, trauma, and epilepsy) to chronic neurodegenerative diseases (e.g., Huntington's disease, Alzheimer's disease, amyotrophic lateral sclerosis, and human immunodeficiency virus [HIV]-associated dementia): glaucoma may possibly be among them (Choi, 1988; Lipton, 1993, 2001, 2003, 2004; Lipton and Rosenberg, 1994; Dreyer and Lipton, 1999). In this chapter, we will

*Corresponding author. Tel.: +1 858 713 6261; Fax: +1 858 713 6262; E-mail: slipton@burnham.org

DOI: 10.1016/S0079-6123(08)01134-5

describe mechanistic insights of excitotoxicity and how excitotoxicity can fit into pathogenesis of glaucoma, at least in part. Thereafter, possible therapeutic interventions to treat glaucoma by interrupting excitotoxic cascades will be discussed.

Channel properties of NMDA receptors correlated with excitotoxicity

The excitatory amino acid, glutamate (glutamic acid), elicits neuronal signaling by binding to glutamate receptors. The glutamate receptors are divided into two major categories, the ionotropic (conducting ions) and metabotropic (triggering biochemical signaling). Excitotoxicity is mediated predominantly through the ionotropic receptors, which comprise of three classes of receptors (α-amino-3-hydroxy-5-methyl-4-isoxazoleproprionic acid [AMPA] receptors, kainate receptors, and N-methyl-D-aspartate [NMDA] receptors). All these are ligand-gated ion channels that normally allow cations to enter into a cell upon ligand binding. Mammalian retinal ganglion cells (RGCs) express all three ionotropic receptors (Aizenman et al., 1988), and expression of NMDA receptors in the retina is found on RGCs and subsets of amacrine cells (Brandstatter et al., 1994; Hartveit et al., 1994; Grunder et al., 2000). Although AMPA and kainate receptors can also contribute to excitotoxicity, NMDA receptors appear to play a prominent role (Hahn et al., 1988; Sucher et al., 1991).

NMDA receptors are probably a tetramer of subunits, consisting of an obligatory NR1 subunit plus NR2A-D subunits and possibly modulatory NR3A-B subunits. NR1 and NR3 subunits have glycine-binding sites. NR2 subunits have glutamate-binding sites. Although the NR1 subunit is mandatory to form functional NMDA receptors, agonist binding to the NR2 subunits is mandatory for functional activity of NMDA receptors responding to glutamate. Upon binding of glycine and glutamate (or NMDA under experimental conditions), a neuron becomes depolarized and consequently allows influx of Ca^{2+} and Na^+ into the cell through the NMDA receptor-operated channels. Among the three classes of ionotropic glutamate receptors, the NMDA receptors are the most permeable to Ca^{2+}, which, if fluxed excessively, signals downstream events leading to cell death (Lipton and Rosenberg, 1994).

NMDA receptors have important modulatory sites that can regulate the ion channel activity

Fig. 1. Illustration of Mg^{2+} block of NMDA receptors, activation modes of NMDA receptors, and pathways to NMDA receptor-mediated toxicity. (A) NMDA receptor activity is strictly controlled under physiological conditions. At the resting membrane potential of healthy neurons, Mg^{2+} blocks NMDA receptor-associated channels. The physiological Mg^{2+} block of NMDA receptors is regulated in a voltage-dependent manner. (B) After binding of agonists (glycine and glutamate or NMDA), neurons become depolarized, and then the Mg^{2+} block is removed to allow Ca^{2+} influx. (C) Under pathological conditions, neurons lose their ionic homeostasis and become depolarized. This voltage change relieves Mg^{2+} block of NMDA receptors, even in the absence of excessive agonist binding. Loss of the Mg^{2+} block results in Ca^{2+} entry. (D) Schema outlining cell injury and death pathways triggered by overactivation of NMDA receptors. An early event after overactivation of NMDA receptors is excessive Ca^{2+} influx. Increase in intracellular Ca^{2+} concentration $[Ca^{2+}]_i$ can trigger downstream signaling cascades, leading to cell death. (1) Influx of excessive Ca^{2+} into mitochondria contributes to loss of mitochondrial membrane potential followed by release of bioactive substances (e.g., cytochrome c, AIF, and ROS) into cytosol. Cytosolic cytochrome c leads to activation of caspases. Active caspases, together with AIF, can contribute to DNA fragmentation and apoptosis. Mitochondria also serve as a major source of ROS. (2) Calmodulin (CaM), potentiated by high $[Ca^{2+}]_i$, triggers NO synthesis by nNOS, which is physically tethered to the NMDA receptor via PSD-95. NO can regulate the activity of a number of proteins by S-nitrosylation. NO also reacts with ROS to form highly toxic peroxynitrite ($ONOO^-$), which injures cells via DNA damage, lipid peroxidation, and protein oxidation/nitration. Pathological activation of PARP after DNA damage leads to disrupted energy metabolism. (3) Activation of Rho GTPase and NO/ROS can link elevation of $[Ca^{2+}]_i$ to activation of p38 MAPK and subsequent cell death. Paradoxically, activation of p38 MAPK can also trigger a survival-promoting pathway through activation of the transcription factor MEF2C. However, if caspases are concurrently activated, they can cleave MEF2C, leading to dominant negative form of this transcription factor, which enters the nucleus and blocks the synthesis of several survival factors, thus contributing to a pro-death pathway. Abbreviations: AIF, apoptosis-inducing factor; CaM, calmodulin; MEF2C, myocyte enhancer factor 2C; NMDAR, N-methyl-D-aspartate receptor; nNOS, neuronal nitric oxide synthase; NO, nitric oxide; p38 MAPK, p38 mitogen-activated protein kinase; PARP, poly(ADP-ribose) polymerase; PSD-95, postsynaptic density-95; ROS, reactive oxygen species. Adapted with permission from Lipton (2006).

of the receptors. Among these are the Mg^{2+} sites within the channel pore and the *S*-nitrosylation (redox) sites (Sullivan et al., 1994; Lipton et al., 1998, 1999; Choi et al., 2000). We have used these targets for therapeutic intervention to block excitotoxicity, as discussed later in this chapter. Mg^{2+} serves as an endogenous negative regulator of NMDA receptors. At resting membrane potentials of healthy neurons (approximately $-70\,mV$), Mg^{2+} blocks NMDA receptor-operated channels (Fig. 1A). This Mg^{2+} block is controlled in a voltage-dependent manner

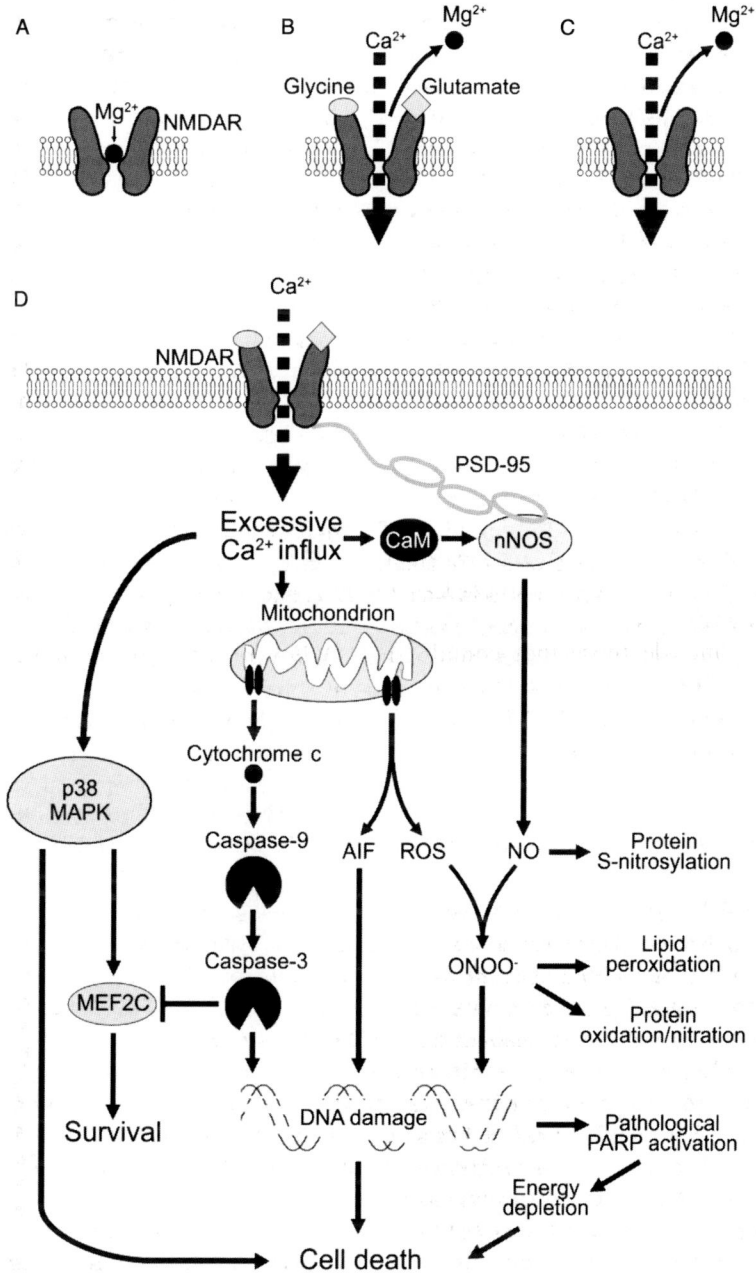

(Mayer et al., 1984; Nowak et al., 1984). If a neuron becomes more positively charged (e.g., depolarized by an excitatory postsynaptic potential [EPSP] to −40 or −30 mV), the Mg^{2+} block is removed to induce consequent Ca^{2+} and Na^+ influx through the now unblocked NMDA receptor-coupled channels (Fig. 1B). Under physiological conditions, the Mg^{2+} block soon recovers during repolarization, thus only allowing the channel to be activated for brief periods of time (milliseconds). However, if the tissue is compromised by ischemia or injury, neurons become depolarized spontaneously, thus removing Mg^{2+} block and rendering NMDA receptors abnormally active, particularly extrasynaptic NMDA receptors because of the location of extracellular glutamate under these pathological conditions (Zeevalk and Nicklas, 1992; Hardingham and Bading, 2003; Lipton, 2006) (Fig. 1C).

Excitotoxicity is a result of overactivation of the glutamate receptors, which can be initiated by the elevated extracellular glutamate concentration or hyperactivity of receptors in the presence of normal glutamate levels. Elevation of extracellular glutamate concentration will occur, for example, after ischemia in the central nervous system, which causes enormous disruption of energy metabolism in both neurons and glia. Under these circumstances, glutamate is not cleared properly by glutamate transporters, which normally take it up mainly into glia, and even reversal of glutamate transport can occur, resulting in increased release (Zeevalk and Nicklas, 1992; Lipton and Rosenberg, 1994; Szatkowski and Attwell, 1994; Billups and Attwell, 1996; Li et al., 1999). Hyperactivity of NMDA receptors in the face of normal glutamate levels can also occur in ischemic nervous tissue because neurons lose their ionic homeostasis and become depolarized. Depolarization relieves Mg^{2+} block of NMDA receptors (Mayer et al., 1984; Nowak et al., 1984; Zeevalk and Nicklas, 1992). Thus, in the absence of physiological Mg^{2+} block, NMDA receptor-coupled channels become abnormally active even in the absence of elevated extracellular glutamate concentration.

Downstream signaling cascades after overactivation of NMDA receptors

Overactivation of NMDA receptors triggers an excessive Ca^{2+} influx into neurons, initiating cell death pathways (Lipton and Rosenberg, 1994) (Fig. 1D). As a consequence of the increase in intracellular Ca^{2+} concentration ($[Ca^{2+}]_i$) and subsequent Ca^{2+} entry into mitochondria, the mitochondrial membrane potential depolarizes (Ankarcrona et al., 1995; Green and Reed, 1998). Depolarized mitochondria release various bioactive substances into the cytosol. Cytochrome c released from mitochondria activates caspases, which play an important role in apoptosis through DNA fragmentation (Green and Reed, 1998). Apoptosis-inducing factor (AIF) is another factor released from mitochondria and also contributes to DNA damage (Yu et al., 2003). Mitochondria are also major sources of reactive oxygen species (ROS), and NMDA stimulation, in fact, causes ROS production in cultured cerebrocortical neurons (Lafon-Cazal et al., 1993; Tenneti et al., 1998).

Calmodulin (CaM) activated by elevation of $[Ca^{2+}]_i$ triggers synthesis of nitric oxide (NO) via neuronal NO synthase (nNOS) (Dawson et al., 1991, 1993; Lipton and Rosenberg, 1994; Lipton, 2006), which is physically tethered to NMDA receptors via interaction with postsynaptic density-95 (PSD-95) protein linked to NR2 subunits, predominantly the NR2B subunit (Kornau et al., 1995; Christopherson et al., 1999; Sattler et al., 1999). NO is involved in many chemical reactions with a great variety of molecules. For example, S-nitrosylation (a chemical reaction representing transfer of NO to the thiol or sulfhydryl group of a critical cysteine residue) regulates the biological activity of many proteins (Lipton et al., 1993; Hess et al., 2005). Depending on the protein, S-nitrosylation may either stimulate or inhibit activity and lead to either neuronal death or survival (Lipton, 1999; Nakamura and Lipton, 2007). Thus, NO can be both neurodestructive and neuroprotective. NO also reacts with ROS to form highly toxic peroxynitrite ($ONOO^-$), which can injure cells by DNA damage, lipid peroxidation, protein oxidation/nitration, and other mechanisms

(Dawson and Dawson, 1996; Lipton, 1999). Additionally, peroxynitrite can elicit Zn^{2+} release from intracellular stores, resulting in loss of mitochondria membrane potential and subsequent neuronal death (Bossy-Wetzel et al., 2004). In the rat retina, intravitreal injection of NMDA increased formation of peroxynitrite, and NMDA-induced RGC death was inhibited by pharmacological inhibition of peroxynitrite formation (El-Remessy et al., 2003). Pathological activation of poly(ADP-ribose) polymerase-1 (PARP-1) after DNA damage is also thought to contribute to cell death by depleting cellular ATP, among other mechanisms, including AIF release (Dawson and Dawson, 1996; Yu et al., 2006). PARP-1 activation occurs in the retina after NMDA injection in the vitreous and is accompanied by a decrease in ATP levels (Goebel and Winkler, 2006). Conversely, a PARP inhibitor protects RGCs from NMDA-induced excitotoxicity (Goebel and Winkler, 2006).

An additional pathway in NMDA excitotoxicity involves p38 mitogen-activated protein kinase (MAPK), which is activated through phosphorylation of its serine and threonine residues. Although initial findings associated p38 MAPK with inflammatory responses (Lee et al., 1994), later studies showed that activation of p38 MAPK can be involved in apoptosis (Xia et al., 1995). Our group has shown that activation of p38 MAPK also occurs in the retina after intravitreal injection of NMDA (Manabe and Lipton, 2003). Moreover, specific p38 MAPK inhibitors can ameliorate NMDA-induced excitotoxicity (Kawasaki et al., 1997; Manabe and Lipton, 2003). A recent study identified Rho GTPase as an essential molecule that links elevation of $[Ca^{2+}]_i$ to p38 MAPK activation and subsequent excitotoxic neuronal death (Semenova et al., 2007). NO or $ONOO^-$ is also capable of activating p38 MAPK upon excitotoxic stimuli (Bossy-Wetzel et al., 2004; Cao et al., 2005). In contrast, depending on the circumstances, p38 MAPK can also trigger a survival-promoting pathway through activation of the downstream transcription factor myocyte enhancer factor 2C (MEF2C) (Mao et al., 1999; Okamoto et al., 2000).

The degree of excitotoxic insult will influence a neuron to undergo apoptosis or necrosis. Exposure to a low concentration of NMDA induces neuronal cell death with apoptotic morphology, whereas a high concentration of NMDA induces predominantly necrosis (Ankarcrona et al., 1995; Bonfoco et al., 1995). The duration of insult is also important. We thus hypothesized that chronic glutamate receptor hyperactivity, even if mild, could trigger apoptotic cell death, possibly following initial synaptic damage (Lipton, 2004). In support of this premise, we and our colleagues reported that mild but prolonged elevation of glutamate levels in the vitreous (30 µM for 3 months, with normal concentrations ~13 µM) resulted in loss of approximately 40% of rat RGCs (Vorwerk et al., 1996). This form of slow and subtle excitotoxicity, leading to oxidative (ROS-related) and nitrosative (NO-related) stress, has been implicated in a variety of chronically progressing neurodegenerative disorders, possibly including glaucoma (Lipton, 2004).

Relevance of excitotoxicity to glaucoma

Whether excitotoxicity participates in the pathophysiology of glaucoma has been a topic of much debate. One contested study detected higher levels of glutamate in the vitreous of glaucoma patients than in controls (Dreyer et al., 1996), while other groups have not replicated this finding (Honkanen et al., 2003). Meanwhile, elevated glutamate levels have been observed in ocular tissues in patients with other retinal diseases in which involvement of glutamate toxicity has been suggested, i.e., in the vitreous of patients with proliferative diabetic retinopathy (Ambati et al., 1997) and in aqueous humor of patients with retinal artery occlusion (Wakabayashi et al., 2006). Importantly, however, one need not have elevated levels of glutamate in order to observe a component of excitotoxicity in pathophysiology of glaucoma. For example, since RGCs that are compromised for almost any reason manifest energy failure, the cells will depolarize as the ionic pumps begin to fail. Therefore, the normal block of NMDA receptor-operated channels by Mg^{2+} decreases (positive charges repel, and, as the intracellular side of the RGC membrane becomes more depolarized or positively charged, it will therefore repel the Mg^{2+} ion normally lodged in the ion channel pore). Hence, sick, depolarized neurons manifest relief from Mg^{2+} block, rendering the cells susceptible to damage by normal levels of glutamate (Zeevalk and Nicklas, 1992; Lipton, 2003).

Neurons and glia contain a high concentration of intracellular glutamate ($\sim 10\,mM$) for both metabolism and neurotransmission (Lipton and Rosenberg, 1994). However, glutamate stored within cells is not harmful. Only "extracellular" glutamate can cause excitotoxicity via the receptor-mediated mechanisms described here. Therefore, localization of glutamate (intracellular or extracellular) is critical, and thus measuring extracellular glutamate levels is much more meaningful than measuring total glutamate contents in tissues. Glutamate levels in the vitreous have been measured under the premise that the vitreous represents the "extracellular space" of the retina. However, the true extracellular space is the "intercellular" space between retinal cells. Because of technical difficulties in measuring such glutamate concentrations in the human retina, no one really knows if there is an elevation of glutamate or not in human patients with glaucoma. Most importantly, however, Hare et al. (2001, 2004a, b) and WoldeMüssie et al. (2002) have shown that an NMDA receptor antagonist can protect RGCs from both histological and electrophysiological correlates of glaucoma in a well-known monkey model of the disease, as discussed below.

How RGCs die in response to elevated intraocular pressure (IOP) in human glaucoma, and whether excitotoxicity is involved, is still a mystery. Furthermore, particularly in Asia, the high prevalence of normal-tension glaucoma, which manifests glaucomatous optic neuropathy without elevated IOP (Shiose et al., 1991; Klein et al., 1992), makes the etiology even more enigmatic. The epicenter of glaucomatous optic neuropathy is proposed to be the optic nerve head or lamina cribrosa, where soft tissues, including RGC axons and blood vessels circulating the optic nerve head, are likely to be compressed as a result of deformation of the laminar structure (Quigley, 1995, 1999). Currently, a dominant hypothesis accounting for RGC loss in glaucoma is that obstructed retrograde axonal transport at the lamina cribrosa deprives RGCs from neurotrophic factors, leading them to die, as proposed by Quigley (1995). Even in this scenario, excitotoxicity may participate in the pathophysiology of glaucoma by causing secondary RGC death because of glutamate leaking out of injured cells, thereby triggering oxidative and nitrosative stress (Fig. 2, dotted arrow). This mechanism has been described as a final common pathway contributing to neuronal degeneration in many neurological disorders (Lipton and Rosenberg, 1994).

Some authorities also opine that the optic nerve head is ischemic in glaucoma (Flammer and Orgul, 1998; Osborne et al., 2001). If this is true, excitotoxicity can almost certainly contribute to the pathophysiology of glaucoma because glutamate clearance by glia decreases under ischemic condition (Lipton and Rosenberg, 1994; Szatkowski and Attwell, 1994; Billups and Attwell, 1996; Li et al., 1999). In this regard, retinas of glaucoma patients showed significantly lower immunoreactivity of the excitatory amino acid transporter-1 (EAAT-1), an enzyme responsible for glutamate clearance (Naskar et al., 2000). Together with inappropriate release of glutamate from metabolically compromised cells (Zeevalk and Nicklas, 1992; Lipton and Rosenberg, 1994; Szatkowski and Attwell, 1994), elevation of extracellular glutamate concentration may occur within the glaucomatous retina (Fig. 2). We emphasize here again that if Mg^{2+} block of NMDA receptors is relieved (as depicted in Fig. 1C), excitotoxicity can come into play in RGCs even without elevated glutamate concentrations (Zeevalk and Nicklas, 1992; Lipton, 2003). During ischemia, disruption of energy metabolism would lead to depolarization of RGCs and relieve Mg^{2+} block of NMDA receptors (Zeevalk and Nicklas, 1992). In addition to depolarization, there are several factors that can impair the Mg^{2+} block, among which is mechanical stress. Sublethal stretch caused almost complete loss of the Mg^{2+} block in cortical neurons (Zhang et al., 1996) and rendered them vulnerable to low concentrations of NMDA (Arundine et al.,

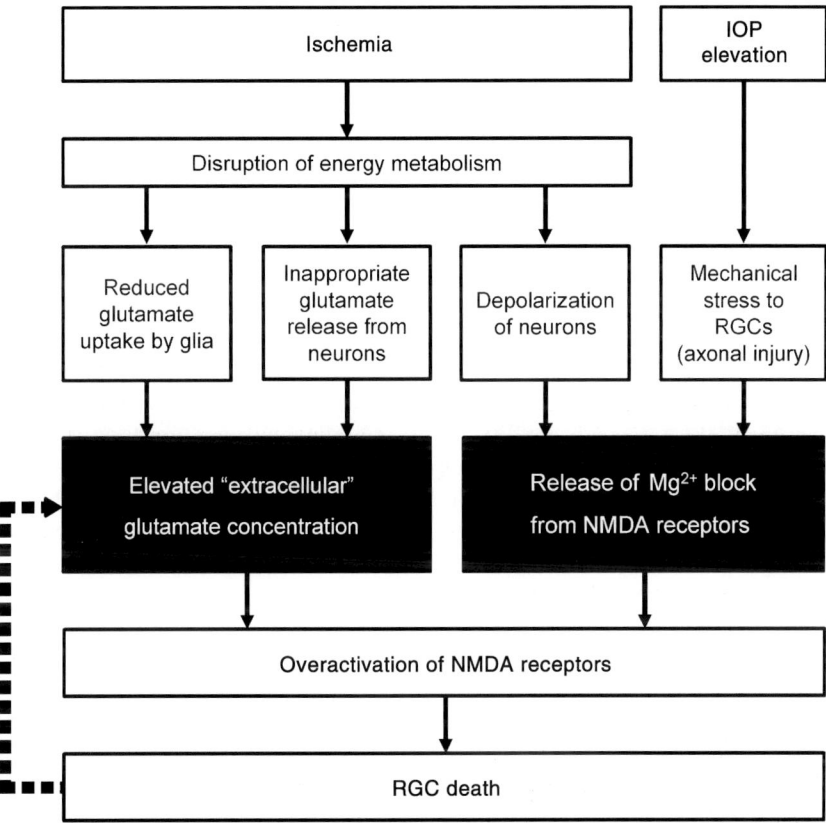

Fig. 2. Hypothetical mechanisms leading to excitotoxicity in glaucoma. The primary site of glaucomatous optic neuropathy is thought to be the optic nerve head, especially at the lamina cribrosa, where RGC axons or blood vessels are likely to be compressed. While two principal hypotheses have been proposed for the pathogenesis of glaucoma (the vascular and mechanical theories), excitotoxicity mediated by the NMDA receptors seems harmonious with both theories. During ischemia, when enormous disruption of energy metabolism occurs, glutamate is not cleared properly by glia and can even be inappropriately released. As a result, the extracellular glutamate concentration may increase. With the loss of energy due to hypoxia-ischemia, neurons lose their ability to maintain energy-dependent ionic homeostasis, and thus neurons become depolarized. This voltage change removes physiological Mg^{2+} block from NMDA receptor-associated channels. As discussed in the text, axonal compression, in this case at the level of the lamina cribrosa, may also relieve Mg^{2+} block, and increased IOP may abnormally increase the activity of NMDA receptor-associated channels. Importantly, either elevated extracellular glutamate concentration or relief from Mg^{2+} block in the face of normal glutamate levels may be sufficient to elicit overactivation of NMDA receptors. In other words, excitotoxicity can play a role in glaucoma even in the absence of elevated extracellular glutamate concentration once Mg^{2+} block of NMDA receptor-associated channels is removed. Glutamate leaking out of dying/dead RGCs or compromised glia may contribute to secondary death of neighboring RGCs via excessive activation of NMDA receptors (dotted arrow). Abbreviations: IOP, intraocular pressure; RGC, retinal ganglion cell; NMDA, *N*-methyl-D-aspartate.

2003), and this affect of stretch in activating NMDA receptors has been replicated on RGCs (R.H. Farkas and S.A. Lipton, unpublished). Other reported factors that impede the Mg^{2+} block of the NMDA receptors are axonal injury (Furukawa et al., 2000) and inflammation (Guo and Huang, 2001). In conclusion, Mg^{2+} blockade of NMDA receptors may be relieved in the face of ischemia, mechanical stress, axonal injury, and inflammation, all of which have been proposed as mechanisms that are associated with high IOP and RGC damage in glaucoma (Quigley, 1995, 1999; Flammer and Orgul, 1998; Osborne et al., 2001; Tezel and Wax, 2004; Burgoyne et al., 2005) (Fig. 2).

Therapeutic approaches to prevent RGC death by targeting the pathways involved in NMDA excitotoxicity

In this section, on the basis of laboratory research and clinical trials, we will discuss if NMDA receptors and downstream signaling molecules contributing to excitotoxic pathways can be targets of therapeutic intervention to prevent RGC death in glaucoma. Importantly, glutamate mediates synaptic transmission, which is essential for the normal function of the nervous system. Hence, complete blockade of NMDA receptor activity can be deleterious because physiological activity is impaired. To be clinically acceptable and well tolerated, anti-excitotoxic therapies must block only excessive activation of the NMDA receptors, while leaving normal function relatively intact to avoid side effects. Although there are a number of potential targets for therapeutic intervention, in this chapter we will focus on molecular pathways that we have been studying in detail.

Drugs targeting NMDA receptors

Kinetics of NMDA receptor antagonists

Various NMDA antagonists have been tested for their therapeutic potential in neurodegenerative disorders. Nevertheless, many of them failed chiefly because of lack of safety and tolerability. If an antagonist binds too tightly (has too high an affinity) and totally blocks activity of the target, it will inhibit the normal activity necessary to maintain physiological function and thus prove to be clinically unacceptable. However, it is important not to confuse affinity with selectivity. If an antagonist acts selectively and specifically on a target, and the therapeutic concentration is achieved, high affinity is not necessary and can even become detrimental (Lipton, 2006). For example, a "neuroprotective" dose of MK-801 (a noncompetitive NMDA receptor antagonist, which blocks the associated ion channel with high affinity) causes drowsiness, hallucinations, and even coma (Lipton, 2004). Indeed, antagonists with high affinity for the NMDA receptors cause psychiatric and other side effects (Domino and Luby, 1981; Leppik, 1988; Javitt and Zukin, 1991; Muir and Lees, 1995).

In addition to affinity, the mechanism of antagonism (competitive, noncompetitive, or uncompetitive) is also important. An uncompetitive antagonist acts at an allosteric site that is different from the agonist-binding site (Lipton, 2004), and its inhibitory action is contingent on prior activation of the receptor by the agonist (Lipton, 2007). Accordingly, an uncompetitive antagonist provides stronger antagonism when an agonist exists at higher concentration, whereas a competitive antagonist exhibits weaker antagonism at higher agonist concentration because it competes with the antagonist for the binding site (Lipton, 2006, 2007) (Fig. 3A). For instance, we have shown that the uncompetitive NMDA receptor antagonist, memantine, inhibits NMDA receptor channel activity only when the receptors are activated by NMDA (Chen et al., 1992). Moreover, an uncompetitive antagonist will block receptor activity more effectively when the receptor is excessively (pathologically) activated, while relatively preserving lower (normal) activity (Lipton, 2004). As a matter of fact, memantine is relatively ineffective at blocking normal synaptic NMDA receptor activity involved in physiological function, but is exceptionally effective when the receptors (usually representing extrasynaptic receptors under pathological conditions) are activated by higher concentration of glutamate (Chen et al., 1992).

Importantly, the off-rate from the NMDA receptor is also a critical factor for developing clinically tolerated drugs. For example, let us contrast MK-801 and memantine. Because MK-801 has a much slower off-rate from NMDA receptor-operated channels (in part accounting for its high affinity), MK-801 resides in the channels for prolonged periods of time and accumulates there. This fact produces prolonged antagonism of NMDA receptors even after attempted washout of MK-801 (Fig. 3B, gray trace), which in turn results in prolonged blockade of not only pathological activity but also normal physiological activity. Conversely, NMDA receptor activity recovers quickly after the washout of memantine owing to its relatively fast off-rate from NMDA-operated channels (Chen et al., 1992; Lipton, 2006) (Fig. 3B,

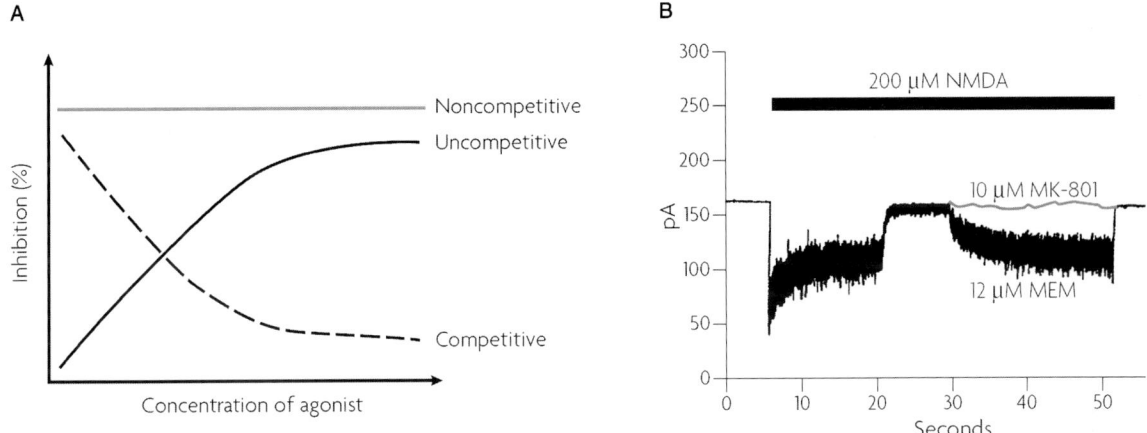

Fig. 3. Characterization of competitive, noncompetititve, and uncompetitive antagonists affecting NMDA receptor-mediated currents. (A) Characteristics of competitive and uncompetitive antagonists. Here, the concentration of agonist is increased while concentration of antagonists is held constant. An uncompetitive antagonist inhibits channel activity contingent on prior activation of the receptor-channel by the agonist. Therefore, a fixed dose of the uncompetitive antagonist manifests a greater degree of antagonism in presence of higher agonist concentration (solid line). In contrast, a competitive antagonist has to outcompete the agonist in order to inhibit channel activity. Thus, a competitive antagonist exhibits less inhibition at higher agonist concentrations (broken line). A noncompetititve antagonist acts at an allosteric site and manifests an equal degree of inhibition irrespective of agonist concentration (gray line). Adapted with permission from Lipton (2007). (B) Inhibition of whole-cell NMDA current recorded with a patch electrode at a holding potential of −50 mV. Action of a noncompetitive NMDA receptor antagonist, MK-801 (gray trace) and an uncompetitive antagonist, memantine (black trace). Importantly, a major difference in action becomes apparent upon washout, reflecting their different off-rates from NMDA receptor-coupled ion channels. MK-801, which possesses a slow off-rate, blocks NMDA current persistently even after washout. In contrast, NMDA current soon recovers after washout of memantine, as this antagonist manifests a relatively rapid off-rate. Abbreviation: MEM, memantine; NMDA, N-methyl-D-aspartate. Adapted with permission from Chen et al. (1992) and Lipton (2006).

black trace). This fast off-rate also allows memantine to block the channels when they are excessively open but leave the channels unblocked when activity returns to normal (Lipton, 2003). Therefore, during prolonged activation of NMDA receptor, as hypothesized to occur in retinal ischemia, glaucoma, and other chronic neurodegenerative disorders, memantine will be a very efficacious blocker (Lipton, 2004). In conclusion, uncompetitive antagonists of NMDA receptors with high selectivity, low affinity, and relatively rapid off-rate can be clinically tolerated (Lipton, 1993, 2001, 2003, 2004, 2006, 2007).

Memantine

Memantine (1-amino-3,4-dimethyladamantane hydrochloride) is an adamantane derivative (an analog of the antiviral drug amantadine) that works as an open-channel blocker of NMDA receptors at or near the Mg^{2+} site within the ion channel (Chen et al., 1992; Chen and Lipton, 1997). What characterizes memantine as unique and safe are the four properties described above (uncompetitive antagonist with high selectivity, low affinity, and fast off-rate) (Chen et al., 1992, 1998). Memantine enters NMDA receptor channels preferentially when the channels are excessively (pathologically) active, while relatively preserving normal channel activity, contributing to the drug's safety and tolerability (Lipton, 2004, 2006). Indeed, clinical trials of memantine for Alzheimer's disease (Reisberg et al., 2003, 2006; Ott et al., 2007) and vascular dementia (Orgogozo et al., 2002; Wilcock et al., 2002) reported no serious adverse events. Similarly, memantine seems less likely to adversely affect visual function because memantine treatment did not alter the electroretinogram (ERG) or visual-evoked potentials (VEPs) of monkeys (Hare et al., 2004a). In addition to proving safe, memantine was also effective in the clinical trials mentioned above and, as a result, has been approved by the European Union and the U.S. Food and Drug Administration (FDA) for treatment of moderate-to-severe Alzheimer's disease. Clinical studies of memantine in the treatment of HIV-associated dementia have also been encouraging (Lipton, 2004; Schifitto et al., 2007). Interestingly, however, the milder the disease, the higher the concentration of memantine needed to combat the damage, since fewer NMDA receptor-operated channels are open in milder disease. Hence, the effective treatment of glaucoma, which often continues for decades in the typical human patient, may necessitate a higher dose of memantine than Alzheimer's disease in which the average patient lives for approximately 7 years (Lipton, 2006, 2007).

At the basic research level, there have been many studies demonstrating therapeutic efficacy of memantine for RGC insults. Our earlier studies have shown that memantine prevents excitotoxic cell death in RGC cultures (Fig. 4A). Additionally, in vivo treatment with memantine inhibited RGC death caused by mild chronic glutamate elevation in rats (Vorwerk et al., 1996) (Fig. 4B) and in EAAT-1 (a glial glutamate transporter) knockout mice, which manifest glaucoma-like pathology (Harada et al., 2007). Memantine also protected the retina, both histologically and functionally, in a rodent model of retinal ischemia caused by raising IOP, which mimics, in part, acute angle closure glaucoma (Osborne, 1999). Additionally, in a monkey model of glaucoma induced by laser coagulation of the trabecular meshwork, memantine successfully inhibited RGC death (Hare et al., 2004b) and shrinkage of neurons in the lateral geniculate nucleus (the major projection site of RGC axons in primates) (Yucel et al., 2006). Encouragingly, memantine treatment improved functional outcomes measured by ERG and VEP in this monkey glaucoma model (Hare et al., 2004a). These results from basic laboratory research imply that memantine may be useful in the treatment of glaucoma. However, a recent

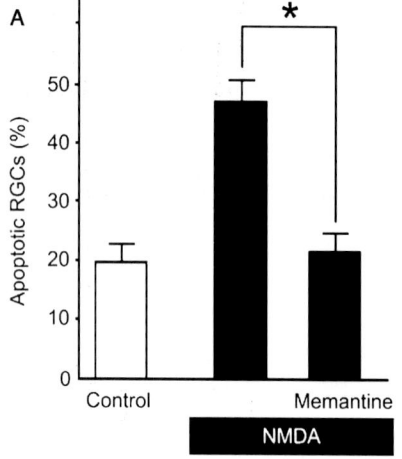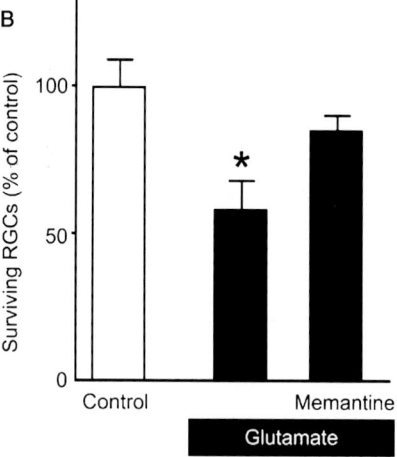

Fig. 4. Neuroprotection of RGCs by memantine. (A) In vitro protection by memantine. Excitotoxicity was induced by exposing rat RGC cultures to NMDA (200 μM, 20 min) in the presence of nominally absent extracellular Mg^{2+} and elevated Ca^{2+} (10 mM) to enhance excitotoxic damage. The percentage of apoptotic RGCs was determined the next day. Simultaneous treatment with memantine (1 μM) significantly prevented apoptosis (*$P<0.01$). Data represent mean ± SEM. (B) In vivo protection by memantine. Rats repeatedly received vitreous injections of glutamate for 3 months, yielding a vitreal concentration of ~30 μM glutamate concentration (cf. normal glutamate level, ~13 μM). Surviving RGCs were counted on retinal whole mounts after retrograde labeling. Whereas eyes with chronic mild elevation of glutamate concentration in the vitreous manifested fewer surviving RGCs than controls (*$P<0.001$), animals intraperitoneally treated with memantine (1 mg/kg) exhibited comparable RGC counts to controls. Data shown represent mean ± SD. Abbreviations: RGC, retinal ganglion cell; NMDA, N-methyl-D-aspartate. Adapted with permission from Vorwerk et al. (1996).

human phase III clinical trial used the same dose of memantine as approved for the treatment of Alzheimer's disease, which for the reasons alluded to above, may represent too low a dose to be effective in glaucoma.

NitroMemantines

NMDA receptors have several cysteine residues whose S-nitrosylation results in downregulation of NMDA receptor activity (Sullivan et al., 1994; Lipton et al., 1998; Lipton, 1999; Choi et al., 2000). Nitroglycerin, which generates NO-related species and is widely used for management of angina pectoris, can decrease NMDA receptor-mediated channel activity (Lipton et al., 1998). Interestingly, nitroglycerin appears to retard progression of glaucomatous changes in primary open-angle glaucoma patients (Zurakowski et al., 1998). Although improved circulation provided by the vasodilating action of nitroglycerin could be the mechanism for this effect, downregulation of excessive NMDA receptor activity by S-nitrosylation of the receptors might have also contributed by attenuating excitotoxicity of RGCs. However, it is not practical to use nitroglycerin for glaucoma treatment because of its significant cardiovascular effects, including hypotension. Therefore, we designed a novel set of dual-acting memantine derivatives, called NitroMemantines, which contain an NO group tethered to memantine in order to target NO to the critical cysteine residues of the NMDA receptor via memantine's specific interaction with overly active NMDA receptor-associated channels (Lipton, 2004, 2006, 2007; Chen and Lipton, 2006). Our preliminary studies have revealed that NitroMemantines display increased neuroprotective properties over memantine, both in culture and in vivo.

Drugs targeting downstream signaling molecules in NMDA-induced cell death pathways

p38 MAPK inhibitors

The p38 MAPK inhibitor, SB203580, can offer neuroprotection. Structural biology studies revealed that SB203580 binds to the ATP-binding site of p38, thus blocking kinase activity without affecting phosphorylation of p38 (Wilson et al., 1997). The neuroprotective potential of SB203580 was first observed in cerebellar granule cells (Kawasaki et al., 1997). We previously showed that SB203580 inhibited NMDA-induced RGC death in culture (Kikuchi et al., 2000) (Fig. 5A). This p38 MAPK inhibitor also protected rat RGCs in vivo from NMDA-induced excitotoxicity (Manabe and Lipton, 2003) (Fig. 5B) and from optic nerve axotomy (Kikuchi et al., 2000) (Fig. 5C). Recent development of improved, clinically tolerated p38 MAPK antagonists for other diseases, including rheumatoid arthritis, suggests the possibility of future clinical application of these drugs to optic neuropathies such as glaucoma.

Averting caspase-mediated neurodegeneration

Caspases are a group of enzymes that cleave their substrates in the execution of apoptotic cell death. These enzymes contribute to apoptotic neuronal cell death observed during mild NMDA-induced excitotoxicity. Thus, a candidate therapeutic strategy in glaucoma is to protect RGCs from excessive caspase activity. Conventional irreversible caspase inhibitors such as *N*-benzyloxycarbonyl-Val-Ala-Asp-fluoromethyl ketone (z-VAD-fmk) work as pseudosubstrates; they bind to caspases like decoys in place of the actual substrate and thus block interaction between caspases and substrates. Because this type of inhibition is irreversible and therefore possibly toxic, these chemicals to date have not proven useful in the clinic for neuroprotection. We and our collaborators have found that a hexapeptide, IQACRG, protects a variety of neuronal cells from apoptotic-like cell death, including rat pheochromocytoma-derived PC12 cell death induced by trophic factor withdrawal and downregulation of superoxide dismutase 1 (Troy et al., 1996), and cultured cerebrocortical neurons exposed to NMDA (Tenneti et al., 1998). The peptide is homologous to the active site of caspases-1, -2, -3, -6, -7, and -14 but is incapable of substrate cleavage. IQACRG is thought to bind to caspase substrates by mimicking the enzyme, thus acting as a pseudoenzyme to protect substrates

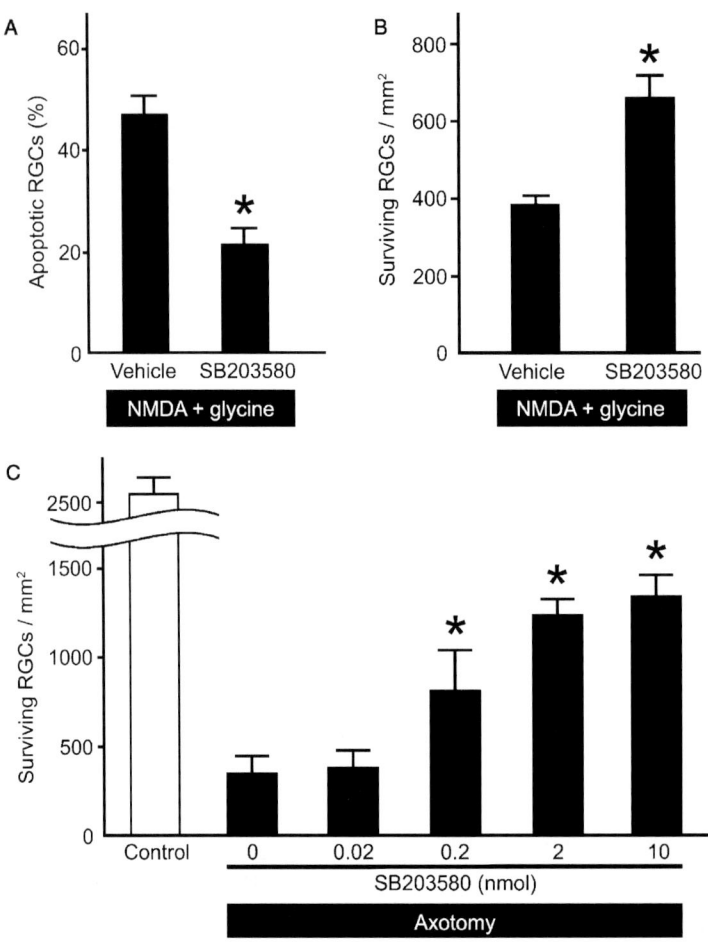

Fig. 5. Neuroprotection of RGCs by the p38 MAPK inhibitor, SB203580, in culture and in vivo. (A) Cultured rat RGCs were stimulated with NMDA (200 μM) and glycine (5 μM). Cells were either treated or not treated with a p38 MAPK inhibitor (SB203580, 1 μM) during NMDA exposure. After an 18-h incubation, RGCs were stained with the RGC marker (Thy-1). After cell permeabilization, nuclei were stained with propidium iodide to judge morphology and scored to determine the percentage of apoptotic RGCs. Treatment with SB203580 significantly inhibited NMDA-induced RGC death ($^*P<0.01$). Data are mean ± SEM. Adapted with permission from Kikuchi et al. (2000). (B) Intravitreal injection of NMDA (200 nmol) and glycine (10 nmol) was performed in rats with or without SB203580 (0.2 nmol). Surviving RGCs were visualized after retrograde labeling with a fluorescent tracer. The p38 MAPK inhibitor protected RGCs from excitotoxicity ($^*P<0.05$). Data are mean ± SEM. Figure modified with permission from Manabe and Lipton (2003). (C) Optic nerve axotomy was performed in rats followed by intravitreal injection of SB203580, with repeated injections on postoperative days 5 and 10. RGC survival was assessed histologically 14 days after axotomy. SB2030580 attenuated RGC death after axotomy at doses of 0.2 nmol or higher ($^*P<0.01$). Data are expressed as mean ± SD. Abbreviations: RGC, retinal ganglion cell; NMDA, N-methyl-D-aspartate; MAPK, mitogen-activated protein kinase. Adapted with permission from Kikuchi et al. (2000).

from caspase cleavage (Troy et al., 1996). We have recently found that the IQACRG peptide protects RGCs from NMDA-induced cell death both in culture and in vivo after intravitreal injection. Future studies investigating the protective properties of IQACRG in more relevant animal models of glaucoma and also exploring clinically applicable modes of drug delivery could make this peptide a viable therapeutic to prevent RGC death in diseases such as glaucoma.

Abbreviations

AIF	apoptosis-inducing factor
CaM	calmodulin
EAAT	excitatory amino acid transporter
HIV	human immunodeficiency virus
IOP	intraocular pressure
MAPK	mitogen-activated protein kinase
MEF2C	myocyte enhancer factor 2C
NMDA	N-methyl-D-aspartate
NMDAR	NMDA receptor
nNOS	neuronal nitric oxide synthase
NO	nitric oxide
PARP	poly(ADP-ribose) polymerase
PSD-95	postsynaptic density-95
RGC	retinal ganglion cell
ROS	reactive oxygen species

Acknowledgments

We thank our collaborators and colleagues, present and former, for their contributions to this work. We are especially grateful to J. Bormann, Y.-B. Choi, H.-S. V. Chen, M. Kikuchi, S. Manabe, C. M. Troy, and M. L. Shelanski. This work was supported in part by National Institutes of Health grants R01 EY05477 and R01 EY09024 (to S.A.L), Allergan, Inc. (to S.A.L.), Astellas Foundation for Research on Metabolic Disorders (to M.S.), and the Japanese Society for the Promotion of Science (JSPS Postdoctoral Fellowship for Research Abroad to M.S).

References

Aizenman, E., Frosch, M.P. and Lipton, S.A. (1988) Responses mediated by excitatory amino acid receptors in solitary retinal ganglion cells from rat. J. Physiol., 396: 75–91.

Ambati, J., Chalam, K.V., Chawla, D.K., D'Angio, C.T., Guillet, E.G., Rose, S.J., Vanderlinde, R.E. and Ambati, B.K. (1997) Elevated gamma-aminobutyric acid, glutamate, and vascular endothelial growth factor levels in the vitreous of patients with proliferative diabetic retinopathy. Arch. Ophthalmol., 115(9): 1161–1166.

Ankarcrona, M., Dypbukt, J.M., Bonfoco, E., Zhivotovsky, B., Orrenius, S., Lipton, S.A. and Nicotera, P. (1995) Glutamate-induced neuronal death: a succession of necrosis or apoptosis depending on mitochondrial function. Neuron, 15(4): 961–973.

Arundine, M., Chopra, G.K., Wrong, A., Lei, S., Aarts, M.M., MacDonald, J.F. and Tymianski, M. (2003) Enhanced vulnerability to NMDA toxicity in sublethal traumatic neuronal injury in vitro. J. Neurotrauma, 20(12): 1377–1395.

Billups, B. and Attwell, D. (1996) Modulation of non-vesicular glutamate release by pH. Nature, 379(6561): 171–174.

Bonfoco, E., Krainc, D., Ankarcrona, M., Nicotera, P. and Lipton, S.A. (1995) Apoptosis and necrosis: two distinct events induced, respectively, by mild and intense insults with N-methyl-D-aspartate or nitric oxide/superoxide in cortical cell cultures. Proc. Natl. Acad. Sci. U.S.A., 92(16): 7162–7166.

Bossy-Wetzel, E., Talantova, M.V., Lee, W.D., Scholzke, M.N., Harrop, A., Mathews, E., Gotz, T., Han, J., Ellisman, M.H., Perkins, G.A. and Lipton, S.A. (2004) Crosstalk between nitric oxide and zinc pathways to neuronal cell death involving mitochondrial dysfunction and p38-activated K^+ channels. Neuron, 41(3): 351–365.

Brandstatter, J.H., Hartveit, E., Sassoe-Pognetto, M. and Wassle, H. (1994) Expression of NMDA and high-affinity kainate receptor subunit mRNAs in the adult rat retina. Eur. J. Neurosci., 6(7): 1100–1112.

Burgoyne, C.F., Downs, J.C., Bellezza, A.J., Suh, J.K. and Hart, R.T. (2005) The optic nerve head as a biomechanical structure: a new paradigm for understanding the role of IOP-related stress and strain in the pathophysiology of glaucomatous optic nerve head damage. Prog. Retin. Eye Res., 24(1): 39–73.

Cao, J., Viholainen, J.I., Dart, C., Warwick, H.K., Leyland, M.L. and Courtney, M.J. (2005) The PSD95-nNOS interface: a target for inhibition of excitotoxic p38 stress-activated protein kinase activation and cell death. J. Cell Biol., 168(1): 117–126.

Chen, H.-S.V. and Lipton, S.A. (1997) Mechanism of memantine block of NMDA-activated channels in rat retinal ganglion cells: uncompetitive antagonism. J. Physiol., 499(Pt 1): 27–46.

Chen, H.-S.V. and Lipton, S.A. (2006) The chemical biology of clinically tolerated NMDA receptor antagonists. J. Neurochem., 97(6): 1611–1626.

Chen, H.-S.V., Pellegrini, J.W., Aggarwal, S.K., Lei, S.Z., Warach, S., Jensen, F.E. and Lipton, S.A. (1992) Open-channel block of N-methyl-D-aspartate (NMDA) responses by memantine: therapeutic advantage against NMDA receptor-mediated neurotoxicity. J. Neurosci., 12(11): 4427–4436.

Chen, H.-S.V., Wang, Y.F., Rayudu, P.V., Edgecomb, P., Neill, J.C., Segal, M.M., Lipton, S.A. and Jensen, F.E. (1998) Neuroprotective concentrations of the N-methyl-D-aspartate open-channel blocker memantine are effective without cytoplasmic vacuolation following post-ischemic administration and do not block maze learning or long-term potentiation. Neuroscience, 86(4): 1121–1132.

Choi, D.W. (1988) Glutamate neurotoxicity and diseases of the nervous system. Neuron, 1(8): 623–634.

Choi, Y.-B., Tenneti, L., Le, D.A., Ortiz, J., Bai, G., Chen, H.-S.V. and Lipton, S.A. (2000) Molecular basis of NMDA receptor-coupled ion channel modulation by S-nitrosylation. Nat. Neurosci., 3(1): 15–21.

Christopherson, K.S., Hillier, B.J., Lim, W.A. and Bredt, D.S. (1999) PSD-95 assembles a ternary complex with the N-methyl-D-aspartic acid receptor and a bivalent neuronal NO synthase PDZ domain. J. Biol. Chem., 274(39): 27467–27473.

Dawson, V.L. and Dawson, T.M. (1996) Nitric oxide neurotoxicity. J. Chem. Neuroanat., 10(3–4): 179–190.

Dawson, V.L., Dawson, T.M., Bartley, D.A., Uhl, G.R. and Snyder, S.H. (1993) Mechanisms of nitric oxide-mediated neurotoxicity in primary brain cultures. J. Neurosci., 13(6): 2651–2661.

Dawson, V.L., Dawson, T.M., London, E.D., Bredt, D.S. and Snyder, S.H. (1991) Nitric oxide mediates glutamate neurotoxicity in primary cortical cultures. Proc. Natl. Acad. Sci. U.S.A., 88(14): 6368–6371.

Domino, E.F. and Luby, E.D. (1981) Abnormal mental states induced by phencyclidine as a model of schizophrenia. In: Domino E.F. (Ed.), PCP (Phencyclidine): Historical and Current Perspectives. NPP Books, Ann Arbor, pp. 401–418.

Dreyer, E.B. and Lipton, S.A. (1999) New perspectives on glaucoma. JAMA, 281(4): 306–308.

Dreyer, E.B., Zurakowski, D., Schumer, R.A., Podos, S.M. and Lipton, S.A. (1996) Elevated glutamate levels in the vitreous body of humans and monkeys with glaucoma. Arch. Ophthalmol., 114(3): 299–305.

El-Remessy, A.B., Khalil, I.E., Matragoon, S., Abou-Mohamed, G., Tsai, N.-J., Roon, P., Caldwell, R.B., Caldwell, R.W., Green, K. and Liou, G.I. (2003) Neuroprotective effect of (-){Delta}9-tetrahydrocannabinol and cannabidiol in N-methyl–aspartate-induced retinal neurotoxicity: involvement of peroxynitrite. Am. J. Pathol., 163(5): 1997–2008.

Flammer, J. and Orgul, S. (1998) Optic nerve blood-flow abnormalities in glaucoma. Prog. Retin. Eye Res., 17(2): 267–289.

Furukawa, Y., Okada, M., Akaike, N., Hayashi, T. and Nabekura, J. (2000) Reduction of voltage-dependent magnesium block of N-methyl-D-aspartate receptor-mediated current by in vivo axonal injury. Neuroscience, 96(2): 385–392.

Goebel, D.J. and Winkler, B.S. (2006) Blockade of PARP activity attenuates poly(ADP-ribosyl)ation but offers only partial neuroprotection against NMDA-induced cell death in the rat retina. J. Neurochem., 98(6): 1732–1745.

Green, D.R. and Reed, J.C. (1998) Mitochondria and apoptosis. Science, 281(5381): 1309–1312.

Grunder, T., Kohler, K., Kaletta, A. and Guenther, E. (2000) The distribution and developmental regulation of NMDA receptor subunit proteins in the outer and inner retina of the rat. J. Neurobiol., 44(3): 333–342.

Guo, H. and Huang, L.Y.M. (2001) Alteration in the voltage dependence of NMDA receptor channels in rat dorsal horn neurones following peripheral inflammation. J. Physiol., 537(1): 115–123.

Hahn, J.S., Aizenman, E. and Lipton, S.A. (1988) Central mammalian neurons normally resistant to glutamate toxicity are made sensitive by elevated extracellular Ca^{2+}: toxicity is blocked by the N-methyl-D-aspartate antagonist MK-801. Proc. Natl. Acad. Sci. U.S.A., 85(17): 6556–6560.

Harada, T., Harada, C., Nakamura, K., Quah, H.M., Okumura, A., Namekata, K., Saeki, T., Aihara, M., Yoshida, H., Mitani, A. and Tanaka, K. (2007) The potential role of glutamate transporters in the pathogenesis of normal tension glaucoma. J. Clin. Invest., 117(7): 1763–1770.

Hardingham, G.E. and Bading, H. (2003) The Yin and Yang of NMDA receptor signalling. Trends Neurosci., 26(2): 81–89.

Hare, W.A., WoldeMüssie, E., Lai, R.K., Ton, H., Ruiz, G., Chun, T. and Wheeler, L. (2004a) Efficacy and safety of memantine treatment for reduction of changes associated with experimental glaucoma in monkey, I: Functional measures. Invest. Ophthalmol. Vis. Sci., 45(8): 2625–2639.

Hare, W., WoldeMüssie, E., Lai, R., Ton, H., Ruiz, G., Feldmann, B., Wijono, M., Chun, T. and Wheeler, L. (2001) Efficacy and safety of memantine, an NMDA-type open-channel blocker, for reduction of retinal injury associated with experimental glaucoma in rat and monkey. Surv. Ophthalmol., 45(Suppl 3): S284–S289.

Hare, W.A., WoldeMüssie, E., Weinreb, R.N., Ton, H., Ruiz, G., Wijono, M., Feldmann, B., Zangwill, L. and Wheeler, L. (2004b) Efficacy and safety of memantine treatment for reduction of changes associated with experimental glaucoma in monkey, II: Structural measures. Invest. Ophthalmol. Vis. Sci., 45(8): 2640–2651.

Hartveit, E., Brandstatter, J.H., Sassoe-Pognetto, M., Laurie, D.J., Seeburg, P.H. and Wassle, H. (1994) Localization and developmental expression of the NMDA receptor subunit NR2A in the mammalian retina. J. Comp. Neurol., 348(4): 570–582.

Hess, D.T., Matsumoto, A., Kim, S.O., Marshall, H.E. and Stamler, J.S. (2005) Protein S-nitrosylation: purview and parameters. Nat. Rev. Mol. Cell. Biol., 6(2): 150–166.

Honkanen, R.A., Baruah, S., Zimmerman, M.B., Khanna, C.L., Weaver, Y.K., Narkiewicz, J., Waziri, R., Gehrs, K.M., Weingeist, T.A., Boldt, H.C., Folk, J.C., Russell, S.R. and Kwon, Y.H. (2003) Vitreous amino acid concentrations in patients with glaucoma undergoing vitrectomy. Arch. Ophthalmol., 121(2): 183–188.

Javitt, D.C. and Zukin, S.R. (1991) Recent advances in the phencyclidine model of schizophrenia. Am. J. Psychiatry, 148(10): 1301–1308.

Kawasaki, H., Morooka, T., Shimohama, S., Kimura, J., Hirano, T., Gotoh, Y. and Nishida, E. (1997) Activation and involvement of p38 mitogen-activated protein kinase in glutamate-induced apoptosis in rat cerebellar granule cells. J. Biol. Chem., 272(30): 18518–18521.

Kikuchi, M., Tenneti, L. and Lipton, S.A. (2000) Role of p38 mitogen-activated protein kinase in axotomy-induced apoptosis of rat retinal ganglion cells. J. Neurosci., 20(13): 5037–5044.

Klein, B.E., Klein, R., Sponsel, W.E., Franke, T., Cantor, L.B., Martone, J. and Menage, M.J. (1992) Prevalence of glaucoma. The beaver dam eye study. Ophthalmology, 99(10): 1499–1504.

Kornau, H.C., Schenker, L.T., Kennedy, M.B. and Seeburg, P.H. (1995) Domain interaction between NMDA receptor subunits and the postsynaptic density protein PSD-95. Science, 269(5231): 1737–1740.

Lafon-Cazal, M., Pietri, S., Culcasi, M. and Bockaert, J. (1993) NMDA-dependent superoxide production and neurotoxicity. Nature, 364(6437): 535–537.

Lee, J.C., Laydon, J.T., McDonnell, P.C., Gallagher, T.F., Kumar, S., Green, D., McNulty, D., Blumenthal, M.J., Keys, J.R., Landvatter, S.W., Strickler, J.E., McLaughlin, M.M., Siemens, I.R., Fisher, S.M., Livi, G.P., White, J.R., Adams, J.L. and Young, P.R. (1994) A protein kinase involved in the regulation of inflammatory cytokine biosynthesis. Nature, 372(6508): 739–746.

Leppik, I.E. (1988) MK-801 for epilepsy: a pilot study. Neurology, 38(Suppl. 1): p. 405.

Li, S., Mealing, G.A.R., Morley, P. and Stys, P.K. (1999) Novel injury mechanism in anoxia and trauma of spinal cord white matter: glutamate release via reverse Na^+-dependent glutamate transport. J. Neurosci., 19(14): 16RC–16RC.

Lipton, S.A. (1993) Prospects for clinically tolerated NMDA antagonists: open-channel blockers and alternative redox states of nitric oxide. Trends Neurosci., 16(12): 527–532.

Lipton, S.A. (1999) Neuronal protection and destruction by NO. Cell Death Differ., 6(10): 943–951.

Lipton, S.A. (2001) Retinal ganglion cells, glaucoma and neuroprotection. Prog. Brain Res., 131: 712–718.

Lipton, S.A. (2003) Possible role for memantine in protecting retinal ganglion cells from glaucomatous damage. Surv. Ophthalmol., 48(Suppl 1): S38–S46.

Lipton, S.A. (2004) Failures and successes of NMDA receptor antagonists: molecular basis for the use of open-channel blockers like memantine in the treatment of acute and chronic neurologic insults. NeuroRx, 1(1): 101–110.

Lipton, S.A. (2006) Paradigm shift in neuroprotection by NMDA receptor blockade: memantine and beyond. Nat. Rev. Drug Discov., 5(2): 160–170.

Lipton, S.A. (2007) Pathologically activated therapeutics for neuroprotection. Nat. Rev. Neurosci., 8(10): 803–808.

Lipton, S.A., Choi, Y.B., Pan, Z.H., Lei, S.Z., Chen, H.-S.V., Sucher, N.J., Loscalzo, J., Singel, D.J. and Stamler, J.S. (1993) A redox-based mechanism for the neuroprotective and neurodestructive effects of nitric oxide and related nitroso-compounds. Nature, 364(6438): 626–632.

Lipton, S.A., Rayudu, P.V., Choi, Y.B., Sucher, N.J. and Chen, H.-S.V. (1998) Redox modulation of the NMDA receptor by NO-related species. Prog. Brain Res., 118: 73–82.

Lipton, S.A. and Rosenberg, P.A. (1994) Excitatory amino acids as a final common pathway for neurologic disorders. N. Engl. J. Med., 330(9): 613–622.

Lucas, D.R. and Newhouse, J.P. (1957) The toxic effect of sodium L-glutamate on the inner layers of the retina. AMA Arch. Ophthalmol., 58(2): 193–201.

Manabe, S. and Lipton, S. (2003) Divergent NMDA signals leading to proapoptotic and antiapoptotic pathways in the rat retina. Invest. Ophthalmol. Vis. Sci., 44(1): 385–392.

Mao, Z., Bonni, A., Xia, F., Nadal-Vicens, M. and Greenberg, M.E. (1999) Neuronal activity-dependent cell survival mediated by transcription factor MEF2. Science, 286(5440): 785–790.

Mayer, M.L., Westbrook, G.L. and Guthrie, P.B. (1984) Voltage-dependent block by Mg^{2+} of NMDA responses in spinal cord neurones. Nature, 309(5965): 261–263.

Muir, K.W. and Lees, K.R. (1995) Clinical experience with excitatory amino acid antagonist drugs. Stroke, 26(3): 503–513.

Nakamura, T. and Lipton, S.A. (2007) S-Nitrosylation and uncompetitive/fast off-rate (UFO) drug therapy in neurodegenerative disorders of protein misfolding. Cell Death Differ., 14(7): 1305–1314.

Naskar, R., Vorwerk, C.K. and Dreyer, E.B. (2000) Concurrent downregulation of a glutamate transporter and receptor in glaucoma. Invest. Ophthalmol. Vis. Sci., 41(7): 1940–1944.

Nowak, L., Bregestovski, P., Ascher, P., Herbet, A. and Prochiantz, A. (1984) Magnesium gates glutamate-activated channels in mouse central neurones. Nature, 307(5950): 462–465.

Okamoto, S., Krainc, D., Sherman, K. and Lipton, S.A. (2000) Antiapoptotic role of the p38 mitogen-activated protein kinase-myocyte enhancer factor 2 transcription factor pathway during neuronal differentiation. Proc. Natl. Acad. Sci. U.S.A., 97(13): 7561–7566.

Olney, J.W. and Ho, O.L. (1970) Brain damage in infant mice following oral intake of glutamate, aspartate or cysteine. Nature, 227(5258): 609–611.

Orgogozo, J.M., Rigaud, A.S., Stoffler, A., Mobius, H.J. and Forette, F. (2002) Efficacy and safety of memantine in patients with mild to moderate vascular dementia: a randomized, placebo-controlled trial (MMM 300). Stroke, 33(7): 1834–1839.

Osborne, N.N. (1999) Memantine reduces alterations to the mammalian retina, in situ, induced by ischemia. Vis. Neurosci., 16(1): 45–52.

Osborne, N.N., Melena, J., Chidlow, G. and Wood, J.P.M. (2001) A hypothesis to explain ganglion cell death caused by vascular insults at the optic nerve head: possible implication for the treatment of glaucoma. Br. J. Ophthalmol., 85(10): 1252–1259.

Ott, B.R., Blake, L.M., Kagan, E. and Resnick, M. (2007) Open label, multicenter, 28 week extension study of the safety and tolerability of memantine in patients with mild to moderate Alzheimer's disease. J. Neurol., 254(3): 351–358.

Quigley, H.A. (1995) Ganglion cell death in glaucoma: pathology recapitulates ontogeny. Aust. N. Z. J. Ophthalmol., 23(2): 85–91.

Quigley, H.A. (1999) Neuronal death in glaucoma. Prog. Retin. Eye Res., 18(1): 39–57.

Reisberg, B., Doody, R., Stoffler, A., Schmitt, F., Ferris, S. and Mobius, H.J. (2003) Memantine in moderate-to-severe Alzheimer's disease. N. Engl. J. Med., 348(14): 1333–1341.

Reisberg, B., Doody, R., Stoffler, A., Schmitt, F., Ferris, S. and Mobius, H.J. (2006) A 24 week open-label extension study of memantine in moderate to severe Alzheimer disease. Arch. Neurol., 63(1): 49–54.

Sattler, R., Xiong, Z., Lu, W.Y., Hafner, M., MacDonald, J.F. and Tymianski, M. (1999) Specific coupling of NMDA receptor activation to nitric oxide neurotoxicity by PSD-95 protein. Science, 284(5421): 1845–1848.

Schifitto, G., Navia, B.A., Yiannoutsos, C.T., Marra, C.M., Chang, L., Ernst, T., Jarvik, J.G., Miller, E.N., Singer, E.J., Ellis, R.J., Kolson, D.L., Simpson, D., Nath, A., Berger, J., Shriver, S.L., Millar, L.L., Colquhoun, D., Lenkinski, R., Gonzalez, R.G. and Lipton, S.A. (2007) Memantine and HIV-associated cognitive impairment: a neuropsychological and proton magnetic resonance spectroscopy study. AIDS, 21(14): 1877–1886.

Semenova, M.M., Maki-Hokkonen, A.M.J., Cao, J., Komarovski, V., Forsberg, K.M., Koistinaho, M., Coffey, E.T. and Courtney, M.J. (2007) Rho mediates calcium-dependent activation of p38alpha and subsequent excitotoxic cell death. Nat. Neurosci., 10(4): 436–443.

Shiose, Y., Kitazawa, Y., Tsukahara, S., Akamatsu, T., Mizokami, K., Futa, R., Katsushima, H. and Kosaki, H. (1991) Epidemiology of glaucoma in Japan — a nationwide glaucoma survey. Jpn. J. Ophthalmol., 35(2): 133–155.

Sucher, N.J., Aizenman, E. and Lipton, S.A. (1991) N-Methyl-D-aspartate antagonists prevent kainate neurotoxicity in rat retinal ganglion cells in vitro. J. Neurosci., 11(4): 966–971.

Sullivan, J.M., Traynelis, S.F., Chen, H.-S.V., Escobar, W., Heinemann, S.F. and Lipton, S.A. (1994) Identification of two cysteine residues that are required for redox modulation of the NMDA subtype of glutamate receptor. Neuron, 13(4): 929–936.

Szatkowski, M. and Attwell, D. (1994) Triggering and execution of neuronal death in brain ischaemia: two phases of glutamate release by different mechanisms. Trends Neurosci., 17(9): 359–365.

Tenneti, L., D'Emilia, D.M., Troy, C.M. and Lipton, S.A. (1998) Role of caspases in N-methyl-D-aspartate-induced apoptosis in cerebrocortical neurons. J. Neurochem., 71(3): 946–959.

Tezel, G. and Wax, M.B. (2004) The immune system and glaucoma. Curr. Opin. Ophthalmol., 15(2): 80–84.

Troy, C.M., Stefanis, L., Prochiantz, A., Greene, L.A. and Shelanski, M.L. (1996) The contrasting roles of ICE family proteases and interleukin-1beta in apoptosis induced by trophic factor withdrawal and by copper/zinc superoxide dismutase down-regulation. Proc Natl. Acad. Sci. U.S.A., 93(11): 5635–5640.

Vorwerk, C.K., Lipton, S.A., Zurakowski, D., Hyman, B.T., Sabel, B.A. and Dreyer, E.B. (1996) Chronic low-dose glutamate is toxic to retinal ganglion cells. Toxicity blocked by memantine. Invest. Ophthalmol. Vis. Sci., 37(8): 1618–1624.

Wakabayashi, Y., Yagihashi, T., Kezuka, J., Muramatsu, D., Usui, M. and Iwasaki, T. (2006) Glutamate levels in aqueous humor of patients with retinal artery occlusion. Retina, 26(4): 432–436.

Wilcock, G., Mobius, H.J. and Stoffler, A. (2002) A double-blind, placebo-controlled multicentre study of memantine in mild to moderate vascular dementia (MMM500). Int. Clin. Psychopharmacol., 17(6): 297–305.

Wilson, K.P., McCaffrey, P.G., Hsiao, K., Pazhanisamy, S., Galullo, V., Bemis, G.W., Fitzgibbon, M.J., Caron, P.R., Murcko, M.A. and Su, M.S.S. (1997) The structural basis for the specificity of pyridinylimidazole inhibitors of p38 MAP kinase. Chem. Biol., 4(6): 423–431.

WoldeMüssie, E., Yoles, E., Schwartz, M., Ruiz, G. and Wheeler, L.A. (2002) Neuroprotective effect of memantine in different retinal injury models in rats. J. Glaucoma, 11(6): 474–480.

Xia, Z., Dickens, M., Raingeaud, J., Davis, R.J. and Greenberg, M.E. (1995) Opposing effects of ERK and JNK-p38 MAP kinases on apoptosis. Science, 270(5240): 1326–1331.

Yu, S.W., Andrabi, S.A., Wang, H., Kim, N.S., Poirier, G.G., Dawson, T.M. and Dawson, V.L. (2006) Apoptosis-inducing factor mediates poly(ADP-ribose) (PAR) polymer-induced cell death. Proc. Natl. Acad. Sci. U.S.A., 103(48): 18314–18319.

Yu, S.W., Wang, H., Dawson, T.M. and Dawson, V.L. (2003) Poly(ADP-ribose) polymerase-1 and apoptosis inducing factor in neurotoxicity. Neurobiol. Dis., 14(3): 303–317.

Yucel, Y.H., Gupta, N., Zhang, Q., Mizisin, A.P., Kalichman, M.W. and Weinreb, R.N. (2006) Memantine protects neurons from shrinkage in the lateral geniculate nucleus in experimental glaucoma. Arch. Ophthalmol., 124(2): 217–225.

Zeevalk, G.D. and Nicklas, W.J. (1992) Evidence that the loss of the voltage-dependent Mg^{2+} block at the N-methyl-D-aspartate receptor underlies receptor activation during inhibition of neuronal metabolism. J. Neurochem., 59(4): 1211–1220.

Zhang, L., Rzigalinski, B.A., Ellis, E.F. and Satin, L.S. (1996) Reduction of voltage-dependent Mg^{2+} blockade of NMDA current in mechanically injured neurons. Science, 274(5294): 1921–1923.

Zurakowski, D., Vorwerk, C.K., Gorla, M., Kanellopoulos, A.J., Chaturvedi, N., Grosskreutz, C.L., Lipton, S.A. and Dreyer, E.B. (1998) Nitrate therapy may retard glaucomatous optic neuropathy, perhaps through modulation of glutamate receptors. Vision Res., 38(10): 1489–1494.

CHAPTER 35

Stem cells for neuroprotection in glaucoma

N.D. Bull[a], T.V. Johnson[a] and K.R. Martin*

Cambridge Centre for Brain Repair, University of Cambridge, Forvie Site, Robinson Way, Cambridge CB2 2PY, UK

Abstract: Stem cell transplantation is currently being explored as a therapy for many neurodegenerative diseases including glaucoma. Cellular therapies have the potential to provide chronic neuroprotection after a single treatment, and early results have been encouraging in models of spinal cord injury and Parkinson's disease. Stem cells may prove ideal for use in such treatments as they can accumulate at sites of injury in the central nervous system (CNS) and may also offer the possibility of targeted treatment delivery. Numerous stem cell sources exist, with embryonic and fetal stem cells liable to be superseded by adult-derived cells as techniques to modify the potency and differentiation of somatic cells improve. Possible neuroprotective mechanisms offered by stem cell transplantation include the supply of neurotrophic factors and the modulation of matrix metalloproteinases and other components of the CNS environment to facilitate endogenous repair. Though formidable challenges remain, stem cell transplantation remains a promising therapeutic approach in glaucoma. In addition, such studies may also provide important insights relevant to other neurodegenerative diseases.

Keywords: glaucoma; stem cell; neuroprotection; transplantation; neurodegeneration; retinal ganglion cell

Introduction

The possible neuroprotective effects of stem cell transplantation are a focus of active investigation in neuroscience at the present time. There are a number of studies where such treatments have shown promise in animal models of neurodegenerative disease and applying such techniques to glaucoma has a number of attractions. Glaucoma is well suited for such investigations as it is possible to directly visualize cellular transplants in vivo and, in addition, techniques to assess small changes in the structure and function of the eye are well developed. Such findings may also apply to other neurodegenerative diseases with similar pathologies. In addition, current treatments for glaucoma are limited mainly to reduction of intraocular pressure, which fails to prevent further deterioration in some glaucomatous eyes and necessitates the development of new treatment approaches. Methods that have the potential for prolonged therapeutic effects after a single treatment are particularly attractive.

In this chapter, we explore the potential use of stem cells for neuroprotection in the central nervous system (CNS) and, in particular, the challenges of using such techniques for glaucoma therapy. We also consider the implications of results obtained in glaucoma models for other neurodegenerative diseases.

*Corresponding author. Tel.: +44 1223 331160; Fax: +44 1223 331174; E-mail: krgm2@cam.ac.uk

[a]NDB and TVJ contributed equally to this chapter

DOI: 10.1016/S0079-6123(08)01135-7

Glaucoma as a model of neurodegenerative disease

Glaucoma is the most common neurodegenerative disease in the world, affecting approximately 80 million people or 1.2% of the global population (Quigley and Broman, 2006). Like many neurodegenerative diseases, glaucoma is characterized by the loss of a discrete neuronal cell population from the CNS. Similarities in the pathophysiological mechanisms of cell death appear to exist between glaucoma and other neurodegenerative diseases. For example, signs of oxidative stress, glutamate toxicity, reactive glial changes, and impaired axonal transport have been identified in glaucoma (Gupta and Yucel, 2007; Kumar and Agarwal, 2007; Nickells, 2007) as well as in other neurodegenerative diseases. As such, progress made in understanding the mechanisms of glaucomatous onset and progression may provide insight into the pathogenesis of other neuropathologies. Furthermore, successful approaches to protecting retinal ganglion cells (RGCs) could potentially translate into therapies for other cell types.

Unlike other neurodegenerative diseases, structural changes associated with glaucomatous progression can be directly visualized in live subjects. In human patients, the optic nerve head and retinal nerve fiber layer can be examined directly and longitudinal changes are routinely documented. In animal models of ocular hypertension and glaucoma, the ability to observe RGC loss directly is even more powerful. RGCs can be visualized at single-cell resolution by scanning laser ophthalmoscopy after fluorescent labeling via retrograde tracing or immunohistochemical identification of apoptotic cells (Cordeiro et al., 2004; Maass et al., 2007). If in vivo imaging of apoptosing RGCs proves to be safe, and to correlate well with RGC death, then such techniques may provide a unique method to visualize cell death longitudinally in a human neurodegenerative disease.

Why use stem cells for neuroprotective therapy?

It is reasonable to ask what advantages there may be in using stem cells, or their various derivatives, as a neuroprotective therapy in glaucoma rather than simply using a pharmacological approach. One distinct advantage cellular therapy could provide is chronic neuroprotection after a single treatment. Once integrated into the host tissue, the ideal cellular therapy would provide lifelong support for RGCs and attenuate visual field loss. A further benefit of a cell-based therapy may be the noted ability of stem cells to migrate or "home" to sites of injury within the CNS and thus deliver support locally, where it is most needed. Such stem cell behavior has been observed in various neuropathological models, and has been particularly well-characterized following stroke (Felling and Levison, 2003; Tai and Svendsen, 2004). Furthermore, some stem cells have the ability to migrate extensively within the CNS, offering the potential for widespread therapeutic activity following focal delivery. Such integration with the host tissue may also allow stem cells to provide support through contact-mediated mechanisms, thereby offering a supporting niche for surviving neurons.

Cell-based therapies may function by a variety of different mechanisms. It is acknowledged that neuroprotection by stem cells involves trophic support of neurons (as discussed below). It seems very likely that such protection is not due to the secretion of single growth factors but is multifactorial. For example, transplantation of neural stem cells into the spinal cord has been shown to delay disease onset and progression in a mouse model of amyotrophic lateral sclerosis (Corti et al., 2007). The protective mechanism in this model was found to involve vascular endothelial growth factor (VEGF)- and insulin-like growth factor 1 (IGF1)-dependent signaling pathways and resulted in improved performance in behavioral tests. In addition, transplantation of immortalized human neural stem cells into the lesioned striatum has been observed to improve function significantly in a rat Parkinson's disease model (Yasuhara et al., 2006). Again, this effect appeared to be due to trophic factor secretion by engrafted cells, although there also was evidence of neuronal differentiation by some of the transplanted cells. Interestingly, the effect of cellular transplantation was greater than that of single injections or continuous infusions of trophic factors. Such a neuroprotective effect might be bolstered by modification or manipulation of stem cells prior to transplantation in order to control the identity and levels of factors supplied.

In addition to supplying trophic factors, transplanted cells may also be able to modify the pathological environment to promote neuronal survival. Neural stem cells appear to possess this ability inherently. As an example, stem cells derived from the subventricular zone have been found to modify the local environment directly through immunomodulatory mechanisms (Pluchino et al., 2005) or by influencing gene expression in surrounding neurons (Madhavan et al., 2008). In addition, it has been demonstrated that the integration of glial precursor cells, which possess active glutamate transporters, into organotypic spinal cord cultures enhanced glutamate uptake and reduced motor neuron cell death (Maragakis et al., 2005). This neuroprotective approach could potentially be extended to other pathologies that may involve glutamate excitotoxicity, such as amyotrophic lateral sclerosis and glaucoma.

Stem cell sources

Stem cells possess the ability to self-renew indefinitely and to differentiate into a variety of different cell types. In contrast, progenitor cells are generally considered a more restricted cell type, often multipotent but usually of limited proliferative capacity. For simplicity, in this chapter, we will use the term "stem cell" in its broadest sense to cover all precursor cell types. Stem cells exist within a wide array of tissues and at all developmental stages from embryogenesis to adulthood. Selection of the optimal stem cell type to be used for therapeutic applications depends upon the ultimate therapeutic goal: cellular replacement and/or neuroprotection of surviving cells. In the case of cell replacement for neurodegenerative disease, the relative levels of lineage potency and commitment are of primary concern. The ideal cell must be capable of differentiating into the specific cell type that has been depleted, but should also be restricted from differentiating into other cell types.

If transplantation is to be used for neuroprotection, on the other hand, grafted cells need only support the surviving tissue and attenuate disease progression, rather than having to acquire neurological function themselves.

Perhaps the most well-known type of stem cell is the embryonic stem (ES) cell, which is derived from the inner cell mass of the developing blastocyst. ES cells are capable of differentiating into any somatic cell type in the body and are of great clinical interest, especially with regard to cell replacement. However, the potential for malignant transformation of transplanted ES cells, concerns regarding the ethics and logistics of obtaining ES cells, and the inherent necessity of allogeneic transplantation leading to the risk of graft rejection may ultimately limit their clinical use.

Neural stem cells are of particular interest in neurodegenerative disease. These cells can self-replicate but are lineage-restricted and capable only of differentiating into neurons and glia. Neural stem cells have been isolated from multiple regions of the CNS, including the brain, spinal cord, and retina. In adults, the most well-characterized neural stem cells are from the lateral wall of the subventricular zone and the subgranular zone of the hippocampal dentate gyrus, two regions where neurogenesis occurs throughout life. Retinal stem cells residing in the ciliary marginal zone have been identified in adult birds, fish, and amphibians, and contribute to retinal repair following injury in these species (Otteson and Hitchcock, 2003; Hitchcock et al., 2004). In mammals, this process does not occur naturally; however, cells isolated from this region can proliferate and differentiate in vitro, and growth-factor induced reactivation of these cells has been reported in vivo (Abdouh and Bernier, 2006). Furthermore, mammalian Müller glia and cells isolated from the iris-pigmented epithelium, as well as the pars plana of the ciliary body, display neural stem cell-like characteristics when isolated in vitro (Asami et al., 2007; Lawrence et al., 2007; MacNeil et al., 2007; Xu et al., 2007).

Somatic cells isolated from nonneural tissue may play an important role in the future of stem cell therapy for neurodegenerative disease. First, many somatic stem cells that are seemingly unrelated to neural tissue reportedly hold neural competence given the proper in vitro differentiation conditions. For instance, some evidence suggests that mesenchymal stem cells (MSCs) isolated from bone marrow, blood, and umbilical cord can transdifferentiate to give rise to neuronal or glial-like cells, though this phenomenon is controversial (Krabbe et al., 2005). Second, many somatic cells have demonstrated benefit in neurodegenerative disease models in the absence of overt differentiation toward a neural phenotype. Potential mechanisms of this neuroprotective effect will be discussed in the following section. Last, very recent evidence suggests that adult human somatic cells, including terminally differentiated non-stem cells such as fibroblasts, can be reprogrammed to achieve a pluripotent, self-renewing stem cell-like state by ectopically inducing the expression of certain genes associated with ES cells (Takahashi et al., 2007; Yu et al., 2007). Importantly, somatic cells that can be easily isolated from adult patients provide the opportunity for autologous cell transplantation therapy, thereby eliminating the risk of graft rejection or lifetime administration of immune suppressant drugs.

Neuroprotection by transplanted stem cells

Numerous published examples exist where stem cell transplantation for neurodegenerative disease has led to structural and functional benefit in the absence of differentiation or functional integration on the part of the engrafted cells. In these cases, the effect is often attributed to neuroprotection of the endogenous surviving tissue. There are a number of mechanisms by which transplantation of stem cells can be neuroprotective, and each may hold potential therapeutic applications for glaucoma. The most widely cited, and perhaps most important mechanism, involves supply of neurotrophic factors and support for surviving neurons. In addition, regulation of immune activity and the promotion of endogenous CNS repair also may play a role.

It has been demonstrated, both in vitro and in vivo, that various types of stem cells synthesize and release neurotrophic factors without forced differentiation or experimental manipulation. Transplantation of these cells into the nervous system is associated with a preservation of surviving neural structure and function in a variety of neurodegenerative conditions including Parkinson's disease, amyotrophic lateral sclerosis, spinal cord injury, and traumatic brain injury. Specifically in the eye, stem cell transplantation has been shown to rescue retinal degeneration in a range of models. For instance, engrafted bone marrow-derived MSCs have preserved photoreceptors in the rhodopsin knockout mouse (Arnhold et al., 2006). Furthermore, transplantation of ES cells (Schraermeyer et al., 2001), Schwann cells (Lawrence et al., 2000), or pigmented iris epithelial cells (Schraermeyer et al., 2000) have all reduced degeneration in the Royal College of Surgeons rat, which displays marked photoreceptor loss. In addition, neuralized ES cells have been found to attenuate retinal degeneration in *mnd* mice (Meyer et al., 2006). The authors of each of these studies suggest that the primary neuroprotective benefit is derived from trophic factors secreted by the grafted cells to support surviving retinal neurons.

Failure of neurotrophic support might be especially important in glaucoma where it has been shown that retrograde axonal transport of brain-derived neurotrophic factor (BDNF) is disrupted (Pease et al., 2000). In fact, an accumulation of motor proteins at the optic nerve head is suggestive of a general failure of axonal transport (Martin et al., 2006), and intraocular supplementation of BDNF has been shown to reduce RGC loss in experimental glaucoma (Martin et al., 2003). Transplantation of MSCs into rats with induced ocular hypertension has also been shown to increase RGC survival (Yu et al., 2006). In this case, the authors demonstrated a concomitant increase in basic fibroblast growth factor, ciliary neurotrophic factor (CNTF), glial cell line-derived neurotrophic factor (GDNF), and BDNF within the retina of treated eyes.

Some stem cells have also revealed an ability to modulate immune cell activity. Immune mechanisms have been implicated in a wide range of neurodegenerative diseases. In multiple sclerosis, autoimmune activity appears critical to the pathogenesis of the disease (McFarland and Martin,

2007). In acute neurodegenerative conditions such as stroke, spinal cord injury, and traumatic brain injury, inflammation may contribute to secondary neuronal degeneration. In more chronic conditions such as Alzheimer's disease and Parkinson's disease, inflammatory "bystander" damage is thought to contribute to disease progression (Lucas et al., 2006). Transplantation of cells able to modulate immune activity has demonstrated benefit in many of these conditions. Bone marrow-derived MSCs are known for being particularly immunomodulatory and are currently the subject of clinical trials for treating multiple sclerosis (Passweg and Tyndall, 2007). Neural stem cells are reportedly capable of downregulating inflammatory processes within the CNS, leading to functional recovery in a range of neurodegenerative diseases (Martino and Pluchino, 2006). Whether these neuroprotective mechanisms are applicable to glaucoma requires further investigation.

Finally, there is some evidence that transplanted stem cells are able to influence local CNS tissue to promote endogenous repair mechanisms. Unlike the peripheral nervous system, the plasticity of the CNS is notoriously restricted and neuritic regrowth following injury is very limited. This lack of regeneration after injury appears to be due to the combination of a lack of trophic cues in the adult CNS, which signal for regenerative activity, and the inhibitory nature of the CNS environment. The production and release of certain neurotrophic factors by neural stem cells has been shown to trigger axonal regrowth in the adult injured spinal cord (Lu et al., 2003) and, therefore, release of neurotrophic factors by engrafted cells might have beneficial consequences beyond neuroprotection alone. The presence of various inhibitory molecules, such as chondroitin sulfate proteoglycans and members of the Nogo family, within the local environment of the CNS also contributes to a general lack of plasticity. Transplantation of neural stem cells has been found to reduce the expression of these inhibitory molecules by increasing the levels of various matrix metalloproteinases within the retina (Zhang et al., 2007). In turn, this has enhanced neurite outgrowth. If neurite sprouting and the formation of local synaptic connections could be triggered in the glaucomatous retina, it is conceivable that the receptive field size of surviving RGCs could be increased in order to improve visual function.

Enhanced neuroprotection by transplantation of modified stem cells

In addition to neurotrophic support by naive transplanted stem cells, the possibility of augmenting the neuroprotective effect by using genetically modified stem cells to deliver specific factors has been explored. This concept has attracted particular attention in stroke research. Many studies have demonstrated that transplantation of various stem cell types into the infarcted brain can ameliorate ischemic damage (Haas et al., 2005). In addition to these earlier studies, a number of groups have now demonstrated substantial protection by engineered stem cells in various models of ischemic disease. For example, a significant improvement in neurological deficiency was observed in a rat transient focal cerebral ischemia model following the engraftment of neural stem cells, modified in vitro to express VEGF, into the perifocal zone (Zhu et al., 2005). The observed improvement was greater than that of control transplantation of naive neural stem cells. Furthermore, the transplantation of human neural stem cells overexpressing VEGF into the cortex overlying an intracerebral hemorrhage lesion has been shown to improve survival of engrafted cells, to stimulate host angiogenesis and to enhance functional recovery in mice (Lee et al., 2007). Similarly, the introduction of MSCs transfected to express BDNF into the brain 6 h after permanent middle cerebral artery occlusion was found to reduce lesion size and improve functional outcome (Nomura et al., 2005). Furthermore, in this model, stem cells engineered to produce BDNF provided greater neuroprotection than that observed following the delivery of naive cells. The same group also reported similar results using MSCs modified to express GDNF (Horita et al., 2006) and placental growth factor (PlGF) (Liu et al., 2006), with PlGF also stimulating angiogenesis. The secretion of GDNF by implanted neural stem cells has also provided neuroprotection in other neurodegenerative disease models, such as Huntington's disease

(Pineda et al., 2007) and Parkinson's disease. In a mouse model of Parkinson's disease, the engraftment of neural stem cells engineered to express GDNF was found to ameliorate the degeneration of dopaminergic neurons upon subsequent exposure to the toxin 6-hydroxydopamine (Akerud et al., 2001). Furthermore, this neuroprotection resulted in a significant alleviation of the behavioral impairment caused by the substantia nigra lesion. In addition to demonstrating neuroprotection, this study confirmed that the engineered stem cells provided a stable, local delivery system with neurotrophic factor production for up to 4 months after transplantation.

While these techniques are yet to be applied to RGC degeneration in glaucoma, it is clear that effects observed in diverse regions of the CNS are likely to be highly transferable to the retina. Furthermore, the retina is more accessible than other regions of the CNS, making it an ideal model for the development of these potential therapies.

Endogenous stem cells

It is now understood that endogenous pools of neural stem cells proliferate in response to brain injury, particularly in response to stroke (Felling and Levison, 2003; Tai and Svendsen, 2004). Furthermore, it has been demonstrated that this proliferation may be enhanced via the exogenous application of growth factors (Nakatomi et al., 2002; Tureyen et al., 2005; Ninomiya et al., 2006) and drugs (Zheng and Chen, 2007). Previous research has primarily focused on the role that enhanced neurogenesis plays in recovery following stroke injury. However, while proliferation of stem cells is upregulated following ischemic stroke injury, only a percentage of the newly generated cells differentiate into neurons and survive in the long term (Naylor et al., 2005). Given that transplantation of undifferentiated neural precursor cells can provide neuroprotection in various neuropathological conditions, it seems entirely conceivable that a similar role may be played by the proliferation of endogenous neural stem cells. However, to date, this concept has received no direct investigation and the contribution of endogenous stem cell proliferation to neuroprotection in disease remains to be elucidated.

Key challenges

Given that many of the cell-based therapies we have discussed are reliant on transplantation, one of the key hurdles to the development of such clinical treatments is finding an acceptable source of stem cells, or their derivatives, for this purpose. This problem has a number of facets, perhaps the most well-publicized of which is the ethical acceptability of using stem cells derived from human embryonic or fetal tissue for clinical therapies — an emotive issue that is unlikely to be resolved anytime soon. Furthermore, we need to consider the safety of using such cells in patients. Most ES cells are cultured in media that contain essential products currently derived from animal sources. Contamination of the therapeutic product by these ingredients is a potential problem, posing a risk of pathogens crossing the species barrier. This may also be exacerbated in patients receiving transplants, as such people are likely to require chronic administration of immune suppression drugs in order to prevent graft rejection. Graft rejection is, of course, a problem in its own right, although the immune-privileged status of the CNS may alleviate this, at least in part. Another safety concern is the propensity of stem cells in general, and ES cells in particular, to generate tumors following transplantation. Finally, if we are to use modified stem cells as a local delivery system, the safety of the transfection system used to manipulate cells in vitro prior to transplantation will need to be verified.

Much of this chapter has considered experiments and findings from investigation within the CNS in general, rather than the eye specifically. This is because very little work examining stem cell-mediated neuroprotection in the retina, and more specifically glaucoma, has been published to date. If these methods are to be translated to a glaucoma therapy, many technical issues will need to be resolved. These include how to deliver stem cells to the retina so their neuroprotection will be most effective. Most grafts are delivered to the brain via injection, which may also work very well for the retina given its accessibility. Within the eye, this method raises the question of whether to deliver the cells intravitreally, close to stressed RGCs in glaucoma, or subretinally, where they may be nourished by the choroidal blood supply. Cells may also be transferred to the host using an artificial scaffold, which has the advantage of protecting the cells during transfer from the in vitro to the in vivo environment. Alternatively, many investigators are now delivering stem cells systemically, via infusion into blood vessels, which may also work for the retina, provided that the cells can cross the blood-retinal barrier and migrate to the site of injury.

A unique consideration for cell transplantation in the eye is whether the introduction of new cells into the retina, or into the vitreous, will impact negatively on vision. Unlike slow-release, biodegradable delivery systems, transplanted cells may have the ability to survive for very long periods, perhaps indefinitely, in vivo and are unlikely to be easily removed once engrafted. Indeed, with a chronic progressive disease like glaucoma, a stem cell-based neuroprotective strategy would need to exhibit very long survival and function to ameliorate visual field loss over the life of the patient. Thus, another hurdle to the development of such therapies will be ensuring transplanted cells survive. In addition, the retina appears to be more resistant to the integration of transplanted cells compared to the brain. Therefore, further research is needed in order to discover how we may manipulate the retinal environment to encourage the integration of engrafted stem cells.

Conclusion

Stem cell transplantation may be a promising approach to human glaucoma treatment if barriers related to the control of differentiation, integration, and long-term survival of grafted cells can be overcome, and if safety and efficacy can be demonstrated. In the short term, however, glaucoma models provide a very useful system in which to explore the neuroprotective potential of cellular transplantation where both degenerating and transplanted cells can be directly visualized.

Abbreviations

CNS	central nervous system
ES	embryonic stem
GDNF	glial cell line-derived neurotrophic factor
IGF1	insulin-like growth factor 1
MSC	mesenchymal stem cell
PlGF	placental growth factor
RGC	retinal ganglion cell
VEGF	vascular endothelial growth factor

Acknowledgments

The funding for our work in this field has been provided by the Gates Cambridge Trust, Fight for Sight, the Glaucoma Research Foundation, and the GSK Clinical Fellowship Program.

References

Abdouh, M. and Bernier, G. (2006) In vivo reactivation of a quiescent cell population located in the ocular ciliary body of adult mammals. Exp. Eye Res., 83(1): 153–164.

Akerud, P., Canals, J.M., Snyder, E.Y. and Arenas, E. (2001) Neuroprotection through delivery of glial cell line-derived neurotrophic factor by neural stem cells in a mouse model of Parkinson's disease. J. Neurosci., 21(20): 8108–8118.

Arnhold, S., Absenger, Y., Klein, H., Addicks, K. and Schraermeyer, U. (2006) Transplantation of bone marrow-derived mesenchymal stem cells rescue photoreceptor cells in the dystrophic retina of the rhodopsin knockout mouse. Graefes. Arch. Clin. Exp. Ophthalmol., 245(3): 414–422.

Asami, M., Sun, G., Yamaguchi, M. and Kosaka, M. (2007) Multipotent cells from mammalian iris pigment epithelium. Dev. Biol., 304(1): 433–446.

Cordeiro, M.F., Guo, L., Luong, V., Harding, G., Wang, W., Jones, H.E., Moss, S.E., Sillito, A.M. and Fitzke, F.W. (2004) Real-time imaging of single nerve cell apoptosis in retinal neurodegeneration. Proc. Natl. Acad. Sci. U.S.A., 101(36): 13352–13356.

Corti, S., Locatelli, F., Papadimitriou, D., Del Bo, R., Nizzardo, M., Nardini, M., Donadoni, C., Salani, S., Fortunato, F., Strazzer, S., Bresolin, N. and Comi, G.P. (2007) Neural stem cells LewisX+ CXCR4+ modify disease progression in an amyotrophic lateral sclerosis model. Brain, 130(Pt 5): 1289–1305.

Felling, R.J. and Levison, S.W. (2003) Enhanced neurogenesis following stroke. J. Neurosci. Res., 73(3): 277–283.

Gupta, N. and Yucel, Y.H. (2007) Glaucoma as a neurodegenerative disease. Curr. Opin. Ophthalmol., 18(2): 110–114.

Haas, S., Weidner, N. and Winkler, J. (2005) Adult stem cell therapy in stroke. Curr. Opin. Neurol., 18(1): 59–64.

Hitchcock, P., Ochocinska, M., Sieh, A. and Otteson, D. (2004) Persistent and injury-induced neurogenesis in the vertebrate retina. Prog. Retin. Eye Res., 23(2): 183–194.

Horita, Y., Honmou, O., Harada, K., Houkin, K., Hamada, H. and Kocsis, J.D. (2006) Intravenous administration of glial cell line-derived neurotrophic factor gene-modified human mesenchymal stem cells protects against injury in a cerebral ischemia model in the adult rat. J. Neurosci. Res., 84(7): 1495–1504.

Krabbe, C., Zimmer, J. and Meyer, M. (2005) Neural transdifferentiation of mesenchymal stem cells: a critical review. APMIS, 113(11–12): 831–844.

Kumar, D.M. and Agarwal, N. (2007) Oxidative stress in glaucoma: a burden of evidence. J. Glaucoma, 16(3): 334–343.

Lawrence, J.M., Sauve, Y., Keegan, D.J., Coffey, P.J., Hetherington, L., Girman, S., Whiteley, S.J., Kwan, A.S., Pheby, T. and Lund, R.D. (2000) Schwann cell grafting into the retina of the dystrophic RCS rat limits functional deterioration. Invest. Ophthalmol. Vis. Sci., 41(2): 518–528.

Lawrence, J.M., Singhal, S., Bhatia, B., Keegan, D.J., Reh, T.A., Luthert, P.J., Khaw, P.T. and Limb, G.A. (2007) MIO-M1 cells and similar Müller glial cell lines derived from adult human retina exhibit neural stem cell characteristics. Stem Cells, 25(8): 2033–2043.

Lee, H.J., Kim, K.S., Park, I.H. and Kim, S.U. (2007) Human neural stem cells over-expressing VEGF provide neuroprotection, angiogenesis and functional recovery in mouse stroke model. PLoS ONE, 2(1): p. e156.

Liu, H., Honmou, O., Harada, K., Nakamura, K., Houkin, K., Hamada, H. and Kocsis, J.D. (2006) Neuroprotection by PlGF gene-modified human mesenchymal stem cells after cerebral ischaemia. Brain, 129(Pt 10): 2734–2745.

Lu, P., Jones, L.L., Snyder, E.Y. and Tuszynski, M.H. (2003) Neural stem cells constitutively secrete neurotrophic factors and promote extensive host axonal growth after spinal cord injury. Exp. Neurol., 181(2): 115–129.

Lucas, S.M., Rothwell, N.J. and Gibson, R.M. (2006) The role of inflammation in CNS injury and disease. Br. J. Pharmacol., 147(Suppl 1): S232–S240.

Maass, A., von Leithner, P.L., Luong, V., Guo, L., Salt, T.E., Fitzke, F.W. and Cordeiro, M.F. (2007) Assessment of rat and mouse RGC apoptosis imaging in vivo with different scanning laser ophthalmoscopes. Curr. Eye Res., 32(10): 851–861.

MacNeil, A., Pearson, R.A., MacLaren, R.E., Smith, A.J., Sowden, J.C. and Ali, R.R. (2007) Comparative analysis of progenitor cells isolated from the iris, pars plana, and ciliary body of the adult porcine eye. Stem. Cells, 25(10): 2430–2438.

Madhavan, L., Ourednik, V. and Ourednik, J. (2008) Neural stem/progenitor cells initiate the formation of cellular networks that provide neuroprotection by growth factor-modulated antioxidant expression. Stem Cells, 26(1): 254–265.

Maragakis, N.J., Rao, M.S., Llado, J., Wong, V., Xue, H., Pardo, A., Herring, J., Kerr, D., Coccia, C. and Rothstein, J.D. (2005) Glial restricted precursors protect against chronic glutamate neurotoxicity of motor neurons in vitro. Glia, 50(2): 145–159.

Martin, K.R., Quigley, H.A., Valenta, D., Kielczewski, J. and Pease, M.E. (2006) Optic nerve dynein motor protein distribution changes with intraocular pressure elevation in a rat model of glaucoma. Exp. Eye Res., 83(2): 255–262.

Martin, K.R., Quigley, H.A., Zack, D.J., Levkovitch-Verbin, H., Kielczewski, J., Valenta, D., Baumrind, L., Pease, M.E., Klein, R.L. and Hauswirth, W.W. (2003) Gene therapy with brain-derived neurotrophic factor as a protection: retinal ganglion cells in a rat glaucoma model. Invest. Ophthalmol. Vis. Sci., 44(10): 4357–4365.

Martino, G. and Pluchino, S. (2006) The therapeutic potential of neural stem cells. Nat. Rev. Neurosci., 7(5): 395–406.

McFarland, H.F. and Martin, R. (2007) Multiple sclerosis: a complicated picture of autoimmunity. Nat. Immunol., 8(9): 913–919.

Meyer, J.S., Katz, M.L., Maruniak, J.A. and Kirk, M.D. (2006) Embryonic stem cell-derived neural progenitors incorporate into degenerating retina and enhance survival of host photoreceptors. Stem Cells, 24(2): 274–283.

Nakatomi, H., Kuriu, T., Okabe, S., Yamamoto, S., Hatano, O., Kawahara, N., Tamura, A., Kirino, T. and Nakafuku, M. (2002) Regeneration of hippocampal pyramidal neurons after ischemic brain injury by recruitment of endogenous neural progenitors. Cell, 110(4): 429–441.

Naylor, M., Bowen, K.K., Sailor, K.A., Dempsey, R.J. and Vemuganti, R. (2005) Preconditioning-induced ischemic tolerance stimulates growth factor expression and neurogenesis in adult rat hippocampus. Neurochem. Int., 47(8): 565–572.

Nickells, R.W. (2007) From ocular hypertension to ganglion cell death: a theoretical sequence of events leading to glaucoma. Can. J. Ophthalmol., 42(2): 278–287.

Ninomiya, M., Yamashita, T., Araki, N., Okano, H. and Sawamoto, K. (2006) Enhanced neurogenesis in the ischemic striatum following EGF-induced expansion of transit-amplifying cells in the subventricular zone. Neurosci. Lett., 403 (1–2): 63–67.

Nomura, T., Honmou, O., Harada, K., Houkin, K., Hamada, H. and Kocsis, J.D. (2005) I.V. infusion of brain-derived neurotrophic factor gene-modified human mesenchymal stem cells protects against injury in a cerebral ischemia model in adult rat. Neuroscience, 136(1): 161–169.

Otteson, D.C. and Hitchcock, P.F. (2003) Stem cells in the teleost retina: persistent neurogenesis and injury-induced regeneration. Vision Res., 43(8): 927–936.

Passweg, J. and Tyndall, A. (2007) Autologous stem cell transplantation in autoimmune diseases. Semin. Hematol., 44(4): 278–285.

Pease, M.E., McKinnon, S.J., Quigley, H.A., Kerrigan-Baumrind, L.A. and Zack, D.J. (2000) Obstructed axonal transport of BDNF and its receptor TrkB in experimental glaucoma. Invest. Ophthalmol. Vis. Sci., 41(3): 764–774.

Pineda, J.R., Rubio, N., Akerud, P., Urban, N., Badimon, L., Arenas, E., Alberch, J., Blanco, J. and Canals, J.M. (2007) Neuroprotection by GDNF-secreting stem cells in a Huntington's disease model: optical neuroimage tracking of brain-grafted cells. Gene Ther., 14(2): 118–128.

Pluchino, S., Zanotti, L., Rossi, B., Brambilla, E., Ottoboni, L., Salani, G., Martinello, M., Cattalini, A., Bergami, A., Furlan, R., Comi, G., Constantin, G. and Martino, G. (2005) Neurosphere-derived multipotent precursors promote neuroprotection by an immunomodulatory mechanism. Nature, 436(7048): 266–271.

Quigley, H.A. and Broman, A.T. (2006) The number of people with glaucoma worldwide in 2010 and 2020. Br. J. Ophthalmol., 90(3): 262–267.

Schraermeyer, U., Kayatz, P., Thumann, G., Luther, T.T., Szurman, P., Kociok, N. and Bartz-Schmidt, K.U. (2000) Transplantation of iris pigment epithelium into the choroid slows down the degeneration of photoreceptors in the RCS rat. Graefes. Arch. Clin. Exp. Ophthalmol., 238(12): 979–984.

Schraermeyer, U., Thumann, G., Luther, T., Kociok, N., Armhold, S., Kruttwig, K., Andressen, C., Addicks, K. and Bartz-Schmidt, K.U. (2001) Subretinally transplanted embryonic stem cells rescue photoreceptor cells from degeneration in the RCS rats. Cell Transplant., 10(8): 673–680.

Tai, Y.T. and Svendsen, C.N. (2004) Stem cells as a potential treatment of neurological disorders. Curr. Opin. Pharmacol., 4(1): 98–104.

Takahashi, K., Tanabe, K., Ohnuki, M., Narita, M., Ichisaka, T., Tomoda, K. and Yamanaka, S. (2007) Induction of pluripotent stem cells from adult human fibroblasts by defined factors. Cell, 131(5): 861–872.

Tureyen, K., Vemuganti, R., Bowen, K.K., Sailor, K.A. and Dempsey, R.J. (2005) EGF and FGF-2 infusion increases post-ischemic neural progenitor cell proliferation in the adult rat brain. Neurosurgery, 57(6): 1254–1263.

Xu, H., Sta Iglesia, D.D., Kielczewski, J.L., Valenta, D.F., Pease, M.E., Zack, D.J. and Quigley, H.A. (2007) Characteristics of progenitor cells derived from adult ciliary body in mouse, rat, and human eyes. Invest. Ophthalmol. Vis. Sci., 48(4): 1674–1682.

Yasuhara, T., Matsukawa, N., Hara, K., Yu, G., Xu, L., Maki, M., Kim, S.U. and Borlongan, C.V. (2006) Transplantation of human neural stem cells exerts neuroprotection in a rat model of Parkinson's disease. J. Neurosci., 26(48): 12497–12511.

Yu, J., Vodyanik, M.A., Smuga-Otto, K., Antosiewicz-Bourget, J., Frane, J.L., Tian, S., Nie, J., Jonsdottir, G.A., Ruotti, V., Stewart, R., Slukvin, I.I. and Thomson, J.A. (2007) Induced pluripotent stem cell lines derived from human somatic cells. Science, 318(5858): 1917–1920.

Yu, S., Tanabe, T., Dezawa, M., Ishikawa, H. and Yoshimura, N. (2006) Effects of bone marrow stromal cell injection in an experimental glaucoma model. Biochem. Biophys. Res. Commun., 344(4): 1071–1079.

Zhang, Y., Klassen, H.J., Tucker, B.A., Perez, M.T. and Young, M.J. (2007) CNS progenitor cells promote a permissive environment for neurite outgrowth via a matrix metalloproteinase-2-dependent mechanism. J. Neurosci., 27(17): 4499–4506.

Zheng, Z. and Chen, B. (2007) Effects of Pravastatin on neuroprotection and neurogenesis after cerebral ischemia in rats. Neurosci. Bull., 23(4): 189–197.

Zhu, W., Mao, Y., Zhao, Y., Zhou, L.F., Wang, Y., Zhu, J.H., Zhu, Y. and Yang, G.Y. (2005) Transplantation of vascular endothelial growth factor-transfected neural stem cells into the rat brain provides neuroprotection after transient focal cerebral ischemia. Neurosurgery, 57(2): 325–333.

CHAPTER 36

The relationship between neurotrophic factors and CaMKII in the death and survival of retinal ganglion cells

N.G.F. Cooper[1,*], A. Laabich[2], W. Fan[1] and X. Wang[1]

[1]Department of Anatomical Sciences and Neurobiology, University of Louisville School of Medicine, Louisville, KY 40292, USA
[2]Acucela Inc., Bothell, WA 98021, USA

Abstract: The scientific discourse relating to the causes and treatments for glaucoma are becoming reflective of the need to protect and preserve retinal neurons from degenerative changes, which result from the injurious environment associated with this disease. Knowledge, in particular, of the signal transduction pathways which affect death and survival of the retinal ganglion cells is critical to this discourse and to the development of a suitable neurotherapeutic strategy for this disease. The goal of this chapter is to review what is known of the chief suspects involved in initiating the cell death/survival pathways in these cells, and what still remains to be uncovered. The least controversial aspect of the subject relates to the potential role of neurotrophic factors in the protection of the retinal ganglion cells. On the other hand, the postulated triggers for signaling cell death in glaucoma remain controversial. Certainly, the restricted flow of neurotrophic factors has been cited as one possible trigger. However, the connections between glaucoma and other factors present in the retina, such as glutamate, long held to be a prospective culprit in retinal ganglion cell death are still being questioned. Whatever the outcome of this particular debate, it is clear that the downstream intersections between the cell death and survival pathways should provide important foci for future studies whose goal is to protect retinal neurons, situated as they are, in the stressful environment of a cell destroying disease. The evidence for CaMKII being one of these intersecting points is discussed.

Keywords: CaMKII-α; CaMKII-α$_B$; calcium/calmodulin-dependent protein kinase II; retinal ganglion cell; amacrine cell; BDNF; brain-derived neurotrophic factor; cell death pathway; cell survival pathway; anti-apoptotic genes; pro-apoptotic genes; glutamate; NMDA; autocamtide-2-related inhibitory peptide; caspase-3

Introduction

"All of the present studies suggest that neuroprotective therapy will probably become the treatment of choice in the near future for glaucomatous optic neuropathy" (Yamamoto, 2001). Neurotrophic factors have the potential to aid recovery from diseases and disorders of the nervous system, including the neural retina and its optic nerve. For this potential to be fully realized, we first need to thoroughly establish the cellular networks, including the genomic and environmental parameters, which regulate such factors. Specifically, we need to map the signal transduction pathways,

*Corresponding author. Tel.: 502-852-1474; Fax: 502-852-3028; E-mail: nigelcooper@louisville.edu

which stimulate and/or inhibit or otherwise modify the cell's regulatory machinery for neurotrophic factors. This review attempts to summarize some of the evidence for the pivotal involvement of Ca^{2+} and calmodulin-dependent protein kinase II (CaMKII) in cell death and survival pathways in the retina. In particular, we review the evidence for the signal transduction pathway leading to the transcription, translation, secretion, and action of brain-derived neurotrophic factor (BDNF) from the retinal ganglion cells (RGCs). These cells are particularly vulnerable in retinal diseases such as retinal ischemia and glaucoma.

Glaucoma and the RGCs

While glaucoma is the leading cause of blindness in the world, it cannot be simply defined, due to its multiple etiologies and its several underlying genetic and environmental causes. However, we can speak of a class of eye diseases in which optic nerve damage is a most frequently observed feature. In addition to optic nerve damage and the consequential visual field defects, increased intraocular pressure (IOP), is often cited as a third correlate in the most common form of the disease, although this may be a coincidental observation or aggravating condition because normal-tension glaucoma results in the same blinding condition without the elevated IOP. The incidence of glaucoma increases with increasing age (Armaly et al., 1980), and other known risk factors including family history, ethnicity, myopia, and diabetes (Leske, 1983).

Once nerve damage occurs there is a functional discontinuity between the retina and the higher order visual processing centers with a consequential degradation and/or loss of vision. This loss is permanent because of the well-established dogmas; (1) the axons of central nervous system neurons in mammals do not re-grow; (2) axonal section results in the death of the cell bodies in the central nervous system of mammals.

Unfortunately, glaucoma is an insidious disease in which visual loss occurs incrementally and patients do not complain of problems until much of the optic nerve damage has already occurred. The exact causes of optic nerve damage from glaucoma are not fully understood, but leading postulates include mechanical compression and/or decreased blood flow to the optic nerve. As to the death of the RGCs, one related hypothesis implicates the loss of neurotrophic factor, due to a block in its retrograde transport from higher visual centers, along the optic nerve to the cell somata of RGCs. Morphological changes of the optic nerve head are some of the earliest macroscopic signs of trouble in the clinic, although it is possible that such signs are not only relatively late indicators of trouble in the visual system, but also distal to the most likely primary causes of the problem. While there are those who strongly advocate a recreation of the condition seen in the clinic in the development of animal and cellular models of this disease, such advocacy may actually impede the exploration and discovery derived from other interesting avenues of research.

Observation of altered structure in the optic nerve head, it is frequently argued, precedes observations of the malaise and/or death of the RGCs. Axonopathy is not unique to glaucoma and perhaps we can learn something from studies and models of other diseases. For example, in a cell culture model for amyotrophic lateral sclerosis, excitotoxicity is known to lead to an accumulation of cytoskeletal proteins in the axons of cultured spinal neurons (King et al., 2007). This may be worthy of further investigation in RGCs. It is most important to recall the dogma, which contends that the cell somata of neurons are the genetic and trophic centers which nourish and support the other parts of the cell including, in the case of glaucoma, the RGC axons present in the optic nerve. Because of this fact, it would seem plausible to examine all models and experimental paradigms which tell us something about the pro-death and pro-survival machinery within these genetic and trophic centers of the RGCs. Thus, the science should look far beyond the account and treatment of symptoms to the underlying and fundamental molecular mechanisms. With this approach, we may be able to determine if the cellular machinery can be manipulated to advantage, to inhibit cell death, to promote cell survival, and even to promote axonal re-growth.

The ultimate and permanent loss of vision in the glaucomatous patient is due to the death of the RGCs. These cells die by apoptosis (Quigley, 1995; Kerrigan et al., 1997) leading some investigators to propose anti-apoptotic therapies for the treatment of glaucoma (Nickells, 1999; Tatton, 1999). Elevated levels of ocular glutamate have been proposed as being responsible for the glaucomatous death of RGCs (Vorverk et al., 1999) and several glutamate receptor antagonists have been tested as potential antiglaucoma agents (Hare et al., 2001; Chaudhary et al., 1998; Pang et al., 1999). Although elevated vitreous glutamate levels have been reported in glaucoma patients and in animal models of glaucoma (Dreyer et al., 1996) other studies were unable to confirm these initial findings (Honkanen et al., 2003).

Are other retinal cells affected in glaucoma?

It is well established that glaucoma describes a condition which impacts the life of the RGCs, and therefore, this condition may be considered as part of a larger class of diseases/disorders in which these particular cells are affected. Although it initially appeared that only RGCs were destroyed in the glaucomatous retina, other retinal cells may also become damaged. Several studies over the last decade or so indicate that there may also be losses among the amacrine cells in the glaucomatous-human, the glaucomatous non-human primate, and in other animal models (Table 1), although the significance of these data with respect to the human condition is less well established.

Kielczewski et al. (2005) used translimbal laser treatment to raise IOP and found that rats with increased IOP did not have significantly reduced numbers of amacrine cells when compared to control eyes. This is in contradiction to many other examples where amacrine cell loss occurs as a result of elevated IOP, and this apparent contradiction may be indicative of the manner in which the elevated IOP is produced. However, the former study did note significant reductions in GABA and glycine containing amacrine cells in eyes when another well established model of glaucoma was used, in which the optic nerve had been transected. Hood et al. (1999) used intravitreal injections of N-methyl-D-aspartate (NMDA) and tetrodotoxin (TTX) in monkeys to examine the altered waveforms of the multifocal electroretinogram (mfERG). These authors noted changes in the ERG waveform contributed by the inner retina (ganglion and amacrine cells) of the treated monkey eyes, which are consistent with records from patients with glaucoma and diabetes. In addition to the RGCs, Diijik et al. (2004) used cell specific transcript to assess cell types lost following acutely elevated IOP, these authors noted that among the amacrine cells, the glycinergic AII amacrine cells were the most affected in this animal model. The authors, May and Mittag (2004), used immunohistochemistry to document the loss of neuronal nitric oxide synthase (nNOS) positive retinal amacrine cells in the DBA/2NNia murine model of angle-closure glaucoma. It should be noted that the possibility of cell death in the displaced amacrine cells of the RGC layer, has not yet been well documented because of the lack of specific markers for such cells. It seems likely that the shared vulnerabilities of the RGCs and the amacrine cells in several of these studies reflect some shared phenotype, some common property or properties in their cell membranes and/or protein compositions. The presence of both ionotropic glutamate receptors (Laabich and Cooper, 1999) and CaMKII (Laabich

Table 1. Observations of amacrine cell losses in glaucoma and other related animal models

Citation	Model	Methods
Dkhissi et al., 1996	a1 mutant quail	immunocytochemistry
Dkhissi et al., 1998	a1 mutant quail	HPLC, immunocytochemistry
Hood et al. (1999)	Human and non-human primates	mfERG
Hood and Zhang, 2000	Human	mfERG
May and Mittag (2004)	DBA/2NNia mouse	immunocytochemistry
Moon et al., 2005	DBA/J2 mouse	immunocytochemistry
Kielczewski et al., 2005	Rats, elevated IOP, nerve transection	immunocytochemistry
Diijik et al., 2004	Rats, acute IOP, ischemia-reperfusion	cell specific transcripts, PCR
Laabich and Cooper, 2000	Rats, excitotoxicity — NMDA	TUNEL-labeled tissue sections

et al., 2000) might represent such a common phenotype.

Retinal ischemia related glaucoma

There is a long history implicating a relationship between ischemia and glaucoma (Posner, 1958). Ocular ischemia can result from a wide variety of systemic (Kaiser et al., 1993; Flammer, 1994; Stroman et al., 1995; Waldmann et al., 1996; Osborne et al., 2004) or local influences, such as trauma, which may then lead to glaucoma, or aggravate an existing glaucomatous condition. Kaiser et al. (1993) noted, for example, that silent myocardial ischemia was correlated with normal-tension glaucoma. The particular forms of ischemia that might be relevant to glaucoma require further investigation and documentation. It is known that high IOP-induced ischemia in rats lead to increased levels of intraretinal glutamate (Nucci et al., 2005). This particular experimental paradigm preferentially affects RGCs (Osborne et al., 1999) and may therefore be suitable for studies of glaucoma. It is also of interest to note in this regard that the RGCs of glaucomatous rats with chronic, moderately elevated IOP, are more susceptible to ischemia reperfusion injury (Kawai et al., 2001).

Excitotoxicity and the retina

We have known for 50 years that an excess of the excitatory transmitter, glutamate, can cause cell death in inner layers of the retina (Lucas and Newhouse, 1957). In hindsight, this is an interesting historical parallel to those studies, which have argued that ischemia as an important risk factor for glaucoma. However, the prospective role of glutamate in glaucoma remains controversial, and recent studies have even questioned the sensitivity of RGCs to glutamate meditated excitotoxicity (Ullian et al., 2004). The sensitivity of RGCs to glutamate receptor agonists may be variable and highly dependent on the experimental conditions. For example, RGC death induced by 20 nmol NMDA is enhanced by the addition of glycine, D-serine, or a competitive glycine transporter-1 inhibitor, sarcosine. Thus, the severity of excitotoxic retinal damage depends on the levels of both glycine and D-serine (Hama et al., 2006). With respect to neurotrophins, it is well established that BDNF can also have profound effects on NMDA receptor currents which may vary depending on the presence of the TrkB and/or the pan-neurotrophin receptor p75 (NTR) (Sandoval et al., 2007). The relationship between the NMDA receptor and BDNF is of particular interest with respect to glaucoma in view of the current clinical trials that are using the NMDA receptor antagonist, memantine (1-amino-3,5-dimethyl-adamantane) to treat glaucomatous patients. Recently, this NMDA receptor antagonist has been shown to upregulate the expression of BDNF in the brain (Meisner et al., 2008). Therefore, it is critical to explore further the signal transduction events and to determine if this is relevant to the RGCs.

Signal transduction

NMDA receptor antagonists and CaMKII

CaMKII is known to be present in the retina (Bronstein et al., 1988a, b, 1989, 1993; Ochiishi et al., 1994; Cooper et al., 1995). The CaMKII is present in amacrine cells of the inner nuclear layer (INL) and displaced amacrine cells and ganglion cells in the ganglion cell layer (GCL) (Ochiishi et al., 1994). The specific effect of NMDA on CaMKII in the retina is known (Laabich et al., 2000). The NMDA subtype of glutamate receptor is also a known substrate of CaMKII (Chen and Huang, 1992; Kitamura et al., 1993) and the CaMKII-mediated phosphorylation of glutamate receptors leads to a positive modulation of receptor function and maintenance of synaptic excitability in other neural systems (Fukunaga et al., 1992; McGlade-McCulloh et al., 1993; Tan et al., 1994). However, NMDA is also a known excitotoxin, and its neurotoxicity in the retina has been demonstrated (Siliprandi et al., 1991; Sabel et al., 1995). The early biochemical feature of NMDA-induced excitotoxicity in neurons is the disturbance in ionic balance triggered by calcium and sodium influx through the NMDA receptor/channel complex (Choi, 1988; Marcoux et al.,

1990; Goldberg and Choi, 1993; Terashima et al., 1994). One consequence of such ionic imbalance is the activation/overactivation of many vital cellular enzymes including protein kinases and phosphatases, phospholipases, and proteases. This is said to lead to a cascade of both biochemical and physical changes (cytoskeletal breakdown) leading to neuronal death (Saido et al., 1994). RGCs express multiple subtypes of ionotropic and metabotropic glutamate receptors (Hamassaki-Britto et al., 1993; Hartveit et al., 1995), yet excitotoxic loss of RGCs is thought to result primarily from glutamate interacting with the NMDA receptor subtype. Administration of NMDA receptor antagonists is an effective method of preventing RGCs loss in various models of RGC death in which excitotoxicity is implicated, such as intraocular injection of glutamate or NMDA (Vorverk et al., 1996; Laabich et al., 2000). If NMDA receptor antagonists similarly reduce RGC loss in models of chronic hypertension, then a role for excitotoxicity in glaucoma would be more acceptable. The first study to address this issue was by Chaudhary et al. (1998), who elevated IOP in rats by cauterizing the episcleral veins, and administered the prototypical, non-competitive NMDA receptor antagonist MK-801. They found a marked preservation of RGC numbers in animals treated with the compound. However, MK-801 is unsuitable for use clinically because it blocks normal glutaminergic neurotransmission and would, therefore give rise to potentially severe adverse effects. Accordingly, other studies have made use of the non-competitive NMDA receptor antagonist, memantine. This compound has recently been shown to be clinically useful in the treatment of moderate to severe Alzheimer's disease (Areosa et al., 2005). A phase III clinical trial is currently in progress to determine the effect of memantine on visual field loss in glaucoma patients. If memantine proves to be clinically useful, this will provide further support for the postulate that excitotoxicity plays a role in the pathogenesis of glaucoma. However, the mechanism by which NMDA receptor antagonists could attenuate glaucoma remains unclear. Therefore, there is an urgent need to study the NMDA-signal transduction pathway in RGCs.

Alteration of CaMKII-α_B and neuroprotective effect of m-AIP in retinal neurons exposed to NMDA

Excessive activation of glutamate receptors mediates neuronal death, but the intracellular signaling pathways that mediate this type of neuronal death are only partly understood. We have demonstrated that the level of the CaMKII-α_B transcript, which contains a nuclear localizing signal (Diagram 1), is

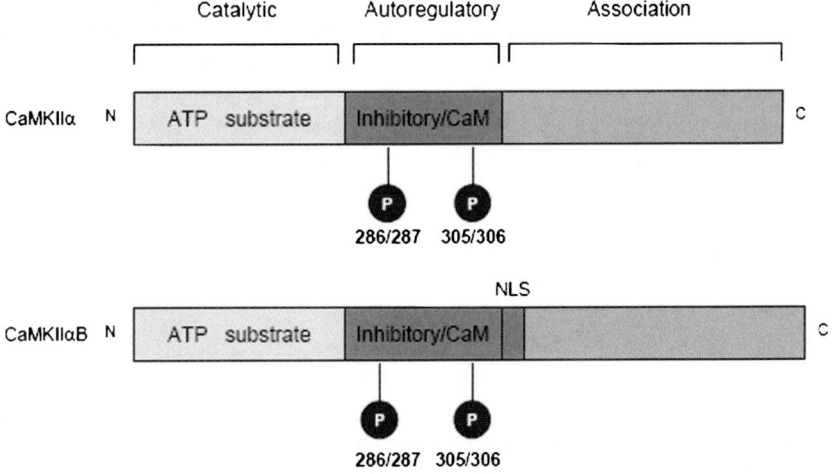

Diagram 1. Linear diagram of prototypical CaMKII-α and α_B transcripts. The α_B transcript contains a 33-nucleotide insert, which comprises the nuclear localizing signal (NLS). The amino acid sequence for the NSL is KRKSSSSVQLM (Li et al., 2001).

altered in retinal neurons exposed to NMDA, whereas the level of the CaMKII-α transcript is not affected by such treatment (Laabich et al., 2000). The CaMKII-$α_B$ gene has been cloned and sequenced from a variety of sources including the human (Li et al., 2001). There is a temporal correlation between the early appearance of the glutamate-stimulated CaMKII-$α_B$ transcript and the later appearance of apoptosis in the retina (Laabich and Cooper, 2000). In these latter studies, the terminal deoxyribonucleotidyl transferase (TdT)-mediated biotin-16-dUTP nick-end labeling (TUNEL) method was used to detect fragmented DNA in fixed tissue sections of rat retina. The TUNEL assay confirmed that cell death is observed in the inner nuclear and GCLs at 24 h following injection of 4 mM NMDA. Previous reports have shown these TUNEL-labeled cells to be ganglion cells and displaced amacrine cells in the GCL, and amacrine cells in the INL, especially cholinergic neurons (Sisk and Kuwabara, 1985; Siliprandi et al., 1991; Gavrielli et al., 1992).

It is suggested that NMDA-induced neuronal cell death is the result of a sustained increase in the intracellular concentration of Ca^{2+}, and over-activation of vital Ca^{2+}-dependent cellular enzymes such as CaMKII (Fukunaga and Soderling, 1990; Fukunaga et al., 1992). While the $α_B$ subunit of CaMKII was shown to be elevated following a single application of NMDA, this response is indicative of a temporal correlation but not necessarily a cause and effect relationship, between the level of such CaMKII transcripts and NMDA-induced cell death, in the retina (Laabich et al., 2000). The use of kinase inhibitors has helped to determine the role of CaMKII. KN-62, an inhibitor of CaMKII, does not provide a complete inhibition of NMDA-mediated neurotoxicity in vitro (Hajimohammadreza et al., 1995). This may be because KN-62 is a competitive inhibitor for calmodulin or that pre-existing CaMKII that is already auto-phosphorylated is not inhibited by this compound. Investigations with m-AIP (myristoylated autocamtide-2-related inhibitory peptide), a non-competitive inhibitor for CaMKII, and reported to be more specific than the KN-series of inhibitors, was evaluated for its putative neuroprotective effect on NMDA-mediated cell death. It is known that m-AIP is specific, and that it completely inhibits CaMKII-mediated phosphorylation of an exogenous substrate in vitro, and does not affect cyclic AMP-dependent protein kinase, protein kinase C, or calmodulin-dependent protein kinase IV (Ishida et al., 1995, 1998).

The previously mentioned study (Laabich et al., 2000) showed an in vivo inhibition of CaMKII activity following an intravitreal injection of NMDA together with m-AIP. A concentration of 500 µM m-AIP resulted in a 52% reduction in CaMKII activity when compared with sham-controls. This reduction in CaMKII activity below basal levels did not reach the zero-threshold, but it was clearly sufficient to inhibit the NMDA-induced cell death. Whereas, a lower concentration of m-AIP (100 µM) decreased the number of TUNEL-labeled cells in the GCL and INL, it did not completely block cell death. At this lower concentration, the NMDA-stimulated enzyme activity appeared to be somewhat reduced, but this reduction did not meet the test of significance, and presumably some residual NMDA-stimulated activity remained. These results compliment other reports of glutamate or NMDA-stimulated CaMKII activity (Fukunaga et al., 1992; Morioka et al., 1995) leading to neuronal cell death. The fact that the CaMKII enzyme activity can be blocked to at least the 50% basal level without causing cell death, indicates that either the increase in CaMKII-$α_B$ level and/or its phosphorylation and/or an increase in the phosphorylation of cytoplasmic CaMKII-α are critical steps in the cell signaling machinery leading to cell death. This is further discussed below.

There is an interesting parallel application of protein kinase inhibitors which have been demonstrated to lower IOP in animal (rabbits and monkeys) studies (Tian et al., 1999; Honjo et al., 2001). Several laboratories have independently discovered that an inhibitor of a new class of kinases, the rho-associated coiled coil-forming kinase (ROCK), is efficacious in lowering IOP. The selective ROCK inhibitor, Y-27632, has been observed to increase outflow facility of aqueous humor in enucleated porcine eyes (Rao et al., 2001) and in the rabbit (Honjo et al., 2001; Waki et al., 2001). It induces change in cell shape of cultured human trabecular meshwork (TM) and Schlemm's canal cells and decreases actin stress fibers and myosin light-chain phosphorylation in these cells, which is associated with widening of the extracellular spaces in the TM, especially of the juxtacanalicular tissue (Honjo et al., 2001; Rao et al., 2001). It is not yet clear if any of these inhibitors affect RGC death or survival.

Caspase-3 activation in NMDA-induced retinal cell death and its inhibition by m-AIP

A number of gene-products may be modulators of neuronal apoptosis resulting from NMDA insult. Caspase-3, a key member of the ICE protease family is implicated in the pathway leading to apoptosis (Chen et al., 1998; Namura et al., 1998). It has been suggested that caspase-3 is activated in neuronal cells that undergo apoptosis but it is not activated during necrosis (Armstrong et al., 1997). In a previous report (Laabich and Cooper, 2000), caspase-3 was found to be expressed at low levels in the adult rat retina However, after 30 min and 2 h post-injection with NMDA (Fig. 1A), the expression of caspase-3 protein was markedly increased. Furthermore, the caspase-3p32 was proteolytically activated, and the caspase-3 active fragment, p17, was markedly increased in the retina. TUNEL-labeled cells started to become apparent at some point after 2 h, but prior to 24 h (Laabich and Cooper, 2000). The increase in caspase-3/p32 protein may be due to an increase in gene expression. The precursor caspase-3 and its activated product returned to baseline 24 h after the intraocular injection with NMDA.

These results demonstrating NMDA-activated caspase in the retina are consistent with other published reports. For example, caspase-3/p32 and the active form p17 were increased in hippocampal neurons following transient global ischemia (Chen et al., 1998). Also, caspase-3 is activated and cleaved in hippocampal neurons 30–60 min of exposure to glutamate (Mattson et al., 1998). Furthermore, caspase-3 protease activation was detected within a similar time frame following the induction of cerebral ischemia (Namura et al., 1998).

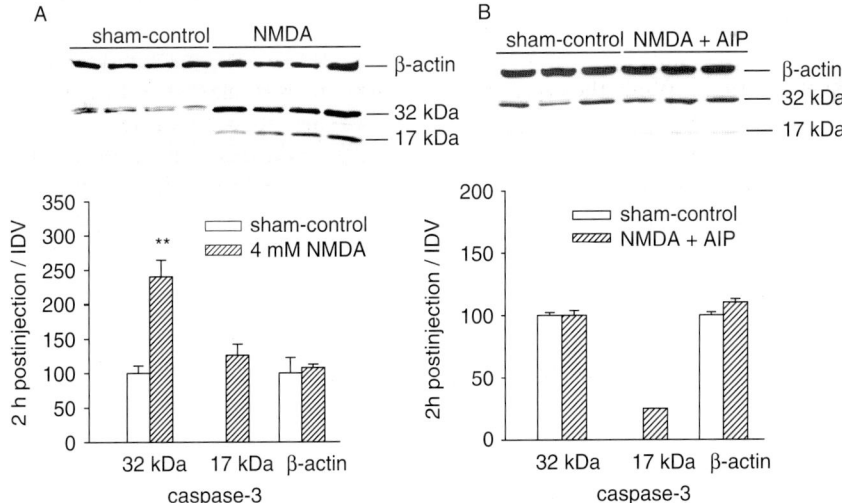

Fig. 1. NMDA-induced regulation of caspase-3 and inhibition by m-AIP. Western blot analysis of the caspase-3 proenzyme, p32, and its cleavage product, p17, in the retina. The activation of caspase-3 was assessed by the observation of the p17 (17 kDa) subunit that was derived from the cleavage of the proenzyme caspase-3p32 (32 kDa). (A) 2 h post-injection with NMDA. Immunolabeling of both caspase-3/p32 (140%) and its larger cleavage form, p17, are increased relative to sham-controls. No significant differences in the density of the β-actin bands were observed. (B) 2 h pretreatment with 500 μM m-AIP and a cotreatment with 4 mM NMDA and 500 μM m-AIP. The caspase-3p/32 remained unchanged and the p17 subunit was significantly reduced relative to the NMDA stimulated condition at this time point. No significant changes in the density of β-actin was observed [values are mean ± SEM ($n = 4$). **$P < 0.01$ vs. sham-control, one-way ANOVA with Bonferroni correction].

A pretreatment with m-AIP, a potent inhibitor for CaMKII, abolished the activity and cleavage of caspase-3 mediated by NMDA in the retina (Fig. 1B). Tenneti and Lipton (2000) had demonstrated that an increase in caspase-3 activation in NMDA-treated cerebrocortical neurons was blocked by MK801 as well as memantine, indicating that its activation there was due specifically to NMDA stimulation. As m-AIP inhibited the cleavage of caspase-3 stimulated by intraocular injections of NMDA, one can predict the activation of the protease, caspase-3, to be downstream of CaMKII. While the possibility that m-AIP may have acted through some other unknown mechanism cannot be eliminated, the results strongly support cell death-effector roles for CaMKII activity and caspase-3 in NMDA-mediated cell death in the retina. The prospective roles of transcription vs. post-translational phosphorylation of CaMKII are the subject of additional studies described below.

It has been suggested that the distribution of NMDA receptors in the retina is not correlated with cells that are killed by NMDA (Hof et al., 1998), the receptors having a wider distribution than the vulnerable cells. The neuronal phenotypes for NMDA-sensitive cells may be extended to include those that contain the NMDA receptor together with CaMKII and caspase-3. Future studies are likely to add other proteins to this prospective profile.

BDNF and neuroprotection of RGCs

While the potential for several neurotrophic factors to protect RGCs from stress has been demonstrated (see for example, Chaum, 2003; Chidlow et al., 2007) the prospective role for BDNF has received the most attention. BDNF is a member of the protein family of neurotrophins (NTs) showing widespread expression in the developing and adult mammalian brain (Lessmann et al., 2003). BDNF plays an important role in neuronal survival, differentiation and synaptic plasticity, as well as being important for protection of neurons in various pathological conditions. The effects of BDNF are mediated through the binding of this factor to its high affinity receptor, TrkB, and a low affinity p75 NTR, respectively. Binding of BDNF to TrkB activates PI3-K/Akt and/or mitogen-activated protein kinase (MAPK) signaling pathways, and thereby mediates numerous cellular functions, including inhibition of apoptosis (Chaum, 2003). The p75 NTR employs distinct signaling pathways to either enhance or suppress TrkB receptor activity, or autonomously activates signaling cascades that result in induction of apoptosis or in the promotion of survival (Roux and Barker, 2002).

In the retina, BDNF has been shown to play critical roles not only in the development and differentiation (Bennett et al., 1999; Bosco and Linden, 1999), but also in survival of retinal neuronal cells of the mature animal both in physiological and pathological conditions (Mey and Thanos, 1993; Unoki and LaVail, 1994; Peinado-Ramon et al., 1996; Kido et al., 2000). The death of RGCs is the hallmark of glaucoma, and the neuroprotective role of BDNF on RGCs has been demonstrated by many studies. For example, administration of exogenous BDNF protects RGCs in various experimental models of glaucoma, including optic nerve axotomy (Mey and Thanos, 1993; Peinado-Ramon et al., 1996), retinal ischemia (Unoki and LaVail, 1994), NMDA-induced neuronal death (Kido et al., 2000), and in eyes with chronic intraocular hypertension (Ko et al., 2000). Transgenic expression of the BDNF gene also prolongs the survival of RGCs in some of the experimental models of glaucoma (Mo et al., 2002; Martin et al., 2003). In the in vitro paradigm, supplements of BDNF in the culture media enhances primary RGC survival (Johnson et al., 1986; Thanos et al., 1989) and has also been shown to rescue transformed RGCs (RGC-5) from cell death following serum deprivation (Krishnamoorthy et al., 2001). Clearly, the mere presence of BDNF in experimental protocols will affect the interpretation of data related to cell death signaling.

There are two sources of BDNF for the RGCs in the retina including that which is retrogradely transported and that which is locally synthesized (Chaum, 2003). The relative contributions of these sources to RGC survival in the in vivo condition remain to be fully elucidated. The retrogradely transported BDNF has been postulated to be an important trophic factor for RGC survival in glaucoma. Thus, RGCs die by apoptosis in models of glaucoma or retinal ischemia, where retrogradely transported BDNF is interrupted (Pease et al., 2000; Quigley et al., 2000; Lambert et al., 2004). BDNF that is locally synthesized in the retina has also been implicated in RGC protection (Gao et al., 1997; Vecino et al., 1998, 1999; Rudzinski et al., 2004). BDNF is expressed by RGCs, amacrine cells, and other neighboring cells such as Müller cells in the retina (Vecino et al., 1998, 2002; Garcia et al., 2003). Therefore, both autocrine and paracrine effects on the RGCs need to be examined, and it seems likely that these local sources have multiple roles in the retina. The high affinity receptor for BDNF, TrkB, is present in the RGCs. The local levels of BDNF mRNA and protein in the retina have been shown to be modulated by injury to the optic nerve (Gao et al., 1997), by ocular hypertension (Rudzinski et al., 2004), by injection of NMDA into the eye (Vecino et al., 2002), and by transient retinal ischemia (Vecino et al., 1998; Lonngren et al., 2006). Taken together, these studies suggest an important paracrine/autocrine mechanism for BDNF action within the retina. However, the mechanism through which BDNF is regulated locally to protect RGCs remains unknown.

Expression of BDNF in RGCs/retina: involvement of nuclear CaMKII-α_β

Recently, the nuclear isoform of CaMKII-α, CaMKII-α_B, has been shown to be involved in a survival response of RGCs (Fan et al., 2007). CaMKII-α_B is a splice variant for CaMKII-α that has a nuclear localization signal (Schulman, 2004) (see Diagram 1), which aids the translocation of the CaMKII-α to the nucleus. Its role in the nucleus remains to be clarified. This variant transcript is particularly interesting because of the NMDA-stimulated increase in the CaMKII-α_B transcript evident in the in vivo rat model (Laabich et al., 2000). This increase also occurs specifically in pan-purified and cultured RGCs isolated from the Sprague–Dawley (SD) rat retina when glutamate is used as the stressor (Fan et al., 2007). Glutamate treatment induced a transient increase in the CaMKII-α_B transcript, which was followed by the increase in CaMKII-α protein detected in the nucleus several hours later. This result is of particular interest because it seems that glutamate stimulation induces an alternative splicing of the α gene whose product is targeted to the nucleus at a later stage. While one might conjecture that this delayed nuclear targeting of CaMKII-α_B may be relevant to the later appearance of cell death, additional experiments suggest otherwise. Specific knockdown of CaMKII-α_B in purified primary RGCs with the aid of RNA interference (RNAi) significantly enhanced glutamate-induced cell death, indicating that CaMKII-α_B is involved in a cell survival signaling pathway in RGCs (Fan et al., 2007).

The precise mechanisms underlying the role of CaMKII-α_B in cell death/survival responses remain unclear. Several reports indicate that CaMKII-α plays a role in Ca^{2+}-mediated transcriptional regulation of genes through phosphorylation of transcription factors such as cAMP response element binding protein (CREB) (Matthews et al., 1994; Sun et al., 1994) activating transcription factor (ATF) (Shimomura et al., 1996; Sun et al., 1996), CCAAT/enhancer-binding protein (C/EBP) (Wegner et al., 1992; Yano et al., 1996), serum response factor (Misra et al., 1994), and NeuroD (Gaudilliere et al., 2004). Thus, it seems likely that the nuclear localized CaMKII-α_B detected after NMDA stimulation is evidence of the regulation of gene expression in RGCs. Our studies have revealed that when CaMKII-α_B was knocked down, there was a corresponding decrease in the level of BDNF mRNA and protein in primary RGCs. Considering that knockdown of CaMKII-α_B also enhanced RGC death, these data may indicate an involvement of CaMKII-α_B in regulating BDNF expression and thus cell survival responses. This may be especially relevant to the in vivo condition, where the micro environment of retinal cells is intact and where locally synthesized BDNF may be of significance for the maintenance of cell survival (Murphy and Clarke, 2006). It seems likely that BDNF is not the only survival gene that is regulated by CaMKII-α_B. Studies in our laboratory also show an increased expression of the anti-apoptotic Bcl-2 gene in RGC-5 cells containing over-expressed CaMKII-α_B (unpublished data). Indeed, the target genes whose expressions are regulated by CaMKII-α_B are the subjects of further investigation.

It should be noted that among the glutamate-responsive transcription factors, CREB (Jiang et al., 2003) and nuclear factor κB (NFκB) (Jiang et al., 2003; Marini et al., 2004) have also been implicated in BDNF expression. NFκB is a critical regulator of many genes involved in inflammatory processes, cell differentiation, and apoptosis. It has been shown that glutamate-induced NFκB is activated in a Ca^{2+}-dependent manner (Ko et al., 1998; Meffert et al., 2003) and cytoplasmic CaMKII-α plays an important role in mediating NFκB activation in neurons, including RGCs in retina (Lilienbaum and Israel, 2003; Meffert et al., 2003; Fan et al., 2007). These studies indicate that the NFκB machinery is a prospective target for CaMKII-α. As both pro- and anti-apoptotic properties have been attributed to NFκB in neurons (Mattson et al., 2000; Pizzi et al., 2002; de Erausquin et al., 2003), the balance between cell death and survival in response to external stimuli most likely relies on the activation of distinct NFκB proteins (Pizzi et al., 2002), as well as the expression of genes that are under the control of the NFκB protein(s). For the RGCs this is an active area of research.

Secretion of BDNF: involvement of cytoplasmic CaMKII-α

While BDNF expression is regulated by multiple transcription factors, including CREB, NFκB, and possibly, the nuclear isoform of CaMKII-α, the mechanism by which BDNF is released is not yet completely understood, and yet this is an important consideration for RGC survival. BDNF, like all other neurotrophins, is generated as pre-pro-BDNF, which is further processed in the endoplasmic reticulum, trans-Golgi network, and secretory vesicles, until they are eventually secreted as mature homodimer proteins into the extracellular space (Lessmann et al., 2003). Secretion is observed in other systems in response to depolarization by K^+ or by glutamate stimulation (Lessmann et al., 2003). To investigate the mechanisms underlying the regulation of BDNF secretion, transformed RGCs (RGC-5) have been used. The RGC-5 cells show most of the characteristics of RGCs, respond to glutamate stimulation (Krishnamoorthy et al., 2001) and can be grown in sufficient quantities to measure secretable proteins such as BDNF. RGC-5 cells not only express and secrete BDNF, but also have the BDNF receptor protein, TrkB, thus providing a valuable in vitro model for studying the modulation of BDNF expression and secretion, as well as signaling pathways and modulatory influences.

Fan et al. (2006) demonstrated that glutamate stimulated a transient increase in BDNF mRNA (0.5–2 h) and protein (6–12 h) in RGC-5 cells, and also stimulated an early release (0.5–2 h) of BDNF into the culture media (Fig. 2A). This early release may be triggered by transmitter dependent depolarization. It is noted that at this early stage, although BDNF mRNA is on the rise, the protein translation is not yet underway. Therefore, the early release of BDNF is most likely derived from pre-existing pools within these cells. The released BDNF may exert some protection for glutamate challenged cells, because blocking antibodies against BDNF or its TrkB receptor led to an elevated level of glutamate-stimulated cell death. However, the protection by this small BDNF release was limited and perhaps insufficient to protect all cells because RGC-5 cells did begin to die within 24 h after exposure to glutamate. Although the level of BDNF protein within the RGC-5 cells started to increase at 6–12 h after exposure to glutamate, there was no corresponding increase in its release at these later time points. This could be a critical point with regard to the eventual cell death.

A specific inhibitor for CaMKII, m-AIP, has been shown to be a neuroprotectant for RGCs treated with NMDA in vivo (Laabich and Cooper, 2000) and glutamate in vitro (Fan et al., 2005). The mechanism for the neuroprotective role of m-AIP remains unclear and may be mediated through multiple signaling pathways. Additional studies have revealed that m-AIP enhanced

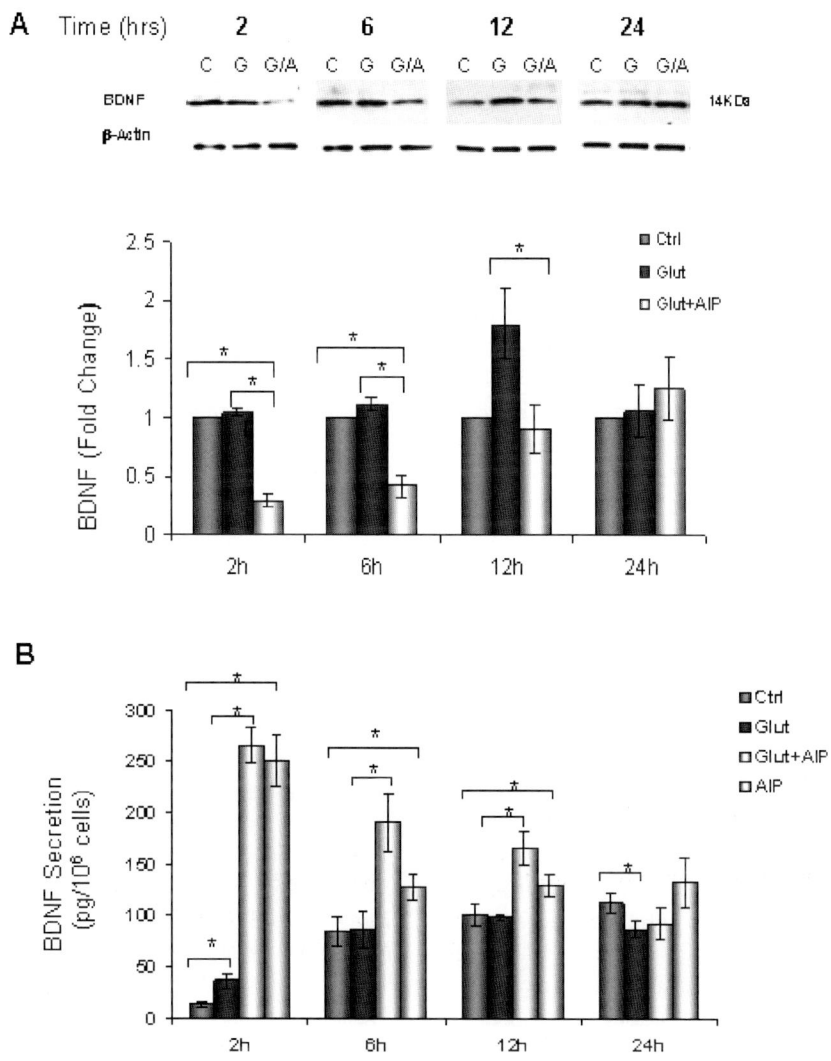

Fig. 2. Intracellular (A) and secreted (B) BDNF protein in response to glutamate and AIP. (A) Analysis of intracellular BDNF (mature proteins, 14 kDa) in RGC-5 cells. Upper panel: Immunoblots of BDNF in RGC-5 cells treated with 1 mM glutamate in the absence or presence of m-AIP (10 μM). Lower panel: The digitized data expressed as fold change in amounts of BDNF. Glutamate treatment caused an increase in the level of BDNF protein in the cells at 6–12 h. Addition of m-AIP in glutamate treated cells led to decrease in the amount of intracellular BDNF from 2–12 h. All data were normalized to β-actin and the values for controls were taken as 1. Control without glutamate (C); glutamate treatment only (G); glutamate treatment in the presence of m-AIP (G/A). Data were presented as means ± S.E.M of triplicate determinations in three independent experiments. One-way ANOVA followed by Newman–Keuls paired comparison was used for statistical analysis. $^*p<0.05$. (B) ELISA analysis of secretion of BDNF by RGC-5 cells treated with glutamate (1 mM) in the absence or presence of m-AIP (10 μM). Glutamate treatment initiated a small increase in BDNF release at 2 h. From 6–12 h, there was no significant difference in the amounts of BDNF being released when compared to the non-treated controls. Application of m-AIP in glutamate treated RGC-5 cells dramatically enhanced the release of BDNF from 2–12 h. AIP alone also increased BDNF release by the cells from 2–12 h. Data were presented as means ± S.E.M of triplicate determinations in three independent experiments. One-way ANOVA followed by Newman–Keuls paired comparison was used for statistical analysis. $^*p<0.05$.

glutamate-stimulated BDNF release in RGC-5 cells at a very early stage (30 min–2 h), when glutamate-induced BDNF production is hardly underway, and also promoted the release of BDNF for a prolonged period, perhaps for even longer than the glutamate-stimulated period of BDNF synthesis (Fig. 2B). These results suggest that m-AIP enhances the release of the pre-existing pool of BDNF, and possibly, of newly synthesized BDNF. Obviously, this relatively fast acting and long lasting role for m-AIP in promoting BDNF release is important with respect to neuroprotection, and could be one of the reasons why cell survival both in vivo and in vitro is evidently enhanced in the presence of m-AIP.

The specific mechanisms underlying the regulation of BDNF release are not yet known, but an increase in the activity of cytoplasmic CaMKII induced by glutamate treatment may be involved (Sucher et al., 1997). CaMKII-α is mainly expressed in neurons, and presynaptically, it is associated with synaptic vesicles (Benfenati et al., 1992). CaMKII-α has been shown to serve as a negative activity-dependent regulator of neurotransmitter release at hippocampal synapses (Hinds et al., 2003). This is possibly the case for BDNF release in RGC-5 cells because they do express CaMKII-α (Fan et al., 2005). Inhibition of CaMKII by m-AIP, led to an enhanced glutamate-stimulated release of BDNF, and m-AIP alone increased BDNF release by cells that are in control conditions (without glutamate). Therefore, CaMKII can inhibit basal levels BDNF release. However, in contradiction with previous results (Hinds et al., 2003; Fan et al., 2005), a recent study has revealed a strong dependence on Ca^{2+} influx and activation of CaMKII for an activity-dependent postsynaptic BDNF secretion (Kolarow et al., 2007). Further studies are warranted to clarify this important topic. Also, it seems likely that in the in vivo condition, the synthesis and release of BDNF, as well as the expression and activation of TrkB receptors, may be more complex, being regulated and influenced by neighboring cells and other factors that are not present in the in vitro models. Additional studies will be needed to show that results seen in the in vitro model apply to RGCs in the in vivo model.

Nuclear CaMKII-α vs. cytoplasmic CaMKII-α

These recent studies have shown the involvement of CaMKII-α in the regulation of both BDNF expression (nuclear CaMKII-$α_B$ isoform) and secretion (cytoplasmic CaMKII-α isoform) in RGCs. Future studies should determine if modulation of BDNF expression and release via CaMKII can protect neurons in glaucomatous animal models. Although application of exogenous BDNF has been shown to be effective, the regulated autocrine/paracrine release of BDNF into the environment of the retina is clearly an important resource for maintaining RGC survival. Thus, the endogenous BDNF and its regulatory machinery should continue to be a target for investigators. Since the application of the CaMKII inhibitor, m-AIP, dramatically enhances BDNF release, this may, in part, explain the prior observations that m-AIP protects neurons in the retina from NMDA-induced cell death in the retina. Thus, modulating CaMKII-α or its nuclear counterpart, CaMKII-$α_B$, to enhance BDNF expression/secretion may be a promising neuroprotective strategy for diseases/disorders such as glaucoma and retinal ischemia where glutamate and excitotoxicity have been implicated.

Patterns of BDNF expression are regulated by the light–dark cycle

Previous studies of BDNF protein in the retina have shown evidence of a diurnal pattern of expression. There is a 1.5-fold higher level of protein at mid-day relative to the mid-night (Pollock et al., 2001). Investigations in our laboratory show a similar trend in mRNA levels with similar mid-day to mid-night ratios of transcribed mRNA, and these observations have been extended to include additional time points. Results of such studies indicate that the peaks of BDNF mRNA expression actually occur shortly after the lights-off condition in the retinas of mice (Fig. 3). This would indicate that the lights-off condition may be a trigger for reducing transcription and the lights-on would be a trigger for ramping up transcription of BDNF. We do know that the levels of CaMKII transcripts and protein

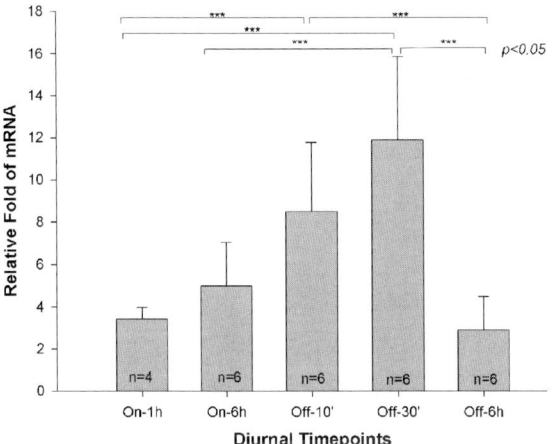

Fig. 3. Real-time PCR of BDNF expression in the diurnal period. Real-time PCR showed that BDNF message increased when light was turned on and reached its maximum 30 min after light was turned off, and then decreased in the dark. Retinas from 30-day-old C57/BL6 mice were used for this study. Real-time PCR was performed using a 7300 real-time PCR system (Applied Biosystems, Foster City, CA) for 50 cycles. The primer set and probe were located within BDNF exon-V (Applied Biosystems). The bracketed time points had significant differences of $P < 0.05$. The ANOVA procedure t-test (LSD) was used. Values represented mean $\pm$ SD (n as labeled on the chart).

in the retina increases in a dark-rearing paradigm (Xue et al., 2001) but the levels of the CaMKII-α_B transcripts, in particular, have not been measured in the diurnal period. It would be useful to see if they correlate with the patterns of BDNF expression. The mRNA levels of both BDNF and its cognate receptor TrkB have also been shown to oscillate during 24 h light–dark cycles in rat hippocampus and the frontal cortex (Bova, 1998). There appears to be a delay in the elevation of TrkB mRNA relative to BDNF in these brain regions (Bova, 1998) consistent with the notion that BDNF may have a role in regulating the transcription of TrkB mRNA (Robinson et al., 1996). While we now know that the expression, translation, and secretion of BDNF in RGCs can be regulated and manipulated via CaMKII-dependent mechanisms, the regulatory mechanisms for BDNF receptors is less clear. However, some evidence indicates that TrkB receptors in the retina are downregulated following application of BDNF (Chen and Weber, 2004). Thus, the application of exogenous BDNF may have limited practical value and other targets may have to be explored for therapeutic intervention.

Summary and conclusions

In summary, data indicate that the cytoplasmic CaMKII-α is likely involved in cell death pathways through activation of caspase-3 and through inhibition of BDNF release. In contrast, the nuclear isoform, CaMKII-α_B seems to be involved in cell survival responses through regulating expression of pro-survival genes including BDNF. The dual action of release inhibition and upregulated expression appears to be related to homeostasis of BDNF levels in retinal cells.

In response to glutamate stimulation, the level of intracellular Ca^{2+} increases and CaMKII-α becomes auto-phosphorylated (T286/287) (see Diagram 1). On the one hand, this leads to the downstream activation of caspase-3, the reduction of BDNF autocrine secretion, and the activation of NFκB (Diagram 2). On the other hand, the translocation of CaMKII-α_B to the nucleus is thought to be inhibited by autophosphorylation (Schulman, 2004) (see Diagram 2). As a result of the actions on these two pathways, the cells die. Thus, m-AIP, a specific inhibitor of CaMKII-α activity, may protect the cells from dying by blocking the cell death pathways at both the cytoplasmic and/or nuclear levels and by enhancing the survival pathways at the nuclear and/or cytoplasmic levels (Diagram 2). In this scenario, CaMKII plays a truly pivotal role. NMDA appears to transmit signals to the nucleus by multiple mechanisms including CaMKII-α_B and NFκB. The relative roles of these two pathways in the cell death and survival still remain to be completely resolved, and the diagrammatic depiction of events (Diagram 2) is intended to provide heuristic value rather than a conclusive summary of the signal transduction pathways.

It is known from in vivo studies that cell death-related genes and gene-products such as TNFR1, FasL, Nur77, GADD45, and GADD153 are also upregulated in the retina following an NMDA

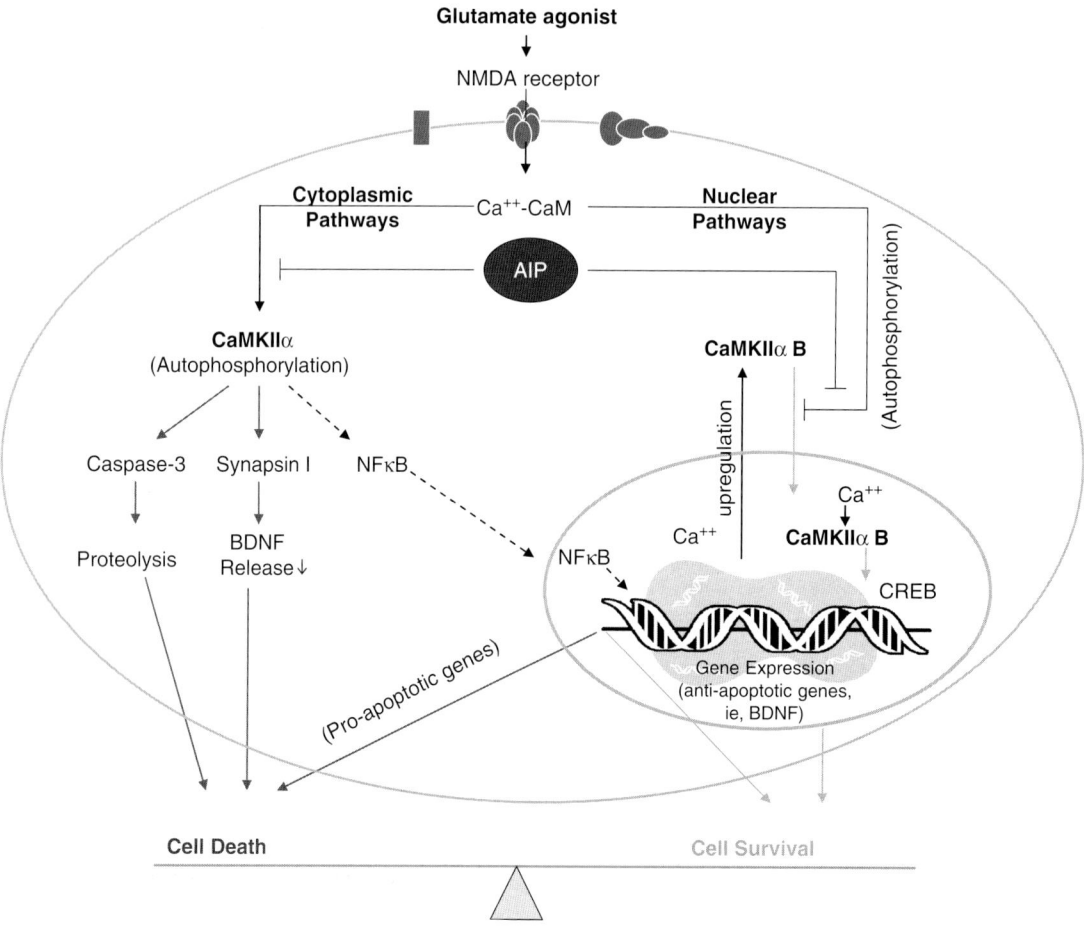

Diagram 2. See-saw model of cell death and survival pathways regulated by CaMKII.

stimulus, and the upregulation of such genes is also blocked by m-AIP (Laabich et al., 2001). It is not yet known if regulation of these "bad" genes is mediated through the cytoplasmic or the nuclear forms of CaMKII. Glutamate agonists appear to stimulate both the death and survival pathways in RGCs, and it appears that the balance is in favor of the cell death pathways. We still do not know why, logically, this would be the case, but clearly the stimulation of the survival pathways is insufficient to meet the challenge in the presence of excess glutamate. However, the use of glutamate receptor antagonists and downstream inhibition of the CaMKII activity appears to redress the balance in favor of survival through inhibiting the transcription and/or activity of the "bad" genes and by enhancing the transcription and/or activity of the "good" genes. The NMDA receptor is a current target for pharmacotherapies. While some details remain to be filled in, CaMKII would appear to be a good second target, downstream of the NMDA receptor, and deserving of further exploration. The prospects for improving the chances of RGC survival under duress are looking better.

Abbreviations

m-AIP   myristoylated autocamtide-2-related inhibitory peptide
BDNF   brain-derived neurotrophic factor

CaMKII   calcium/calmodulin-dependent
             protein kinase II
GCL   ganglion cell layer
INL   inner nuclear layer
NMDA   N-methyl-D-aspartate
RGC   retinal ganglion cell
CREB   cAMP response element binding
             protein
C/EBP   CCAAT/enhancer-binding protein
nNOS   neuronal nitric oxide synthase
NTs   neurotrophins
ATF   activating transcription factor
NFκB   nuclear factor-κB
TTX   tetrodotoxin
MAPK   mitogen-activated protein kinase
RNAi   RNA interference

Acknowledgments

This work is supported in part by NIH grants: R01EY17594; P20RR16481; P30ES014443.

References

Areosa, S.A., Sherriff, F. and McShane, R. (2005) Memantine for dementia. Cochrane Database Syst. Rev., 20(3): p. CD003154.

Armaly, M.F., Krueger, D.E., Maunder, L., Becker, B., Hetherington, J., Jr., Kolker, A.E., Levene, R.Z., Maumenee, A.E., Pollack, I.P. and Shaffer, R.N. (1980) Biostatistical analysis of the collaborative glaucoma study. I. Summary report of the risk factors for glaucomatous visual-field defects. Arch. Ophthalmol., 98: 2163–2171.

Armstrong, R.C., Aja, T.J., Hoang, K.D., Gaur, S., Bai, X., Alnemri, E.S., Litwack, G., Karanewsky, D.S., Fritz, L.C. and Tomaselli, K.J. (1997) Activation of the CED3/ICE-related protease CPP32 in cerebellar granule neurons undergoing apoptosis but not necrosis. J. Neurosci., 17: 553–562.

Benfenati, F., Valtorta, F., Rubenstein, J.L., Gorelick, F.S., Greengard, P. and Czernik, A.J. (1992) Synaptic vesicle-associated Ca^{2+}/calmodulin-dependent protein kinase II is a binding protein for synapsin I. Nature, 359: 417–420.

Bennett, J.L., Zeiler, S.R. and Jones, K.R. (1999) Patterned expression of BDNF and NT-3 in the retina and anterior segment of the developing mammalian eye. Invest. Ophthalmol. Vis. Sci., 40: 2996–3005.

Bosco, A. and Linden, R. (1999) BDNF and NT-4 differentially modulate neurite outgrowth in developing retinal ganglion cells. J. Neurosci. Res., 57: 759–769.

Bova, R. (1998) BDNF and TrkB mRNAs oscillate in rat brain during the light-dark cycle. Brain Res. Mol. Brain Res., 57: 321–324.

Bronstein, J.M., Farber, D.B. and Wasterlain, C.G. (1993) Regulation of type-II calmodulin kinase: functional implications. Brain Res. Revs., 18: 135–147.

Bronstein, J.M., Wasterlain, C.G., Bok, D., Lasher, R. and Farber, B.D. (1988a) Localization of retinal calmodulin kinase. Exp. Eye Res., 47: 391–402.

Bronstein, J.M., Wasterlain, C.G. and Farber, B.D. (1988b) A retinal calmodulin-dependent kinase: calcium/calmodulin-stimulated and inhibited states. J. Neurochem., 50: 438–446.

Bronstein, J.M., Wasterlain, C.G., Lasher, R. and Farber, B.D. (1989) Dark-induced changes in activity and compartmentalization of retinal calmodulin kinase in the rat. Brain Res., 495: 83–88.

Chaudhary, P., Ahmed, F. and Sharma, S.C. (1998) MK801-a neuroprotectant in rat hypertensive eyes. Brain Res., 792: 154–158.

Chaum, E. (2003) Retinal neuroprotection by growth factors: a mechanistic perspective. J. Cell Biochem., 88: 57–75.

Chen, H. and Weber, A.J. (2004) Brain-derived neurotrophic factor reduces TrkB protein and mRNA in the normal retina and following optic nerve crush in adult rats. Brain Res., 1011: 99–106.

Chen, J.C., Nagayama, T., Jin, K., Stetler, R.A., Zhu, R.L., Graham, S.H. and Simon, R.P. (1998) Induction of caspase-3-like protease may mediate delayed neuronal death in the hippocampus after transient cerebral ischemia. J. Neurosci., 18: 4914–4928.

Chen, L. and Huang, L.Y.M. (1992) Protein kinase C reduces magnesium block of NMDA-receptor channels as a mechanism of modulation. Nature, 356: 521–523.

Chidlow, G., Wood, J.P.M. and Casson, R.J. (2007) Pharmacological neuroprotection for glaucoma. Drugs, 67: 725–759.

Choi, D.W. (1988) Glutamate neurotoxicity and diseases of the nervous system. Neuron, 1: 623–634.

Cooper, N.G.F., Wei, X. and Liu, N. (1995) Onset of expression of the alpha subunit of Ca^{2+}/calmodulin-dependent protein kinase II and a novel related protein in the developing retina. J. Mol. Neurosci., 6: 75–89.

de Erausquin, G.A., Hyrc, K., Dorsey, D.A., Mamah, D., Dokucu, M., Masco, D.H., Walton, T., Dikranian, K., Soriano, M., Verdugo, J.M.G., Goldberg, M.P. and Dugan, L.L. (2003) Nuclear translocation of nuclear transcription factor-kappa B by alpha-amino-3-hydroxy-5-methyl-4-isoxazolepropionic acid receptors leads to transcription of p53 and cell death in dopaminergic neurons. Mol. Pharmacol., 63: 784–790.

Diijik, F., van Leeuwen, S. and Kamphius, W. (2004) Differential effects of ischemia/reperfusion on amacrine cell subtype-specific transcript levels in the rat retina. Brain Res., 1026: 194–204.

Dkhissi, O., Chanut, E., Versaux-Botteri, C., Minvielle, F., Trouvin, J.H. and Nguyen-Legros, J. (1996) Changes in retinal dopaminergic cells and dopamine rhythmic metabolism during the development of a glaucoma-like disorder in quails. Invest. Ophthalmol. Vis. Sci., 37: 2335–2344.

Dkhissi, O., Chanut, E., Versaux-Botteri, C., Trouvin, J.H., Reperant, J. and Nguyen-Legros, J. (1998) Day and night dysfunction in intraretinal melatonin and related indoleamines metabolism, correlated with the development of glaucoma-like disorder in an avian model. J. Neuroendocrinol., 10: 863–869.

Dreyer, E.B., Zurakowski, D., Schumer, R.A., Podos, S.M. and Lipton, S.A. (1996) Elevated glutamate levels in the vitreous body of humans and monkeys with glaucoma. Arch. Ophthalmol., 114: 299–305.

Fan, W., Agarwal, N. and Cooper, N.G.F. (2006) The role of CaMKII in BDNF-mediated neuroprotection of retinal ganglion cells (RGC-5). Brain Res., 1067: 48–57.

Fan, W., Agarwal, N., Kumar, M.D. and Cooper, N.G.F. (2005) Retinal ganglion cell death and neuroprotection: involvement of the CaMKII-alpha gene. Brain Res. Mol. Brain Res., 139: 306–316.

Fan, W., Li, X. and Cooper, N.G.F. (2007) CaMKII-α_B mediates a survival response in retinal ganglion cells subjected to a glutamate stimulus. Invest. Ophthalmol. Vis. Sci., 48: 3854–3863.

Flammer, J. (1994) The vascular concept of glaucomaSurv. Ophthalmol., 38(Suppl.): S3–S6. Review

Fukunaga, K. and Soderling, T.R. (1990) Activation of calcium/calmodulin-dependent kinase II in cerebellar granule cells by N-methyl-D-aspartate receptor activation. Mol. Cell. Neurosci., 1: 133–138.

Fukunaga, K., Soderling, T.R. and Miyamoto, E. (1992) Activation of Ca^{2+}/calmodulin-dependent protein kinase II and protein kinase C by glutamate in cultured rat neurons. J. Biol. Chem., 267: 22527–22533.

Gao, H., Qiao, X., Hefti, F., Hollyfield, J.G. and Knusel, B. (1997) Elevated mRNA expression of brain-derived neurotrophic factor in retinal ganglion cell layer after optic nerve injury. Invest. Ophthalmol. Vis. Sci., 38: 1840–1847.

Garcia, M., Forster, V., Hicks, D. and Vecino, E. (2003) In vivo expression of neurotrophins and neurotrophin receptors is conserved in adult porcine retina in vitro. Invest. Ophthalmol. Vis. Sci., 44: 4532–4541.

Gaudilliere, B., Konishi, Y., de la Iglesia, N., Yao, G.l. and Bonni, A. (2004) A CaMKII-NeuroD signaling pathway specifies dendritic morphogenesis. Neuron, 41: 229–241.

Gavrielli, Y., Sherman, Y. and Ben-Sasson, S.A. (1992) Identification of programmed cell death in situ via specific labeling of nuclear DNA fragmentation. J. Cell Biol., 119: 493–501.

Goldberg, M.P. and Choi, D.W. (1993) Combined oxygen and glucose deprivation in cortical cell culture: calcium-dependent and calcium-independent protein kinase mechanism of neuronal injury. J. Neurosci., 13: 3510–3524.

Hajimohammadreza, I., Probert, A.W., Coughenour, L.L., Borosky, S.A., Marcoux, F.W., Boxer, P.A. and Wang, K.W. (1995) A specific inhibitor of calcium/calmodulin-dependent protein kinase II provides neuroprotection against NMDA and Hypoxia/Hypoglycemia induced cell death. J. Neurosci., 15: 4093–4101.

Hama, Y., Katsuki, H., Tochikawa, Y., Suminaka, C., Kume, T. and Akaike, A. (2006) Contribution of endogenous glycine site NMDA agonists to excitotoxic retinal damage in vivo. Neurosci. Res., 56: 279–285.

Hamassaki-Britto, D.E., Hermans-Borgmeyer, I., Heinemann, S. and Hughes, T.E. (1993) Expression of glutamate receptors genes in the mammalian retina: the localization of GluR7 mRNAs. J. Neurosci., 13: 1888–1898.

Hare, W., WoldeMussie, E., Lai, R., Ton, H., Ruiz, G., Feldmann, B., Wijono, M., Chun, T. and Wheeler, L. (2001) Efficacy and safety of memantine, an NMDA-type open-channel blocker, for reduction of retinal injury associated with experimental glaucoma in rat and monkey. Surv. Ophthalmol., 45(suppl. 3): S284–S289.

Hartveit, E., Brandstatter, J.H., Enz, R. and Wassle, H. (1995) Expression of the mRNA of seven metabotropic glutamate receptors (mGluR1 to 7) in the rat retina. An in situ hybridization study on tissue sections and isolated cells. Eur. J. Neurosci., 7: 1472–1483.

Hinds, H.L., Goussakov, I., Nakazawa, K., Tonegawa, S. and Bolshakov, V.Y. (2003) Essential function of alpha-calcium/calmodulin-dependent protein kinase II in neurotransmitter release at a glutamatergic central synapse. Proc. Natl. Acad. Sci. U.S.A., 100: 4275–4280.

Hof, P.R., Lee, P.Y., Yeung, G., Wang, R.F., Podos, S.M. and Morrison, J.H. (1998) Glutamate receptor subunit GluR2 and NMDAR1 immunoreactivity in the retina of macaque monkeys with experimental glaucoma does not identify vulnerable neurons. Exp. Neur., 153: 234–241.

Honjo, M., Tainihara, H., Inatani, M., Kido, N., Sawamura, T., Yue, B.Y., Narumiya, S. and Honda, Y. (2001) Effects of Rho-associated protein kinase inhibitor Y-27632 on intraocular pressure and outflow facility. Invest. Ophthalmol. Vis. Sci., 42: 137–144.

Honkanen, R.A., Baruah, S., Zimmerman, M.B., Khanna, C.L., Weaver, Y.K., Narkiewicz, J., Waziri, R., Gehrs, K.M., Weingeist, T.A., Boldt, H.C., Folk, J.C., Russell, S.R. and Kwon, Y.H. (2003) Vitreous amino acid concentrations in patients with glaucoma undergoing vitrectomy. Arch. Ophthlmol., 121: 183–188.

Hood, D.C., Greenstein, V., Frishman, L., Holopigian, K., Viswanathan, S., Seiple, W., Ahmed, J. and Robson, J.G. (1999) Identifying inner retinal contributions to the human multifocal ERG. Vis. Res., 39: 2285–2291.

Hood, D.C. and Zhang, X. (2000) Multifocal ERG and VEP responses and visual fields: comparing disease-related changes. Doc. Ophthalmol., 100: 115–137.

Ishida, A., Kameshita, I., Okuno, S., Kitani, T. and Fujisawa, H. (1995) A novel highly specific and potent inhibitor of calmodulin-dependent protein kinase II. Biochem. Biophys. Res. Comm., 212: 806–812.

Ishida, A., Shigeri, Y., Tatsu, Y., Uegaki, K., Kameshita, I., Okuno, S., Kitani, T., Yumoto, N. and Fujisawa, H. (1998) Critical amino acid residues of AIP, a highly specific inhibitory peptide of calmodulin-dependent protein kinase II. FEBS Lett., 427: 115–118.

Jiang, X., Zhu, D., Okagaki, P., Lipsky, R., Wu, X., Banaudha, K., Mearow, K., Strauss, K.I. and Marini, A.M. (2003)

N-methyl-D-aspartate and TrkB receptor activation in cerebellar granule cells: an in vitro model of preconditioning to stimulate intrinsic survival pathways in neurons. Ann. N. Y. Acad. Sci., 993: 134–145.

Johnson, J.E., Barde, Y.A., Schwab, M. and Thoenen, H. (1986) Brain-derived neurotrophic factor supports the survival of cultured rat retinal ganglion cells. J. Neurosci., 6: 3031–3038.

Kaiser, H.J., Flammer, J. and Burckhardt, D. (1993) Silent myocardial ischemia in glaucoma patients. Ophthalmologica, 207: 6–7.

Kawai, S.I., Vora, S., Das, S., Gachie, E., Becker, B. and Neufeld, A.H. (2001) Modeling of risk factors for the degeneration of retinal ganglion cells after ischemia/reperfusion in rats: effects of age, caloric restriction, diabetes, pigmentation, and glaucoma. FASEB J., 15: 1285–1287.

Kerrigan, L.A., Zack, D.J., Quigley, H.A., Smith, S.D. and Pease, M.E. (1997) TUNEL-positive ganglion cells in human primary open-angle glaucoma. Arch. Ophthalmol., 115: 1031–1035.

Kido, N., Tanihara, H., Honjo, M., Inatani, M., Tatsuno, T., Nakayama, C. and Honda, Y. (2000) Neuroprotective effects of brain-derived neurotrophic factor in eyes with NMDA-induced neuronal death. Brain Res., 884: 59–67.

Kielczewski, J.L., Pease, M.E. and Quigley, H.A. (2005) The effect of experimental glaucoma and optic nerve transection on amacrine cells in the rat retina. Invest. Ophthalmol. Vis. Sci., 46: 3188–3196.

King, A.E., Dickson, T.C., Blizzard, C.A., Foster, S.S., Chung, R.S., West, A.K., Chuah, M.I. and Vickers, J.C. (2007) Excitotoxicity mediated by non-NMDA receptors causes distal axonopathy in long-term cultured spinal motor neurons. Eur. J. Neurosci., 26: 2151–2159.

Kitamura, Y., Miyazaki, A., Yamanaka, Y. and Nomura, Y. (1993) Stimulatory effects of protein kinase C and calmodulin kinase II on N-methyl-D-aspartate receptor/channels in the postsynaptic density of rat brain. J. Neurochem., 61: 100–109.

Ko, H.W., Park, K.Y., Kim, H., Han, P.L., Kim, Y.U., Gwag, B.J. and Choi, E.J. (1998) Ca^{2+}-mediated activation of c-Jun N-terminal kinase and nuclear factor kappa B by NMDA in cortical cell cultures. J. Neurochem., 71: 1390–1395.

Ko, M.L., Hu, D.N., Ritch, R. and Sharma, S.C. (2000) The combined effect of brain-derived neurotrophic factor and a free radical scavenger in experimental glaucoma. Invest. Ophthalmol. Vis. Sci., 41: 2967–2971.

Kolarow, R., Brigadski, T. and Lessmann, V. (2007) Postsynaptic secretion of BDNF and NT-3 from hippocampal neurons depends on calcium calmodulin kinase II signaling and proceeds via delayed fusion pore opening. J. Neurosci., 27: 10350–10364.

Krishnamoorthy, R.R., Agarwal, P., Prasanna, G., Vopat, K., Lambert, W., Sheedlo, H.J., Pang, I.H., Shade, D., Wordinger, R.J., Yorio, T., Clark, A.F. and Agarwal, N. (2001) Characterization of a transformed rat retinal ganglion cell line. Brain Res. Mol. Brain Res., 86: 1–12.

Laabich, A. and Cooper, N.G.F. (1999) Regulation of calcium/calmodulin-dependent protein kinase II in the adult rat retina is mediated by ionotropic glutamate receptors. Exp. Eye Res., 68: 703–713.

Laabich, A. and Cooper, N.G.F. (2000) Neuroprotective effect of AIP on N-methyl-D-aspartate-induced cell death in retinal neurons. Brain Res. Mol. Brain Res., 85: 32–40.

Laabich, A., Li, G.Y. and Cooper, N.G.F. (2000) Calcium/calmodulin-dependent protein kinase II containing a nuclear localizing signal is altered in retinal neurons exposed to N-methyl-D-aspartate. Brain Res. Mol. Brain Res., 76: 253–265.

Laabich, A., Li, G. and Cooper, N.G.F. (2001) Characterization of apoptosis-genes associated with NMDA mediated cell death in the adult rat retina. Mol. Brain Res., 91: 34–42.

Lambert, W.S., Clark, A.F. and Wordinger, R.J. (2004) Neurotrophin and Trk expression by cells of the human lamina cribrosa following oxygen-glucose deprivation. BMC Neurosci., 5: 51–65.

Leske, M.C. (1983) The epidemiology of open-angle glaucoma: a review. Am. J. Epidemiol., 118: 166–191.

Lessmann, V., Gottmann, K. and Malcangio, M. (2003) Neurotrophin secretion: current facts and future prospects. Prog. Neurobiol., 69: 341–374.

Li, G., Laabich, A., Liu, L.O., Xue, J. and Cooper, N.G. (2001) Molecular cloning and sequence analyses of calcium/calmodulin-dependent protein kinase II from fetal and adult human brain. Sequence analyses of human brain calcium/calmodulin-dependent protein kinase II. Mol. Biol. Rep., 28: 35–41.

Lilienbaum, A. and Israel, A. (2003) From calcium to NF-{kappa}B signaling pathways in neurons. Mol. Cell. Biol., 23: 2680–2698.

Lonngren, U., Napankangas, U., Lafuente, M., Mayor, S., Lindqvist, N., Vidal-Sanz, M. and Hallbook, F. (2006) The growth factor response in ischemic rat retina and superior colliculus after brimonidine pre-treatment. Brain Res. Bull., 71: 208–218.

Lucas, D.R. and Newhouse, J.P. (1957) The toxic effect of sodium l-glutamate on the inner layers of the retina. AMA Arch. Ophthalmol., 58: 193–201.

Marcoux, F.W., Probert, A.W. and Weber, M.L. (1990) Hypoxic neuronal injury in tissue culture is associated with delayed calcium accumulation. Stroke, 21(Suppl. III): 71–74.

Marini, A.M., Jiang, X., Wu, X., Tian, F., Zhu, D., Okagaki, P. and Lipsky, R.H. (2004) Role of brain-derived neurotrophic factor and NF-kappaB in neuronal plasticity and survival: From genes to phenotype. Restor. Neurol. Neurosci., 22: 121–130.

Martin, K.R., Quigley, H.A., Zack, D.J., Levkovitch-Verbin, H., Kielczewski, J., Valenta, D., Baumrind, L., Pease, M.E., Klein, R.L. and Hauswirth, W.W. (2003) Gene therapy with brain-derived neurotrophic factor as a protection: retinal ganglion cells in a rat glaucoma model. Invest. Ophthalmol. Vis. Sci., 44: 4357–4365.

Matthews, R.P., Guthrie, C.R., Wailes, L.M., Zhao, X., Means, A.R. and McKnight, G.S. (1994) Calcium/calmodulin-dependent protein kinase types II and IV differentially regulate CREB-dependent gene expression. Mol. Cell Biol., 14: 6107–6116.

Mattson, M.P., Culmsee, C., Yu, Z. and Camandola, S. (2000) Roles of nuclear factor kappaB in neuronal survival and plasticity. J. Neurochem., 74: 443–456.

Mattson, M.P., Keller, J.N. and Begley, J.G. (1998) Evidence of synaptic apoptosis. Exp. Neurol., 153: 35–48.

May, C.A. and Mittag, T. (2004) Neuronal nitric oxide synthase (nNOS) positive retinal amacrine cells are altered in the DBA/2NNia mouse, a murine model for angle-closure glaucoma. J. Glaucoma., 13: 496–499.

McGlade-McCulloh, E., Yamamoto, H., Tan, S.E., Brickey, D.A. and Soderling, T.R. (1993) Phosphorylation and regulation of glutamate receptors by calcium/calmodulin-dependent protein kinase II. Nature, 362: 640–642.

Meffert, M.K., Chang, J.M., Wiltgen, B.J., Fanselow, M.S. and Baltimore, D. (2003) NF-kappa B functions in synaptic signaling and behavior 73. Nat. Neurosci., 6: 1072–1078.

Meisner, F., Scheller, C., Kneitz, S., Sopper, S., Neuen-Jacob, E., Riederer, P., Meulen, V.T. and Koutsilieri, E. (2008) Memantine upregulates BDNF and prevents dopamine deficits in SIV-infected macaques: a novel pharmacological action of memantine. Neuropsychopharmacology, 33: 2228–2236.

Mey, J. and Thanos, S. (1993) Intravitreal injections of neurotrophic factors support the survival of axotomized retinal ganglion cells in adult rats in vivo. Brain Res., 602: 304–317.

Misra, R.P., Bonni, A., Miranti, C.K., Rivera, V.M., Sheng, M. and Greenberg, M.E. (1994) L-type voltage-sensitive calcium channel activation stimulates gene expression by a serum response factor-dependent pathway. J. Biol. Chem., 269: 25483–25493.

Mo, X., Yokoyama, A., Oshitari, T., Negishi, H., Dezawa, M., Mizota, A. and Adachi-Usami, E. (2002) Rescue of axotomized retinal ganglion cells by BDNF gene electroporation in adult rats. Invest. Ophthalmol. Vis. Sci., 43: 2401–2405.

Moon, J.I., Kim, I.B., Gwon, J.S., Park, M.H., Kang, T.H., Lim, E.J., Choi, K.R. and Chun, M.H. (2005) Changes in retinal neuronal populations in the DBA/2J mouse. Cell Tissue Res., 320(1): 51–59.

Morioka, M., Fukunaga, K., Nagahiro, S., Kurino, M., Ushio, Y. and Miyamoto, E. (1995) Glutamate-induced loss of Ca^{2+}/calmodulin-dependent protein kinase II activity in cultured rat hippocampal neurons. J. Neurochem., 64: 2132–2139.

Murphy, J.A. and Clarke, D.B. (2006) Target-derived neurotrophins may influence the survival of adult retinal ganglion cells when local neurotrophic support is disrupted: implications for glaucoma. Med. Hypotheses, 67: 1208–1212.

Namura, N., Zhu, J., Fink, K., Endres, M., Srinivasan, A., Tomaselli, K.J., Yuan, J. and Moskowitz, M.A. (1998) Activation and cleavage of caspase-3 in apoptosis induced by experimental cerebral ischemia. J. Neurosci., 18(10): 3659–3668.

Nickells, R.W. (1999) Apoptosis of retinal ganglion cells in glaucoma: an update of the molecular pathways involved in cell death. Surv. Ophthalmol., 43: 151–161.

Nucci, C., Tartaglione, R., Rombola, L., Morrone, L.A., Fazzi, E. and Bagetta, G. (2005) Neurochemical evidence to implicate elevated glutamate in the mechanisms of high intraocular pressure (IOP)-induced retinal ganglion cell death in rat. Neurotoxicology, 26: 935–941.

Ochiishi, T., Terashima, T., Sugiura, H. and Yamauchi, T. (1994) Immunohistochemical localization of Ca^{2+}/calmodulin-dependent protein kinase II in the rat retina. Brain Res., 634: 257–265.

Osborne, N.N., Casson, R.J., Wood, J.P.M., Chidlow, G., Graham, M. and Melena, J. (2004) Retinal ischemia: mechanisms of damage and potential therapeutic strategies. Prog. Retin. Eye Res., 23: 91–147.

Osborne, N.N., Ugarte, M., Chao, M., Bae, J.H., Wood, J.P. and Nash, M.S. (1999) Neuroprotection in relation to retinal ischemia and relevance to glaucoma. Surv. Ophthalmol., 43(Suppl 1): S102–S128.

Pang, I.H., Wexler, E.M., Nawys, S., Desantis, L. and Kapin, M.A. (1999) Protection by eliprodil against excitotoxicity in cultured rat retinaganglion cells. Invest. Ophthalmol. Vis. Sci., 40: 1170–1176.

Pease, M.E., McKinnon, S.J., Quigley, H.A., Kerrigan-Baumrind, L.A. and Zack, D.J. (2000) Obstructed axonal transport of BDNF and its receptor TrkB in experimental glaucoma. Invest. Ophthalmol. Vis. Sci., 41: 764–774.

Peinado-Ramon, P., Salvador, M., Villegas-Perez, M.P. and Vidal-Sanz, M. (1996) Effects of axotomy and intraocular administration of NT-4, NT-3, and brain-derived neurotrophic factor on the survival of adult rat retinal ganglion cells: a quantitative in vivo study. Invest. Ophthalmol. Vis. Sci., 37: 489–500.

Pizzi, M., Goffi, F., Boroni, F., Benarese, M., Perkins, S.E., Liou, H.C. and Spano, P. (2002) Opposing roles for NF-kappa B/Rel factors p65 and c-Rel in the modulation of neuron survival elicited by glutamate and interleukin-1beta. J. Biol. Chem., 277: 20717–20723.

Pollock, G.S., Vernon, E., Forbes, M.E., Yan, Q., Ma, Y.T., Hsieh, T., Robichon, R., Frost, D.O. and Johnson, J.E. (2001) Effects of early visual experience and diurnal rhythms on BDNF mRNA and protein levels in the visual system, hippocampus, and cerebellum. J. Neurosci., 21: 3923–3931.

Posner, A. (1958) Ischemia and the glaucoma syndrome. The Eye, Ear, Nose and Throat Monthly, 37: p. 528.

Quigley, H.A. (1995) Ganglion cell death in glaucoma: pathology recapitulates ontogeny. Aust. N.Z. J. Ophthalmol., 23: 85–91.

Quigley, H.A., McKinnon, S.J., Zack, D.J., Pease, M.E., Kerrigan-Baumrind, L.A., Kerrigan, D.F. and Mitchell, R.S. (2000) Retrograde axonal transport of BDNF in retinal ganglion cells is blocked by acute IOP elevation in rats. Invest. Ophthalmol. Vis. Sci., 41: 3460–3466.

Rao, P.V., Deng, P.F., Kumar, J. and Epstein, D.L. (2001) Modulation of aqueous humor outflow facility by the Rho kinase-specific inhibitor Y-27632. Invest. Ophthalmol. Vis. Sci., 42: 1029–1037.

Robinson, M., Adu, J. and Davies, A.M. (1996) Timing and regulation of TrkB and BDNF mRNA expression in placode-derived sensory neurons and their targets. Eur. J. Neurosci., 8: 2399–2406.

Roux, P.P. and Barker, P.A. (2002) Neurotrophin signaling through the p75 neurotrophin receptor. Prog. Neurobiol., 67: 203–233.

Rudzinski, M., Wong, T.P. and Saragovi, H.U. (2004) Changes in retinal expression of neurotrophins and neurotrophin receptors induced by ocular hypertension. J. Neurobiol., 58: 341–354.

Sabel, B.A., Sautter, J., Stoehr, T. and Siliprandi, R.A. (1995) A behavioral model of excitotoxicity: retinal degeneration loss of vision, and subsequent recovery after intraocular NMDA-administration in adult rats. Exp. Brain. Res., 106: 93–105.

Saido, T.C., Sorimachi, H. and Suzuki, K. (1994) Calpain: new perspectives in molecular diversity and physiological-pathological involvement. FASEB J., 8: 814–822.

Sandoval, M., Sandoval, R., Thomas, U., Spilker, C., Small, K.H., Falcon, R., Marengo, J.J., Calderón, R., Saavedra, V., Heumann, R., Bronfman, F., Garner, C.C., Gundelfinger, E.D. and Wyneken, U. (2007) Antagonistic effects of TrkB and p75(NTR) on NMDA receptor currents in post-synaptic densities transplanted into Xenopus oocytes. J. Neurochem., 101: 1672–1684.

Schulman, H. (2004) Activity-dependent regulation of calcium/calmodulin-dependent protein kinase II localization. J. Neurosci., 24: 8399–8403.

Shimomura, A., Ogawa, Y., Kitani, T., Fujisawa, H. and Hagiwara, M. (1996) Calmodulin-dependent protein kinase II potentiates transcriptional activation through activating transcription factor 1 but not cAMP response element-binding protein. J. Biol. Chem., 271: 17957–17960.

Siliprandi, R., Canella, R., Carmignoto, G., Schiavo, N., Zanellato, N., Zanoni, R. and Vantini, G. (1991) N-methyl-D-aspartate-induced neurotoxicity in the adult rat retina. Vis. Neurosci., 8: 567–573.

Sisk, D.R. and Kuwabara, T. (1985) Histologic changes in the inner retina of albino rats following intravitreal injection of monosodium l-glutamate. Graefes Arch. Clin. Exp. Ophthalmol., 223: 250–258.

Stroman, G.A., Stewart, W.C., Golnik, K.C., Curé, J.K. and Olinger, R.E. (1995) Magnetic resonance imaging in patients with low-tension glaucoma. Arch. Ophthalmol., 113: 168–172.

Sucher, N.J., Lipton, S.A. and Dreyer, E.B. (1997) Molecular basis of glutamate toxicity in retinal ganglion cells. Vis. Res., 37: 3483–3493.

Sun, P., Enslen, H., Myung, P.S. and Maurer, R.A. (1994) Differential activation of CREB by Ca^{2+}/calmodulin-dependent protein kinases type II and type IV involves phosphorylation of a site that negatively regulates activity. Genes Dev., 8: 2527–2539.

Sun, P., Lou, L. and Maurer, R.A. (1996) Regulation of activating transcription factor-1 and the cAMP response element-binding protein by Ca/calmodulin-dependent protein kinases type I, II, and IV. J. Biol. Chem., 271: 3066–3073.

Tan, S.E., Robert, J.W. and Soderling, T.R. (1994) Phosphorylation of AMPA-type glutamate receptors by calcium/calmodulin-dependent protein kinase II and protein kinase C in cultured hippocampal neurons. J. Neurosci., 14: 1123–1129.

Tatton, W.G. (1999) Apoptosis mechanisms in neurodegeneration: possible relevance to glaucoma. Eur. J. Ophthalmol., 9(suppl 1): S22–S29.

Tenneti, L. and Lipton, S.A. (2000) Involvement of activated caspase-3-like proteases in N-methyl-D-aspartate-induced apoptosis in cerebrocortical neurons. J. Neurosci., 74: 134–142.

Terashima, T., Ochiishi, T. and Yamauchi, T. (1994) Immunocytochemical localization of calcium/calmodulin-dependent protein kinase II isoforms in the ganglion cells of the rat retina: immunofluorescence histochemistry combined with a fluorescent retrograde tracer. Brain Res., 650: 133–139.

Thanos, S., Bahr, M., Barde, Y.A. and Vanselow, J. (1989) Survival and axonal elongation of adult rat retinal ganglion cells. In vitro effects of lesioned sciatic nerve and brain derived neurotrophic factor. Eur. J. Neurosci., 1: 19–26.

Tian, B., Gabelt, B.T., Peterson, J.A., Kiland, J.A. and Kaufman, P.L. (1999) H7 increases trabecular facility and facility after ciliary muscle disinsertion in monkeys. Invest. Ophthalmol. Vis. Sci., 40: 239–242.

Ullian, E.M., Barkis, W.B., Chen, S., Diamond, J.S. and Barres, B.A. (2004) Invulnerability of retinal ganglion cells to NMDA excitotoxicity. Mol. Cell. Neurosci., 26: 544–557.

Unoki, K. and LaVail, M.M. (1994) Protection of the rat retina from ischemic injury by brain-derived neurotrophic factor, ciliary neurotrophic factor, and basic fibroblast growth factor. Invest. Ophthalmol. Vis. Sci., 35: 907–915.

Vecino, E., Caminos, E., Ugarte, M., Martin-Zanca, D. and Osborne, N.N. (1998) Immunohistochemical distribution of neurotrophins and their receptors in the rat retina and the effects of ischemia and reperfusion. Gen. Pharmacol., 30: 305–314.

Vecino, E., Garcia-Grespo, D., Garcia, M., Martinez-Millan, L., Sharma, S.C. and Carrascal, E. (2002) Rat retinal ganglion cells co-express brain derived neurotrophic factor (BDNF) and its receptor TrkB. Vis. Res., 42: 151–157.

Vecino, E., Ugarte, M., Nash, M.S. and Osborne, N.N. (1999) NMDA induces BDNF expression in the albino rat retina in vivo. Neuroreport, 10: 1103–1106.

Vorverk, C.K., Gorla, M.S. and Dreyer, E.B. (1999) An experimental basis for implicating excitotoxicity in glaucomatous optic neuropathy. Surv. Ophthalmol., 43(suppl 1): S142–S150.

Vorverk, C.K., Lipton, S.A., Zurakowski, D., Hyman, B.T. and Sabel, B.A. (1996) Chronic low-dose glutamate is toxic to retinal ganglion cells: toxicity blocked by memantine. Invest. Ophthalmol. Vis. Sci., 37: 1618–1624.

Waldmann, E., Gasser, P., Dubler, B., Huber, C. and Flammer, J. (1996) Silent myocardial ischemia in glaucoma and cataract patients. Graefes. Arch. Clin. Exp. Ophthalmol., 234: 595–598.

Waki, M., Yoshida, Y., Oka, T. and Azuma, M. (2001) Reduction of intarocular pressure by topical administration of an inhibitor of the Rho-associated protein kinase. Curr. Eye Res., 22: 470–474.

Wegner, M., Cao, Z. and Rosenfeld, M.G. (1992) Calcium-regulated phosphorylation within the leucine zipper of C/EBP beta. Science, 256: 370–373.

Xue, J., Li, G., Laabich, A. and Cooper, N.G. (2001) Visual-mediated regulation of retinal CaMKII and its GluR1 substrate is age-dependent. Brain Res. Mol. Brain Res., 93: 95–104.

Yamamoto, T. (2001) The dawn of neuroprotective therapy for glaucomatous optic neuropathy. Nippon Ganka Gakkai Zasshi, 105: 866–883.

Yano, S., Fukunaga, K., Takiguchi, M., Ushio, Y., Mori, M. and Miyamoto, E. (1996) Regulation of CCAAT/Enhancer-binding protein family members by stimulation of glutamate receptors in cultured rat cortical astrocytes. J. Biol. Chem., 271: 23520–23527.

Evidence of the neuroprotective role of citicoline in glaucoma patients

Vincenzo Parisi[1,*], Giovanni Coppola[2], Marco Centofanti[1,3], Francesco Oddone[1], Anna Maria Angrisani[2], Lucia Ziccardi[1], Benedetto Ricci[2], Luciano Quaranta[4] and Gianluca Manni[1,3]

[1] *G.B. Bietti Eye Foundation-IRCCS, Rome, Italy*
[2] *Ophthalmology Unit, Association Columbus Clinic, Catholic University of Rome, Rome, Italy*
[3] *Department of Ophthalmology, University of Rome Tor Vergata, Rome, Italy*
[4] *Department of Ophthalmology, University of Brescia, Brescia, Italy*

Abstract: The glaucomatous disease is currently considered a disease involving ocular and visual brain structures. This new approach to glaucoma introduces the possibility of inducing an improvement by means of a pharmacological approach similar to that used in different degenerative brain disorders. In line with this hypothesis, we studied the effects of oral (1600 mg/die, Cebrolux®, Tubilux Pharma, Pomezia, Rome, Italy) or intramuscular (1000 mg/die, Cebroton®, Tubilux Pharma) cytidine-5′-diphosphocholine (citicoline) treatment on retinal function and neural conduction in the visual pathways of glaucoma patients with moderate visual defects. Improvement of retinal function (objectively evaluated by pattern electroretinogram recordings) and of neural conduction along visual pathways (objectively evaluated by visual evoked potential recordings) were observed in glaucoma patients after two 60-day periods of oral or intramuscular treatment with citicoline. However, partial regression of this improvement was detected after two 120-day periods of washout. This suggests that the beneficial effects observed are in part treatment-dependent. The extension of citicoline treatment up to a period of 8 years lead to the stabilization or improvement of the glaucomatous visual dysfunction. These results suggest potential neuroprotective effects of citicoline in the glaucomatous disease.

Keywords: citicoline; glaucoma; visual evoked potentials; pattern electroretinogram; innermost retinal layers; visual pathways; visual function

Introduction

The glaucomatous disease open angle glaucoma (OAG) is commonly characterized by the presence of increased intraocular pressure (IOP), typical optic nerve head cupping, and visual field defects, in particular, evaluated by Humphrey Field Analyzer (HFA) (Graham et al., 1996).

HFA perimetry, however, does not selectively reveal which structures contribute to the impairment of the visual system observed in glaucoma. Alternatively, electrophysiological methods may allow us to explore the different structures that contribute to visual function.

The function of retinal preganglionic elements can be objectively evaluated by recording electroretinographic signals evoked by flash stimuli (flash

*Corresponding author. Tel.: +390685356727; Fax: +390686216880; E-mail: vparisi@tin.it

electroretinogram, ERG) (Armington, 1974). Studies in animals (Maffei and Fiorentini, 1982) and humans (Holder, 1997; Parisi et al., 1999d; Parisi, 2003) suggest that the function of retinal ganglion cells (RGCs) and their fibers can be assessed by electroretinographic signal recordings evoked by pattern stimuli (pattern ERG, PERG). The function of the entire visual pathway can be assessed by recording cortical potentials evoked by patterned stimuli (visual evoked potentials, VEPs) (Celesia et al., 1993).

Several studies performed in groups of patients with ocular hypertension without visual field defects (OHT patients) or in groups of patients with glaucoma showed the presence of normal or impaired flash ERG responses (Gur et al., 1987; Vaegan et al., 1995), impaired PERG (Porciatti et al., 1987; Watanabe et al., 1989; Parisi, 1997, 2001; Parisi et al., 1997, 1999c, 2001, 2006; Salgarello et al., 1999; Bach, 2001; Garway-Heath et al., 2002; Ventura et al., 2005), and VEP (Parisi and Bucci, 1992; Greenstein et al., 1998; Martus et al., 1998; Klistorner and Graham, 1999; Parisi, 1997, 2001; Horn et al., 2002; Parisi et al., 1997, 1999c, 2001, 2006) responses when compared to responses obtained in groups of normal subjects.

The results provided by electrophysiological studies suggest that the natural history of glaucoma involves the early impairment of the innermost retinal layers (which may precede the onset of visual field defects) (Parisi et al., 2006), which may be followed by an impairment, due to transynaptic degeneration, in postretinal visual pathways and in particular at the level of the lateral geniculate nucleus (LGN) (Chaturvedi et al., 1993; Weber et al., 2000; Yucel et al., 2000, 2003; Gupta and Yucel, 2003). These observations lead us to believe that the glaucomatous disease must not be considered as a disease exclusively involving ocular structures, but is a pathology in which visual brain structures are also compromised.

Thus, since 1996, we are paying our attention on the possibility of inducing an improvement of visual function using a pharmacological approach similar to that used in different brain disorders ascribed to vascular, traumatic, or degenerative processes (Boismare et al., 1978; Serra et al., 1981; Agnoli et al., 1985; Zappia et al., 1985; Kakihana et al., 1988; Cacabelos et al., 1996).

In particular, in our two previously published studies, we observed that the glaucomatous visual impairment may be improved by treatment with nicergoline (Parisi et al., 1999a) or with cytidine-5′-diphosphocholine (citicoline) (Parisi et al., 1999b). Nevertheless, the beneficial effects of these treatments are treatment-dependent. In particular, 45 and 300 days (respectively for nicergoline and citicoline) (Parisi et al., 1999a, b) after the end of the treatment, no differences were detected with respect to pretreatment conditions. When a second period of citicoline treatment was performed, we observed that even after a long period of washout (120 days), it is possible to once again detect visual function improvement, suggesting that repeated treatments may inhibit the development of the visual impairment (Parisi et al., 1999b).

In the above-mentioned study (Parisi et al., 1999b), citicoline treatment was performed by intramuscular injection. As this administration route requires suitable paramedical abilities, it could represent a real problem for patients' self-administration and thus patient compliance.

At present, citicoline is available as an oral formulation. This can represent an advantage for many patients whose visual dysfunction, as observed in previously published studies, can improve after citicoline treatment.

This chapter presents data regarding studies that evaluated the effects of oral or intramuscular treatment with citicoline on the function of the retina and visual pathways. In these studies, the effects of citicoline on visual field sensitivity were purposely not considered because any detectable improvement could be ascribed to the associated effects on consciousness level and attention (Zappia et al., 1985).

In addition, this chapter discusses whether citicoline effects may be considered as "neuroprotective", which is able to prevent the development of the glaucomatous disease.

Patients: selection and recruitment criteria

Seventy eyes of 60 patients (range: 38–62 years, mean age 52.77 ± 5.28 years) affected by OAG with moderate visual field defects [HFA with mean deviation (MD) between -2 and $-14\,dB$; corrected pattern standard deviation (CPSD) between $+2$ and $+12\,dB$] were studied. OAG patients were selected from a very large population (172 OAG patients) on the basis of the inclusion criteria extensively reported in our previously published study (Parisi et al., 2006).

Because it is known that PERG responses can be modified by the pharmacological reduction of IOP (Colotto et al., 1995; Ventura and Porciatti, 2005), we only enrolled OHT and OAG patients with IOP values less than 18 mmHg on beta-blocker monotherapy maintained during the 8 months preceding the electrophysiological evaluation.

All OAG patients were randomly divided on the basis of age and visual field defects into three groups (see below):

- Group NT-OAG (not treated open-angle glaucoma): 20 OAG patients (range: 39–61 years, mean age 51.64 ± 4.68 years, 20 eyes) in which no additional pharmacological treatment was performed.
- Group TI-OAG (glaucoma patients treated with intramuscular citicoline): 20 OAG patients (range: 41–62 years, mean age 52.54 ± 5.43 years, 20 eyes) treated with a daily intramuscular dose of 1000 mg citicoline (Cebroton 1000®, Tubilux Pharma, Pomezia, Rome, Italy).
- Group TO-OAG (glaucoma patients treated with oral citicoline): 20 OAG patients (range: 38–60 years, mean age 50.84 ± 6.62 years, 20 eyes) treated with a daily oral dose of 1600 mg citicoline (Cebrolux®, Tubilux Pharma).

The research followed the tenets of the Declaration of Helsinki. The protocol was approved by the local Institutional Review Board (IRB). Upon recruitment, each patient gave informed consent.

Pharmacological treatment protocol

The pharmacological treatment was performed according to the following schedule.

First period — A daily intramuscular dose of 1000 mg citicoline or oral dose of 1600 mg citicoline was prescribed according to the following protocol:

- 0–60 days: First period of pharmacological treatment with oral or intramuscular citicoline.
- 61–180 days: First period of washout and follow-up at the sixth month.

Second period — A daily intramuscular dose of 1000 mg citicoline or oral dose of 1600 mg citicoline was prescribed according to the following protocol:

- 181–240 days: Second period of pharmacological treatment with oral or intramuscular citicoline.
- 241–360 days: Second period of washout and follow-up at the twelfth month.

Methodology of visual function evaluation: electrophysiological examinations

PERG and VEP recordings were performed using the following methods (Parisi, 1997, 2001; Parisi et al., 1997, 1999c, 2001, 2006).

Briefly, subjects were seated in a semidark, acoustically isolated room in front of the display surrounded by a uniform field of luminance of $5\,cd/m^2$. Prior to the experiment, each subject was adapted to the ambient room light for 10 min and pupil diameter was approximately 5 mm. Mydriatic or miotic drugs were never used. Stimulation was monocular after occlusion of the other eye. Visual stimuli were checkerboard patterns (contrast 80%, mean luminance $110\,cd/m^2$) generated on a TV monitor and reversed in contrast at the rate of 2 reversals per second; at the viewing distance of 114 cm, the check edges subtended 15 minutes (15′) of visual angle. The monitor screen subtended 18°. PERG and VEP recordings were performed with full correction of refraction at the viewing distance. A small red fixation target, subtending a visual angle of approximately 0.5 degrees (estimated after taking into account spectacle-corrected individual refractive errors) was placed at the center of the pattern stimulus. At every PERG and VEP examination, each patient positively reported that he/she could clearly perceive the fixation target. The refraction of all subjects was corrected for viewing distance.

PERG recordings

The bioelectrical signal was recorded by a small Ag/AgCl skin electrode placed over the lower eyelid. PERGs were derived bipolarly between the stimulated (active electrode) and the patched (reference electrode) eye using a previously described method (Fiorentini et al., 1981). As the recording protocol was extensive, the use of skin electrodes with interocular recording represented a good compromise between signal-to-noise ratio (SNR) and signal stability. The ground electrode was in Fpz. Interelectrode resistance was lower than 3000 ohms. The signal was amplified (gain 50,000), filtered (band pass 1–30 Hz), and averaged with automatic rejection of artifacts (200 events free from artifacts were averaged for every trial) by BM 6000 (Biomedica Mangoni, Pisa, Italy). Analysis time was 250 ms. The transient PERG response is characterized by a number of waves with three subsequent peaks, of negative, positive, negative polarity, respectively. In normal subjects, these peaks have the following implicit times: 35, 50, and 95 ms (N35, P50, N95).

VEP recordings

Cup-shaped electrodes of Ag/AgCl were fixed with collodion in the following positions: active electrode in Oz, reference electrode in Fpz, and ground in the left arm. Interelectrode resistance was kept below 3000 ohms. The bioelectric signal was amplified (gain 20,000), filtered (band pass 1–100 Hz), and averaged (200 events free from artifacts were averaged for every trial) by BM 6000. Analysis time was 250 ms. The transient VEP response is characterized by a number of waves with three subsequent peaks, of negative, positive, negative polarity, respectively. In normal subjects, these peaks have the following implicit times: 75, 100, and 145 ms (N75, P100, N145).

During a recording session, simultaneous VEPs and PERGs were recorded at least twice (2–6 times) and the resulting waveforms were superimposed to check the repeatability of the results. All control, OHT, and OAG eyes underwent at least two recording sessions, 1–7 days apart, to determine test–retest variability.

In each subject or patient, the SNR of PERG and VEP responses was assessed by measuring a "noise" response, while the subject fixated at an unmodulated field of the same mean luminance as the stimulus. At least two "noise" records of 200 events each were obtained and the resulting grand average was considered for measurement. The peak-to-peak amplitude of this final waveform (i.e., average of at least two replications) was measured in a temporal window corresponding to that at which the response component of interest (i.e., VEP N75-P100, PERG P50-N95) was expected to peak. SNRs for this component were determined by dividing the peak amplitude of the component by the noise in the corresponding temporal window. An electroretinographic noise $<0.1\,\mu V$ (mean: $0.085\,\mu V$, range: 0.065–$0.095\,\mu V$, resulting from the grand average of 400–1200 events) and an evoked potential noise $<0.15\,\mu V$ (mean: $0.093\,\mu V$, range: 0.072–$0.112\,\mu V$, resulting from the grand average of 400–1200 events) were observed in all tested subjects. In all subjects and patients, we accepted VEP and PERG signals with an SNR >2.

Implicit time and peak amplitude of each wave were measured directly on the displayed records by means of a pair of cursors for all VEPs and PERGs. Simultaneous recordings of VEPs and PERGs allow us to derive Retinocortical Time (RCT) as the difference between VEP P100 and PERG P50 peak latencies (Celesia and Kaufmann, 1985).

At baseline, all OAG patients underwent at least two simultaneous recordings of PERG and VEP 1–7 days apart to determine test–retest variability. During the follow-up assessment (60, 180, 240, and 360 days), PERGs and VEPs were performed in OAG patients at least three times and the resulting waveforms were superimposed to check the repeatability of the results. The recording with the highest PERG P50-N95 amplitude was considered in the statistical analysis (see below). During all follow-up examinations, PERG and VEP recordings were performed in a condition of pupil diameter equal to that measured in baseline conditions (see above).

Statistic evaluation of electrophysiological results

Sample size estimates were obtained from pilot evaluations performed in 20 eyes from 20 OAG patients and 20 eyes from 20 control subjects, other than those included in the current study. Interindividual variability, expressed as data standard deviation (SD), was estimated for PERG P50-N95 amplitude and VEP P100 implicit time measurements. It was found that data SDs were significantly higher for patients when compared to controls (about 35% vs. 15%). It was also established that, assuming the above between-subjects SD in the current study, sample sizes of control subjects and patients belonging to OAG group provided a power of 90%, at an alpha = 0.05, for detecting a between-group difference of 55% or greater in PERG P50-N95 amplitude and VEP P100 implicit time measurements. These differences were preliminarily observed by comparing OAG and control data (see above). They were also expected to be clinically meaningful when comparing results of treated OAG eyes observed in baseline conditions versus those observed at 60, 180, 240, and 360 days.

Test–retest data of PERG and VEP results were expressed as the mean difference between two recordings obtained in separate sessions plus/minus the SD of this difference. Ninety-five percent confidence limits of test–retest variability in normal subjects and patients were established assuming a normal distribution.

The differences of PERG and VEP responses between groups (NT-OAG, TO-OAG, and TI-OAG eyes) were evaluated by one-way analysis of variance (ANOVA). Changes in PERG and VEP responses observed in NT-OAG, TO-OAG, and TI-OAG groups with respect to baseline were evaluated by one-way repeated measures ANOVA. The differences observed in individual TO-OAG eyes after observation or citicoline treatment with respect to baseline values were calculated by performing a logarithmic transformation to better approximate a normal distribution. In all analyses, a conservative p value less than 0.01, to compensate for multiple comparisons, was considered as statistically significant.

Electrophysiological (PERG and VEP) responses in OAG patients after the first period of evaluation

Examples of simultaneous recordings of VEP and PERG before and after treatment with oral citicoline are displayed in Fig. 1. Individual changes observed in TO-OAG eyes are shown in Fig. 2. Mean values observed in NT-OAG, TI-OAG, and TO-OAG eyes are presented in Figs. 3 and 4.

A decrease in PERG P50 implicit time, VEP P100 implicit time, and RCT, and an increase in PERG P50-N95 and VEP N75-P100 amplitudes ($p<0.01$) were found in TO-OAG and TI-OAG patients after 60 days of treatment (day 60) with respect to baseline values. TO-OAG and TI-OAG patients displayed shorter PERG P50 implicit time, VEP P100 implicit time, and RCT, and greater PERG and VEP amplitudes with respect to NT-OAG patients ($p<0.01$).

Increased PERG P50 implicit time, VEP P100 implicit time, and RCT, and decreased PERG P50-N95 and VEP N75-P100 amplitudes were found at day 180 with respect to values observed at 60 days. PERG and VEP parameters observed were still shorter (P50 implicit time, P100 implicit time, RCT) and still greater (P50-N95 and N75-P100 amplitudes) with respect to those observed in baseline conditions ($p<0.01$) and with respect to those observed in NT-OAG patients ($p<0.01$).

No changes ($p>0.01$) in PERG and VEP responses were observed in NT-OAG patients after 60 and 180 days with respect to baseline conditions.

Electrophysiological (PERG and VEP) responses in OAG patients after the second period of evaluation

A further decrease in PERG P50 implicit time, VEP P100 implicit time, and RCT, and a further increase in PERG P50-N95 and VEP N75-P100 amplitudes ($p<0.01$) were observed in TO-OAG

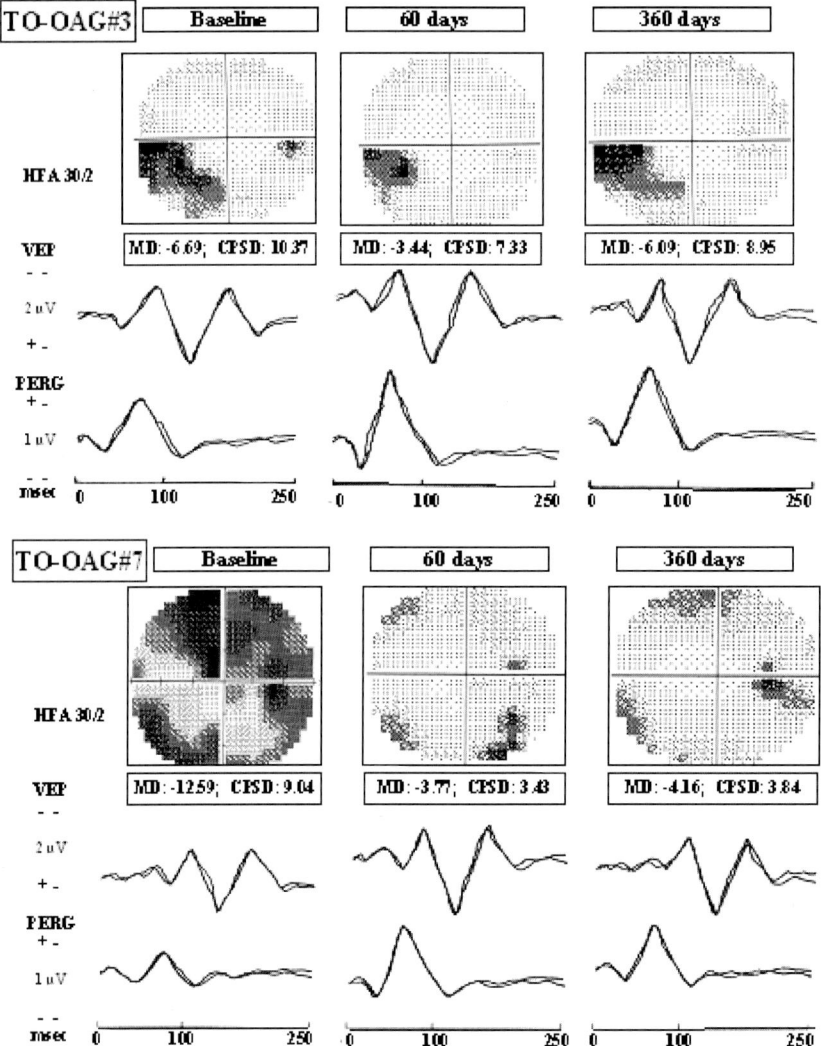

Fig. 1. Examples of visual field (Humphrey 30-2) and layouts of simultaneous visual evoked potentials (VEPs) and pattern electroretinograms (PERGs) recorded in two patients affected by open-angle glaucoma treated with oral citicoline (TO-OAG#3, TO-OAG#7). Electrophysiological examinations were assessed at baseline conditions and 60, 180, 240, and 360 days after medical treatment with oral citicoline. Oral citicoline treatment was performed in two different 60-day periods (0–60 and 181–240 days), each followed by a period of washout (61–180 and 241–360 days). TO-OAG patients showed a decrease in implicit times and an increase in amplitude after the first period of citicoline treatment (60 days) when compared to baseline conditions. At the end of the second period of washout (360 days), one OAG patient (TO-OAG#3) showed electrophysiological and visual field parameters similar to those observed in baseline conditions, while in the other example (TO-OAG#7), the visual field and electrophysiological improvements observed after oral citicoline treatment remained stable even after the washout period.

and TI-OAG patients, at day 240, with respect to values observed at day 180. PERG and VEP implicit times and RCT were still shorter and PERG and VEP amplitudes were still greater when compared to NT-OAG patients ($p < 0.01$). At day 360, TO-OAG and TI-OAG patients showed an increase in VEP and PERG implicit times and in RCT, and a decrease in amplitudes with respect to values observed at 240 days. Implicit time and amplitude values were respectively still shorter and

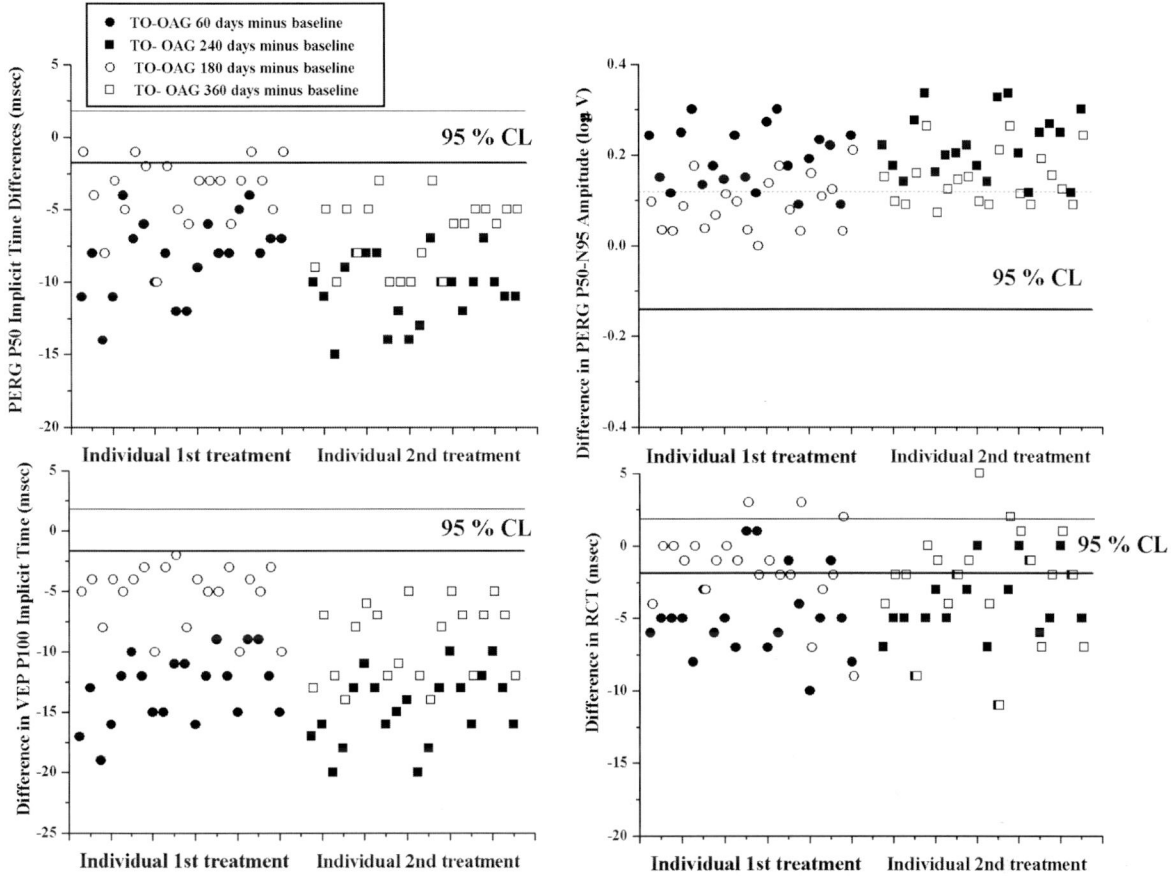

Fig. 2. Individual changes in pattern electroretinogram (PERG), visual evoked potential (VEP) responses, and retinocortical time (difference between VEP P100 and PERG P50 implicit times and RCT) observed in eyes affected by open-angle glaucoma treated with oral citicoline (TO-OAG). Values refer to the difference observed after 60, 180, 240, and 360 days with respect to baseline conditions. Medical treatment with citicoline was performed in TO-OAG eyes over a 60-day period followed by 120 days of washout. At day 180, a second 60-day period of citicoline treatment followed by a second period of 180 days of washout was performed. Solid and dashed lines refer respectively to the upper and lower 95% confidence limit of the intraindividual variability resulting from test–retest analysis.

still greater when compared to baselines ($p<0.01$) and when compared to NT-OAG patients ($p<0.01$).

No changes ($p>0.01$) in PERG and VEP responses were observed in NT-OAG patients, after 240 and 360 days, with respect to baseline conditions.

Considering electrophysiological responses observed after the two different periods of treatment or washout, no significant differences ($p>0.05$) were found between TO-OAG and TI-OAG patients when comparing VEP and PERG changes in implicit times or amplitudes (see Fig. 4).

Adverse side effects were not reported by any of the patients enrolled in the study during the entire period of treatment. Significant changes in IOP were not found in any of the subjects tested.

Effects of citicoline on retinal function in glaucoma patients: neurophysiological implications

Oral or intramuscular treatment with citicoline induces an improvement of retinal bioelectric responses in our glaucomatous patients, as suggested by the increase in amplitudes and shortening in implicit times of PERG recordings.

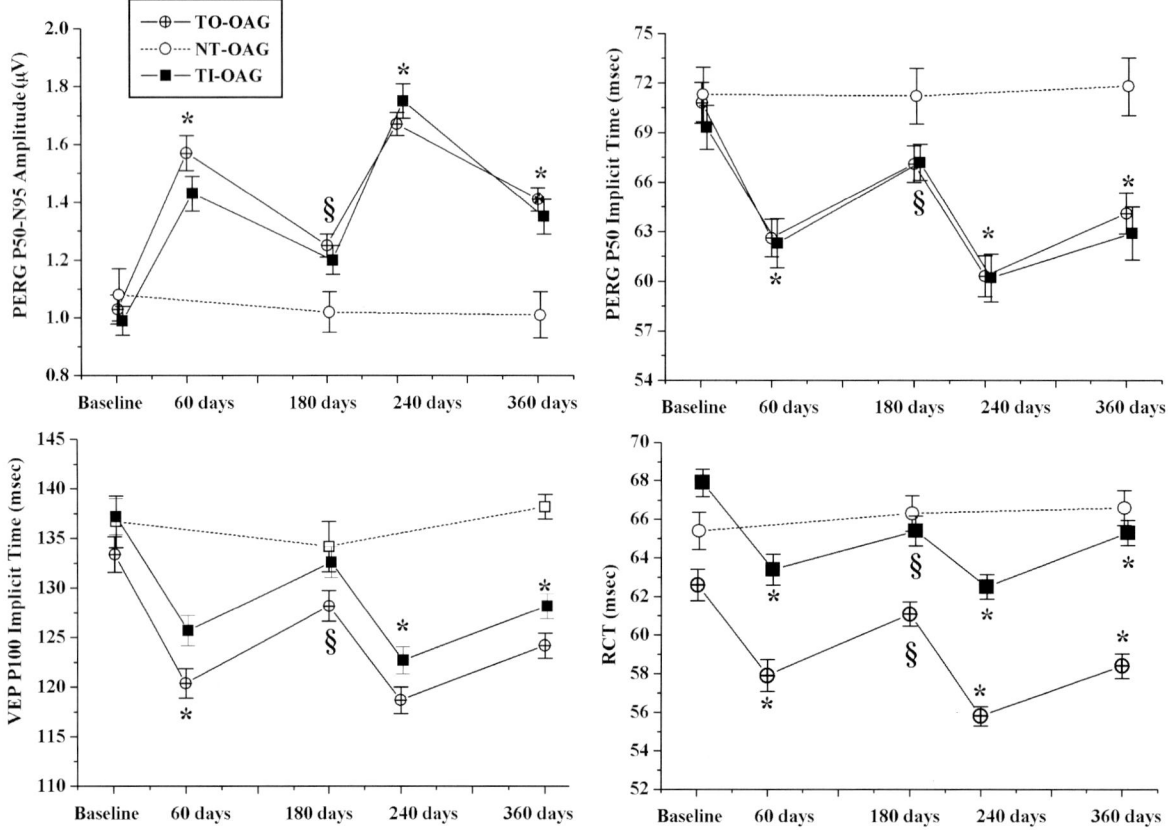

Fig. 3. Graphic representation of mean values of PERG P50 implicit time and P50-N95 amplitude, VEP P100 implicit time, and RCT observed in patients affected by open-angle glaucoma without additional treatments (NT-OAG) and in patients with glaucoma treated with oral (TO-OAG) or intramuscular (TI-OAG) citicoline. Solid lines indicate periods of medical treatment, while dashed lines indicate periods of washout. Vertical lines represent one standard error of the mean. Statistical analysis (ANOVA) evaluating the differences between TO-OAG and TI-OAG groups versus baseline conditions and with respect to NT-OAG group: $^*p<0.01$; $^§p>0.01$.

In our studies we found that there were no differences between oral and intramuscular treatment with citicoline. The results obtained with intramuscular treatment are consistent with those observed in our previous study (Parisi et al., 1999b).

Although in this study, just as in our previous study (Parisi et al., 1999b), our results clearly suggest a positive effect of citicoline in improving the retinal and postretinal glaucomatous dysfunction, the mechanism of action of citicoline in the visual system is only in part understood.

Citicoline is an endogenous substance that acts as intermediary in the synthesis of phosphatidylcholine (a major phospholipid in the neuronal membrane) (Kennedy, 1957; Goracci et al., 1985; Secades and Frontera, 1995; Weiss, 1995) through the activation of the biosynthesis of structural phospholipids in neuronal membranes. Citicoline increases the metabolism of cerebral structures (Secades and Frontera, 1995) and inhibits phospholipid degradation (Weiss, 1995). It may therefore have potential neuroprotective and neuromodulator roles as demonstrated in conditions of cerebral hypoxia and ischemia (Secades and Frontera, 1995; Weiss, 1995) and by the evidence that it induces an increase in the levels of different neurotransmitters and neuromodulators, including noradrenaline, in the central nervous

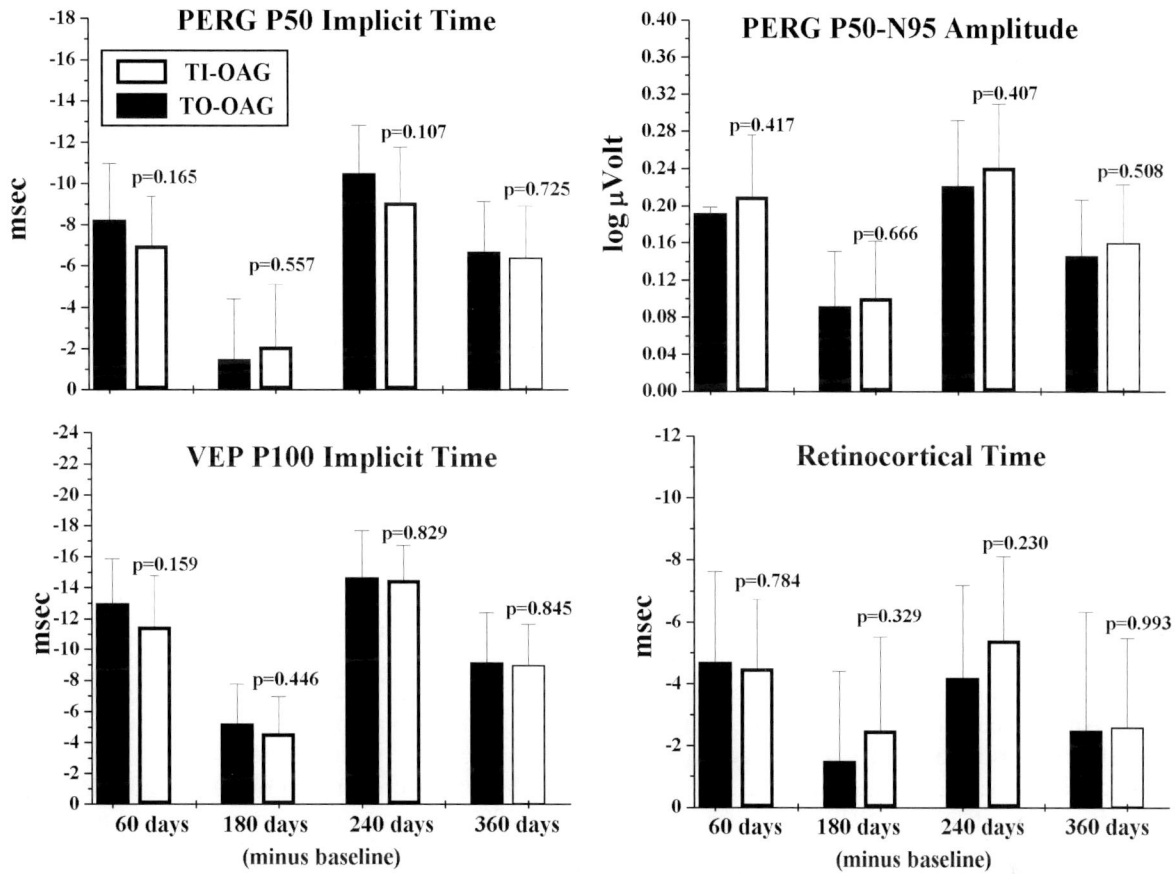

Fig. 4. Bar graphs of the mean differences observed after oral or intramuscular citicoline treatment (60 and 240 days) or after washout (180 and 360 days) with respect to baseline conditions. Means refer to changes in PERG P50 implicit time and P50-N95 amplitude, VEP P100 implicit time, and RCT observed in patients affected by open-angle glaucoma treated with oral (TO-OAG) or intramuscular (TI-OAG) citicoline. Vertical lines represent one standard deviation of the mean. The p value refers to a statistical analysis (ANOVA) evaluating differences between TI-OAG and TO-OAG groups.

system. In addition, several studies suggest that citicoline successfully increases the level of consciousness in different brain disorders ascribed to vascular, traumatic, or degenerative processes (Boismare et al., 1978; Serra et al., 1981; Agnoli et al., 1985; Zappia et al., 1985; Kakihana et al., 1988; Cacabelos et al., 1996).

Previous studies report that treatment with citicoline may induce an improvement of the glaucomatous visual field defects (Pecori-Giraldi et al., 1989; Virno et al., 2000). In our study, the effects of citicoline on visual field sensitivity were purposely not considered because any detectable improvement could be ascribed to the associated effects on consciousness level and attention (Boismare et al., 1978; Serra et al., 1981; Agnoli et al., 1985; Zappia et al., 1985; Kakihana et al., 1988; Cacabelos et al., 1996).

A dopaminergic-like activity of citicoline may also be involved in the improvement of PERG responses after oral treatment; in fact, levodopa was found to increase retinal function in humans treated with this substance (Gottlob et al., 1989), and our results could therefore be explained by a similar neuromodulator activity.

In our study, we did not perform any morphological examination, and thus, although our results indicate that oral citicoline improves bioelectrical retinal activity, we were not able to demonstrate whether there were other effects on retinal fibers (i.e., an increase in retinal nerve fiber layer thickness).

Effects of citicoline on neural conduction along the visual pathways in glaucoma patients: neurophysiological implications

Oral or intramuscular treatment with citicoline induced an improvement of visual cortical bioelectric responses (VEPs with increased amplitudes and shortened times to peak) and an improvement of the index of neural conduction in postretinal visual pathways (reduced RCT) in our glaucomatous patients. This is in agreement with a similar previous study (Rejdak et al., 2003).

VEP abnormalities observed in patients with glaucoma have recently been ascribed to impaired neural conduction along postretinal visual pathways related to a dysfunction of the innermost retinal layers (ganglion cells and their fibers) (Parisi, 2001). An independent effect on neural conduction in postretinal visual pathways or in visual cortical cells could also be hypothesized in order to explain the influence of citicoline in VEP responses.

The effects of citicoline in the neural visual system were revealed by the improvement of visual acuity (Campos et al., 1995; Porciatti et al., 1998), VEP responses, and contrast sensitivity (Porciatti et al., 1998) in amblyopic subjects after treatment with this substance. Because similar results were obtained in amblyopic subjects after treatment with levodopa (Gottlob and Stangler-Zuschrott, 1990; Gottlob et al., 1992; Leguire et al., 1993) and studies performed in patients with Parkinson's disease recommended the use of citicoline as a complement to levodopa therapy (Birbamer et al., 1990), a dopaminergic-like activity could once again be suggested to explain VEP and RCT results after treatment with citicoline.

Because there are no clear or conclusive experimental or published data to support the hypothesis of an independent effect on neural conduction in postretinal visual pathways, the changes in neural conduction along the visual pathways following citicoline treatment could be dependent on the reduced dysfunction of the innermost retinal layers.

Possibility of neuroprotective role of citicoline in glaucoma patients

Our results advocate an interesting question: can oral or intramuscular citicoline effects be considered as "neuroprotective," preventing the development of the glaucomatous disease?

For citicoline effects to be considered "neuroprotective," glaucomatous visual defects should undergo stabilization or improvement long after treatment termination.

When considering that after the first period of washout there were no differences with respect to pretreatment conditions, this leads to the conclusion that one cycle of treatment with citicoline is not sufficient to induce changes in the natural history of glaucoma. On the other hand, we observed that the second treatment period with oral citicoline was able to induce an improvement, which persisted after 120 days of washout.

These observations suggest that the improvement in retinal function and postretinal neural conduction along the visual pathways could be considered as "treatment-dependent."

The results obtained in our first study (Parisi et al., 1999b) suggested us to perform in a restrict cohort of selected patients (12 OAG patients only) a series of 60-day periods of treatment each followed by 120 days of washout during a total period of 8 years (Parisi, 2005). This study showed that after 8 years, glaucomatous patients subjected to citicoline treatment displayed a stable or improved electrophysiological and visual field impairment with respect to pretreatment conditions (8 years before), while in similar glaucomatous patients not subjected to citicoline treatment, there was a worsening of the electrophysiological and visual field impairment with respect to pretreatment conditions (8 years before). The results of this study are summarized in Fig. 5.

Indeed, the data observed in our OAG patients treated with beta-blockers plus several periods of treatment with intramuscular citicoline with respect to those found in OAG patients treated with beta-blockers only may suggest the potential use of citicoline in order to obtain the stabilization or improvement of the glaucomatous visual dysfunction.

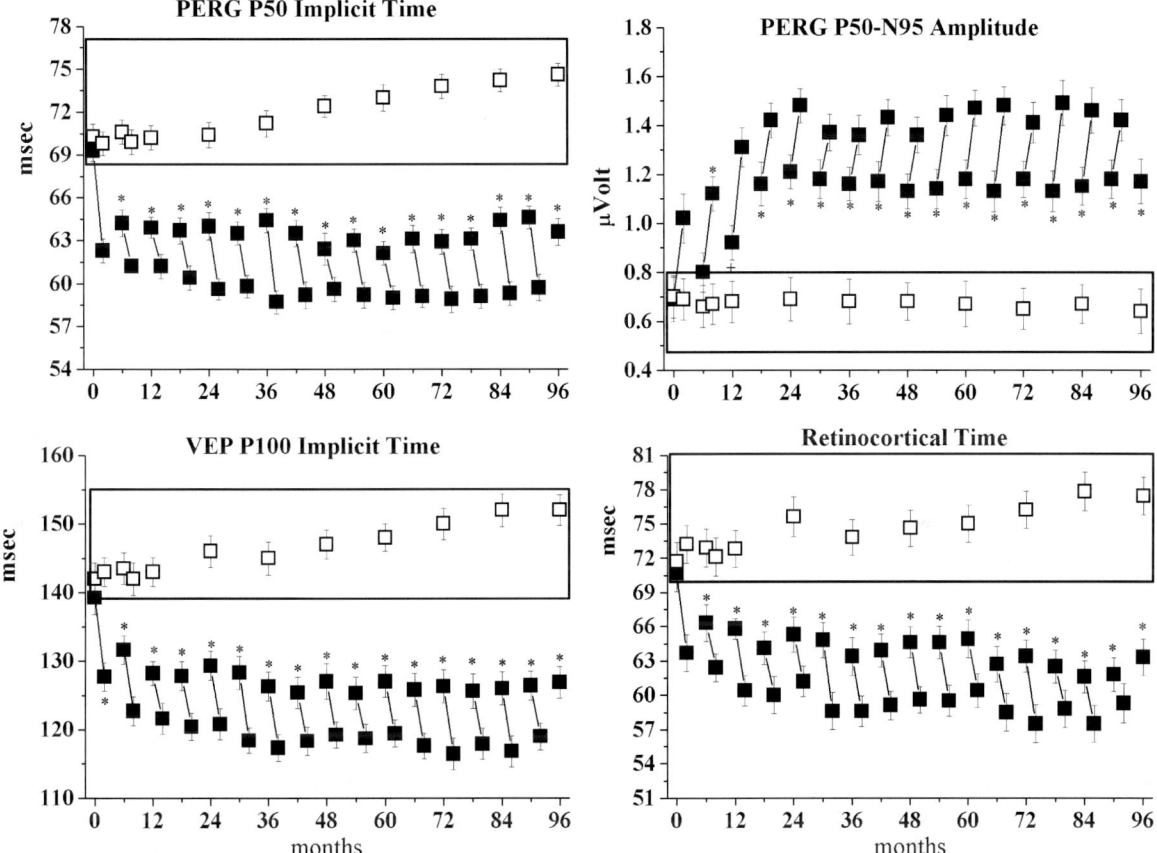

Fig. 5. Graphic representation of mean values of PERG P50 implicit time, PERG P50-N95 amplitude, VEP P100 implicit time, and retinocortical time (difference between VEP P100 and PERG P50 implicit times, an electrophysiological index of neural conduction along postretinal visual pathways) observed in glaucoma patients at baseline conditions and without additional treatment (hollow bar) or treated with intramuscular citicoline (solid bar). Medical treatment with citicoline was performed over several 2-month periods followed by several 4-month periods of washout during an entire period of 96 months. The solid line within solid bars indicates periods of treatment. The absence of a line within a solid bar indicates washout periods. Vertical lines represent one standard error of the mean. We concentrated on the long-term effects of citicoline treatment by comparing the differences observed at the end of each period of washout with respect to baseline conditions through ANOVA. *$p < 0.01$ versus baseline and untreated glaucoma patients. Adapted with permission from Parisi, 2005 (Figs. 2–5, pp. 98–99).

These observations suggest that citicoline could be considered a neuroprotective drug. Our study showed that the neuroprotective effects of citicoline may be achieved in a restricted cohort of patients (see inclusion criteria, Parisi et al., 2006) and must be considered "treatment-dependent," leading to the possibility that repeated treatments may determine possible neuroprotective effects.

In agreement with the reported observations of similar studies (Kennedy, 1957; Campos et al., 1995; Secades and Frontera, 1995; Weiss, 1995; Porciatti et al., 1998; Parisi et al., 1999b; Parisi, 2005), an important aspect of this study is the lack of adverse pharmacological side effects in all participating subjects, even after long-term administration of the drug.

Conclusive remarks

The results provided by our studies suggest that oral and intramuscular citicoline treatment significantly improves retinal and cortical responses in glaucoma patients. This indicates the potential use of this substance in the medical treatment of glaucoma, as a complement to hypotensive therapy, with a possible direct neuroprotective effect.

Abbreviations

CPSD	Corrected Pattern Standard Deviation
ERG	electroretinogram
HFA	Humphrey Field Analyzer
IOP	intraocular pressure
IRB	Institutional Review Board
LGN	lateral geniculate nucleus
MD	mean deviation
NT-OAG	not treated open-angle glaucoma
OAG	open-angle glaucoma
OHT	ocular hypertension
PERG	pattern electroretinogram
RCGs	retinal ganglion cells
RCT	retinocortical time
SD	standard deviation
SNR	signal-to-noise ratio
TI-OAG	glaucoma patients treated with intramuscular citicoline
TO-OAG	glaucoma patients treated with oral citicoline
VEPs	visual evoked potentials

References

Agnoli, A., Fioravanti, M. and Lechner, H. (1985) Efficacy of CDP–choline in chronic cerebral vascular diseases (CCVD). In: Novel Biochemical, Pharmacological and Clinical Aspects of Cytidinediphosphocholine. *Proceedings of the International Meeting*, Sorrento, Italy, 1984. New York: Elsevier, pp. 305–308.

Armington, J.C. (1974) The Electroretinogram. Academic Press, New York.

Bach, M. (2001) Electrophysiological approaches for early detection of glaucoma. Eur. J. Ophthalmol., 11(Suppl 2): S41–S49.

Birbamer, G., Gesterbrand, E., Rainer, J. and Eberhardt, R. (1990) CDP-choline in the treatment of Parkinson's disease. New Trends Clin. Pharmacol., 4: 1–6.

Boismare, F., Le Poncin, M., Lefrancois, J. and Lecordier, J.C. (1978) Action of cytidine diphosphocholine on functional and hemodynamic effects of cerebral ischemia in cats. Pharmacology, 17: 15–20.

Cacabelos, R., Caamano, J., Gomez, M.J., Fernandez-Novoa, L., Franco-Maside, A. and Alvarez, X.A. (1996) Therapeutic effects of CDP-choline in Alzheimer's disease. Cognition, brain mapping, cerebrovascular hemodynamics, and immune factors. Ann. N. Y. Acad. Sci., 777: 399–403.

Campos, E.C., Schiavi, C., Benedetti, P., Bolzani, R. and Porciatti, V. (1995) Effect of citicoline on visual acuity in amblyopia: preliminary results. Graefes. Arch. Clin. Exp. Ophthalmol., 233: 307–312.

Celesia, G.C. and Kaufmann, D. (1985) Pattern ERG and visual evoked potentials in maculopathies and optic nerve disease. Invest. Ophthalmol. Vis. Sci., 26: 726–735.

Celesia, G.G., Bodis-Wollner, I., Chatrian, G.E., Harding, G.F., Sokol, S. and Spekreijse, H. (1993) Recommended standards for electroretinograms and visual evoked potentials. Report of an IFCN committee. Electroencephalogr. Clin. Neurophysiol., 87: 421–436.

Chaturvedi, N., Hedley-Whyte, E.T. and Dreyer, E.B. (1993) Lateral geniculate nucleus in glaucoma. Am. J. Ophthalmol., 116: 182–188.

Colotto, A., Salgarello, T., Giudiceandrea, A., De Luca, L.A., Coppè, A., Buzzonetti, L. and Falsini, B. (1995) Pattern electroretinogram in treated ocular hypertension: a cross-sectional study after timolol maleate therapy. Ophthalmic Res., 27: 168–177.

Fiorentini, A., Maffei, L., Pirchio, M., Spinelli, D. and Porciatti, V. (1981) The ERG in response to alternating gratings in patients with diseases of the peripheral visual pathway. Invest. Ophthalmol. Vis. Sci., 21: 490–493.

Garway-Heath, D.F., Holder, G.E., Fitzke, F.W. and Hitchings, R.A. (2002) Relationship between electrophysiological, psychophysical, and anatomical measurements in glaucoma. Invest. Ophthalmol. Vis. Sci., 43: 2213–2220.

Goracci, G., Francescangeli, E., Mozzi, R., Porcellati, S. and Porcellati, G. (1985) Regulation of phospholipid metabolism by nucleotides in brain and transport of CDP-choline into brain. In: Zappia V., Kennedy E.P., Nilsson B.I. and Galletti P. (Eds.), Novel Biochemical, Pharmacological and Clinical Aspects of Cytidinediphosphocholine. Elsevier, New York, pp. 105–109.

Gottlob, I., Charlier, J. and Reinecke, R.D. (1992) Visual acuities and scotomas after one week levodopa administration in human amblyopia. Invest. Ophthalmol. Vis. Sci., 33: 2722–2728.

Gottlob, I. and Stangler-Zuschrott, E. (1990) Effect of levodopa on contrast sensitivity and scotomas in human amblyopia. Invest. Ophthalmol. Vis. Sci., 31: 776–780.

Gottlob, I., Weghaupt, H., Vass, C. and Auff, E. (1989) Effect of levodopa on the human pattern electroretinogram and pattern visual evoked potentials. Graefes. Arch. Clin. Exp. Ophthalmol., 227: 421–427.

Graham, S.L., Drance, S.M., Chauhan, B.C., Swindale, N.V., Hnik, P., Mikelberg, F.S. and Douglas, G.R. (1996) Comparison of psychophysical and electrophysiological testing in early glaucoma. Invest. Ophthalmol. Vis. Sci., 37: 2651–2662.

Greenstein, V.C., Seliger, S., Zemon, V. and Ritch, R. (1998) Visual evoked potential assessment of the effects of glaucoma on visual subsystems. Vision Res., 38: 1901–1911.

Gupta, N. and Yucel, Y.H. (2003) Brain changes in glaucoma. Eur. J. Ophthalmol., 13(Suppl 3): S32–S35.

Gur, M., Zeevi, Y.Y., Bielik, M. and Neumann, E. (1987) Changes in the oscillatory potentials of the electroretinogram in glaucoma. Curr. Eye Res., 6: 457–466.

Holder, G.E. (1997) The pattern electroretinogram in anterior visual pathway dysfunction and its relationship to the pattern visual evoked potential: a personal clinical review of 743 eyes. Eye, 11: 924–934.

Horn, F.K., Jonas, J.B., Budde, W.M., Junemann, A.M., Mardin, C.Y. and Korth, M. (2002) Monitoring glaucoma progression with visual evoked potentials of the blue-sensitive pathway. Invest. Ophthalmol. Vis. Sci., 43: 1828–1834.

Kakihana, M., Fukuda, N., Suno, M. and Nagaoka, A. (1988) Effects of CDP-choline on neurologic deficits and cerebral glucose metabolism in a rat model of cerebral ischemia. Stroke, 19: 217–222.

Kennedy, E.P. (1957) Biosynthesis of phospholipids. Fed. Proc., 16: 847–853.

Klistorner, A.I. and Graham, S.L. (1999) Early magnocellular loss in glaucoma demonstrated using the pseudorandomly stimulated flash visual evoked potential. J. Glaucoma, 8: 140–148.

Leguire, L.E., Rogers, G.L., Bremer, D.L., Walson, P.D. and McGregor, M.L. (1993) Levodopa/carbidopa for childhood amblyopia. Invest. Ophthalmol. Vis. Sci., 34: 3090–3095.

Maffei, L. and Fiorentini, A. (1982) Electroretinographic responses to alternating gratings in the cat. Exp. Brain Res., 48: 327–334.

Martus, P., Korth, M., Horn, F., Junemann, A., Wisse, M. and Jonas, J.B. (1998) A multivariate sensory model in glaucoma diagnosis. Invest. Ophthalmol. Vis. Sci., 39: 1567–1574.

Parisi, V. (1997) Neural conduction in the visual pathways in ocular hypertension and glaucoma. Graefes. Arch. Clin. Exp. Ophthalmol., 235: 136–142.

Parisi, V. (2001) Impaired visual function in glaucoma. Clin. Neurophysiol., 112: 351–358.

Parisi, V. (2003) Correlation between morphological and functional retinal impairment in patients affected by ocular hypertension, glaucoma, demyelinating optic neuritis and Alzheimer's disease. Semin. Ophthalmol., 18: 50–57.

Parisi, V. (2005) Electrophysiological assessment of glaucomatous visual dysfunction during treatment with cytidine-5′-diphosphocholine (citicoline): a study of 8 years of follow-up. Doc. Ophthalmol., 110: 91–102.

Parisi, V. and Bucci, M.G. (1992) Visual evoked potentials after photostress in patients with primary open-angle glaucoma and ocular hypertension. Invest. Ophthalmol. Vis. Sci., 33: 436–442.

Parisi, V., Colacino, G., Milazzo, G., Scuderi, A.C. and Manni, G. (1999a) Effects of nicergoline on the retinal and cortical electrophysiological responses in glaucoma patients: a preliminary open study. Pharmacol. Res., 40: 249–255.

Parisi, V., Manni, G., Centofanti, M., Gandolfi, S.A., Olzi, D. and Bucci, M.G. (2001) Correlation between optical coherence tomography, pattern electroretinogram, and visual evoked potentials in open-angle glaucoma patients. Ophthalmology, 108: 905–912.

Parisi, V., Manni, G., Colacino, G. and Bucci, M.G. (1999b) Cytidine-5′-diphosphocholine (citicoline) improves retinal and cortical responses in patients with glaucoma. Ophthalmology, 106: 1126–1134.

Parisi, V., Manni, G., Gandolfi, S.A., Centofanti, M., Colacino, G. and Bucci, M.G. (1999c) Visual function correlates with nerve fiber layer thickness in eyes affected by ocular hypertension. Invest. Ophthalmol. Vis. Sci., 40: 1828–1833.

Parisi, V., Manni, G., Spadaro, M., Colacino, G., Restuccia, R., Marchi, S., Bucci, M.G. and Pierelli, F. (1999d) Correlation between morphological and functional retinal impairment in multiple sclerosis patients. Invest. Ophthalmol. Vis. Sci., 40: 2520–2527.

Parisi, V., Miglior, S., Manni, G., Centofanti, M. and Bucci, M.G. (2006) Clinical ability of pattern electroretinograms and visual evoked potentials in detecting visual dysfunction in ocular hypertension and glaucoma. Ophthalmology, 113: 216–228.

Parisi, V., Pernini, C., Guinetti, C., Neuschuler, R. and Bucci, M.G. (1997) Electrophysiological assessment of visual pathways in glaucoma. Eur. J. Ophthalmol., 7: 229–235.

Pecori-Giraldi, J., Virno, M., Covelli, G., Grechi, G. and De Gregorio, F. (1989) Therapeutic value of citicoline in the treatment of glaucoma (computerized and automated perimetric investigation). Int. Ophthalmol., 13: 109–112.

Porciatti, V., Falsini, B., Brunori, S., Colotto, A. and Moretti, G. (1987) Pattern electroretinogram as a function of spatial frequency in ocular hypertension and early glaucoma. Doc. Ophthalmol., 65: 349–355.

Porciatti, V., Schiavi, C., Benedetti, P., Baldi, A. and Campos, E.C. (1998) Cytidine-5′-diphosphocholine improves visual acuity, contrast sensitivity and visually-evoked potentials of amblyopic subjects. Curr. Eye Res., 17: 141–148.

Rejdak, R., Toczolowski, J., Kurkowski, J., Kaminski, M.L., Rejdak, K., Stelmasiak, Z. and Grieb, P. (2003) Oral citicoline treatment improves visual pathway function in glaucoma. Med. Sci. Monit., 9: PI24–PI28.

Salgarello, T., Colotto, A., Falsini, B., Buzzonetti, L., Cesari, L., Iarossi, G. and Scullica, L. (1999) Correlation of pattern electroretinogram with optic disc cup shape in ocular hypertension. Invest. Ophthalmol. Vis. Sci., 40: 1989–1997.

Secades, J.J. and Frontera, G. (1995) CDP-choline: pharmacological and clinical review. Methods Find Exp. Clin. Pharmacol., 17(Suppl B): 1–54.

Serra, I., Alberghina, M., Viola, M., Ristretta, A. and Giuffrida, A.M. (1981) Effect of CDP-choline on the biosynthesis of nucleic acids and proteins in brain regions during hypoxia. Neurochem. Res., 6: 607–618.

Vaegan, G.S.L., Goldberg, I., Buckland, L. and Hollows, F.C. (1995) Flash and pattern electroretinogram changes with optic atrophy and glaucoma. Exp. Eye Res., 60: 697–706.

Ventura, L.M. and Porciatti, V. (2005) Restoration of retinal ganglion cell function in early glaucoma after intraocular pressure reduction: a pilot study. Ophthalmology, 112: 20–27.

Ventura, L.M., Porciatti, V., Ishida, K., Feuer, W.J. and Parrish, R.K., 2nd (2005) Pattern electroretinogram abnormality and glaucoma. Ophthalmology, 112: 10–19.

Virno, M., Pecori-Giraldi, J., Liguori, A. and De Gregorio, F. (2000) The protective effect of citicoline on the progression of the perimetric defects in glaucomatous patients (perimetric study with a 10-year follow-up). Acta Ophthalmol. Scand., 232(Suppl): 56–57.

Watanabe, I., Iijima, H. and Tsukahara, S. (1989) The pattern electroretinogram in glaucoma: an evaluation by relative amplitude from the Bjerrum area. Br. J. Ophthalmol., 73: 131–135.

Weber, A.J., Chen, H., Hubbard, W.C. and Kaufman, P.L. (2000) Experimental glaucoma and cell size, density, and number in the primate lateral geniculate nucleus. Invest. Ophthalmol. Vis. Sci., 41: 1370–1379.

Weiss, G.B. (1995) Metabolism and actions of CDP-choline as an endogenous compound and administered exogenously as citicoline. Life Sci., 56: 637–660.

Yucel, Y.H., Zhang, Q., Gupta, N., Kaufman, P.L. and Weinreb, R.N. (2000) Loss of neurons in magnocellular and parvocellular layers of the lateral geniculate nucleus in glaucoma. Arch. Ophthalmol., 118: 378–384.

Yucel, Y.H., Zhang, Q., Weinreb, R.N., Kaufman, P.L. and Gupta, N. (2003) Effects of retinal ganglion cell loss on magno-, parvo-, koniocellular pathways in the lateral geniculate nucleus and visual cortex in glaucoma. Prog. Retin. Eye Res., 22: 465–481.

Zappia V., Kennedy P., Nilsson B.I. and Galletti P. (Eds.), (1985). Novel Biochemical, Pharmacological and Clinical Aspects of Cytidine-diphosphocholine. Elsevier, New York.

CHAPTER 38

Neuroprotection: VEGF, IL-6, and clusterin: the dark side of the moon

S. Pucci[1,*,a], P. Mazzarelli[1], F. Missiroli[2], F. Regine[2] and F. Ricci[2,**,a]

[1]*Department of Biopathology, Institute of Anatomic Pathology, University of Rome "Tor Vergata", 00133 Rome, Italy*
[2]*Section of Ophthalmology, Department of Biopathology, University of Rome "Tor Vergata", 00133 Rome, Italy*

Abstract: Growth factors and their respective receptors are key regulators in development and homeostasis of the nervous system, and changes in the function, expression, or downstream signaling of growth factors are involved in many neuropathological disorders.

Recently, research has yielded a rich harvest of information about molecules and gene, and currently the assumption "a gene–a protein", where each gene encodes the structure of a single protein, is becoming a paradox. In the past years, the discovery of synergic or antagonistic proteins deriving from the same gene is a novelty upsetting. In some way, the conventional function of proteins involved in DNA repair, cell death/growth induction, vascularization, and metabolism is inhibited or shifted toward other pathways by soluble mediators that orchestrate such change depending on the microenvironment conditions. In this chapter, we focus on the antithetic properties that proteins could exert, depending on the microenvironment that orchestrates the complex networks among proteins and their respective partners.

Keywords: VEGF; interleukin-6; clusterin; glaucoma

Neuroprotection: VEGF-A, a shared growth factor

Recently, the characterization of the molecular interactions among soluble factors in the microenvironment points out the action of some neuronal growth factors on the neovascular system (i.e., NGF, nerve growth factor; BDNF, brain-derived neurotrophic factor). In the same way, some vascular endothelial growth factors (VEGFs) first identified as endothelial specific mitogens exert a specific action on neuronal cells. The vascular and nervous systems are functionally and structurally different but they share some multiple similarities in the development, construction, and function. During development, the gradient of growth factors leads the long-distance final target. This chapter aims to point out the growing field of common growth factors, receptors, and inflammatory mediators within different networks that cooperate in the neuronal survival in stress conditions. The knowledge of the molecular interactions of these shared growth factors, pleiotropic but specific for the vascular and neural cells, is relevant for therapies addressed to the vascular and neurological diseases.

*Corresponding author. Tel.: +39-06-20903953; Fax: +39-06-20902209; E-mail: sabinapuc@yahoo.it
*Co-Corresponding author. Tel.: +39-06-23188334; Fax: +39-06-20902209; E-mail: federico.ricci@uniroma2.it

[a]These authors contributed equally to this work.

DOI: 10.1016/S0079-6123(08)01138-2

Development and homeostasis of vascular and neuronal systems were believed to be controlled by different, specialized growth factors and their receptors that, in some cases, are the key regulatory steps by their own. Actually, different studies have shed light on the shared overlapping repertoire of growth factors that affect the development and homeostasis of the nervous and vascular systems. This suggests the cooperative evolution of molecular mechanisms to control path finding, spatial patterning proliferation, and survival. These include members of VEGF, ephrin, and neutrophin growth factor families. One prominent member among these is vascular endothelial growth factor-A (VEGF-A), an important multifunctional molecule with several important biological activities that depend on both the stage of development and physiological function of the organ in which it is expressed. The VEGF family comprehends six different homologous factors: VEGF-A–E and placenta growth factor (PlGF). All the components of this family are involved in the development of the vasculature and display multiple functions on endothelial cells; in particular, VEGF-A–C exert their specific action on neuronal cells. VEGF-A was first identified as a homodimeric protein of 32–42 kDa that increased vascular permeability in the skin. The observation that tumor growth can be accompanied by increased vascularity was reported more than a century ago (see Ferrara, 2002). It was identified in the supernatant of a guinea pig tumor cell line and proposed as specific mediator of the hyperpermeability of tumor blood vessels, involved also in the formation of tumor-associated ascites. This protein-enhancing vascular permeability was named vascular permeability factor (VPF) (Senger et al., 1983). Further characterization of the bioactivity of this growth factor reveals that it exerts mitogenic effects also on vascular endothelial cells isolated from several districts. Due to its apparent target cell selectivity, this factor was renamed vascular endothelial growth factor (VEGF). By the end of 1989, Ferrara and coworkers reported the isolation of cDNA clones for bovine VEGF164 and three human VEGF isoforms: VEGF121, VEGF165, and VEGF189 (Leung et al., 1989). Additional reports showed the high conservation of VEGF across species, with approximately 85% homology between human and rat VEGF underlying the crucial role of this factor in the bioevolution (Conn et al., 1990).

VEGF-A isoforms

The human VEGF-A gene is organized in eight exons, separated by seven introns, and is localized in chromosome 6p21.3 (Vincenti et al., 1996). Alternative splicing of the human VEGF-A gene gives rise to at least six different transcripts, encoding isoforms of 121-, 145-, 165-, 183-, 189-, and 206-amino-acid residues (Fig. 1; Tischer et al., 1991). Multiple protein forms are encoded through alternative exon splicing. All transcripts contain exons 1–5, which codify for the signal sequence and core VEGF binding or VEGF/PDGF homology domain, and exon 8, with diversity generated through the alternative splicing of exons 6 and 7. Exon 6 encodes a heparin-binding domain, whereas exons 7 and 8 encode a domain that mediates binding to neuropilin-1 (NP1) and heparin exons (Senger et al., 1983; Leung et al., 1989). Several additional minor splice variants also have been described including VEGF145 (Poltorak et al., 1997), VEGF162 (Lange et al., 2003), and VEGF165b, a variant reported to have an antagonistic effect on VEGF165a-induced mitogenesis (Bates et al., 2002). Recently identified VEGF165b displays different activities in respect to its isoform VEGF165a; it is not mitogenic and does not increase proliferation, but its functions are still not well characterized. Human VEGF-A165, the most abundant and biologically active form, and VEGF-A121 are secreted as covalently linked homodimeric proteins, whereas the larger isoforms VEGF-A189 and VEGF-A206 are not readily diffusible and may remain sequestered in the extracellular matrix. Similar to VEGF165, VEGF121 that lacks heparin-binding properties is a freely diffusible protein (Plouet et al., 1989). VEGF189 and VEGF206 bind to heparin with particularly high affinity due to the presence of a highly basic 24-amino-acid insertion, and they are almost completely sequestered in the extracellular

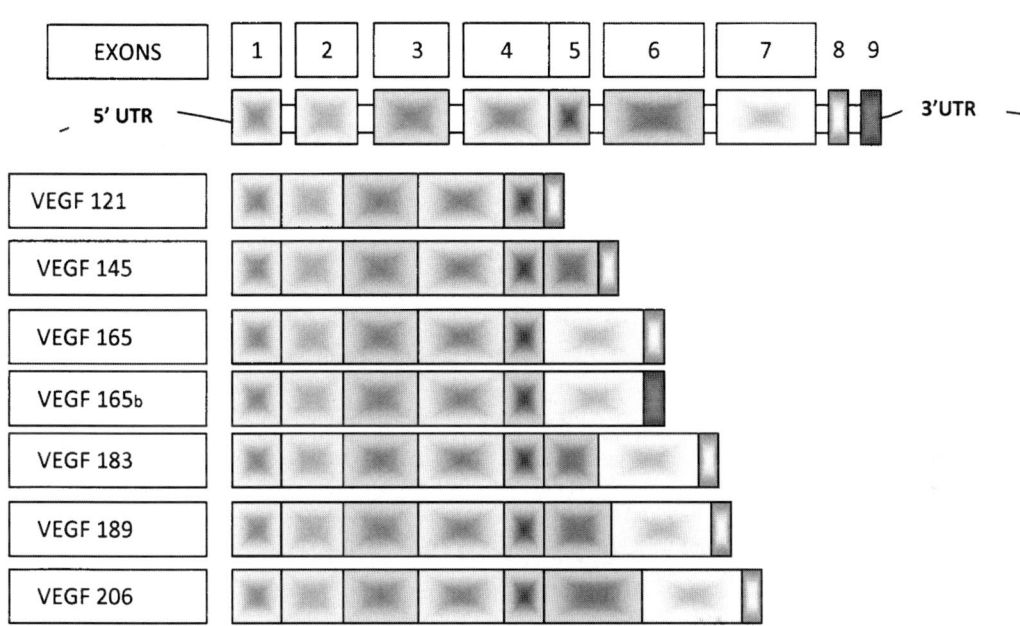

Fig. 1. VEGF-A isoforms. There are six different isoforms of VEGF-A arising by alternative exon splicing. VEGF165 induces and VEGF165b inhibits angiogenesis. All isoforms contain exons 1–5.

matrix (Houck et al., 1992; Park et al., 1993). Not only alternative RNA splicing but also extracellular proteolysis regulates the activity of VEGF. Proteolytic mechanisms are known to be important during pathologic angiogenesis, suggesting that non–heparin-binding cleaved forms of VEGF-A may play an especially significant role (Pepper, 2001). The extracellular matrix–bound isoforms may be released in a diffusible form by plasmin cleavage at the C terminus, which generates a bioactive fragment (VEGF110), consisting of the first 110 N-terminal amino acids. VEGF165 may also undergo a similar processing by plasmin (Houck et al., 1992). The C-terminal domain (111–165) of VEGF is critical for its mitogenic potency. More recent studies have shown that various matrix metalloproteases (MMPs), especially MMP-3, can also cleave VEGF165 to generate diffusible, non–heparin-binding fragments. Proteolytic processing of VEGF165 by MMP-3 takes place in sequential steps, with initial cleavage occurring at residues VEGF135, VEGF120, and, finally, VEGF113 (Lee et al., 2005). Thus, the final product of MMP-3 processing, VEGF113, is expected to be biologically and biochemically very similar to the VEGF110, the plasmin-generated fragment. According to recent studies, levels of plasminogen and MMP-3 in the vitreous of patients affected by proliferative retinopathies are higher than those in controls, providing evidence for a prodegradative environment that may locally generate non–heparin-binding VEGF fragments.

VEGF-A receptors

The transduction pathway of VEGF-A is exploited by the binding of VEGF to its receptors. VEGF-A isoforms VEGF121, VEGF165a and VEGF165b, VEGF189, and VEGF206 are the products of alternative splicing where VEGF165 is the prevalent one. The differential splicing is influenced by the external growth factors and environmental conditions that guide the production of the predominant form needed in a particular district. The molecular factors that could influence this posttranscriptional modification that generates preferentially a specific isoform still need to be investigated. Every VEGF isoform displays a particular affinity for a class of receptors, which in turn activate preferentially a cascade of transduction pathways that lead to proliferative impulse, anti–cell death signals, and survival. The biological functions of VEGF are mediated via tyrosine kinase receptors, VEGF receptor-1 (Ftl-1), and VEGF receptor-2 (KDR/Flk-1). Additionally, certain VEGF isoforms bind to NPs, non-tyrosine transmembrane receptors (Fig. 2A).

In particular, NP1 is a non-tyrosine kinase receptor for VEGF165, the heparin-binding PlGF-2 isoform, VEGF-B, and VEGF-E. NP1 was first identified as a receptor for semaphorin-3A, a member of a family of polypeptides involved in axonal guidance and patterning; moreover, it is

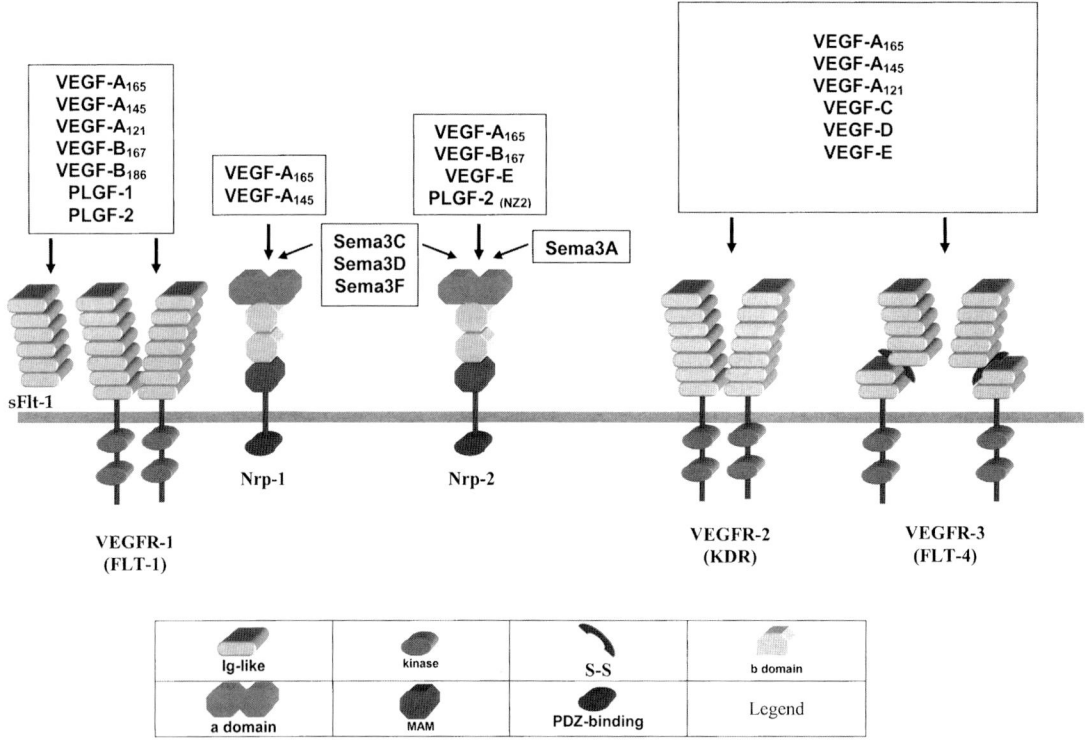

Fig. 2. (A) VEGF ligands and their receptors. VEGFR-1 (Flt-1) binds VEGF-A165, VEGF-A145, VEGF-A121, VEGF-B186, VEGF-B167, PLGF-1, and PLGF-2 (human isoforms). The extracellular domain of VEGFR-1 is independently expressed as a soluble protein (sFlt-1 or sVEGFR-1), with predicted ligand specificity identical to that of the complete receptor. VEGFR-2 (KDR) is a receptor for VEGF-A165, VEGF-A145, VEGF-A121, the processed forms of VEGF-C and VEGF-D, and the viral VEGF-Es. Unprocessed VEGF-C and VEGF-D bind VEGFR-3 (Flt-4) with higher affinity. NP1, a non-tyrosine kinase receptor for semaphorin 3A (Sema-3A), is also a co-receptor for VEGF-A165, PLGF-2, VEGF-B167, and VEGF-E(NZ2). NP2 binds only VEGF-A165 and VEGF-A145. (B) Neuronal receptors for VEGF and downstream signaling mechanisms. The major receptors for VEGF in neuronal cells are NP1 and VEGFR-2/KDR. VEGF-A165 also competes with Sema-3A for binding to NP1. (Left) Associations between NP1 and the transmembrane proteins, plexin A1/2 and L1, are required for Sema-3A-induced growth cone collapse. L1 binds to both actin and ankyrin, which associates with spectrin and the associated actin cytoskeleton. Recruitment of the small Rho-like GTPase, Rnd, by the plexin A1 intracellular domain is sufficient to induce growth cone collapse and this is antagonized by RhoD. NP1 also associates with the PDZ domain protein, synectin/NIP-1/GPIC. VEGF competition with Sema-3A may regulate associations between NP1 and either plexin A1/2 or L1, but this remains to be determined. Plexins have a large extracellular domain containing short cysteine-rich motifs; the plexin intracellular domain contains two separate regions of homology highly conserved within plexins but sharing little homology with other proteins. (Right) It is unknown whether VEGF binding to NP1 is sufficient to trigger intracellular signaling independently of Sema-3A. VEGF would stimulate activation of PI3K/Akt, PLC-γ, and MEK/ERK pathways via VEGFR-2 in a variety of neuronal cells.

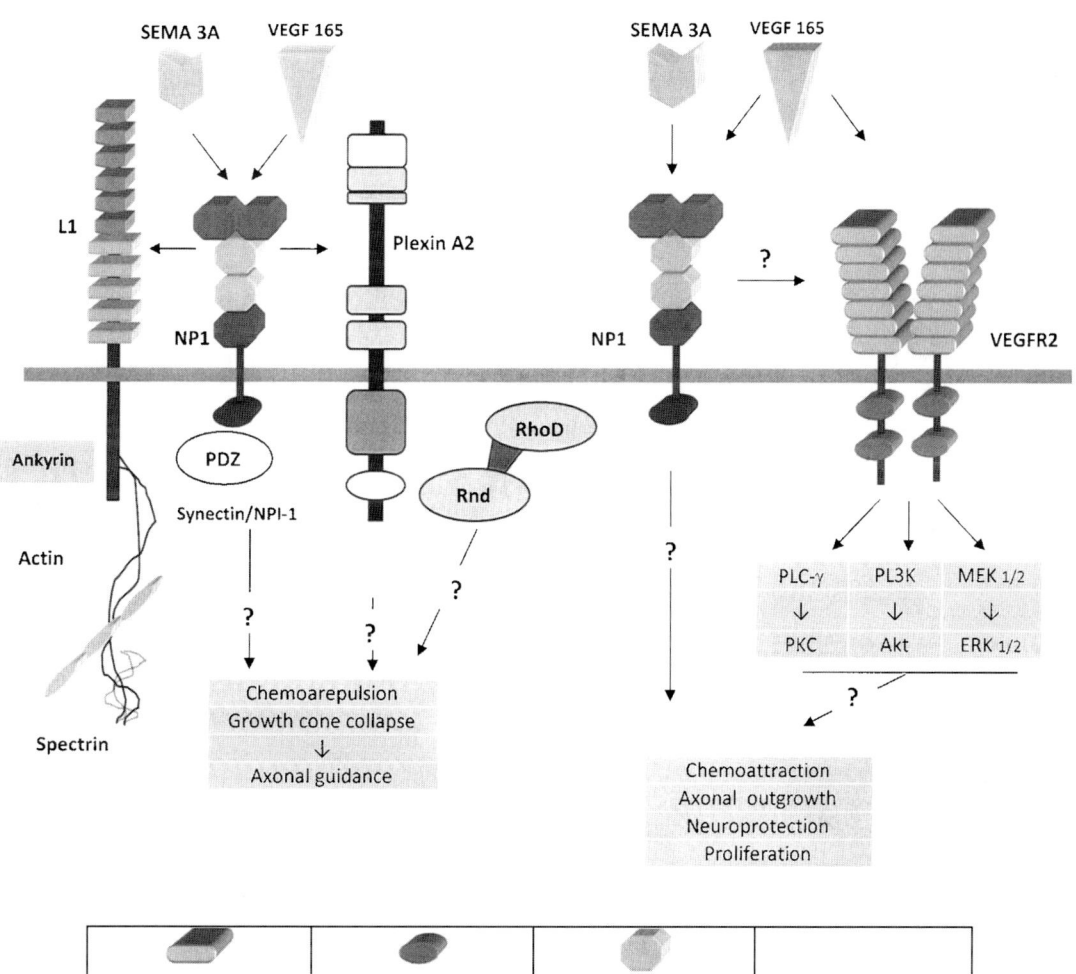

Fig. 2. (Continued).

expressed by endothelial cells, several tumor cell types, and different types of sensory neurons, including dorsal root ganglion, olfactory and optic nerves, as well as some sympathetic neurons. VEGF165 also binds specifically to NP2 that has domain similar to that of NP1 displaying a 44% of amino acid identity and distinct expression pattern in the developing nervous system. NP2 is also a receptor for the VEGF isoform 145 (Fig. 2B).

VEGF-A and its receptors are of utmost significance for many diseases accompanied by vascular pathology and inflammation, such as cancer stroke, intraocular neovascular syndromes, inflammatory disorders, and brain edema. Although vascular endothelial growth factor receptor-1 (VEGFR-1) has a higher affinity to VEGF-A than VEGFR-2, VEGFR-1 autophosphorylation upon VEGF-A binding is weak in comparison to VEGFR-2. This particular effect could be linked to the fine control of VEGF-A activity during development. In fact, early embryonic lethality in knockout (VEGFR-1 −/−) mice has been demonstrated; conversely, mice lacking only the intracellular domain of the VEGFR-1 develop almost normal and reach adulthood. Therefore, it seems that VEGFR-1 mainly functions as a "decoy receptor" during development by regulating VEGF availability. Besides membrane-bound VEGFR-1, a soluble form exists (sFlt-1) that has also the capacity to perform decoy functions. Hence, the VEGF action could be different depending on the dominant isoform produced in a particular district. Therefore, alternative splicing, posttranslational modification, and receptor and antagonistic receptor transduction pathway are the key regulators of this pleiotropic growth and survival factor.

Angiogenesis, mitogenesis, and endothelial survival

During the development, blood vessels are generated by two processes: vasculogenesis and angiogenesis. Vasculogenesis is the process that defines the de novo formation of blood vessels from undifferentiated endothelial precursor cells. During development, the extra- and intra-embryonic primary vascular plexus, the dorsal aorta, and the primitive heart are formed by vasculogenesis. In angiogenesis, the formation of new blood vessels from preexisting vessels gives rise to a continuous expansion of the primitive vasculature. In these processes, the VEGF-A action is pointed out by knockout mice where the loss of one allele leads to embryonic lethality at midgestation due to defective vasculogenesis in the entire embryo. In addition, the specific inactivation of VEGF-A in brain resulted in growth retardation and neural apoptosis. VEGF-A stimulates proliferation of endothelial cells, and the newly formed blood vessels are guided by VEGF gradient to their target tissues. A specialized, nonproliferating cell at the apical part of a navigating vessel extends filopodia that recognize VEGF-A gradient by its VEGF receptors and guides the vessel toward its target area. Gradients are formed by the higher expression of VEGF in the targeted area and by the different binding capacities of VEGF-A isoforms to the extracellular matrix. The isoforms differ also for the diverse binding capacity to the heparan sulfate proteoglycan and therefore the diffusion within the tissue. VEGF supports in vitro growth of vascular endothelial cells derived from arteries, veins, and lymphatics (Ferrara, 2002). In particular, VEGF promotes angiogenesis in vitro in three-dimensional models, inducing confluent microvascular endothelial cells to invade collagen gels and form capillary-like structures (Pepper et al., 1992) Furthermore, VEGF induces sprouting from rat aortic rings embedded in a collagen gel and also elicits a pronounced angiogenic response *in different* in vivo models such as the chick chorioallantoic membrane (Leung et al., 1989), stimulates direct angiogenesis in the rabbit cornea (Phillips et al., 1994), is able to induce iris neovascularization, and stimulates direct angiogenesis in the rabbit cornea (Tolentino et al., 1996). VEGF is also a survival factor for endothelial cells, both in vitro and in vivo (Alon et al., 1995). In vitro, VEGF prevents endothelial apoptosis induced by serum starvation. Such activity is mediated by the phosphatidylinositol 3-kinase (PI3K)/Akt pathway (Gerber et al., 1998). Moreover, VEGF induces the expression of the antiapoptotic proteins Bcl-2 and A1 in vascular endothelial cells. In vivo, the prosurvival effects of VEGF are developmentally regulated. In fact, VEGF inhibition results in extensive apoptotic changes in the vasculature of neonatal, but not adult mice. Furthermore, a strong VEGF dependence has been demonstrated within tumors in endothelial cells of newly formed vessels but not of established ones. Although endothelial cells are the primary targets of VEGF, several studies have reported mitogenic effects also on certain nonendothelial cell types, such as retinal pigment epithelial cells, pancreatic duct cells, and Schwann cells.

Neurotrophic and neuroprotective effect

Recent evidence suggests that VEGF-A directly affects neurons. In particular, VEGF is a relevant key factor in the developing and in the adult nervous system, displaying a relevant role in angiogenesis, neurogenesis, cell survival, and neuronal migration. During the early development,

VEGFR-2 is expressed in neural progenitor cells and in some differentiated cells in the developing mouse retina, at this stage still avascular. In this tissue, VEGF-A is expressed concomitantly in VEGFR-2 expressing cell, demonstrating the direct action of VEGF-A and VEGFR-2 in retinal neurogenesis and development (Raab and Plate, 2007). In vitro studies demonstrate that VEGF-A, transducing its action by VEGFR-2, regulates retinal progenitor cell proliferation and neuronal differentiation in chick retina (Hashimoto et al., 2006). The direct effect of VEGF-A on retinal cells was demonstrated in experimental model using isolated cells from newborn rats, by increasing the number of photoreceptors and amacrine cells. Therefore, its action is regulated by the expression of specific receptors alternatively by the action of its antagonistic soluble receptor. An example of this fine regulatory system is the corneal avascularity, required for optical clarity and optimal vision. The molecular basis of corneal avascularity has been for a long time unclear. Recently, it was found that the cornea contains VEGF-A but nearly all of it was bound to sVEGFR-1, also known as sFlt-1, the soluble receptor, recognized as essential factor for preserving the avascular ambit of the cornea.

VEGF-A exerts its action also on axonal outgrowth, on differentiation, and on the survival of superior cervical and dorsal root ganglion neurons and mesencephalic neurons in organotypic explant cultures. In vitro, VEGF-A stimulates axonal outgrowth, improves the survival of superior cervical and dorsal-root ganglion neurons, enhances the survival of mesencephalic neurons in organotypic explant cultures, and can rescue HN33 hippocampal cells from apoptosis induced by serum withdrawal (Jin et al., 2000). In vivo, VEGF-A coordinates the migration of motor neuron soma (Schwarz et al., 2004), whereas local delivery of VEGF-A prolongs motor neuron survival (Storkebaum et al., 2004). Conversely, low VEGF-A levels have been linked to motor neuron degeneration in both animal models and human disease (Oosthuyse et al., 2001).

In addition to a cellular role in regulating proliferation, migration, and permeability, the most recent function assigned to VEGF is that of a neuroprotective agent. A neuroprotective effect has been demonstrated also in various brain diseases: Parkinson's disease, stroke, and amyotrophic lateral sclerosis (ASL). The neurotrophic and neuroprotective effect of VEGF-A is the result of the complex cross talk among different soluble factors and involves a combination of direct neuroprotective effects and stimulation of angiogenesis, implying an enhanced supply of oxygen and nutrients that indirectly exert a neuroprotective impact.

Therefore, VEGF-mediated neuroprotection in the central nervous system has been attributed to both indirect and direct effects of VEGF-A on neuron survival. VEGF exerts neuroprotective actions indirectly through multiple mechanisms such as stimulation of angiogenesis, increasing blood flow, and enhancing brain barrier permeability for glucose and antioxidant activation (Góra-Kupilas and Josko, 2005).

Furthermore, it has been demonstrated that VEGF-A has a direct survival effect on neuronal cells of the retina, independent of blood flow, and functional analyses showed that VEGFR-2, detected in several neuronal cell layers of the retina, was involved in retinal neuroprotection.

Mechanisms implicated in *direct* neuroprotection involve the inhibition of the programmed cell death of neuronal cells under hypoxic conditions and stimulation of neurogenesis.

Under critical conditions for neuronal cells (hypoxia, glucose deprivation, oxidative stress), VEGF-A becomes a mediator of multiple molecular reactions leading to the inhibition of programmed cell death–inducing Bcl-2 antiapoptotic factor, inhibiting the activation of the proapoptotic factor Bax, and the stimulating neurogenesis (Góra-Kupilas and Josko, 2005). A significant regulatory step of apoptosis is the activation of the proapoptotic factor Bax. The overexpression of Bcl-2 induced by VEGF-A inhibits the activation of Bax. The mechanism by which Bax activation is regulated is still unclear, although several downstream events have been elucidated.

Following its activation, Bax homodimerizes and translocates into the mitochondria membrane, conferring permeability and leading to the release of several death-promoting factors (cytochrome *c*) in the cytoplasmic compartment. Therefore, the regulation of its activation is clearly relevant in normal cell survival under physiological conditions and still conflicting in tumor insurgence as well as in tumor progression. Recently, some information was published on its activity regulation. Bax is localized physiologically inactive in the cytoplasm in normal undamaged cells interacting with the Ku70 protein C terminus; this status determines its inability to homodimerize and give rise to apoptotic key events. Overexpression of Ku70 in vitro blocks the Bax-induced apoptosis under some variety of stimuli in epithelial cells (Leskov et al., 2003). Following UV-induced DNA damage, the DNA double-strand breaks sensor Ku70 translocates to the nucleus, allowing Bax release and its translocation to the mitochondria. This important function of Ku as regulator of Bax-mediated releases of several death-promoting factors is in agreement with its role as caretaker in the nucleus. Microenvironmental factors, such as VEGF or IL-6, are strictly connected to the Ku70–Bax-binding induction to prevent Bax activation and translocation in the mitochondria (Sawada et al., 2003). This scenario strongly influences cell survival increasing Bcl-2, Bax antagonist, when the cell death program is redirected into survival (Pucci et al., personal communication). These molecular interactions are still obscure, but underline the relevant role of VEGF and other microenvironmental factors in the complicated cross talk among molecules that could effectively turn the cell fate.

Moreover, VEGF exerts a *direct* effect on neurons, as demonstrated in vitro. VEGF-A has a neurotrophic effect on cultured neural cells of the peripheral and central nervous systems by stimulating the axonal outgrowth and by protecting neural cells from serum deprived-glutamate-induced or hypoxia-induced cell death. The neuroprotective function of VEGF includes also the *indirect* actions that affect neuronal survival under critical conditions and may be related to (i) increased blood flow and oxygen supply, (ii) enhanced endothelial permeability with increased glucose transport across blood–brain barrier (BBB), (iii) the antioxidative function of VEGF and heme oxygenases, and (iv) angiogenesis.

i. It has been demonstrated that VEGF-A can increase tissue blood flow via nitric oxide induction and that increased flow (and oxygen supply) in many different ischemic settings has been shown to preserve tissue integrity and function (Endres et al., 2004). In a model of ischemia–reperfusion injury (I/R injury model), it has been demonstrated that VEGF-A injection can increase volumetric blood flow also in the retina provoking a dose-dependent reduction in retinal neuron apoptosis. The role of the incremental increase in blood flow on neuroprotection was assessed by inhibiting the activity of iNOS in the I/R injury model, which led to a significant reduction (but not complete suppression) of the neuroprotective effects of VEGF-A, suggesting that VEGF-A-induced volumetric blood flow to the retina may be at least in part responsible for the neuroprotection (Nishijima et al., 2007).

ii. Glucose, the most important energetic substrate for the brain, must pass the BBB to obtain access to the brain. There are studies indicating VEGF as a candidate for directly regulating brain glucose uptake under critical conditions (Dorle et al., 2002).

iii. In fact, VEGF mediates the enhancement of glucose transport through the BBB by increasing GLUT-1 gene expression and translocation of cytosolic GLUT-1 to the plasma membrane surface. Furthermore, VEGF mediates induction of endothelial fenestrations, thereby increasing the transport of small molecules (Sone et al., 2000). In conclusion, VEGF can increase brain glucose uptake by enhancing the transport via GLUT-1, dilating cerebral arterioles, and inducing endothelial fenestrations under conditions of brain glucose deprivation.

iv. Moreover, VEGF-A also exerts its action on oxidative stress. Human heme oxygenase (HO) exists in three isoforms: HO-1 (inducible), HO-2, and HO-3 (constitutive). HO-1 is widely distributed and induced by a range of stimuli including shear stress, oxidative stress, nitric oxide (NO), and hypoxia (Maines, 1997). HO-1 is a cytoprotective molecule that acts, in the presence of a variety of noxious stimuli, to exert antiapoptotic,

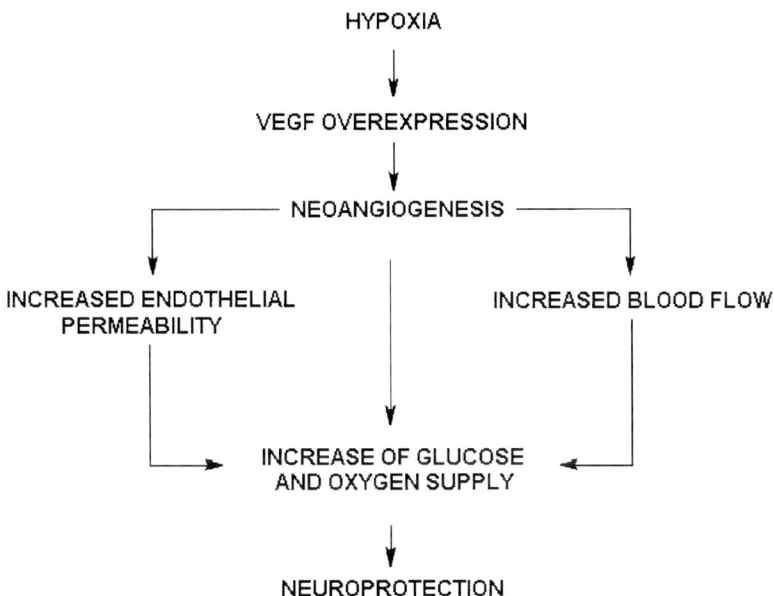

Fig. 3. Indirect neuroprotective effect of VEGF. VEGF regulates brain glucose uptake under critical conditions.

anti-inflammatory, and antiproliferative effects. HO-1 is rapidly induced during hypoxia, ischemia–reperfusion, hyperthermia, and endotoxic shock, providing cytoprotection during the resolution of stress-induced inflammatory injury (Fig. 3).

v. Recent in vivo studies report that VEGF regulates HO-1 expression and activity in vascular endothelial cells. The inhibition of HO-1 abrogates VEGF-induced endothelial activation and subsequent angiogenesis but promotes VEGF-induced monocyte recruitment and inflammatory angiogenesis. HO-1 may also regulate the synthesis and activity of VEGF, resulting in a positive-feedback loop (Józkowicz et al., 2003). HO-1 participates in heme degradation. Carbon monoxide (CO), ferrous ions (Fe^{2+}), and biliverdin are the products of this reaction, which prevent cells from oxidative damage caused by free radicals. Ferritin, biliverdin, and bilirubin are physiological antioxidants in serum and extravascular space, which protect neurons from damage due to the production of reactive oxygen species. Hypoxia by inducing HOs contributes to the increase in the concentration of heme degradation products. These products exert their neuroprotective actions through vessel dilatation and antioxidation. Moreover, HO-1 induces expression hypoxia-inducible factor-1α (HIF-1α), including the gene encoding VEGF. VEGF induces backward HO-1 and, through this positive biofeedback, it mediates in antioxidation (Fig. 4) (Motterlini et al., 2000).

vi. Many studies on cerebral ischemia in animals reveal that hypoxia induced by occlusion of the cerebral middle artery contributes to overexpression of VEGF and angiogenesis. Other observations on rats show that overexpression of VEGF occurs also after subarachnoid hemorrhage (SAH). Increased expression of this factor is particularly detected in some regions of the brain hippocampus, thalamus, lateral ventricles, the fourth ventricle, smooth muscle cells of subarachnoid vessels, and cerebellar cortex. The potential, therapeutic usefulness of VEGF in cerebral ischemia is also limited by the fact that its angiogenic effect is delayed in onset and, therefore, presumably too late to rescue many vulnerable neurons. The

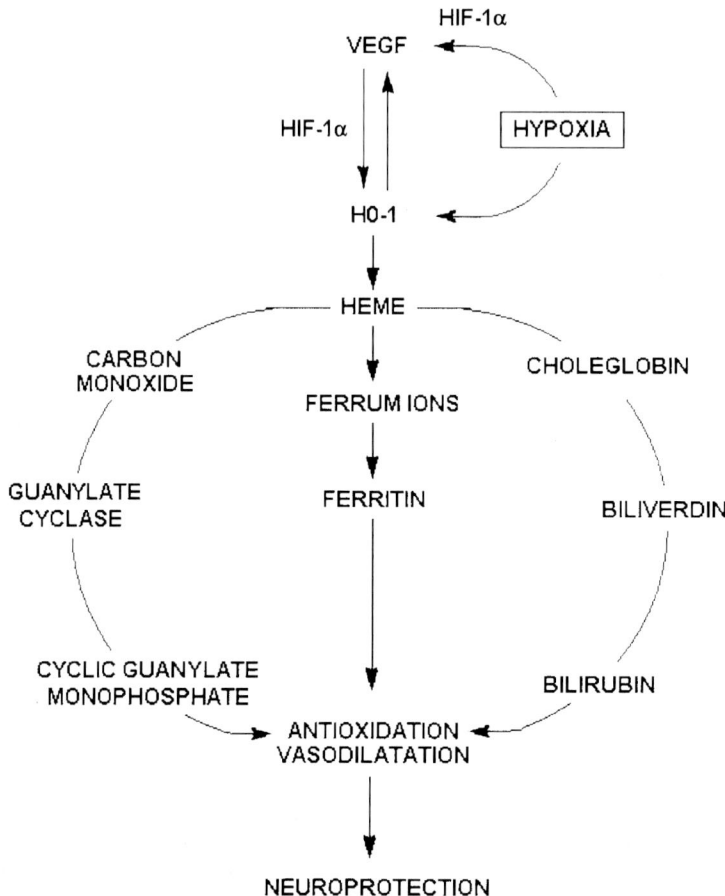

Fig. 4. Effect of VEGF on the oxidative stress and its downstream cascade.

possibility of direct neuroprotective VEGF actions in ischemic tissue in the interval preceding angiogenesis may help prolong cell survival until angiogenesis can occur.

VEGF neuroprotection is exerted mainly by the activation of the VEGFR-2. Additional evidence for direct neuronal protection by VEGF is that VEGF inhibits the death of cultured hippocampal neurons from glutamate and N-methyl-d-aspartate toxicity and that deletion of the hypoxia-response element from the VEGF promoter causes motor neuron degeneration in mice.

Because neuroprotective effects of other growth factors are associated with improved outcome from stroke, this could be true for VEGF as well. Finally, VEGF has been implicated as a factor that promotes neurogenesis in the adult brain. One possibility is that this occurs through the establishment of a vascular niche that favors the proliferation and differentiation of neuronal precursors, perhaps by the release of BDNF from endothelial cells. Alternatively, VEGF may exercise a direct mitogenic effect on neuronal precursors.

Thus, VEGF expression in the ischemic brain could contribute to ischemia-induced neurogenesis and modify outcome in that way as well.

Intravitreal VEGF inhibition therapy and neuroretina toxicity

The neuroprotective function of VEGF-A in the retina has not been characterized despite the fact that VEGF antagonists are being applied widely to combat retinal vascular disease. A recent study compared the antiproliferative and cytotoxic properties of bevacizumab (Avastin), pegaptanib (Macugen), and ranibizumab (Lucentis) on human retinal pigment epithelium (ARPE19) cells, rat retinal ganglion cells (RGC5), and pig choroidal endothelial cells (CECs). Ranibizumab reduced CEC proliferation by 44.1%, bevacizumab by 38.2%, and pegaptanib by 35.1% when the drugs were used at their established clinical doses. However, the differences between the three drugs in respect to cell growth inhibition were not statistically significant. Only a mild antiproliferative effect of bevacizumab or pegaptanib on ARPE19 cells could be observed. Ranibizumab did not alter ARPE19 cell proliferation. No cytotoxicity on CEC, RGC5, and ARPE19 cells could be seen. In this view, bevacizumab, pegaptanib, and ranibizumab significantly suppress CEC proliferation. However, when used at the currently established doses none of the drugs was superior over the others in respect to endothelial cell growth inhibition. The biocompatibility of all three drugs — including the off-label bevacizumab — seems to be excellent when used at the currently recommended intravitreal dose (Spitzer et al., 2007).

Neuroprotection: clusterin, a multifunctional protein

Clusterin/ApoJ: a debated physiological role

Clusterin/Apolipoprotein J (Clu/ApoJ) is a heterodimeric, highly conserved, secreted glycoprotein being expressed in a wide variety of tissues and found in all human fluids. The protein has been reportedly implicated in several diverse physiological processes such as sperm maturation, lipid transportation, complement inhibition, tissue remodeling, membrane recycling, cell–cell and cell–substratum interactions, and promotion or inhibition of apoptosis.

The human clusterin comprises 449 amino acids, generating an unglycosylated holoprotein (precursor form, pCLU) with a predicted molecular mass of 60 kDa that can be proteolytically cleaved into α- and β-subunits held together by disulfide bonds.

Mature clusterin is glycosylated and secreted as a protein of 76–80 kDa (Wong et al., 1993; Lakins et al., 1998) depending on the degree of glycosylation, which appears like a 40-kDa α- and β-protein smear, by polyacrylamide gel under reducing conditions (Leskov et al., 2003). Its high degree of sequence conservation, almost ubiquitous tissue distribution, and the absence of functional clusterin polymorphisms in humans suggest that the protein accomplishes a function of fundamental biological importance.

Two different mRNA transcripts for clusterin, derived from an alternative splicing, have been identified, one coding for the secreted form of clusterin (40 kDa) and the other one coding for the 50- to 55-kDa nuclear form, without leader peptide. The presence of these two different isoforms and their function within the cell is a much-debated question. One form was found expressed preferentially in the nucleus (nClu) of different cell types, while the highly glycosylated form (sClu) was found in the cytoplasm and secreted in biological fluids. The activation of nuclear clusterin form seems to induce cell cycle arrest and cell death. Leskov et al. (2003) demonstrated that clusterin nuclear form is unglycosylated and is involved in apoptosis induction. Its function is related to the activation of Ku70 and Bax cascade as previously described as one of VEGF antiapoptotic direct functions. In normal cells after irreversible cell damage, nClu cooperates with Ku70 to induce apoptotic death activating Bax translocation to mitochondria. In tumors, its expression is inhibited favoring the presence of the antiapoptotic form (sClu) that inhibits cell death and favors the aberrant neoplastic cell clone expansion. In fact, it has been demonstrated that the overexpression of the secreted isoforms in the cytoplasm correlated with the apoptotic cell death resistance (Leskov et al., 2003; Pucci et al., 2004). Paradoxically, cell survival and cell death are finely regulated by the balanced expression of different isoforms of the same protein and by the intriguing pathways of growth- or death-promoting factors of the microenvironment.

In response to cellular stress conditions, including heat shock, UV radiation, and oxidative stress, and in a wide array of pathological conditions, such as neurodegenerative disorders, multiple sclerosis, atherosclerosis, myocardial infarction, and cancer, clusterin expression is markedly upregulated in vitro and in vivo. Evidence suggests that sClu exerts a cytoprotective function under these stress conditions, which may be mediated by protection against oxidative stress (i), inhibition of apoptosis and complement-mediated cell lysis (ii, iii), promotion of cellular contacts (iv), and protection of cell membranes (v) (Zenkel et al., 2006).

Clusterin and diseases

Clusterin gene is differentially regulated by cytokines, growth factors, and stress-inducing agents; upregulation of clusterin mRNA and protein levels is also detected in neurodegenerative conditions related to aging (Trougakos and Gonos, 2002). nClu accumulates during the growth-arrested cellular state of senescence that is thought to contribute to aging and suppression of tumorigenesis (Kyprianou et al., 1991; Bettuzzi et al., 2002); however, the secreted isoform sClu is also upregulated in several cases of in vivo cancer progression and tumor formation (Pucci et al., 2004). Pucci et al. (2004) have demonstrated that the progression toward high-grade and metastatic colorectal carcinomas correlates to the overexpression of the glycosylated cytoplasmic form of clusterin (sClu) and to a complete loss of the proapoptotic nuclear form (nClu). Thus, the increased level of the secreted form and the disappearance of the nuclear unglycosylated one are directly connected to increased cell survival, aggressiveness, and enhanced metastatic potential.

Therefore, clusterin can act either to promote or to inhibit cell death, depending on the cellular context or molecular species. The question of whether clusterin is a multifunctional protein, or deploys a single primary function influenced by cellular context, remains a central issue continuing to stimulate interest in this enigmatic molecule (Jones and Jomary, 2002).

More recently, it has been proposed that clusterin represents a new class of highly efficient chaperones, similar to the small heat-shock proteins; however, the heat-shock proteins act in the intracellular space, whereas for sClu the main site of action is the extracellular space. Molecular chaperones serve as a cellular rescue system by interacting with partially folded or unfolded stressed proteins to prevent their aggregation and precipitation and to promote protein refolding. Misfolding of proteins is a result of structural modification typically caused by oxidative stress or mutation. sClu binds to exposed hydrophobic regions of a broad range of partially unfolded proteins producing solubilized high-molecular-weight complexes, thereby preventing their aggregation and precipitation (Poon et al., 2000; Wilson and Easterbrook-Smith, 2000). Moreover, sClu has been shown to preferentially interact with slowly aggregating target proteins (Poon et al., 2002). Clusterin does not affect the refolding of destabilized proteins, but it stabilizes them in a state competent for refolding into functional proteins by other chaperones (such as heat-shock protein 70).

Clusterin and the nervous system

The normal adult brain is a major site of Clu/ApoJ mRNA synthesis in several mammalian species (Collard and Griswold, 1987; Duguid et al., 1989). High levels of clusterin mRNA were found in several neuron-rich layers and nuclei, including the pyramidal and granular cell layers of the hippocampal formation, the hypothalamus, and some motor neurons (Danik et al., 1999). In the cerebral cortex, a homogeneous pattern of hybridization was observed in which neurons as well as astrocytes expressed clusterin mRNA. In contrast with the high levels and ubiquitous distribution of clusterin mRNA, only a few structures showed clusterin immunoreactivity by immunohistochemistry (Pasinetti et al., 1994; Calero et al., 2000). Overall, we can distinguish two general mechanisms exerted by clusterin that could induce a *direct* neuroprotection, through the triggering of signaling cascades that affect neurodegenerative pathways and apoptotic processes, and an *indirect* neuroprotection, through (i) its interaction with amyloid-β (Aβ) peptides, by its role of chaperone; (ii) the inhibition of the formation of the membrane-attack complex of the complement system; and (iii) neurotrophic functions, promoting a general recovery from neuronal injury via lipid transport or membrane recycling.

In the Alzheimer's disease (AD), the most frequent form of amyloidosis in humans and the major cause of dementia in the elderly, clusterin is found associated with Aβ plaques and cerebrovascular deposits. Clusterin coprecipitates with Aβ from cerebrospinal fluid, suggesting a physiological interaction with Aβ. The presence of sClu in parenchymal and vascular amyloid deposits may result as a consequence of the inflammatory process and complement activation occurring in the lesions rather than constituting a specific mechanism for the disease. Current knowledge suggests that sClu has a protective function in AD: mammalian recombinant clusterin obtained from stable transfection of hamster kidney fibroblasts as well as a commercial native clusterin, purified from serum, provides dose-dependent neuroprotection against Aβ (Boggs et al., 1996). Thus, the interaction of clusterin with Aβ could modulate Aβ neurotoxicity. Furthermore, in primary neuronal cell cultures, clusterin prevents Aβ toxicity (Calero et al., 2000).

In biological fluids, clusterin maintains sAβ solubility and modulates its uptake by the brain across the BBB. Moreover, clusterin promotes the Aβ catabolism through the formation of soluble complexes that may follow catabolic pathways common to the lipoprotein particles.

As reported above, clusterin is markedly induced after tissue injury. In the nervous system, clusterin upregulation may represent a defense response against local nerve damage. Moreover, as a chaperone protein, clusterin could be involved in the protection of the nervous system from potential toxic debris.

Dati et al. (2007) have investigated the effects of recombinant human clusterin (r-h-clusterin) in different experimental models of peripheral neuropathies. Daily treatment with clusterin accelerated the recovery of nerve motor–evoked potential parameters, after nerve injury. Treatment of experimental autoimmune neuritis rats with clusterin accelerated the rate of recovery from the disease, associated with remyelination of demyelinated nerve fibers. These data demonstrate that clusterin is capable of ameliorating clinical, neurophysiological, and pathological signs in models of peripheral neuropathies. Taking into account the interplay between inflammation and neurodegeneration in the peripheral nervous system, the effects of r-h-clusterin could be both immunomodulatory and neurotrophic on myelinating Schwann cells. Recombinant clusterin may prevent apoptosis in these experimental models. These authors concluded that r-h-clusterin does not exert immunomodulatory effect, based on the absence of significant changes in blood and nerve cytokine concentrations or in situ macrophage infiltration. However, r-h-clusterin could modify the cytokine-network abnormalities known to occur in inflammatory neuropathies. For example, it has been known that transforming growth factor-β1 (TGF-β1) controls clusterin expression (Morgan et al., 1995). This observation emphasizes the importance of early modulation of the inflammatory microenvironment.

Interestingly, investigations of knockout (Clu–/–) mice have complicated the picture, highlighting the two faces of clusterin. On the one hand, absence of clusterin reduces cell death in hypoxia-ischemia-induced brain damage, suggesting that it normally functions to exacerbate neuronal damage in these circumstances (Han et al., 2001), whereas in the same knockout strain, autoimmune myocardial damage is increased, implying a normally protective role (McLaughlin et al., 2000). Moreover, inducing brain injury by middle cerebral artery occlusion in clusterin gene knockout (Clu–/–) mice identical to those used by Han et al., in transgenic mice overexpressing clusterin and wild-type mice, Wehrli et al. (2001) demonstrated a 30–50% increase in clusterin mRNA levels in the ischemic mouse brains, as compared with the control hemisphere of wild-type mouse brains. This area sustains the inflammation and contains cells that retain the potential for salvage. At 14 days post ischemia, wild-type mice displayed a much higher number of inflammatory cells than clusterin-overexpressing mice, and Clu–/– mice showed the worst morphological resolution. Overall, an inverse correlation between clusterin expression and ischemic brain damage was observed, showing that clusterin protects against ischemic brain damage (Wehrli et al., 2001).

Cortical spreading depression (CSD) is a powerful benign stimulus that is capable of providing long-lasting ischemic tolerance. It is classically characterized as a slowly propagating wave (1–6 mm/min) of neuronal and astrocytic depolarization that results in transient suppression of electroencephalographic activity. According to previous studies that have implicated clusterin as a protective molecule assisting in neuronal survival, the clusterin mRNA levels significantly increase 1–14 days after CSD. These findings, along with other studies in clusterin-transgenic mice, suggest that increased clusterin production and secretion, particularly by astrocytes, could play a role in CSD-induced conditioning/protection of cortex, perhaps via one or more of its actions, such as inhibition of complement activation and cytolysis, effects on chemotaxis and apoptosis, or as an antistress protein chaperone (Wiggins et al., 2003).

Moreover, HIF-1α activation, influenced by VEGF-A, acts on clusterin gene induction and could be directly related to the preferential clusterin isoforms produced. Moreover under stress conditions, many stress-derived factors and cytokines (i.e., IL-6, TGF-β) could induce this prosurvival protein, as demonstrated by another enigmatic link among clusterin, the function of VHL and its target HIF (Nakamura et al., 2006). Therefore, sClu could be induced in the presence of HIF-1α, condition necessary but not sufficient for its production.

Neuroprotection: IL-6, VEGF, clusterin, and glaucoma

Glaucoma represents a group of neurodegenerative diseases characterized by anatomical damage to the optic nerve and slow, progressive death of RGCs.

Elevated intraocular pressure (IOP) is considered to be the most important risk factor for glaucoma, and treatment options for the disease have been traditionally limited to its reduction. However, visual field loss and RGC death continue to occur in patients with well-controlled IOP normal tension glaucoma and thus the concept that treatment strategies alternative to IOP lowering are needed has recently emerged.

This concept emphasizes that several pressure-independent mechanisms are involved in the development and progression of glaucomatous neuropathy and that high IOP and optic nerve vascular deficiency could be necessary but not sufficient in the pathogenesis of glaucomatous damage.

An advocated strategy concerns the development of neurotrophic agents that would interact with neuronal or glial elements within the retina and/or optic nerve head allowing the RGCs survival.

The advent of animal models of chronic glaucoma has greatly enhanced the understanding of many of the processes occurring in glaucoma pathogenesis and addressed "theoretical" targets for pharmacological intervention. Such targets include glutamate receptors, autoimmune elements, neurotrophin deprivation, nitric oxide synthesis, oxidative stress products, sodium and calcium channels, heat-shock proteins, and apoptotic pathways.

For example, the death of RGC could be the result of a biphasic process: a primary injury responsible for damage induction that is followed by a slower secondary degeneration, related to noxious environment surrounding the degenerating cells. Retinal ischemia may provoke a cascade of events that result in cell death: hypoxia leads to cytotoxic levels of glutamate that cause a rise in intracellular calcium, which in turn leads to neuronal death, due to apoptosis or necrosis. Neuroprotective agents would attempt to protect the cells that were spared during the initial insult but are still vulnerable to environmental damage.

Many downstream factors are involved in neuroprotection after damage or inflammatory disorders. Usually, for example, IL-6 involved in some pro-inflammatory diseases or tumorigenesis is retained as a negative prognostic factor, but the presence of this cytokine in the acute phase of cerebral ischemia plays a critical role in preventing damaged neurons from undergoing apoptosis and its role may be mediated by Stat3 activation.

Moreover, IL-6 determines an increased neuroprotection inducing VEGF, clusterin, and hypoxia-1α activation.

The increased amount of this pro-inflammatory cytokine after neuronal damage orchestrates the cascade of events that turn death into survival.

IL-6, which has been identified as B-cell-stimulating factor, belongs to the subfamily of cytokines, including leukemia inhibitory factor and ciliary neurotrophic factor (CNTF) that share the gp130 subunit as a common receptor subunit.

IL-6 binds to IL-6 receptor α that is composed of a membrane-bound and soluble form. The association of the IL-6/IL-6 receptor α with gp130 causes the dimerization of gp130, followed by the activation of the Janus kinase–signal transducer and activator of transcription signal (JAK–STAT) pathway. In addition, gp130 in turn activates the extracellular signal–regulated kinase (ERK) and the PI3K/Akt pathway.

Clinical studies on ischemic stroke have previously demonstrated a correlation between IL-6 serum levels, body temperature, expression of inflammatory proteins (C-reactive protein), infarct volume, and severity of paresis. Previous studies in rat brain and glia demonstrated that IL-6 is expressed by neurons and glia. Neuroprotective effect of exogenous IL-6 has been observed in PC12 cell cultures in a hypoxia–reoxygenation paradigm, in peripheral nerves after trauma, and in ischemic neurons. Although the role of IL-6 in stroke is still a matter of controversy, identifying extracellular signals that modulate RGC survival in glaucoma and determining whether these signals depend on pressure are essential for delineating the mechanisms of disease and for defining new targets for its treatment. Astrocytes and microglia represent a source of these signals.

The concept that the glia in the optic nerve could influence the survival of RGCs in glaucoma is widely accepted; however, the contribution of the retinal glia to pathological processes associated with glaucoma is less clear. Thus, glia-derived IL-6 could modulate RGC survival in glaucoma, particularly in the retina, where astrocytes and microglia are in proximity to RGC somas and axons. Astrocytes and microglia are subjected to elevated pressure, affecting the survival of RGCs, and IL-6 is a key component of microglial response to elevated pressure serving as a neuroprotectant for RGCs challenged by pressure (Orellana et al., 2005; Yamashita et al., 2005).

In addition to IL-6 itself, proteins related to IL-6, such as CNTF and neuropoietin, may also influence the development and survival of retinal cells. Many antiapoptotic and survival factors are induced by IL-6. Overall, the presence of IL-6 induced VEGF by the activation of nuclear factor-kappa B (NF-κB) and activator protein 1 (AP-1). The AP-1 activation resulted in the increased expression of both, VEGF and IL-6 regulated in paracrine and in autocrine fashion. Concomitantly, it was found that increased levels of IL-6 and VEGF are closely linked to increased nuclear protein levels of HIF-1α and enhanced nuclear transcription factor DNA-binding activity to a hypoxia-inducing element located in the VEGF promoter. The characterization of the VEGF165 and VEGF120 variants is a relevant improvement in the discovery of the regulatory pathway of this abundant growth factor. In particular, the neuroprotective action of VEGF120 has been reported as a survival factor for retinal neurons and a critical neuroprotectant during adaptive response to ischemic injury (Nishijima et al., 2007). It resulted that treatment with VEGF might provide neuroprotection in the retina, particularly during ischemic diseases. The use of receptor-specific ligand confirmed that VEGF2 activation is sufficient to trigger retinal neuroprotection and the role of the NP-1 coreceptor appears to be not essential in mediating neuroprotection effects. In fact, blocking VEGF165 with pegaptanib, which does not bind to VEGF120, there is no decrease in retinal RGC viability. As VEGF120 does not bind to NP1, it is unlikely that NP1 is required for this effect. These data demonstrated the important role of VEGFR-2 bound in transducing signals related only to a specific function of this neuroprotectant factor.

Concomitantly to the releases of VEGF in stress condition, there is also the activation of HIF-1α factor that leads to cascade of prosurvival pathways. Although the role of HIF-1α in glaucoma was rarely examined due to the problems of collecting intraocular samples from the retina and the optic nerve head, there is little evidence to demonstrate hypoxia in the human glaucomatous retina (Tezel and Wax, 2004). Some data on immunolocalization of cellular HIF-1α in donor eyes from cadaver patients with a diagnosis of glaucoma were obtained by immunoperoxidase and doubleimmunofluorescence experiments. These data showed that the immunostaining for HIF-1α in the retina and optic nerve head of glaucomatous eyes was higher than that in control eyes, the immunostaining being most prominent in the retinal regions corresponding to the areas of decreased light sensitivity, as determined by visual field testing. Interestingly, HIF-1α could be localized by immunostaining from the eyes of patients with normal tension glaucoma with a similar spatial relationship with visual field damage (Tezel and Wax, 2004). Further studies are warranted, revealing a possible role for HIF-1α in the neuroprotection of the glaucomatous retina (Naskar and Dreyer, 2001).

The presence of network of shared neuroprotectant pathways seems to be due to IL-6 that orchestrates the connection among prosurvival factors. Hence, IL-6 induces VEGF that directly influences HIF-1α activation and clusterin anti-apoptotic isoform production (sClu), Bax activity inhibition and Bcl-2 overexpression: exactly as a tumor cell behaves to survive (Fig. 5).

In the experimental model of the oxidative stress, neuroprotection may be mediated by IL-6 via its activation of gp130, the transmembrane

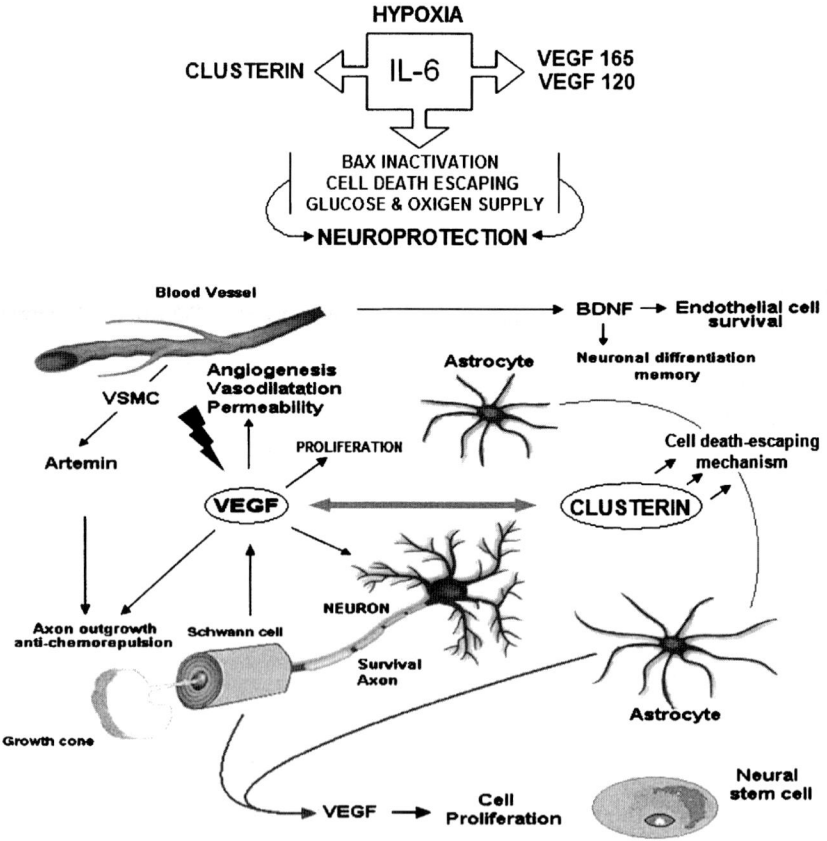

Fig. 5. Downstream factors involved in neuroprotection after damage or oxidative stress. IL-6 is a key component of survival especially in microglial response to elevated pressure. The induction of IL-6 after stress orchestrates the survival pathways. The IL-6 release induces the expression of VEGF that activates HIF-1α and the prosurvival sClu. The overexpression of sClu and the disappearance of the proapoptotic nuclear isoform (nClu) cooperate with Ku70 to inhibit Bax-dependent cell death.

glycoprotein involved in many cytokine-mediated responses (Van Lenten et al., 2001). This experimental model, mimicking the oxidative stress, demonstrated a direct connection between IL-6 production and the increased clusterin mRNA levels, as we have found in colon tumorigenesis (Pucci et al., 2000, and confidential data). Therefore, the important aspect of secreted clusterin as antiapoptotic stress folding protein in this context assumes a key role in neuronal cell survival.

In fact, clusterin plays an important role as a neuroprotectant agent, during different sources of cellular stress as an antiapoptotic signal. In particular, the cytoplasmic glycosylated isoform would play the antiapoptotic role, promoting cell survival, displaying a protection against oxidative stress; inhibiting the membrane-attack complex of complement proteins, locally activated as a result of inflammation (Jenne and Tschopp, 1989); and binding to hydrophobic regions of unfolded, stressed proteins, and therefore avoiding aggregation in a chaperone-like manner (Poon et al., 2000). Although there is good laboratory evidence for glaucoma neuroprotection by several drugs, proof of clinical efficacy for any neuroprotective agent has not yet been obtained in glaucoma. As a consequence, new drugs are needed for testing in appropriate randomized clinical trials for glaucoma neuroprotection. This striking result on clusterin, VEGF-A120, and VEGF-A165 suggests that future therapeutic strategies may need to refocus on the challenge of normalizing rather than abrogating VEGF-A responses.

Taking into account all of the experimental evidences reported above about the role of clusterin, VEGF-A120, and VEGF-A165 in enhancing neuronal survival and in decreasing or preventing neuronal damage, these factors could be considered as new potential glaucoma neuroprotective agents.

References

Alon, T., Hemo, I., Itin, A., Pe'er, J., Stone, J. and Keshet, E. (1995) Vascular endothelial growth factor acts as a survival factor for newly formed retinal vessels and has implications for retinopathy of prematurity. Nat. Med., 1: 1024–1028.

Bates, D.O., Cui, T.G., Doughty, J.M., Winkler, M., Sugiono, M., Shields, J.D., Peat, D., Gillatt, D. and Harper, S.J. (2002) VEGF(165)b, an inhibitory splice variant of vascular endothelial growth factor, is down-regulated in renal cell carcinoma. Cancer Res., 62: 4123–4131.

Bettuzzi, S., Scorcioni, F., Astancolle, S., Davalli, P., Scaltriti, M. and Corti, A. (2002) Clusterin (SGP-2) transient overexpression decreases proliferation rate of SV40-immortalized human prostate epithelial cells by slowing down cell cycle progression. Oncogene, 21(27): 4328–4334.

Boggs, L.N., Fuson, K.S., Baez, M., Churgay, L., McClure, D., Becker, G. and May, P.C. (1996) Clusterin (Apo J) protects against in vitro amyloid-β(1-40) neurotoxicity. J. Neurochem., 67(3): 1324–1327.

Calero, M., Rostagno, A., Matsubara, E., Zlokovic, B., Frangione, B. and Ghiso, J. (2000) Apolipoprotein J (clusterin) and Alzheimer's disease. Microsc. Res. Tech., 50: 305–315.

Collard, M.W. and Griswold, M.D. (1987) Biosynthesis and molecular cloning of sulfated glycoprotein 2 secreted by rat Sertoli cells. Biochemistry, 26(12): 3297–3303.

Conn, G., Bayne, M.L., Soderman, D.D., Kwok, P.W., Sullivan, K.A., Palisi, T.M., Hope, D.A. and Thomas, K.A. (1990) Amino acid and cDNA sequences of a vascular endothelial cell mitogen that is homologous to platelet-derived growth factor. Proc. Natl. Acad. Sci. U.S.A., 87: 2628–2632.

Danik, M., Champagne, D., Petit-Turcotte, C., Beffert, U. and Poirier, J. (1999) Brain lipoprotein metabolism and its relation to neurodegenerative disease. Crit. Rev. Neurobiol., 13(4): 357–407.

Dati, G., Quattrini, A., Bernasconi, L., Malaguti, M.C., Antonsson, B., Nicoletti, F., Alliod, C., Di Marco, R., Sagot, Y., Vitte, P.A., Hiver, A., Greco, B., Roach, A. and Zaratin, P.F. (2007) Beneficial effects of r-h-CLU on disease severity in different animal models of peripheral neuropathies. J. Neuroimmunol., 190: 8–17.

Dorle, D., Bewersdorf, J., Fruehwald-Schultes, B., Kern, W., Jelkmann, W., Born, J., Fehm, H.L. and Peters, A. (2002) Vascular endothelial growth factor: a novel endocrine defensive response to hypoglycemia. J. Clin. Endocrinol. Metab., 87: 835–840.

Duguid, J.R., Bohmont, C.W., Liu, N.G. and Tourtellotte, W.W. (1989) Changes in brain gene expression shared by scrapie and Alzheimer disease. Proc. Natl. Acad. Sci. U.S.A., 86(18): 7260–7264.

Endres, M., Laufs, U., Liao, J.K. and Moskowitz, M.A. (2004) Targeting eNOS for stroke protection. Trends Neurosci., 27: 283–289.

Ferrara, N. (2002) VEGF and the quest for tumour angiogenesis factors. Nat. Rev. Cancer, 2: 795–803.

Gerber, H.P., Mc Murtrey, A., Kowalski, J., Yan, M., Keyt, B.A., Dixit, V. and Ferrara, N. (1998) VEGF regulates endothelial cell survival by the 1/KDR activation. J. Biol. Chem., 273: 30336–30343.

Góra-Kupilas, K. and Josko, J. (2005) The neuroprotective function of vascular endothelial growth factor (VEGF). Folia Neuropathol., 43(1): 31–39.

Han, B.H., DeMattos, R.B., Dugan, L.L., Kim-Han, J.S., Brendza, R.P., Fryer, J.D., Kierson, M., Cirrito, J., Quick, K., Harmony, J.A.K., Aronow, B.J. and Holtzman, D.M. (2001) Clusterin contributes to caspase-3-independent brain injury following neonatal hypoxia-ischemia. Nat. Med., 7(3): 338–343.

Hashimoto, T., Zhang, X.M., Chen, B.Y. and Yang, X.J. (2006) VEGF activates divergent intracellular signaling components to regulte retinal progenitor cell proliferation and neuronal differentiation. Development, 133: 2201–2210.

Houck, K.A., Leung, D.W., Rowland, A.M., Winer, J. and Ferrara, N. (1992) Dual regulation of vascular endothelial growth factor bioavailability by genetic and proteolytic mechanisms. J. Biol. Chem., 267: 26031–26037.

Jenne, D.E. and Tschopp, J. (1989) Molecular structure and functional characterization of a human complement cytolysis inhibitor found in blood and seminal plasma: identity to sulfated glycoprotein 2, a constituent of rat testis fluid. Proc. Natl. Acad. Sci. U.S.A., 86(18): 7123–7127.

Jones, S.E. and Jomary, C. (2002) Clusterin. Int. J. Biochem. Cell Biol., 34: 427–431.

Józkowicz, A., Huk, I., Nigisch, A., Weigel, G., Dietrich, W., Motterlini, R. and Dulak, J. (2003) Heme oxygenase and angiogenic activity of endothelial cells: stimulation by carbon monoxide and inhibition by tin protoporphyrin-IX. Antioxid. Redox Signal., 5: 155–162.

Kyprianou, N., Martikainen, P., Davis, L., English, H.F. and Isaacs, J.T. (1991) Programmed cell death as a new target for prostatic cancer therapy. Cancer Surv., 11: 265–277

Lakins, J., Bennett, S.A., Chen, J.H., Arnold, J.M., Morrissey, C., Wong, P., O'Sullivan, J. and Tenniswood, M. (1998) Clusterin biogenesis is altered during apoptosis in the regressing rat ventral prostate. J. Biol. Chem., 273(43): 27887–27895.

Lange, T., Guttmann-Raviv, N., Baruch, L., Machluf, M. and Neufeld, G. (2003) VEGF162, a new heparin-binding vascular endothelial growth factor splice form that is expressed in transformed human cells. J. Biol. Chem., 278: 17164–17169.

Lee, S., Jilani, S.M., Nikolova, G.V., Carpizo, D. and Iruela-Arispe, M.L. (2005) Processing of VEGF-A by matrix metalloproteinases regulates bioavailability and vascular patterning in tumors. J. Cell Biol., 169: 681–691.

Leskov, K.S., Klokov, D.Y., Li, J., Kinsella, T.J. and Boothman, D.A. (2003) Synthesis and functional analyses of nuclear clusterin, a cell death protein. J. Biol. Chem., 278(13): 11590–11600.

Leung, D.W., Cachianes, G., Kuang, W.J., Goeddel, D.V. and Ferrara, N. (1989) Vascular endothelial growth factor is a secreted angiogenic mitogen. Science, 246: 1306–1309.

Maines, M.D. (1997) The heme oxygenase system: a regulator of second messenger gases. Annu. Rev. Pharmacol. Toxicol., 37: 517–554.

McLaughlin, L., Zhu, G., Mistry, M., Ley-Ebert, C., Stuart, W.D., Florio, C.J., Groen, P.A., Witt, S.A., Kimball, T.R., Witte, D.P., Harmony, J.A. and Aronow, B.J. (2000) Apolipoprotein J/clusterin limits the severity of murine autoimmune myocarditis. J. Clin. Invest., 106(9): 1105–1113.

Morgan, T.E., Laping, N.J., Rozovsky, I., Oda, T., Hogan, T.H., Finch, C.E. and Pasinetti, G.M. (1995) Clusterin expression by astrocytes is influenced by transforming growth factor beta 1 and heterotypic cell interactions. J. Neuroimmunol., 58(1): 101–110.

Motterlini, R., Foresti, R., Bassi, R., Calabrese, V., Clark, J.E. and Green, C.J. (2000) Endothelial heme oxygenase-1 induction by hypoxia. Modulation by inducible nitric-oxide synthase and S-nitrosothiols. J. Biol. Chem., 275: 13613–13620.

Nakamura, E., Abreu-e-Lima, P., Awakura, Y., Inoue, T., Kamoto, T., Ogawa, O., Kotani, H., Manabe, T., Zhang, G.J., Kondo, K., Nosé, V. and Kaelin, W.G., Jr. (2006) Clusterin is a secreted marker for a hypoxia-inducible factor-independent function of the von Hippel-Lindau tumor suppressor protein. Am. J. Pathol., 168(2): 574–584.

Naskar, R. and Dreyer, E.B. (2001) New horizons in neuroprotection. Surv. Ophthalmol., 45(Suppl 3): S250–S255.

Nishijima, K., Ng, Y.S., Zhong, L., Bradley, J., Schubert, W., Jo, N., Akita, J., Samuelsson, S.J., Robinson, G.S., Adamis, A.P. and Shima, D.T. (2007) Vascular endothelial growth factor-A is a survival factor for retinal neurons and a critical neuroprotectant during the adaptive response to ischemic injury. Am. J. Pathol., 171(1): 53–67.

Oosthuyse, B., Moons, L., Storkebaum, E., Beck, H., Nuyens, D., Brusselmans, K., Van Dorpe, J., Hellings, P., Gorselink, M., Heymans, S., Theilmeier, G., Dewerchin, M., Laudenbach, V., Vermylen, P., Raat, H., Acker, T., Vleminckx, V., Van Den Bosch, L., Cashman, N., Fujisawa, H., Drost, M.R., Sciot, R., Bruyninckx, F., Hicklin, D.J., Ince, C., Gressens, P., Lupu, F., Plate, K.H., Robberecht, W., Herbert, J.M., Collen, D. and Carmeliet, P. (2001) Deletion of the hypoxia-response element in the vascular endothelial growth factor promoter causes motor neuron degeneration. Nat. Genet., 28: 131–138.

Orellana, D.I., Quintanilla, R.A., Gonzalez-Billault, C. and Maccioni, R.B. (2005) Role of the JAKs/STATs pathway in the intracellular calcium changes induced by interleukin-6 in hippocampal neurons. Neurotox. Res., 8: 295–304.

Park, J.E., Keller, G.-A. and Ferrara, N. (1993) The vascular endothelial growth factor isoforms (VEGF): differential deposition into the subepithelial extracellular matrix and bioactivity of extracellular matrix-bound VEGF. Mol. Biol. Cell, 4: 1317–1326.

Pasinetti, G.M., Johnson, S.A., Oda, T., Rozovsky, I. and Finch, C.E. (1994) Clusterin (SGP-2): a multifunctional glycoprotein with regional expression in astrocytes and neurons of the adult rat brain. J. Comp. Neurol., 339(3): 387–400.

Pepper, M.S. (2001) Extracellular proteolysis and angiogenesis. Thromb. Haemost., 86: 346–355.

Pepper, M.S., Ferrara, N., Orci, L. and Montesano, R. (1992) Potent synergism between vascular endothelial growth factor and basic fibroblast growth factor in the induction of angiogenesis in vitro. Biochem. Biophys. Res. Commun., 189: 824–831.

Phillips, G.D., Stone, A.M., Jones, B.D., Schultz, J.C., Whitehead, R.A. and Knighton, D.R. (1994) Vascular endothelial growth factor (rhVEGF165) stimulates direct angiogenesis in the rabbit cornea. In Vivo, 8(6): 961–965.

Plouet, J., Schilling, J. and Gospodarowicz, D. (1989) Isolation and characterization of a newly identified endothelial cell mitogen produced by AtT20 cells. EMBO J., 8: 3801–3808.

Poltorak, Z., Cohen, T., Sivan, R., Kandelis, Y., Spira, G., Vlodavsky, I., Keshet, E. and Neufeld, G. (1997) VEGF145, a secreted vascular endothelial growth factor isoform that binds to extracellular matrix. J. Biol. Chem., 272: 7151–7158.

Poon, S., Easterbrook-Smith, S.B., Rybchyn, M.S., Carver, J.A. and Wilson, M.R. (2000) Clusterin is an ATP-independent chaperone with very broad substrate specificity that stabilizes stressed proteins in a folding-competent state. Biochemistry, 39(51): 15953–15960.

Poon, S., Treweek, T.M., Wilson, M.R., Easterbrook-Smith, S.B. and Carver, J.A. (2002) Clusterin is an extracellular chaperone that specifically interacts with slowly aggregating proteins on their off-folding pathway. FEBS Lett., 513(2–3): 259–266.

Pucci, S., Bonanno, E., Pichiorri, F., Angeloni, C. and Spagnoli, L.G. (2004) Modulation of different clusterin isoforms in human colon tumorigenesis. Oncogene, 23(13): 2298–2304.

Raab, S. and Plate, K.H. (2007) Different networks, common growth factors: shared growth factors and receptors of the vascular and the nervous system. Acta Neuropathol., 113: 607–626.

Redondo, M., Villar, E., Torres-Muñoz, J., Tellez, T., Morell, M. and Petito, C.K. (2000) Overexpression of clusterin in human breast carcinoma. Am. J. Pathol., 157(2): 393–399.

Sawada, M., Sun, W., Hayes, P., Leskov, K., Boothman, D.A. and Matsuyama, S. (2003) Ku70 suppresses the apoptotic translocation of Bax to mitochondria. Nat. Cell Biol., 5(4): 320–329.

Schwarz, Q., Gu, C., Fujisawa, H., Sabelko, K., Gertsenstein, M., Nagy, A., Taniguchi, M., Kolodkin, A.L., Ginty, D.D., Shima, D.T. and Ruhrberg, C. (2004) Vascular endothelial growth factor controls neuronal migration and cooperates with Sema3A to pattern distinct compartments of the facial nerve. Genes Dev., 18: 2822–2834.

Senger, D.R., Galli, S.J., Dvorak, A.M., Perruzzi, C.A., Harvey, V.S. and Dvorak, H.F. (1983) Tumor cells secrete a vascular permeability factor that promotes accumulation of ascites fluid. Science, 219: 983–985.

Sone, H., Deo, B.K. and Kumagai, A.K. (2000) Enhancement of glucose transport by vascular endothelial growth factor in retinal endothelial cells. Invest. Ophthalmol. Vis. Sci., 41: 1876–1884.

Spitzer, M.S., Yoeruek, E., Sierra, A., Wallenfels-Thilo, B., Schraermeyer, U., Spitzer, B., Bartz-Schmidt, K.U. and Szurman, P. (2007) Comparative antiproliferative and cytotoxic profile of bevacizumab (Avastin), pegaptanib (Macugen) and ranibizumab (Lucentis) on different ocular cells. Graefes Arch. Clin. Exp. Ophthalmol., 245(12): 1837–1842.

Storkebaum, E., Lambrechts, D. and Carmeliet, P. (2004) VEGF: once regarded as a specific angiogenic factor, now implicated in neuroprotection. Bioessays, 26(9): 943–954.

Tezel, G. and Wax, M.B. (2004) Hypoxia-inducible factor 1alpha in the glaucomatous retina and optic nerve head. Arch. Ophthalmol., 122: 1348–1356.

Tischer, E., Mitchell, R., Hartman, T., Silva, M., Gospodarowicz, D., Fiddes, J.C. and Abraham, J.A. (1991) The human gene for vascular endothelial growth factor. Multiple protein forms are encoded through alternative exon splicing. J. Biol. Chem., 266(18): 11947–11954.

Tolentino, M.J., Miller, J.W., Gragoudas, E.S., Chatzistefanou, K., Ferrara, N. and Adamis, A.P. (1996) Vascular endothelial growth factor is sufficient to produce iris neovascularization and neovascular glaucoma in a nonhuman primate. Arch. Ophthalmol., 114: 964–970.

Trougakos, I.P. and Gonos, E.S. (2002) Clusterin/apolipoprotein J in human aging and cancer. Int. J. Biochem. Cell Biol., 34: 1430–1448.

Van Lenten, B.J., Wagner, A.C., Navab, M. and Fogelman, A.M. (2001) Oxidized phospholipids induce changes in hepatic paraoxonase and ApoJ but not monocyte chemoattractant protein-1 via interleukin-6. J. Biol. Chem., 276(3): 1923–1929.

Vincenti, V., Cassano, C., Rocchi, M. and Persico, G. (1996) Assignment of the vascular endothelial growth factor gene to human chromosome 6p21.3. Circulation, 93: 1493–1495.

Wehrli, P., Charnay, Y., Vallet, P., Zhu, G., Harmony, J., Aronow, B., Tschopp, J., Bouras, C., Viard-Leveugle, I., French, L.E. and Giannakopoulos, P. (2001) Inhibition of post-ischemic brain injury by clusterin overexpression. Nat. Med., 7(9): 977–979.

Wiggins, A.K., Shena, P.J. and Gundlacha, A.L. (2003) Delayed, but prolonged increases in astrocytic clusterin (ApoJ) mRNA expression following acute cortical spreading depression in the rat: evidence for a role of clusterin in ischemic tolerance. Mol. Brain Res., 114: 20–30.

Wilson, M.R. and Easterbrook-Smith, S.B. (2000) Clusterin is a secreted mammalian chaperone. Trends Biochem. Sci., 25(3): 95–98.

Wong, P., Pineault, J., Lakins, J., Taillefer, D., Léger, J., Wang, C. and Tenniswood, M. (1993) Genomic organization and expression of the rat TRPM-2 (clusterin) gene, a gene implicated in apoptosis. J. Biol. Chem., 268(7): 5021–5031.

Yamashita, T., Sawamoto, K., Suzuki, S., Suzuki, N., Adachi, K., Kawase, T., Mihara, M., Ohsugi, Y., Abe, K. and Okano, H. (2005) Blockade of interleukin-6 signaling aggravates ischemic cerebral damage in mice: possible involvement of Stat3 activation in the protection of neurons. J. Neurochem., 94: 459–468.

Zenkel, M., Kruse, F.E., Jünemann, A.G., Naumann, G.O. and Schlötzer-Schrehardt, U. (2006) Clusterin deficiency in eyes with pseudoexfoliation syndrome may be implicated in the aggregation and deposition of pseudoexfoliative material. Invest. Ophthalmol. Vis. Sci., 47(5): 1982–1990.

elevated IOP, may adversely affect neuronal survival by inducing cell death (Kitano et al., 1996; Rosenbaum et al., 1997; Gross et al., 1999; Tezel and Wax, 1999, 2000; Osborne et al., 2004). Oxidative stress causes degeneration of trabecular meshwork (Tamm et al., 1996; Izzotti et al., 2003; Saccà et al., 2005) leading to alterations in the aqueous outflow pathway, ultimately increasing IOP that, in turn, damages RGCs. Free radical species can cause neuronal cell death by inhibition of key enzymes of the tricarboxylic acid cycle, the mitochondrial electron transport chain, and the mitochondrial calcium homeostasis, leading to defective energy metabolism (Patel et al., 1996; Duchen, 2000; Beal, 2005).

Growing evidence also supports the hypothesis that oxidative stress is the leading mechanism of excitotoxic, glutamate-mediated RGC loss in vitro (Luo et al., 2001) and in vivo (Nucci et al., 2005). Accumulation of free radical species may damage glutamate transporter proteins (Muller et al., 1998), decreasing the capacity of neurons and astrocytes to metabolize glutamate, resulting in abnormal synaptic concentration of glutamate and excitotoxicity (Trotti et al., 1996; Sandhu et al., 2003). Moreover, glutamine synthetase is also oxidatively modified in ocular hypertensive eyes (Tezel et al., 2005). In addition, retinal glutamate-induced damage has been shown to be mediated in part through nitric oxide (NO), a highly reactive oxidant (Nucci et al., 2005).

It is well established that free radical scavengers are useful pharmacological tools to improve mitochondrial function and prevent neuronal cell death under excitotoxic conditions (Lipton and Rosenberg, 1994). Free radical scavengers when combined with trophic factors result in increased survival of RGCs in ocular hypertensive eyes (Ko et al., 2000) and in combination with a nitric oxide synthase inhibitor potentiate the neurotrophic effect of brain-derived neurotrophic factor on axotomized RGCs (Klocker et al., 1998).

Coenzyme Q10 (CoQ10) is an important component of the mitochondrial electron transport chain, and it is also a potent antioxidant that has been shown to afford neuroprotection in Huntington's and other neurodegenerative diseases (Beal and Matthews, 1997; Matthews et al., 1998; Beal, 1999, 2004; Ferrante et al., 2002; Shults et al., 2002, 2004). Neuroprotection by CoQ10 has in part been attributed to its free radical scavenger ability and to a specific inhibition of the mitochondrial permeability transition pore (PTP), a channel whose opening causes the mitochondrial membrane potential collapse that leads to apoptosis (Papucci et al., 2003).

In this chapter, using an established animal model of retinal ischemia/reperfusion in rat (Osborne et al., 2004), we will be focusing on assessment of neuroprotective properties of CoQ10 in RCG damage in rat retina.

Ischemia model

High IOP-induced ischemia is a frequently used animal model for retinal ischemia research (see Osborne et al., 1999, 2003, 2004) that produces pathological features almost identical to those seen in patients after central retinal artery or ophthalmic artery occlusion but may also represent a model of acute-angle closure glaucoma. The latter hypothesis is based on the observation that in animal subjected to ischemia caused by elevated IOP, a retinal damage similar to that seen in human glaucoma is produced, whereas occlusion of retinal blood vessels (e.g., occluding the carotid or vertebral arteries) causes a different pattern of damage. In particular, clamping of the carotids showed that opsin mRNA levels were decreased dramatically, indicating that the photoreceptors were preferentially affected. In contrast, ischemia induced by elevated IOP induces a large decrease in the RGC surface-associated antigen, Thy-1, whereas opsin levels were unaffected, indicating that the ganglion cells were preferentially affected (Osborne et al., 1999).

Neuroprotective effect of Coenzyme Q10 against cell loss yielded by transient ischemia in the RGC layer

Retinal ischemia for 45 min followed by 24 h reperfusion causes retinal damage and cell death (Nucci et al., 2005). Under these experimental conditions, reduction in the number of cells in the

RGC layer is also observed. The loss of RGC seems to occur via an apoptotic mechanism and this event seems to be initiated early after the ischemic insult. In fact, DNA fragmentation, evaluated by the occurrence of positive cells by terminal transferase (TdT) dvDP nick end-labeling (TUNEL) technique (Gavrieli et al., 1992), in the RGC layer was observed as early as 6 h after reperfusion (Nucci et al., 2005). The mechanism underlying cell loss implicates overactivation of N-methyl-D-aspartate (NMDA) and non-NMDA subtypes of glutamate receptors and consequent accumulation of NO (Nucci et al., 2005). Topical treatment with antioxidant agents such as CoQ10 (solution of CoQ10 0.1% + Vit-E 0.5% + saline + EDTA 0.1%) and vitamin E (Vit-E) minimizes retinal damage and cell death (Nucci et al., 2007; Fig. 1).

The mechanism underlying neuroprotection afforded by CoQ10 is not known, although it is conceivable that a free radical scavenging mechanism may play a minor role. In fact, neuroprotection afforded by CoQ10 is far greater than that provided by treatment with Vit-E alone. In addition, further support to our hypothesis comes from the neuroprotection afforded by CoQ10 in the course of experimental neurodegenerative disorders such as Alzheimer's and Parkinson's diseases whose pathogenesis implicates failure of mitochondrial energy metabolism (Beal and Matthews, 1997; Matthews et al., 1998; Beal, 1999, 2004; Ferrante et al., 2002; Shults et al., 2002, 2004; Cleren et al., 2008). Accordingly, an alternative hypothesis would be that CoQ10 reduces the detrimental action of ischemia/reperfusion on mitochondrial energy metabolism and, consequently, on the function of glutamate transporters, thus limiting accumulation of extracellular glutamate and preventing apoptotic death of RGCs. In fact, it is well documented that excessive activation of glutamate receptors via the excitotoxic cascade leads to the PTP formation and cytochrome c, a member of the mitochondrial electron transport chain, to be released from the mitochondrial intermembrane space into the cytosol, where it functions as a pro-apoptotic factor committing the RGC to death (see Kroemer and Reed, 2000). Incidentally, CoQ10 has been shown to inhibit apoptosis by maintaining the PTP in the

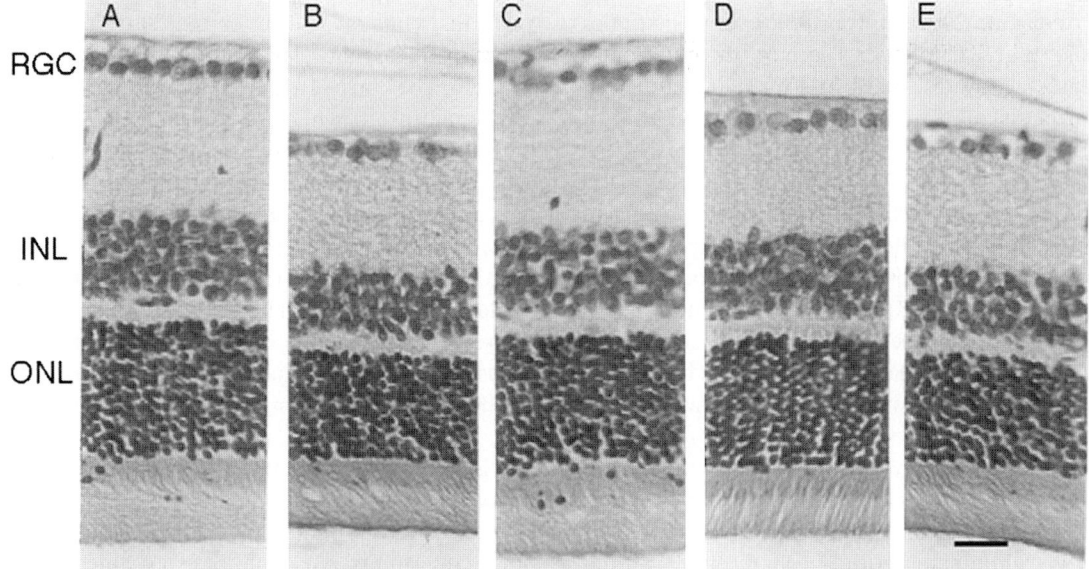

Fig. 1. Retinal ischemia for 45 min followed by 24 h reperfusion reduces the number of cells in the retinal ganglion cell layer (B) as compared to sham-operated eyes (A). Topical treatment with CoQ10 (C), or with Vit-E (D), prevents the tissue damage observed in (B). Treatment with vehicle does not prevent RGC layer loss (E). Hematoxylin and eosin staining. RGC, retinal ganglion cell layer; INL, inner nuclear layer; and ONL, outer nuclear layer. Scale bar, 50 μm. Adapted with permission from Nucci et al. (2007).

closed conformation via a mechanism independent from free radical scavenging (Papucci et al., 2003).

Retinal ischemia and glutamate

Glutamate is the major excitatory neurotransmitter in the retina and is removed from the extracellular space by an energy-dependent process involving neuronal and glial cell transporters (Barnett et al., 2001). Control of extracellular glutamate levels under physiological and pathophysiological conditions is a prerequisite for the prevention of neurodegeneration, particularly in the retina where elevated glutamate levels are associated with experimental glaucoma (Dreyer and Grosskreutz, 1997; Osborne et al., 1999, 2004; Nucci et al., 2005). Clinical studies also support a role for excitotoxicity in the development of the RGC damage in subjects suffering from glaucoma (Manabe et al., 2005). Excessive activation of glutamate receptors, mainly of the NMDA subtype, results in calcium overload, free radical formation, and activation of proteases, including caspases, and transcription factors, leading to excitotoxic neuronal death (Joo et al., 1999; Osborne et al., 2004; Lipton, 2006). Several studies suggest that excitotoxicity occurs during retinal ischemia with subsequent loss of RGCs and that this process plays a role in the pathogenesis of ischemic retinopathy (Sucher et al., 1997). In vitro exposure of the retina to the ionotropic glutamate receptor agonist NMDA produces histological changes similar to those caused by experimental ischemia (see Sucher et al., 1997). Subcutaneous injections of glutamate into young mice cause severe damage of the inner layers of the retina, most notably in the ganglion cell layer (Lucas and Newhouse, 1957). A similar pattern of damage was observed when an acute bolus of glutamate was injected intravitreally to adult rats (Sisk and Kuwabara, 1985) or following chronic administration (Samy et al., 1994). The pivotal role of excessive glutamate in the mechanisms of retinal damage is also documented by the evidence that NMDA and non-NMDA receptor antagonists afford protection in experimental models of RGC death, both in vitro (see Joo et al., 1999; Osborne et al., 1999) and in vivo (Adachi et al., 1998; Nucci et al., 2005). Under normal conditions, in the retina glutamate is removed from the extracellular space by an energy-dependent process involving neuronal and glial cell transporters (excitatory amino acid transporters, EAATs), thus terminating the excitatory signal and preventing excitotoxic neuronal damage. During ischemia, glutamate levels are thought to rise by a combination of vesicular release and reversal of the EAATs (Billups and Attwell, 1996; Maguire et al., 1998; Jabaudon et al., 2000; Rossi et al., 2000). It has been reported that glutamate levels are elevated in the vitreous humor of glaucoma patients (Dreyer et al., 1996; but see also Honkanen et al., 2003) and the progressive deterioration of visual field in glaucomatous patients is a consequence of progressive RGC death (Osborne et al., 1999).

Coenzyme Q10 minimizes glutamate increase induced by ischemia/reperfusion

However, direct evidence of accumulation of toxic levels of glutamate in experimental glaucoma has only recently been gathered by using a microdialysis technique in high IOP-induced retinal ischemia in living rat (Nucci et al., 2005). These data demonstrate that elevation of glutamate occurs during the first 10 min after induction of the ischemic insult, while a significant increase is seen during reperfusion (Fig. 2). A similar observation has been reported in the rabbit and cat retina where a much larger increase in extracellular glutamate was also seen during the reperfusion as compared to the ischemic phase (Louzada-Junior et al., 1992; Adachi et al., 1998). The lower levels of glutamate observed during the period of ischemia might reflect the tolerant nature of the retina to the ischemic insult as compared to the brain (Neal et al., 1994). In addition, the prolonged accumulation of glutamate that occurs after the ischemic insult may stem from a defective glutamate uptake system in the retinal neurons known to be devoid of high affinity for glutamate (White and Neal, 1976). Previous microdialysis experiments carried out in anesthetized rats undergone high IOP-induced ischemia have documented that glutamate levels increase in the retina and these are reversed

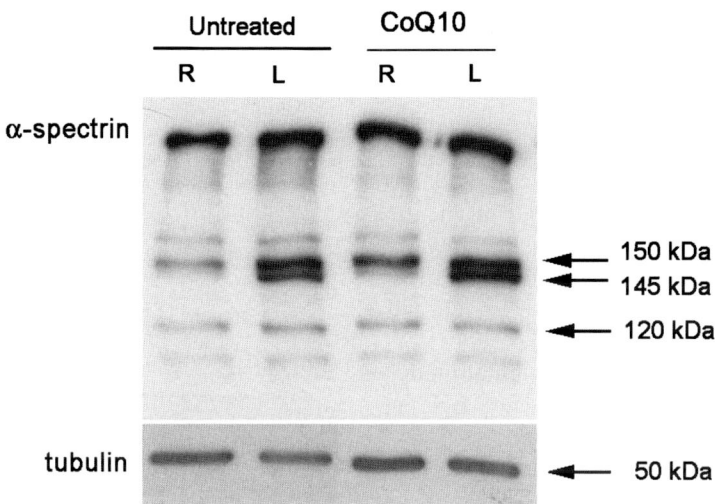

Fig. 2. High IOP-induced ischemia enhances the activity of calpain, a calcium-dependent protease, and this is unaffected by CoQ10 in the retina of rat. Immunoblots show increases of 150/145 kDa fragments specific for calpain-mediated cleavage of the substrate α-spectrin observed after 50 min of retinal ischemia in vehicle-treated rat and in rat treated intravitreally with CoQ10 (dose/volume).

by neuroprotective doses of the NMDA receptor antagonist, MK801 (Nucci et al., 2005).

To evaluate the potential role of CoQ10 in modulating glutamate neurotoxicity, this was administered 30 min before high IOP-induced ischemia in rat. Interestingly, intravitreal CoQ10 failed to counteract glutamate increase typically observed during ischemia (Nucci et al., 2007). This treatment also failed to counteract the activation of calpain, a calcium-dependent protease highly active under excitotoxic conditions (Branca, 2004), measured as cleavage of the calpain substrate α-spectrin, confirming the lack of CoQ10 effect on the release of glutamate during the ischemic phase (unpublished observation; Fig. 2).

Astrocytes and Müller cells are involved in mediating the early death of RGCs upon elevated IOP (Lam et al., 2003) and functional disorders of the glutamate uptake in Müller cells may be one of the causes of glaucoma, especially in patients with satisfactory control of IOP (Dreyer et al., 1996; Kawasaki et al., 2000).

Altogether, these data make it possible for the hypoxic nature of glutamate increase (Napper et al., 2001). At variance with these negative results, pretreatment with CoQ10, but not with Vit-E (data not shown), inhibited the increase of glutamate observed during reperfusion, suggesting that mechanisms other than the antioxidant may account for it (Nucci et al., 2007).

Accordingly, an attractive hypothesis would be that CoQ10 reduces the detrimental action of ischemia/reperfusion on mitochondrial energy metabolism and, consequently, on the function of glutamate transporters, thus limiting accumulation of extracellular glutamate and preventing apoptotic death of RGC in rat (Nucci et al., 2007).

Furthermore, pretreatment with CoQ10 slightly seems to potentiate the activation of the pro-survival self-defense pathway mediated by the serine/threonine kinase PKB/Akt (see Franke et al., 2003) normally observed after 1 h of reperfusion. This latter effect is probably a consequence of the prevention, mediated by the CoQ10 treatment, of the energy failure and the reduction of the oxidative stress associated with the retina reperfusion (unpublished observation).

Summary

Glaucoma is a neurodegenerative disease characterized by progressive death of the RGCs. Neuronal cell death may be the consequence of energy impairment that triggers secondary excitotoxicity and free radical generation. Accordingly, improvement of energy production by sustaining mitochondrial function may contribute to neuroprotection in the course of glaucoma. CoQ10, which serves as the electron acceptor for complexes I and II of the mitochondrial electron transport chain and also acts as an antioxidant, has the potential to be a beneficial agent in neurodegenerative diseases, including Parkinson's and Alzheimer's diseases, in which impaired mitochondrial function and/or excessive oxidative damage has been reported.

Here we report that the use of an animal model of RGC death induced by acute rise of IOP in combination with a neurochemical approach allows to gain more insight in the neuroprotective profile of CoQ10 in glaucoma.

Under these conditions, CoQ10, but not Vit-E, is able to reduce significantly the pathological increase of glutamate observed during reperfusion and this may contribute to the neuroprotection afforded in rats undergone transient ischemia. In conclusion, minimization of energy failure may underlie the observed neuroprotection afforded by CoQ10 in rat undergone transient ischemia and this lend support to the concept that mitochondrial dysfunction may be a risk factor for patients with primary open-angle glaucoma (Abu-Amero et al., 2006).

Acknowledgment

Financial support from the Italian Ministry of Health (Rome) is gratefully acknowledged.

References

Abu-Amero, K.K., Morales, J. and Bosley, T.M. (2006) Mitochondrial abnormalities in patients with primary open angle glaucoma. Invest. Ophthalmol. Vis. Sci., 47: 2533–2541.

Adachi, K., Kashii, S., Masai, H., Ueda, M., Morizane, C., Kaneda, K., Kume, T., Akaike, A. and Honda, Y. (1998) Mechanism of the pathogenesis of glutamate neurotoxicity in retinal ischemia. Graefes Arch. Clin. Exp. Ophthalmol., 236: 766–774.

Barnett, N.L., Pow, D.V. and Bull, N.D. (2001) Differential perturbation of neuronal and glial glutamate transport systems in retinal ischaemia. Neurochem. Int., 39: 291–299.

Beal, M.F. (1999) Coenzyme Q10 administration and its potential for treatment of neurodegenerative diseases. Biofactors, 9: 261–266.

Beal, M.F. (2004) Therapeutic effects of coenzyme Q10 in diseases. Methods Enzymol., 382: 473–487.

Beal, M.F. (2005) Mitochondria take center stage in aging and neurodegeneration. Ann. Neurol., 58: 495–505.

Beal, M.F. and Matthews, R.T. (1997) Coenzyme Q10 in the central nervous system and its potential usefulness in the treatment of neurodegenerative diseases. Mol. Aspects Med., 18(Suppl): 169–179.

Billups, B. and Attwell, D. (1996) Modulation of non-vesicular glutamate release by pH. Nature, 379: 171–174.

Branca, D. (2004) Calpain-related diseases. Biochem. Biophys. Res. Commun., 322: 1098–1104.

Cleren, C., Yang, L., Lorenzo, B., Calingasan, N.Y., Schomer, A., Sireci, A., Wille, E.J. and Beal, M.F. (2008) Therapeutic effects of coenzyme Q10 (CoQ10) and reduced CoQ10 in the MPTP model of Parkinsonism. J. Neurochem., 104: 1613–1621.

Dreyer, E.B. and Grosskreutz, C.L. (1997) Excitatory mechanisms in retinal ganglion cell death in primary open angle glaucoma (POAG). Clin. Neurosci., 4: 270–273.

Duchen, M.R. (2000) Mitochondria and calcium: from cell signaling to cell death. J. Physiol., 529: 57–68.

Ferrante, R.J., Andreassen, O.A., Dedeoglu, A., Ferrante, K.L., Jenkins, B.G., Hersch, S.M. and Beal, M.F. (2002) Therapeutic effects of coenzyme Q10 and remacemide in transgenic mouse models of Huntington's disease. J. Neurosci., 22: 592–599.

Franke, T.F., Hornik, C.P., Segev, L., Shostak, G.A. and Sugimoto, C. (2003) PI3K/Akt and apoptosis: size matters. Oncogene, 22: 8983–8998.

Gavrieli, Y., Sherman, Y. and Ben-Sasson, S.A. (1992) Identification of programmed cell death in situ via specific labeling of nuclear DNA fragmentation. J. Cell Biol., 119: 493–501.

Gross, R.L., Hensley, S.H., Gao, F. and Wu, S.M. (1999) Retinal ganglion cell dysfunction induced by hypoxia and glutamate: potential neuroprotective effects of beta-blockers. Surv. Ophthalmol., 43(Suppl 1): S162–S170.

Honkanen, R.A., Baruah, S., Zimmerman, M.B., Khanna, C.L., Weaver, Y.K., Narkiewicz, J., Waziri, R., Gehrs, K.M., Weingeist, T.A., Boldt, H.C., Folk, J.C., Russell, S.R. and Kwon, Y.H. (2003) Vitreous amino acid concentrations in patients with glaucoma undergoing vitrectomy. Arch. Ophthalmol., 121: 183–188.

Izzotti, A., Saccà, S.C., Cartiglia, C. and De Flora, S. (2003) Oxidative deoxyribonucleic acid damage in the eyes of glaucoma patients. Am. J. Med., 114: 638–646.

Jabaudon, D., Scanziani, M., Gähwiler, B.H. and Gerber, U. (2000) Acute decrease in net glutamate uptake during energy deprivation. Proc. Natl. Acad. Sci. U.S.A., 97: 5610–5615.

Joo, C., Cho, K., Kim, H., Choi, J.S. and Oh, Y.J. (1999) Protective role for bcl-2 in experimentally induced cell death of bovine corneal endothelial cells. Ophthalmic Res., 31: 287–296.

Kawasaki, A., Otori, Y. and Barnstable, C.J. (2000) Muller cell protection of rat retinal ganglion cells from glutamate and nitric oxide neurotoxicity. Invest. Ophthalmol. Vis. Sci., 41: 3444–3450.

Kitano, S., Morgan, J. and Caprioli, J. (1996) Hypoxic and excitotoxic damage to cultured rat retinal ganglion cells. Exp. Eye Res., 63: 105–112.

Klocker, N., Cellerino, A. and Bahr, M. (1998) Free radical scavenging and inhibition of nitric oxide synthase potentiates the neurotrophic effects of brain-derived neurotrophic factor on axotomized retinal ganglion cells in vivo. J. Neurosci., 18: 1038–1046.

Ko, M.L., Hu, D.N., Ritch, R. and Sharma, S.C. (2000) The combined effect of brain-derived neurotrophic factor and a free radical scavenger in experimental glaucoma. Invest. Ophthalmol. Vis. Sci., 41: 2967–2971.

Kroemer, G. and Reed, J.C. (2000) Mitochondrial control of cell death. Nat. Med., 6: 513–519.

Lam, T.T., Kwong, J.M. and Tso, M.O. (2003) Early glial responses after acute elevated intraocular pressure in rats. Invest. Ophthalmol. Vis. Sci., 44: 638–645.

Lipton, S.A. (2006) Paradigm shift in neuroprotection by NMDA receptor blockade: memantine and beyond. Nat. Rev. Drug Discov., 5: 160–170.

Lipton, S.A. and Rosenberg, P.A. (1994) Excitatory amino acids as a final common pathway for neurologic disorders. N. Engl. J. Med., 330: 613–622.

Louzada-Júnior, P., Dias, J.J., Santos, W.F., Lachat, J.J., Bradford, H.F. and Coutinho-Netto, J. (1992) Glutamate release in experimental ischemia of the retina: an approach using microdialysis. J. Neurochem., 59: 358–363.

Lucas, D.R. and Newhouse, J.P. (1957) The toxic effect of sodium L-glutamate on the inner layers of the retina. AMA Arch. Ophthalmol., 58: 193–201.

Luo, X., Lambrou, G.N., Sahel, J.A. and Hicks, D. (2001) Hypoglycemia induces general neuronal death, whereas hypoxia and glutamate transport blockade lead to selective retinal ganglion cell death in vitro. Invest. Ophthalmol. Vis. Sci., 42: 2695–2705.

Maguire, G., Simko, H., Weinreb, R.N. and Ayoub, G. (1998) Transport-mediated release of endogenous glutamate in the vertebrate retina. Pflugers Arch., 436: 481–484.

Manabe, S., Gu, Z. and Lipton, S.A. (2005) Activation of matrix metalloproteinase-9 via neuronal nitric oxide synthase contributes to NMDA-induced retinal ganglion cell death. Invest. Ophthalmol. Vis. Sci., 46: 4747–4753.

Matthews, R.T., Yang, L., Browne, S., Baik, M. and Beal, M.F. (1998) Coenzyme Q10 administration increases brain mitochondrial concentrations and exerts neuroprotective effects. Proc. Natl. Acad. Sci. U.S.A., 95: 8892–8897.

Muller, A., Maurin, L. and Bonne, C. (1998) Free radicals and glutamate uptake in the retina. Gen. Pharmacol., 30: 315–318.

Napper, G.A., Pianta, M.J. and Kalloniatis, M. (2001) Localization of amino acid neurotransmitters following in vitro ischemia and anoxia in the rat retina. Vis. Neurosci., 18: 413–427.

Neal, M.J., Cunningham, J.R., Hutson, P.H. and Hogg, J. (1994) Effects of ischemia on neurotransmitter release from the isolated retina. J. Neurochem., 62: 1025–1033.

Nucci, C., Tartaglione, R., Cerulli, A., Mancino, R., Spanò, A., Cavaliere, F., Rombolà, L., Bagetta, G., Corasaniti, M.T. and Morrone, L.A. (2007) Retinal damage caused by high intraocular pressure-induced transient ischemia is prevented by coenzyme Q10 in rat. Int. Rev. Neurobiol., 82: 397–406.

Nucci, C., Tartaglione, R., Rombolà, L., Morrone, L.A., Fazzi, E. and Bagetta, G. (2005) Neurochemical evidence to implicate elevated glutamate in the mechanisms of high intraocular pressure (IOP)-induced retinal ganglion cell death in rat. Neurotoxicology, 26: 935–941.

Osborne, N.N., Casson, R.J., Wood, J.P., Childlow, G., Graham, M. and Melena, J. (2004) Retinal ischemia: mechanisms of damage and potential therapeutic strategies. Prog. Retin. Eye Res., 23: 91–147.

Osborne, N.N., Chidlow, G., Wood, J. and Casson, R. (2003) Some current ideas on the pathogenesis and the role of neuroprotection in glaucomatous optic neuropathy. Eur. J. Ophthalmol., 13: s19–s26.

Osborne, N.N., Ugarte, M., Chao, M., Childlow, G., Bae, J.H., Wood, J.P. and Nash, M.S. (1999) Neuroprotection in relation to retinal ischemia and relevance to glaucoma. Surv. Ophthalmol., 43(Suppl 1): s102–s128.

Papucci, L., Schiavone, N., Witort, E., Donnini, M., Lapucci, A., Tempestini, A., Formigli, L., Zecchi-Orlandini, S., Orlandini, G., Carella, G., Brancato, R. and Capaccioli, S. (2003) Coenzyme Q10 prevents apoptosis by inhibiting mitochondrial depolarization independently of its free radical scavenging property. J. Biol. Chem., 278: 28220–28228.

Patel, M., Day, B.J., Crapo, J.D., Fridovich, I. and McNamara, J.O. (1996) Requirement for superoxide in excitotoxic cell death. Neuron, 16: 345–355.

Rosenbaum, D.M., Rosenbaum, P.S., Gupta, A., Michaelson, M.D., Hall, D.H. and Kessler, J.A. (1997) Retinal ischemia leads to apoptosis which is ameliorated by aurintricarboxylic acid. Vis. Res., 37: 3445–3451.

Rossi, D.J., Oshima, T. and Attwell, D. (2000) Glutamate release in severe brain ischaemia is mainly by reversed uptake. Nature, 403: 316–321.

Saccà, S.C., Pascotto, A., Camicione, P., Capris, P. and Izzotti, A. (2005) Oxidative DNA damage in the human trabecular meshwork: clinical correlation in patients with primary open angle glaucoma. Arch. Ophthalmol., 123: 458–463.

Samy, C.N., Lui, C.J., Kaiser, P.K., Lipton, S.A. and Dreyer, E.B. (1994) Toxicity of chronic glutamate administration to the retina. Invest. Ophthalmol. Vis. Sci., 35: p. 497.

Sandhu, J.K., Pandey, S., Ribecco-Lutkiewicz, M., Monette, R., Borowy-Borowski, H., Walker, P.R. and Sikorska, M. (2003) Molecular mechanisms of glutamate neurotoxicity in mixed cultures of NT2-derived neurons and astrocytes: protective effects of coenzyme Q10. J. Neurosci. Res., 72: 691–703.

Shults, C.W., Beal, M.F., Song, D. and Fontaine, D. (2004) Pilot trial of high dosages of coenzyme Q10 in patients with Parkinson's disease. Exp. Neurol., 188: 491–494.

Shults, C.W., Oakes, D., Kieburtz, K., Beal, M.F., Haas, R., Plumb, S., Juncos, J.L., Nutt, J., Shoulson, I., Carter, J., Kompoliti, K., Perlmutter, J.S., Reich, S., Stern, M.,

Watts, R.L., Kurlan, R., Molho, E., Harrison, M. and Lew, M. (2002) Effects of coenzyme Q10 in early Parkinson disease: evidence of slowing of the functional decline. Arch. Neurol., 59: 1541–1550.

Sisk, D.R. and Kuwabara, T. (1985) Histologic changes in the inner retina of albino rats following intravitral injection of monosodium L-glutamate. Graefes Arch. Clin. Exp. Ophthalmol., 223: 250–258.

Sucher, N.J., Lipton, S.A. and Dreyer, E.B. (1997) Molecular basis of glutamate toxicity in retinal ganglion cells. Vision Res., 37: 3483–3493.

Tamm, E.R., Russell, P., Johnson, D.H. and Piatigorsky, J. (1996) Human and monkey trabecular meshwork accumulate alpha B-crystallin in response to heat shock and oxidative stress. Invest. Ophthalmol. Vis. Sci., 37: 2402–2413.

Tezel, G. (2006) Oxidative stress in glaucomatous neurodegeneration: mechanisms and consequences. Prog. Retin. Eye Res., 25: 490–513.

Tezel, G. and Wax, M.B. (1999) Inhibition of caspase activity in retinal cell apoptosis induced by various stimuli in vitro. Invest. Ophthalmol. Vis. Sci., 40: 2660–2667.

Tezel, G. and Wax, M.B. (2000) Increased production of tumor necrosis factor-alpha by glial cells exposed to simulated ischemia or elevated hydrostatic pressure induces apoptosis in cocultured retinal ganglion cells. J. Neurosci., 20: 8693–8700.

Tezel, G., Yang, X. and Cai, J. (2005) Proteomic identification of oxidatively modified retinal proteins in a chronic pressure-induced rat model of glaucoma. Invest. Ophthalmol. Vis. Sci., 46: 3177–3187.

Trotti, D., Rossi, D., Gjesdal, O., Levy, L.M., Racagni, G., Danbolt, N.C. and Volterra, A. (1996) Peroxynitrite inhibits glutamate transporter subtypes. J. Biol. Chem., 271: 5976–5979.

White, R.D. and Neal, M.J. (1976) The uptake of L-glutamate by the retina. Brain Res., 111: 79–93.

CHAPTER 40

17β-Estradiol prevents retinal ganglion cell loss induced by acute rise of intraocular pressure in rat

Rossella Russo[1], Federica Cavaliere[1], Chizuko Watanabe[2], Carlo Nucci[3,4], Giacinto Bagetta[1,5], Maria Tiziana Corasaniti[4,6], Shinobu Sakurada[2] and Luigi Antonio Morrone[1,5,*]

[1] *Department of Pharmacobiology, University of Calabria, 87036 Arcavacata di Rende, Italy*
[2] *Department of Physiology and Anatomy, Tohoku Pharmaceutical University, Sendai, Japan*
[3] *Physiopathological Optics, Department of Biopathology, University of Rome "Tor Vergata," 00133 Rome, Italy*
[4] *"Mondino-Tor Vergata" Center for Experimental Neurobiology, University of Rome "Tor Vergata," 00133 Rome, Italy*
[5] *Section of Neuropharmacology of Normal and Pathological Neuronal Plasticity, University Center for Adaptive Disorders and Headache (UCADH), University of Calabria, 87036 Arcavacata di Rende, Italy*
[6] *Department of Pharmacobiological Sciences, University "Magna Graecia" of Catanzaro, 88100 Catanzaro, Italy*

Abstract: Glaucoma, is a progressive optic neuropathy often associated with increased intraocular pressure (IOP) and characterized by progressive death of retinal ganglion cells (RGCs). High acute rise of IOP is a model for retinal ischemia and may represent a model of acute angle closure glaucoma. Here we have used this experimental model in combination with a neurochemical and neuropathological approach to gain more insight in the neuroprotective profile of 17β-estradiol (E2), a steroid hormone, which has been shown to increase the viability, survival, and differentiation of primary neuronal cultures from different brain areas including amygdala, hypothalamus, and neocortex. Our data demonstrate that systemic administration of E2 significantly reduces RGC loss induced by high IOP in rat. In addition, pretreatment with E2, 30 min before ischemia, minimizes the elevation of glutamate observed during the reperfusion period. These effects seem to be in part mediated by the activation of the estrogen receptor, since a pretreatment with ICI 182-780, a specific estrogen receptor antagonist, partially counteracts the neuroprotection afforded by the estrogen.

Keywords: glaucoma; excitotoxicity; oxidative stress; estrogens; microdialysis; neuroprotection

Introduction

Glaucoma, one of the leading causes of blindness in the world (Quigley, 1996), is a progressive optic neuropathy often associated with increased intraocular pressure (IOP) that is characterized by progressive death of retinal ganglion cells (RGCs) and consequent deterioration of the visual field. Several studies suggest that excitotoxicity occurs during retinal ischemia with subsequent loss of RGCs and that this process plays a role in the pathogenesis of ischemic retinopathy (Nucci et al., 2005; Casson, 2006). Glutamate functions as the major excitatory amino acid neurotransmitter

*Corresponding author. Tel.: +39 0984 493054; Fax: +39 0984 493462; E-mail: luigimorron@libero.it

in the retina, but at high concentrations it becomes neurotoxic. The pivotal role of excessive glutamate in the mechanisms of retinal damage is also documented by the evidence that NMDA and non-NMDA receptor antagonists afford protection in experimental models of RGC death, both in vitro and in vivo (Adachi et al., 1998; Joo et al., 1999; Osborne et al., 1999; Nucci et al., 2005). Growing evidence also supports that oxidative stress is the leading mechanism of excitotoxic, glutamate-induced RGC loss in vitro (Luo et al., 2001) and in vivo (Nucci et al., 2005). Not only the interference with glutamate transmission may stem from the decrease of the retinal glutamate transport (Muller et al., 1998), but also glutamine synthetase, which converts glutamate to glutamine, is oxidatively modified in ocular hypertensive eyes (Tezel et al., 2005). In addition, retinal glutamate damage has been shown to be mediated in part through nitric oxide, a highly reactive radical species (Nucci et al., 2005). Based on previous observations, neuroprotection of RGC from excitotoxicity is becoming an important approach of glaucoma therapy. Interestingly, it has been recently demonstrated that 17β-estradiol (E2) minimizes RGC loss in DBA/2J mouse, an in vivo model of an inherited (pigmentary) glaucoma (Zhou et al., 2007), shows neuroprotective effect on axotomy-induced RGC death (Nakazawa et al., 2006), and protects RGC against glutamate cytotoxicity (Kumar et al., 2005). In addition, several studies showed that estrogens, a family of cholesterol-derived steroid hormones, are protective against various oxidative stress insults including excitotoxicity (Goodman et al., 1996; Singer et al., 1996, 1999; Weaver et al.,1997; Zaulyanov et al., 1999). Cumulative evidence from basic science and clinical research suggests that estrogens play a significant neuromodulatory and neuroprotective role in the brain, and this underlies their ability to ameliorate symptoms and decrease the risk of neurodegenerative conditions such as cerebrovascular stroke, Alzheimer's disease, and Parkinson's disease (see Amantea et al., 2005). The mechanisms underlying estrogen neuroprotection have not been completely elucidated and several mechanisms have been proposed to explain the neurotrophic and neuroprotective actions of estrogens, including modulation of synaptogenesis, protection against apoptosis, anti-inflammatory activity, and increased cerebral blood flow (Garcia-Segura et al., 2001; Wise, 2003; Maggi et al., 2004).

Using high IOP experimental model in rat, in combination with a neurochemical and neuropathological approach, we now report the neuroprotection afforded by systemic treatment with E2, showing that this hormone is able to prevent the glutamate-induced loss of RGC under these experimental conditions.

Methods

Male, Wistar rats (250–300 g) (Charles River, Lecco, Italy) were maintained on a 12-h light–dark cycle. Before ischemia was induced, animals were anesthetized with chloral hydrate (400 mg kg^{-1}, intraperitoneally (i.p.)). Corneal analgesia was achieved using topical drops of oxibuprocaine 0.4% (Novesina, Novartis Farma, Italy). Pupillary dilation was maintained using 0.5% tropicamide (Visumidriatic 0.5%, Visufarma, Italy). The anterior chamber of the right eye was cannulated with a 27-gauge infusion needle connected to a 500 mL plastic container of sterile saline, the IOP was raised to 120 mmHg for 50 min by elevating the saline reservoir. Retinal ischemia was confirmed by observing whitening of the iris and loss of the red reflex of the retina. Sham procedure was performed without the elevation of the bottle in control rats.

Morphometric analysis

Rats receiving ischemic insult or sham procedure were anaesthetized (chloral hydrate 400 mg kg^{-1}, i.p.) and perfused through the left ventricle of the heart with 50 mL of heparinized phosphate buffered (pH 7.4) saline followed by 50 mL of 4% paraformaldehyde in phosphate buffered saline at 6 h, 12 h, 24 h, 48 h, and 7 days after reperfusion ($n = 6$ per group). Two hours after the perfusion procedure had been completed, the eyes were enucleated and post-fixed in 4% paraformaldehyde for 72 h. Serial coronal sections, cut along the vertical meridian of the eye passing through the optic nerve head, were stained with hematoxylin and eosin (H&E). The number of RGC was counted in six areas (25 μm × 25 μm each) of each section ($n = 6$ per eye) at 300 μm from the optic nerve head on the superior and inferior hemisphere, using light microscopy (40 × magnification). The data were expressed as mean ± SEM per area, and were evaluated statistically for differences using the Student's t-test.

Microdialysis

Extracellular glutamate was monitored in the retina of anesthetized rats (urethane, 1500 mg kg^{-1}, i.p.) during and after pressure-induced ischemia using a microdialysis technique. For implantation, a microdialysis probe (concentric design, 2 mm regenerated cellulose membrane, molecular weight cutoff 5 kDa) was implanted into vitreous chamber through the nonvascular pars plana region of the sclerotic coat after it had been punctured with a surgical needle (23 gauge). The surface of the dialysis membrane was secured perpendicularly to the retina for stable sampling during the experiment. Superfusion medium was continuously delivered via the probe at a rate of 2 μL min^{-1}. The composition of the medium (in mM) was as follows: NaCl, 125; KCl, 2.5; MgCl$_2$, 1.18; CaCl$_2$, 1.26; NaH$_2$PO$_4$, 0.2; pH adjusted to 7.0. After 2 h stabilization period, dialysate samples (20 μL) were collected at 10 min intervals before, during, and after ischemia. For analysis, the dialysate samples were derivatized with o-phthalaldehyde (OPA) and the concentration of glutamate determined as previously reported (Richards et al., 2000) by means of a high-performance liquid chromatography (HPLC) equipped with a fluorescence detector. Briefly, separation was achieved with a Hypersil ODS column (5 mm, 150 mm × 3 mm, Chrompack, Milan, Italy) using a short methanol gradient (7–14% over 15 min) in 50 mM sodium acetate buffer, pH 6.95, followed by elution of remaining peaks with 95% methanol. Total run time was 17 min. The baseline concentration of glutamate was the mean concentration obtained by averaging the six samples collected consecutively at 10 min intervals immediately prior to the onset of ischemia and was used as control.

All experiments were carried out in accordance with the European Community Council Directive of November 24, 1986 (86/609/EEC). All efforts were made to minimize animal suffering and to use only the number of animals necessary to produce reliable results.

Drug application

For neuropathological studies, control animals ($n = 6$) received injections of saline (1 mg kg^{-1}, given i.p. twice daily), whereas test group received E2 (i.p., 0.2 mg kg^{-1}; $n = 3$) 30 min before ischemia or the estrogen receptor antagonist, ICI 182-780 (i.p., 0.2 and 2 mg kg^{-1}; $n = 3$ per group) 1 h before injection of E2.

For neurochemical studies, animals received systemic administration of E2 (i.p., 0.2 mg kg^{-1}; $n = 5$) and the 17α-isomer of estradiol (E2α, i.p., 0.2 mg kg^{-1}; $n = 3$) 30 min before ischemia. ICI 182-780 (i.p., 0.2 and 2 mg kg^{-1}; $n = 5$ and $n = 3$, respectively) was administered 1 h before injection of E2.

E2, E2α, and ICI 182-780 were purchased from SIGMA (Italy).

Statistical analysis

All numerical data are expressed as the mean ± SEM. Data were tested for statistical significance with paired Student's t-test or by ANOVA followed by Dunnett's test for multiple comparisons.

Results

17β-Estradiol pretreatment minimizes RGC loss

As shown in Table 1, 50 min of IOP-induced ischemia followed by 24 h of reperfusion caused a reduction in the number of RGCs by 28.03%. Systemic administration of E2 (0.2 mg kg^{-1}), 30 min before ischemia, protected against RGC damage observed 24 h after delivery of the ischemic insult (Fig. 1) and significantly reduced the percentage loss of RGC to 7.14% (Table 1). A pretreatment with ICI 182-780, a specific estrogen receptor antagonist, failed to abrogate the neuroprotection afforded by E2 (−6.63%) at the doses of 0.2 mg kg^{-1} (Table 1, Fig. 1), whereas it partially counteracted (−15.18%) the effect of E2 at a dose of 2 mg kg^{-1} (Table 1).

Table 1. Retinal ganglion cell (RGC) loss induced by acute high intraocular pressure is prevented by systemic treatment with 17β-estradiol

Experimental conditions	Number of RGC	Percentage vs. control
Control (sham-operated)	35.43 ± 0.08	
Ischemia	25.50 ± 0.29[#]	−28.03
17β-Estradiol (0.2 mg kg^{-1}) + ischemia	32.90 ± 0.82*	−7.14
ICI 182-780 (0.2 mg kg^{-1}) + 17β-estradiol + ischemia	33.08 ± 0.32*,§	−6.63
ICI 182-780 (2 mg kg^{-1}) + 17β-estradiol + ischemia	30.05 ± 0.22	−15.18

Elevated IOP-induced ischemia for 50 min was followed by a 24-h reperfusion period. Control animals ($n = 6$) received injections of saline (1 mg kg^{-1}, given i.p. twice daily), whereas test group received i.p. 17β-estradiol (E2, $n = 3$) 30 min before ischemia or the estrogen receptor antagonist, ICI 182-780 (i.p., $n = 3$ per group), 1 h before injection of E2.
Cell counting was performed in the ganglion cell layer of ischemic/reperfused and sham-operated rat retinas stained with hematoxylin and eosin. The number of RGC was counted in six areas of each section ($n = 6$ for eye) using light microscopy. The data were expressed as mean ± SEM per area, and were evaluated statistically for differences using the Student's t-test.
[#]$p = 0.000$ vs. control.
*$p < 0.01$ vs. ischemia.
§$p = 0.8$ vs. E2 + ischemia

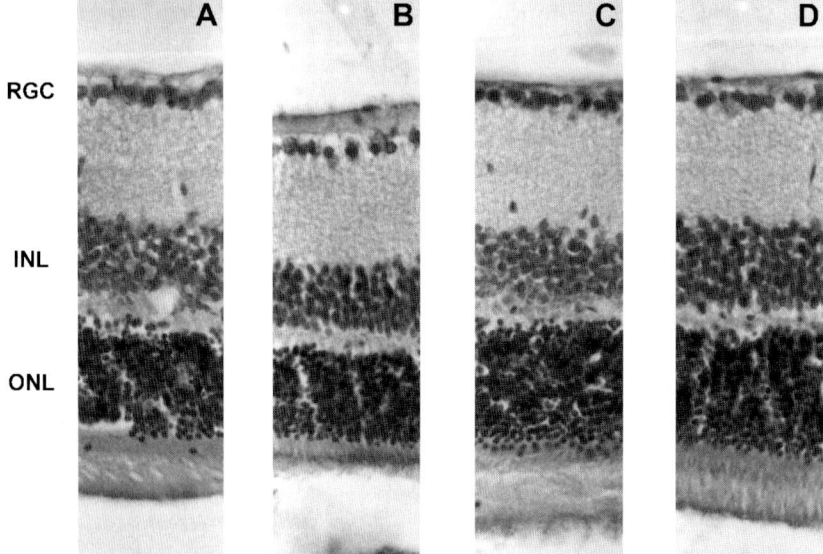

Fig. 1. Retinal ischemia for 50 min followed by 24 h reperfusion reduces the number of cells in the retinal ganglion cell layer (B, $n = 6$ rats) as compared to sham-operated rats (A, $n = 6$). Systemic treatment with 17β-estradiol (C, i.p., 0.2 mg kg^{-1}, $n = 3$ rats) prevents the tissue damage observed in (B). A pretreatment with ICI 182-780 (D, i.p., 0.2 mg kg^{-1}, $n = 3$ rats), a specific estrogen receptor antagonist, failed to abrogate the neuroprotection afforded by 17β-estradiol. H&E staining. RGC: retinal ganglion cell layer; INL: inner nuclear layer; ONL: outer nuclear layer.

High IOP-induced ischemia enhances extracellular glutamate in the retina: effect of 17β-estradiol

The time course of changes in extracellular glutamate during ischemia and reperfusion in rat ($n = 6$) is illustrated in Fig. 2. The extracellular level of glutamate from the retina (1.089 ± 0.160 μM) increased after the first 10 min of ischemia (2.032 ± 0.258 μM) with a larger and statistically significant increase observed at 10 and 150 min after the reperfusion had started (4.465 ± 0.746, $p < 0.001$ and 3.683 ± 1.158 μM, $p < 0.05$, respectively).

Systemic administration of E2 (0.2 mg kg^{-1} given i.p.; $n = 5$ rats), 30 min before ischemia, did not

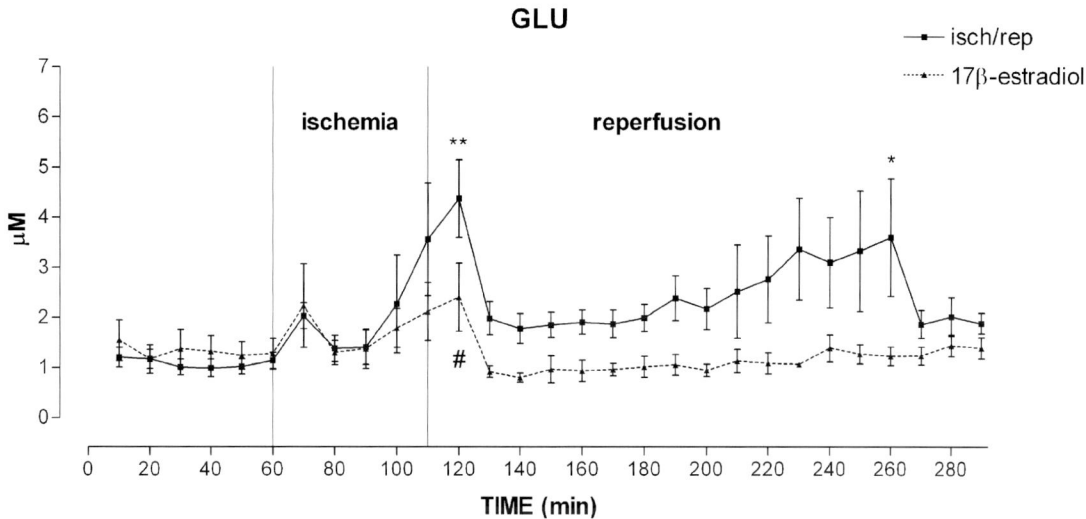

Fig. 2. Neurochemical data obtained by intraocular microdialysis experiments in anesthetized rats ($n = 6$) demonstrate that ischemia/reperfusion insult increases intraretinal glutamate. The extracellular level of glutamate (GLU, solid line) from the retina tended to increase after the first 10 min of ischemia with a larger and statistically significant increase observed at 10 and 150 min after the reperfusion had started. Systemic administration of 17β-estradiol (i.p., $0.2\,mg\,kg^{-1}$, $n = 5$ rats, dashed line), 30 min before ischemia, did not significantly affect the GLU peak increase observed at 10 min after ischemia, whereas it minimized the elevation of GLU observed during the reperfusion period. The baseline concentrations of GLU were the mean concentrations obtained by averaging the six samples collected consecutively at 10 min intervals immediately prior to the onset of ischemia and were used as basal values. Glutamate values (μM) are expressed as mean ± SEM. Statistical significance was assessed by ANOVA followed by Dunnett's test for multiple comparisons [#] and [*]$p < 0.05$ vs. basal values; [**]$p < 0.001$ vs. basal values.

significantly affect the glutamate peak increase observed at 10 min after ischemia whereas it minimized the elevation of glutamate observed during the reperfusion period (Fig. 2). More importantly, E2 counteracted the glutamate increase observed after 10 and 150 min of reperfusion ($2.406 \pm 0.681\,\mu M$ vs. basal levels $1.318 \pm 0.307\,\mu M$, $p < 0.05$ and 1.224 ± 0.183 vs. basal levels $1.318 \pm 0.307\,\mu M$, respectively) (Figs. 2 and 3). Pretreatment with the estrogen receptor antagonist ICI 182-780 ($0.2\,mg\,kg^{-1}$ given i.p.; $n = 5$ rats) failed to counteract the effect on extracellular glutamate levels by E2 during reperfusion (Fig. 3), whereas at the dose of $2\,mg\,kg^{-1}$ ($n = 3$) it counteracted the effect of E2 at 10 min after reperfusion (data not shown). Interestingly, systemic administration, 30 min before ischemia, of E2α ($0.2\,mg\,kg^{-1}$, given i.p.; $n = 3$ rats), which weakly binds to estrogen receptors, does not affect the glutamate peak observed at 10 min of reperfusion but, likewise to E2, counteracted the glutamate increase in the late reperfusion phase (Fig. 4).

Discussion

High IOP-induced ischemia is an established animal model to study the mechanisms underlying RGC death that also recapitulates features of acute angle closure glaucoma (Osborne et al., 2004). Recently, under these experimental conditions, we have reported that a delayed and progressive loss of viable cells in the RGC layer is observed starting from 6 h after the beginning of the reperfusion to peak at 7 days (Nucci et al., 2005). The mechanism underlying cell loss implicates overactivation of NMDA and non-NMDA subtypes of glutamate receptors and consequent accumulation of nitric oxide, being the loss minimized by systemic pretreatment with antagonists of the NMDA and non-NMDA receptors and by systemic pretreatment with l-NAME, an inhibitor of nitric oxide synthase (Nucci et al., 2005). The excitotoxic, glutamate-mediated, nature of the underlying mechanism of RGC death has also been confirmed by neurochemical data demonstrating that, during

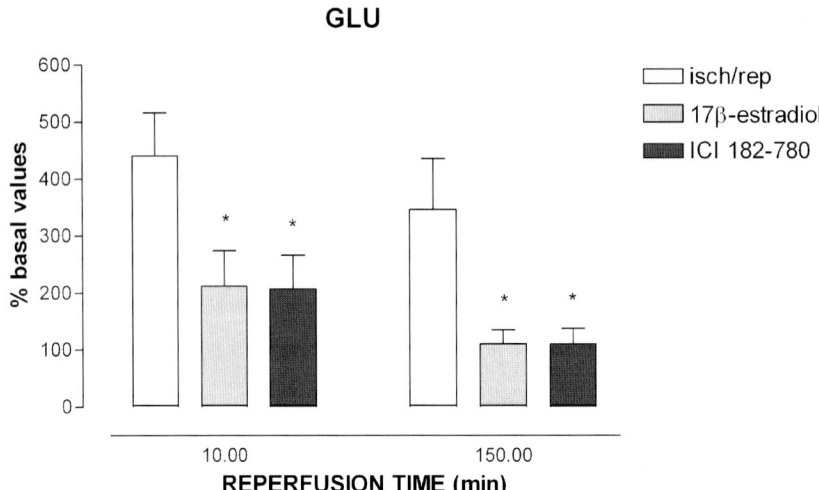

Fig. 3. Effect of 17β-estradiol (i.p., 0.2 mg kg^{-1}, $n = 5$ rats) and of the estrogen receptor antagonist ICI 182-780 (0.2 mg kg^{-1}, $n = 5$ rats) on levels of glutamate (GLU) observed at 10 and 150 min after reperfusion had started. Administration of 17β-estradiol (gray columns) significantly reduced the GLU increase observed after 10 and 150 min of reperfusion. ICI 182-780 (black columns) failed to counteract the effect on extracellular GLU levels afforded by 17β-estradiol. The white columns show the extracellular GLU levels obtained in ischemia/reperfusion (isch/rep) experiments ($n = 6$). Data are expressed as mean ± SEM percentage of basal values of GLU. The baseline concentrations of glutamate were the mean concentrations obtained by averaging the six samples collected consecutively at 10 min intervals immediately prior to the onset of ischemia. Data were tested for statistical significance with paired, two-tailed, Student's t-test. $^{*}p < 0.05$ vs. isch/rep.

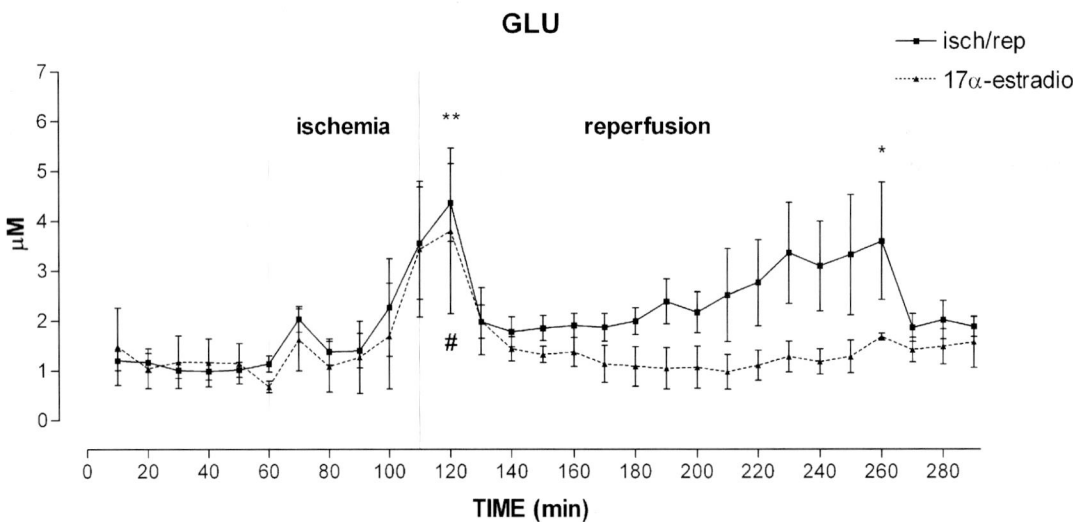

Fig. 4. Neurochemical data obtained by intraocular microdialysis experiments in anaesthetized rats ($n = 6$) demonstrate that the extracellular level of glutamate (GLU, solid line) from the retina tended to increase after the first 10 min of ischemia with a larger and statistically significant increase observed at 10 and 150 min after the reperfusion had started. Systemic administration of 17α-estradiol (i.p., 0.2 mg kg^{-1}, $n = 3$ rats, dashed line), 30 min before ischemia, did not significantly affect the GLU peak increase observed at 10 min after ischemia and after 10 min of reperfusion, whereas it minimized the elevation of GLU observed during the late reperfusion period. The baseline concentrations of GLU were the mean concentrations obtained by averaging the six samples collected consecutively at 10 min intervals immediately prior to the onset of ischemia and were used as basal values. Glutamate values (μM) are expressed as mean ± SEM. Statistical significance was assessed by ANOVA followed by Dunnett's test for multiple comparisons $^{\#}$ and $^{*}p < 0.05$ vs. basal values; $^{**}p < 0.001$ vs. basal values.

reperfusion, extracellular glutamate increases significantly in the retina of the ischemic eye and this is sensitive to the prevention afforded by systemic MK801 (Nucci et al., 2005).

The present study shows that systemic pretreatment with E2, 30 min before ischemia, prevents glutamate increase during reperfusion and this is accompanied by minimization of RGC death at 24 h of reperfusion. Accordingly, recent data demonstrated that E2 minimizes RGC loss in DBA/2J mouse, an in vivo model of an inherited glaucoma (Zhou et al., 2007), has neuroprotective effect on axotomy-induced RGC death (Nakazawa et al., 2006), and protects RGC against glutamate cytotoxicity (Kumar et al., 2005). Here we monitored in vivo extracellular glutamate levels from rat retina during and after pressure-induced ischemia using an established microdialysis technique (Nucci et al., 2005). An increase of dialysate glutamate levels occurred during ischemia, followed by a more pronounced and statistically significant increase during the reperfusion phase.

Interestingly, neurochemical data show that pretreatment with E2 fails to counteract glutamate increase typically observed during ischemia, while it significantly inhibits the subsequent increase observed during reperfusion.

The mechanism underlying neuroprotection afforded by E2 is not known; however, under our experimental conditions, the effect of E2 on RGC seems to be mediated in part by the activation of the estrogen receptors. In fact, a pretreatment with high concentrations of a specific estrogen receptor antagonist, the compound ICI 182-780, partially counteracts the neuroprotection afforded by the estrogen. Interestingly, this high dose of ICI 182-780 is able to counteract partially the effect of E2 on extracellular levels of glutamate during reperfusion period, whereas a lower dose of ICI 182-780 fails to counteract any effect of E2 on extracellular glutamate levels during reperfusion. Particularly, the higher dose of ICI 182-780 prevents the effect of E2 in the early reperfusion phase, but not during the late phase. This experimental observation seems to suggest that in our experimental conditions, the increase of glutamate observed in the early phase of reperfusion is under the control of estrogen receptors whereas the accumulation of glutamate in the late reperfusion phase seems to stem from an estrogen receptor-independent mechanism. Accordingly, we demonstrate that the E2α, a weak estrogen receptor agonist (Clark and Markaverich, 1983; Lubahn et al., 1985, Yang et al., 2003), does not affect the glutamate increase observed in the early phase of reperfusion but it is effective as E2 in minimizing the glutamate increase observed in the late reperfusion period. The latter result strengthens the hypothesis that this estrogenic effect is mediated via a mechanism that does not require binding to the cytosolic estrogen receptor.

Interestingly, although many of the estrogen neuroprotective effects appear to be mediated through the activation of intracellular estrogen receptors (ER), ERα and ERβ, several observations support the possibility that estrogens exert their potent neuroprotective effects through a mitochondrial mechanism (see Simpkins and Dykens, 2008). Increasing experimental evidence implicates failure of mitochondrial energy metabolism in the pathogenesis of neurodegenerative disorders such as Alzheimer's and Parkinson's diseases (see Tezel, 2006). Following a detrimental stimulus, estrogens preserve ATP production, prevent production of ROS, moderate excessive cellular and mitochondrial Ca^{2+} loading, and preserve mitochondrial membrane potential (see Simpkins and Dykens, 2008). Accordingly, the estrogen can prevent glutamate-related cell death by decreasing extracellular glutamate levels through an increased glutamate uptake capacity by astrocytes (Pawlak et al., 2005). The reported effects might be responsible for the ability of estrogen in reducing the detrimental action of ischemia/reperfusion on glutamate transporters thus limiting accumulation of extracellular glutamate and preventing the death of RGC.

Acknowledgment

Financial support from the Italian Ministry of Health (Rome) is acknowledged.

References

Adachi, K., Kashii, S., Masai, H., Ueda, M., Morizane, C., Kaneda, K., Kume, T., Akaike, A. and Honda, Y. (1998) Mechanism of the pathogenesis of glutamate neurotoxicity in retinal ischemia. Graefes Arch. Clin. Exp. Ophthalmol., 236: 766–774.

Amantea, D., Russo, R., Bagetta, G. and Corasaniti, M.T. (2005) From clinical evidence to molecular mechanisms underlying neuroprotection afforded by estrogens. Pharmacol. Res., 52: 119–132.

Casson, R.J. (2006) Possible role of excitotoxicity in the pathogenesis of glaucoma. Clin. Exp. Ophthalmol., 34: 54–63.

Clark, J.H. and Markaverich, B.M. (1983) The agonistic and antagonistic effects of short acting estrogens: a review. Pharmacol. Ther., 21: 429–453.

Garcia-Segura, L.M., Azcoitia, I. and DonCarlos, L.L. (2001) Neuroprotection by estradiol. Prog. Neurobiol., 63: 29–60.

Goodman, Y., Bruce, A.J., Cheng, B. and Mattson, M.P. (1996) Estrogens attenuate and corticosterone exacerbates excitotoxicity, oxidative injury, and amyloid beta-peptide toxicity in hippocampal neurons. J. Neurochem., 66: 1836–1844.

Joo, C., Cho, K., Kim, H., Choi, J.S. and Oh, Y.J. (1999) Protective role for bcl-2 in experimentally induced cell death of bovine corneal endothelial cells. Ophthalmic Res., 31: 287–296.

Kumar, D.M., Perez, E., Cai, Z.Y., Aoun, P., Brun-Zinkernagel, A.M., Covey, D.F., Simpkins, J.W. and Agarwal, N. (2005) Role of nonfeminizing estrogen analogues in neuroprotection of rat retinal ganglion cells against glutamate-induced cytotoxicity. Free Radic. Biol. Med., 38: 1152–1163.

Lubahn, D.B., McCarty, K.S., Jr. and McCarty, K.S., Sr. (1985) Electrophoretic characterization of purified bovine, porcine, murine, rat, and human uterine estrogen receptors. J. Biol. Chem., 260: 2515–2526.

Luo, X., Lambrou, G.N., Sahel, J.A. and Hicks, D. (2001) Hypoglycemia induces general neuronal death, whereas hypoxia and glutamate transport blockade lead to selective retinal ganglion cell death in vitro. Invest. Ophthalmol. Vis. Sci., 42: 2695–2705.

Maggi, A., Ciana, P., Belcredito, S. and Vegeto, E. (2004) Estrogens in the nervous system: mechanisms and nonreproductive functions. Annu. Rev. Physiol., 66: 291–313.

Muller, A., Maurin, L. and Bonne, C. (1998) Free radicals and glutamate uptake in the retina. Gen. Pharmacol., 30: 315–318.

Nakazawa, T., Takahashi, H. and Shimura, M. (2006) Estrogen has a neuroprotective effect on axotomized RGCs through ERK signal transduction pathway. Brain Res., 1093: 141–149.

Nucci, C., Tartaglione, R., Rombolà, L., Morrone, L.A., Fazzi, E. and Bagetta, G. (2005) Neurochemical evidence to implicate elevated glutamate in the mechanisms of high intraocular pressure (IOP)-induced retinal ganglion cell death in rat. Neurotoxicology, 26: 935–941.

Osborne, N.N., Casson, R.J., Wood, J.P., Childlow, G., Graham, M. and Melena, J. (2004) Retinal ischemia: mechanisms of damage and potential therapeutic strategies. Prog. Retin. Eye Res., 23: 91–147.

Osborne, N.N., Ugarte, M., Chao, M., Childlow, G., Bae, J.H., Wood, J.P. and Nash, M.S. (1999) Neuroprotection in relation to retinal ischemia and relevance to glaucoma. Surv. Ophthalmol., 43(Suppl 1): s102–s128.

Pawlak, J., Brito, V., Kuppers, E. and Beyer, C. (2005) Regulation of glutamate transporter GLAST and GLT-1 expression in astrocytes by estrogen. Brain Res. Mol. Brain Res., 138: 1–7.

Quigley, H.A. (1996) Number of people with glaucoma worldwide. Br. J. Ophthalmol., 80: 389–393.

Richards, D.A., Morrone, L.A., Bagetta, G. and Bowery, N.G. (2000) Effects of α-dendrotoxin and dendrotoxin k on extracellular excitatory amino acids and on electroencephalograph spectral power in the hippocampus of anaesthetised rats. Neurosci. Lett., 293: 183–186.

Simpkins, J.W. and Dykens, J.A. (2008) Mitochondrial mechanisms of estrogen neuroprotection. Brain Res. Rev., 57: 421–430.

Singer, C.A., Figueroa-Masot, X.A., Batchelor, R.H. and Dorsa, D.M. (1999) The mitogen-activated protein kinase pathway mediates estrogen neuroprotection after glutamate toxicity in primary cortical neurons. J. Neurosci., 19: 2455–2463.

Singer, C.A., Rogers, K.L., Strickland, T.M. and Dorsa, D.M. (1996) Estrogen protects primary cortical neurons from glutamate toxicity. Neurosci. Lett., 212: 13–16.

Tezel, G. (2006) Oxidative stress in glaucomatous neurodegeneration: mechanisms and consequences. Prog. Retin. Eye Res., 25: 490–513.

Tezel, G., Yang, X. and Cai, J. (2005) Proteomic identification of oxidatively modified retinal proteins in a chronic pressure-induced rat model of glaucoma. Invest. Ophthalmol. Vis. Sci., 46: 3177–3187.

Weaver, C.E., Jr., Park-Chung, M., Gibbs, T.T. and Farb, D.H. (1997) 17beta-Estradiol protects against NMDA-induced excitotoxicity by direct inhibition of NMDA receptors. Brain Res., 761: 338–341.

Wise, P. (2003) Estradiol exerts neuroprotective actions against ischemic brain injury: insights derived from animal models. Endocrine, 21: 11–15.

Yang, S.H., Liu, R., Wu, S.S. and Simpkins, J.W. (2003) The use of estrogens and related compounds in the treatment of damage from cerebral ischemia. Ann. N.Y. Acad. Sci., 1007: 101–107.

Zaulyanov, L.L., Green, P.S. and Simpkins, J.W. (1999) Glutamate receptor requirement for neuronal death from anoxia–reoxygenation: an in vitro model for assessment of the neuroprotective effects of estrogens. Cell. Mol. Neurobiol., 19: 705–718.

Zhou, X., Li, F., Ge, J., Sarkisian, S.R., Jr., Tomita, H., Zaharia, A., Chodosh, J. and Cao, W. (2007) Retinal ganglion cell protection by 17-beta-estradiol in a mouse model of inherited glaucoma. Dev. Neurobiol., 67: 603–616.

Subject Index

acetazolamide, to treat acute glaucoma, 189
achromatic automated perimetry, for management of glaucoma, 139
achromatopsia, 174
ACTSEB, see anterior chamber tube shunt to encircling band
acute primary angle closure, 35
AD, see Alzheimer's disease
Adamantane, for glaucoma, 398
adeno-associated virus, ability of transfecting RGCs using, 283
Advanced glaucoma intervention study (AGIS), for glaucoma examination, 79
age
   and gender in risk factors of AC, 33
   in risk factors of glaucoma development and progression, 19
   in treatment of POAG, 183–184
aging, and TM cell loss, 389
agiogenesis, of VEGF-A, 560
agnosia, 174
Ahmed glaucoma valve, 265
   comparative data between Baerveldt implant and, 268
   implantation of, 267
akinotopsia, 174
alpha-agonists, for POAG, 188–189
alpha-agonists-beta-blockers, for POAG, 192
ALPI, see argon laser peripheral iridoplasty
ALS, see amyotrophic lateral sclerosis
ALT, see argon laser trabeculoplasty
Alzheimer's disease (AD), 211, 437, 440, 467, 472–474, 495, 515, 525, 567, 575, 577, 580, 584, 589
   ADCS-ADLsev, for dementia, 329
   amyloid-β neurotoxicity in, 442–443
   CBs therapeutic effects in, 452
   due to mutations of APP, 312
   mechanisms of, 354

   memantine clinical trials for, 503–505
   memantine for, 328
   NMDA receptor in, involvement of, 329
   therapy of, 323, 324
   XFG associated with, 215
amacrine, 398
   cells, 481, 484, 524
      ChAT associated with, 345
   reduction of, 345
α-amino-3-hydroxy-5-methyl-4-isoxazoleproprionic acid [AMPA] receptors, 496
amniotic membrane transplantation, uses of, 238
amyloid-β (A-β)
   abnormal, in AD, 312
   drugs, 438
   implication in glaucoma, 443
   neurotoxicity in AD, 442–443
   pathway in experimental glaucoma, 443–445
   pathway targeting with DARC, 442–445
amyloidosis, 567
amyloid precursor protein (APP), 378, 443
   mutations of, AD due to, 312
   proteolytic cleavage of, 443
β-amyloid protein, 378
amyotrophic lateral sclerosis (ALS), 512, 561
   creatine supplementation for, 345
   Kaplan-Meier plots of survival in patients with, 329
   mechanisms of, 354
   riluzole for, 328
   therapy of, 323, 324, 327
anandamide (N-arachidonoylethanolamine, AEA), 452
anemia, 206
angina
   prinzmetal, 184
   XFG associated with, 215

angiography
  capabilities of, 146
  for detection of ocular ischemia, 214
angiotensin-converting enzyme (ACE) inhibitors, in NTG, effects of, 207
angle-closure, 31–32
  acute, primary, 39
    ALPI for management of, 39–40
    fellow eye of, 41
    laser PI for management of, 40
    lens extraction for management of, 40
    management of, 41
    medical therapy for management of, 39
    monitoring for IOP rise in eyes with, 40–41
  acute management of, 42
  chronic, primary, 41
    combined lens extraction and trabeculectomy for management of, 43
    goniosynechialysis for management of, 43
    laser iridoplasty for management of, 41
    laser PI for management of, 41
    lens extraction for management of, 41–42
    medical therapy for management of, 41
    trabeculectomy for management of, 41
  diagnosis of, 35
    acute primary angle closure, 35
    angle assessment in, 35–36
    anterior segment OCT for, 37–38
    gonioscopy technique for, 36
    SPAC for, 38
    UBM for, 36–37
    visual-field loss for, 38
  management of, 39
    acute primary angle closure in, 39–41
    chronic primary angle closure in, 41–43
  mechanism of, 32–33
  primary, classification of, 8
  risk factors of, 33
    age and gender in, 33
    ethnicity in, 33–34
    genetics in, 34
    ocular biometry in, 34
angle-closure glaucoma
  due to XFS, 211
  surgical iridectomy for, 225
animal models, for POAG, use of, 286–287
aniridia, 232
annexin 5-labeled apoptosis, 438–439

anoxia, GABA-immunoreactive neurons induced by, 346
anterior chamber, high-frequency ultrasound study of, 294
anterior chamber depth (ACD), for angle-closure, 34
anterior chamber tube shunt to encircling band (ACTSEB), for aqueous drainage, 264
anterior segement dysgenesis (ASD), causes of, 308
antioxidant
  enzymes, in glaucoma, 367
  glutathione (GSH), production of, 366
  property, of EGCG, 348
APAC, see acute primary angle-closure
aphakia, 190
apolipoprotein J (ApoJ), 565
apoptosis, 208, 409–410, 412, 415, 417
  annexin 5-labeled, 438–439
  ganglion cell, 339
  in glaucoma, 343
  intrinsic and extrinsic, 424–425
  methodology for, 348
  mitochondrial functions and, 343
  pathways of, 424–425
  retinal ganglion cell death by, 423, 425
  RGC, 309, 312
    EPO for, 315
  in RGCs, 480–481
  signaling of ganglion cell, 426, 429–433
apoptosis inducing factor-1 (Apaf-1), 424
apoptosome, 424
apoptotic death, 306
apoptotic pathways, mitochondrial-dependent and -independent, 344
apoptotic signals, in glaucoma, 378
APP, see amyloid precursor protein
aqueous drainage, ACTSEB for, 264
aqueous humor (AqH), 388, see also trabecular meshwork (TM)
  ascorbic acid in, 389–390
  drainage, 255
    molecular understanding of, 304
  external filtration of, 256
  GSH in, 389, 390
  outflow obstruction
    elevated IOP due to, 292, 294
    feature of rat glaucoma models produced by, 296

outflow pathways, for producing elevated IOP, 293
production, reduction of, 186
TGF in, 394
aqueous hyposecretion, 270–271
aqueous outflow
  for management of exfoliative glaucoma, 218–219
  obstruction of, 290
aqueous shunt implantation, modern, 263–264
  current shunts and factors affecting their function, 264
  future challenges of, 272
    cataract formation in, 273
    long-term effect on cornea in, 273–275
    predictability in, 272–273
  patient and ocular factors for, 269
    aqueous hyposecretion, 270–271
    previous ocular surgery, 271
    scleral thinning, 271–272
    severity of glaucoma damage, 269–270
    tolerance of potential slit-lamp interventions, 272
    tolerance of topical ocular hypotensive medications, 270
  shunt-related factors in, 264
    commercially available devices, 267–268
    comparative studies of, 268–269
    plate material, 267
    surface area, 264–266
    valved vs non-valved, 267
argon laser peripheral iridoplasty (ALPI)
  for eyes with plateau iris, 40
  for management of APAC, 39–40
argon laser trabeculoplasty (ALT)
  for eyes with XFS, 217
  preparation for, 226
arteritis, temporal, 206
arthritis, juvenile idiopathic, uveitis associated with, 271
ascorbate, regeneration of, 346
ascorbic acid, 398–390
  and IOP, 394–395
Asian racial group, glaucoma prevalence studies in, 5
astrocytes, 569, 579
  columnar structure of, 287
  migration, in glaucomatous OHN, 360–363
  in optic nerve, 312
  within rat ONH, predominance of, 296
astrocytes, in glaucomatous optic neuropathy, 353–354
  cell adhesion of ONH, 363–364
  cell-cell communication in ONH, 355–356
  connective tissue changes in, 364
    extracellular matrix degradation by reactive astrocytes, 364–365
    extracellular matrix synthesis by ONH astrocytes, 364
    TGFβ signaling in ONH astrocytes in, 365–366
  migration in glaucomatous OHN, 360–363
  oxidative stress in ONH, 366–368
  quiescent, 354
  reactive, 354–355
  signal transduction in glaucomatous, 356
    G protein-coupled receptors, 360
    MAPKs, 357–360
    PTKs in, 356–357
    ras superfamily of small G proteins, 360
astrocytomas, 364
astrogliosis, role of endothelin-1 in, 360
ataxia, optic, 174
athalamia, 226
atrophy
  of glaucomatous nerve, 145
  of iris, 40
  of LGN, 102
  of optic nerve, 125, 166, 309
  of parapapillary, 185
  of peripapillary, 200
  of RNFL, 142
Avastin, 565
Axenfeld-Rieger syndrome, *Col4a1* associated with, 308
axonal injury
  cellular mechanisms of, 287
  elevation of IOP results in, 297
axonopathy, 522
axotomy, 280, 283
  lesion, 423

elevated, as risk factors for glaucoma, 303
elevated, due to aqueous humor outflow
  obstruction, 292, 294
elevation, magnitude of, 306
elevation of, 182
excessive elevation of, 340
Gaussian distribution of, 196
immune mechanisms for mechanisms of, 205
induced ischemia, 586–588
induced loss of RGCs, 583–589
local vascular factors for mechanisms of,
  204–205
lowering, non-penetrating glaucoma surgery for,
  255, 259
lowering, viscocanalostomy for, 258
lowering effects of endocannabinoid, 451–452,
  454–455
measurement for early diagnosis of
  glaucoma, 48
mechanisms of, 204
for NTG, reduction of, 207
OHTS for reduction of, 191
predictability of early control of, 272–273
in rats, general considerations for measurment
  of, 289–291
in rats, methods for measurment of, 288–289
reduction of, 219
as risk factor associated with glaucoma, 181
in risk factors of glaucoma development and
  progression, 17–18
in risk factors of NTG, 200–203
transient elevation of, 229
intraocular pressure (IOP), and glaucoma, 375,
  388
  free radicals, 389–390
intraocular pressure (IOP) fluctuations
  in treatment of POAG, 184
intraocular pressure (IOP) producing elevated
  episcleral vein cautery for, 294–295
  experimental methods of, 293
  hypertonic saline injection of aqueous humor
    outflow pathways for, 293
  laser treatment of limbal tissues for, 293–294
intravitreal inhibition therapy, of VEGF-A,
  564–565
intrinsic apoptosis, 424–425
IOP, see intraocular pressure
iridectomy, hyphema from, 218

iridocorneal endothelial syndrome, 33
iridoplasty, see laser iridoplasty
iridotomy, see also laser iridotomy
  laser peripheral, for management of APAC, 40
iris
  atrophy in iridoplasty complications, 40
  cystis, 190
  dysgenesis, 308
  neovascularization of, 226
  plateau, for cause of CPAC, 41
  plateau, UBM image of, 38
  and pupil, 32–33
iritis, laser-induced, 227
ischemia, 500–501, 567–568
  CoQ10 and, 576–578
  glaucoma and, 524
  increased ROS production in, 343
  induced glutamate and CoQ10, 578–579
  reperfusion injury model, 562
ischemic attacks, XFG associated with, 215
ischemic-optic neuropathy, 423
  nonarteritic anterior, patients with, 324
ischemic optic neuropathy decompression trial
  (IONDT)
  to assess neuroprotective therapy, 330
ischemic retinopathy, 583
ischemic stroke
  acute, neuroprotective therapy for, 328
  MRI for investigation of, 165
8-iso-prostaglandin F_{2a}
  for free radical-induced oxidative damage,
    214

juvenile glaucomas, 228
  effect of viscocanalostomy in, 258
juvenile idiopathic arthritis, uveitis associated
  with, 271
juvenile open-angle glaucoma, cause of, 305

kainate receptors, 496
ketoacidosis in diabetics, glycerol for, 39
α-ketoglutarate, oxidative decarboxylation
  of, 346
kidney aggregates of XFM in, 215
koniocells, 103
koniocellular LGN layers, neural degeneration in,
  468, 470–471
koniocellular neurons, 467

lamina, in glaucoma
    insult to axons of RGCs within, 311–312
        nature of, 312–313
lamina cribrosa, 310–311, 393
    cribriform plates of, 357
    in ONH, 353
lamina of optic nerve, 309
laminin
    deposition of, 365
    expression of, 354
laser cyclophotocoagulation, for glaucoma, 231
    complications, 233–234
    efficacy of, 234
    ELT, 234–235
    endoscopic, 232
    indications and contraindications, 231–232
    patient preparation, 232
    transpupillary, 232
    transscleral, 232
    transscleral contact, 233
    transscleral noncontact, 233
laser Doppler flowmetry (LDF)
    for blood flow changes in ONH, 160
    functional, of optic nerve, 149–150
    of optic nerve, 150–151
laser iridoplasty
    complications of, 231
    contraindications for, 230
    efficacy of, 231
    for glaucoma, 230
    indications for, 230
    for management of CPAC, 41
    treatment technique of, 230–231
laser iridotomy
    argon, 227
    complications in, 227
    contraindications of, 226
    efficacy of, 227
    for glaucoma, 225
    indications for, 225–226
    Nd:YAG, 226
    patient preparation for, 226
    technique of, 226
laser ophthalmoscopy
    confocal scanning, and scanning laser
        polarimetry uses in glaucoma, 125–126
    confocal scanning, and SLP-VCC for
        monitoring glaucoma progression, 95

confocal scanning, for optic disc abnormalities,
        157
    confocal scanning, for optic nerve
        imaging, 145
    confocal scanning, for visual function, 49
    scanning, for evaluation of RNFL, 142
laser peripheral iridoplasty (LPI)
    argon, for management of APAC, 39–40
laser peripheral iridotomy (LPI)
    for management of APAC, 40
    for management of CPAC, 41
    plateau iris after, 38
laser polarimeter
    GDx scanning, for glaucoma, 126–128
    GDx scanning, nerve fiber analysis in,
        128–131
    GDx scanning, serial analysis in, 131
laser polarimetry
    scanning, and CSLO uses in glaucoma,
        125–126
    scanning, for evaluation of RNFL, 142
    scanning, for visual function, 49
laser surgery for management of exfoliative
        glaucoma, 217
laser therapy for IOP, 185
laser trabeculoplasty
    argon, for eyes with XFS, 217
    complications in, 229
    contraindications for, 228
    efficacy of, 229
    for glaucoma, 227–228
    indications for, 228
    mechanism of, 228
    patient preparation and postoperative follow-up
        in, 228–229
    treatment technique of, 228
laser treatment
    of limbal tissues for producing elevated IOP,
        293–294
latanoprost-timolol for glaucoma treatment, 191
lateral geniculate nucleus (LGN), 466–467,
        481
    transsynaptic degeneration in glaucoma,
        467–468
    neural degeneration, 468, 470–471
    visual cortex, 471
LDF, see laser Doppler flowmetry (LDF)
learning effect from SWAP printouts, 113

lens
   dislocation, risk of, 214
   enlargement of, 293
   subluxation of, 33
      due to XFS, 212
lens extraction
   combined, and trabeculectomy for management of CPAC, 43
   for management of APAC, 40
   for management of CPAC, 41–42
lentivirus
   ability of transfecting RGCs using, 283
light microscopy (LM)
   grading system for optic nerve damage, 292
   for identification of normal axons, 291
limbal tissues
   for producing elevated IOP, laser treatment of, 293–294
lipid peroxidation, 387
α-lipoic acid
   for mitochondrial function enhancement and attenuation of ganglion cell death, 346–347
   for treatment of glaucoma, 350
liver, aggregates of XFM in, 215
LM, see light microscopy
low-pressure glaucoma treatment, study of, 200
low-tension glaucoma (LTG)
   Friedman analyzer for identification of, 4
   in Japanese population, 9
LTG, see low-tension glaucoma
Lucentis, 565
lung, aggregates of XFM in, 215
lymphocytes in NTG, studies of, 208
lysyl oxidase, for high concentrations of homocysteine, 220
lysyl oxidase-like (LOXL), decreased levels of, 364
lysyl oxidase-like 1 (LOXL1) gene
   on chromosome 15, location of, 216
   exfoliative glaucoma due to, 212

machine learning classifiers (MLC)
   to optimize VF assessment, 94
   use of, 96
MacKay-Marg, principles of, 288
Macugen, 565
macular degeneration, visual function questionnaire in, 332

magnetic resonance imaging (MRI)
   conventional, and visual pathways for glaucoma, 166
   diffusion, for neuroimaging of visual pathways, 166
   diffusion tensor (DT), for optic nerves degeneration, 165
   for evaluation of microstructural integrity, 165
   functional, for neuroimaging of visual pathways, 173–175
   for patients with NTG, 205
   SAP and, for monitoring glaucoma progression, 96–97
magnetic resonance spectroscopy, proton, for neuroimaging of visual pathways, 175–176
magnocellular LGN layers, neural degeneration in, 468, 470–471
m-AIP, inhibition of caspase-3 activation, 527–528
MAP, see motion automated perimetry
MAPKs, see mitogen-activated kinases (MAPKs)
marijuana, glaucoma treatment by, 454–455
matrix metalloproteinases (MMPs)
   expression of, 360
   inhibitors, uses of, 244
   role in glaucoma, 413
   tissue inhibitors of, 364
MBP, see myelin basic protein (MBP)
medical therapy
   for glaucoma, 183
   for management of exfoliative glaucoma, 216–217
memantine, 437–438, 441, 445, 472, 495, 502–505, 524–525, 528
   for AD, 328
   for glaucoma, 398
memory loss, 187
meninges, aggregates of XFM in, 215
mesenchymal stem cells (MSCs), 514–515
metabotropic glutamate receptor (mGluR) agonist, 441
metalloproteinases (MMP), 395
methanandamide, 457–458
microdialysis technique, 456, 578, 585–589
microglia in optic nerve, 312
microtubule-associated protein 2 (MAP2), involvement of, 282
migraine, 184
   in NTG, 204

mitochondria, RGCs and, 340–341
mitochondrial function enhancement
   and attenuation of ganglion cell death, 343–344
     creatine in, 344–345
     epigallocatechin gallate in, 348–350
     α-lipoic acid in, 346–347
     nicotinamide in, 347–348
mitochondrial functions and apoptosis, 343
mitogen-activated kinases (MAPKs)
   signaling pathways, 528
   in signal transduction in glaucomatous
      astrocytes, 357–360
p38 mitogen-activated protein kinase (p38MAPK)
   involvement of, 243
mitogenesis, of VEGF-A, 560
mitomycin-C (MMC)
   antiscarring effects of, 244
   effect of, 238, 239
   mechanism of, 243
   trabeculectomy
     risks of, 272
     study of, 274
MLC, see machine learning classifiers
MMP, see metalloproteinases (MMP)
modified stem cells, neuroprotection by
     transplantation of, 515–516
Molteno implants, 268
   principles of, 263
monocytes, 380
Moorfields regression analysis (MRA)
   for volume comparison in stereometric
     parameters, 133
motion automated perimetry (MAP)
   for detection of VF loss, 94
mouse models
   to characterize processes involved in
     glaucomatous neurodegeneration,
     310–315
   to develop neuroprotective strategies
     axonal protection, 315
     erythropoietin (EPO) administration, 315
     radiation-based treatment, 315–316
     somal protection, 315
   of glaucoma
     developmental glaucoma, 308–309
     experimentally induced models, 309–310
     myocilin gene, MYOC, 305–306
     pigmentary glaucoma (PG), 309

POAG, 305
   POAG, strategies for developing new models
     of, 307–308
MRI, see magnetic resonance imaging
Muller cells, 481, 579
Müller glia
   and ceruloplasmin, 314
   and heat shock proteins, 314–315
mydriasis, 231, 232
myelin basic protein (MBP), and RGC, 379
MYOC, 305–306
   associated with POAG, mutations of, 183
   effects of mutations in, 308
   glaucoma caused by mutants of, 307
   in patients with PACG, analysis of, 34
myocardial dysfunction, XFG associated with, 215
myocardial infarction (MI), XFG associated with,
   215
myocardial ischemia, 460
   asymptomatic, in NTG, 204
myocilin gene, see *MYOC*
myocilin (MYOC), see trabecular meshwork-
     inducible glucocorticoid response (TIGR)
myopia, 132, 184
   as risk factor for vision loss, 340
myopic refractive error, 48
myotics, for POAG, 189–190

N-acetyl aspartate (NAA), 175
   proton MR spectroscopy for concentration of,
     176
nasal defect, 201
N-benzyloxycarbonyl-Val-Ala-Asp-fluoromethyl
     ketone (z-VAD-fmk), 505
NCAM, see neural cell adhesion molecules
nephrolithiasis, 189
nerve conduction, energy requirement
   for, 339
nerve fiber analysis (NFA), in GDx scanning laser
     polarimeter, 128–131
nerve fiber index (NFI), for glaucomatous damage,
   127
nerve fiber layer (NFL), thinning of, 185
nerve fiber layer polarimetry (GDx NFA), for
     glaucoma, 206
neural activity, retinal
   assessement of, 152–153
   changes of, 158

neural cell adhesion molecules (NCAM), in CNS, 363
neural rim loss, ODDSS for, 60
neural stem cells, 512–513
neural tissues, in ONH, homeostasis of, 353
neurites, RGC-5, 280, 282
neurodegeneration
   in glaucoma, 409
   in POAG, progression of, 368
neurodegenerative disease
   fatal, 328
   glaucoma as, see glaucoma, neuronal loss in
neurofilament light (NF-L) protein, Western blotting for, 345
neuroimaging of visual pathways, advances in, 165–166
neuronal apoptosis, associated with Parkinson's disease, 339
neuronal injury
   attenuate, 347
   counteracts inflammation-mediated, 348
   direct, 309
neuronal loss, in glaucoma, see Glaucoma, neuronal loss in
neuronal nitric oxide synthase (nNOS), 523
neuro-ophthalmic disorders, detection of, 161
neuropathy
   glaucomatous, associated with visual-field loss, 53
   glaucomatous, identification of, 47
   optic, HIOP associated with, 4
neuroplasticity and adaptive changes, 488
neuroprotection
   clinical trials of, 323–324, 327–329
      endpoints in, 330–332
      and glaucoma, 332
      issues in design and conduct of, 326–327
      methods of, 324–326
      in ophthalmology, 329–330
   clusterin, 565–571
   creatine supplementation for, 345
   by endocannabinoid, 455–460
   for glaucoma, 378–379
   glaucoma and, 568–571
   IL-6, 568–571
   for NTG, 207–208
   in retina with DARC, 437–446
   by transplantation of modified stem cells, 515–516
   by transplanted stem cells, 514–515
   VEGF-A, 555–565, 568–571
neuroprotection and glaucoma, pathogenesis of ganglion cells in, 339–340
   causes for ganglion cell death in, 341–343
   mitochondrial function enhancement and attenuation of ganglion cell death in, 343–344
      creatine, 344–345
      epigallocatechin gallate, 348–350
      α-Lipoic acid, 346–347
      nicotinamide, 347–348
   mitochondrial functions and apoptosis in, 343
   RGCs and mitochondria in, 340–341
neuroprotective agents, 437, 465, 488–489, 568, 571
neuroretinal rim
   CSLO for measurements of, 95
   in NTG, thinning of, 196
   in NTG patient, sloping of, 198
   in NTG patient, thinning of, 201
neuroretina toxicity, of VEGF-A, 564–565
neurotrophins, in glaucoma, 377–378
neurovascular coupling
   FLDF and, in humans, 158–159
   in humans, 155–156
   of ONH, loss of, 158
NF-L, see neurofilament light
NF-L protein
   in ischemia/reperfusion, decrease of, 349
   in optic nerve, drastic loss of, 348
nicergoline, 542
nicotinamide, for treatment of glaucoma, 350
nifedipine, against glaucoma progression, benefits of, 207
nimodipine against glaucoma progression, benefits of, 207
nitric oxide (NO), 390–391, 409, 413–414, 456, 460, 472, 480, 496, 498, 523, 563, 568, 575–576, 584, 587
   elevation of, 342
   NOS enzyme, 390
   and vitamin C, 391

nitric oxide synthase (NOS), 480
  for glaucoma, 377
  induction of, 356
  localization of, 348
nitroglycerin, 505
NitroMemantine, 505
NMDA, see N-methyl-D-aspartate
N-methyl-D-aspartate (NMDA) receptors, 396, 441, 452, 523–525, 575
  in AD, involvement of, 328
  antagonists
    and CaMKII, 524–525
    kinetics of, 502–503
  averting caspase-mediated neurodegeneration, 505–506
  caspase-3 activation in, 527–528
  downstream signaling cascades, 498–499
  drugs targeting, 502–506
    downstream signaling molecules, 505–506
    memantine, 503–505
    nitro-memantine, 505
    p^{38} MAPK inhibitors, 505
  induced
    cell death pathways, 505–506
    excitotoxicity, 495–498, 505
    retinal cell death, 527–528
    RGC death, 499
  intraocular injection of, 345
  neuroprotective effect of m-AIP in retinal neurons exposed to, 525–527
  overactivation, 498–499
  targeting pathways involved in excitotoxicity, 502–506
nomogram, uses of, 50
nonarteritic anterior ischemic optic neuropathy (NAION)
  patients with, 324, 330
noncompliance for NTG, 208
non-NMDA glutamate receptors, 452
nonsteroidal anti-inflammatory drugs (NSAIDs)
  application of, 229
  effects of, 239
normal-tension glaucoma (NTG), 375, 452
  clinical features of, 196
    CCT in, 198
    disease course in, 198–200
    optic disk in, 196
    visual field in, 196–198
  differential diagnosis for, 205
    diagnostic evaluation in, 205–207
  epidemiology of, 196
  genetics of, 208–209
  IOP-independent mechanisms, 204
    immune mechanisms, 205
    local vascular factors, 204–205
  risk factors of, 200
    IOP in, 200–203
  therapy for, 207
    IOP reduction, 207
    neuroprotection, 207–208
    noncompliance, 208
    systemic medications, 207
  treatment of, 195–196
NOS, see nitric oxide synthase (NOS)
NTG, see normal-tension glaucoma
nuclear CaMKII-α_β, involvement in BDNF expression in RGCs, 529–530, 532

OAG, see open angle glaucoma
OCT, see optical coherence tomography (OCT)
octopus indices, for monitoring glaucoma progression, 80
Octopus program peritrend, uses of, 89
ocular and systemic abnormalities, in treatment of POAG, 184
ocular associations for exfoliative glaucoma, 214
ocular biometry in risk factors of AC, 34
ocular factors, of aqueous shunt implantation, 269
  aqueous hyposecretion in, 270–271
  previous ocular surgery in, 271
  scleral thinning in, 271–272
  severity of glaucoma damage in, 269–270
  tolerance of potential slit-lamp interventions in, 272
  tolerance of topical ocular hypotensive medications in, 270
ocular hypertension (HIOP)
  application of FLDF in patients with, 149
  associated with optic neuropathy, 4
  and corneal thickness, relationship between, 19
  and glaucoma, risk assessment in, 16
  glaucoma in patients with, development of, 19
  medical therapy for, 183
  SAP for, 77

ocular hypertension (OHT), 388
  patients with, 157
  RF_{onh} in, 157–158
ocular hypertension treatment (OHT)
  CCT for glaucoma development from, 27
  CCT in management of, 25
  progression of, 29
ocular hypertension treatment study (OHTS)
  criteria for SAP VF defects assessment, 84–85
  for evaluating risk factors of glaucoma
      development, 25
    interocular asymmetries, for predictor of
        POAG, 88
  for IOP reduction, 191
  for prediction of glaucoma development, 17
ocular hypotensive medications
  topical, tolerance of, 270
ocular inflammation, glaucoma associated
    with, 269
ocular ischemia, angiography for detection of, 214
ocular ischemic syndrome, 271
ocular neuroprotection strategy, 474
ocular perfusion, 23
ocular response analyzer (ORA)
  for measurement of corneal response, 28
ocular surgery previous, in ocular factors of
    aqueous shunt implantation, 271
OD, see optic disc
ODDSS, see optic disc damage staging system
OHT, see ocular hypertension (OHT)
OHTS, see ocular hypertension treatment study
oligodendrocyte death, and TNF-α, 412
oligodendrocytes
    loss, TNF for, 313
    in optic nerve, 312
ONH, see optic nerve head
open-angle glaucoma (OAG), 473, 479, 505, 541,
    543, 547–549
  chronic, early diagnosis of, 59
  chronic, SLT for, 230
  development of, 84
  due to XFS, 211
  early, associated with nerve fiber layer
      loss, 161
  ELT for treatment of, 234
  incidence of, 19
  primary, aqueous shunts in, 272
  primary, deep sclerectomy for, 257

  primary, fMRI studies of patients with, 165
  primary, patients with, 88
open angle glaucoma (OAG), primary, overview
    on medical therapy for, 181
  fixed combinations for, 190–191
    beta-blockers-alpha-agonists, 192
    beta-blockers-CAIs, 191
    beta-blockers-prostaglandins, 191–192
  how to treat, 185–186
    alpha-agonists, 188–189
    beta-blockers, 186–187
    CAIs, 189
    myotics, 189–190
    prostaglandins, 187–188
  when to treat, 182–183
  whom to treat, 183
    age, 183–184
    genetics, 183
    IOP fluctuations, 184
    ocular and systemic abnormalities, 184
    optic nerve condition and visual field stage,
        184–185
    race, 184
    tonometry and pachymetry, 184
operative surgery
  for management of exfoliative glaucoma,
      217–218
ophthalmology
  clinical trials of neuroprotection in,
      329–330
  neuroprotective therapies in, 331
ophthalmoloscopy
  and annexin 5-labeled apoptosis, 438–439
ophthalmoscopy
  indirect, for ocular analysis, 304
optical coherence tomograph (OCT)
  for assessing glaucomatous damage, 331
  and SAP for monitoring glaucoma progression,
      95–96
  for visual function, 49
optical coherence tomography (OCT)
  anterior segment, 274
  anterior segment, for diagnosis of AC, 37–38
  for glaucoma, 206
  in glaucoma management, 144–146
  for measurements of RNFL thickness, 95
  for optic disc damage, 96
  performance of, 141

for vitreoretinal diseases, 66
works of, 140–141
optical heterodyne detection, 140
optical interferometry tonometer (OIT)
for IOP measurements, 304
optic ataxia, 174
optic axons, 488
optic cup, deepening of, 144
optic disc damage staging system (ODDSS)
for neural rim loss, 60
optic disc (OD)
abnormalities of, 157
cupping, clinical findings of, 139
damage, OCT for, 96
examination of, 7
GDx and HRT for morphology of, 125, 126
for glaucoma management, evaluation of, 143–144
hemorrhages, 185
morphological alterations of, 59
neuro-retinal rim of, 160
slit lamp biomicroscopy for assessment of, 60
topography, measurement of, 144
optic disk hemorrhage
associated with NTG, 196
development of, 201
in NTG patient, 199
as predictor and progression for POAG, 204
optic disk in clinical features of NTG, 196
optic nerve
absence of enhancement of, 168
anatomy of, 166
atrophy of, 125
Col4A1 mutation in abnormalities of, 309
computerized analysis of, 77
DT fiber tractography of, 171
functional hyperemia measured in, 161
intraorbital portion of, 167
LDF of, 150–151
neurodegenerative disease of, 16
in NTG patient, 197, 198
prechiasmatic segment of, 169
vascular insufficiency of, 204
vertical cup/disc ratio of, 17
optic nerve, FLDF of, 149–150
clinical application of, 152–153
conclusion and future aspect of, 161
discussion for, 158–161

physiology of, 153–156
technology for, 150
LDF in, 150–151
retinal neural activity assessed from ERG in, 152–153
visual stimulation in, 151–152
optic nerve and RGC damage, 482
optic nerve atrophy, progression of, 309
optic nerve axon, TEM for counts of, 291
optic nerve condition and visual field stage, in treatment of POAG, 184–185
optic nerve damage
assessment of, 291–293
in glaucoma, mechanisms of, 203
LM grading system for, 292
for POAG, rat models of, 287–288
in POAG, rat models for, 293
optic nerve decompression surgery
safety and efficacy of, 330
optic nerve degeneration, RGC death associated with, 309
optic nerve head (ONH)
appearance of, 119
astrocytes
cell-cell communication in, 355–356
exposure of, 361
extracellular matrix synthesis by, 364
migration of, 362
TGFβ signaling in, 365–366
cell adhesion of astrocytes of, 363–364
characteristic cupping of, 285
color image of, 129
for early diagnosis of glaucoma, 48
evaluation of, 205, 206, 309
glaucomatous, astrocyte migration in, 360–363
GPS analysis for 3-D shape of, 135
hemodynamic response of, 149
HRT for analysis of, 131
HRT stereometric parameters of, 133
HRT three-dimensional image of, 132
LDF for blood flow changes in, 160
loss of neurovascular coupling of, 158
monkey, expression of connexin-43 in, 357
monkey, expression of SOD-2 in, 367
neural tissues in, homeostasis of, 353
oxidative stress in astrocytes of, 366–368
perfusion of, 207
rat, longitudinal section of, 287

rat eye, longitudinal section of, 295
remodeling of, 340
structure of, 286
superior, TEM of, 297
twodimensional map of, 144
optic nerve head (ONH) topography
correlations between CCT and, 27
optic nerves
DT MR imaging for degeneration of, 165
injury
mechanisms of, 295
in POAG, mechanisms of, 285
protein levels in, 347
optic neuritis, 157, 169, 175
optic neuropathies (ON)
clinical relevance of, 480–481
glaucomatous, 279
glaucomatous, evidence of, 32
glaucomatous, pathophysiology of, 161
glaucomatous, SAP in patients with, 86
HIOP associated with, 4
ischemic, 205
nonarteritic anterior ischemic, patients with, 324
pathogenesis of, 341
optic radiation, DT fiber tractography of, 173
optic tectum, 479, 482, 485–486
optic tract, pregeniculate segment of, 169
optineurin gene, mutations in glaucoma, 412–413
OPTN, see optineurin gene
OPTN gene, effects of mutations in, 308
orphan trabeculectomy, 270
oxidation, defined, 386
oxidative damage
free radical-induced, 8-*iso*-prostaglandin F_{2a} for, 214
oxidative stress, 409, 413, 445, 467, 512, 561, 563–566, 568, 570–571, 575–576, 579, 583–584
chronic, 367
in ONH astrocytes, 366–368
oxidative stress, and glaucoma, 386–387, see also Free radicals, in glaucoma
animal metabolism, 386
effect on TM, see Trabecular meshwork (TM), in glaucoma
oxidation, 386
oxidative stress, in glaucoma, 377
oxygen, in animal metabolism, 386

PACG, see primary angle-closure glaucoma
pachymetry and tonometry, in treatment of POAG, 184
Panax ginseng, see Ginseng
parapapillary atrophy, 185
Parkinson's disease (PD), 440, 512, 514–516, 561, 575, 577, 580, 584, 589
CBs therapeutic effects in, 452
creatine supplementation for, 345
L-Dopa for, 328
mechanisms of, 354
neuronal apoptosis associated with, 339
therapy of, 323, 324, 327
Parvocellular LGN layers, neural degeneration in, 468, 470–471
PAS, see peripheral anterior synechiae
pattern electroretinogram (PERG)
for assessing glaucomatous damage, 305
mechanisms of, 313
pattern standard deviation (PSD)
in risk factors of glaucoma development and progression, 19
visual-field index, 17
Pegaptanib (Macugen), 565
peripapillary atrophy, 200
peripheral anterior synechiae (PAS)
causes of, 33
gonioscopy for, 36
goniosynechialysis for stripping of, 43
in iridoplasty complications, 40
peripheral iridoplasty (PI)
argon laser, for management of APAC, 39–40
peripheral iridotomy (PI)
laser, for management of APAC, 40
laser, for management of CPAC, 41
peripheral nerve, ischemic-reperfusion injuries in, 346
peroxisomal targeting signal type 1 receptor (PTS1R)
cryptic, binding site of, 306
peroxynitrite, scavenging of, 345
phacotrabeculectomy, complication rates of, 43
phosphatidylserine (PS), 438
photodynamic therapy for control healing, 244
physiology of optic nerve, 153–156
PI, see peripheral iridotomy
pigmentary glaucoma (PG), 309
pigment dispersion syndrome, 226

pilocarpine
    for APAC, 39
    for eyes with XFS, 217
    for glaucoma, 326
    for glaucoma treatment, 190
placenta growth factor (PlGF), 515, 556
plasminogen activator inhibitor-1 (PAI-1), 394, 395
plasminogen activators, effect of, 281
plateau iris, for cause of CPAC, 41
plateau iris syndrome, 230, 231
platelet-derived growth factor (PDGF)
    activation of, 243
    expression of, 354
p^{38} Mitogen-activated protein kinase (p^{38}MAPK) inhibitors, 495–496, 499, 505
pneumatonometer for measuring IOP, 289
POAG, see primary open-angle glaucoma (POAG)
pointwise linear regression analysis (PLRA)
    peridata softwares based on, 89
polarimetry techniques (GDx)
    for morphometry and morphology of RNFL, 125, 126
poly-ADP-ribose polymerase (PARP), activation of, 347
polyglactin mesh in GFS, uses of, 239
polypropylene with silicone end plates, influence of, 267
polytetrafluoroethylene (PTFE), in GFS, uses of, 239
Polyunsaturated fatty acids, and free radicals, 387
Posner-Schlossman syndrome, 35
potassium, elevation of, 342
primary angle-closure glaucoma (PACG)
    for cause of blindness, 31
    development of, 39
    MYOC gene in patients with, analysis of, 34
    prevalence of, 3, 8–9
    visual-field loss in, 38
primary angle closure (PAC)
    acute, 39
        ALPI for management of, 39–40
        fellow eye of, 41
        laser PI for management of, 40
        lens extraction for management of, 40
        medical therapy for management of, 39
        monitoring for IOP rise in eyes with, 40–41
        symptoms of, 32

    chronic, 41
        combined lens extraction and trabeculectomy for management of, 43
        goniosynechialysis for management of, 43
        laser iridoplasty for management of, 41
        laser PI for management of, 41
        lens extraction for management of, 41–42
        medical therapy for management of, 41
        trabeculectomy for management of, 41
    classification of, 8, 32
primary angle-closure suspect (PACS), 31
primary open-angle glaucoma (POAG), 305, 388
    aqueous shunts in, 272
    deep sclerectomy for, 257
    diagnosis of, 125
    fMRI studies of patients with, 165
    frequency of, 10
    IOP associated with incidence of, 18
    mechanisms of optic nerve injury in, 285
    nerve fiber indicator for presence of, 130
    neurodegeneration in, progression of, 368
    OPTN mutations cause, 306
    pachymetry and tonometry in treatment of, 184
    patients with, 88
    predictor and progression of, 204
    predictor of, 88
    prevalence of, 3, 4, 8
    proportion of, 196
    rat models for optic nerve damage in, 293
    strategies for developing new models of, 307–308
    suitability of rat models of optic nerve damage in, 287–288
    TM cells in, 389, 390
    use of animal models for, 286–287
    visual-field loss in, 38
primary open angle glaucoma (POAG), overview on medical therapy for, 181
    fixed combinations for, 190–191
        beta-blockers-alpha-agonists, 192
        beta-blockers-CAIs, 191
        beta-blockers-prostaglandins, 191–192
    how to treat, 185–186
        alpha-agonists, 188–189
        beta-blockers, 186–187
        CAIs, 189
        myotics, 189–190
        prostaglandins, 187–188

when to treat, 182–183
whom to treat, 183
  age, 183–184
  genetics, 183
  IOP fluctuations, 184
  ocular and systemic abnormalities, 184
  optic nerve condition and visual field stage, 184–185
  race, 184
  tonometry and pachymetry, 184
Programmed cell death (PCD), 438
prosopagnosia, 174
prostaglandins
  analogs for treatment of XFG, 216
  for POAG, 187–188
prostaglandins-beta-blockers for POAG, 191–192
"protective autoimmunity", concept of, 379
proteins
  abnormal accumulation of, 378
  phosphatase calcineurin, domain of, 309
protein tyrosine kinases (PTKs)
  in signal transduction in glaucomatous astrocytes, 356–357
proteoglycans, expression of, 354
proton MR spectroscopy for neuroimaging of visual pathways, 175–176
prototype instruments for measuring IOP, 289
PSD, see pattern standard deviation
pseudoexfoliation glaucoma, ONHs of patients with, 364
pseudomyopia, 190
psychophysical tests
  for magnocellular pathways in glaucoma, 54
PTFE, see polytetrafluoroethylene
PTKs, see protein tyrosine kinases (PTKs)
ptosis, 53
  congenital or acquired, 87
pupil dysgenesis, 308
pupillary block, 32
  angleclosure glaucoma associated with, 225
pyridoxine in extracellular matrix integrity, role of, 219
pyruvate, oxidative decarboxylation of, 346

quiescent astrocytes in glaucomatous optic neuropathy, 354

race in treatment of POAG, 184
Ranibizumab (Lucentis), 565
rat eye ONH
  longitudinal section of, 295
rat models
  for glaucoma research, 285–286
  of optic nerve damage for POAG, suitability of, 287–288
  for optic nerve damage in POAG, 293
rat ONH
  longitudinal section of, 287
Raynaud's phenomenon, 460
  in NTG, 204
reactive astrocytes
  extracellular matrix degradation by, 364–365
  in glaucomatous optic neuropathy, 354–355
reactive glial changes, 512
reactive oxidative stress (ROS)
  generation of, 341
  production of, 343
  on RGC axons, effects of, 366
reactive oxygen species (ROS), 377, 410–411, 424, 472, 480, 495–496, 498, 563
recombinant human clusterin (r-h-clusterin), 567
rectus muscles
  Baerveldt glaucoma implant in, placement of, 272
red blood cells (RBCs), velocity of, 150, 160
regulators of G protein signaling (RGS)
  in human ONH, expression of, 361
reperfusion induced glutamate, and Coenzyme Q10, 578–579
retina
  damage, assessment of, 291–293
  evaluation of, 309
  and excitotoxicity, 524
  neural activity changes in, 158
  protein levels in, 347
  scanning laser polarimetry study of, 126
retina in glaucoma
  dendrites and neurofilament accumulation, 314
  Müller glia
    and ceruloplasmin, 314
    and heat shock proteins, 314–315
  PERG and complement, 313
retinal artery, occlusion of, 157
retinal circuitry, remodeling of, 481–484

retinal ganglion cell (RGC-5)
  neurites, 280, 282
  in RGCs, 281
  in RGCs, differentiation of, 281–282
retinal ganglion cells (RGC), 409–410, 437, 439, 465–466, 479, 501, 504, 506, 512, 521, 577, 583, 586
  advantages and disadvantages of culture models of, 282–283
  anatomy and function of, 102–104
  apoptosis, 309, 312, 392–393, 437–439, 443, 445, 451, 480–481
  assessment of loss of, 295
  BDNF and neuroprotection of, 528–533
  CaMKII role in death and survival pathways of, 521–534
  cell death pathways in, 378
  in culture, experimental study of, 279
  in DBA/2J mice, age-dependent death of, 316
  death
    by apoptosis, 423, 437–439, 443, 445, 480–481
    BAX protein for, 426–429
    Bcl2 gene family role, 423–433
    in glaucoma, 410–413, 415
    TNF-α induced, 410–413
  death of, 310
  degeneration of, 304
  development of, 309
  dysfunction and death of, 303
  and free radicals, 377
  functional features of, 116
  and glaucoma, 522–523, see also glaucoma
  in glaucoma, degeneration of, 315
  Glial cultures of, 280–281
  and glutamate, 377
  immune system, see Immune system, and RGC
  intrinsic changes in, 312
  IOP induced loss, 583–589
  isolation of, 104
  and lamina cribrosa, 393
  loss and 17β-estradiol, 583–589
  loss of, 292
  loss of axons of, 285
  mechanisms of injury in glaucoma, 467
  microdialysis, 585
  and mitochondria, 340–341
  mixed, in culture, 279–280
  morphometric analysis, 584–585
  neuroprotection by
    memantine, 504
    p^{38} MAPK inhibitor, 506
  neurotrophic factors role in protection of, 521–534
  and neurotrophin, 377–378
  in NTG, 375
  OPTN expression in, 306
  phospho-JNK immunolabeling, 412
  progressive loss of, 340
  purified, 280
  receptive field structure and redundancy of, 104–105
  reduction of, 294
  retinal explants in, 280
  RGC-5 cell neurites, 282
  RGC-5 cells in, 281
  RGC-5 cells in, differentiation of, 281–282
  segregation of, 104
  soma death BAX for, 426–429
  subtypes and features of, 102
  therapeutic approaches to prevent death of, 502–506
  vision loss by death of, 465
retinal ganglion cell (RGC) apoptosis
  EPO for, 315
retinal ischemia, 524, 568, 583
  and glutamate, 578–579
retinal nerve fiber layer (RNFL)
  analysis of, 61
  computerized analysis of, 77
  damage, GDx-VCC for, 131
  defects, patient with, 116
  for diagnosis of visual field, 54
  for early diagnosis of glaucoma, 48
  evaluation of, 205
  GDx and HRT for morphology of, 125, 126
  for glaucomatous progression, 62
  GPS analysis for 3-D shape of, 135
  images of, 127
  loss, associated with visual-field loss, 47
  morphological alterations of, 59
  thickness
    evaluation of, 141–143
    OCT for measurement of, 139
retinal neural activity from ERG, assessement of, 152–153

retinal pathology for perimetry, 49
retinal protection, CoQ10 as neurotherapeutic
 agent for, 575–580
retinal stem cells, 513
retinal vein occlusion
 associated with XFS, 214
 due to XFS, 211
 XFS with, 212
retinitis cytomegalovirus, visual function
 questionnaire in, 332
retinitis pigmentosa
 vitamin A and vitamin E supplementation for,
 331
retino-geniculate neurons, 466
RGC, see retinal ganglion cell (RGC)
Rho-associated kinase (ROCK)
 application of inhibitors of, 243
 effectiveness of, 242
riluzole for ALS, 328
RNFL, see retinal nerve fiber layer
ROS, see reactive oxidative stress (ROS)

saline injection, hypertonic, of aqueous humor
 outflow pathways, 293
SAP, see standard automated perimetry
scanning laser ophthalmoscopy (SLO)
 for evaluation of RNFL, 142
scanning laser polarimetry (SLP)
 and CSLO uses in glaucoma, 125–126
 for evaluation of RNFL, 142
 for optic nerve imaging, 145
 for visual function, 49
scanning laser polarimetry (SLP)-VCC
 for measurements of RNFL thickness, 95
scanning peripheral anterior chamber depth
 analyzer (SPAC)
 for diagnosis of AC, 38
scarring
 in glaucoma surgery, 238
 induction of, 239
scarring activity, IFN-α for, 239
schizophrenia, 175
Schlemm's canal, removal of inner wall
 of, 256
Schwann cells, 514, 567
scleral thinning
 in ocular factors of aqueous shunt implantation,
 271–272

sclerectomy, deep, in surgical alternative to
 trabeculectomy, 256–257
sclerosis
 amyotrophic lateral, riluzole for, 328
 amyotrophic lateral, therapy of, 323, 324
 amyotropic lateral, creatine supplementation
 for, 345
 DTI for brain lesions in, 169
 manifestation of, 175
 multiple, mechanisms of, 354
scotoma, 479, 482–484
secondary degeneration of neurons, in glaucoma,
 376
 BDNF and, 378
β-Secretase, 443
selective laser trabeculoplasty (SLT)
 for glaucoma, 229–230
 results of, 230
seprafilm in GFS, uses of, 239
serine, elevation of, 342
servo-null micropipette system (SNMS)
 for IOP measurements, 304
short-wavelength automated perimetry (SWAP)
 for detection of VF loss, 94
 for early diagnosis of glaucoma, 54
 and FDT clinical data comparison in glaucoma,
 114–116
 and FDT perimetry in glaucoma, 101–102
 in glaucoma
 clinical data of, 111–114
 rationale and perimetric techniques of,
 108–110
 for monitoring glaucoma progression, 94–95
 for OHT patients with thin corneas, 19
 for visual field testing, 331
 for visual function, 49
SITA, see Swedish interactive threshold algorithm
SLDT technique
 GJIC measured by, effects of HP on, 358
sleep disturbances
 as risk factor for vision loss, 340
SLT, see selective laser trabeculoplasty
S-nitrosylation, 495–498, 505
sodium nitroprusside, intraocular injection of, 349
somatic cells, 513–514
SPAC, see scanning peripheral anterior chamber
 depth analyzer
spatial frequency, 486–487

spinal cord injury (SCI)
  role for scar production after, 365
standard automated perimetry (SAP)
  advantages of, 91
  and algorithms for monitoring glaucoma
      progression, 77–79
  for detecting functional damage, 59
  for early diagnosis of glaucoma, 49
  and FDP-matrix for monitoring glaucoma
      progression, 94
  and functional MRI for monitoring glaucoma
      progression, 96–97
  for glaucoma management, 101
  for measurement of VF defects, 96
  for monitoring glaucoma progression, 94–95
  and optical coherence tomography for
      monitoring glaucoma progression, 95–96
  and SLP-VCC for monitoring glaucoma
      progression, 95
  for VF damage, 125
  and VF progression for monitoring glaucoma
      progression, 88–94
  for VF testing in glaucoma, 66
  visual function test by, 104
standard automated perimetry (SAP): interocular
    asymmetries in OHTS
  for monitoring glaucoma progression, 88
standard automated perimetry (SAP): relationship
    between function and structure
  for monitoring glaucoma progression, 95
standard automated perimetry (SAP) VF
    assessment
  glaucoma staging system for, 86–88
  for monitoring glaucoma progression, 79
standard automated perimetry (SAP) VF
    assessment: full-threshold strategy
  for monitoring glaucoma progression, 82–84
standard automated perimetry (SAP) VF defects
    assessment
  AGIS criteria for, 85
  CIGTS for, 85
  OHTS criteria for, 84–85
staurosporine differentiation of RGC-5, 281
stem cells
  based neuroprotective therapy, 512–513
  neuroprotection in glaucoma, 511–517
    challenges, 516–517
    edogenous stem cells, 516

  modified stem cells transplantation, 515–516
  transplanted stem cells, 514–515
  sources, 513–514
  transplantation of, 512, 514–516
  transplant strategies, 474
stereophotography, for representation of OD, 60
Stereovision, 473
stress, RF_{onh} alteration in, 150
stroke
  acute ischemic, neuroprotective therapy for, 328
  diffusion MR imaging for evaluation of, 168
  ischemic, MRI for investigation of, 165
  XFG associated with, 215
superior colliculus, 482–486
superoxide
  generation of, 343
  scavenging of, 345
superoxide dismutase (SOD), 387
super oxide dismutase (SOD-2), in monkey ONH,
      expression of, 367
surgery for IOP, 185
SWAP, see short-wavelength automated perimetry
Swedish interactive threshold algorithm (SITA)
  for monitoring glaucoma progression, 86
  uses of, 6
α-synuclein protein, 378
syphilis, 206
systemic associations
  for exfoliative glaucoma, 214–215
systemic medications for NTG, 207

T cells, and RGC, 379–380
  cytokines, 380
  growth factors, 380
TEM, see transmission electron microscopy
tenascin C, expression of, 354
Tetrodotoxin (TTX), 523
TGF, see transforming growth factors (TGF)
TGF-α, expression of, 354
TGFβ signaling in ONH astrocytes, 365–366
TGM2, expression of, 366
thrombocytopenia, 189
TIGR, see Trabecular meshwork-inducible
      glucocorticoid response (TIGR)
timolol-bimatoprost for glaucoma treatment, 192
timolol-brimonidine for glaucoma treatment, 192
timolol-dorzolamide for glaucoma treatment, 191
timolol-latanoprost for glaucoma treatment, 191

timolol on IOP, effect of, 332
timolol-travoprost for glaucoma treatment, 191–192
TNF, see tumor necrosis factor
TNF-α, elevation of, 342
TNF receptor–associated death domain protein (TRADD), 416–417
α-tocopherol, regeneration of, 346
TonoLab tonometers for measuring IOP, 289
tonometer, optical interferometry and I/R, for IOP measurements, 304
tonometry
  applanation, IOP measurement with, 205
  and pachymetry, in treatment of POAG, 184
  rebound, principle of, 289
Tono-Pen tonometer for measuring IOP, 288
topographic change analysis (TCA)
  for glaucomatous structural damage, 61
topography of optic nerve head, 206
Trabecular meshwork-inducible glucocorticoid response (TIGR), 390
Trabecular meshwork (TM), 388–389
  in aqueous humor (AH) regulation, 388
  and extracellular matrix (ECM), see extracellular matrix (ECM)
  in POAG, 389, 390
  TIGR, 390
trabeculectomy
  benefits of, 185
  combined lens extraction and, for management of CPAC, 43
  consequences of, 273
  for management of CPAC, 41
  mitomycin, risks of, 272
trabeculectomy, surgical alternative to, 255–256
  deep sclerectomy in, 256–257
  viscocanalostomy in, 257–259
trabeculoplasty, see also laser trabeculoplasty; selective laser trabeculoplasty
  argon laser, for eyes with XFS, 217
tractography
  to deduce pathways of nerve fiber tracts, 169
  DT, for disclosing of complex fiber tracts, 170
  DT fiber, of optic nerve, 171
  DT fiber, of optic radiation, 173
  of visual pathways, 172
transforming growth factors (TGF), 394

transforming growth factor-β(TGF-β)
  effect of, 239
  on elastic tissue formation, influence of, 220
  inhibitors of, 242
transient receptor potential vanilloid type 1 (TRPV1), 452–453
transmission electron microscopy (TEM)
  for optic nerve axon counts, 291
  of superior ONH, 297
transplantation therapy, 512, 514
  neuroprotection by
    modified stem cells, 515–516
    stem cells, 514–515
transpupillary cyclophotocoagulation for glaucoma, 232
transscleral contact and noncontact cyclophotocoagulation for glaucoma, 233
transscleral cyclophotocoagulation for glaucoma, 232
transsynaptic neural degeneration, disease model for, 465–474
traumatic brain injury
  acute, 327
  creatine supplementation for, 345
travoprost-timolol for glaucoma treatment, 191–192
trichostatin (TSA)
  neuritogenesis in RGC-5 cells due to, 281
TSNIT double hump graph, 129
tubulin protein in ischemia/reperfusion, decrease of, 349
tumor necrosis factor-alpha (TNF-α)
  consequences of signaling in glaucoma, 417–418
  cytotoxicity, 409, 411, 418
  functions during Wallerian degeneration, 415
  glial production of, 409–410, 414–415, 417–419
  induced RGCs death in glaucoma, 410–413
  mediated neurotoxicity, 418
  neurodestructive and neuroprotective effects, 416–417
  oligodendrocyte death and, 412
  for oligodendrocyte loss, 313
  pathogenesis of, 409
  signaling with glaucomatous neurodegeneration, 413–416
tyrosinase, genetic deficiency of, 308

UBM, see ultrasound biomicroscopy
ultrasound, high-frequency, study of anterior chamber, 294
ultrasound biomicroscopy (UBM)
   for diagnosis of AC, 36–37
   for rotation of ciliary processes, 34
   studies of anterior chamber angle in rat eyes, 296
   use of, 257
uveitic glaucomas
   effect of deep sclerectomy in, 257
   effect of viscocanalostomy in, 258
uveitis, 190
   anterior, 33
   associated with juvenile idiopathic arthritis, 271

vascular endothelial growth factors (VEGFs), 555–556, 555–571, see also Vascular endothelial growth factor-A (VEGF-A)
   angiogenesis, mitogenesis, and endothelial survival, 560
   dependent signaling pathways, 512, 515
   effect on oxidative stress and downstream cascade, 564
   and glaucoma, 568–571
   intravitreal inhibition therapy, 564–565
   isoforms, 556–557
   neuroprotection, 555–565, 568–571
   neuroretina toxicity, 564–565
   neurotrophic and neuroprotective effect, 560–564
   receptors, 557–560
vasculogenesis, 560
vasospasm as risk factor for vision loss, 340
vein cautery episcleral, for producing elevated IOP, 294–295
verapamil against glaucoma progression, benefits of, 207
VF, see visual field
vimentin
   expression of, 354
   localization of, 363
viral infection, variability of, 306
viscocanalostomy
   in surgical alternative to trabeculectomy, 257–259

vision
   function tests for early glaucoma, 47
   intermittent blurring of, 32
   loss
      in glaucoma, visual-field test for, 51
      in patients with glaucoma, 331
      in POAG, 286
      symptomatic, risk of, 48, 55
   preservation of, 332
visual acuity, 479, 486–488, 550
   cause of, 229
visual cortex, 471, 488–489
visual deficits
   fMRI for brain function in patients with, 174
visual dysfunction
   severe, and blindness, 16
visual evoked potentials, 541–542, 546
visual field (VF)
   assessment
      MLC to optimize, 94
      SAP, for monitoring glaucoma progression, 79
      SAP, full-threshold strategy for, 82–84
      SAP, glaucoma staging system for, 86–88
   in clinical features of NTG, 196–198
   damage, SAP for, 125
   defect
      classification of, 86
      EGS criteria for, 69
      GDx-VCC for, 131
      patient with, 68
      P-ERG response for development of, 157
   defects assessment
      SAP, AGIS criteria for, 85
      SAP, CIGTS for, 85
      SAP, OHTS criteria for, 84–85
   examination for glaucoma management, 144
   FDT for changes in, 6
   loss
      clinical findings of, 139
      for diagnosis of AC, 38
      fMRI for visual stimulation in patients with, 175
      glaucomatous, 204
      glaucomatous neuropathy associated with, 53
      GSS for, 88
      identification of patterns of, 96
      risk of, 184

RNFL loss associated with, 47
SWAP, FDP, MAP and HPRP for detection of, 94
test
in glaucoma, SAP for, 66
for vision loss in glaucoma, 51
visual function
measures of, 331
selective tests of, 54
test by SAP, 104
visual pathways
conventional MR imaging and, for glaucoma, 166
glaucoma neuropathology in, 471–472
neuroimaging of, advances in, 165–166
tractography of, 172
visual receptive fields
changes in experimental glaucoma, 479–489
expansion in glaucoma, 484–486
visual stimulation
hemodynamic response of ONH elicited by, 149
of optic nerve, 151–152
vitamin A
and vitamin E supplementation for retinitis pigmentosa, 331
vitamin E
and vitamin A supplementation for retinitis pigmentosa, 331
vitamins B_6
decreased, in patients with XFS, 219
vitamins B_{12}
decreased, in patients with XFS, 219

vitreoretinal diseases, OCT for, 66
vitreous loss, risk of, 214

Wallerian degeneration
protection of, 315
TNF-α functions during, 415
WDR36 gene, effects of mutations in, 308
Western blotting, for NF-L protein, 345
whitematter fibers, diffusivity of, 168
White racial group, glaucoma prevalence studies in, 6
wound healing, 237–238
after GFS, modulation of, 246–250
during and after glaucoma surgery, modulation of, 237
blood clotting and fibrin formation in, 239
cell motility, matrix contraction and synthesis in, 244
cellular proliferation and vascularization in, 243–244
drug delivery in, 245
growth factors in, 239–243
inflammatory cells and mediators in, 239
using surgical and anatomical principles to modify therapy in, 238–239

XFM, see exfoliation material
XFS, see exfoliation syndrome

zonular dialysis, 212
risk of, 214

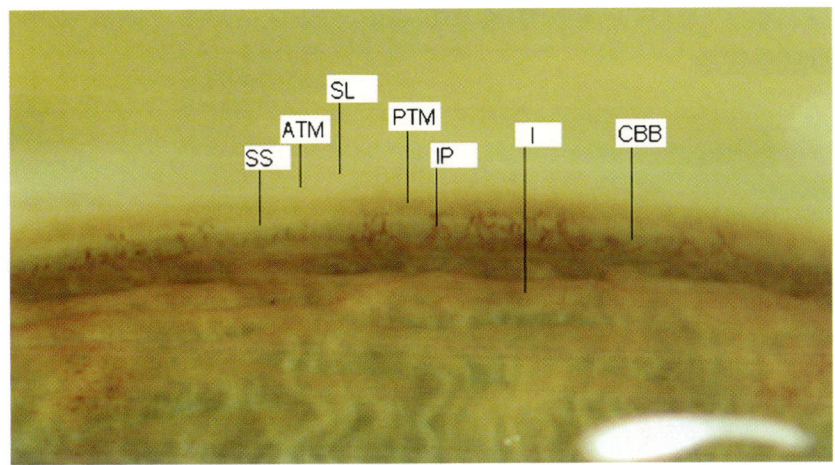

Plate 4.1. Gonioscopic view of normal angle anatomy, showing iris (I), ciliary body band (CBB), scleral spur (SS), posterior trabecular meshwork (PTM), anterior trabecular meshwork (ATM), and Schwalbe's line (SL). Iris processes can also be clearly seen (IP). (Courtesy of Lisandro Sakata, MD, PhD, University of Alabama, Birmingham, USA.)

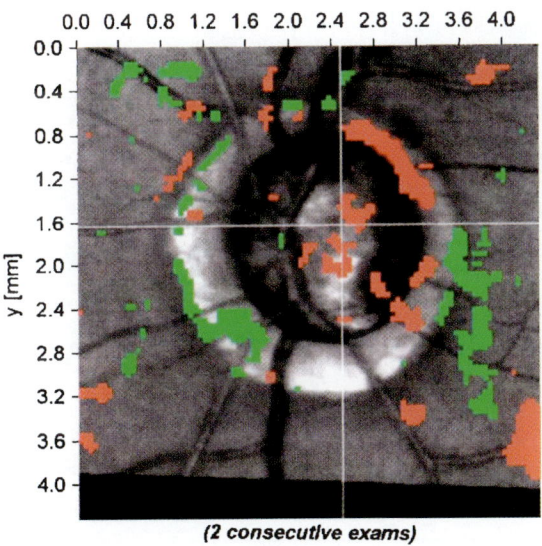

Plate 6.2. HRT 2 change probability map.

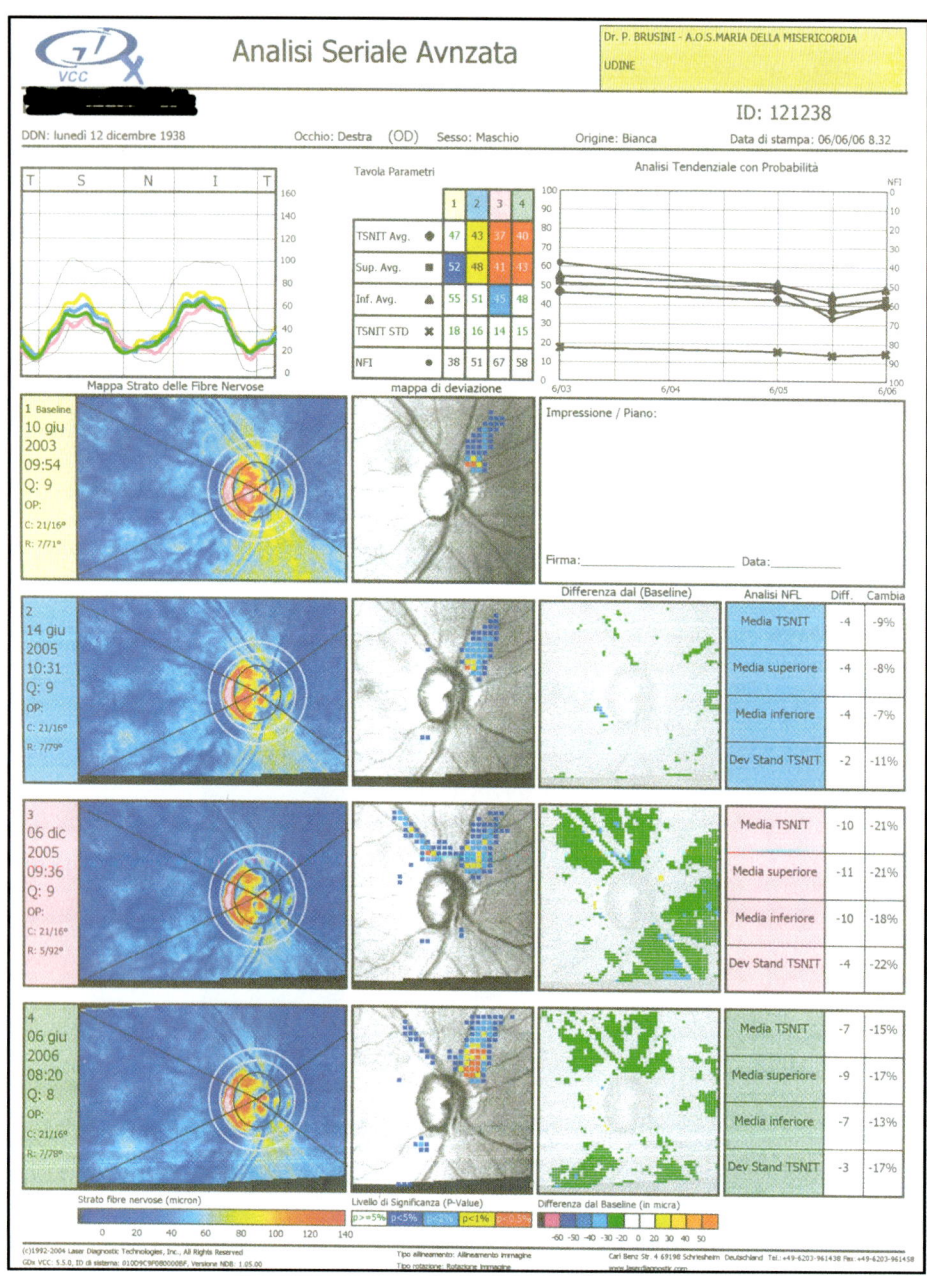

Plate 6.4. GDx VCC advanced serial analysis. This glaucomatous patient shows a progressive worsening and expansion of the superior fascicular defect over time.

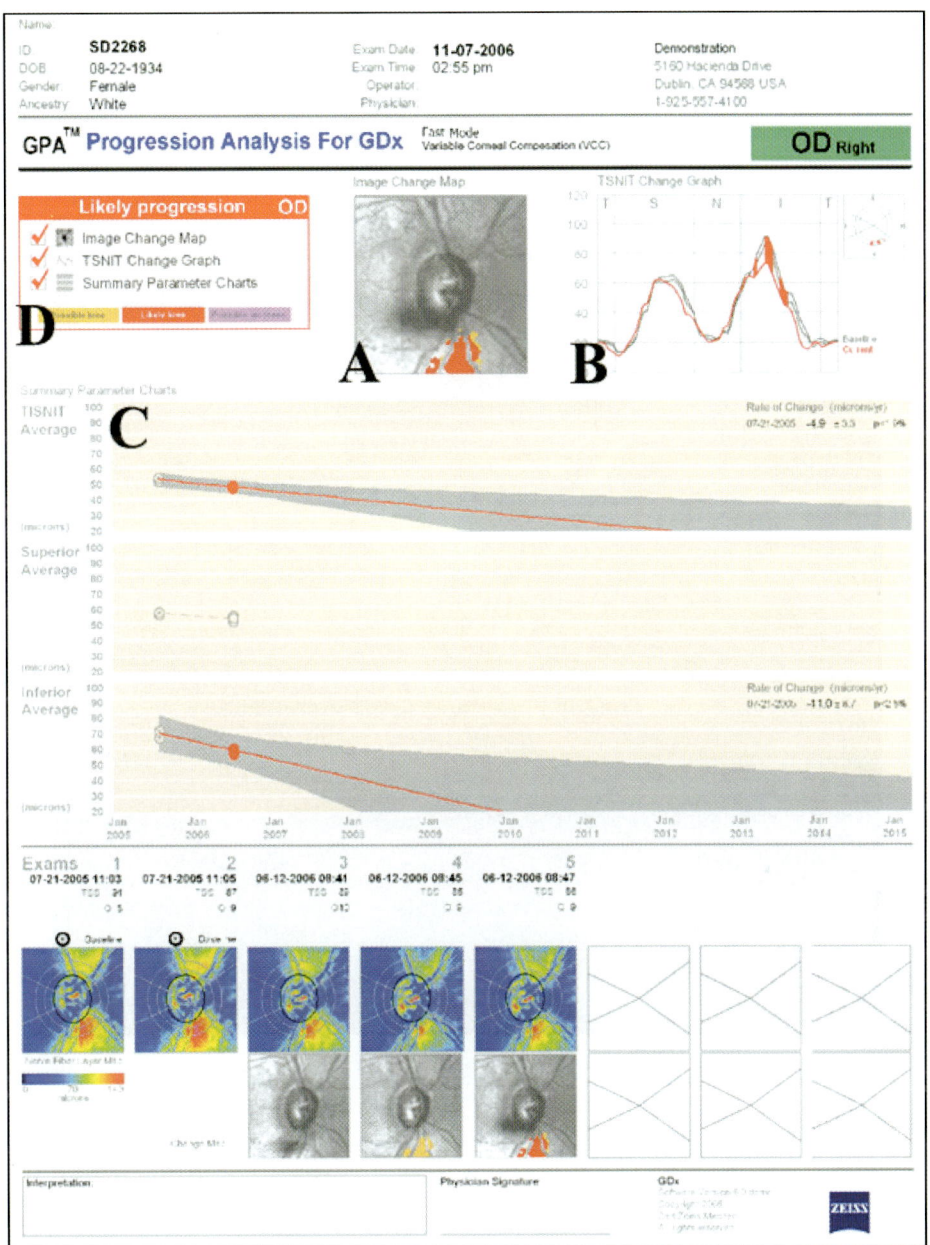

Plate 6.5. GDx VCC guided progression analysis (GPA). Significant RNFL thinning over time can be seen in the inferior sector. (A) Image change map; (B) TSNIT change graph; (C) Summary parameter charts; (D) Global progression assessment.

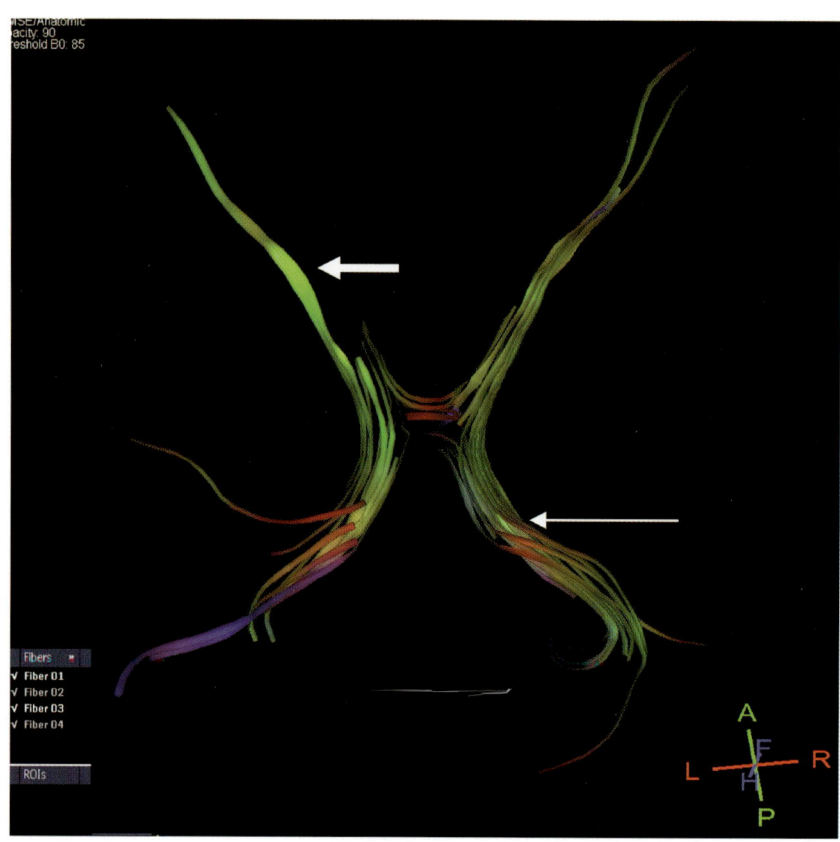

Plate 12.6. Reconstructed diffusion tensor fiber tractography of the optic nerves (short arrow), chiasm and the optic tracts (long arrow).

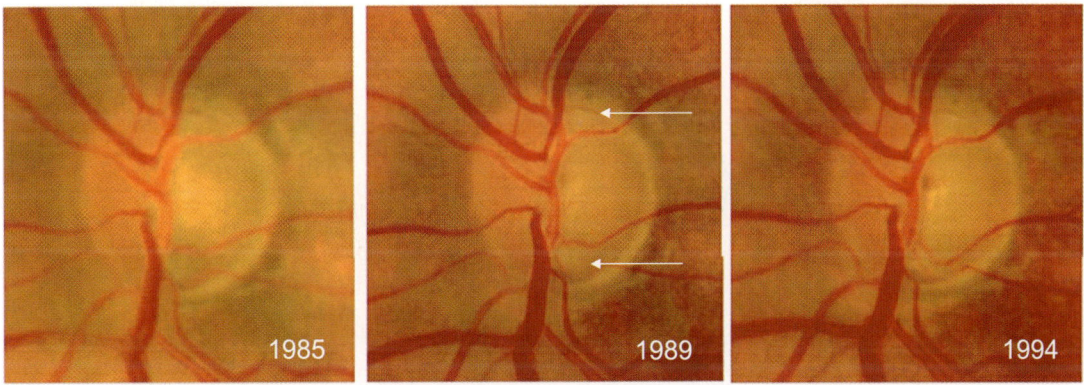

Plate 14.2. The progression of both a superior and inferior acquired pit of the optic nerve (APON) in an NTG patient.

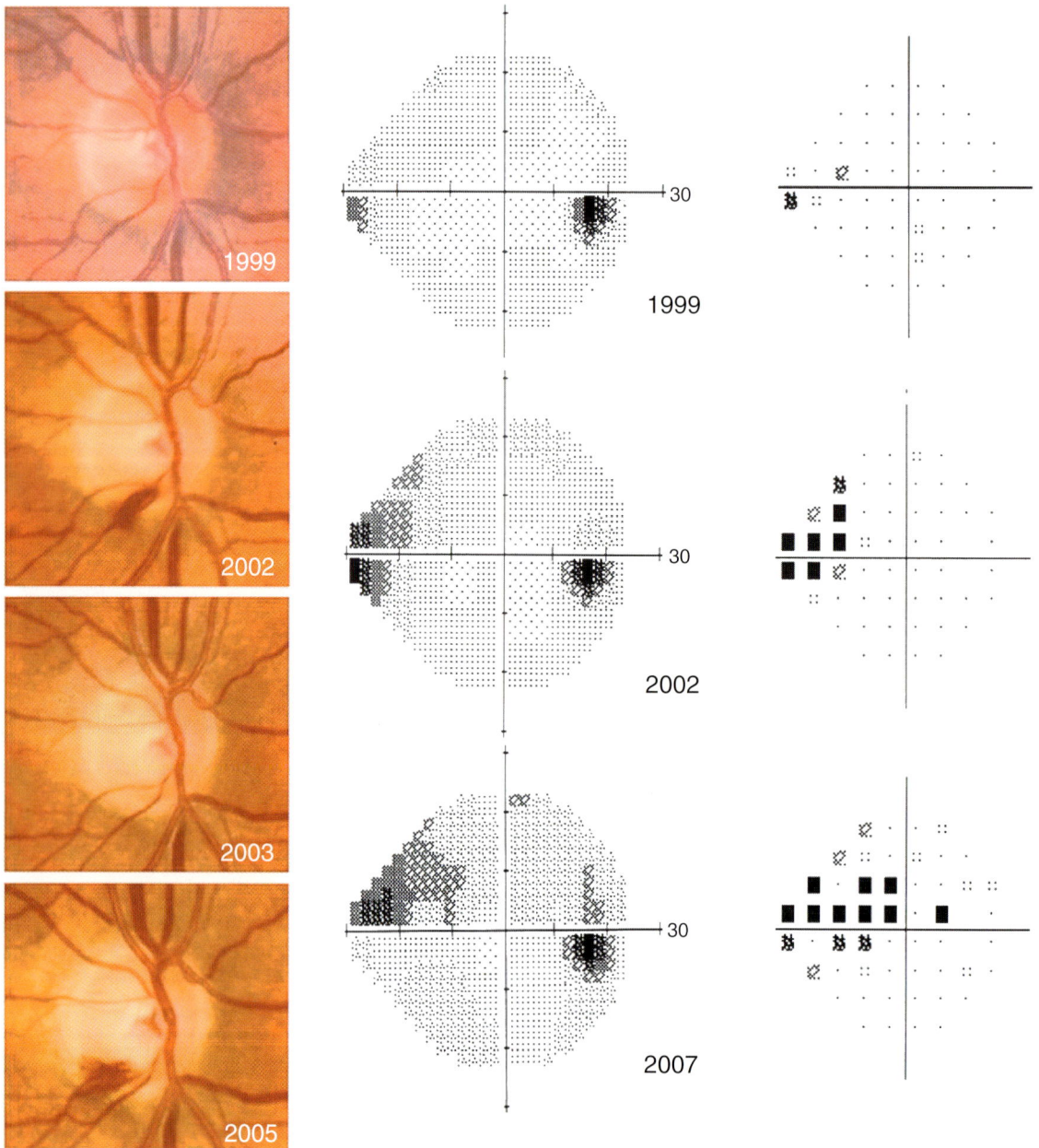

Plate 14.5. Disk photos document recurrent disk hemorrhage in an NTG patient. The visual fields show new focal nasal step defect after the first disk hemorrhage in 2002 and then progression of this defect to encroach on fixation after second disk hemorrhage.

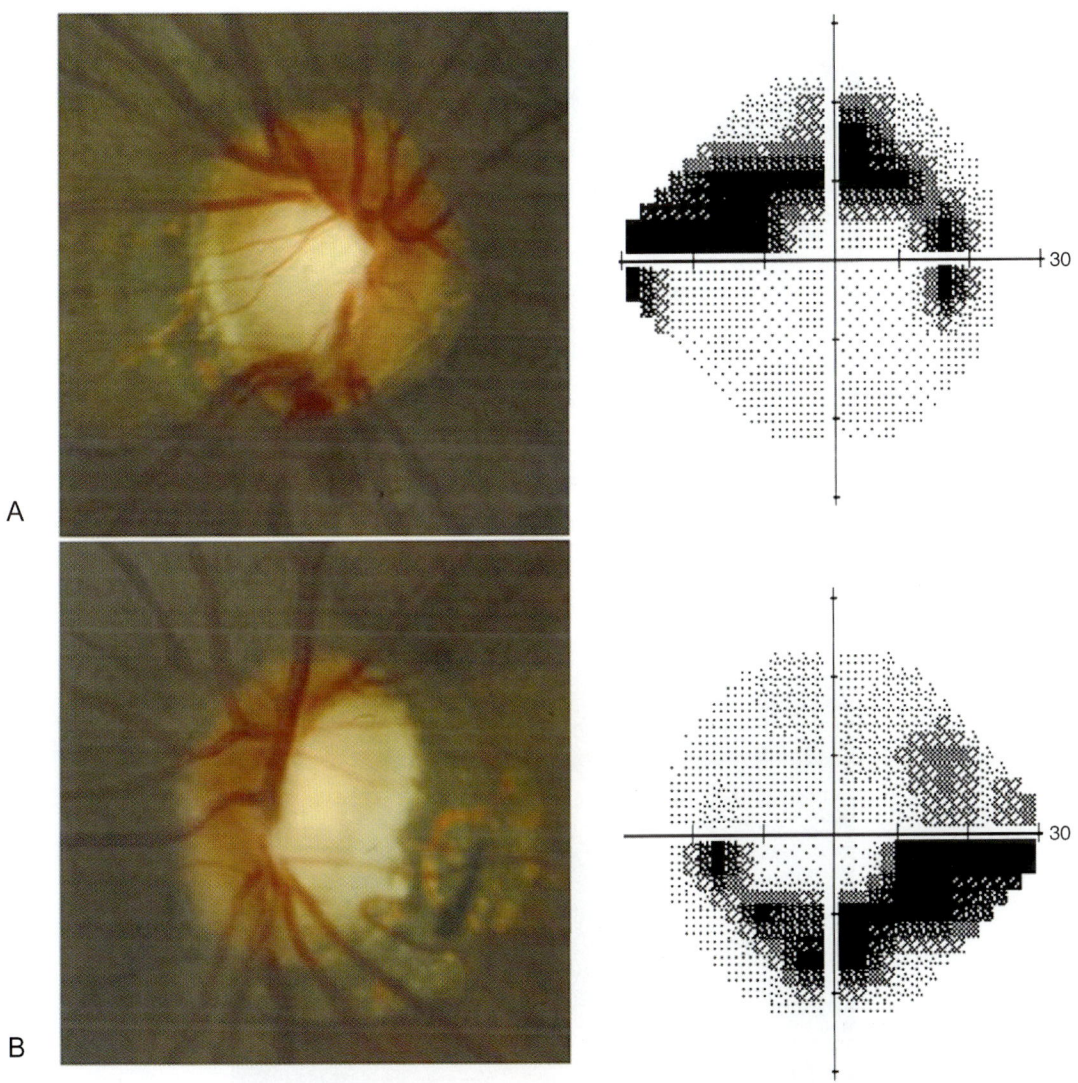

Plate 14.6. Disk photos of an NTG patient that show tilted disks with inferior and temporal peripapillary atrophy and corresponding visual fields. (A) Note the inferior notch on the right nerve. (B) Note the superior notch on the left nerve.

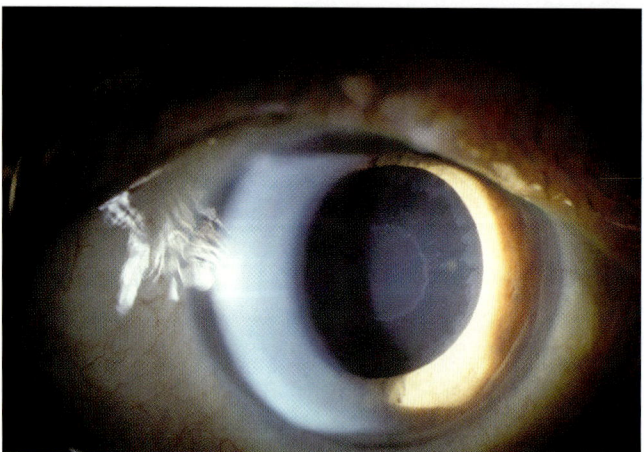

Plate 15.1. Classic appearance of XFM on the anterior lens surface consists of a central disc, intermediate clear zone, and granular peripheral zone.

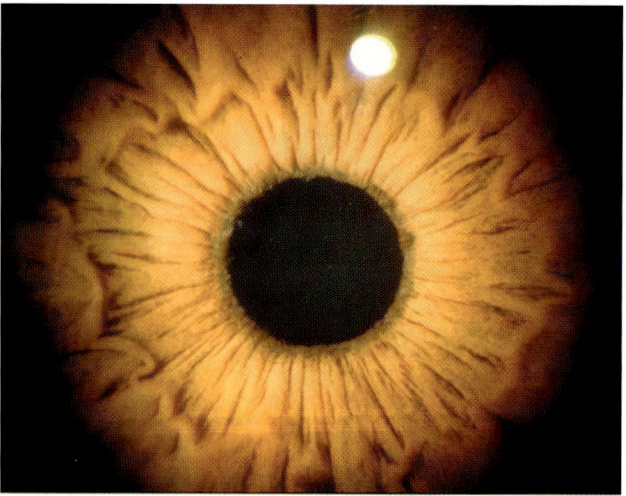

Plate 15.2. Loss of iris pigment ruff in XFS. Exfoliation material is present on the pupillary border.

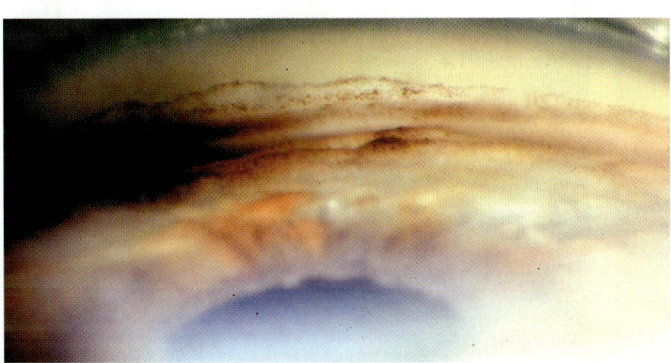

Plate 15.3. Hyperpigmentation of the trabecular meshwork. Pigment is also found on Schwalbe's line and on the peripheral corneal shelf (Sampaolesi line).

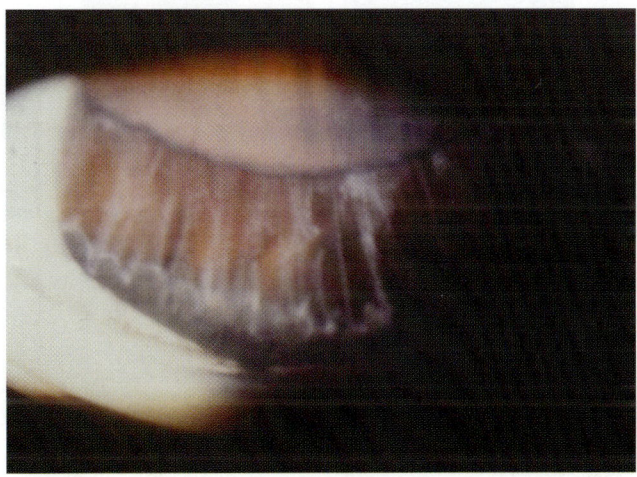

Plate 15.4. Fragmented zonules in XFS.

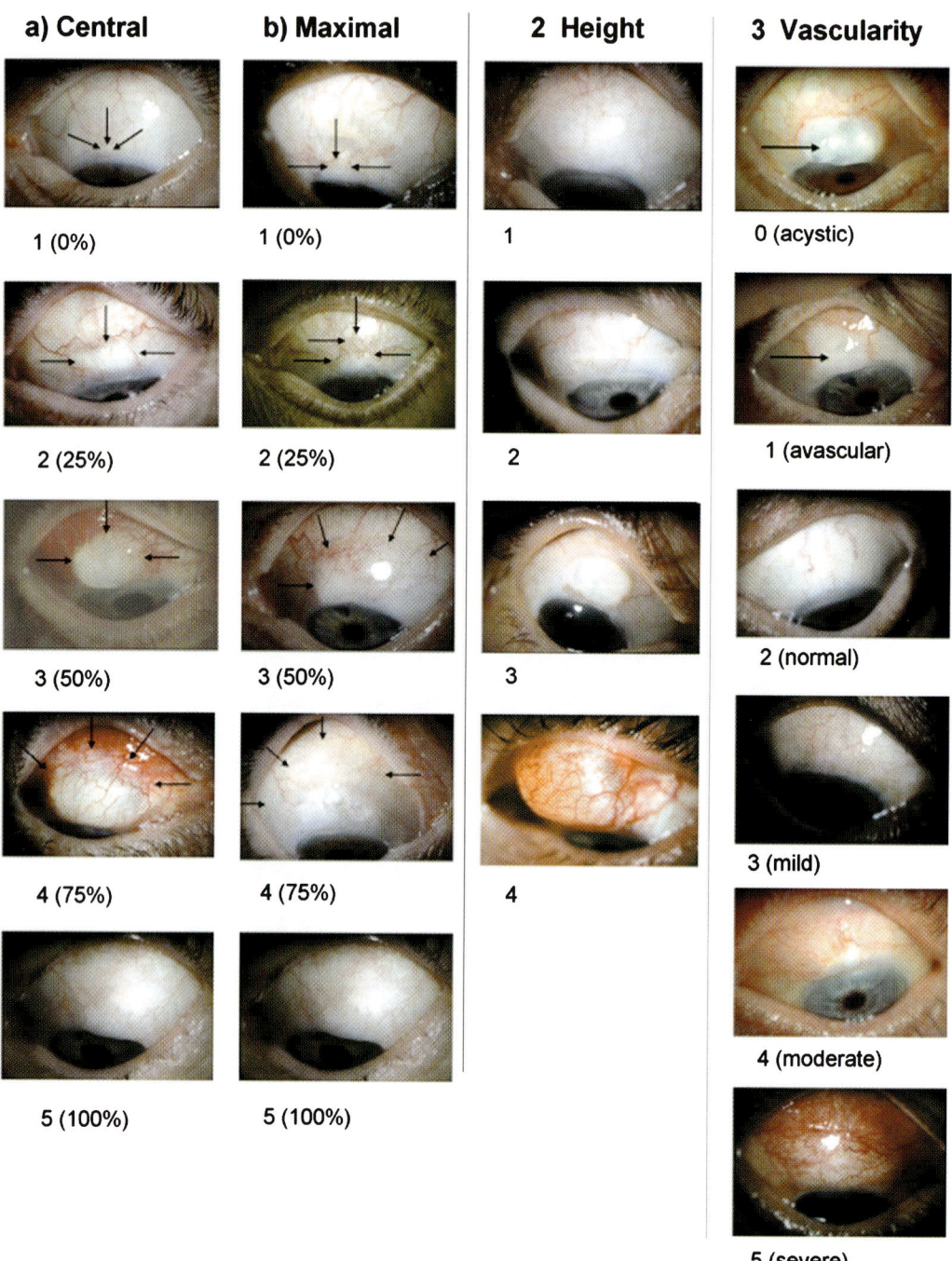

Plate 17.4. Moorfields bleb grading system.

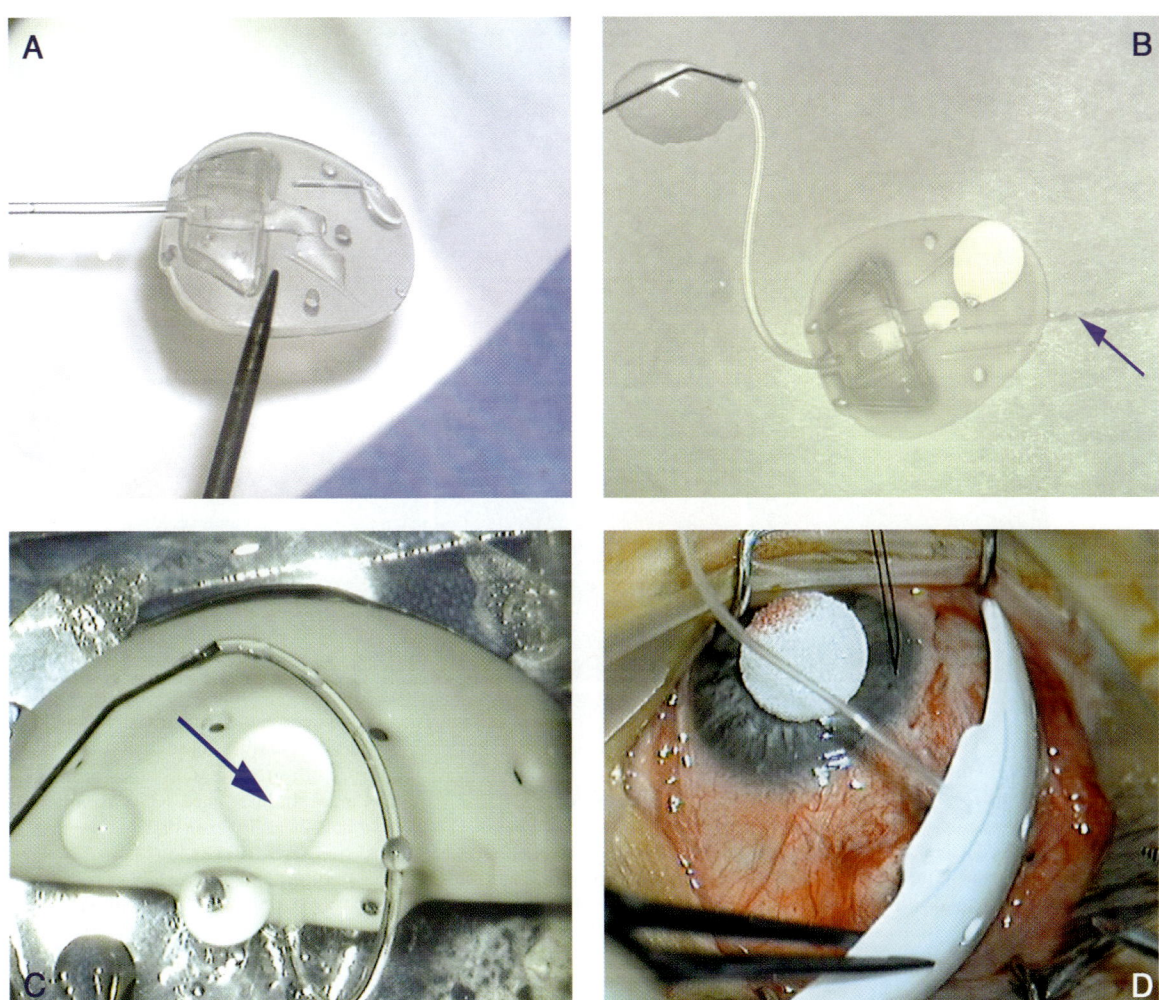

Plate 19.2. The Ahmed Glaucoma Valve (A) contains a valve mechanism on the upper surface, which must be primed (B) prior to insertion. The arrow shows balanced salt solution (BSS) emerging from the valve mechanism on priming. (C) The arrow shows infused BSS flowing across the upper surface of the Baerveldt 350 Glaucoma Implant and (D) demonstrates the large plate surface area in comparison with the globe, as well as the curved thin profile on implantation. Adapted with permission from Shaarawy et al. (2008). Courtesy of Moorfields Eye Hospital.

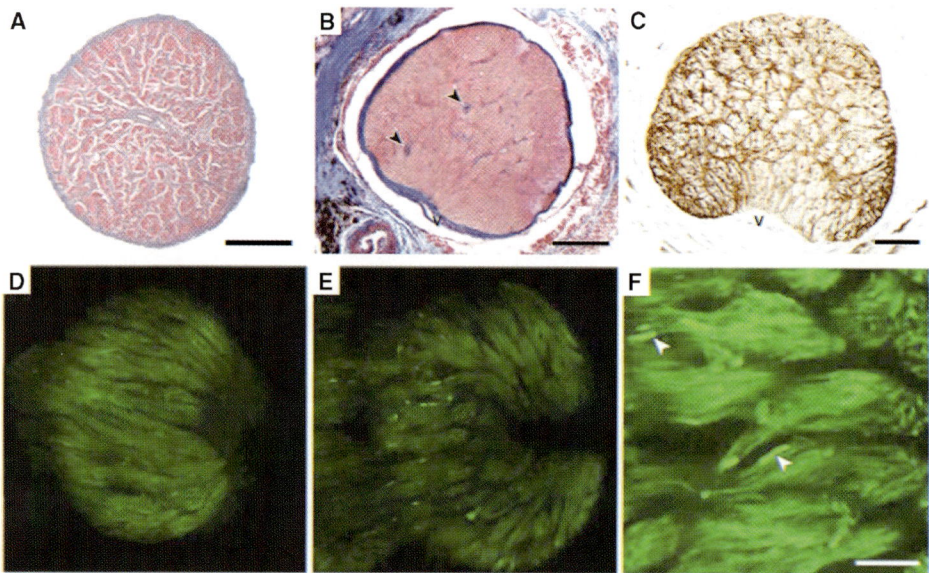

Plate 22.2. (A–C) The lamina cribrosa has robust ECM plates (A, shown in blue, Masson's Trichrome) through which bundles of axons pass. In contrast, the mouse glial lamina has no collagenous plates (B), although collagens are clearly visible in blood vessel walls (arrowheads). Similar to the human lamina, the mouse glial lamina has an extensive meshwork of astrocytes (C, stained positive for GFAP). (D–F) Early focal damage at the glial lamina is visualized using DBA/2J.*Thy1*-CFP mice. In these mice, RGC axons appear green when viewed by confocal microscopy. Because axonal contents accumulate in regions of damage, early axon damage is evident as very brightly fluorescent axon segments (arrows). (D) No damage is seen in preglaucomatous young eyes. (E) Obvious axonal swellings were evident specifically in the lamina of eyes that were at early stages of glaucoma. (F) Focal regions of damage are clearly visible as bright, slightly swollen axonal regions (arrowheads). Scale bars: (A) 1 mm; (B, C) 50 μm; (D–F) 20 μm. (A) Adapted with permission from Karim et al. (2004). (B–F) Adapted with permission from Howell et al. (2007).

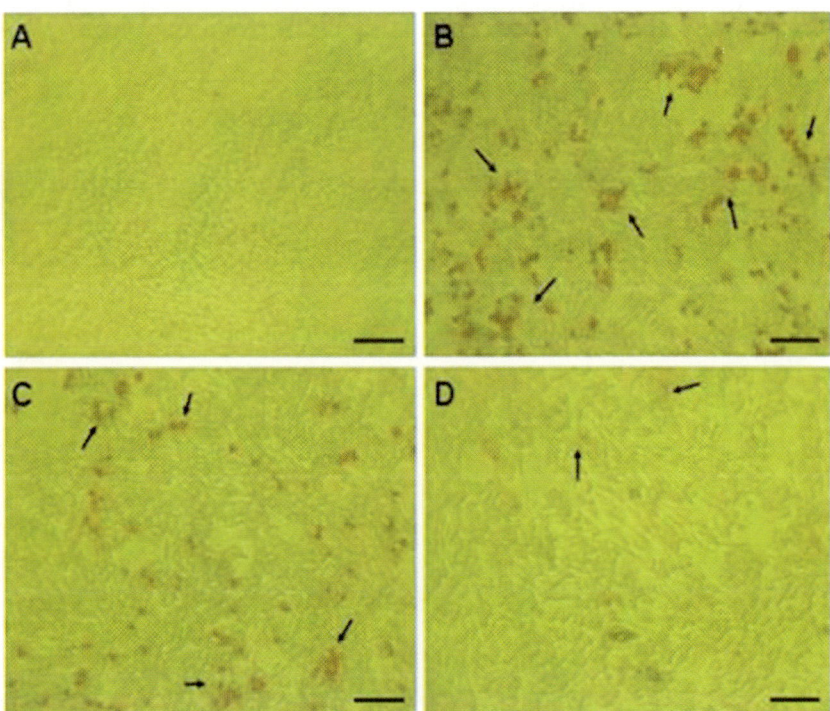

Plate 24.7. Staining of RGC-5 cells exposed to light for 96 h (B) compared with cells maintained in the dark (A) by the APOP*ercentage*^{TM} methodology for apoptosis (arrows). Inclusion of 0.5 mm nicotinamide (C) or 100 μM NU1025, a known PARP inhibitor (D), clearly reduced the positive staining of cell membranes as an indication for apoptosis induced by light. Scale bar = 20 μm.

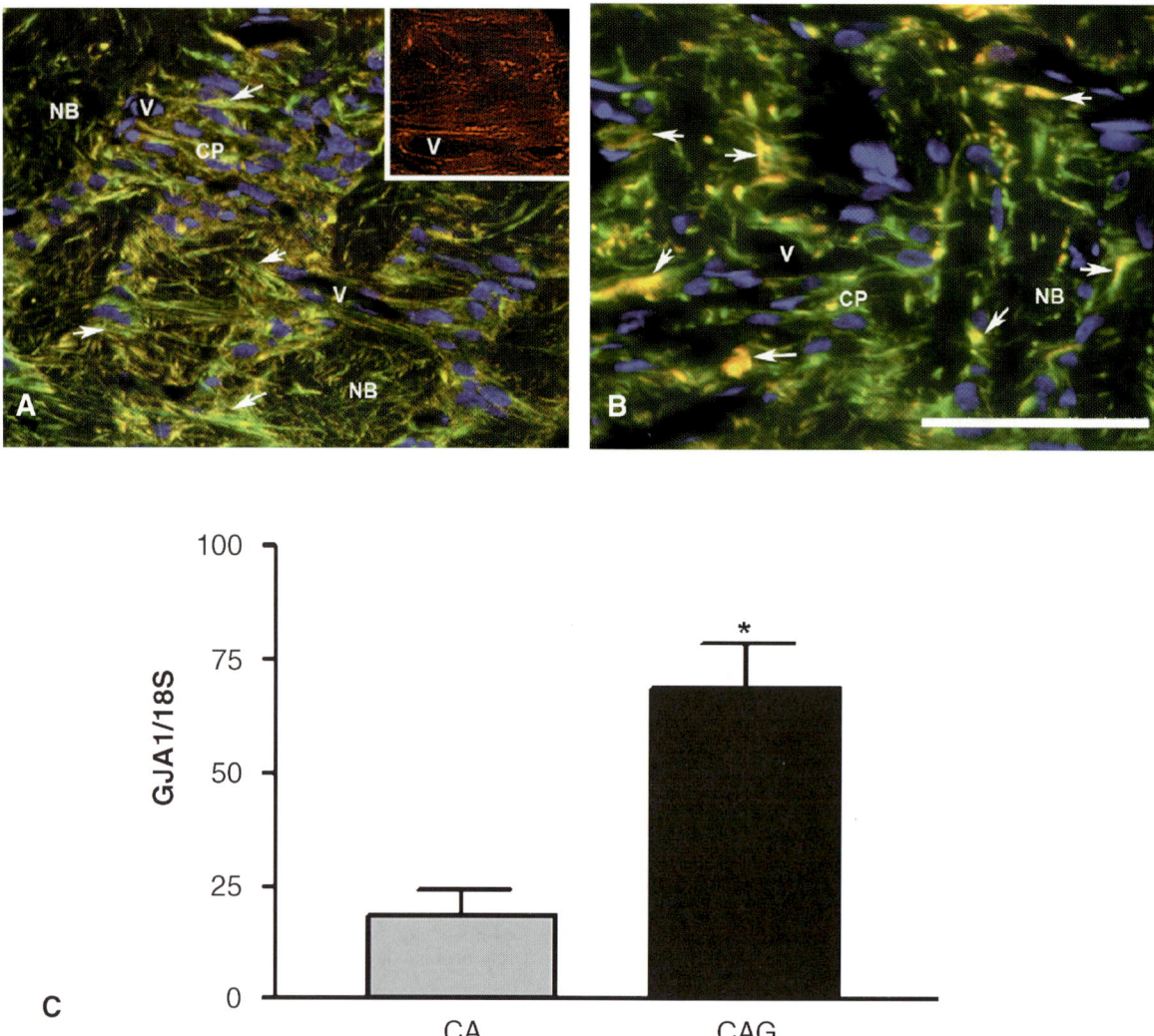

Plate 25.2. Expression of connexin-43 in the monkey optic nerve head. (A) Immunofluorescent staining for GFAP shows astrocytes (green) forming lamellae in the cribriform plates of the lamina cribrosa (CP) stained with Cx43 (red). Fine lines of colocalization of Cx43 on astrocytes membranes (yellow, arrows), V: blood vessel, NB: nerve bundles. Inset: Cx43 immunoreactivity following the fine lamellar processes of astrocytes in the normal lamina cribrosa. (B) In the contralateral eye with experimental glaucoma, astrocytes (green) are disorganized and appear rounded with intracellular Cx43 immunoreactivity (yellow, arrows) in the lamina cribrosa. A and B: Magnification bar = 35 μm. For methods see (Agapova et al., 2006a). (C) Relative amount of Cx43 (GJA1) mRNA in human normal and glaucomatous ONH astrocytes measured by quantitative RT-PCR. Bar graphs represent relative expression of Cx43 mRNA normalized to 18S in normal ($n = 8$) and glaucomatous ($n = 8$) ONH astrocyte cultures. Two-tailed t test was used. *Indicates $p < 0.0002$.

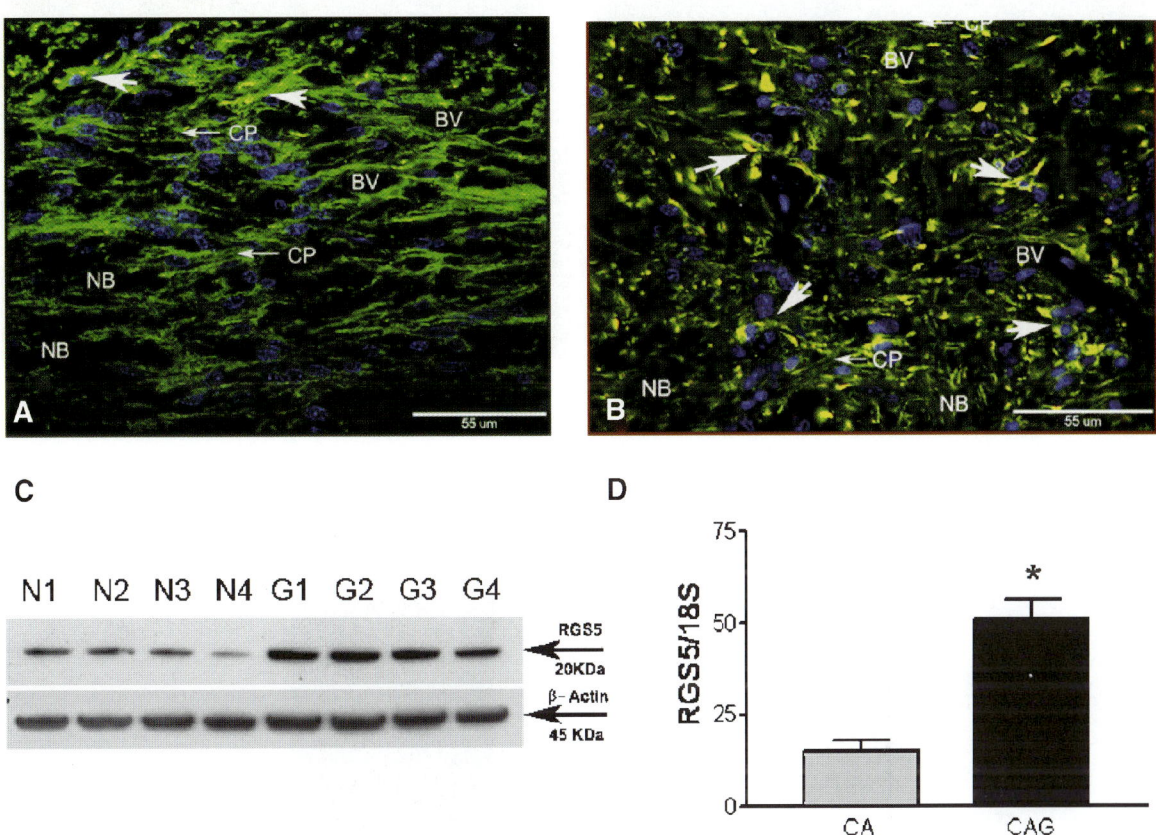

Plate 25.5. Expression of RGS5 in the human optic nerve head. (A) Immunofluorescent staining for GFAP shows astrocytes (green) forming lamellae in the cribriform plates of the lamina cribrosa (CP) (yellow, arrows), few astrocytes stained with RGS5 antibody (yellow) V: blood vessel, NB: nerve bundles. (B) In an eye with POAG, there is an increase in RGS5 immunoreactivity in laminar astrocytes (yellow, arrows). A and B: Magnification bar = 35 μm. For methods see (Agapova et al., 2006a). (C) Western blot analysis. (D) Relative amount of RGS5 mRNA in human normal and glaucomatous ONH astrocytes measured by quantitative RT-PCR. Bar graphs represent relative expression of SOD2 mRNA normalized to 18S in normal ($n = 8$) and glaucomatous ($n = 8$) ONH astrocyte cultures.

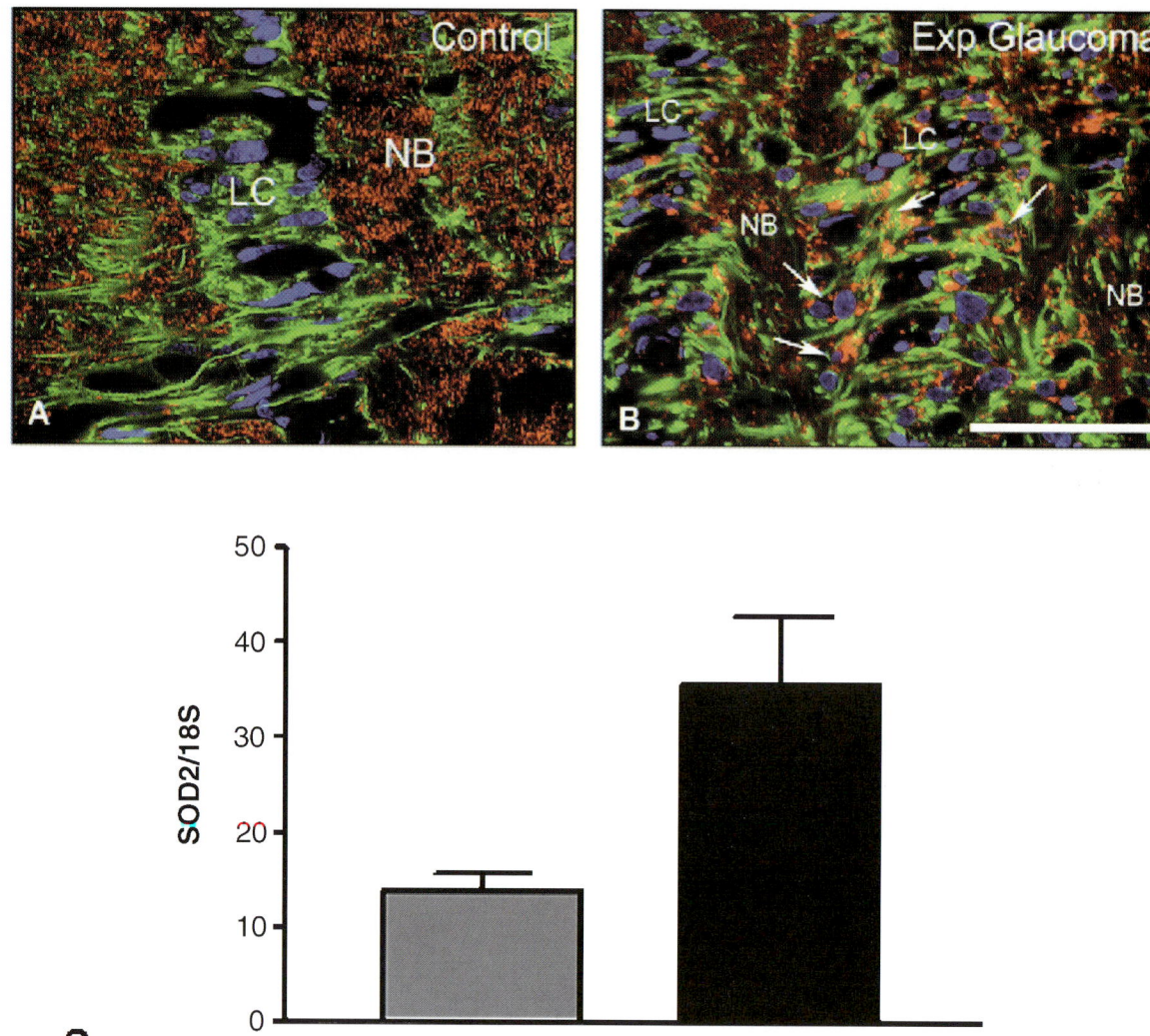

Plate 25.8. Expression of SOD-2 in the monkey optic nerve head. (A) Immunofluorescent staining shows SOD-2 (red) localized to abundant mitochondria in the axons of the RGC in the lamina cribrosa (LC) (arrows) and very few in GFAP+ astrocytes (green) in a normal monkey eye. (B) In experimental glaucoma (ExpG), there is an increase in SOD-2 immunoreactivity in astrocytes (arrows) and a marked decrease in axons in the nerve bundles. A and B: Magnification bar = 35 μm. For methods see (Agapova et al., 2003a). (C) Relative amount of SOD2 mRNA in normal and glaucomatous ONH astrocytes measured by quantitative RT-PCR. Bar graphs represent relative expression of SOD2 mRNA normalized to 18S in normal ($n = 8$) and glaucomatous ($n = 8$) ONH astrocyte cultures ($n = 8$, respectively, two-tailed t test was used. *Indicates $p<0.05$).